T0306080

LIVING AROUND ACTIVE STARS

IAU SYMPOSIUM 328

IAU SYMPOSIUM PROCEEDINGS SERIES

Chief Editor
PIERO BENVENUTI, IAU General Secretary
IAU-UAI Secretariat
98-bis Blvd Arago
F-75014 Paris
France
iau-general.secretary@iap.fr

Editor
MARIA TERESA LAGO, IAU Assistant General Secretary
Universidade do Porto
Centro de Astrofísica
Rua das Estrelas
4150-762 Porto
Portugal
mtlago@astro.up.pt

INTERNATIONAL ASTRONOMICAL UNION
UNION ASTRONOMIQUE INTERNATIONALE

International Astronomical Union

LIVING AROUND ACTIVE STARS

PROCEEDINGS OF THE 328TH SYMPOSIUM OF THE INTERNATIONAL ASTRONOMICAL UNION HELD IN MARESIAS, BRAZIL OCTOBER 17–21, 2016

Edited by

DIBYENDU NANDY

Center of Excellence in Space Sciences India and Department of Physical Sciences, Indian Institute of Science Education and Research Kolkata, India

ADRIANA VALIO

Center for Radio Astronomy and Astrophysics, Mackenzie Presbyterian University, São Paulo, Brazil

and

PASCAL PETIT

Observatoire Midi-Pyrénées Toulouse, France

CAMBRIDGE
UNIVERSITY PRESS

CAMBRIDGE
UNIVERSITY PRESS

University Printing House, Cambridge CB2 8BS, United Kingdom

One Liberty Plaza, 20th Floor, New York, NY 10006, USA

477 Williamstown Road, Port Melbourne, VIC 3207, Australia

314-321, 3rd Floor, Plot 3, Splendor Forum, Jasola District Centre, New Delhi - 110025, India

103 Penang Road, #05-06/07, Visioncrest Commercial, Singapore 238467

Cambridge University Press is part of the University of Cambridge.

It furthers the University's mission by disseminating knowledge in the pursuit of education, learning and research at the highest international levels of excellence.

www.cambridge.org
Information on this title: www.cambridge.org/9781107170056

First published 2017

A catalogue record for this publication is available from the British Library

ISBN   978-1-107-17005-6   Hardback
ISSN   1743-9213

Table of Contents

Preface: Living Around Active Stars   x

The Organizing Committee.   xiii

Conference Photograph   xiv

Participants   xv

The Puzzling Dynamos of Stars: Recent Progress With Global Numerical Simula-
    tions   1
    A. Strugarek, P. Beaudoin, P. Charbonneau & A. S. Brun

On the connections between solar and stellar dynamo models.   12
    L. Jouve & R. Kumar

Starspot Activity and Superflares on Solar-type Stars   22
    H. Maehara

Magnetic field generation in PMS stars with and without radiative core   30
    B. Zaire, G. Guerrero, A. G. Kosovichev, P. K. Smolarkiewicz
    & N. R. Landin

Estimating the chromospheric magnetic field from a revised NLTE modeling: the
    case of HR 7428.   38
    I. Busá

CARMENES – M Dwarfs and their Planets: First Results   46
    A. Quirrenbach, P. J. Amado, I. Ribas, A. Reiners, J. A. Caballero,
    W. Seifert, M. Zechmeister & the CARMENES Consortium

Magnetic activity of interacting binaries   54
    C. A. Hill

Are tachoclines important for solar and stellar dynamos? What can we learn from
    global simulations.   61
    G. Guerrero, P. K. Smolarkiewicz, E. M. de Gouveia Dal Pino,
    A. G. Kosovichev, B. Zaire & N. N. Mansour

Starspots properties and stellar activity from planetary transits.   69
    A. Valio

Dynamo action and magnetic activity during the pre-main sequence: Influence of
    rotation and structural changes.   77
    C. Emeriau-Viard & A. S. Brun

The Suppression and Promotion of Magnetic Flux Emergence in Fully Convective
    Stars   85
    M. A. Weber, M. K. Browning, S. Boardman, J. Clarke, S. Pugsley
    & E. Townsend

Beyond sunspots: Studies using the McIntosh Archive of global solar magnetic field
    patterns...   93
    S. E. Gibson, D. Webb, I. M. Hewins, R. H. McFadden, B. A. Emery,
    W. Denig & P. S. McIntosh

Magnetic fields of weak line T-Tauri stars...............................   101
    C. A. Hill & the MaTYSSE Collaboration

Differential rotation of stars with multiple transiting planets   107
    Y. Netto & A. Valio

Studies of synoptic solar activity using Kodaikanal Ca K data   110
    K. P. Raju

Solar Radius at Sub-Terahertz Frequencies   113
    F. Menezes & A. Valio

Exploring shallow sunspot formation by using Implicit Large-eddy simulations .   117
    F. J. Camacho, G. Guerrero, P. K. Smolarkiewicz, A. G. Kosovichev
    & N. N. Mansour

Contribution of energetic ion secondary particles to solar flare radio spectra.....   120
    J. Tuneu, S. Szpigel, G. G. de Castro & A. MacKinnon

Low-mass eclipsing binaries in the WFCAM Transit Survey   124
    P. Cruz, M. Diaz, D. Barrado & J. Birkby

Evolution of the Active Region NOAA 12443 based on magnetic field extrapola-
    tions: preliminary results   127
    A. Chicrala, R. S. Dallaqua, L. E. A. Vieira, A. D. Lago, J. M. R. Gómez,
    J. Palacios, T. R. C. Stekel, J. E. R. Costa & M. da Silva Rockenbach

Deriving the solar activity cycle modulation on cosmic ray intensity observed by
    Nagoya muon detector from October 1970 until December 2012   130
    R. R. S. de Mendonça, C. R. Braga, E. Echer, A. D. Lago, M. Rockenbach,
    N. J. Schuch & K. Munakata

Coherent Synchrotron Radiation in Laboratory Accelerators and the Double-Spectral
    Feature in Solar Flares.......................................   134
    W. Cruz, S. Szpigel, P. Kaufmann, J.-P. Raulin & M. Klopf

The behavior of the spotless active regions during the solar minimum 23-24....   137
    A. J. de Oliveira e Silva & C. L. Selhorst

Analysis Of Kepler-71 Activity Through Planetary Transit..................   140
    E. A. Gusmão, C. L. Selhorst & A. S. Oliveira

A large catalog of young active RAVE stars in the Solar neighborhood   143
    M. Žerjal, T. Zwitter, G. Matijevič & RAVE Collaboration

Rossby numbers of fully convective and partially convective stars   146
    N. R. Landin & L. T. S. Mendes

Modelling coronal electron density and temperature profiles of the Active Region
    NOAA 11855.   149
    J. M. R. Gómez, L. E. A. Vieira, A. D. Lago, J. Palacios,
    L. A. Balmaceda & T. Stekel

Using planetary transits to estimate magnetic cycles lengths in Kepler stars.... 152
R. Estrela & A. Valio

Modelling coronal electron density and temperature profiles based on solar magnetic field observations................................. 159
J. M. R. Gómez, L. E. A. Vieira, A. D. Lago, J. Palacios,
L. A. Balmaceda & T. Stekel

Solar and stellar coronae and winds................................. 162
M. Jardine

The Influences of Stellar Activity on Planetary Atmospheres 168
C. P. Johnstone

Solar activity forcing of terrestrial hydrological phenomena................. 180
P. J. D. Mauas, A. P. Buccino & E. Flamenco

On the influence of magnetic fields in neutral planetary wakes 192
C. V. D'Angelo, M. Schneiter & A. Esquivel

Hunting for Stellar Coronal Mass Ejections............................ 198
H. Korhonen, K. Vida, M. Leitzinger, P. Odert & O. E. Kovács

Carrington Class Solar Events and How to Recognize Them................. 204
C. T. Russell, J. G. Luhmann & P. Riley

Space Weather Storm Responses at Mars: Lessons from A Weakly Magnetized Terrestrial Planet 211
J. G. Luhmann, C. F. Dong, Y. J. Ma, S. M. Curry, Y. Li, C. O. Lee,
T. Hara, R. Lillis, J. Halekas, J. E. Connerney, J. Espley, D. A. Brain,
Y. Dong, B. M. Jakosky, E. Thiemann, F. Eparvier, F. Leblanc, P. Withers
& C. T. Russell

Coronal Mass Ejections travel time 218
C. R. Braga, R. R. S. de Mendonça, A. D. Lago & E. Echer

Nickel-Phosphorous Development for Total Solar Irradiance Measurement 221
F. Carlesso, L. A. Berni, L. E. A. Vieira, G. S. Savonov, M. Nishimori,
A. D. Lago & E. Miranda

Preliminary Design of the Brazilian's National Institute for Space Research Broadband Radiometer for Solar Observations 224
L. A. Berni, L. E. A. Vieira, G. S. Savonov, A. D. Lago, O. Mendes,
M. R. Silva, F. Guarnieri, M. Sampaio, M. J. Barbosa, J. V. V. Boas,
R. H. F. Branco, M. Nishimori, L. A. Silva, F. Carlesso, J. M. R. Gómez,
L. R. Alves, B. V. Castilho, J. Santos, A. S. Paula & F. Cardoso

Ground-based observations of the [SII] 6731 Å emission lines of the Io plasma torus 227
F. P. Magalhães, W. Gonzalez, E. Echer, M. P. Souza-Echer, R. Lopes,
J. P. Morgenthaler & J. Rathbun

A study on Electron Oscillations in the Magnetosheath of Mars with Mars Express observations 230
A. M. de Souza, E. Echer, M. J. A. Bolzam & M. Fränz

Extreme solar-terrestrial events   233
  A. D. Lago, L. E. A. Vieira, E. Echer, L. A. Balmaceda, M. Rockenbach &
  W. D. Gonzalez

How to make the Sun look less like the Sun and more like a star?   237
  A. A. Vidotto

The X-ray Light-Curves and CME onset of a M2.5 flare of July 6, 2006   240
  J. E. Mendoza-Torres & J. E. Pérez-León

A Framework for Finding and Interpreting Stellar CMEs   243
  R. A. Osten & S. J. Wolk

The long-term evolution of stellar activity.   252
  S. G. Gregory

Stellar Midlife Crises: Challenges and Advances in Simulating Convection and Differential Rotation in Sun-like Stars.   264
  N. J. Nelson, C. Payne & C. M. Sorensen

Improved rotation-activity-age relations in Sun-like stars   274
  J. Meléndez, L. A. dos Santos & F. C. Freitas

Hunting for hot Jupiters around young stars.   282
  L. Yu & the MaTYSSE collaboration

Observable Impacts of Exoplanets on Stellar Hosts – An X-Ray Perspective . . .   290
  S. J. Wolk, I. Pillitteri & K. Poppenhaeger

Possible effects on Earth's climate due to reduced atmospheric ionization by GCR
  during Forbush Decreases   298
  W. Portugal, E. Echer, M. P. de Souza Echer & A. A. Pacini

Interaction of extra solar planets with their host star.   301
  D. Silva & A. Valio

The influence of eclipses in the stellar radio emission   305
  C. L. Selhorst & A. Valio

Tidal effects on stellar activity   308
  K. Poppenhaeger

The Environment of the Young Earth in the Perspective of An Young Sun   315
  V. S. Airapetian

Evolution of Long Term Variability in Solar Analogs   329
  R. Egeland, W. Soon, S. Baliunas, J. C. Hall & G. W. Henry

The solar proxy κ^1 Cet and the planetary habitability around the young Sun. . .   338
  J.-D. do Nascimento, Jr., A. A. Vidotto, P. Petit, C. Folsom,
  G. F. P. de Mello, S. Meibom, X. C. Abrevaya, I. Ribas, M. Castro,
  S. C. Marsden, J. Morin, S. V. Jeffers, E. Guinan & Bcool Collaboration

The Faint Young Sun and Faint Young Stars Paradox   350
  P. C. Martens

High Energy Exoplanet Transits.   356
  J. Llama & E. L. Shkolnik

Detection of secondary eclipses of WASP-10b and Qatar-1b in the Ks band and
   the correlation between Ks-band temperature and stellar activity.   363
   P. Cruz, D. Barrado, J. Lillo-Box, M. Diaz, M. López-Morales,
   J. B. Jonathan, J. Fortney & S. Hodgkin

Atmospheric Parameters and Luminosities of Nearby M Dwarfs − Estimating Habitable Exoplanet Detectability with the E-ELT   371
   G. F. P. de Mello, R. E. Giribaldi, D. Lorenzo-Oliveira & N. M. P. Leme

Author index   374

Preface: Living Around Active Stars

The variable activity of stars such as the Sun is mediated via stellar magnetic fields, radiative and energetic particle fluxes, stellar winds and magnetic storms. This activity influences planetary atmospheres, climate and habitability. Studies of this intimate relationship between the parent star, its astrosphere (i.e., the equivalent of the heliosphere) and the planets that it hosts have reached a certain level of maturity within our own Solar System – fuelled both by advances in theoretical modelling and a host of satellites that observe the Sun-Earth system. Based on this understanding the first attempts are being made to characterize the interactions between stars and planets and their coupled evolution, which have relevance for habitability and the search for habitable planets. In this scientific context, the International Astronomical Union Symposium 328 "Living around Active Stars" was organized in Maresias, Brazil from 17–21 October, 2016. The symposium brought together scientists from diverse, interdisciplinary scientific areas such as solar, stellar and planetary physics, atmospheric and climate physics and astrobiology to review the current state of our understanding of solar and stellar environments and its relevance for life and society.

The magnetic fields of stars originate deep in their interior via a magnetohydrodynamic dynamo mechanism that relies on interactions between plasma flows and magnetic fields. These fields manifest as star spots and small-scale magnetic features on the surface and are dispersed in the surrounding space through stellar winds. While some stars such as the Sun exhibit magnetic cycles, others display a diverse variety of activity output. Variation in this magnetic output contributes to radiative flux, particle flux and stellar wind modulation whose impacts are relevant throughout the astrosphere of the star. Energetic, transient events, e.g., flares and coronal mass ejections originate in magnetic structures in stellar atmospheres and generate extreme conditions that persist over several days in the stellar environment. Slower, long-term variation in the stellar radiative output forces planetary atmospheres and climate. Thus, planets and their parent star(s) share a physical bond across space and over a broad range of timescales that is not only important for forcing of planetary atmospheres but is also relevant for coupled star-planet evolution and habitability.

The advent of the space age and the consequent deployment of a host of satellites have led to the appreciation that there is a variable environment or weather in space. The ambient stellar wind forms the background environment. Although steady, the speed of the wind is coupled to the activity of the star and younger, more active stars are known to have stronger winds. It is also known that low-mass M-type dwarf stars have relatively stronger winds. These winds play an important role in atmospheric evaporation through sputtering processes, especially in the absence of a magnetosphere. Magnetic storms which carry significant amount of magnetized plasma perturb planetary magnetospheres and create ionospheric disturbances. Solar flares, which are accompanied by intense high energy radiation, impact the state of the Earth's upper atmosphere. While causal connections between an event in the Sun and its heliospheric and planetary impact are being characterized and modelled routinely, to what extent significant, long-term perturbations of these nature impact the evolution of planetary atmospheres remains an open question.

In contrast to hundreds of stars, for which we have only a few years or decades of systematic observations, the Sun is the only star whose activity can be traced back for tens of thousands of years using diverse proxies (e.g., cosmogenic isotopes). This gives a unique opportunity to study solar activity on the time scale of thousands of the prominent cycle, thus providing clues to deciphering long-term solar variability and an assessment of extreme events. A hotly debated topic of current interest is the possibility of intense stellar super-flares and their impact on planets and life. How strong can stellar flares be in solar-like planet hosting stars, can observations of flares in other stars constrain the extremity of events in our solar system? Questions such as what role the presence of planetary magnetospheres play in moderating this interaction are just beginning to be asked.

The Sun's radiative output is the primary natural driver of planetary climates. Recently it is being appreciated that it may be more important to account for spectral irradiance variability and study the impact of UV radiation in climate dynamics (although the solar cycle variation of total solar irradiance is small, the variability in higher energy radiation is much stronger). The young Sun, and rapidly rotating stars would have had much higher levels of magnetic activity which would lead to stronger high energy radiative fluxes due to heating of stellar atmospheres. This leads one to conjecture on the role of changing solar radiative variability in shaping planetary climates and in governing the conditions in which life may have emerged.

Sun-like stars have a typical lifetime of 10 billion years in the main sequence stage. Stars are born as rapid rotators, plausibly with strong differential rotation and therefore stronger magnetic activity, extreme and frequent flares, stronger fluxes of high energy radiation and energetic particles. Thus, a young planet in a young stellar system would likely be subject to extreme conditions; conditions which would also influence habitability and the evolution of life. With age, and wind mediated loss of angular momentum, the young star would spin down gradually. The magnetic dynamo would become less efficient and so will stellar magnetic output reduce with time. Stellar models indicate the Sun was much fainter in total luminosity compared to the present. This leads to the faint young Sun paradox, which raises the enigmatic question, how life could have emerged in a planet that would, plausibly, have been frozen given the lower solar luminosity? What was the activity of the young Sun like, how did its activity evolve in time until the present, and how is it expected to evolve as the Sun ages? What can observations and modelling of other Sun-like stars tell us about long-term solar evolution? How does the habitable zone in an astrosphere evolve with an evolving parent star? These are open questions that have a direct bearing on the question of life around an active star.

Understanding the impact of an active star on its environment would help us not only to understand habitability but would also be useful in characterizing exoplanetary systems. A case in point is that planetary radio emission similar to that of Jupiter is expected from the interaction between stellar wind particles and exoplanetary magnetic fields. Searches are underway to detect such effects since they can characterise planetary magnetic fields thus setting constraints on exoplanetary dynamos and by extension the internal structure of planets. The detection of transit asymmetries – which are potentially linked to the interaction of the planet with its stellar environment – may also shed light on the loss of atmospheres and magnetic fields of exoplanets. Magnetic and tidal interactions between planets and host stars can also be explored through numerical modelling and observable signatures of anomalous stellar activity enhancement and spin-up that may show up in new observation campaigns.

Multiple current and planned future missions and instruments such as Kepler, COROT, LOFAR, JWST, CHEOPS, PLATO, SPIRou, HARPS, TESS, EXPRESSO (VLT) and HIRES (E-ELT) are seeking to characterize exoplanets and aid in the search for terrestrial planets orbiting within habitable zones. These instruments can also help us characterize the activity of host stars and their magnetism and through a synergy between observations and theoretical modelling, constrain their internal properties.

We are, therefore, standing at an epoch with the promise of unparalleled opportunities towards characterizing and understanding the environment and habitability around active stars. These considerations inspired this Symposium which was a functional outcome of the International Astronomical Union Working Group on "Impact of Magnetic Activity on Solar and Stellar Environments". The papers presented in this symposium and collated in this proceedings deals with many of these outstanding scientific questions – which are multi-disciplinary in character and immense in their scope.

On the one hand, this symposium sought to bridge the boundaries across these diverse disciplines which are each, independently important. On the other hand, the symposium sought to address those outstanding questions that can only be explored through an interdisciplinary approach that brings together diversity in expertise. We are hopeful that the deliberations in this symposium and the scientific papers in this proceedings will motivate coordinated research to explore the coupled evolution of star-planet systems and understand astrophysical conditions that have a direct bearing on living around active stars.

Dibyendu Nandy, Editor and SOC Chair
Adriana Valio, Co-Editor and LOC Chair
Pascal Petit, Co-Editor and SOC Co-Chair
Kolkata, India, 24 May 2017

THE ORGANIZING COMMITTEE

Scientific

Dibyendu Nandi (chair, India)
Sarah Gibson (co-chair, USA)
Emre Isik (Turkey)
Kanya Kusano (Japan)
Cristina Mandrini (Argentina)
Adriana Valio (Brazil)
David Webb (USA)

Pascal Petit (co-chair, France)
Margit Haberreiter (Switzerland)
Heidi Korhonen (Finland)
Duncan Mackay (UK)
Allan Sacha Brun (France)
Aline Vidotto (Ireland)

Local

Adriana Valio (chair)
Alisson Dal Lago
Emilia Correia

Gustavo Guerrero (co-chair)
Jorge Melendez
Caius L. Selhorst

Acknowledgements

The symposium was organized by the IAU Working Group "Impact of Magnetic Activity on Solar and Stellar Environments" and coordinated and supported by the IAU Division E (Sun and Heliosphere), Division F (Planetary Systems and Bioastronomy), Division G (Stars and Stellar Physics) and many of their associated Commissions, namely 10 (Solar Activity), 12 (Solar Radiation and Structure), 25 (Astronomical Photometry and Polarimetry) and 49 (Interplanetary Plasma and Heliosphere).

The Local Organizing Committee operated under the auspices of the Center for Radio Astronomy and Astrophysics, Mackenzie Presbyterian University. Logistical support was provided by the Brazilian Astronomical Society (SAB).

Funding by the
International Astronomical Union (IAU),
Solar Physics Division-American Astronomical Society (SPD-AAS),
Scientific Committee on Solar-Terrestrial Physics (SCOSTEP),
FAPESP,
CAPES,
and
CNPq,
is gratefully acknowledged.

CONFERENCE PHOTOGRAPH

Participants

Ximena C. **Abrevaya**, Inst. Astronomia y Fisica del Espacio, UBA, Buenos Aires, Argentina — abrevaya@iafe.uba.ar

Vladimir **Airapetian**, NASA GSFC, MA, USA — vladimir.airapetian@nasa.gov

Livia **Alves**, Space Geophysics Division, INPE, Brazil — liviarib@gmail.com

Sudeshna **Boro Saikia**, Inst. Astrophysics, University of Goettingen, Germany — sudeshna@astro.physik.uni-goettingen.de

Carlos Roberto **Braga**, Space Geophysics Division, INPE, Brazil — carlos.braga@inpe.br

Allan Sacha **Brun**, CEA-Saclay, France — sacha.brun@cea.fr

Innocenza **Busa**, INAF -OA, Catania, Italy — busainnocenza@gmail.com

Yong-Ik **Byun**, Yonsei University, Seul, South Korea — yongikbyun@gmail.com

Francisco **Camacho**, Universidade Federal de Minas Gerais, Brazil — pachocamacho@gmail.com

Andrew **Collier Cameron**, School of Physics and Astronomy, University of St Andrews, UK — acc4@st-andrews.ac.uk

Franciele **Carlesso**, Space Geophysics Division, INPE, Brazil — fccarlesso@gmail.com?

Andre **Chicrala**, Space Geophysics Division, INPE, Brazil — andre.chicrala@inpe.br

Patricia **Cruz**, IAG, Universidade de São Paulo, Brazil — patricia.cruz@usp.br

Wellington **Cruz**, CRAAM, Universidade Presbiteriana Mackenzie, Brazil — wellington.cruz@gmail.com

Ligia Alves **da Silva**, Space Geophysics Division, INPE, Brazil — ligia.silva@inpe.br

Alisson **Dal Lago**, Space Geophysics Division, INPE, Brazil — alisson.dallago@inpe.br

Rafael R. S. **de Mendona**, Space Geophysics Division, INPE, Brazil — rafael.mendonca@inpe.br

Adriane **de Souza**, Space Geophysics Division, INPE, Brazil — adriane.souza@inpe.br

Fabio **Del Sordo**, Nordita, Sweden — fabio.delsordo@yale.edu

Mekhi **Dhesi**, Blue Skies Space Ltd. / University College London, UK — mekhi@bssl.space

Jose-Dias **do Nascimento Jr.**, Universidade Federal do Rio Grande do Norte, Brazil — jdonascimento@fisica.ufrn.br

Ricky **Egeland**, High Altitude Observatory/Montana State University, USA — egeland@ucar.edu

Constance **Emeriau-Viard**, CEA-Saclay, France — constance.emeriau@cea.fr

Raissa **Estrela**, CRAAM, Universidade Presbiteriana Mackenzie, Brazil — rlf.estrela@gmail.com

Fabricio **Freitas**, IAG, Universidade de São Paulo, Brazil — fabricio.freitas@usp.br

Sarah E. **Gibson**, National Center for Atmospheric Research, Boulder, CO, USA — sgibson@ucar.edu

Scott G. **Gregory**, University of St Andrews, UK — sg64@st-andrews.ac.uk

Manuel **Guedel**, University of Vienna, Austria — manuel.guedel@univie.ac.at

Gustavo **Guerrero**, Universidade Federal de Minas Gerais, Brazil — guerrero@fisica.ufmg.br

Colin A. **Hill**, Institut de Recherche en Astrophysique et Plan.tologie, Toulouse, France — chill@irap.omp.eu

Moira **Jardine**, University of St Andrews, UK — mmj@st-andrews.ac.uk

Colin **Johnstone**, Dept. Astrophysics, University of Vienna, Austria — colin.johnstone@univie.ac.at

Lourene **Jouve**, Institut de Recherche en Astrophysique et Plan.tologie, Toulouse, France — ljouve@irap.omp.eu

Pierre **Kaufmann**, CRAAM, Universidade Presbiteriana Mackenzie, Brazil — luciola@craam.mackenzie.br

Kristina **Kislyakova**, Space Research Inst., Austrian Academy of Sciences, Austria — kristina.kislyakova@oeaw.ac.at

Heidi **Korhonen**, Niels Bohr Institute, University of Copenhagen, Denmark — heidi.korhonen@nbi.ku.dk

Raju Paul **Kuttickat**, Indian Institute of Astrophysics, Bangalore, India — kpr@iiap.res.in

Natalia R. **Landin**, Universidade Federal de Viçosa, Brazil — nlandin@ufv.br

J. **Llama**, Lowell Observatory, Flagstaff, AZ, USA — joe.llama@lowell.edu

Theresa **Lueftinger**, Department of Astrophysics, University of Vienna, Austria — theresa.rank-lueftinger@univie.ac.at

Janet **Luhmann**, Space Sciences Laboratory, University of California at Berkeley, CA, USA — jgluhman@ssl.berkeley.edu

Hiroyuki **Maehara**, National Astronomical Observatory, Okayama, Japan — h.maehara@oao.nao.ac.jp

Fabiola P. **Magalhães**, Space Geophysics Division, INPE, Brazil — fabiola.magalhaes@inpe.br

Sushant **Mahajan**, Georgia State University, USA — mahajan@astro.gsu.edu

Stephen **Marsden**, University of Southern Queensland, Australia — stephen.marsden@usq.edu.au

Petrus **Martens**, Dept. of Physics and Astronomy, Georgia State University, USA — martens@astro.gsu.edu

Christopher **Marvin**, IAG, University of Goettingen, Germany — cmarvin@astro.physik.uni-goettingen.de

Pablo **Mauas**, Instituto de Astronomia y Fisica del Espacio, UBA, Buenos Aires, Argentina — pablo@iafe.uba.ar

Jorge **Melendez**, IAG, Universidade de São Paulo, Brazil — jorge.melendez@iag.usp.br

J.E. **Mendoza-Torres**, Instituto Nacional de Astrofisica, Optica y Electronica, Mexico — mend@inaoep.mx

Fabian **Menezes**, CRAAM, Universidade Presbiteriana Mackenzie, Brazil — fabianme17@gmail.com

Subhanjoy **Mohanty**, Imperial College London, London, UK — smohanty@ic.ac.uk

Clarisse **Monteiro Fernandes**, Observatorio Nacional, Rio de Janeiro, Brazil — clarisse.fernandes206@gmail.com

Dibyendu **Nandi**, Center of Excellence in Space Sciences India, IISER Kolkata, India — dnandi@iiserkol.ac.in

Nicholas J. **Nelson**, California State University, Chico, CA, USA — njnelson@csuchico.edu

Alexandre Jose **Oliveira e Silva**, Universidade do Vale do Paraiba, Brazil — ajoliveiraesilva@gmail.com

Rachel **Osten**, Space Telescope Science Institute, Baltimore, MD, USA — osten@stsci.edu

Pascal **Petit**, Toulouse University, France — ppetit@irap.omp.eu

Katja **Poppenhaeger**, Queen's University Belfast, UK — k.poppenhaeger@qub.ac.uk

Gustavo F. **Porto Mello**, Obs. Valongo, Universidade Federal do Rio de Janeiro, Brazil — gustavo@astro.ufrj.br

Williamary **Portugal**, Space Geophysics Division, INPE, Brazil — williamary@uol.com.br

Andreas **Quirrenbach**, Landessternwarte Heildelberg, Germany — a.quirrenbach@lsw.uni-heidelberg.de

Rosa Alejandra **Reyes**, Universidad Nacional de Colombia, Colombia — rareyesq@unal.edu.co

Marlos **Rockenbach da Silva**, Space Geophysics Division, INPE, Brazil — marlosrs@gmail.com

Jenny Marcela **Rodriguez Gomez**, Space Geophysics Division, INPE, Brazil — jemfisi@gmail.com

Rachael **Roettendahl**, University of Stockholm, Sweden — rmroett@umich.edu

Christopher **Russell**, University of California at Los Angeles, CA, USA — ctrussel@igpp.ucla.edu

J.C. **Santos**, Universidade Tecnologica Federal do Parana, Brazil — jeansanth@utfpr.edu.br

Caius L. **Selhorst**, Universidade Cruzeiro do Sul, Brazil — caiuslucius@gmail.com

Alexander **Shapiro**, Max-Planck-Institut f.r Sonnensystemforschung, Germany — shapiroa@mps.mpg.de

Suzana S. A. **Silva**, Space Geophysics Division, INPE, Brazil — suzana.seas@gmail.com

Dirceu Yuri **Simplicio Netto**, CRAAM. Universidade Presbiteriana Mackenzie, Brazil — dirceuyuri@hotmail.com

Antoine **Strugarek**, Universite de Montreal, Canada — strugarek@astro.umontreal.ca

Jordi Tuneu **Serra**, CRAAM, Universidade Presbiteriana Mackenzie, Brazil — tuneu.jordi@gmail.com

Adriana **Valio**, CRAAM, Universidade Presbiteriana Mackenzie, Brazil — avalio@craam.mackenzie.br

Aline **Vidotto**, Trinity College Dublin, University of Dublin, Ireland — aline.vidotto@tcd.ie

Luis Eduardo **Vieira**, Space Geophysics Division, INPE, Brazil — luis.vieira@inpe.br

Carolina **Villarreal D'Angelo**, IATE-CONICET, Argentina — carolina.villarreal@oac.edu.ar

M.R. **Voelzke**, Universidade Cruzeiro do Sul, Brazil — mrvoelzke@hotmail.com

Maria **Weber**, University of Exeter, UK — weber@astro.ex.ac.uk

J.S. **Wolk**, Harvard-Smithsonian Center for Astrophysics, USA — swolk@cfa.harvard.edu

Rakesh **Yadav**, Harvard-Smithsonian Center for Astrophysics, USA — rakesh.yadav@cfa.harvard.edu

Louise **Yu**, Institut de Recherche en Astrophysique et Plan.tologie, Toulouse, France — louise.yu@irap.omp.eu

Bonnie **Zaire**, Universidade Federal de Minas Gerais, Brazil — zaire@fisica.ufmg.br

Marusa **Zerjal**, University of Ljubljana, Slovenia — marusa.zerjal@fmf.uni.lj.si

Living Around Active Stars
Proceedings IAU Symposium No. 328, 2016
D. Nandy, A. Valio & P. Petit, eds.

© International Astronomical Union 2017
doi:10.1017/S1743921317003957

The Puzzling Dynamos of Stars: Recent Progress With Global Numerical Simulations

Antoine Strugarek[1,2], Patrice Beaudoin[2], Paul Charbonneau[2] and Allan S. Brun[1]

[1] Laboratoire AIM Paris-Saclay,
CEA/Irfu Université Paris-Diderot CNRS/INSU,
F- 91191 Gif-sur-Yvette
email: `antoine.strugarek@cea.fr`
[2] Département de physique, Université de Montréal,
C.P. 6128 Succ. Centre-Ville, Montréal, QC H3C-3J7, Canada

Abstract. The origin of magnetic cycles in the Sun and other cool stars is one of the great theoretical challenge in stellar astrophysics that still resists our understanding. Ab-initio numerical simulations are today required to explore the extreme turbulent regime in which stars operate and sustain their large-scale, cyclic magnetic field. We report in this work on recent progresses made with high performance numerical simulations of global turbulent convective envelopes. We rapidly review previous prominent results from numerical simulations, and present for the first time a series of turbulent, global simulations producing regular magnetic cycles whose period varies systematically with the convective envelope parameters (rotation rate, convective luminosity). We find that the fundamentally non-linear character of the dynamo simulated in this work leads the magnetic cycle period to be inversely proportional to the Rossby number. These results promote an original interpretation of stellar magnetic cycles, and could help reconcile the cyclic behaviour of the Sun and other solar-type stars.

Keywords. Sun: magnetic fields, Sun: interior, stars: magnetic fields, (magnetohydrodynamics:) MHD

1. Introduction

Cool stars such as the Sun possess a convective envelope that is thought to be at the origin of their internal large-scale flows (differential rotation, see Brun & Toomre 2002, meridional circulation, see Featherstone & Miesch 2015) and dynamic magnetic fields (see, e.g. Brun *et al.* 2015). Many possible mechanisms sustaining a dynamo in a convective envelope have been invoked in the literature (see, e.g. the review Charbonneau 2010), most of which rely on some parametrization of the magnetohydrodynamical turbulence that animates the convective layer. The magnetism of solar-like stars hence is a formidable theoretical challenge due to the extreme parameter regime in which interiors of stars operate. The magnetic field of a star furthermore plays a crucial role during its life, by shaping its wind and astrosphere (Cranmer 2012), determining its rotational braking (Réville *et al.* 2015), and even affecting its (exo)planets habitability (Lammer *et al.* 2009; Gallet *et al.* 2016).

In particular, the cyclic aspect of solar magnetism (and stellar magnetism, see e.g. Noyes *et al.* 1984; Baliunas *et al.* 1995; Saar & Brandenburg 1999; Bohm Vitense 2007; Egeland *et al.* 2015; Metcalfe *et al.* 2016) remains as of today one of the great mysteries of stellar astrophysics. Indeed, if the dynamo process sustaining the large-scale cyclic magnetic field of the Sun is rooted in the convective turbulence itself, how can a characteristic time-scale of eleven years emerge since all hydrodynamical time-scales are significantly different? In other words, what sets the magnetic cycle period of the Sun and of other cool stars? Observational constraints on the magnetic cycle of distant stars are difficult to acquire because they require repeated monitoring of stars over long (at least decadal) periods of time. As a result, only a few observational programs (see previously cited efforts) have as of today been able to provide very useful observations of magnetic cycles in stars different than the Sun. Numerical simulations of stars hence provide as of today promising laboratories to explore the possible drivers of cyclic magnetic activity in stars.

In the past decade, significant progress has been made in understanding the sustainment of large-scale magnetic fields in turbulent convection zones of stars thanks to global, 3D numerical simulations. Only a handful simulations, though, present today a cyclic behaviour. We quickly review the current state of the art for modelling magnetic cycles in stars with global turbulent simulations in Section 2 (the reader may find a more detailed review in the Section 2.3 of Brun *et al.* 2015). In Section 3 and 4 we will present for the first time a series of numerical simulations exhibiting a cyclic behaviour, with a period that systematically varies with the rotation and luminosity of the modelled star.

2. Cycles in global turbulent numerical simulations

Large-scale magnetic fields in turbulent global numerical simulations. The basis of 3D global modelling of stellar interior flows and magnetism was first developed in the pioneering work of Gilman & Miller (1981); Gilman (1983); Glatzmaier (1984, 1985); Brun *et al.* (2004). At that time, global scale magnetism was already achieved in turbulent simulations using enhanced diffusivity coefficients (which are still widely used today) to model the contribution of the scales unresolved by the numerical model. In those models the large-scale field was nonetheless less energetic than the small scale magnetic field, which is ubiquitously produced by the turbulent convective motions under the influence of rotation as long as the magnetic Reynolds number exceeds the critical onset of dynamo action. Many teams have since then attempted to carry out such simulations, mainly in the solar/stellar physics and planetary dynamo communities (a convective dynamo benchmark conducted by Jones *et al.* 2011). One very intriguing regime was in particular discovered by Brown *et al.* (2010), in which persistent wreaths of magnetic field are sustained near the bottom of the convection zone, inside the turbulent region itself.

Reversing magnetic fields in turbulent global numerical simulations. Many simulations then started to show quasi-regular magnetic field reversals, using the ASH code (Brown *et al.* 2011; Augustson *et al.* 2013; Nelson *et al.* 2013), the PENCIL code (Käpylä *et al.* 2013), the MagIC code (Gastine *et al.* 2012), or the PaRoDy code (see Schrinner *et al.* 2012, 2014). In addition, evidence of buoyantly rising magnetic wreaths was found by Nelson *et al.* (2011). Finally, an almost cyclic activity was also found in simulations of the fully convective M-star Proxima Centauri by Yadav *et al.* (2016). In all these simulations, the large-scale magnetic field is observed to invert quasi-regularly on a yearly timescale, over a period of several hundreds of rotation periods of the star. Nevertheless, the detailed mechanism setting the exact inversion timing remains to be clarified as of today.

Cyclic magnetic fields in turbulent global numerical simulations. Only a handful of global turbulent simulations successfully produced truly regular magnetic cycles. The first solar-like cycles were arguably obtained by Ghizaru *et al.* (2010), using the EULAG code with an implicit large-eddy simulation approach (see Section 3.1). These results were subsequently analyzed (Beaudoin *et al.* 2013; Lawson *et al.* 2015) and compared to classical mean-field theory (Racine *et al.* 2011; Simard *et al.* 2013; Beaudoin *et al.*

2016; Simard *et al.* 2016), revealing an $\alpha^2 - \Omega$ like behaviour with significant departure from isotropic mean field α and β tensors. In the meantime, Käpylä *et al.* (2012) also reported on cyclic solutions in global spherical wedges using the PENCIL code. In their solution, the cycle period was found to be much shorter (of the order of a few years), but presented a clear equatorial propagation of what would be equivalent to a solar activity band. In the simulations of Gastine *et al.* (2012), almost cyclic solutions are reported but exhibit a polar propagation and tend to show strong hemispheric decoupling. Finally, Augustson *et al.* (2015) also found a cyclic solution (also with a short cycle-period of a few years) using a hybrid dynamic-Smagorinski approach as a sub-grid scale model. They interestingly found a cyclic solution showing self-consistently a long period of minimal activity, that may relate to epochs like the so-called Maunder minimum of the Sun. So it is now clear that 3D global simulations can yield cyclic magnetic behaviour but such solutions are quite sensitive to the global parameters of the simulations, making it impossible to study their properties through an exploration of the parameter space and to deduce systematic trends.

3. An implicit large-eddy simulation of a convection zone exhibiting cyclic magnetism

3.1. *Numerical model*

The numerical simulations presented in this work are based on the hydrodynamical simulations presented in Strugarek *et al.* (2016). They consist in a spherical shell with a solar-like aspect ratio ($R_{\mathrm{bottom}} = 0.7 R_{\mathrm{top}}$) subject to a convective instability. The EULAG code (Smolarkiewicz & Charbonneau 2013) is used to solve the Lipps-Helmer set of ideal MHD equations, written in the stellar rotating frame $\mathbf{\Omega}_\star$ as

$$\nabla \cdot (\bar{\rho}\mathbf{u}) = 0 \,, \tag{3.1}$$

$$\mathrm{D}_t \mathbf{u} = -\nabla \left(\frac{p}{\bar{\rho}} \right) - \frac{S}{c_p}\mathbf{g} - 2\mathbf{\Omega}_\star \times \mathbf{u} \,, \tag{3.2}$$

$$\mathrm{D}_t S = -\left(\mathbf{u} \cdot \nabla \right) S_a - \frac{S}{\tau} \,, \tag{3.3}$$

$$\mathrm{D}_t \mathbf{B} = \left(\mathbf{B} \cdot \nabla \right) \mathbf{u} - \mathbf{B} \left(\nabla \cdot \mathbf{u} \right) \,. \tag{3.4}$$

In these anelastic equations the variables are perturbed quantities around a background isentropic state denoted with bars. In addition, an ambient hydrostatic state, denoted with the subscript $_a$, was substracted from the equations. D_t is the material derivative. We recall that we use standard notation for the basic fluid quantities, *i.e.* $\mathbf{u}$ is the fluid velocity, ρ its density, p its pressure, S its specific entropy, $\mathbf{B}$ the magnetic field, and $c_p = 3.4\,10^8$ erg/g/K the specific heat at constant pressure. The background state is chosen to cover 3.22 density scale heights, which allows moderately small-scale convective structures at the top of the convective layer while retaining non-negligible stratification throughout the domain. The grid is held constant in all the simulations presented here, and is chosen relatively coarse ($51 \times 64 \times 128$) to allow the exploration of decadal time-scale phenomena over a parameter space exploration.

As we consider a spherical, fully convective shell, boundary conditions are extremely important. The top and bottom boundaries are classically assumed to be stress-free walls. The magnetic boundary conditions are more delicate and were reported in equivalent setups (*e.g.* Brown *et al.* 2010) to significantly change the simulated dynamo. Here we use a radial field top boundary condition, to mimic the connection of the top of our convective envelope to a chromosphere and lower corona. At the bottom of our domain, we use a perfect conductor boundary condition to resemble the connection to an underlying conductive layer such as a stably stratified zone.

The convective instability is forced by the conjunction of the advection of the unstable ambient entropy profile S_a, and a Newtonian cooling term of characteristic timescale τ (for details, see Prusa *et al.* 2008; Smolarkiewicz & Charbonneau 2013). The ambient entropy profile S_a is defined in Strugarek *et al.* (2016) and is controlled by an entropy contrast ΔS throughout the modelled convective shell (in the model shown here, $\Delta S \in [8, 15] \times 10^3$ erg/g/K). The Newtonian cooling in Equation 3.3 damps entropy perturbations over the timescale τ which is always chosen to exceed the convective overturning time. It ensures that on long time-scales, the model mimics a stellar convection zone remaining in thermal equilibrium (*e.g.* Cossette *et al.* 2016).

The resulting convective turbulence organizes such as to transport heat outward and is characterized by a convective luminosity calculated *a posteriori* and defined as

$$L_c = 4\pi\bar{\rho}c_P \left\langle v_r T \right\rangle_{t,\varphi} , \qquad (3.5)$$

where T is the temperature perturbation and c_P the specific heat at constant pressure, and $\langle\rangle_{t,\varphi}$ stands for the average over time and the azimuthal angle φ. Due to the particular convection forcing in our setup, and the existence of radial wall boundaries, the convective luminosity vary with radius in our models. Here we choose to estimate the convective luminosity L_{bc} by averaging it over the $[0.75\,R_\odot, 0.8\,R_\odot]$, which is safely away from the lower boundary but close to the bottom part of the domain where the dynamo action primarily takes place (see below).

The EULAG code solves the ideal set of MHD equations, and as a result do not resolves any explicit dissipative process. Note that explicit dissipative process can be solved in the EULAG code, the results presented here nonetheless do not take them into account. The approach used here is a so-called *implicit large-eddy simulation*, in which the advective scheme *MPDATA* (Prusa *et al.* 2008; Smolarkiewicz & Charbonneau 2013) adds up the necessary numerical dissipation to ensure the stability of the numerical scheme. This results in a very time- and space-dependant effective dissipation that occurs only in strong gradient regions. This approach has the advantage that the largest scales in the domain are saturated and sustained by inviscid non-linear processes as in real stellar convection zones, while the numerical dissipation acts mostly at small-scale. This scheme was shown to be compatible with classical turbulent cascades in 3D isotropic and homogeneous turbulence (Domaradzki *et al.* 2003) and with standard laplacian viscosity and heat diffusion (at small scale) in spherical convective shells (Strugarek *et al.* 2016).

3.2. *Cyclic magnetic fields in turbulent convection zones*

We consider a set of 7 numerical simulations covering about a factor of 2 in rotation rate and 3 in convective luminosity. After a transient phase of exponential growth of the magnetic field, the magnetic energy saturates in all the models at about 10% of the total kinetic energy.

We display the time evolution of the energies in a subset of three of our models in Figure 1. The three simulations exhibit a cyclic magnetic energy (black, 'ME') and the cyclic behaviour is found in both the mean toroidal (magenta, 'TME') and fluctuating (yellow, 'FME') components of the magnetic energy. The three magnetic energies oscillate in phase, and in these simulations the fluctuating magnetic energy dominates over the mean toroidal energy. The same cycle is also found in the differential rotation energy (green, 'DRKE', see also the differential rotation pattern on the right panels) and in the total kinetic energy (blue, 'KE'). Epochs of maximum DRKE correlate very well with minima of magnetic energy, suggesting an energy beating between the magnetic

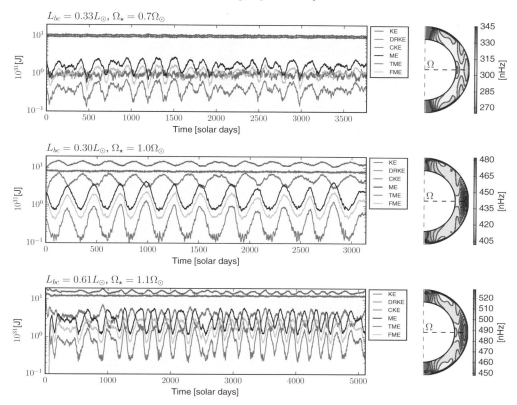

Figure 1. Magnetic cycles in three representative models. The left panels show energies as a function of time, integrated over the all convective envelope. The total kinetic energy (KE) is shown in blue, the differential rotation kinetic energy (DRKE) in green, and the convective kinetic energy (CKE) in red. The total magnetic energy (ME) is shown in black, the toroidal magnetic energy (TME) in magenta and the turbulent magnetic energy (FME) in yellow. The global parameters of the models (convective luminosity L_{bc}, rotation rate $\Omega_\star$) are indicated in each plot. The right panel show the differential rotation profile on the meridional plane, color-centered on the model rotation rate Ω.

and differential rotation energy reservoirs along the observed cycles. Supporting this interpretation, the convective kinetic energy (red, 'CKE') does not show any significant modulation with the magnetic cycle, which suggests a dynamo scenario in which energy is exchanged between the large-scale differential rotation and the small and large-scale magnetic energy, and points to a significant back-reaction of the Lorentz force on the balance sustaining the large-scale differential rotation profile (right panels in Figure 1). Indeed, when the magnetic energy decreases the differential rotation is able to grow back until the magnetic back-reaction sets in again. The differential rotation profile remains *solar-like* (fast equator, slow poles) in the whole set of simulations presented here, but its amplitude and detailed latitudinal profile change significantly in our simulation set. As a result, the dynamo mechanism at the heart of these simulations seems to be robust with respect to the detailed profile of the differential rotation.

The azimuthal component of the cyclic large-scale magnetic field is shown in Figure 2 in a time-latitude diagram at depth $r = 0.75\,R_\odot$, near the bottom of our domain. Three snapshots of the azimutal magnetic field on the meridional plane are shown, showing the reversal of the large-scale field. The azimuthal field is primarily located in the bottom half

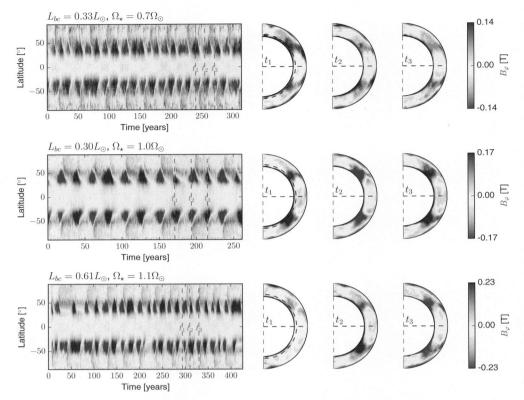

Figure 2. Azimuthal component of the mean magnetic field $\langle B_\varphi \rangle_\varphi$ in the same series of models as in Figure 1. $\langle B_\varphi \rangle_\varphi$ is shown in a time-latitude diagram at $r = 0.75\,R_\odot$ on the left panel. The three meridional plane snapshots in the right panels are taken at the time labeled by the dashed magenta lines.

of our convection zone and at mid-latitude, right outside of the inner tangent cylinder. A clear equatorial propagation of the azimuthal field is observed on the left panels, confined at mi-latitude near the maximum of the latitudinal shear. In all models, the magnetic field reaches a few tenths of a Tesla at the base of the convective envelope.

We readily see in Figure 2 that the large-scale azimuthal field is either symmetric (*i.e.* quadrupolar) or anti-symmetric (*i.e.* dipolar) with respect to the equator depending on the model and on the epoch of the model. We show in Figure 3 the dipolar/quadrupolar ratio at $R = 0.75\,R_\odot$ in the first model ($L_\star = 0.33\,L_\odot$, $\Omega = 0.7\,\Omega_\odot$). The magnetic cycle clearly appears (over-plotted in grey in Figure 3), on top of which the large-scale field oscillates between periods of predominantly dipolar and predominantly quadrupolar geometries, with no clear regularity (longer time integrations would be required with this simulation to confirm this lack or regularity). The beating between the two topologies suggests a coupling between the dynamo families in our models (McFadden *et al.* 1991; Gubbins & Zhang 1993; Knobloch *et al.* 1998), which is often a trace of a strongly non-linear dynamo regime. In both situations, the dominant polarity is maximized (resp. minimized) during epochs of maxima (resp. minima) of magnetic energy, which differs from the situation observed in the Sun (DeRosa *et al.* 2012).

The dynamo acting in our simulations operates for differential rotations of various strength and profile (see the right panels of Figure 1). This warrants further

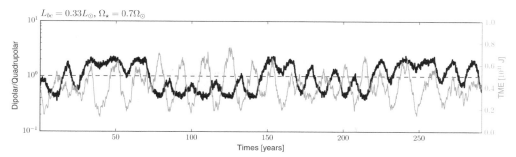

Figure 3. Dipolar/quadrupolar ratio (thick black line) as a function of time at $r = 0.75\,R_\odot$, for the first models shown in Figures 1 and 2. The volume-integrated toroidal magnetic energy (TME) is overlaid as a thin grey line.

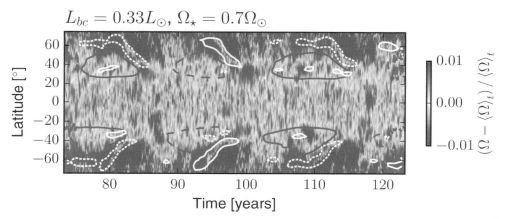

Figure 4. Toroidal field inversion in our modelled dynamo. The color map represents the temporal fluctuation of the differential rotation, defined as $\delta\Omega = \Omega - \langle\Omega\rangle_t$. Red tones denote an acceleration with respect to the mean differential rotation, while blue tones denotes a deceleration. The $[-0.1\,\mathrm{T}, 0.1\,\mathrm{T}]$ contours of $\langle B_\varphi\rangle_\varphi$ are overlaid in grey. Contours of the action of the differential rotation on the mean magnetic field are shown in white, and correlate very well with the cyclic destruction of $\langle B_\varphi\rangle_\varphi$. Positive contours are shown with solid lines, and negative contours with dashed lines.

investigation on the origin of the magnetic reversals in our models. We display in Figure 4 iso-contours of the azimuthal component of the mean magnetic field at 0.1 T in grey on a time-latitude diagram zoomed over a particular subset of cycles. We overplot the contribution of the differential rotation shear to the time-evolution of $\langle B_\varphi\rangle_\varphi$ as white contours (for all contours, plain line denote positive value and dashed lines negative values). We observe that the mean component of the induction associated with the mean azimuthal flow correlates very well with the destruction of the mean azimuthal magnetic field. The background colormap actually reveals the origin of the reversals. It shows the deviation from the mean differential rotation over time. Red threads originating at high latitudes and propagating poleward trace acceleration epochs in the differential rotation that trigger the polarity reversal of the mean magnetic field. These modulations of the differential rotation take their roots in the non-linear feedback of the magnetic field on the balance establishing the differential rotation through the Lorentz force. As result, the reversals observed in our simulations originate from a non-linear modulation of the differential rotation due to the presence of a large-scale magnetic field. Taken at face value, these considerations suggest a dynamo mechanism in which the cycle period, which is a tracer of the strength of dynamo action, is expected to decrease as the differential rotation weakens (since the non-linear magnetic feedback is then able to impact more efficiently the balance establishing the differential rotation). This is indeed what we observe, and we now turn to our full set of numerical simulations to determine what are the control parameters setting the cycle period in our non-linear, global dynamo.

4. Modulation of the cycle period in global dynamo simulations

We saw in Figure 2 that the cycle period of our simulated dynamos varies when the luminosity or the rotation rate of the model changes. In order to robustly estimate the cycle period in our models, we consider the azimuthally averaged azimuthal component of the magnetic field $b_\varphi(r, \theta, t) = \langle B_\varphi \rangle_{t,\varphi}$. At each (r, θ) point on the meridional plane, we compute the Fourier transform of b_φ. We automatically identify the peak in the Fourier spectrum, and calculate the width of the peak (defined as the width for which the Fourier spectrum decreases to 10% of its peak value) to be used as a proxy for an error bar. We then compute the probability density function of the cycle periods obtained for each (r, θ), and define the cycle period of the model as the peak in this distribution function. The error-bar is then also averaged over the (r, θ) points close to the peak cycle period.

We further define the fluid Rossby number in our models as

$$R_{of} = \frac{|\nabla \times \mathbf{u}|}{2\Omega_\star}. \tag{4.1}$$

We calculate the Rossby number R_{of} as function of radius and latitude, and define the average Rossby number R_o inside the white wedge defined in the right panel of Figure 5, which corresponds to the region where the dynamo action is strong (*i.e.* near the bottom of the convection zone, excluding the polar regions). We show on the left panel the cycle period realized in each of our seven models as a function of the Rossby number (in log scale). The cycle period is found to be inversely proportional to the Rossby number, $P_{\rm cyc} \propto R_o^{-1.1\pm0.2}$. This result may seem counter-intuitive, as standard and fully linear α-Ω dynamo models are expected to exhibit a cycle period which is directly proportional to the Rossby number (*e.g.* Noyes *et al.* 1984). This apparent contradiction is not so surprising, as we showed in Figure 4 that the dynamo operating in our set of simulations is different from the standard α-Ω dynamo: temporal fluctuations of the large-scale differential rotation associated to the back-reaction from the Lorentz force play a dominant role in the polarity reversal. The dynamo at stake here is consequently fundamentally non-linear (*i.e.* non kinematic, as in Augustson *et al.* 2015), and subsequently exhibits a different scaling law compared to kinematic models of stellar dynamos. We are currently investigating in more details the exact dynamo process presented here, and in particular the dynamics of the poloidal field reversal, the role played by the small scale turbulence, and the effect of the cyclic modulations on the meridional circulation pattern (see also Beaudoin *et al.* 2013).

5. Conclusions

In this work we have shown for the very first time a series of global, 3D turbulent numerical simulations of stellar convective envelopes producing regular stellar magnetic cycles, which period varies with the rotation rate of the star and its convective luminosity. The dynamo sustaining the cyclic field is found to be fully non-linear and be able to generate large scale fields of both dipolar and quadrupolar families. The non-linearity of

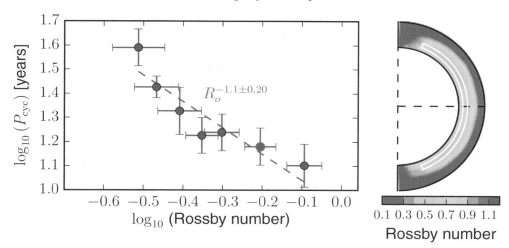

Figure 5. Left panel: cycle period as a function of the Rossby number in log scale. Right panel: Rossby number in the meridional plane of the model shown in Figures 3 and 4 ($L_{bc} = 0.33L_\odot$, $\Omega_\star = 0.7\Omega_\odot$) and highlighted in red in the left panel. The white wedge labels the region in which the representative Rossby number of the simulation is evaluated.

the dynamo strikingly appears in the predominant role played by the temporal fluctuations of the differential rotation driven by the feedback of the magnetic field through the Lorentz force.

The boundary conditions, especially the lower magnetic boundary, play a dominant role in the dynamo state achieved in the simulations. Here we considered reasonable choices given the simplicity of our numerical setup. We also tested changing the lower magnetic boundary condition from perfect conductor to radial field, which completely changed the dynamo state and shut down the cycle. It seems that being able to store horizontal fields in an underlying region is henceforth important in developing a magnetic cycle, albeit the detailed fate of such a stored field appear not to matter in our set simulations.

We also explored various density contrasts N_ρ (not shown here) from 2 to 4, which effectively changes the Rossby number of the simulated convective envelope. All these simulations fall on the same trend identified in Figure 5, namely the cycle period is found to be robustly inversely proportional to the Rossby number when changing the rotation rate, the luminosity, or the stratification of the modelled convection zone.

Several aspects require further investigations which are currently being pursued. Simulations at higher resolution are currently underway, in order to assess the robustness of the dynamo mechanism identified in our simulations with respect to the dissipative properties of the plasma. Furthermore, the complete dynamo loop still eludes our understanding and dedicated analysis are still required to fully assess the mode of operation of the dynamo in the simulations. Finally, we are also investigating twin simulations where a stably stratified zone is added to the bottom of the convective envelope to assess the importance of an underlying radiative zone in setting the cycle period (*e.g.* Browning *et al.* 2006; Lawson *et al.* 2015; Guerrero *et al.* 2016). Ultimately, we aim at reproducing the benchmarking exercise carried out in (see Strugarek *et al.* 2016) for the MHD models presented here, and evaluate the robustness of our dynamo mechanism for various small and sub-grid scale behaviours. The dynamo presented in this work opens new possibilities in interpreting the magnetic cycle of solar-like stars (Bohm Vitense 2007; Metcalfe *et al.*

2016), and could help reinstate the Sun as a truly typical solar-type star (Salabert *et al.* 2016).

References

Augustson, K., Brun, A. S., Miesch, M., & Toomre, J. 2015, *ApJ*, 809, 149

Augustson, K. C., Brun, A. S., & Toomre, J. 2013, *ApJ*, 777, 153

Baliunas, S. L., Donahue, R. A., Soon, W. H., *et al.* 1995, *ApJ*, 438, 269

Beaudoin, P., Charbonneau, P., Racine, E., & Smolarkiewicz, P. K. 2013, *Sol. Phys.*, 282, 335

Beaudoin, P., Simard, C., Cossette, J.-F., & Charbonneau, P. 2016, *ApJ*, 826, 138

Bohm Vitense, E. 2007, *ApJ*, 657, 486

Brown, B. P., Browning, M. K., Brun, A. S., Miesch, M. S., & Toomre, J. 2010, *ApJ*, 711, 424

Brown, B. P., Miesch, M. S., Browning, M. K., Brun, A. S., & Toomre, J. 2011, *ApJ*, 731, 69

Browning, M. K., Miesch, M. S., Brun, A. S., & Toomre, J. 2006, *ApJ*, 648, L157

Brun, A. S., Garcia, R. A., Houdek, G., Nandy, D., & Pinsonneault, M. 2015, *Space Sci Rev*, 196, 303

Brun, A. S., Miesch, M. S., & Toomre, J. 2004, *ApJ*, 614, 1073

Brun, A. S. & Toomre, J. 2002, *ApJ*, 570, 865

Charbonneau, P. 2010, *LRSP*, 7, 3

Cossette, J.-F., Charbonneau, P., Smolarkiewicz, P. K., & Rast, M. P. 2016, Submitted to ApJ

Cranmer, S. R. 2012, *Space Sci Rev*, 172, 145

DeRosa, M. L., Brun, A. S., & Hoeksema, J. T. 2012, *ApJ*, 757, 96

Domaradzki, J. A., Xiao, Z., & Smolarkiewicz, P. K. 2003, *PoF, 15, 3890*

Egeland, R., Metcalfe, T. S., Hall, J. C., & Henry, G. W. 2015, *ApJ*, 812, 12

Featherstone, N. A. & Miesch, M. S. 2015, *ApJ*, 804, 67

Gallet, F., Charbonnel, C., Amard, L., *et al.* 2016, To appear in *A&A*, 1608

Gastine, T., Duarte, L., & Wicht, J. 2012, *A&A*, 546, 19

Ghizaru, M., Charbonneau, P., & Smolarkiewicz, P. K. 2010, *ApJL*, 715, L133

Gilman, P. A. 1983, *ApJ Supp. Series*, 53, 243

Gilman, P. A. & Miller, J. 1981, *ApJ Supp. Series*, 46, 211

Glatzmaier, G. A. 1984, J. Comp. Phys., 55, 461

—. 1985, *ApJ*, 291, 300

Gubbins, D. & Zhang, K. 1993, *Physics of the Earth and Planetary Interiors, 75, 225*

Guerrero, G., Smolarkiewicz, P. K., de Gouveia Dal Pino, E. M., Kosovichev, A. G., & Mansour, N. N. 2016, *ApJ*, 819, 104

Jones, C. A., Boronski, P., Brun, A. S., *et al.* 2011, *Icarus, 216, 120*

Käpylä, P. J., Mantere, M. J., & Brandenburg, A. 2012, *ApJ*, 755, L22

Käpylä, P. J., Mantere, M. J., Cole, E., Warnecke, J., & Brandenburg, A. 2013, *ApJ*, 778, 41

Knobloch, E., Tobias, S. M., & Weiss, N. O. 1998, *MNRAS*, 297, 1123

Lammer, H., Bredehöft, J. H., Coustenis, A., *et al.* 2009, *The Astron. and Astrophys. Rev.*, 17, 181

Lawson, N., Strugarek, A., & Charbonneau, P. 2015, *ApJ*, 813, 95

McFadden, P. L., Merrill, R. T., McElhinny, M. W., & Lee, S. 1991, *J. of Geophys. Res.*, 96, 3923

Metcalfe, T. S., Egeland, R., & van Saders, J. 2016, *ApJL*, 826, L2

Nelson, N. J., Brown, B. P., Brun, A. S., Miesch, M. S., & Toomre, J. 2011, *ApJL*, 739, L38

—. 2013, *ApJ*, 762, 73

Noyes, R. W., Weiss, N. O., & Vaughan, A. H. 1984, *ApJ*, 287, 769

Prusa, J. M., Smolarkiewicz, P. K., & Wyszogrodzki, A. A. 2008, *Computers & Fluids, 37, 1193*

Racine, É., Charbonneau, P., Ghizaru, M., Bouchat, A., & Smolarkiewicz, P. K. 2011, *ApJ*, 735, 46

Réville, V., Brun, A. S., Matt, S. P., Strugarek, A., & Pinto, R. F. 2015, *ApJ*, 798, 116

Saar, S. H. & Brandenburg, A. 1999, *ApJ*, 524, 295

Salabert, D., Garcia, R. A., Beck, P. G., *et al.* 2016, *A&A*, 596, A31

Schrinner, M., Petitdemange, L., & Dormy, E. 2012, *ApJ*, 752, 121

Schrinner, M., Petitdemange, L., Raynaud, R., & Dormy, E. 2014, *A&A*, 564, A78

Simard, C., Charbonneau, P., & Bouchat, A. 2013, *ApJ*, 768, 16

Simard, C., Charbonneau, P., & Dubé, C. 2016, *Adv. Spa. Res.*, 58, 1522

Smolarkiewicz, P. K. & Charbonneau, P. 2013, *J. Comp. Phys.*, *236, 608*

Strugarek, A., Beaudoin, P., Brun, A. S., *et al.* 2016, *Adv. Spa. Res.*, 58, 1538

Yadav, R. K., Christensen, U. R., Wolk, S. J., & Poppenhaeger, K. 2016, To appear in *A&A*, 1610.02721

Living Around Active Stars
Proceedings IAU Symposium No. 328, 2016
D. Nandy, A. Valio & P. Petit, eds.

© International Astronomical Union 2017
doi:10.1017/S1743921317004124

On the connections between solar and stellar dynamo models

Laurène Jouve[1,2] and Rohit Kumar[1,2]

[1]Université de Toulouse, UPS-OMP, Institut de Recherche en Astrophysique et Planétologie, 31028 Toulouse Cedex 4, France

[2]CNRS, Institut de Recherche en Astrophysique et Planétologie, 14 avenue Edouard Belin, 31400 Toulouse, France

Abstract. We here discuss the various dynamo models which have been designed to explain the generation and evolution of large-scale magnetic fields in stars. We focus on the models that have been applied to the Sun and can be tested for other solar-type stars now that modern observational techniques provide us with detailed stellar magnetic field observations. Mean-field flux-transport dynamo models have been developed for decades to explain the solar cycle and applications to more rapidly-rotating stars are discussed. Tremendous recent progress has been made on 3D global convective dynamo models. They do not however for now produce regular flux emergence that could be responsible for surface active regions and questions about the role of these active regions in the dynamo mechanism are still difficult to address with such models. We finally discuss 3D kinematic dynamo models which could constitute a promising combined approach, in which data assimilation could be applied.

Keywords. dynamo, magnetic fields, convection, rotation, numerical simulations

1. Introduction

The magnetic activity of our Sun has been monitored for centuries through the recording of the number of spots regularly emerging at the solar surface. The first telescope observations date back from the beginning of the 17th century when the attention of famous astronomers like Galileo and Thomas Harriott started to be drawn by those spots present at the surface of our star. It was only much later that those spots were associated with the presence of a magnetic field with a Zeeman signature discovered in a sunspot by Hale(1908). We now know that the solar magnetic field is not only responsible for the periodic appearance of sunspots at the surface but is also the triggering mechanism for powerful flares and coronal mass ejections that may strongly interact with our own terrestrial magnetosphere. Understanding the various consequences of the solar magnetic field requires to investigate its origins. Quite rapidly, the dynamo mechanism was invoked to be the process through which the Sun would maintain its magnetic field. A need to develop models and later numerical simulations appeared inevitable, to fully understand how the plasma flows would organize themselves in the solar interior to produce a magnetic field with a 22-yr period and all the associated phenomena at the solar photosphere and higher in its atmosphere. Two different approaches were adopted for the simulation of the solar dynamo, as we will develop later: mean-field dynamo models (Moffatt(1978); Krause & Rädler(1980)) which deal only with the large-scale magnetic field, assuming some parameterization of the underlying small-scale turbulence and magnetism and 3D global models which solve the full set of magnetohydrodynamical (MHD) equations, allowing to self-consistently produce the flow and magnetic field structures which will non-linearly interact (see reviews of Miesch & Toomre(2009) and Brun *et al.*(2015)).

Various models and associated numerical simulations have thus been developed to explain the characteristics of the solar magnetic field, with different levels of success (see review by Charbonneau(2014)). Despite all theses efforts, the community still does not have at its disposal a global model of the Sun which reproduces self-consistently the cyclic appearance of sunspots at the solar surface, showing the extraordinary complexity of our nearest star and the difficulties to model it accurately. Tremendous progress has been made recently on the observations of magnetic field on stars other than the Sun. It was indeed known from observations of the emissions in Ca lines H and K at the Mount Wilson Observatory that some stars possess chromospheric activity cycles on time-scales of years to decades (Baliunas *et al.*(1995)). Even if the relationship with a magnetic cycle is not necessarily direct (See *et al.*(2016)), it implied that other stars could also exhibit cyclic reversals of their magnetic field. Zeeman-Doppler Imaging (ZDI) now provides a way to monitor the topology and intensity of stellar magnetic fields in time. This tomographic technique is capable of reconstructing large-scale magnetic field topologies at stellar surfaces by inverting a series of spectropolarimetric observations (Donati & Brown(1997)). This technique revealed fascinating features about the magnetism of cool stars (Folsom *et al.*(2016), Morin *et al.*(2010)). It was for example found that fully convective stars appear to mostly host strong dipolar fields (Gregory *et al.*(2012)) or that rapidly rotating partly convective stars tend to produce strong toroidal structures (Morin *et al.*(2010), Petit *et al.*(2008)). Repeated observations of individual targets have now even revealed polarity reversals (Fares *et al.*(2009), Morgenthaler *et al.*(2011), Boro Saikia *et al.*(2016)), suggesting the presence of magnetic cycles.

We are thus now reaching a point where models which have been applied to the Sun for years or decades should now be tested on other cool stars and compared to stellar observations. We discuss the connections between solar and stellar dynamo models in 4 different sections. In section 1, 2D mean field dynamo models are discussed. In section 2, we rapidly review the great recent progress of 3D global dynamo models. In the last 2 sections, we focus on magnetic flux emergence: section 3 is dedicated to the detailed simulations of the particular process of flux emergence through the stellar interior and the production of spots and section 4 discusses the use of a combined approach: 3D kinematic dynamo models.

2. Flux transport dynamo models : applying the solar paradigm to stellar dynamos

2.1. *The Babcock-Leighton dynamo model applied to the Sun*

The past few decades have seen the advent of multidimensional numerical simulations to better understand the generation and the intricate non-linear evolution of the solar magnetic field. In order to model and understand the large-scale magnetic field, a useful approach has been to make use of the mean field dynamo theory. Among the various mean field dynamo models, the Babcock-Leighton (Babcock(1961), Leighton(1969)) flux transport dynamo models have recently been massively applied to the Sun. In the Babcock-Leighton (BL) model, the toroidal magnetic field owes its origin to the differential rotation at play in the stellar convection zone, while the poloidal field originates from the decay of active regions popping up at the stellar surface with a particular field strength and tilt angle. If we then add a large-scale meridional circulation, whose role is to advect the magnetic field concentrations inside the convection zone, the model is called a flux-transport model. This model demonstrated great success at reproducing some solar observations such as the 11-yr cycle, the mid-latitude activity belt, the

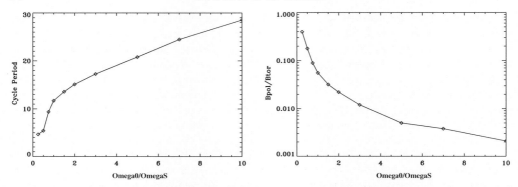

Figure 1. Period of the dynamo cycle (left panel) and poloidal to toroidal field ratio (right) as a function of the stellar rotation rate, normalized to the solar rotation rate. From simulations of Jouve *et al.*(2010).

phase relationship between toroidal and poloidal fields or the equatorward propagation of sunspot emergence (see Living Review by Charbonneau(2005)). These models even start to be used now to give tentative predictions of the next solar cycle, using data assimilation techniques which have been commonly used in meteorology for decades (Jouve *et al.*(2011), Dikpati & Anderson(2014), Hung *et al.*(2015)). It is then legitimate to ask if such models can be applied to other stars and lead to the same agreement with available observations. In particular, the observations seem to indicate that rapidly rotating solar-type stars tend to possess shorter magnetic cycles and exhibit stronger toroidal field components. Such a study was performed by Jouve *et al.*(2010) and we here summarize their findings.

2.2. *The Babcock-Leighton dynamo model applied to the rapidly-rotating stars*

The peculiarity of flux-transport models is that they indeed produce a magnetic field regularly reversing its polarity, with a cycle period extremely sensitive to the meridional flow amplitude v_0. Dikpati & Charbonneau(1999) or Jouve & Brun(2007) report scalings such as $P_{cyc} \propto v_0^{-0.83}$. It is thus necessary to have a hint of the amplitude of the meridional circulation in stars other than the Sun, which is a difficult problem. Indeed, already in the Sun are the characteristics of the meridional flow poorly constrained by helioseismology (see review by Gizon *et al.*(2010)). In other stars, this flow is not observed at all. However, 3D MHD numerical simulations of rapidly rotating stars do exist, showing that the meridional flow amplitude decreases with increasing rotation rate (Ballot *et al.*(2007), Brown *et al.*(2008), Augustson *et al.*(2012)) with a typical scaling such that $v_0 \propto \Omega_0^{-0.45}$. The same 3D numerical simulations also tend to show that the meridional flow becomes more and more multicellular as the rotation rate increases.

Applying BL flux-transport dynamo models with faster rotation and thus slower meridional circulation, Jouve *et al.*(2010) thus found that the magnetic cycle period in those stars should be much longer than suggested by the observations. Figure 1 illustrates those results by showing the period of the mean-field dynamo cycle with respect to the stellar rotation rate (left panel). The period is clearly increased when the rotation is increased because of the decrease in amplitude of the meridional flow speed. We note however that more toroidal field is produced compared to the poloidal field when the rotation is increased (right panel), in agreement with the spectropolarimetric observations of Petit *et al.*(2008) for example. It is however still possible to reconcile the models with observations by reducing the dependency of the magnetic cycle period with respect to v_0, for example, by invoking other transport processes such as turbulent diffusion or magnetic pumping (Guerrero & de Gouveia Dal Pino(2008), Do Cao & Brun(2011), Hazra *et al.*(2014)). Nevertheless, applying a model well-calibrated for the Sun to other stars here proved this model had to be severely modified to fit with stellar observations.

3. 3D global MHD models

3.1. *Hydrodynamical models*

As already stated above, the full set of MHD equations is solved in 3D MHD global models in spherical geometry. Those models are then necessarily much more costly than 2D mean-field models, but have the decisive advantage of self-consistently computing the flows and magnetic fields which will non-linearly interact. Tremendous progress has been done in the past decade on these 3D models and several properties about stellar convection, large-scale flows and dynamos have been found to be quite robust in the various simulations performed with such models (see review in Brun *et al.*(2015)). For instance, they all show that the differential rotation profile is directly linked to the Rossby number of the simulation, which is a measure of the importance of the inertia term compared to the Coriolis term in the Navier-Stokes equation. Indeed, it was found by several authors that anti-solar differential rotation state (slow equator, fast poles) occur at large Rossby number whereas solar-like differential rotation state (fast equator, slow poles) occurs at low Rossby number (see synthesis of large number of numerical simulations by Gastine *et al.*(2014)). Those calculations also agree on the fact that fast rotation (i.e. low Rossby numbers) implies a decrease in amplitude of the meridional circulation (as stated above) and a more complex structure, with several circulation cells appearing both in latitude and radius (Featherstone & Miesch(2015)). This again may have strong consequences on flux-transport dynamo models (Jouve & Brun(2007)).

3.2. *Magnetic models of low-mass stars*

As far as magnetic fields are concerned in those 3D MHD models, robust features have also been recovered by various groups. In particular, a wide variety of dynamo behaviours are found, from steady to irregular to well-defined cyclic magnetic activity. For low-mass stars (spectral type M), numerical simulations tend to demonstrate the ordering role of the Coriolis force, also seen in planetary dynamos (Christensen & Aubert(2006)). More specifically, when the Rossby number is increased, the magnetic field switches from being mostly dipolar to mostly multipolar. However, it has also been found that the low Rossby number regime could maintain both a dipolar solution and a multipolar solution depending on the initial magnetic conditions. This interesting bistability was also seen in observations (Morin *et al.*(2010)) where two stars with very similar rotation rates and masses (thus probably similar Rossby numbers) exhibit very different magnetic fields (strong and dipolar for one and weak and multipolar for the other). However, with a stronger stratification, this bistable behaviour seems to disappear in simulations. More computations are thus needed to further investigate this property. For instance, a recent simulation of a fully convective star by Yadav *et al.*(2015) with a reasonable degree of stratification (a density ratio of 150) was shown to possess both large-scale (mostly dipolar) and small-scale magnetic fields. A ZDI reconstruction was then applied to the simulation to see how well this analysis technique was able to recover the magnetic field content. As expected, the large-scale strong polar spot was perfectly recovered but not the smaller-scale features, which represent most of the magnetic flux in the simulation.

3.3. *Magnetic models of solar-type stars*

Simulations of solar-type stars with high rotation rates were also performed, showing strong belts of toroidal field in the convection zone which could start to undergo cyclic reversals as the level of turbulence is increased (Brown *et al.*(2011), Käpylä *et al.*(2013)). Some Maunder minima-like periods were even found in some simulations of F stars (Augustson *et al.*(2013)). It is still not entirely clear in those models possessing cyclic reversals what sets the cycle period. From published results, the meridional circulation amplitude does not seem to play a key role in establishing the time scale for the magnetic cycle, contrary to what is assumed in a Babcock-Leighton flux-transport dynamo model. However, those 3D dynamos do not produce spots at their surfaces and may thus be difficult to reconcile with BL models. We will come back on this in the next section.

Most of the simulations cited above do not possess a tachocline and a stable layer beneath. Only recently have some simulations been performed considering a tachocline and comparisons with convective shells having similar properties (Guerrero *et al.*(2016)). It is reported that a tachocline helps to organise the magnetic field by building strong concentrations of large-scale field. However, the influence of the tachocline on the cyclic behaviour of the solution still needs to be clarified.

4. Flux emergence : what properties of solar flux emergence coud be applied to other stars ?

4.1. *Spot formation in 3D global models*

In both mean-field dynamo models and global MHD models discussed above, the process of magnetic flux emergence through the stellar convection zone is crucial. In the Sun, it is the strong toroidal structures built at the base of the convection zone which are assumed to be unstable to a buoyancy instability and rise through the convection zone to produce sunspots (Parker(1955)). This particular step is crucial for BL models since the source of poloidal field is directly linked to the presence of active regions. In 3D models, the strong toroidal structures built in rapidly-rotating stars can become buoyant (Nelson *et al.*(2013), Fan & Fang(2014)) but rarely rises all the way to the top of the computational domain and those models consequently do not produce spots. It is thus still an open question if we can really rely on *spotless* dynamo models to reproduce what could happen in stars or in the Sun. The particular step of flux emergence being potentially important for the whole dynamo mechanism, detailed numerical simulations of such a process are thus needed. It has to be noted that other theories exist which do not rely so strongly on the presence of strong toroidal structures built in the tachocline and then becoming unstable. Some authors (Stein & Nordlund(2012), Brandenburg *et al.*(2013)) have argued that local flux concentrations by convective motions or by instabilities appearing in very strongly stratified zones could also lead to the formation of active regions in the Sun. We here concentrate only on the first picture of flux tubes rising from the base of the convection zones where they are produced, to the surface where they emerge as spots.

4.2. *Simulations of flux emergence in the Sun and possible applications to other stars*

Numerous numerical simulations of flux emergence have been performed for the Sun for which detailed observations of active region formation and evolution exist (see Living Review by Fan(2004)). An illustration of a numerical simulation of a buoyant loop rising in a convective shell is shown in Fig. 2. These simulations are able to reproduce several features of solar active regions: their morphology, their tilt angle due both to the twist of

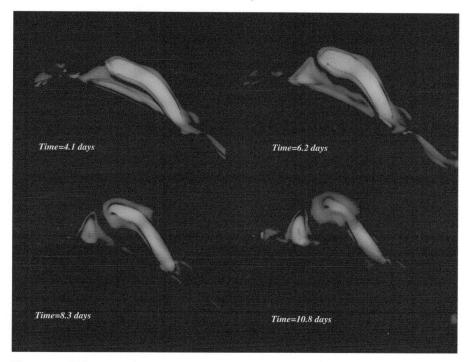

Figure 2. Volume rendering of the toroidal magnetic field of a loop rising through a convective layer. From simulations of Jouve *et al.*(2013).

the field lines and the Coriolis force acting on the rising tubes, the asymmetry between the leading and trailing spots or the interplay with the local convective flows. However, very few investigations have been conducted on similar processes of flux emergence in other stars. We note that thin flux tube calculations have been conducted by for giants (Holzwarth & Schüssler(2001)) and rapidly rotating stars (Holzwarth *et al.*(2006)), and simulations of thin flux tubes evolving in a fully convective star performed recently by Weber & Browning(2016). In this last article, they were particularly interested in the latitude of emergence of starspots, which they found to be strongly related to the stellar rotation and to the thermodynamical characteristics of their thin flux tubes.

Indications exist today of spots on other stars, with various degrees of surface coverage and magnetic fluxes (see review by Berdyugina(2005)). These properties have strong implications for potential eruptive activity on those stars and consequences on surrounding planets. Some properties found in simulations of large-scale flux emergence in the Sun could be easily applied to other stars. In particular, the rise trajectory of field concentrations from the base of the convection zone to the surface is strongly influenced by the Coriolis force, with a tendency of flux tubes in a fast rotating environment to rise parallel to the rotation axis and thus emerge at high latitudes. We thus expect rapidly rotating stars to exhibit spots at high latitude, which seem to be in reasonable agreement with observations. The typical size of active regions is then another question which could be addressed through numerical simulations of flux emergence in a convective domain. Is this size determined by the mean size of convective cells (which is different in stars rotating at different rates) or by the typical length-scale of the buoyancy instability, this question remains to be answered in detail.

5. A combined approach: 3D kinematic dynamo models

5.1. *Modelling buoyancy in kinematic models*

The Babcock-Leighton mechanism relies on the decay of active regions to reverse the polar field. In such kinds of models, the production of bipolar magnetic regions (BMR) at the stellar surface is thus essential to sustain dynamo action. This mechanism is supported by various solar observations: firstly, BMR are seen to continuously emerge and diffuse at the solar surface with a sufficient flux to be able to reverse the polar field. Secondly, correlations exist between the strength of the BL mechanism in one cycle and the amplitude (i.e. the number of sunspots) of the next cycle (Dasi-Espuig *et al.*(2010)). However, as stated before, the 3D global models exhibiting dynamo cycles do not produce well-defined BMRs and thus can not address the question of the role of the BL mechanism in the full solar and stellar dynamo loop. Motivated by solar observations and lack of self-consistent models to produce spots at the solar/stellar surface, some research groups started to develop an intermediate approach between the 2D BL flux-transport dynamo models which take for granted that the BL mechanism is the main player in the dynamo loop and 3D global models which do not capture the spot-producing mechanism. In this type of models, called 3D kinematic dynamo models, a velocity field is prescribed (as in 2D mean-field kinematic models) and a "buoyancy algorithm" is implemented to extract toroidal flux concentrations produced at the base of the convection zone and to translate them into BMRs at the surface. Two different models were developed so far, differing in this "buoyancy algorithm": Miesch & Dikpati(2014) and then Miesch & Teweldebirhan(2016) use a version of the "double-ring" algorithm (Durney(1997), Nandy & Choudhuri(2001)) to place BMRs at the solar surface in response to the dynamo-generated field at the base of the convection zone. Yeates & Muñoz-Jaramillo(2013) use a more self-consistent model in which an additional velocity is applied to the toroidal structures to make them rise to the surface.

5.2. *Promising models for data assimilation?*

Kumar & Jouve (in prep.) are currently building a model similar to the one of Yeates & Muñoz-Jaramillo(2013), by adapting the pseudo-spectral code MagIC (Wicht(2002)). For their simulation, they use an initial magnetic field which has two components: an equatorially antisymmetric strong toroidal field at the base of the convection zone, and a relatively weaker large-scale poloidal field. The magnetic buoyancy along with a helical flow give rise to tilted BMRs at the outer surface; eight such BMRs are illustrated in Figure 3(a). Note that the tilt angle is large for the BMRs at higher latitudes. Later, the decay and dispersal of these tilted BMRs lead to the polarity reversal of the large-scale polar field (see Figure 3(b)). These models are very appealing since they provide a way to directly test the ability of BMRs to reverse the polar fields, taking into account the main ingredients thought to be important in the dynamo mechanism: the meridional circulation and magnetic diffusion at the solar surface and the building of toroidal flux at the base of the convection zone by the differential rotation. Moreover, several prescriptions from 3D models of flux emergence can be easily reintroduced in these models, such as the field strength-dependent rise time of the flux tubes or the latitude of emergence which will be modified when faster rotation is considered. Such models will thus be able to be designed not only for the Sun but for other stars which produce spots. Finally, those models will be very well-adapted to the application of data assimilation techniques to predict the future solar activity. Data assimilation has been used for decades for weather forecasting on Earth. The idea is to combine time-dependent models and data available over a certain time interval so that the misfit between the outputs of the model and

$B_r(r = 0.95 R_\odot)$ (a)

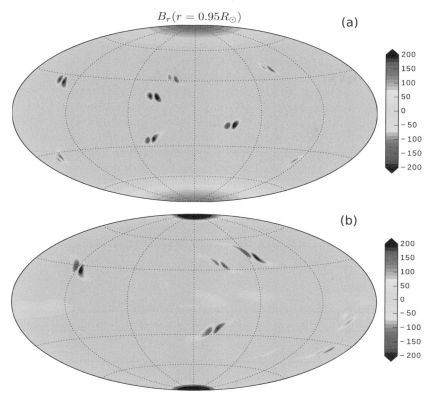

(b)

Figure 3. Snapshots of the non-dimensional radial magnetic field (B_r) at the surface corresponding to the radius $r = 0.95 R_\odot$ showing (a) the initial large-scale poloidal field along with eight tilted BMRs, and (b) after the polarity reversal, the poloidal field and BMRs of the next cycle (note the polarity switch). From Kumar & Jouve (in prep.).

the observational data is minimized. After this "learning phase", the model is optimized to fit previous observations and is ready to provide forecast for the future behavior of the system. Attempts to apply data assimilation to solar physics have already been performed in 2D mean-field dynamo models (Hung *et al.*(2015), Dikpati *et al.*(2016) for the most recent) but only on synthetic data. The direct comparisons between BMRs produced in 3D kinematic models and solar active regions, between surface observations of the meridional flow and the one used in the simulation and between the observed and simulated polar flux could be very promising for the application of data assimilation in such models. This will be the subject to future work.

6. Conclusion

We rapidly reviewed here some features of dynamo models used for the Sun and the possibilities to apply such models to other stars. We are indeed reaching a fascinating time when detailed observations of magnetic fields on stars other than the Sun become available and where confronting solar models to stellar observations can be performed. Some attempts have been made to apply Babcock-Leighton flux-transport dynamo models to rapidly-rotating solar like stars. In those models, the magnetic cycle period is typically set by the amplitude of the meridional flow, which is still not well known in the solar interior and completely unknown for other stars. However, 3D global models of rapidly-rotating stars tend to show that the meridional flow speed decreases when the rotation rate is increased, thus producing a longer magnetic cycle period for more rapidly rotating stars. This does not seem to agree with observations of chromospheric activity cycles. Moreover, recent 3D dynamo simulations which exhibit cyclic reversal do not find any correlations between the meridional flow amplitude and the magnetic cycle period, contrary to what is expected in flux-transport models. We note however that those 3D more complete and self-consistent models do not produce a regular emergence of bipolar magnetic regions as seen in the Sun. This may be only a question of time before these global models do produce starspots but in the meantime, a useful approach is to consider 3D kinematic dynamo models. They represent an intermediate approach between the 2D dynamo models where the BL mechanism is crucial for the dynamo and 3D global models which do not capture the spot-producing mechanism. Application of this kind of models to other stars would also be possible and particularly interesting, to investigate the effects of faster rotation, larger convection zones on the production of spots and on the BL mechanism. We anticipate that those models will be very promising for the application of data assimilation techniques, with the aim to produce reliable forecast of future solar (and possibly stellar?) activity.

Acknowledgements

LJ and RK acknowledge support from the Indo-French Centre for the Promotion of Advanced Research (CEFIPRA), via the collaborative project grant 5004-B.

References

Augustson, K. C., Brown, B. P., Brun, A. S., Miesch, M. S., & Toomre, J. 2012, *Astrophysical Journal*, 756, 169
Augustson, K. C., Brun, A. S., & Toomre, J. 2013, *Astrophysical Journal*, 777, 153
Babcock, H. W. 1961, *Astrophysical Journal*, 133, 572
Baliunas, S. L., Donahue, R. A., Soon, W. H., et al. 1995, *Astrophysical Journal*, 438, 269
Ballot, J., Brun, A. S., & Turck-Chièze, S. 2007, *Astrophysical Journal*, 669, 1190
Berdyugina, S. V. 2005, *Living Reviews in Solar Physics*, 2, 8
Boro Saikia, S., Jeffers, S. V., Morin, J., et al. 2016, *Astronomy & Astrophysics*, 594, A29
Brandenburg, A., Kleeorin, N., & Rogachevskii, I. 2013, *Astrophysical Journal, Letters*, 776, L23
Brown, B. P., Browning, M. K., Brun, A. S., Miesch, M. S., & Toomre, J. 2008, *Astrophysical Journal*, 689, 1354-1372
Brown, B. P., Miesch, M. S., Browning, M. K., Brun, A. S., & Toomre, J. 2011, *Astrophysical Journal*, 731, 69
Brun, A. S., Browning, M. K., Dikpati, M., Hotta, H., & Strugarek, A. 2015, *Space Science Reviews*, 196, 101
Charbonneau, P. 2005, *Living Reviews in Solar Physics*, 2, 2
Charbonneau, P. 2014, *Annual Review of Astron and Astrophys*, 52, 251
Christensen, U. R., & Aubert, J. 2006, *Geophysical Journal International*, 166, 97
Dasi-Espuig, M., Solanki, S. K., Krivova, N. A., Cameron, R., & Peñuela, T. 2010, *Astronomy & Astrophysics*, 518, A7
Dikpati, M., & Charbonneau, P. 1999, *Astrophysical Journal*, 518, 508
Dikpati, M., & Anderson, J. L. 2014, *AGU Fall Meeting Abstracts*
Dikpati, M., Anderson, J. L., & Mitra, D. 2016, *Astrophysical Journal*, 828, 91
Guerrero, G., & de Gouveia Dal Pino, E. M. 2008, *Astronomy & Astrophysics*, 485, 267
Do Cao, O., & Brun, A. S. 2011, *Astronomische Nachrichten*, 332, 907
Donati, J.-F., & Brown, S.F. 1997, *Astronomy & Astrophysics*, 326, 1135
Durney, B. R. 1997, *Astrophysical Journal*, 486, 1065
Fan, Y. 2004, *Living Reviews in Solar Physics*, 1, 1

Fan, Y., & Fang, F. 2014, *Astrophysical Journal*, 789, 35

Fares, R., Donati, J.-F., Moutou, C., *et al.* 2009, *Monthly Notices of the Royal Astronomical Society*, 398, 1383

Featherstone, N. A., & Miesch, M. S. 2015, *Astrophysical Journal*, 804, 67

Folsom, C.P., Petit, P., Bouvier, J., *et al.* 2016, *Monthly Notices of the Royal Astronomical Society*, 457, 580

Gastine, T., Yadav, R. K., Morin, J., Reiners, A., & Wicht, J. 2014, *Monthly Notices of the Royal Astronomical Society*, 438, L76

Gizon, L., Birch, A. C., & Spruit, H. C. 2010, *Annual Review of Astron and Astrophys*, 48, 289

Gregory, S. G., Donati, J.-F., Morin, J., *et al.* 2012, *Astrophysical Journal*, 755, 97

Guerrero, G., & de Gouveia Dal Pino, E. M. 2008, *Astronomy & Astrophysics*, 485, 267

Guerrero, G., Smolarkiewicz, P. K., de Gouveia Dal Pino, E. M., Kosovichev, A. G., & Mansour, N. N. 2016, *Astrophysical Journal*, 819, 104

Hale, G. E. 1908, *Astrophysical Journal*, 28, 315

Hazra, S., Passos, D., & Nandy, D. 2014, *Astrophysical Journal*, 789, 5

Holzwarth, V., & Schüssler, M. 2001, *Astronomy & Astrophysics*, 377, 251

Holzwarth, V., Mackay, D. H., & Jardine, M. 2006, *Monthly Notices of the Royal Astronomical Society*, 369, 1703

Hung, C. P., Jouve, L., Brun, A.-S., Fournier, A., & Talagrand, O. 2015, *EGU General Assembly Conference Abstracts*, 17, 10832

Jouve, L. & Brun, A. S. 2007, *Astronomy & Astrophysics*, 474, 239

Jouve, L., Brown, B. P., & Brun, A. S. 2010, *Astronomy & Astrophysics*, 509, A32

Jouve, L., Brun, A. S., & Talagrand, O. 2011, *Astrophysical Journal*, 735, 31

Jouve, L., Brun, A. S., & Aulanier, G. 2013, *Astrophysical Journal*, 762, 4

Käpylä, P.J., Mantere, M.J., Cole, E., Warnecke, J., & Brandenburg, A. 2013, *Astrophysical Journal*, 778, 41

Krause, F. & Rädler, K.-H. 1980, *Organic Photonics and Photovoltaics*

Leighton, R. B. 1969, *Astrophysical Journal*, 156, 1

MacGregor, K. B. & Charbonneau, P. 1997, *Astrophysical Journal*, 486, 484

Miesch, M. S., & Toomre, J. 2009, *Annual Review of Fluid Mechanics*, 41, 317

Miesch, M. S. & Dikpati, M. 2014, *Astrophysical Journal, Letters*, 785, L8

Miesch, M. S. & Teweldebirhan, K. 2016, *Advances in Space Research*, 58, 1571

Moffatt, H. K. 1978, Cambridge, England, Cambridge University Press, 1978. 353 p.,

Morgenthaler, A., Petit, P., Morin, J., *et al.* 2011, *Astronomische Nachrichten*, 332, 866

Morin, J., Donati, J.-F., Petit, P., *et al.* 2010, *Monthly Notices of the Royal Astronomical Society*, 407, 2269

Nandy, D. & Choudhuri, A. R. 2001, *Astrophysical Journal*, 551, 576

Nelson, N. J., Brown, B. P., Brun, A. S., Miesch, M. S., & Toomre, J. 2013, *Astrophysical Journal*, 762, 73

Nelson, N. J., Brown, B. P., Sacha Brun, A., Miesch, M. S., & Toomre, J. 2014, *Solar Physics*, 289, 441

Parker, E. N. 1955, *Astrophysical Journal*, 122, 293

Parker, E. N. 1993, *Astrophysical Journal*, 408, 707

Petit, P., Dintrans, B., Solanki, S. K., *et al.* 2008, *Monthly Notices of the Royal Astronomical Society*, 388, 80

See, V., Jardine, M., Vidotto, A.A., *et al.* 2016, *Monthly Notices of the Royal Astronomical Society*, 462, 4442

Stein, R. F. & Nordlund, Å. 2012, *Astrophysical Journal, Letters*, 753, L13

Weber, M. A. & Browning, M. K. 2016, *Astrophysical Journal*, 827, 95

Yadav, R. K., Christensen, U. R., Morin, J., *et al.* 2015, *Astrophysical Journal, Letters*, 813, L31

Wicht, J. 2002, *Physics of the Earth and Planetary Interiors*, 132, 281

Yeates, A. R., & Muñoz-Jaramillo, A. 2013, *Monthly Notices of the Royal Astronomical Society*, 436, 3366

Living Around Active Stars
Proceedings IAU Symposium No. 328, 2016
D. Nandy, A. Valio & P. Petit, eds.

© International Astronomical Union 2017
doi:10.1017/S1743921317003945

Starspot Activity and Superflares on Solar-type Stars

Hiroyuki Maehara

email: h.maehara@oao.nao.ac.jp

Okayama Astrophysical Observatory, National Astronomical Observatory of Japan, 3037-5 Honjo, Kamogata, Asakuchi, Okayama, Japan, 719-0232

Abstract. Recent high-precision photometry from space (e.g., Kepler) enables us to investigate the nature of "superflares" on solar-type stars. The bolometric energy of superflares detected by Kepler ranges from 10^{33} erg to 10^{36} erg which is 10-10,000 times larger than that released by a typical X10 class solar flare. The occurrence frequency (dN/dE) of superflares as a function of flare energy (E) shows the power-law distribution with the power-law index of ~ -1.8 for $10^{34} < E < 10^{36}$ erg. Most of superflare stars show quasi-periodic light variations which suggest the presence of large starspots. The bolometric energy released by flares is consistent with the magnetic energy stored near the starspots. The occurrence frequency of superflares increases as the rotation period decreases. However, the energy of the largest flares observed in a given period bin does not show any clear correlation with the rotation period. These results suggest that superflares would occur on the slowly-rotating stars.

Keywords. stars:activity, stars:flare, stars:solar-type

1. Introduction

Solar flares are sudden and rapid releases of magnetic energy stored near the sunspots caused by the magnetic reconnection (e.g., Shibata & Magara 2011). The typical solar flare releases the order of $10^{29} - 10^{32}$ erg of energy with the time scale of minutes to hours. The occurrence frequency of solar-flares (dN/dE) as a function of flare energy (E) can be well represented by a power-law function ($dN/dE \propto E^{-\alpha}$) with the power-law index of $\alpha = 1.5 - 1.9$ in the wide energy range from 10^{24} erg to 10^{32} erg (e.g., Crosby *et al.* 1993, Shimizu 1995, Aschwanden *et al.* 2000). The energy of the largest solar flares observed so far is the order of 10^{32} erg (e.g., Emslie *et al.* 2012) and such solar flares occur approximately once in 10 years. It is unclear that whether more energetic solar flares would occur on our Sun because the history of the modern solar flare research is less than 100 years.

More energetic flares, called "superflares", have been observed on solar-type stars other than the Sun (e.g., Weaver & Naftilan 1973, Landini *et al.* 1986, Schaefer 1989, Schaefer *et al.* 2000). Recent space-based, high-precision photometry of large number of stars by the Kepler space telescope discovered many superflares on on solar-type stars (G-type main sequence stars; e.g., Maehara *et al.* 2012, Shibayama *et al.* 2013). The bolometric energy released by the superflares observed with the Kepler ranges from 10^{33} to 10^{36} erg, which is $10 - 10^4$ times more energetic than that of the largest solar flares observed so far ($\sim 10^{32}$ erg).

2. Data set of superflares on solar-type stars

Maehara *et al.* (2012) and Shibayama *et al.* (2013) searched for superflares on solar-type stars from the long-cadence (~ 30 min interval) data observed with the Kepler

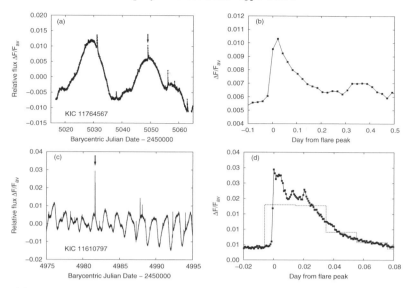

Figure 1. (a): Long-term light variations of the G-dwarf KIC 11764567 from the long-cadence data. The vertical axis means the relative difference between observed brightness of the star and the average brightnes during the observation period. The horizontal axis means the times of the observations in Barycentric Julian Date. (b): Enlarged light curve of a superflare on KIC 11764567 indicated by the down arrow in panel (a) from the long-cadence data. (c): Same as (a), but for KIC 11610797 from the short-cadence data. KIC 11610797 from the short-cadence data (c). (d): Same as (c), but for a superflare on KIC 11610797 indicated by the down arrow in panel (c). Filled-squares with solid-lines and dashed-lines represent the light curves from short- and long-cadence data, respectively.

space telescope from 2009 April (quarter 0: Q0) to 2010 September (Q6). The number of superflares detected in Shibayama *et al.* (2013) is 1547 on 279 stars. Maehara *et al.* (2015) searched for superflares from short-cadence (~1 min interval) data observed between 2009 April (Q0) and 2013 May (Q17) and found 187 superflares on 23 stars.

3. Statistical properties of superflares

3.1. *Light curves of superflares on solar-type stars*

The light curves of typical superflares on solar-type stars are presented in figure 1. As shown figure 1 (a) and (c), most of the stars with superflares show the quasi-periodic brightness modulations with the period and amplitude of ~0.5 − ~30 days, and ~0.1 − ~5 %, respectively. These light variations can be explained by the rotation of the star with spotted surface (e.g., Notsu *et al.* 2013). Figure 1 (b) and (d) show the enlarged light curve of a "spike" (as indicated by a down arrow in figure 1 (a) and (c)) which are thought to be a "superflare". The normalized amplitude of the detected superflares ranges from $\sim 10^{-3}$ to $\sim 8 \times 10^{-2}$. The flare light curves from short-cadence (e.g., 1 (d)) show that some of energetic superflares exhibit flare oscillations with the peak separation of the order of 100-1000 seconds during the decay phase. The duration of typical superflares detected from long-cadence data is ~0.1 day and the duration of superflares from short-cadence data ranges from a few minutes to ~100 minutes. The bolometric energy of superflares detected from the long-cadence data ranges from the order of 10^{33} to 10^{36} erg.

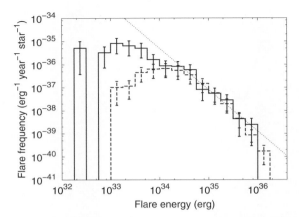

Figure 2. Bold solid and dashed lines represent the occurrence frequency of superflares on all solar-type stars from short- (Maehara *et al.* 2015) and long-cadence data (Shibayama *et al.* 2013) as a function of the bolometric energy of superflares. The vertical axis indicates the number of observed superflares per star, per year, and per unit energy. The thin-dotted line indicates the power-law fit to the frequency distribution in the energy range between 10^{34} and $3 \times 10^{36} \, erg$. The power-law index is -1.8 ± 0.2.

3.2. *Occurrence frequency of superflares*

Figure 2 shows the occurrence frequency distribution of superflares on solar-type stars as a function of flare energy. As mentioned in Maehara *et al.* (2012), Shibayama *et al.* (2013), and Maehara *et al.* (2015), the frequency-energy distribution of superflares on solar-type stars can be represented by a power-law function in the large energy regime ($E_{\text{flare}} > 10^{34}$ erg). The power-law index of frequency distribution is -1.8 ± 0.2. This value is consistent with the power-law indexes of frequency distributions of solar-flares (e.g., Aschwanden *et al.* 2000) and stellar flares on M-dwarfs (e.g., Shakhovskaia 1989).

Shibata *et al.* (2013) found that the frequency-energy distribution of superflares on solar-type stars and that of solar-flares are roughly on the same power-law line. Figure 3 shows the comparison between the frequency-energy distribution of superflares on solar-type stars and those of solar-flares, micro-flares, and nano-flares. All of them are roughly on the same power-law line with the power-law index of -1.8 (thin-solid line in figure 3).

The frequency of superflares depends on the rotation period and effective temperature (e.g., Maehara *et al.* 2012, Notsu *et al.* 2013, Candelaresi *et al.* 2014). As shown in figure 4, the frequency of superflares decreases as the rotation period increases in the long-period regime ($P_{\text{rot}} > 3$ days). On the other hand, in the short-period regime ($P_{\text{rot}} < 3$ days), the frequency is roughly constant. The similar "saturation" of the energy release rate by flares in the short-period (small Rossby number) regime was reported by Davenport (2016).

3.3. *Bolometric energy of superflares*

Solar and stellar flares are thought to be rapid releases of magnetic energy stored near sunspots and starspots. Therefore the total energy released by solar and stellar flares (E_{flare}) must be limited by the magnetic energy (E_{mag}). According to Shibata *et al.* (2013), the flare energy can be written as

$$E_{\text{flare}} \approx f E_{\text{mag}} \approx f \frac{B^2 L^3}{8\pi} \approx f \frac{B^2 A_{\text{spot}}^{3/2}}{8\pi}, \tag{3.1}$$

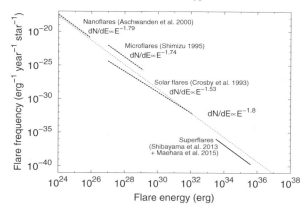

Figure 3. Comparison between the energy-frequency distribution of superflares on Sun-like stars and those of solar flares. Bold-solid line represents the power-law frequency distribution of superflares on Sun-like stars (early G-dwarfs with $P_{\mathrm{rot}} > 10$ days) taken from Shibayama *et al.* (2013) and Maehara *et al.* (2015). Bold-dashed lines indicate the power-law frequency distribution of solar flares observed in hard X-ray (Crosby *et al.* 1993), soft X-ray microflares (Shimizu 1995), and EUV nanoflaers (Aschwanden *et al.* 2000). Occurrence frequency distribution of superflares on Sun-like stars and those of solar flares are roughly on the same power-law line with an index of -1.8 (thin-solid line; Shibata *et al.* 2013).

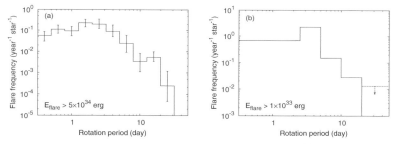

Figure 4. (a): Frequency of the superflares with $E_{\mathrm{flare}} > 5 \times 10^{34}$ erg as a function of the rotation period derived from the long-cadence data (Notsu *et al.* 2013). (b): Same as (a), but for the superflares with $E_{\mathrm{flare}} > 1 \times 10^{33}$ erg derived from the short-cadence data (Maehara *et al.* 2015) The dashed line and down-arrow indicate the upper limit of the flare frequency in the period bin ($1/80$ year^{1}).

where f is the fraction of magnetic energy released by a flare, B, L, and A_{spot} are the magnetic field strength, the scale length of active region, and the area of sunspots/starspots. According to Aschwanden *et al.* (2014), $f < 0.1$. The equation 3.1 suggests that the upper limit of energy released by flares is proportional to $A_{\mathrm{spot}}^{3/2}$. Figure 5 shows the scatter plot of the energy of superflares and solar flares as a function of the area of sunspots / starspots. The area of starspots on superflare stars was estimated from the amplitude of rotational brightness variations. Solid line in figure 5 indicates the analytic relation between flare energy and spot area based on the equation 3.1 for $f = 0.1$ and $B = 3000$ G. Majority of superflares and almost all of solar flares are below the analytic line. This result suggests that the energy released by a flare is basically consistent with the magnetic energy stored near sunspots / starspots and the large starspots whose area is larger than ~ 1 % of the area of solar hemisphere are required to produce superflares with $E_{\mathrm{flare}} > 10^{34}$ erg.

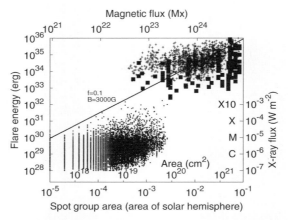

Figure 5. Scatter plot of flare energy as a function of the area of starspots from Maehara *et al.* (2015). The lower horizontal axis indicates the area of starspot group in the unit of the area of solar hemisphere (3×10^{22} cm^2). The upper horizontal axis indicates the magnetic flux if we assume that the magnetic field strength around spots (B) is 3000 G. The vertical axis represents the energy released by each flare. Filled-squares and small-crosses indicate superflares on solar–type stars from the short- and long-cadence data, respectively. Small filled-circles represent solar flares based on the data retrieved from the website of the National Geophysical Data Center of the National Oceanic and Atmospheric Administration (NOAA/NGDC), Solar-Terrestrial Physics Division at http://www.ngdc.noaa.gov/stp/ (Ishii *et al.*, private communication).

However, some superflares occurred on the stars with small amplitude light variations (superflares located above the analytic line in figure 5). The energy released by such superflares is larger than that estimated from equation (3.1). Notsu *et al.* (2015) measured the rotational velocity ($v \sin i$) of superflare stars by high-dispersion spectroscopy and found that the superflares occurred on the stars with low inclination angle (i) are located above the analytic line. In the case of the star with low inclination angle, the amplitude of rotational brightness modulations caused by starspots becomes smaller than that expected from the area of starspots.

Figure 6 shows the scatter plot of flare energy as a function of the rotation period based on the data set of superflares taken from Shibayama *et al.* (2013) and Maehara *et al.* (2015). Maehara *et al.* (2012) and Notsu *et al.* (2013) pointed out that the energy of the largest superflares observed in a given period bin does not depend on the rotation period. This result indicates that superflares can occur on not only the rapidly-rotating stars but also slowly-rotating stars. As shown in figure 4, the frequency of superflares decreases as the rotation period increases in the period range of $P_{\rm rot} > 3$ days. On the other hand, the energy of the largest superflares depends on the area of starspots (figure 5) and does not depend on the rotation period. These results imply that the appearance frequency of large starspots which could produce superflares may decreases as the rotation period increases.

3.4. *Decay timescale of superflares vs. flare energy*

Maehara *et al.* (2015) found a clear correlation between the duration of superflares and the energy of superflares. Figure 7 shows the scatter plot of the flare duration (*e*-folding time) as a function of the bolometric energy of superflares. A linear fit for the duration (τ) and energy ($E_{\rm flare}$) of superflares on the log-log plot yields the following correlation:

$$\tau \propto E_{\rm flare}^{0.39 \pm 0.03}. \tag{3.2}$$

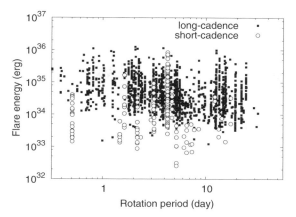

Figure 6. Scatter plot of flare energy as a function of the rotation period. The horizontal and vertical axis indicate the rotation period of superflare stars and the bolometric energy of superflares. Filled squares and open circles represent each superflares detected from the long-cadence data (taken from Shibayama *et al.* 2013) and short-cadence data (from Maehara *et al.* 2015), respectively.

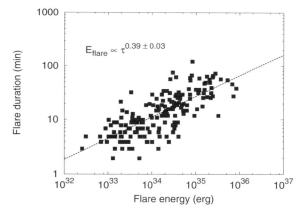

Figure 7. Scatter plot of the flare duration (*e*-folding time) as a function of flare energy. Filled squares indicate superflares on solar-type stars from short-cadence data (taken from Maehara *et al.* 2015). Dotted line represents the linear regression to the data set of superflares. The power-law slope of the line is 0.39 ± 0.03.

Veronig *et al.* (2002) found a similar correlation between the decay time and X-ray fluence of solar flares observed with GOES which shows the power-law slope of $\sim 1/3$. Moreover, Christe *et al.* (2008) reported that the correlation between the duration and peak flux of solar flares observed with RESSI also shows similar power-law slope of ~ 0.2

As discussed in the previous subsection, the flare energy is a part of magnetic energy stored near the starspots. From equation (3.1), if the magnetic field strength of starspots (B) is almost the same on all the solar-type stars, the flare energy would be proportional to L^3,

$$E_{\text{flare}} \propto L^3, \qquad (3.3)$$

where L is the scale length of the active region. On the other hand, the duration of white-light flares is roughly comparable to the reconnection time (τ_{rec}) that can be written as

Table 1. Comparison between the fraction of planet-hosting stars among all solar-type stars and that among superflare stars.

	Number of planet-hosting stars[1]	Total number of stars	Fraction
All solar-type stars	1029	102251	~1.0 %
Superflare stars	1	279	~0.4 %

Note:
[1] The data set of exoplanets were retrieved from the NASA Exoplanet Archive (http://exoplanetarchive.ipac.caltech.edu/) on Dec 14th, 2016.

follows by using the Alfvén time ($\tau_A = L/v_A$):

$$\tau_{\text{flare}} \sim \tau_{\text{rec}} \sim \tau_A/M_A \sim L/v_A/M_A, \tag{3.4}$$

where v_A is the Alfvén velocity, and M_A is the non-dimensional reconnection rate (0.1-0.01 for the fast reconnection; Shibata & Magara 2011). From equation (3.3) and (3.4), if we assume that v_A is roughly the same on all the solar-type stars, the duration of flares can be written as

$$\tau_{\text{flare}} \propto E_{\text{flare}}^{1/3}. \tag{3.5}$$

The power-law slope for the correlation between the flare duration and flare energy derived from the reconnection model ($\sim 1/3$) is comparable to the observed value of 0.39 ± 0.03.

3.5. *Superflares and close-in planet*

It has been discussed that close-in giant planets such as hot Jupiters can affect the stellar magnetic activity (e.g., Cuntz *et al.* 2000, Ip *et al.* 2004). Rubenstein & Schaefer (2000) proposed that superflares are caused by the magnetic reconnection between magnetic fields of the primary star and a close-in Jovian planet, and the superflares on solar-type stars occur only on the stars with hot Jupiters on the basis of an analogy with the RS CVn binaries. However, Maehara *et al.* (2012) found no hot Jupiters orbiting around 148 solar-type stars showing superflares. Table 1 shows the comparison between the fraction of planet hosting stars among all solar-type stars in the Kepler field and that among superflare stars. Only 1 confirmed exoplanet (Kepler-491 b) was discovered around 279 superflare stars listed in Shibayama *et al.* (2013). The fraction of planet-hosting stars is not significantly different between in all solar-type stars and superflare stars. This result suggests that hot Jupiters are not necessary to produce superflares on solar-type stars.

4. Summary

Statistical studies of superflares on solar-type stars by using the Kepler data reveal the following results:

• The frequency-energy distribution of superflares on solar-type stars can be represented by a power-law function with the power-law index of ~ -1.8. The frequency distribution of superflares on solar-type stars and those of solar flares are almost on the same power-law line.

• The average frequency of superflares depends on the rotation period (P_{rot}). In the short-period regime ($P_{\text{rot}} < 3$ days), the flare frequency is almost constant. On the other hand, in the long-period regime, ($P_{\text{rot}} > 3$ days), the flare frequency decreases as the rotation period increases.

• The upper limit of flare energy depends on the area of starspots. The relation between the flare energy and the area of starspots (A_{spot}) suggests that the largest flare energy is proportional to $A_{\mathrm{spot}}^{3/2}$. However, there is no clear correlation between the energy of largest flares observed in a given period bin and the rotation period. These results indicate that slowly-rotating solar-type stars can produce superflares if they have large starspots.

• The flare duration (τ_{flare}) increases as the the energy of flares (E_{flare}) increases. The data set of superflares from the short-cadence data yields the following correlation:

$$\tau_{\mathrm{flare}} \propto E_{\mathrm{flare}}^{0.39\pm0.03}.$$

This suggests that the time-scale of flares is determined by the Alfvén time.

• There is no significant difference between the fraction of planet-hosting stars among all solar-type stars in the Kepler field and that among superflare stars. Hot Jupiters may not be necessary to produce superflares on solar-type stars.

References

Aschwanden, M. J., Tarbell, T. D., Nightingale, R. W., *et al.* 2000, *ApJ*, 535, 1047

Aschwanden, M. J., Xu, Y., & Jing, J. 2014, *ApJ*, 797, 50

Brown, T. M., Latham, D. W., Everett, M. E., & Esquerdo, G. A. 2011, *ApJ*, 142, 112

Candelaresi, S., Hillier, A., Maehara, H, Brandenburg, A., & Shibata, K. 2014, *ApJ*, 792, 67

Christe, S., Hannah, I. G., Krucker, S., *et al.* 2008, *ApJ*, 677, 1385

Crosby, N. B., Aschwanden, M. J., & Dennis, B. R. 1993, *Solar Phys.*, 143, 275

Cuntz, M., Saar, S. H., & Musielak, Z. E. 2000, *ApJ*, 533, L151

Davenport, J. R. A. 2016, *ApJ*, 829, 23

Emslie, A. G., Dennis, B. R., Shih, A. Y., *et al.* 2012, *ApJ*, 759, 71

Gilliland, R. L., Jenkins, J. M., Borucki, W. J., *et al.* 2010, *ApJ* (Letters), 713, L160

Huber, D., Silva Aguirre, V., Matthews, J. M., *et al.* 2014, *ApJS*, 211, 2

Ip, W.-H., Kopp, A., & Hu, J.-H. 2004, *ApJ*, 602, L53

Landini, M., Monsignori Fossi, B. C., Pallavicini, R., & Piro, L. 1986, *A&A*, 157, 217

Maehara, H., Shibayama, T., Notsu, S., *et al.* 2012, *Nature*, 485, 478

Maehara, H., Shibayama, T., Notsu, Y., *et al.* 2015, *Earth, Planet and Space*, 67, 59

McQuillan, A., Mazeh, T., & Aigrain, S. 2014, *ApJS*, 211, 24

Notsu, Y., Shibayama, T., Maehara, H., *et al.* 2013, *ApJ*, 771, 127

Notsu, Y., Honda, S., & Maehara, H., *et al.* 2015, *PASJ*, 67, 33

Rubenstein, E. P. & Schaefer, B. E. 2000, *ApJ*, 529, 1031

Sammis, I., Tang, F., & Zirin, H. 2000, *ApJ*, 540, 583

Shakhovskaia, N. I. 1989, *Solar Physics*, 121, 375

Schaefer, B. E. 1989, *ApJ*, 337, 927

Schaefer, B. E., King, J. R., & Deliyannis, C. P. 2000, *ApJ*, 529, 1026

Shibata, K. & Magara, T. 2011, *Living Reviews in Solar Physics*, 8, 6

Shibata, K., Isobe, H., Hillier, A., *et al.* 2013, *PASJ*, 65, 49

Shibayama, T., Maehara, H., Notsu, S., *et al.* 2013, *ApJS*, 209, 5

Shimizu, T. 1995, *PASJ*, 47, 251

Veronig, A., Temmer, M., Hanslmeier, A., *et al.* 2002, *A&A* 382, 1070

Weaver, W. B. & Naftilan, S. A. 1973, *PASP*, 85, 213

Living Around Active Stars
Proceedings IAU Symposium No. 328, 2016
D. Nandy, A. Valio & P. Petit, eds.

© International Astronomical Union 2017
doi:10.1017/S1743921317003970

Magnetic field generation in PMS stars with and without radiative core

B. Zaire[1,*], G. Guerrero[1,†], A. G. Kosovichev[2], P. K. Smolarkiewicz[3] and N. R. Landin[4]

[1] Physics Department, Universidade Federal de Minas Gerais
Belo Horizonte, MG 31270-901, Brazil
[*] e-mail: zaire@fisica.ufmg.br, [†] e-mail: guerrero@fisica.ufmg.br
[2] New Jersey Institute of Technology
Newark, NJ 07103, USA
[3] European Centre for Medium-Range Weather Forecasts
Reading RG2 9AX, UK
[4] Campus UFV Florestal, Universidade Federal de Viosa
Florestal, MG 35690-000, Brazil

Abstract. Recent observations of the magnetic field in pre-main sequence stars suggest that the magnetic field topology changes as a function of age. The presence of a tachocline could be an important factor in the development of magnetic field with higher multipolar modes. In this work we performed MHD simulations using the EULAG-MHD code to study the magnetic field generation and evolution in models that mimic stars at two evolutionary stages. The stratification for both stellar phases was computed by fitting stellar structure profiles obtained with the ATON stellar evolution code. The first stage is at 1.1Myr, when the star is completely convective. The second stage is at 14Myrs, when the star is partly convective, with a radiative core developed up to 30% of the stellar radius. In this proceedings we present a preliminary analysis of the resulting mean-flows and magnetic field. The mean-flow analysis shown that the star rotate almost rigidly on the fully convective phase, whereas at the partially convective phase there is differential rotation with conical contours of iso-rotation. As for the mean magnetic field both simulations show similarities with respect to the field evolution. However, the topology of the magnetic field is different.

Keywords. Star: interior — star: dynamo — T Tauri

1. Introduction

Magnetic field plays an important role in the evolution of pre-main sequence (PMS) stars. At this phase, solar type stars are called T Tauri stars. They have a disk which will dissipate within the first ten millions of years of their life. The star-disc interaction is mediated by the action of a large-scale magnetic field, which governs the accretion process. Both, the disk and the magnetic field strength are key factors to understand the angular momentum evolution of these objects (Gallet *et al.* 2013).

Recently Donati *et al.* (2007, 2008, 2010, 2011, 2012) derived profiles of the magnetic field for a small sample of T Tauri stars. The differences in those maps suggest that a dynamo mechanism operates in the stellar interior generating and sustaining a large-scale magnetic field. For the observed T Tauri stars, Gregory *et al.* (2012) found a relation between the large-scale magnetic field and the position in the H-R diagram. They found that the magnetic field gains complexity with the evolution of the star. For instance, when the star is completely convective, the magnetic field is mainly dipolar. After the development of the radiative core the dipolar component looses power compared to high order components of the mutipole expansion.

According to the mean-field theory the dynamo depends on the large scale motions, differential rotation and meridional circulation, and on turbulent convection. All these properties vary with the age of the star. For example, the results of Vidotto *et al.* (2014) indicate that for T Tauri stars the period of rotation decreases with age. Unfortunately, apart from very recent observations (e.g., Donati *et al.* 2000, 2010), little is known about the differential rotation of these objects. As for the meridional circulation the only hint we have is provided by global numerical simulations (e.g, Guerrero *et al.* 2013; Gastine *et al.* 2012). Note, however, that these models correspond to main sequence stars. Finally, for the convective motions we rely on the results of stellar structure and evolution models based on the mixing length theory (MLT).

Previous numerical results of rotating turbulent convection in fully convective models have shown that as the density stratification progressively increases, it weakens the dipole-dominant magnetic field topology (Gastine *et al.* 2012). As an illustration, Browning *et al.* (2008) carried out a highly-stratified simulation for a M dwarf star obtaining a weak dipolar field. More recently, Yadav *et al.* (2015) carried out a numerical simulation obtaining a consistent magnitude and morphology, with a dipolar-dominant surface magnetic field. However, as pointed by Bessolaz *et al.* (2011), this picture is different for the stratification of T Tauri stars once the gravitational contraction is still present and the nuclear reactions are absent. Bessolaz *et al.* simulated the interior of BP Tau star after the development of a radiative core covering the inner 14% of the stellar radius. Although their results provided important insights on the convective motions, the amplitude of the large-scale magnetic field was weaker than expected by the observations.

Here we perform 3D numerical simulations for the target star BP Tau in two different phases of evolution. The first one corresponds to the fully convective phase, corresponding to BP Tau today. The second one corresponds to the same star once its radiative core has developed until 30% of the stellar radius. Our main goal is to explore the generation of large-scale flows and magnetic field in this type of stars and study the role of the tachocline in the dynamo mechanism. Our results will allow to verify through numerical experiments the hypothesis presented in Gregory *et al.* (2012) regarding the topological differences between the stellar magnetic field of stars at different ages. Besides, this study will allow to establish a better connection between the models and the observations for this kind of objects.

2. Numerical model

We adopt a full spherical shell, $0 \leqslant \phi \leqslant 2\pi$, $0 \leqslant \theta \leqslant \pi$. In the radial direction the bottom boundary is located at $r_{\rm b} = 0.10R_{\rm s}$ and the upper boundary is at $r_{\rm t} = 0.95R_{\rm s}$, where $R_{\rm s}$ is the stellar radius. Similarly to Guerrero *et al.* (2016a) we solve the set of anelastic MHD equations:

$$\nabla \cdot (\rho_{\rm s} \boldsymbol{u}) = 0, \tag{2.1}$$

$$\frac{{\rm D}\boldsymbol{u}}{{\rm D}t} + 2\boldsymbol{\Omega} \times \boldsymbol{u} = -\boldsymbol{\nabla}\left(\frac{p'}{\rho_{\rm s}}\right) + \mathbf{g}\frac{\Theta'}{\Theta_{\rm s}} + \frac{1}{\mu_0 \rho_{\rm s}}(\boldsymbol{B} \cdot \nabla)\boldsymbol{B}, \tag{2.2}$$

$$\frac{{\rm D}\Theta'}{{\rm D}t} = -\boldsymbol{u} \cdot \boldsymbol{\nabla}\Theta_{\rm e} - \frac{\Theta'}{\tau}, \tag{2.3}$$

$$\frac{{\rm D}\boldsymbol{B}}{{\rm D}t} = (\mathbf{B} \cdot \nabla)\boldsymbol{u} - \boldsymbol{B}(\nabla \cdot \boldsymbol{u}), \tag{2.4}$$

where $D/Dt = \partial/\partial t + \boldsymbol{u}\cdot\nabla$ is the total time derivative, $\boldsymbol{u}$ is the velocity field in a rotating frame with $\boldsymbol{\Omega} = \Omega_0(\cos\theta, -\sin\theta, 0)$, p' is a pressure perturbation variable that accounts for both the gas and magnetic pressure, $\boldsymbol{B}$ is the magnetic field, and Θ' is the potential temperature perturbation with respect to an ambient state Θ_{e} (as explained in Guerrero *et al.* 2013). The term Θ'/τ maintains a steady axisymmetric solution of the stellar structure against the action of convective turbulent motions by restoring the ambient state with potential temperature Θ_{e} within a time scale given by τ. Furthermore, ρ_{s} and Θ_{s} are the density and potential temperature of the reference state which is chosen to be isentropic (i.e., $\Theta_{\mathrm{s}} = $ const) and in hydrostatic equilibrium; $\boldsymbol{g}$ is the gravity acceleration and μ_0 is the magnetic permeability. The potential temperature, Θ, is related to the specific entropy: $s = c_p \ln\Theta + $ const.

The equations are solved numerically using the EULAG-MHD code (Guerrero *et al.* 2013; Ghizaru *et al.* 2010; Racine *et al.* 2011; Smolarkiewicz *et al.* 2013), a spin-off of the hydrodynamical model EULAG predominantly used in atmospheric and climate research (Prusa *et al.* 2008). The ambient and isentropic states, as well as the gravity acceleration, have been computed by fitting the stellar structure profiles obtained with the ATON stellar evolution code (Landin *et al.* 2006) to simple hydrostatic polytropic models. For the velocity field we use impermeable, stress-free conditions at the top and bottom surfaces of the shell; whereas the magnetic field is assumed to be radial at these boundaries. Finally, for the thermal boundary condition we consider zero divergence of the convective flux at the bottom and zero flux at the top surface.

3. Results

3.1. *Convective structures*

Figure 1 shows instantaneous snapshots of the radial velocity, u_r, for models TT01 (**a**) and TT14 (**b**). The three upper panels in each side represent the vertical velocity in the Mollweide projection at 30%, 60% and 90% of the stellar radius, respectively. The bottom panels shows the same component of the velocity in a longitudinal plane at the stellar equator. For model TT01, at the top of the domain small scales of convection can be observed at hight latitudes, however, the convective patters are less evident at low-latitudes. In the deeper stellar interior the convective patterns become columnar, exhibiting the so-called "banana cells". For the partially convective star we can identify, as expected, the existence of the radiative core in the longitudinal cut. Since this layer is stably stratified, the vertical motions only slightly overshoot in this region. An important feature observed is the development of large vertical upflows and downflows at equatorial latitudes in the upper part of the domain. These motions contribute to the mean profile of the meridional circulation, as it will be presented below. The banana cells are also evident in this models at $r = 0.35 r_s$ and $r = 0.6 r_s$.

3.2. *Large-scale flows*

The Figure 1 shows the final moments of the simulations. Representing the vertical velocity, they show that the flow patterns are distributed uniformly around the rotation axis. Therefore, the models are suitable for mean-field analysis. We are interested specially in the large-scale flows and magnetic fields (see next section).

Figures 2 and 3 present the angular velocity and the meridional circulation for the models TT01 and TT14, respectively. It is possible to observe that for model TT01 the star rotates almost uniformly, with the exception of a prominent shear region in the upper part of the domain at lower latitudes (Figure 2 **a**). The model TT14 exhibits a conical profile of differential rotation and it is possible to notice the presence of the tachocline,

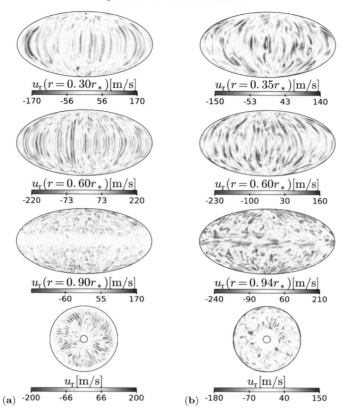

Figure 1. Panels (**a**) and (**b**) shows instantaneous snapshots of the radial velocity in the MHD simulations TT01 and TT14, respectively. The three upper panels in each side are Mollweide projections of u_r at 30%, 60% and 90% of the stellar radius, respectively. The bottom panel is a longitudinal, $r - \phi$, cut at the stellar equator.

although it is not sharply defined (Figure 3 **a** and **b**). Both models develop solar-like rotation, which means that the Coriolis force dominates over the buoyant convective motions. As for the meridional circulation, in both models we observe the existence of circulation cells. These cells are better structured in the model TT14 than in the model TT01 (compare Figures 2 **c** and with 3 **c**). Furthermore, in model TT14 the circulation cells are clearly symmetric across the equator.

3.3. *Mean-field magnetic fields*

Figure 4 presents the evolution of the large-scale magnetic field obtained in model TT01. The upper (lower) panel is a time-latitude (time-radius) butterfly diagram for the toroidal (left) and radial (right) components of the magnetic field. For all component of the magnetic field we notice the existence of a linear phase during which the magnetic field grows until reaching a magnetic energy of the same order than the kinetic one. The magnetic field then reaches a non-linear stage when it inhibits fluid motions via the Lorentz force. At the beginning of this phase the dynamo is oscillatory and anti-symmetric across the equator, with dynamo waves that develop near the surface and propagate upwards, and from mid latitudes towards the equator. After roughly 3 years the magnetic field becomes steady for approximately more 3 yr. Afterwards a new topological change

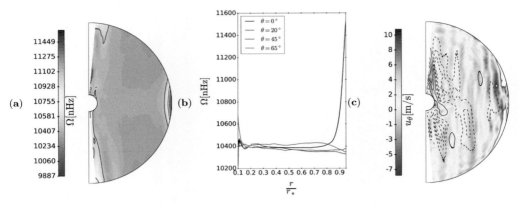

Figure 2. Panels (**a**) and (**b**) present the angular velocity for model TT01. Panel (**a**) shows the differential rotation profile in the meridional plane, and panel (**b**) exhibits the radial distribution of the angular velocity at different latitudes. Panel (**c**) presents the meridional circulation profile in the meridional plane. The colored contours show $\overline{u}_\theta$. The contour lines show the stream function Ψ, computed from $\rho\mathbf{u} = \nabla \times \Psi$. Solid (dashed) lines correspond to clockwise (counter-clockwise) circulation. These profiles correspond to azimuthal and temporal averages considered at the final stages of the simulation.

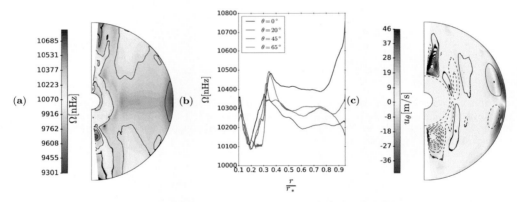

Figure 3. The same as in Figure 2, but for model TT14.

is observed. This indicates that the simulation has not achieved yet its final steady state and should be ran longer. In the non-linear phase of the dynamo, both components of the magnetic field are initially generated in the upper part of the domain. As time evolves, the magnetic field develops in the entire convective envelope. We hypothesize that this occurs because of the back-reaction of the magnetic field on the flow, which gradually establishes new mean-flow profiles (see above) until a final steady state (probably not yet reached) is obtained.

The evolution of the mean magnetic field developed in model TT14 is presented in Figure 5. Similar to the previous case we notice a linear phase when the magnetic field is amplified until its magnetic energy is comparable to the kinetic energy. Following this exponential growing a non-linear phase is observed. In this stage, both the toroidal and radial components of the magnetic field are initially oscillatory and symmetric across the equator. The dynamo waves during the first years of the non-linear phase propagate

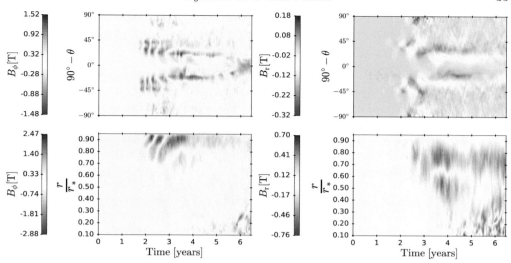

Figure 4. Butterfly diagrams for the fully convective model, TT01. The upper panels show the evolution of the toroidal (left) and radial (right) field components in time and latitude at $r = 0.93R_s$. The lower panels show the time-radius evolution for these quantities at 20° degrees latitude.

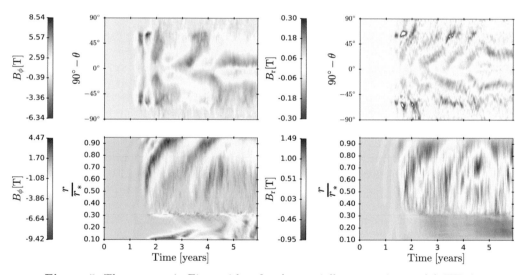

Figure 5. The same as in Figure 4 but for the partially convective model, TT14.

upwards from the bottom of the convection zone. No latitudinal migration of the fields is initially observed, however this migration appears after about one year of evolution. The dynamo is then oscillatory and propagates poleward. Both components are initially symmetric across the equator but this parity changes after one cycle of evolution. A stationary-phase begins at $t \sim 4.5$ years with an anti-symmetric magnetic field. The radial large-scale magnetic field corresponds to a quadrupolar topology. However, as for case TT01, we still need to run the simulation further to be sure that this is the final morphology of the magnetic field. It is worth mentioning that in the non-linear phase the magnetic field is generated also in the radial shear layer and the radiative core. However, the shear does not seem to be sufficient to generate a strong layer of magnetic fields.

4. Conclusions

Recent observations of the magnetic field in low mass stars have identified different magnetic morphologies (Vidotto *et al.* 2014). The topology of the large-scale magnetic field of these stars seems increase in complexity as a function of the stellar age (Gregory *et al.* 2012). It has been hypothesized that the appearance of a radiative core changes an initially dipolar dominant magnetic field (observed in several fully convective stars), towards a field with higher multipolar orders. With the aim to verify this hypothesis, here we studied the generation of magnetic field in PMS stars through global MHD simulation. Models of stars in two different evolutionary stages are presented. The first one is at 1.1Myr, when the star is completely convective. The second stage is at 14Myrs, when the star is partially convective, with a radiative core developed until 30% of the stellar radius. For this work we performed 3D numerical simulations using the anelastic EULAG-MHD code. The initial seed magnetic field is random and imposed on the steady state HD solutions. The background and environmental stratifications, necessary in our set of anelastic equations, are polytropic atmospheres that fit the profiles obtained with the ATON stellar evolution code.

The results are successful in developing large-scale motions and magnetic fields from rotating turbulent convection in stars with convective envelopes of different sizes. For the fully convective model these turbulent motions exhibit small convection cells at the upper layers at higher latitudes and less pronounced cells at the equator. For the model with the radiative core the convective structures represent strong downflows in the convection layer. In the stable layer, the vertical motions only slightly overshoot. In deeper layers for both models the convective pattern becomes cylindrical with respect to the rotation axis and displays convective "banana cells". In both models the convective structures are distributed uniformly in the azimuthal direction. This allows us to perform a mean-field analysis and compute the large scale motions and magnetic fields.

For the fully convective model the final rotation is almost rigidly, with the existence of shear only in the outer layers and at equatorial latitudes. For the partially convective case the differential rotation is slightly conical with a radial shear layer at $r \simeq 0.3 r_s$. The meridional circulation cells that arise in both simulations seemed to be sensitive to the size of the convective domain and to the presence of a large-scale magnetic field. For model TT01 the cells are unorganized (this could be an evidence that the model has not reached a fully saturated state). For the partially convective model there is a symmetric multicellular structure across the equator. The mean-flows drastically change due to the mean-magnetic field and we expect to reach a saturated state when both the mean flows and the mean magnetic field reach equilibrium.

The magnetic field evolution in the fully convective model, TT01, is clearly different from the partially convective model, TT14, which naturally develops a tachocline. Both models have a linear phase, with exponential growing of **B**, as it is characteristic of a mean-field dynamo. Compared to solar dynamo models, for which non-linear saturation is fast (Guerrero *et al.* 2016), the models presented here have a long non-linear phase when both, mean flows and fields mutually adjust. The propagation of dynamo waves differs among the models with convection zones of different thickness. At the current state of the simulations, the field is steady in both cases, however, their topology is different. It is still necessary to run the simulations further before inferring conclusive results. A more quantitative analysis, regarding angular momentum transport as well as studying the source terms of the magnetic field will be presented in forthcoming work (Zaire *et al.*, 2017, in preparation).

Acknowledgements

This work was partially supported by FAPEMIG grant APQ-01168/14 (BZ and GG) and by NASA grant NNX14AB70G. We thank the scientific organizers for the wonderful meeting and for the opportunity to present this work. BZ acknowledges the IAU, the Physics Department of UFMG for travel support, and NASA Ames Heliophysics Summer Program where part of this work was developed. PKS is supported by funding received from the European Research Council under the European Union's Seventh Framework Programme (FP7/2012/ERC Grant agreement no. 320375). The simulations were performed in the NASA cluster Pleiades and in the LNCC cluster SDumont.

References

Browning, M. K. 2008, The Astrophysical Journal, 676, 1262

Bessolaz, N. & Brun, A. 2011, Astronomische Nachrichten, 332, 1045

Donati, J.-F., Mengel, M., Carter, B. D., *et al.* 2000, Monthly Notices of the Royal Astronomical Society, 316, 3

Donati, J.-F., Jardine, M., Gregory, S., *et al.* 2007, Monthly Notices of the Royal Astronomical Society, 380, 1297

—. 2008, Monthly Notices of the Royal Astronomical Society, 386, 1234

Donati, J.-F., Skelly, M., Bouvier, J., *et al.* 2010, Monthly Notices of the Royal Astronomical Society, 402, 1426

Donati, J.-F., Bouvier, J., Walter, F., *et al.* 2011, Monthly Notices of the Royal Astronomical Society, 412, 2454

Donati, J.-F., Gregory, S., Alencar, S., *et al.* 2012, Monthly Notices of the Royal Astronomical Society, 425, 2948

Gallet, F. & Bouvier, J. 2013, Astronomy & Astrophysics, 556, A36

Gastine, T., Duarte, L., & Wicht, J. 2012, Astronomy & Astrophysics, 546, A19

Ghizaru, M., Charbonneau, P., & Smolarkiewicz, P. K. 2010, The Astrophysical Journal Letters, 715, L133

Gregory, S. G., Donati, J.-F., Morin, J., *et al.* 2012, The Astrophysical Journal, 755, 97

Guerrero, G., Smolarkiewicz, P. K., Kosovichev, A. G., & Mansour, N. N. 2013, The Astrophysical Journal, 779, 176

Guerrero, G. and Smolarkiewicz, P. K. and de Gouveia Dal Pino, E. M., & Kosovichev, A. G. and Mansour, N. N. 2016, The Astrophysical Journal Letters, 828, L3

Guerrero, G., Smolarkiewicz, P. K., de Gouveia Dal Pino, M., Kosovichev, A. G., & Mansour, N. N. 2016, The Astrophysical Journal, 819, 104

Johns-Krull, C. M. 1996, Astronomy and Astrophysics, 306, 803

Landin, N. R., Ventura, P., D'Antona, F., Mendes, L. T. S., & Vaz, L. P. R. 2006, Astronomy & Astrophysics, 456, 269

Prusa, J. M., Smolarkiewicz, P. K., & Wyszogrodzki, A. A. 2008, Comput. Fluids, 37, 1193

Racine, É., Charbonneau, P., Ghizaru, M., Bouchat, A., & Smolarkiewicz, P. K. 2011, The Astrophysical Journal, 735, 46

Smolarkiewicz, P. K. & Charbonneau, P. 2013, J. Comput. Phys., 236, 608

Vidotto, A., Gregory, S., Jardine, M., *et al.* 2014, Monthly Notices of the Royal Astronomical Society, 441, 2361

Yadav, R. K., Christensen, U. R., Morin, J., *et al.* 2015, The Astrophysical Journal Letters, 813, L31

Living Around Active Stars
Proceedings IAU Symposium No. 328, 2016
D. Nandy, A. Valio & P. Petit, eds.

© International Astronomical Union 2017
doi:10.1017/S1743921317003684

Estimating the chromospheric magnetic field from a revised NLTE modeling: the case of HR 7428

Innocenza Busá[1]

[1]INAF - Catania Astrophysical Observatory,
Via S. Sofia, 78, 95123
- Catania - Italy
email: `innocenza.busa@oact.inaf.it`

Abstract. Semi-empirical atmospheric modeling is here used to obtain the chromospheric magnetic field distribution versus height in the K2 primary component of the RS CVn binary system HR 7428. The chromospheric magnetic field estimation versus height comes from considering the possibility of not imposing hydrostatic equilibrium in the atmospheric modeling. The stability of the best Non-hydrostatic equilibrium model, implies the presence of and additive (toward the center of the star) pressure, that decrease in strength from the base of the chromosphere toward the outer layers. Interpreting the additive pressure as magnetic pressure and I derive a magnetic field intensity of about 500 Gauss at the base of the chromosphere.

Keywords. magnetic field, stars: chromospheres, stars: modeling atmosphere, stars: activity

---

1. Introduction

HR 7428 (=V 1817 CYgni) is a bright (V = 6.3) long-period (108.578d) spectroscopic RS CVn binary composed by a K2 II-III star and a main sequence A2 star (Parsons & Ake 1987). The magnetic activity of the system is well known: Ca II H & K emission was firstly reported by Gratton (1950), by a detailed analysis of photometric observations Hall *et al.* (1990) were able to detect starspot signatures on the K2 primary star. Is now well established that stellar atmospheres of cool stars are characterized by a temperature gradient inversion. That is explained inside a magnetic activity theory, but not yet definitively understood. Late-type stars with H-α in emission show a fairly stable chromospheric emission outside flares (see. e.g., Byrne *et al.* 1998). This reinforces the hypothesis that chromospheres are globally in a quasi-stationary state, modulated mainly by the stellar activity cycle, whose temperature-density structure results from the balance between global dissipation of non-radiative energy and radiative cooling (see Kalkofen *et al.* 1999).

Most of what we know about stars and systems of stars is derived from an analysis of their radiation and this knowledge will be secure only as long as the analytical technique is physically reliable. A well tested technique to get information on physical properties of chromospheric layers of active stars is the NLTE radiative transfer semi-empirical modeling: for different temperature vs.height distributions, the NLTE populations for hydrogen are computed, solving simultaneously the equations of hydrostatic equilibrium, radiative transfer and statistical equilibrium. The emerging profiles for some chromospheric lines and continua are computed and compared to the observations. Then, the modeling is iterated until a satisfactory match is found. (see, e.g.,Vernazza *et al.* 1981; Fontenla *et al.* 1993). These models are built to match the observations in different spectral features, and make no assumption about the physical processes responsible for the heating of the chromosphere, but can be used as constraints for these processes. The obtained models describe the variations of the essential physical parameters, in particular the temperature, pressure and electron density across the outer atmosphere, and give information on its "mean" state, both temporally and spatially.

The most important problem of this approach lies in the uniqueness of the solution. In fact, knowing that a particular atmosphere would emit a line profile like the one we observe for a given star does not imply that the star has indeed this atmospheric structure, since we do not know whether some other atmospheres would produce the same line profiles. To solve, or at least to reduce, this problem, the modeling has to be based on several spectral features, with different regions of formation. The amount and the kind of diagnostics used to build an atmospheric model is in fact, very important, by combining several spectral lines that are formed at different but overlapping depths in the atmosphere, we can obtain a more reliable model (Mauas *et al.* 2006).

Certainly the best known semiempirical model is the one for the average Quiet-Sun, Model C by Vernazza *et al.* (1981). Semiempirical modeling was successfully applied also to the atmospheres of cool stars. An extensive modeling of dM stars, has been done, starting with the work by Cram & Mullan (1979), Short & Doyle (1998), and by Mauas & Falchi (1994), and Mauas *et al.* (1997). In cool stars the application of NLTE semi-empirical chromospheric modelling can be based on optical and ultraviolet (UV) observations. This is because lines such as the H-α Na I D Ca II IRT become dominated by electron-collision excitation processes, which make them effective chromospheric diagnostics (Houdebine 1996). The possibility of using H-α profile as a diagnostic of stellar chromospheres was discussed in detail by Cram & Mullan (1979) and Mullan & Cram (1982) in terms of control of the source function by collisional processes. H-α is observed in active stars, in a wide variety of shapes and sizes; when the effective temperature is low enough, the collisional control of the H-α source function become possible over a wide range of chromospheric pressures. Under conditions of collisional control, H-α can be a good chromospheric pressure diagnostic. Mg II h&k UV lines, due to their large opacity, provide excellent diagnostic over a wide range of heights of the upper chromospheric layers (Uitenbroek 1992), and Ca II IRT triplet is a constraint for the shape of the middle chromosphere from the temperature minimum up to the plateau (Andretta *et al.* 2005). Here I applied the NLTE semi-empirical chromospheric modelling to the K2 star of HR 7428 binary system basing the analysis on the H-α H-β Na I D Ca II IRT triplet and Mg II h&k lines and UV continuum diagnostics.

2. Data acquisition and reduction

H-α Na I D H-β Ca II IRT spectroscopic observations of HR 7428 were carried out at the 91-cm telescope of Catania Astrophysical Observatory, "M. G. Fracastoro" station (Serra La Nave, Mt. Etna, Italy), using the new Catania Astrophysical Observatory Spectropolarimeter (CAOS) which is a fiber fed, high-resolution, cross-dispersed echelle spectrograph (Leone *et al.* 2016, Spanó *et al.* 2004, Spanó *et al.* 2006.)

The spectra were obtained in July 2015. Exposure times have been tuned in order to obtain a signal-to-noise ratio of at least 200 in the continuum in the 390-900 nm, with a resolution of $R = \frac{\Delta(\lambda)}{\lambda} = 45,000$, as measured from ThAr and telluric lines.

Echelle IRAF packages have been used for data reduction, following the standard steps: bias subtraction, background subtraction, trimming, flat-fielding, scattered light subtraction and order extraction. Several ThAr lamp exposures were obtained during each night and then used to provide a wavelength calibration of the observations.

Table 1. HR 7428 K2II-III and A2 components as determined by Marino *et al.* (2001)

Element	Primary (cooler K2II-III)	Secondary (hotter A2)
R	$40.0 \pm 6.5 R_\odot$	$2.25 \pm 0.5 R_\odot$
T_{eff}	$4400\ K \pm 150\ K$	$9000\ K \pm 200\ K$
$\log g$	2.0 ± 0.5	4.0 ± 0.5

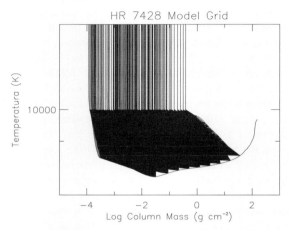

Figure 1. Grid of 15691 models used for the study of HR7428 atmosphere. The grid has been built s described in the text.

Each spectral order was normalized by a polynomial fit to the local continuum. Mg II h&k spectroscopic observations have been obtained, in 1997, by the IUE satellite. IUE spectra have been corrected for interstellar extinction. The typical value of 1 magnitude per kilo-parsec for the interstellar extinction and the Hipparcos distance (d = 323 pc), adopted for the computations, lead to an extinction magnitude A(V) = 0.32. Assuming the standard reddening law $A(V)=3.1\times$E(B-V), a color excess E(B-V) = 0.10 has been derived. IUE spectra have been de-reddened according to the selective extinction function of Cardelli *et al.* (1989). The spectral resolution is about 0.2 Å for the Mg II region.

3. The model grid

The atmospheric model of the K2 primary magnetic active component has been built computing a photospheric model, a chromospheric model and a transition region model and joining the three together.

The Photospheric Model

Taking into account the Marino *et al.* (2001) HR 7428 physical parameters (Table 1) we computed the A2 secondary component atmospheric model, by the ATLAS9 code (Kurucz 1993) using the parameters $\log g = 4.0$, $T_{\text{eff}} = 9000$ K and solar metalicity and the K2 primary component photospheric model using the parameters $\log g = 2.0$, $T_{\text{eff}} = 4400$ K and solar metalicity. The A star dominate the continuum emission for $\lambda \leqslant 3200$ Å, while for wavelength longer than $\lambda =3200$ Å, the continuum is dominated by the K star, therefore I neglected the contribution of the A2 when computing optical line profiles while I took it into account in the computation of the Mg II h&k lines.

The Transition Region Model

A first estimation of a plane-parallel model for the lower transition region of the HR 7428 K2 primary component, has been built using the method of the Volumetric Emission Measure (Jordan & Brown 1981, Harper 1992). The flux at the star, for lines forming at temperature $T_e \approx 10^5$ are, dominated by collisions and this results in emission lines that are optically thin and with a contribution function sharply picked in temperature, that is, typically formed over a temperature range of $\Delta \log(T_e) = 0.30$. The above conditions allow to derive the temperature gradient as a function of the averaged Emission Measure over $\Delta \log(T_e) = 0.30$ that we indicate as $EM_{0.3}$.

By imposing hydrostatic equilibrium, including turbulent pressure, the transition region model can be obtained by combining the temperature gradient as a function of $EM_{0.3}$ and the pressure variation from the equation of hydrostatic equilibrium (see, e.g., Harper 1992). The expression for the temperature gradient combined with the equation of hydrostatic equilibrium gives the relationship:

$$P_T^2(T_2) - P_T^2(T_1) = = 2 * (1.4)^2 m_p g k \int_{T_1}^{T_2} [EM_{0.3}(1 + 1.1x) + \tfrac{1.4 x m_p \eta^2 EM_{0.3}}{2kT}]dT \qquad (3.1)$$

where m_p is the H^+ mass, k is the Boltzmann constant and x= N_H/N_e.

The Eq. 3.1 together with an estimate of electron density and turbulent velocity in a layer, allows to find the pressure as a function of temperature for the TR model. The total particle density (N_{tot}), gas pressure (P_G), turbulence pressure (P_{turb}) and electron density (N_e) are then obtained according to the following relations

$$\begin{aligned}
N_{tot} &= \sqrt{P_T}\,(k\,T + 0.5 m_p \eta^2) \\
P_G &= N_{tot}\,k\,T \\
P_{turb} &= 0.5 N_{tot}\,m_p\,\mu\,\eta^2 \\
N_e &= N_{tot}/(1. + 1.1\,x)
\end{aligned} \qquad (3.2)$$

I used the equations above in order to build plane-parallel models for the lower transition region of the HR 7428 K2 primary component. As an estimation of the Volumetric Emission Measure (VEM) vs T_{eff} of the HR 7428 K2 primary component I used the measured by Griffiths & Jordan 1998 for the RS CVn system, HR 1099 primary component ($R_{HR1099K1IV}$ =3.9 $R_\odot$), opportunely scaling them, for the bigger radius (R_{K2} =40 $R_\odot$) of HR 7428 K2 star, according the formula $EM_{0.3} = VEM/(4\pi R^2)$ (Brown *et al.* 1991) and the parameters of Table 1.

From this estimated $EM_{0.3}$ a grid of 160 transition region models has been built by means of equations 3.1 and 3.2 using a grid of 23 values of electron density at the fixed temperature T_e =50000 K obtained scaling of a factor from 0.5 up to 30 the values of electron density measured in HR 1099 ($N_e = 5 \cdot 10^{+11}$). For each of these 23 electron density values, a grid of seven values of the turbulent velocity in the layer with $T_0 = 10^4$ $v_{turb}(T_0 = 10^4)$= 10, 20, 30, 40, 50, 60, 70 kms^{-1} has been considered for the calculation of turbulent velocity distribution according the empirical law by Griffiths & Jordan98 (1998) $v_{turb}(T) = v_{turb}(T_0) * (T/T_0)^{1/4}$ between log (T)=4.0 and log (T)=5.3.

These transition region models provide the upper boundaries for the radiative transfer calculations of the chromospheric models, while the adopted photospheric model provides the lower boundaries of the chromospheric models.

The Chromospheric Model

For each one of the 160 transition region models and for each of nine values of T_{min} in the range between ~2800 K and 4200 K that we have chosen with a step of less than 200 K as points where to cut the photospheric Kurucz model, a grid of 25 chromospheric models are generated by a smooth spline interpolation between the photosphere and the transition region using as free parameters a grid of 5×5 interpolation knots. We impose

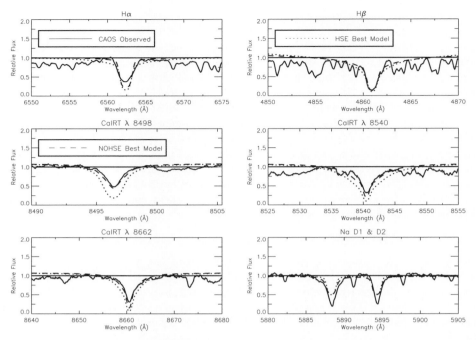

Figure 2. NLTE H-α, H-β, Ca II IRT, Na I D normalized profiles computed for the best HSE (dotted line) and the best $NOHSE$ (dashed line) models compared with observations.

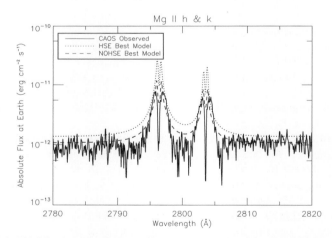

Figure 3. Mg II h&k absolute flux profiles computed from the best HSE and the best $NOHSE$ atmospheric models compared with the IUE observation.

the chromospheric structures to have a monotonic temperature dependence on column mass $(dT/dm \leqslant 0)$.

The final grid of models includes 160*9*25=36225 models, and only 15691 satisfy the $dT/dm \leqslant 0)$ constraint and are shown in Fig. 1.

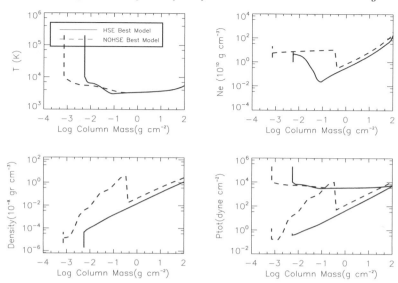

Figure 4. Temperature, electron density, density mass and pressure versus Column Mass for the best *HSE* and *NOHSE* models of the K2 primary component of the RS CVn system HR 7428. In the last plot, the two best models are shown for references.

4. Computation

The coupled equations of radiative transfer and statistical equilibrium were solved using the version 2.2 of the code *Multi* (Carlsson 1986), for the H, Ca, Na, Mg atomic models. The H atomic model incorporates 16 states of H, with 84 b-b and 9 b-f transitions. The Ca atomic model incorporates 8 states of CaI, the lowest 5 states of CaII and the ground state of CaIII, 9 b-b and 13 b-f transitions are treated in detail. The Mg II atomic model is made of 3 states MgI, the lowest 6 states of MgII and the ground state of mgIII, 9 b-b and 9 b-f transitions are treated in detail. The Na atomic model incorporates 12 levels: 11 levels of NaI and the ground state of NaII and 29 b-b and 11 b-f transitions are treated in detail. The opacity package included in the code takes into account free-free opacity, Rayleigh scattering, and bound-free transitions from hydrogen and metals, I included the line blanketing contribution to the opacity using the method described in Busá *et al.* (2001).

Imposing Hydrostatic Equilibrium - HSE Models

As a first step I imposed hydrostatic equilibrium to the hydrogen. In detail, starting with the LTE hydrogen populations for the electron pressure and density calculation, hydrogen is iterated to convergence. Then Hydrostatic equilibrium $dP_{Tot} = \rho g dh$ is solved and electron pressure and hydrogen populations are updated, the loop continues until we obtain a convergence. Theobtained electron density is then used to solve the Ca, Mg, and Na radiative transfer and statistical equilibrium equations. The population densities obtained from the H calculation are used to obtain the background NLTE source function in the Ca, Mg, and Na calculations. This computation modifies the initial grid of Fig. 1 because a new column of electron density is obtained, and we obtain a new grid of only 2052 models that we call HSE models.

Not imposing Hydrostatic Equilibrium - NOHSE Models

I also considered not imposing hydrostatic equilibrium that means fixing the electron densities to the ones of the original grid of Fig. 1. In this case we consider all the 15691 models that only satisfy the $dT/dm \leqslant 0$ constraint and we call these, *NOHSE* models.

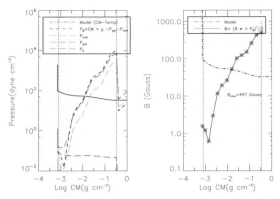

Figure 5. Chromospheric magnetic field distribution obtained considering the whole lacking pressure as magnetic pressure.

This approach takes into account the possibility that some pressure contributions are neglected and becomes a method to derive an estimate of the lacking pressure component. In such an approach I use the electron density as a free parameter, looking for the electron density distribution that best fit the data. I accept as best solution, also a distribution whose total pressure is not balancing the gravity, accepting the hypothesis that a new pressure component could be considered for achieving the equilibrium.

Implications of HSE - NOHSE Models

Both the HSE grid of 2052 models and the initial grid of 15691 model have been considered in the radiative transfer calculations for H, Na, Ca and Mg. For each grid, the models that converge to solution for all the atoms have been taken into account for the comparison with the observed spectrum. The computed profiles of H-α, H-β, Ca II IRT, Na I D, have been convolved for a rotational profile with $v \sin i$=17 kms^{-1} (Marino *et al.* 2001) and an instrumental profile with $\frac{\Delta(\lambda)}{\lambda} = 45,000$, normalized and then compared with observations. The computed profiles of Mg II h&k have been obtained by the weighted sum of the K2 star profile and the A2 star profile and weighting for the d^2/R^2 factor. Wavelength shifts, to account for orbital velocities, are applied to synthetic spectra.

5. Results

I used a χ^2 minimization procedure for the selection of the model that best describes the mean outer atmosphere of HR 7428. This procedure has been applied both to the HSE and $NOHSE$ set of models. We find that the best $NOHSE$ model has a χ^2_{Tot}=1.22 while the best HSE model has a χ^2_{Tot}=2.60. This result let me conclude that the $NOHSE$ best model distribution of temperature, gas pressure, electron and population densities is the best description of the mean outer atmosphere of the K2 star of the binary system HR7428. In Fig. 3, the HSE and the $NOHSE$ best models are compared in temperature, electron density and mass density versus Column Mass (first three plots). But the most important difference between the two models is shown in the last plot of Fig. 3 where the total pressure is plotted versus Column Mass. While, in the HSE model, the total chromospheric pressure is imposed to be equal to the Column Mass multiplied for the gravity, that is PTot= Column Mass × g, and we obtain a straight line, in the $NOHSE$ model, the total chromospheric pressure exceed the gravity pressure (dashed line is not a straight line but, in the chromospheric layers lays well above of the gravity pressure).

This exceeding pressure has to be balanced by an equal and opposite pressure otherwise we cannot have a stable star. Therefore, I assumed the difference P_{Tot}-CM×g as equal, with opposite sign, to the lacking pressure in our calculations, that is: $P_{Totnew} = P_{turb} + P_{gas} + P_{new} = CM \times g$ and therefore $P_{new} = CM \times g \text{-} P_{Tot}$.

I make the hypothesis that the additive pressure, could be a magnetic pressure that from the base of the chromosphere decrease toward the outer layers. This is quite probable in the case of cool active stars. The atmospheres of cool active stars are, in fact, permeated by magnetic fields that emerge from deeper layers. With increasing height we might expect the structure to be more greatly influenced by magnetic fields, since the energy density of the magnetic fields should fall off more slowly than the energy density of the gas (this is the case of solar atmosphere where $\beta = 8\pi N K T_e / |(B)|^2$ is $\geqslant 1$ in photosphere and $\ll 1$ in transition region layers). In Fig. 4 the new pressure component is shown in comparison to electron pressure, gas pressure, turbulent pressure. It is clear that this pressure component is not negligible being of the same order of strength of the gas pressure. In the hypothesis that the additive pressure, could be a magnetic pressure, I calculated a ≈ 500 gauss magnetic field corresponding magnetic field that, from the base of the chromosphere decrease toward the outer layers. It is worthwhile to notice that the *NOHSE* refers only to the chromospheric layers, therefore this new component pressure estimation refers only to this layer.

References

Andretta, V., Busá I., Gomez, M. T., & Terranegra, L. 2005 *Astronomy & Astrophysics* 430, 669

Byrne, P. B., Abdul Aziz, H., Amado, P. J., *et al.* 1998 *A&AS* 127, 505

Brown, A. *et al.* 1991 *Astrophysical Journal* 373, 614B

Busá I., Andretta, V., Gomez, M. T., & Terranegra, L. 2001 *Astronomy & Astrophysics* 373, 993

Carlsson, M. 1986, *Technical report 33, Uppsala Astronomical Observatory*

Cram, L. E. & Mullan, D. J. 1979, *Astrophysical Journal* 234, 579

Fontenla, J. M., Avrett, E. H., & Loeser, R. 1993, *Astrophysical Journal* 406, 319

Gratton, L. 1950, *Astrophysical Journal* 111, 31

Griffiths, N. W. & Jordan, C. 1998, *Astrophysical Journal* 497, 883

Jordan, C. & Brown, A. 1981 SPSS, 199J

Hall, D. S., Gessner, S. E. Lines, H. C., & Lines, R. D. 1990 *Astronomical Journal* 100, 2017

Harper, G. M. 1992 *Monthly Notices of the Royal Astronomical Society* 256, 37

Houdebine, E. R. 19960 *IAUS* 176, 547H

Kalkofen, W., Ulmschneider, P., & Avrett, E. H. 1999 *Astrophysical Journal* 521, 141

Kurucz, R. L. 1993, *IAU Coll. 138, ASP Conf. Ser.* 44, 87

Kurucz, R. L. & Avrett, E. H. 1981, *SAOSR*, 391

Leone, F., Avila, G., Bellassai, G., *et al.* 2016, *AJ*, 151, 116L

Marino, G., Catalano, S., Frasca, A., & Marilli, E. 2001, *Astronomy & Astrophysics* 375, 100

Mauas, P. J. D. & Falchi, A. 1994, *Astronomy & Astrophysics* 281, 129

Mauas, P. J. D., Falchi, A., Pasquini, L., & Pallavicini, R. 1997, *Astronomy & Astrophysics* 326, 249

Mauas, P. J. D., Cacciari, C., & Pasquini, L. 2006, *Astronomy & Astrophysics* 454, 609

Cram, L. E. & Mullan, D. J. 1979, *Astrophysical Journal* 234, 579

Parsons, S. B. & Ake, T. B. 1987, *Bull. Am. Astron. Soc.* 19, 708

Short, C. I. & Doyle, J. G. 1998, *Astronomy & Astrophysics* 336, 613

Spanó P., Leone, F., Bruno, P., Catalano, S., Martinetti, E., Scuderi, S. 2006, *MSAIS* 9, 481S

Spanó P., Leone, F., Scuderi, S., Catalano, S., & Zerbi, F. M. 2004, *SPIE*, 5492, 373S

Uitenbroek, H. 1992, *ASPC* 26, 546

Vernazza, J. E., Avrett, E. H., & Loeser, R. 1973, *Astrophysical Journal* 184,605V

Vernazza, J. E., Avrett, E. H., & Loeser, R. 1981, *Astrophysical Journal Supplement* 45,635

Living Around Active Stars
Proceedings IAU Symposium No. 328, 2016
D. Nandy, A. Valio & P. Petit, eds.

© International Astronomical Union 2017
doi:10.1017/S1743921317004070

CARMENES – M Dwarfs and their Planets: First Results

A. Quirrenbach[1], P.J. Amado[2], I. Ribas[3], A. Reiners[4], J.A. Caballero[1], W. Seifert[1], M. Zechmeister[4] and the CARMENES Consortium[1,2,3,4,5,6,7,8,9,10,11]

[1]Landessternwarte, Zentrum für Astronomie der Universität Heidelberg, Königstuhl 12, D-69117 Heidelberg, Germany
[2]Instituto de Astrofísica de Andalucía (CSIC), Glorieta de la Astronomía s/n, E-18008 Granada, Spain
[3]Institut de Ciències de l'Espai (CSIC-IEEC), Campus UAB, Facultat Ciències, Torre C5 - parell - 2a planta, E-08193 Bellaterra, Barcelona, Spain
[4]Institut für Astrophysik (IAG), Friedrich-Hund-Platz 1, D-37077 Göttingen, Germany
[5]Max-Planck-Institut für Astronomie, Königstuhl 17, D-69117 Heidelberg, Germany
[6]Centro de Astrobiología (CSIC-INTA), Campus ESAC, Camino Bajo del Castillo, s/n, E-28691 Villanueva de la Cañada, Madrid, Spain
[7]Calar Alto Observatory (MPG-CSIC), Centro Astronómico Hispano-Alemán, Jesús Durbán Remón, 2-2, E-04004 Almería, Spain
[8]Departamento de Astrofísica, Facultad de Física, Universidad Complutense de Madrid, E-28040 Madrid, Spain
[9]Thüringer Landessternwarte Tautenburg, Sternwarte 5, D-07778 Tautenburg, Germany
[10]Instituto de Astrofísica de Canarias, Vía Láctea s/n, E-38205 La Laguna, Tenerife, Spain, and Dept. Astrofísica, Universidad de La Laguna, E-38206 La Laguna, Tenerife, Spain
[11]Hamburger Sternwarte, Gojenbergsweg 112, D-21029 Hamburg, Germany

Abstract. CARMENES is a pair of high-resolution ($R \gtrsim 80,000$) spectrographs covering the wavelength range from 0.52 to 1.71 μm with only small gaps. The instrument has been optimized for precise radial velocity measurements. It was installed and commissioned at the 3.5 m telescope of the Calar Alto observatory in Southern Spain in 2015. The first large science program of CARMENES is a survey of ∼300 M dwarfs, which started on Jan 1, 2016. We present an overview of the instrument, and provide a few examples of early science results.

Keywords. planetary systems, stars: late-type, surveys, instrumentation: spectrographs, techniques: radial velocities, techniques: spectroscopic

1. Introduction

CARMENES is a new radial-velocity facility for the 3.5 m telescope of Calar Alto Observatory (CAHA) close to Almería, Spain (see also Quirrenbach *et al.* 2010, 2012, 2014, 2016). The main scientific objective of CARMENES is carrying out a survey of M-type main sequence stars, and the instrument has been optimized solely for this purpose. The CARMENES survey will characterize the population of planets around these stars, and detect low-mass planets in their habitable zones (HZs). In the focus of the project are very cool stars of spectral type M4 V and later, and moderately active stars, but the target list also comprises earlier and therefore brighter M dwarfs. In particular, we aim at being able to detect 2 $M_\oplus$ planets in the HZs of M5 V stars. A long-term radial velocity precision of ∼1 m/s per measurement will permit to attain this goal. The CARMENES survey will also produce a unique data base of high-resolution spectra of M dwarfs, enabling studies of stellar activity and improved determinations of stellar parameters. These data will thus be of high scientific value by themselves, and they will also be needed for disentangling the signatures of planetary companions from activity-induced radial-velocity variations.

2. The CARMENES Instrument

The CARMENES instrument has been optimized for obtaining precise radial velocities of cool stars. In the front end attached to the Cassegrain focus of the 3.5 m telescope, the light is separated by a dichroic beam splitter at 0.96 μm. The spectral ranges shortward and longward of this wavelength are sent to two separate spectrographs, which are mounted on benches inside vacuum tanks located in the coudé laboratory of the 3.5 m dome. The main instrument components of CARMENES are the following:

- *Front End.* The front end is attached to the Cassegrain focus of the 3.5 m telescope and contains a camera for acquisition and guiding, an atmospheric dispersion compensator, the dichroic beam splitter, a shutter in the visible channel, input selectors to switch between the sky and calibration light, and fiber heads (Seifert *et al.* 2012). The first mirror in the front end is motorized; when it is detracted the light passes straight through to a separate focus so that it is possible to switch rapidly between CARMENES and another Cassegrain instrument.

- *Fibers.* The optical fibers transporting the light from the front end to the spectrographs also fulfill the important task of "scrambling", i.e., of reducing the jitter at the spectrograph pseudo-slit with respect to guiding errors and seeing at the fiber input. For improved scrambling, the long circular fibers leading from the telescope to the coudé room are connected to shorter fiber sections with an octagonal diameter (Stürmer *et al.* 2014). The fiber diameter has been chosen to provide a 1″.5 acceptance angle on the sky, matched to somewhat worse than median seeing on Calar Alto.

- *Visible-Light Spectrograph.* The visible-light échelle spectrograph covers the wavelength range from 0.52 μm to 1.05 μm with a resolving power of $R = 94,600$ and a mean sampling of 2.8 pixels per resolution element. It accepts light from two fibers; the first fiber carries the light from the target star, while the second fiber can either be used for simultaneous wavelength calibration or for monitoring the sky. The optical design is a grism cross-dispersed, white pupil, échelle spectrograph working in quasi-Littrow mode using a two-beam, two-slice, image slicer. The spectrograph is housed in a vacuum vessel and operated at room temperature. The detector is a back-side illuminated 4112×4096 pixel CCD (model e2v CCD231-84).

- *Near-Infrared Spectrograph.* The design of the near-IR spectrograph is very similar to that of its visible counterpart. It provides $R = 80,400$ over the range 0.95 μm to 1.71 μm with a mean sampling of 2.5 pixels per resolution element. It is cooled to 138 K with a continuous flow of gaseous nitrogen. The detector is a mosaic of two 2048×2048 pixel HAWAII-2RG infrared arrays with a long-wavelength cutoff at 2.5 μm (see also Amado *et al.* 2012). The near-IR cooling system employs an external heat exchanger / evaporator unit that is fed by liquid nitrogen and provides a continuous flow of gaseous nitrogen to the near-IR spectrograph (Becerril *et al.* 2012).

- *Calibration Units.* CARMENES uses hollow-cathode emission line lamps and Fabry-Pérot etalons for spectral calibration. For each spectrograph, the arc lamps as well as quartz lamps for flat-fielding are housed in a calibration unit that is connected to the front end with a fiber link.

- *Exposure Meters.* The zeroth-order light from the échelle gratings in the two spectrographs is routed to photomultiplier tubes, which monitor the received intensity with high time resolution. This information is needed for an accurate conversion of the

Table 1. CARMENES installation and commissioning milestones

Date	Milestone
December 22, 2014	3.5 m coudé room refurbishment for CARMENES complete
April 23, 2015	Front end arrives at CAHA
April to June 2015	Commissioning of front end, first ICS tests
July 6, 2015	VIS vacuum tank arrives at CAHA
August 17, 2015	VIS spectrograph arrives at CAHA
September 1, 2015	Installation of VIS calibration unit at CAHA
October 3, 2015	First Light for the VIS spectrograph
October 20, 2015	NIR spectrograph arrives at CAHA
November 7, 2015	First Light for NIR spectrograph
November 9, 2015, 20:20:51UT	**CARMENES First Light** (VIS and NIR simultaneously)
November and December 2015	Commissioning of the complete instrument
December 30, 2015	Provisional Acceptance complete
January 1, 2016	**Start of CARMENES Survey**

observed radial velocity to the barycenter of the Solar System. It can also be used to make real-time adjustments to the integration time depending on atmospheric conditions.

• *Instrument Control System.* The coordination and management of the sub-systems of CARMENES is handled by the instrument control system (ICS), which provides a tool to operate the instrument in an integrated manner (Colomé *et al.* 2016). The ICS includes a scheduler that can autonomously prioritize and select targets for observation.

• *Infrastructure.* The CARMENES spectrographs and ancillary equipment are located in the coudé room of the 3.5 m telescope dome. Each spectrograph is placed within a temperature-controlled chamber, providing shielding from annual temperature variations and from heat sources such as electronics, pumps, and calibration lamps. An intelligent interlock system monitors the status of the instrument and of the auxiliary systems, and organizes information about their overall status and health (Helmling *et al.* 2016).

3. Installation and Commissioning

The subsystems of CARMENES were moved to Calar Alto and installed at the 3.5 m telescope in the course of 2015 (see Tab. 1). The front end was mounted at the Cassegrain flange in April, followed by extensive testing of the acquisition and guiding procedures and the software interfaces with the telescope control system. The optical fibers connecting the front end to the spectrographs were routed through the telescope fork at the same time. The visible-light spectrograph was shipped to the observatory in July. The optical bench and the vacuum system had been separately pre-integrated at Landessternwarte Heidelberg and at the Max-Planck-Institut für Astronomie, respectively; they were first integrated with each other on site. The near-infrared spectrograph was fully integrated at the Instituto de Astrofísica de Andalucía and moved to Calar Alto in October. The calibration system and the Fabry-Pérot etalons were installed in parallel. CARMENES had "First Light" – defined as taking stellar spectra with both spectrographs simultaneously – on Nov 9, 2015. This event marked the beginning of the commissioning, in which the whole instrument was tested and characterized. The CARMENES M dwarf survey started on Jan 1, 2016, after the instrument passed its provisional acceptance tests.

4. Calibration Strategy

Precision spectroscopy at red optical and infrared wavelengths requires a novel strategy for   wavelength   calibration.   In   a   spectrograph   like   HARPS,   for   example,

Figure 1. Exposures of hollow-cathode lamps (from left to right: Th-Ne, U-Ne, U-Ar) with the visible-light spectrograph. The spectra are rich in calibration information, but suffer from very bright noble gas lines.

Th-Ar hollow-cathode lamps (HCLs) provide a dense forest of emission lines that are not severely affected by noble gas emission (Ar) because it is strong only at wavelengths on the red side of the HARPS wavelength cutoff (680 nm). In contrast, CARMENES operates in the region where all HCL fill gases emit very bright lines (see Fig. 1). Some of these lines are much stronger than typical Th (or other cathode material) lines and can saturate the detectors. Furthermore, Th emits most of its lines at optical wavelengths but not so many in the infrared. In preparation for CARMENES, we investigated different HCLs and constructed new line lists (Sarmiento *et al.* 2014). In CARMENES, we are using three different types of emission lamps (Th-Ne, U-Ar, and U-Ne, see Fig. 1) to provide optimal coverage. In addition, we operate two passively stabilized Fabry-Pérot etalons (FPs) optimized for the two spectrographs in order to cover the entire CARMENES wavelength with dense emission lines. With more than 10^4 FP emission lines, we can construct a precise wavelength solution for the FP comb that is incorporated in our wavelength calibration scheme (Bauer *et al.* 2015). The FPs are also used during the night to monitor short-term spectrograph drifts; long-term stability is ensured by comparing the FPs to HCL exposures taking during daytime.

5. Data Reduction

During standard operation at night, each CARMENES spectrograph simultaneously receives light from a target in the first (science) fiber and the corresponding FP etalon in the second (calibration) fiber. (For faint targets, it is also possible to use the second fiber as a sky fiber.) The extraction of the spectra follows the reduction procedure described in Baranne *et al.* (1996). The data reduction software was built on the basis of RE-DUCE, a package for cross-dispersed échelle spectra reduction written in IDL (Piskunov & Valenti 2002). The key feature is the optimal extraction that we use together with a new algorithm optimized for stabilized spectrographs (Zechmeister *et al.* 2014). The measurement of radial velocities from the spectra is carried out following two complementary approaches, the method of least squares fitting (a detailed description of our algorithms and their applicability to M dwarfs is given in Anglada-Escudé & Butler (2012), see also Caballero *et al.* 2016), and the cross-correlation method using a consistent flux weighting algorithm (Pepe *et al.* 2002).

6. Stellar Spectra

In this section we show a few examples demonstrating the capabilities of CARMENES for stellar astrophysics. The CARMENES spectra cover the wavelength range from 0.52 to 1.71 µm with only minor gaps; an example in which some important chromospheric

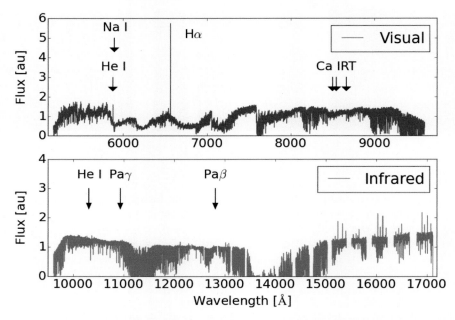

Figure 2. CARMENES spectrum of YZ CMi. Important chromospheric lines are identified. Courtesy S. Czesla.

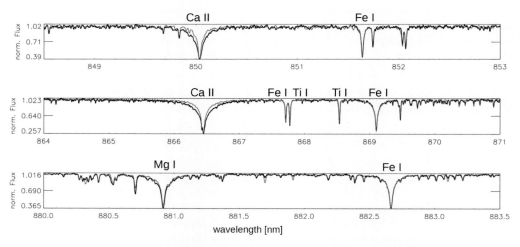

Figure 3. Section of the CARMENES spectrum of an M0.5 V star (black) and the best fit model (blue: model outside fit region, red: model inside fit regions for χ^2-minimization. Courtesy V. Passegger.

lines have been identified is shown in Fig. 2. The high signal-to-noise (typically 100 or more) that is needed for measuring precise radial velocities makes the spectra also very well suited for determining stellar parameters such as effective temperature, gravity, and metallicity. The result from fitting PHOENIX-ACES models (Husser *et al.* 2013) with a downhill simplex methods to a section of the spectrum of an M0.5 V star is shown in Fig. 3.

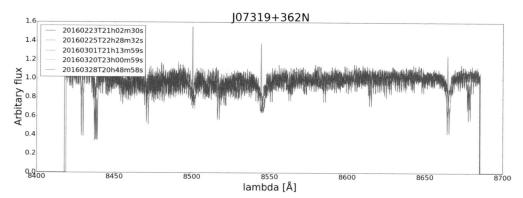

Figure 4. Sections of five CARMENES spectra of the active star BL Lyn around the Ca II infrared triplet taken in February / March 2016. The infrared triplet lines were in absorption during four epochs, but showed narrow emission components on March 1. Adapted from Brinkmöller (2016).

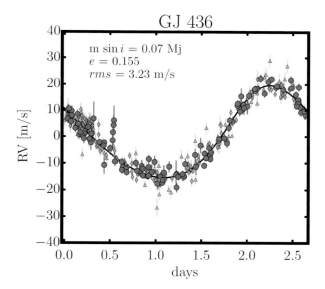

Figure 5. Phase-folded radial-velocity data and orbital fit for GJ 436. The grey triangles are literature data from Keck-HIRES (Maness *et al.* 2007) and HARPS (Lanotte *et al.* 2014), the red dots are data from the CARMENES visual spectrograph. Courtesy T. Trifonov.

In Fig. 4 we show sections of five spectra of the active star BL Lyn covering the Ca II infrared triplet. While the triplet lines are in absorption during four of the epochs, sharp emission components are apparent in the spectrum taken on March 1. Time series of spectra taken during the CARMENES survey will enable analyses of flaring activity, and of correlations between activity indicators and radial velocities.

7. Radial Velocities

CARMENES was designed with the goal of achieving a stability of 1 m/s for radial velocity measurements of late-type stars. An initial look at the data from the visual spectrograph shows that an r.m.s. velocity precision of a few m/s has in fact been achieved for most stars observed during the first few months of operation; several stars show a velocity scatter less than 2 m/s. Since this figure includes a contribution from stellar "jitter", it places an upper limit on the intrinsic stability of the spectrograph. CARMENES data on GJ 436, which is known to harbor a planet, are shown in Fig. 5 along with measurements from Keck-HIRES and HARPS, showing that CARMENES delivers data that are comparable to those from other state-of-the-art instruments.

Getting precise radial velocities from the NIR spectrograph is much more complicated for several reasons: The spectrograph needs to be actively cooled and stabilized (whereas the visible-light spectrographs rely on passive stabilization), the NIR detectors are more difficult to characterize and calibrate than CCDs, and the spectra are much more heavily contaminated by telluric absorption. Work on optimizing the calibration and stabilization of the NIR spectrograph is still ongoing, but observations covering several nights with high-cadence sampling and RV scatter of a few m/s have already been realized.

8. The CARMENES Survey

To define the CARMENES survey sample, we have ranked the M dwarfs with declination $\delta > -23°$ by apparent magnitude within each spectral subtype, and selected the brightest stars in each subclass (not considering binaries with separation $< 5''$). This creates a sliding magnitude cut-off that helps biasing the sample towards later spectral subtypes, while maintaining a simple selection criterion that can be modeled easily in statistical analyses. We thus obtain a sample that takes advantage of the "sweet spot" for CARMENES in the M3 V to M4 V spectral range: Earlier M subtypes can be observed quite efficiently with spectrographs working at bluer wavelengths; reaching larger numbers of later stars requires near-IR spectrographs at larger telescopes.

During the first year of observations, more than 5,000 visible-light and 4,500 NIR spectra on 330 individual M dwarfs were taken. The first data product from CARMENES to be released in mid 2017 will be a library of single-epoch spectra of these stars.

Acknowledgements

CARMENES is an instrument for the Centro Astronómico Hispano-Alemán de Calar Alto (CAHA, Almería, Spain). CARMENES is funded by the German Max-Planck-Gesellschaft (MPG), the Spanish Consejo Superior de Investigaciones Científicas (CSIC), the European Union through FEDER/ERF funds, and the members of the CARMENES Consortium (see author list), with additional contributions by the Spanish Ministry of Economy, the state of Baden-Württemberg, the German Science Foundation (DFG), the Klaus Tschira Foundation (KTS), and by the Junta de Andalucía.

References

Amado, P.J., Lenzen, R., Cárdenas, M.C., et al. (2012). *CARMENES. V: non-cryogenic solutions for YJH-band NIR instruments.* In *Modern technologies in space- and ground-based telescopes and instrumentation II.* SPIE 84501U

Anglada-Escudé, G., & Butler, R.P. (2012). *The HARPS-TERRA project. I. Description of the algorithms, performance, and new measurements on a few remarkable stars observed by HARPS.* ApJS 200, 15

Baranne, A., Queloz, D., Mayor, M., et al. (1996). *ELODIE: A spectrograph for accurate radial velocity measurements.* A&AS 119, 373

Bauer, F.F., Zechmeister, M., & Reiners, A. (2015). *Calibrating echelle spectrographs with Fabry-Péerot etalons.* A&A 581, A117

Becerril, S., Lizon, J.L., Sánchez-Carrasco, M.A., *et al.* (2012). *CARMENES. III: an innovative and challenging cooling system for an ultra-stable NIR spectrograph*. In *Modern technologies in space- and ground-based telescopes and instrumentation II*. SPIE 84504L

Brinkmöller, M. (2016). *Analysis of the activity of M dwarfs observed by CARMENES assessed on the calcium infrared triplet*. BSc Thesis, Univ. Heidelberg

Caballero, J.A., Guárdia, J., López del Fresno, M., *et al.* (2016). *CARMENES: data flow*. In *Observatory operations: strategies, processes, and systems VI*. SPIE 99100E

Colomé, J., Guàrdia, J., Hagen, H.J., *et al.* (2016). *CARMENES: The CARMENES instrument control software suite*. In *Software and cyberinfrastructure for astronomy IV*. SPIE 991334

Helmling, J., Wagner, K., Hernández Castaño, L., *et al.* (2016). *CARMENES: interlocks or the importance of process visualization and system diagnostics in complex astronomical instruments*. In *Ground-based and Airborne Instrumentation for Astronomy VI*. SPIE 990890

Husser, T.O., Wende-von Berg, S., Dreizler, S., *et al.* (2013). *A new extensive library of PHOENIX stellar atmospheres and synthetic spectra*. A&A 553, A6

Lanotte, A.A., Gillon, M., Demory, B.O., *et al.* (2014). *A global analysis of Spitzer and new HARPS data confirms the loneliness and metal-richness of GJ 436 b*. A&A 572, A73

Maness, H.L., Marcy, G.W., Ford, E.B., *et al.* (2007). *The M Dwarf GJ 436 and its Neptune-Mass Planet*. PASP 119, 90

Pepe, F., Mayor, M., Galland, F., *et al.* (2002). *The CORALIE survey for southern extra-solar planets VII. Two short-period Saturnian companions to HD 108147 and HD 168746*. A&A 388, 632

Piskunov, N.E., & Valenti, J.A. (2002). *New algorithms for reducing cross-dispersed echelle spectra*. A&A 385, 1095

Quirrenbach, A., Amado, P.J., Caballero, J.A., *et al.* (2014). *CARMENES instrument overview*. In *Ground-based and airborne instrumentation for astronomy V*. SPIE 91471F

Quirrenbach, A., Amado, P.J., Caballero, J.A., *et al.* (2016). *CARMENES: an overview six months after first light*. In *Ground-based and Airborne Instrumentation for Astronomy VI*. SPIE 990812

Quirrenbach, A., Amado, P.J., Mandel, H., *et al.* (2010). *CARMENES: Calar Alto high-Resolution search for M dwarfs with Exo-earths with Near-infrared and optical Echelle Spectrographs*. In *Ground-based and airborne instrumentation for astronomy III*. SPIE 773513

Quirrenbach, A., Amado, P.J., Seifert, W., *et al.* (2012). *CARMENES. I: Instrument and survey overview*. In *Ground-based and airborne instrumentation for astronomy IV*. SPIE 84460R

Sarmiento, L.F., Reiners, A., Seemann, U., *et al.* (2014). *Characterizing U-Ne hollow cathode lamps at near-IR wavelengths for the CARMENES survey*. In *Ground-based and airborne instrumentation for astronomy V*. SPIE 914754

Seifert, W., Sánchez Carrasco, M.A., Xu, W., *et al.* (2012). *CARMENES. II: optical and opto-mechanical design*. In *Ground-based and airborne instrumentation for astronomy IV*. SPIE 844633

Stürmer, J., Stahl, O., Schwab C., *et al.* (2014). *CARMENES in SPIE 2014. Building a fibre link for CARMENES*. In *Advances in optical and mechanical technologies for telescopes and instrumentation*. SPIE 915152

Zechmeister, M., Anglada-Escudé, G., & Reiners, A. (2014). *Flat-relative optimal extraction. A quick and efficient algorithm for stabilised spectrographs*. A&A 561, A59

Living Around Active Stars
Proceedings IAU Symposium No. 328, 2016
D. Nandy, A. Valio & P. Petit, eds.
© International Astronomical Union 2017
doi:10.1017/S1743921317004112

Magnetic activity of interacting binaries

Colin A. Hill

Université de Toulouse / CNRS-INSU, IRAP / UMR 5277, F-31400, Toulouse, France
email: chill@irap.omp.eu

Abstract. Interacting binaries provide unique parameter regimes, both rapid rotation and tidal distortion, in which to test stellar dynamo theories and study the resulting magnetic activity. Close binaries such as cataclysmic variables (CVs) have been found to differentially rotate, and so can provide testbeds for tidal dissipation efficiency in stellar convective envelopes, with implications for both CV and planet-star evolution. Furthermore, CVs show evidence of preferential emergence of magnetic flux tubes towards the companion star, as well as large, long-lived prominences that form preferentially within the binary geometry. Moreover, RS CVn binaries also show clear magnetic interactions between the two components in the form of coronal X-ray emission. Here, we review several examples of magnetic interactions in different types of close binaries.

Keywords. stars: novae, cataclysmic variables, stars: activity, stars: spots, stars: imaging, stars: magnetic fields, stars: binaries (including multiple): close, techniques: spectroscopic

1. Introduction

Interacting binaries provide unique parameter regimes in which to test stellar dynamo theories, and to study the resulting magnetic activity. Cataclysmic variables (CVs) are semi-detached binaries (typically) consisting of a lower main-sequence secondary star that is overflowing its Roche-lobe, transferring material to a white dwarf (WD) primary star. These systems typically have orbital periods of $P_{orb} < 10$ hr, with the tidal distortion (due to the companion star) forcing the secondary to synchronously rotate (on average) with the binary (i.e. $P_{rot} = P_{orb}$). This co-rotation means that magnetic braking (where charged particles stream along open field lines, removing angular momentum) does not slow the rotation period, but instead acts to shrink the orbit of the system, driving it to shorter orbital periods and sustaining the mass transfer. Furthermore, as the secondary stars may be massive enough to possess large radiative cores, they also have very low Rossby numbers (Ro $= P_{rot}/\tau_c < 0.003$) despite being significantly evolved. Moreover, Doppler imaging of the secondary stars has revealed that they are highly magnetically active, with starspots covering a large fraction of their surfaces, with evidence to suggest that magnetic flux tubes may preferentially emerge facing the companion star. The secondary stars in this class of binary have also been found to harbour large prominences (extending out to several stellar radii), that may also be influenced by the close companion star.

Thus, close binaries such as CVs (and related objects) provide excellent laboratories in which to test stellar dynamo theories, to study the resulting magnetic activity, and to observe the magnetic interactions between the two components. Indeed, these interactions may mirror those found between a planet and its host star, and so the phenomena observed in stellar binaries has far reaching applications.

2. Magnetic activity of the cataclysmic variable AE Aqr

As an example of the type of magnetic activity that may be observed in close binaries, we present our studies on the CV, AE Aqr. This bright CV (V mag $\sim$11.6) consists of a WD primary star with a K4V-type secondary, with P_{rot} = 0.41 d (1.5% that of the Sun). By using the technique of Roche tomography (analogous to Doppler imaging, but adapted specifically for close binary systems) we have constructed brightness maps of the surface, showing the highly spotted nature of this rapidly rotating star (see Figure 1).

Comparing two spot maps taken 9 days apart (labelled 2009a and 2009 in Figure 1), we determined that the surface of the secondary star in AE Aqr is differentially rotating, with a shear rate $d\Omega$ around 40% that of the Sun (Hill *et al.* 2014). This surprising result overturned the decades-old assumption held by theorists and observers (e.g., Scharlemann 1982) that tides raised on the secondary star by the WD act to suppress differential rotation in tidally locked systems (such as CVs), forcing the stellar envelope to co-rotate on average, and shows that CV secondaries are not necessarily tidally locked.

Given that differential rotation is thought to play a crucial role in amplifying and transforming initially poloidal magnetic field into toroidal field through dynamo processes (the so-called Ω effect), one may test dynamo theories in such stars with further measurements of differential rotation in tidally distorted systems, thus disentangling the most important parameters (period, stellar type, mass-ratio, Roche-lobe filling-factor) driving the stellar dynamo and influencing tidal dissipation efficiency. Furthermore, such studies may also lead to a better understand of why close-in exoplanets are preferentially misaligned with the stellar rotation axis around hot stars ($>$6250 K), and are aligned around cool stars ($<$6250 K, see Brothwell *et al.* 2014). Given that the outer convective envelope is responsible for tidal interactions, is a change in internal structure the cause of this transition? Tests of tidal dissipation efficiency in interacting binaries (of different spectral types) may help to answer this question.

As well as studying its short-term behaviour, we have also observed the long-term trends of the magnetic activity on the secondary star in AE Aqr. For this purpose, we constructed 7 brightness maps of the surface over an 8 yr period (see Figure 1). This is the first time such a campaign has carried out for a CV secondary, and our maps of AE Aqr have given us a unique insight into its magnetic activity (Hill *et al.* 2016). The first map, created with data taken in 2001, shows a high spot coverage around the polar region (due to a large high-latitude spot), as well as an increasing fractional spot coverage towards lower latitudes ($\sim$20°, see Figure 1). However, there is a clear paucity of spots at around 45° latitude. In contrast, the 6 maps made with data taken in the years since then show a clear increase in fractional spot coverage at around 45° latitude, with a similarly high concentration of spots at 20° (see Figure 1). The emergence of this band of spots at 45° latitude in the 8 years between observations may be indicative of a solar-like magnetic activity cycle in operation in AE Aqr, where the latitude of spot formation may change in a manner that mimics the butterfly-diagram for the Sun. Furthermore, the increase in fractional spot coverage around 20° latitude may be a second band of spots that form part of a previous cycle. In the case of the Sun, the latitude of emergence of flux tubes gradually moves towards the equator over the course of an activity cycle, taking $\sim$11 yr, with little overlap between consecutive cycles of flux tube emergence. However, simulations by Işık *et al.* 2011 show that stronger dynamo excitation may cause a larger overlap between consecutive cycles, and given the strong dynamo excitation in AE Aqr, the presence of two prominent bands of spots may be indicative of such an overlap between cycles. Over

C. A. Hill

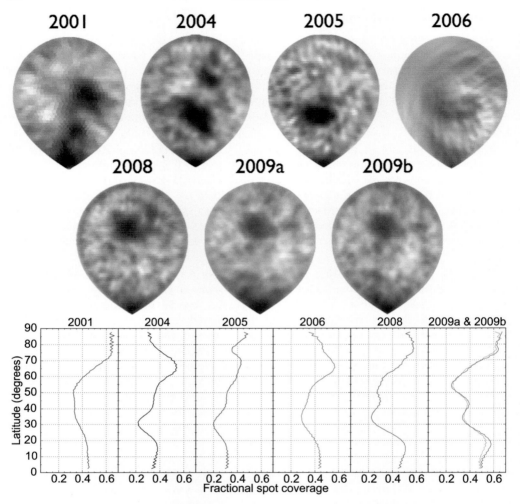

Figure 1. Top panel: Roche tomograms of AE Aqr, shown pole-on, where dark features are starspots, with some irradiation and gravity darkening around the L1 point (also appearing dark). Bottom panel: Fractional spot coverage as a function of latitude for the Northern hemisphere of AE Aqr for each data set (year above plot), normalized by the surface area at that latitude. Note, the absolute level of fractional spot coverage is not directly comparable between maps.

the course of such a cycle, we would expect to see the higher-latitude peak move towards lower latitudes and, as we do not clearly see this, any solar-like activity cycle in AE Aqr must take place over a timescale longer than the 8 years between our observations. Indeed, using the correlations between the duration of the magnetic activity cycle and the rotation period by Saar & Brandenburg 1999, we estimate that AE Aqr would have a magnetic activity cycle lasting $\sim$16–22 yr. If these correlations are also true for AE Aqr, then we may have observed less than half of an activity cycle.

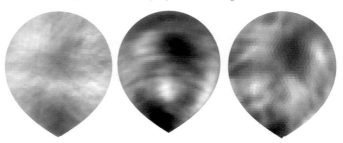

Figure 2. Roche tomograms of 3 CVs, all showing an increase in fractional spot coverage on the hemisphere facing the WD. Left to right, the stars are V426 Oph (Hill *et al.* in prep.), SS Cyg (Hill *et al.* in prep.) and BV Cen (Watson *et al.* 2007).

3. Tidal effects on magnetic ativity

As well as testing stellar dynamo theories in these rapidly-rotating, tidally-distorted stars, we may also study how the nearby companion star may influence its magnetic activity.

Figure 2 shows Roche tomograms of 3 CVs, namely V426 Oph, SS Cyg and BV Cen. For these objects, as well as for AE Aqr, we find a significant increase in spot coverage on the hemisphere facing the WD, even after accounting for the increased area (of the significantly distorted hemisphere) and the affects of gravity darkening and irradiation (both appearing dark in the maps, mimicking spots). This increase in spot coverage may suggest that magnetic flux tubes are forced to emerge at preferred longitudes, as predicted by Holzwarth & Schüssler (2003), and may possibly be related to the impact of tidal forces from the nearby compact object. If these particular spot distributions are confirmed to be long-lasting features, they would require explanation by stellar dynamo theory, and would provide evidence for the impact of tidal forces on magnetic flux emergence. In addition, since the number of starspots should change dramatically over the course of an activity cycle, the density of spots around the mass transfer nozzle may also vary (providing an explanation for the extended high and low accretion states seen in polar type CVs such as AM Her). However, further observations are required to confirm whether this higher concentration of spots is indeed altered over the course of an activity cycle.

As well as surface features, one may observe clear signs of magnetic interaction between binary components in the form of slingshot prominences. Here, a magnetic structure (rooted on the secondary) is pulled towards the WD, causing the loop to expand. Plasma is confined at the top of the loop by magnetic tension, and is illuminated by the disk and secondary star. This results in an emission source that co-rotates with, but is well separated from, the secondary star, and is seen in emission at 0 kms^{-1}. Figure 3 shows several examples of this phenomena, as seen in the CVs BV cen (Watson *et al.* 2007), IP Peg and SS Cyg (Steeghs *et al.* 1996). Although prominences may (in principle) appear anywhere on the secondary, the material is loosely bound at the inner hemisphere, and as the effective potential decreases as one approaches the WD, one only requires surface fields of ~1 kG to produce such prominences seen in these systems.

In other work by Parsons *et al.* 2016, large prominences were also seen in the pre-CV, QS Vir ($P_{orb} = 3.62$ h). Here, the highly magnetically-active M dwarf (that almost fills its Roche lobe) features long-lived spots that remain in fixed locations, preferentially found on the hemisphere facing the WD (see Figure 4). The system also displays three large prominences that cross the line of sight, with one passing in from of both stars (allowing is position to be well constrained, see Figure 5). Furthermore, despite showing small variations on a time-scale of days, they persist for more than ~1 yr. Indeed, one of

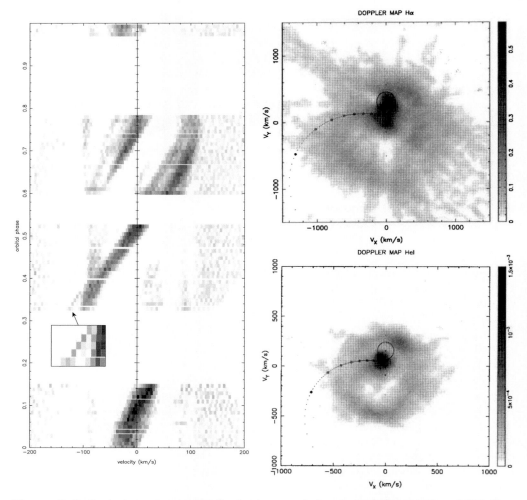

Figure 3. Left: A dynamic spectra the least-squared deconvolved line profiles of BV Cen, where the orbital motion has been removed, and the profiles have had a theoretical line profile subtracted. Features due to starspots and irradiation appear dark, with a slingshot prominence appearing as a narrow feature (indicated with an arrow and shown enlarged) lying off the blue edge of the stellar limb at phases 0.328–0.366. Top-right: A Doppler map of the Hα emission in IP Peg. Bottom-right: A Doppler map of the He I (6678 Å) emission in SS Cyg. For both Doppler maps, the Roche lobe and gas stream are additionally plotted, the centre of mass is denoted by a cross, and the WD is plotted as a point. For both IP Peg and SS Cyg there is significant emission at zero velocity, indicative of slingshot prominences.

these prominences may have been detected (as a sharp absorption feature) in a spectra taken ∼11 yr previously, and so these features may last decades. Moreover, the heavily spotted regions may well be related to these prominences; since the prominences appear to be stable on a time-scale of years, they must still be anchored to the surface of the M star by its magnetic field, and starspots are likely to form near these anchor regions, as is seen in solar coronal loops.

Lastly, magnetic interactions between binary stars have also been observed in the RS CVn, AR Lac (Siarkowski *et al.* 1996). Here, coronal X-ray emission was found to be concentrated on the hemispheres facing the companion star, with evidence that extended

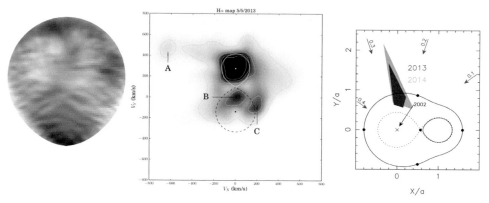

Figure 4. Left: A Roche tomogram of QS Vir, where dark grey scales indicate the presence of starspots or the impact of irradiation. Centre: A Doppler map of the Hα line, where the Roche lobes of the WD and M star are highlighted with a black dashed line and solid white line, respectively. Contours are also plotted to highlight the additional emission features against the strong emission from the M star. Three emission features that do not originate from either star are also labelled. Right: Top-down view of the binary indicating the location of the large prominence feature in 2013 and 2014 (orbital phases shown in blue). Dotted lines indicate the Roche lobes, a solid line shows the M star radius. The arrow indicates the viewing angle to the WD during a 2002 observation in which prominence material was also observed. The WD is marked with a cross and the black dots indicate the 5 Lagrange points. The outer solid black line indicates the effective 'co-rotation' radius of the binary.

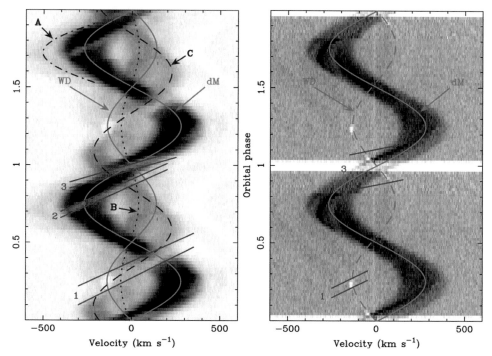

Figure 5. Trailed spectrogram of the Ca II 3934 Å (left) and Hα Å (right) lines with the main features labelled. The velocities of the two stars are indicated by the bright red sinusoids. In blue, we highlight three clear absorption features that cross one of both of the stars (only two are visible in the Ca II trail). We also highlight three other emission components in the Hα trail with dark red lines.

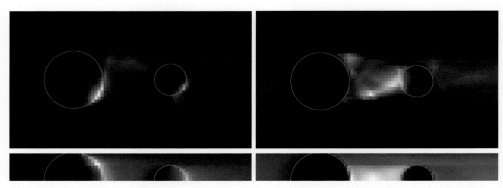

Figure 6. Comparison of the coronal structure (determined from X-ray emission) of AR Lac at two epochs, separated by an interval of 9 yr. The left panel from Siarkowski 1992, right panel from Siarkowski *et al.* 1992. Each panel shows a top view and a side view (at phase 0.25)

regions may connect the K and G-type stars (see Figure 6), with the pattern of the coronal emission showing clear variation over the 9 yr separation between observations.

4. Conclusions

There is clear evidence of magnetic interactions between the components in close binaries, such as preferential emergence of flux tubes, hot spots on facing hemispheres and extended coronal X-ray emission in RS CVn binaries, and large prominences that preferentially form due to the binary geometry, perhaps lasting over a decade. Furthermore, as stellar binaries show scaled-up interactions that may also take place in a planet-star system, as has been seen for hot Jupiters (see other author's work in this volume), understanding these systems may have far reaching consequences. In particular, one could test tidal dissipation efficiency in star-star binaries, with implications for both CV and planetary evolution.

References

Brothwell, R. D., *et al.*, 2014, *MNRAS*, 440, 3392
Hill, C. A., *et al.*, 2016, *MNRAS*, 459, 1858
Işık, E., & and Schmitt, D. and Schüssler, M., 2011, *A&A*, 528, 135
&Saar, S. H. and Brandenburg, A., 1999, *ApJ*, 524, 295
Scharlemann, E. T, 1982, *ApJ*, 253, 298
Siarkowski, M., 1992, *MNRAS*, 259, 453
Siarkowski, M., *et al.*, 1996, *ApJ*, 473, 470
Steeghs, D. T. H., *et al.*, 1996, *MNRAS*, 281, 626
Watson, C. A., *et al.*, 2007, *MNRAS*, 382, 1105

Living Around Active Stars
Proceedings IAU Symposium No. 328, 2016
D. Nandy, A. Valio & P. Petit, eds.

© International Astronomical Union 2017
doi:10.1017/S1743921317003982

Are tachoclines important for solar and stellar dynamos? What can we learn from global simulations

G. Guerrero[1], P. K. Smolarkiewicz[2], E. M. de Gouveia Dal Pino[3], A. G. Kosovichev[4], B. Zaire[1] and N. N. Mansour[5]

[1]Physics Department, Universidade Federal de Minas Gerais, Av. Presidente Antonio Carlos 6627, Belo Horizonte, MG, 31270-901, Brazil, email: **guerrero@fisica.ufmg.br**
[2]European Centre for Medium-Range Weather Forecasts, Reading RG2 9AX, UK
[3]Astronomy Department, IAG-USP, Rua do Matão, 1226, SP, 05508-090, Brazil
[4]New Jersey Institute of Technology, Newark, NJ 07103, USA
[5]NASA, Ames Research Center, Moffett Field, Mountain View, CA 94040, USA

Abstract. The role of tachoclines, the thin shear layers that separate solid body from differential rotation in the interior of late-type stars, in stellar dynamos is still controversial. In this work we discuss their relevance in view of recent results from global dynamo simulations performed with the EULAG-MHD code. The models have solar-like stratification and different rotation rates (i.e., different Rossby number). Three arguments supporting the key role of tachoclines are presented: the solar dynamo cycle period, the origin of torsional oscillations and the scaling law of stellar magnetic fields as function of the Rossby number. This scaling shows a regime where the field strength increases with the rotation and a saturated regime for fast rotating stars. These properties are better reproduced by models that consider the convection zone and a fraction of the radiative core, naturally developing a tachocline, than by those that consider only the convection zone.

Keywords. Stars; rotation; Stars: magnetism; Stars: dynamo

1. Introduction

In stellar interiors, tachoclines are interface layers where the solid body rotation of the stars interior becomes differential rotation. It is also associated with the mechanism by which the star transfers energy from the core to the photosphere, i.e., the interface between a radiative and a convective layers where rotating turbulence drives the transport of energy and angular momentum. Thus, the location of the tachocline depends on the evolutionary stage of the star. In the case of the Sun it is located at about 71% of the solar radius.

Since strong gradients of angular velocity occur at these layers, it is believed that they are relevant for the stellar dynamos. For the solar case, particularly, some dynamo models (Dikpati & Charbonneau 1999; Chatterjee *et al.* 2004; Guerrero & de Gouveia Dal Pino 2008) assume that the large-scale toroidal magnetic field generated at the tachocline directly manifests as sunspots, the regions of strong radial magnetic field observed at the surface. A buoyancy process is invoked as the transport agent of magnetic flux ropes. This scenario of solar dynamo is controversial and matter of debate in the literature Guerrero & Käpylä (2011). However, the same mechanism has already been considered for stars with deeper convection zones.

As for solar-type stars in different evolutive stages, observations of X-ray luminosities, a proxy of stellar activity (Pizzolato *et al.* 2003; Wright *et al.* 2011), or magnetic fields (and references therein Vidotto *et al.* 2014), indicate two regimes of activity as function of the Rossby number, Ro. For higher values of the Rossby number (buoyancy dominated convection) the stellar activity follows a exponential relation ($Lx \propto \text{Ro}^{-2.7}$, $\langle B \rangle \propto \text{Ro}^{-1.39}$). For low Rossby numbers (rotation dominated convection) the stellar activity shows a saturated regime. In the sample studied by Wright *et al.* (2011), the stars in the saturated regime are either partly or fully convective, whereas in the buoyancy dominated regime the stars have already developed a radiative core. This suggested a relation between photospheric and coronal magnetic field and the field developed at the tachocline. Most recently, however, Wright & Drake (2016) reported new observations of fully convective stars that fall within the rotation dominated regime and follow the $Lx \propto \text{Ro}^{-2.7}$ scaling. This result questions the relevance of the tachocline in the generation of the large-scale magnetic field observed in stars. It suggest that in partly convective stars the field does not develop in the shear region but in the convection zone. A distributed dynamo process, with a turbulent α-effect (Parker 1955; Steenbeck *et al.* 1966), is then invoked as the field generating mechanism. The field observed in the Sun, in the form of sunspots, and in other solar-type stars, via Zeeman-Doppler Imaging (Donati & Brown 1997; Petit *et al.* 2008), might be of shallow origin as proposed one decade ago by Brandenburg (2005).

To understand the physics behind the stellar dynamo mechanism in a self-consistent form nowadays we draw upon global MHD models. Although these simulations are still far from the numerical resolution required to capture the full dynamics of the stellar interior, they have been successful in reproducing some key ingredients of the dynamo. For instance Browning (2008) found dynamo action in fully convective models mimicking M dwarfs while Brown *et al.* (2008) obtained steady magnetic field solutions, in simulations of rapidly rotating convection with solar-like stratification. More recently Ghizaru *et al.* (2010) reported oscillatory dynamo solutions in ILES simulations of convection rotating at the solar rate. Thereafter different groups have found dynamos with periodic magnetic fields in simulations with different Rossby numbers (Käpylä *et al.* 2012; Guerrero *et al.* 2016a; Augustson *et al.* 2015). From these models, the natural development of a tachocline is reproduced only by Ghizaru *et al.* (2010) and (Guerrero *et al.* 2016a) who used the EULAG-MHD code. In fact, Guerrero *et al.* (2016a) presented a comparison between models with and without a stable layer at the bottom of the domain. Their results remark important differences between both kind of models performed with the same code, as well as with models performed with different codes that only consider the convection zone.

In this work we present what we consider the most relevant features observed in simulations of solar-type stars that include a fraction of the radiative zone (and therefore a tachocline). The first two sections extrapolate the results of the models to the solar dynamo case. They discuss how MHD instabilities at the tachocline might set the dynamo period, and how the so-called torsional oscillations might develop in the same region before propagating upwards up to the solar surface. The third section discusses the scaling of the dynamo generated magnetic field with the Rossby number. We present the results of a large number of simulations with (RC models) and without tachoclines (CZ models) which indicate that the models that consider the stable radiative zone are the ones that better reproduce the observational laws discussed above.

2. The solar dynamo period

One of the most relevant quantities defining the evolution of a dynamo generated magnetic field is the turbulent magnetic diffusivity, η_t. Given the values for temperature and density in the convection zone, the molecular, Ohmic, magnetic diffusivity is small

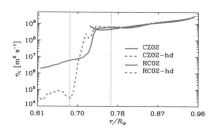
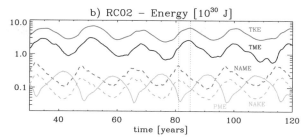

Figure 1. Left: Turbulent magnetic diffusivity of models without (blue) and with radiative zone (red). Solid an dashed lines correspond to MHD and HD cases, respectively. Rigth: Volume integrated energies of the different quantities in the simulation; TKE, TME and PME stand for toroidal kinetic and magnetic energies, and for the poloidal magnetic energy. NAME and NAKE correspond to the non-axisymmetric magnetic and kinetic energies, respectively. Adapted from Guerrero *et al.* (2016a).

enough ($\sim 10^4$ m^{2}s^{-1}) to allow a oscillatory dynamo with 22 yr period. According to dynamo theory the rate at which magnetic field is diffusing is competing with the amount of magnetic field being generated and advected. This competition defines what parameter influences more in setting the oscillation period. The value of η_t has been estimated from the mixing length theory (e.g., Muñoz-Jaramillo *et al.* 2011) and also from the properties of the observed butterfly diagram (Cameron & Schüssler 2016). Both estimations agree with values between $10^8 - 10^9$ m^{2}s^{-1}. If this is the case the magnetic diffusivity must be controlling the cycle period of the solar dynamo. Turbulence diffusivity quenching, due to the non-linear interaction between the large scale magnetic field and the turbulent flows, has proven to be inefficient for the period to be controlled by, for example, the meridional circulation (Guerrero *et al.* 2009; Muñoz-Jaramillo *et al.* 2011).

What is puzzling about this estimative for the value of η_t is that it will diffuse the magnetic field in the convection zone in about 1-2 years, but not in 11. The simulation without tachocline, CZ02, presented in Guerrero *et al.* (2016a) results in a value of the turbulent diffusivity, $\eta_t = 1.09 \times 10^9$ m^{2}s^{-1} (see blue line in Fig. 1(left)) and a full-cycle period of 2.21 yr. The model, including radiative zone, RC02, results in $\eta_t = 1.12 \times 10^9$ m^{2}s^{-1} (see red line in 1(left)) and a cycle period of full-cycle period of 34.5 yr. By comparing the solid (MHD case) and the dashed (HD case) red lines in Fig. 1(left), it is possible to notice that η_t is not quenched by the large-scale magnetic field in the convection zone. However, it is quenched in the range $0.7R_\odot \lesssim r \lesssim 0.73R_\odot$, where the stronger magnetic field is generated. Interestingly, the evolution of the magnetic field in the rest of the computational domain seems to be governed by the deep seated magnetic field.

According to the results presented in Guerrero *et al.* (2016a), the cycle period is set by the value of the turbulent diffusivity at the bottom of the convection zone and the radiative layer, $\eta_t \simeq 5 \times 10^6$ m^{2}s^{-1}. This value is the result of turbulent motions that develop at the tachocline and the stable layer due to current-driven instabilities (e.g., Cally *et al.* 2003; Miesch *et al.* 2007). Because of the shear profile and the configuration of toroidal magnetic field an instability develops. It grows, exchanging energy with the magnetic field (see black and green solid lines in 1(right)). When the magnetic field is weak, the energy of the turbulent motions and magnetic field decays (see dashed lines in Fig. 1(right)). Thus, both the shear and the magnetic field adjust themselves to a equilibrium. Similar results have been obtained for a different global simulation performed with the EULAG-MHD code (Lawson *et al.* 2015).

3. Torsional oscillations

The slow-down and speed-up of the angular velocity observed at the solar surface are called torsional oscillations (TO). When their latitudinal distribution is plotted against time, two branches can be observed. One of them, with the largest amplitude, migrates towards the poles, the second one migrates equatorwards. Since these branches oscillate with the same periodicity than sunspots, it is believed that the TO are correlated with the large-scale magnetic field. In fact, some correlation is observed between the equatorial branch of TO and the sunspot latitudes of activity. However, the speed-up of the angular velocity starts few years before than the sunspot cycle.

The model RC02 of Guerrero *et al.* (2016a) shows variations of the angular velocity that resemble quite well the TO pattern, i.e, there are two brances, one polar and one quatorial, that oscillate with the dynamo period (see the upper panel of Fig. 2(a)). The origin of these oscillations was studied in Guerrero *et al.* (2016b). It was found that the pattern observed at the surface of the model ($r = 0.95R_\odot$) does not form locally but it is the result of the strong axial torque carried on by the large-scale magnetic tension at the tachocline. Therefore, the morphology of the TO does not corresponds to the magnetic field itself, but to the divergence of the correlations $\overline{B_\phi B_r}$ and $\overline{B_\phi B_\theta}$ at the bottom of the convection zone. This perturbation propagates upwards as can be seen in Fig. 3 where the time evolution of the axial torques is compared with the angular velocity perturbation, $\delta\Omega(r,\theta,t) = (2\pi\varpi)^{-1}(u_\phi(r,\theta,t) - \overline{u}_\phi(r,\theta))$, where $\varpi = r\sin\theta$ and $(2\pi\varpi)^{-1}\overline{u}_\phi$ is the zonal and temporal average of the angular velocity. In this figure the computational domain has been divided in 6 regions over which volume averages of the quantities are performed. Regions R31, R21 and R11 correspond to the polar, middle latitudes and equatorial tachocline; R32, R22 and R12 to the same latitudes and the bulk of the convection zone; and R33, R23 and R13 to the top of the domain. It can be seen that in the tachocline region, the shape of the TO closely resembles that of the magnetic tension torque with a slight delay. This resemblance disappears in the other regions but the shape of the TO propagates upwards while loosing some amplitude.

From this study, Guerrero *et al.* (2016b) found that, in general, the meridional circulation appears as a large-scale motion that compensates for the axial torque due to the Reynolds stresses. This is the so-called gyroscopic pumping mechanism (e.g., Miesch & Hindman 2011) for a quase-steady system (i.e., while full stationarity is never obtained due to the cyclic magnetic field, the establishment of the meridional circulation occurs in a faster time-scale). The steady profile of the meridional circulation varies in a time-scale compared with the magnetic cycle. While at the tachocline, this change is the response to the net magnetic tension torque, at the surface, where the magnetic tension is weaker, the variation is a local response to the TO. This can be seen in panel (b) of 2 where $\delta u_\theta = u_\theta(r,\theta,t) - \overline{u}_\theta(r,\theta)$ is presented in time-latitude (at $r = 0.95R_\odot$, upper panel) and time-radius (at 30° latitude, bottom) butterfly diagrams are presented. The correlation between δu_ϕ and δu_θ is evident at the surface with a residual poleward (equatorward) flow appearing when the rotation speeds-up (slows-down). The bottom panel of the same figure indicates that this correlation does not exist all depths, as expected. The correlation between the TO (black line) and a negative axial torque due to the meridional circulation (red line) can be seen in the top and bottom rightmost panels of Fig. 3. It is worth noticing that a similar correlation between δu_ϕ and δu_θ has been reported by Komm *et al.* (2015). These result ultimately indicates that the meridional circulation must be correlated with the magnetic field at the bottom of the convection zone.

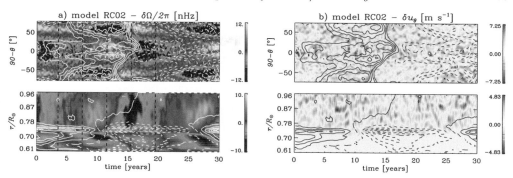

Figure 2. Time-latitude diagram at $r = 0.95R_\odot$ (upper panel), and time-radius diagram at 30° latitude (bottom panel) of $\delta\Omega(r,\theta,t)/2\pi = (2\pi\varpi)^{-1}(u_\phi(r,\theta,t) - \overline{u}_\phi(r,\theta))$, left, and $\delta u_\theta = u_\theta(r,\theta,t) - \overline{u}_\theta(r,\theta)$, right. The continuous (dashed) line contours depict the positive (negative) toroidal magnetic field shown in Fig. 6 of Guerrero *et al.* (2016a).

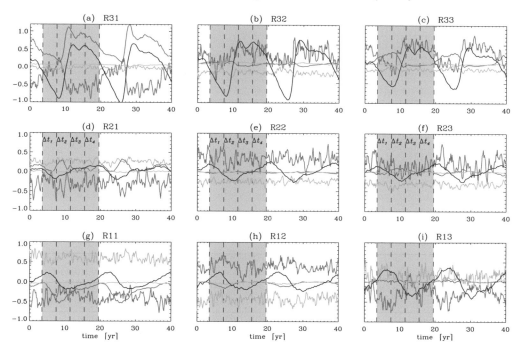

Figure 3. Time evolution of the axial torques computed for different latitudinal and radial regions (see the text). Red, orange, blue and green lines correspond to the meridional circulation (MC), Reynolds stresses (RS), magnetic tension (MT) and Maxwell stresses (MS) axial torques normalized to the local maximum value of $\langle \nabla \cdot \mathcal{F}_{MC} \rangle$. The black line shows the evolution of $\langle \delta\Omega \rangle$ normalized to 10^{-7}. The angular brackets mean volume averages over each region. Adapted from Guerrero *et al.* (2016b).

4. Saturation of magnetic field in solar-type stars

In this section we present results from dynamo simulations of solar-type stars with different Rossby numbers. From these models only six cases were studied in detail in Guerrero *et al.* (2016a). Here we present eight simulations which include a fraction of the radiative zone (RC cases) and seven cases which consider only the convection zone (CZ cases). In Fig. 4 we present the amplitude of the magnetic field as a function of Ro. The

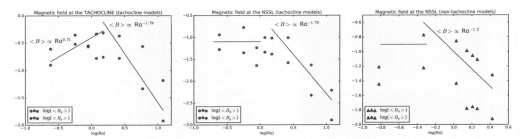

Figure 4. Magnetic field strength as a function of Ro for global simulations that include the radiative zone (left and middle panels) and for models that consider the convection zone only (right). In the left panel the field is measured at the shear region, $0.72R_\odot \leqslant r \leqslant 0.75R_\odot$. In the middle and right panels the field is measured at surface levels, $0.92R_\odot \leqslant r \leqslant 0.96R_\odot$. Red and blue symbols correspond to toroidal and poloidal field components, respectively.

left and middle panels corresponds to RC models while the right panel shows the results for CZ models. In left panel the rms field is computed in the tachocline region where the shear is stronger ($0.72R_\odot \leqslant r \leqslant 0.75R_\odot$). In the middle panel the field is computed in the near-surface shear layer (NSSL, $0.92R_\odot \leqslant r \leqslant 0.96R_\odot$). The Rossby number has been computed here by using the definition used by Landin *et al.* (2010): Ro $= P_{\mathrm{rot}}/\tau$, where P_{rot} is the rotation period and $\tau = H_p/u_{\mathrm{rms}}$ is the convective turnover time computed one pressure scale height above the tachocline.

Starting from the slow rotation regime, at tachocline levels, the toroidal magnetic field (red dots) increases with the decreasing of the Rossby number until Ro $\gtrsim 1$, for lower values of Ro, it exhibits a decay. This happens because faster rotating models result in less shear at the tachocline. The latitudinal shear also decreases and the rotations turns more and more homogeneous. The continuous lines show scaling laws for the magnetic field varying as Ro$^{0.71}$ and Ro$^{-1.79}$ for lower and larger values of the Rossby number, respectively. The amplitude of the poloidal magnetic field (blue dots) also increases with the decreasing of the Ro for slow rotating models. For $1 \lesssim$ Ro $\lesssim 3$, the $\langle B_p \rangle$ shows a platoo, afterwards it increases and then decays again.

The scaling is different when the rms magnetic field is computed at the NSSL levels. Both, the toroidal and the poloidal components increase with the decrease of Ro for slow rotating models, but then they reach a saturation level. It seems, however, that the Rossby number of saturation is different for $\langle B_\phi \rangle$ (Ro ~ 2.5) than for $\langle B_p \rangle$ (Ro ~ 1). The continuous line shows a scaling law with Ro$^{-1.79}$. Although the shear is decreasing for fast rotating models, the amplitude of the toroidal field remains constant. It could be argued that the α effect increases with the rotation so that α^2 dynamos are operating in these cases. However, the same analysis for models without the stable layer (right panel of Fig. 4) indicate that this might not be the case. For these models the rms large-scale magnetic field is computed only at the NSSL because there is not a radial shear layer at the bottom of the domain. From right to left, the magnetic field increases for decreasing Ro (the scaling law in this case fits better for Ro$^{-1.2}$) but then decreases quickly for the faster rotating models. Similar behaviour is observed for both, toroidal and poloidal components of the large-scale magnetic field.

This result indicates that the presence of the tachocline is relevant for reproducing the observed saturation of the magnetic field mentioned above for partly convective stars. Nevertheless, the lhs panel of Fig. 4 suggest that the observed magnetic field might not come from the tachocline but is the surface magnetic field generated by a distributed dynamo. The evolution of this large-scale field is governed by the strong field anchored in the tachocline and the stable layer underneath. The dynamo ingredients, naturally occurring in a global MHD simulation, will be presented in a forthcoming paper (Guerrero *et al.* 2017, in preparation).

5. Discussion

Global dynamo simulations have become an essential tool to understand the physics of the solar and stellar dynamos. The evolution of rotating turbulent convection of a plasma in hydrostatic equilibrium results in large-scale motions and magnetic fields. For the solar case, the models capture some of the observed characteristics of both, motions and magnetism. From the variety of global dynamo models in the literature, the ILES performed with the EULAG-MHD code have proven successful in reproducing two important features: the thin shear layer at the interface between the radiative and the convective zones, and the reversals of the large scale magnetic field in a time scale similar to the solar cycle. Based on the results of these models, in this proceeding we discussed the relevance of tachoclines for the generation of solar/stellar magnetic fields.

Three arguments are presented supporting the relevance of tachoclines. (i) the solar cycle period: according to estimations the value of the turbulent magnetic diffusivity in the bulk of the convection zone must be between $10^8 - 10^9$ m^{2}s^{-1}. These value means a fast decay of the large-scale magnetic field and therefore a cycle period of about 2 years. In the global dynamo models the strong radial shear at the tachocline generates a strong toroidal magnetic field which keeps stored in the stable layer underneath. The co-existence of a shear profile together with a large-scale field is unstable and results in non-axisymmetric field and motions and therefore in a turbulence diffusivity even in the stable layer. This shear current instability develops obtaining energy from the magnetic field, when the magnetic field is not sufficiently enough, the instability decays. The time-scale of this balance sets the period of the oscillatory dynamo. (ii) The torsional oscillations: Guerrero *et al.* (2016b) reported a pattern of TO that resembles the observed pattern of speed-up and slow-down of the zonal flow observed in the solar surface. They demonstrated that in the simulation this pattern is formed in the tachocline due to the large-scale magnetic tension. The TO propagate upwards up to the surface of the model. The simulation also exhibit changes in the meridional flow pattern with the same period of the magnetic cycle. These changes are also observed at the solar surface. We found that the surface variation of the meridional circulation is a response to the TO, i.e., whenever the zonal flow speeds-up (slows-down), the residual of the meridional circulation is poleward (equatorward) such as it has been observed (Komm *et al.* 2015). (iii) Scaling of stellar magnetic fields with the Rossby number: the X-ray luminosity as well as the magnetic field show a well defined scaling with the Rossby number. For large values of Ro, buoyancy dominated convection, the magnetic field increases with the decrease of Ro; for smaller values of Ro, rotation dominated convection, there is a saturated regime. Global simulations of partly convective stars, with a solar-like stratification and in a wide range of Rossby numbers are able to reproduce a similar scaling profile with a Ro$^{-1.79}$ for slow rotation rates and saturated for fast rotation rates. Models with radiative cores are necessary to reproduce this law. The observed magnetic field does not correspond to the magnetic field at the tacholine but to the surface magnetic field generated by a distributed dynamo. However, the large-scale field evolution is governed by the deep seated field anchored in the stable layer. On the other hand, solar-like models which consider only the convection zone do not reproduce the saturated regime but scale with Ro$^{-1.2}$ in the buoyancy dominated regime. Further work is still necessary to determine if the same scaling law is reproduced in simulations of fully convective stars. Initial efforts have started in this sense in EULAG-MHD simulations (see Zaire *et al.*, in this proceedings).

Acknowledgements

This work was partly funded by FAPEMIG grant APQ-01168/14 (GG and BZ), FAPESP grant 2013/10559-5 (EMGDP), CNPq grant 306598/2009-4 (EMGDP), NASA grants NNX09AJ85g and NNX14AB70G. PKS is supported by funding received from the European Research Council under the European Union's Seventh Framework Programme (FP7/2012/ERC Grant agreement no. 320375). The simulations were performed in the NASA cluster Pleiades and Brazilian super computer SDumont of the National Laboratory of Scientific Computation (LNCC).

References

Augustson, K., Brun, A. S., Miesch, M., & Toomre, J. 2015, *ApJ*, 809, 149
Brandenburg, A. 2005, *ApJ*, 625, 539
Brown, B. P., Browning, M. K., Brun, A. S., Miesch, M. S., & Toomre, J. 2008, *ApJ*, 689, 1354
Browning, M. K. 2008, *ApJ*, 676, 1262
Cally, P. S., Dikpati, M., & Gilman, P. A. 2003, *ApJ*, 582, 1190
Cameron, R. H. & Schüssler, M. 2016, *A&A*, 591, A46
Chatterjee, P., Nandy, D., & Choudhuri, A. R. 2004, *A&A*, 427, 1019
Dikpati, M. & Charbonneau, P. 1999, *ApJ*, 518, 508
Donati, J.-F. & Brown, S. F. 1997, *A&A*, 326, 1135
Ghizaru, M., Charbonneau, P., & Smolarkiewicz, P. K. 2010, *ApJL*, 715, L133
Guerrero, G. & de Gouveia Dal Pino, E. M. 2008, *A&A*, 485, 267
Guerrero, G., Dikpati, M., & de Gouveia Dal Pino, E. M. 2009, *ApJ*, 701, 725
Guerrero, G. & Käpylä, P. J. 2011, *A&A*, 533, A40
Guerrero, G., Smolarkiewicz, P. K., de Gouveia Dal Pino, E. M., Kosovichev, A. G., & Mansour, N. N. 2016a, *ApJ*, 819, 104
—. 2016b, *ApJL*, 828, L3
Guerrero, G., Zaire, B., Smolarkiewicz, P. K., *et al.* 2017
Käpylä, P. J., Mantere, M. J., & Brandenburg, A. 2012, *ApJL*, 755, L22
Komm, R., González Hernández, I., Howe, R., & Hill, F. 2015, *Sol. Phys.*, 290, 3113
Landin, N. R., Mendes, L. T. S., & Vaz, L. P. R. 2010, *A&A*, 510, A46
Lawson, N., Strugarek, A., & Charbonneau, P. 2015, ArXiv e-prints
Miesch, M. S., Gilman, P. A., & Dikpati, M. 2007, *ApJ*, 168, 337
Miesch, M. S. & Hindman, B. W. 2011, *ApJ*, 743, 79
Muñoz-Jaramillo, A., Nandy, D., & Martens, P. C. H. 2011, *ApJL*, 727, L23
Parker, E. N. 1955, *ApJ*, 122, 293
Petit, P., Dintrans, B., Solanki, S. K., *et al.* 2008, *MNRAS*, 388, 80
Pizzolato, N., Maggio, A., Micela, G., Sciortino, S., & Ventura, P. 2003, *A&A*, 397, 147
Steenbeck, M., Krause, F., & Rädler, K.-H. 1966, *Zeitschrift Naturforschung Teil A*, 21, 369
Vidotto, A. A., Gregory, S. G., Jardine, M., *et al.* 2014, *MNRAS*, 441, 2361
Wright, N. J. & Drake, J. J. 2016, *Nature*, 535, 526
Wright, N. J., Drake, J. J., Mamajek, E. E., & Henry, G. W. 2011, *ApJ*, 743, 48

Living Around Active Stars
Proceedings IAU Symposium No. 328, 2016
D. Nandy, A. Valio & P. Petit, eds.

© International Astronomical Union 2017
doi:10.1017/S1743921317004094

Starspots properties and stellar activity from planetary transits

Adriana Valio

Center for Radio Astronomy and Astrophysics (CRAAM)
Mackenzie Presbyterian University, Sao Paulo, Brazil
email: `avalio@craam.mackenzie.br`

Abstract. Magnetic activity of stars manifests itself in the form of dark spots on the stellar surface. This in turn will cause variations of a few percent in the star light curve as it rotates. When an orbiting planet eclipses its host a star, it may cross in front of one of these spots. In this case, a "bump" will be detected in the transit lightcurve. By fitting these spot signatures with a model, it is possible to determine the spots physical properties such as size, temperature, location, magnetic field, and lifetime. Moreover, the monitoring of the spots longitude provides estimates of the stellar rotation and differential rotation. For long time series of transits during multiple years, magnetic cycles can also be determined. This model has been applied successfully to CoRoT-2, CoRoT-4, CoRot-5, CoRoT-6, CoRoT-8, CoRoT-18, Kepler-17, and Kepler-63.

Keywords. Stars: activity, stars: rotation stars: spots

1. Introduction

Very likely, all cool stars with a convective envelope like the Sun will have spots on their surfaces. So far, a few thousand planets are confirmed to eclipse their host star, as observed by the CoRoT and the Kepler satellites. During one of these transits, the planet may pass in front of a spot group and cause a detectable signal in the light curve of the star.

To better study these stars spots, I developed a model that simulates planetary transits in front of stars (Silva 2003). In this model, the planet is used as a probe do determine the spots physical characteristics such as

- Size (area coverage)
- Intensity → temperature → magnetic field
- Location (longitude & latitude)
- Lifetime

From the intensity of the spot with respect to the stellar center disc intensity, it is possible to estimate the spots temperature assuming that both the spot and the stellar photosphere irradiates as a black body. Once the temperature is known, the magnetic field of the spot can be inferred considering a solar like behavior and a temperature-magnetic field relation such as the one given in Dicke (1970). According to this author the temperature varies with the magnetic field squared.

This model has already been successfully applied to stars CoRoT-2 (Silva-Valio *et al.* 2010, Silva-Valio & Lanza 2011), Kepler-17 (Valio *et al.* 2017), and and Kepler-63 (Estrela & Valio 2016). Other authors have also studied starspots properties. For example, the effect of the spots not occulted by the planetary transit was studied by Pont *et al.* (2013). Also, these same effects can influence the transmission spectroscopy measurements of the transit (Oshagh *et al.* 2014). Moreover, the detection of the same spot in consecutive transits can also provide information about the spin-orbit alignment of the planetary system (Sanchis & Winn 2011, Sanchis *et al.* 2012).

From spot detection on different transits, some stellar properties may also be determined, for example:

- Rotation period
- Differential rotation
- Activity cycle

The rotation period is calculated the same way as Thomas Harriot did for the Sun in 1610, that is, by tracing the sunspots "move" across the solar disc. This is done by detecting the same spot on a consecutive transit and noting how much its longitude has changed. Silva-Valio (2008) estimated the rotation period of HD 209458 applying this method.

Just as the Sun has an activity cycle of 11 years, so do other stars as shown by the studies of Baliunas *et al.* (1995), Saar & Brandenburg (1999), Olah *et al.* (2009), Messina & Guinan (2002). For long enough monitoring of stars with transits, it is possible to estimate the number of spots and its variation with the years. Since Kepler did observe thousands of stars for a period of 4 yeas, we may identify short duration cycles for some active stars (Estrela & Valio 2016). For more details see article by Estrela & Valio in this volume.

The next section presents the spot model used in this work, whereas the following section describes the studied stars. The results of the model such as spots physical parameters and the stellar rotational profile are described in Sections 4 and 5, respectively. Finally, the last section lists the conclusions.

2. Planetary Transit Model

The model used in this work simulates the passage of a planet (dark disc) in front of a star. For the star we can use an image of the Sun or a synthesised image of a star with limb darkening (see Figure 1):

$$I = I_0[1 - u_1(1 - \mu) - u_2(1 - \mu)^2] \tag{2.1}$$

where $\mu = \cos(\theta)$, and θ is the heliocentric angle (or longitude if we consider the brightness variation along the equator). The parameters u_1 and u_2 define the decrease in brightness from the center to the limb. In the case of the Sun, $u_1 = 0.59$ and $u_2 = 0$, so the limb darkening is of linear form. For $u_2 \neq 0$ then the stellar limb darkening is said to be quadratic.

The planet is an opaque circular disc of radius r given in units of the stellar radius R_s (black disc in the center of the star on Figure 1). In the case of a Jupiter-size planet orbiting a solar like star, $r/R_s = 0.1$. The orbit of the planet is calculated using the period, P_{orb}, semi-major axis, a (also given in units of R_s), and the inclination angle, i. The orbit is considered circular (that is, zero eccentricity). If the orbital plane is parallel to the stellar equator, then the obliquity is null, otherwise, the obliquity parameter, λ, can be set in the model with the appropriate value. Once the orbit is set, then every 2 min (or the desired time interval), the planet is centred at its calculated position and all the pixels intensity in the image star + planet are summed, yielding the light curve. Thus the transit light curve of a spotless star is obtained, this is shown as the blue curve in the top panel of Figure 2.

An important feature of this planetary model is that it allows for the inclusion of spots on the stellar surface. These spots are described by three parameters:

- Intensity: measured with respect to stellar maximum intensity, I_0 (at the center);
- Size: measured in units of planetary radius (R_s);

Table 1. Stellar and Planetary Parameters

Star	CoRoT-2	CoRoT-4	CoRoT-5	CoRoT-6	CoRoT-8	CoRoT-18	Kepler-17	Kepler-63
Spectral type	G7V	F8V	F9V	F9V	K1V	G9V	G2V	G8V
Mass ($M_\odot$)	0.97	1.10	1.0	1.055	0.88	0.95	1.16	0.984
Radius ($R_\odot$)	0.902	1.17	1.19	1.025	0.77	1.0	1.05	0.901
Prot (d)	4.54	8.87	26.6	6.35	21.7	5.4	12.28	5.4
Teff (K)	5625	6190	6100	6090	5080	5440	5781	5576
Age (Gyr)	0.13-0.5	0.7-2.0	5.5-8.3	1.0-3.3	2.0-3	?	>1.78	0.2
Planet								
Mass (M_{Jup})	3.31	0.72	0.467	2.96	0.22	3.47	2.45	–
Radius(R_{star})	0.172	0.107	0.120	0.117	0.090	1.31	1.312	0.0662
Porb (d)	1.743	9.203	4.038	8.886	6.212	1.90	1.49	9.434
a (R_{star})	6.7	17.47	9.877	17.95	17.61	6.35	5.31	19.55

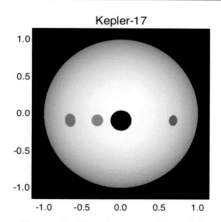

Figure 1. Planetary transit model.

• Position: Longitude and Latitude (restricted to the transit band).

The effect of foreshortenning as the spots approach the limb, a projection effect, is taken into account. An example of a star, Kepler-17, with three spots is shown in the top panel of Figure 1.

3. Stellar observations

Thus far we have analysed a set of 8 stars observed by the CoRoT and Kepler missiona, all of them solar like stars of spectral types F, G, and K, with varying ages. These stars have a Hot-Jupiter in close orbit around them with semi-major axis less than 20 stellar radii. The planetary mass vary from 0.22 to 3.5 Jupiter mass. The basic data of the stars and their planets are listed on Table 1.

4. Spots characteristics

The model described in Section 2 was applied to the 8 stars listed on Table 1. First the transits were cut from the continuous lightcurve, then we subtract a spotless star model from the light curve yielding the residuals. Figure 2 shows the transit lightcurve with the spotless model (blue curve) in the top panel, whereas the residuals after the subtraction is displayed in the bottom panel. The spots signature are clearly seen as peaks in the residuals. Each of these peaks identified in the transit light curve was modelled using a χ^2-minimisation routine (AMOEBA). Only the peaks that exceed a certain threshold

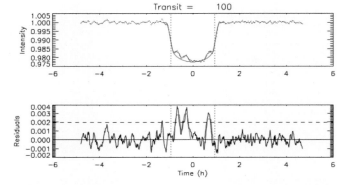

Figure 2. Top: Transit light curve with spotless star model overploted (red curve). Bottom: Residuals after spotless model subtraction from transit light curve, emphasising the spots signatures. The spot model is also shown.

were modelled. In the case of the CoRoT stars the threshold was 2 times the rms of the noise, whereas for the Kepler stars, this value equals 10 times the CDPP (Combined Differential Photometric Precision, Christiansen *et al.* 2012).

Each of the peaks, assumed to be caused by the passage in front of a spot, was modelled with the three parameters mentioned in Section 2, namely, size, intensity and position. The best fit of the model is shown as a red curve on Figure 2. The true size of the spot can be obtained once the size of the star, R_s is known. The average size of spots (in millions of meters) are given in the first line of Table 2. For comparison the values for sunspots are also given in the last column.

Then the area covered by spots within the band of the stellar surface occulted by the planet during its transit can be measured. Usually, for the planets studied here the transit band has a latitudinal width of about $10°$. The second line of Table 2 lists the average area covered by spots within the transit band.

The temperature of the spot may also be determined from the effective temperature of the star, T_{eff}, assuming both the spot and the surface to emit as black bodies. For this purpose, the spot temperature, T_0, is calculated from the spot intensity, f_i, using:

$$T_0 = \frac{K_b}{h\nu \, \ln\left(f_i \left(e^{h\nu/KT_{eff}} - 1\right) + 1\right)} \tag{4.1}$$

where K_b and h are the Boltzmann and Planck constants, respectively, T_{eff} is the photospheric temperature given in the fourth line of Table 2 for each star, ν is the observation frequency corresponding to a wavelength of $600nm$, and f_i is the ratio of spot intensity with respect to the central stellar intensity I_c. The spot temperature calculated using Eq. 4.1 and averaged for all the spots of a given star is listed on the third line of Table 2.

An interesting result from the spot temperature that we found is that the ratio of the spot temperature to the effective temperature of the stellar photosphere is basically constant and equal to 0.84. This temperature ratio is presented in the last line of Table 2, and also shown in the plot of Figure 3.

5. Stellar rotation and differential rotation

If the same spot is detected on a later transit, then the stellar rotation can be estimated by measuring its longitudinal displacement in time. Actually what is measured by the starspot movement is the rotation period at the transit latitude, $P_{rot}(\alpha)$, which may

Table 2. Starspots Parameters

Star	CoRoT-2	CoRoT-4	CoRoT-5	CoRoT-6	CoRoT-8	CoRoT-18	Kepler-17	Kepler-63	Sun
Radius (Mm)	55±19	51±14	75±17	48±14	82±21	65±19	80±50	32±14	12±10
Area (%)	13	6	13	9	29	13	6±4	5±2	< 1
Tspot (K)	4600±700	5100±500	5100±600	4900±600	4400±600	4800±600	5100±500	4700±400	4800±400
Teff (K)	5625	6190	6100	6090	5080	5440	5780	5580	5780
Tspot/Teff	0.818	0.824	0.836	0.804	0.866	0.882	0.875	0.846	0.830

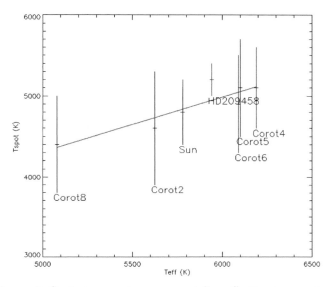

Figure 3. Spots temperature versus stellar effective temperature.

be different from the average rotation period of the star, P_{rot} obtained from the whole light curve, neglecting the transits. To estimate this period, $P_{rot}(\alpha)$, we need the relation between the central time of the "bump" and the longitude on the surface of the star, as seen from Earth. The central time of the "bump", t_s (in hours) is related to the longitude of the spot, lg_{spot} by:

$$lg_{spot} = asin\left(\frac{acos(90\,^\circ - 360\,^\circ)(t_s/24)/P_{orb}}{cos(lat_{spot})}\right) \qquad (5.1)$$

where P_{orb} is the orbital period and lat_{spot} is the latitude of the spot in the stellar surface. The spot latitude is determined by the projection of the planetary transit onto the stellar disc, and is given by:

$$lat_{spot} = -\arcsin[a\,\cos(i)] \qquad (5.2)$$

where a is the planetary semi major axis and i is the inclination angle of the orbit. The minus sign is arbitrarily chosen so that the planetary transit crosses the Southern hemisphere of the star.

Once the longitudes of the spots, lg_{topo}, as seen from Earth, on different transits is determined, we transform the longitudes of these spots to ones in the reference system that rotates with the star, lg_{ror}:

$$lg_{rot} = lg_{topo} - 360^\circ\,\frac{nP_{orb}}{P_{rot}(\alpha)} \qquad (5.3)$$

Table 3. Stellar rotation pattern

Star	CoRoT-2	CoRoT-4	CoRoT-5	CoRoT-6	CoRoT-8	CoRoT-18	Kepler-17	Kepler-63	Sun
P_{rot} (d)	4.54	8.87	26.63	6.35	21.7	5.4	12.28	5.4	27.6
Latitude, α (°)	-14.6	0	-47.2	-16.4	-29.4	-22.8	-4.6	-48.5	0
$P_{rot}(\alpha)$ (d)	4.48	8.71	26.49	6.08	21.42	4.68	11.4	–	24.7
$\Delta\Omega$ (rd/d)	0.042	0.026	0.103	0.101	0.014	0.45	0.077	0.133	0.050
$\Delta\Omega/\bar{\Omega}$ (%)	3	3.6	43	10	5	38	15	11	22

where n is the transit number, P_{orb} the orbital period, and $P_{rot}(\alpha)$ the rotation period of the star at the transit latitude. The main task here is to determine this rotation period, that is done by trying to find many spots on different transits with the same lg_{rot}. For this purpose, we create a flux deficit function with respect to the spot longitude calculated for a certain P_{star} according to Eq. 5.3. Then we calculate the auto correlation of this flux deficit function parametrised by P_{star}. The chosen value for the rotation period of the star at the transit latitude is that of the thinnest auto correlation function. This procedure is explained in detail in Valio (2013).

Table 3 list the stellar rotation period for the 8 stars studied and also for the Sun. The first line displays the average rotation period of the star that is usually obtained from a Lomb-Scargle periodogram, and is considered here to be an average period over all latitudes of the star. The latitude of planet center crossing, determined from Eq. 5.2 is given in the second line of Table 3. The stellar rotation period at the latitude of the shadow of the planet as it transits, obtained as described above, is given in the third line of Table 3.

The differential rotation of the star can be obtained by assuming a solar like (or anti-solar like) rotation model of the type:

$$\Omega(\alpha) = A - B\sin(\alpha) \tag{5.4}$$

$$A = \Omega_{eq} \tag{5.5}$$

$$B = \Delta\Omega = \Omega_{eq} - \Omega_{pole} \tag{5.6}$$

where α is the stellar latitude, whereas the constants A and B are the rotation at the equator, Ω_{eq}, and the rotational shear (or differential rotation), $\Delta\Omega$, respectively. To determine the two constants we use the two measured values of stellar rotation, $P_{rot}(\alpha)$ and P_{rot}, the averaged rotation values over all latitudes.

The rotational shear, $\Delta\Omega$, calculated from Eq. 5.6, is given in the fourth line of Table 3, whereas the relative differential rotation, given by dividing the shear by the average rotation, $\bar{\Omega}$, is listed in the last line of Table 3.

We note here that except for Kepler-63, all the planets are considered to orbit in a plane aligned with their host star equator, having null or very small obliquity as measured from the Rossiter-MacLaughin effect. This is not the case for Kepler-63 that is known to have an almost poleward orbit (Sanchis & Winn 2011, Sanchis et al. 2012). For this star, the way to calculate the rotation differential rotation is more complicated and is explained in an article by Netto & Valio in this volume.

6. Conclusions

Here we applied a planetary transit model that allows stellar spot modelling to 8 stars observed by the CoRoT and Kepler satellites (listed on Table 1). The model fits the peaks in the residuals of the transit light curves that are assumed to be the signatures of the spots when occulted by the planet. Three parameters are used to model each spot, namely, the size (in units of the stellar radius), intensity (with respect to the stellar central intensity) and position. From the intensity of the spots it is possible to infer their temperature using Eq. 4.1.

The results of the physical characteristics of the spots are displayed on Table 2. By comparing the average sizes of starspots with those of the Sun, we found these to be 4 to 7 times larger than sunspots. We believe that what we are measuring is actually active regions, or spot groups, and not individual spots, since the precision is not enough for that. After estimating the spots temperature from its intensity, an interesting result found was that the ratio between the mean spot temperature and the stellar photospheric temperature was basically constant and equal to ~84%. That is, hotter stars have, on average, hotter spots.

The rotation period of the stellar surface within the latitudes occulted by the planet was determined, and since this was different from the mean stellar rotation period estimated from the variations in the out–of–transit lightcurve (see Table 3), we concluded that the stars rotated differentialy. An estimate of this differential rotation may be done by considering a solar-like rotation profile, given by Eq. 5.5. The results yield stars with a shear comparable to the Sun, such as CoRoT-2 and Kepler-17, and others that show twice this shear such as CoRoT-5, CoRoT-6 and Kepler-63. As for the relative differential rotation, only CoRoT-5 and CoRoT-18 showed larger values than the solar ones.

With this work, we hope to have shown the many possibilities of the spot model presented here, developed in Silva (2003). That is, by modelling the small variations observed in the transit light curves one can determine:

• Spots physical characteristics: such as size, temperature, location, evolution or life-time, surface area coverage, etc.

• Stellar rotation profile with differential rotation

• Magnetic activity cycles

From the known location of recurrent spots, one may infer the presence of active longitudes, a key ingredient to stellar dynamos. From the temperature of the spots it is also possible to estimate their average magnetic field, if we assume a relation similar to that of the Sun.

If the same spots are detected on multiple transits, then the rotation profile of the star can be constructed, as described in detail by Silva-Valio (2008), Silva-Valio *et al.* (2010), Silva-Valio & Lanza (2011), and Valio (2017). For longer observing period, stellar activity cycles can be determined. Short magnetic cycles were estimated for two Kepler stars by Estrela & Valio (2016) applying this method.

The results obtained from this model, such as spots characteristics, active longitudes, spots lifetime, and differential rotation, are crucial ingredients to the dynamo mechanisms believed to be at work on solar-like stars.

References

Baliunas, S. A., Donahue, R., Soon, W., Horne, J., Frazer, J., Woodard-Eklund, L., Bradford, M., Rao, L., Wilson, O., Zhang, Q., *et al.* 1995, *ApJ*, 438, 269

Christiansen, J. L., Jenkins, J. M., Caldwell, D. A., Burke, C. J., Tenenbaum, *et al.* 2012, *PASP* 124, 1279

Dicke, R. H. 1970, *ApJ*, 159, 25

Estrela, R. & Valio, A. 2016, *ApJ*, 831, 57

Messina, S. & Guinan, E. F. 2002, *A&A*, 393, 225

Olah, K., Kollath, Z., Granzer, T., Strassmeier, K. G., Lanza, A. F. *et al.* 2009, *A&A*, 501, 703

Oshagh, M., Santos, N. C., Ehrenreich, D., Haghighipour, N., Figueira, P., Santerne, A., &
    Montalto, M., 2014, *A&A*, 568, A99.

Pont, F. , Sing, D. K., Gibson, N. P., Aigrain, S., Henry, G., & Husnoo, N. 2013, *MNRAS*, 432,
    2917

Saar, S. H. & Brandenburg, A. 1999, *ApJ*, 524, 295

Sanchis-Ojeda, R. & Winn, J. N., 2011, *ApJ*, 743, 61

Sanchis-Ojeda, R., Fabrycky, D. C. , Winn, J. N., Barclay, T. , Clarke, B. D. , Ford, E. B. *et al.*
    2012, *Nature*, 487, 449

Silva, A. V. R. 2003, *ApJ* (Letters), 585, L147

Silva-Valio, A. 2008, *ApJ* (Letters), 683, L179

Silva-Valio, A., Lanza, A. F., Alonso, R., & Barge, P. 2010, *A&A*, 510, 25

Silva-Valio, A. & Lanza, A. F. 2011, *A&A* 529, 36

Valio, A. 2013, *Astronomical Society of the Pacific Conference Series*, 472, 239

Valio, A., Estrela, R. Netto, Y., Bravo, J. P., & de Medeiros, J. R. 2017, *ApJ*, 835, 294

Living Around Active Stars
Proceedings IAU Symposium No. 328, 2016
D. Nandy, A. Valio & P. Petit, eds.

© International Astronomical Union 2017
doi:10.1017/S174392131700429X

Dynamo action and magnetic activity during the pre-main sequence : Influence of rotation and structural changes

Constance Emeriau-Viard & Allan Sacha Brun

Laboratoire AIM Paris-Saclay,
CEA/DSM - CNRS - Université Paris Diderot,
IRFU/SAp Centre de Saclay,
F-91191 Gif-sur-Yvette Cedex, France
email: `constance.emeriau@cea.fr`
email: `sacha.brun@cea.fr`

Abstract. During the PMS, structure and rotation rate of stars evolve significantly. We wish to assess the consequences of these drastic changes on stellar dynamo, internal magnetic field topology and activity level by mean of HPC simulations with the ASH code. To answer this question, we develop 3D MHD simulations that represent specific stages of stellar evolution along the PMS. We choose five different models characterized by the radius of their radiative zone following an evolutionary track, from 1 Myr to 50 Myr, computed by a 1D stellar evolution code. We introduce a seed magnetic field in the youngest model and then we spread it through all simulations. First of all, we study the consequences that the increase of rotation rate and the change of geometry of the convective zone have on the dynamo field that exists in the convective envelop. The magnetic energy increases, the topology of the magnetic field becomes more complex and the axisymmetric magnetic field becomes less predominant as the star ages. The computation of the fully convective MHD model shows that a strong dynamo develops with a ratio of magnetic to kinetic energy reaching equipartition and even super-equipartition states in the faster rotating cases. Magnetic fields resulting from our MHD simulations possess a mixed poloidal-toroidal topology with no obvious dominant component. We also study the relaxation of the vestige dynamo magnetic field within the radiative core and found that it satisfies stability criteria. Hence it does not experience a global reconfiguration and instead slowly relaxes by retaining its mixed poloidal-toroidal topology.

Keywords. Convection,Dynamo,Magnetohydrodynamics,Stars interiors

1. Introduction

During the pre-main sequence phase (PMS), stellar structure and rotation rate change drastically. As the star contracts, the rotation rate also evolves. It was modeled by Mac-Gregor & Brenner (1991) and Gallet & Bouvier (2013) by three different phases : at the very beginning of the PMS the rotation rate remains constant as the star is still locked with its protostellar disk, then it increases as the radius of the star decreases, finally the star reaches a stable outer radius and the rotation rate decreases because of the wind-braking Réville (2015) The internal structure also significantly changes. At the very beginning of the PMS, stars are fully convective as there are no long-term thermo-nuclear reactions in their interior. As the star ages, hydrogen burning develops, the opacity of the core decreases and a stable radiative core appears and grows. These major changes impact the star's properties, especially their internal rotation and magnetic field.

77

Stellar rotation rate and magnetic field are strongly linked through a feedback loop created by complex physical processes. Indeed stellar magnetic field impacts the transport of angular momentum (Brun *et al.* 2004; Strugarek 2017) trough the Maxwell stresses and also the braking of the star by the wind while stellar rotation influences the magnetic field through dynamo process, and especially through Ω effect. This mutual influence was shown observationally for instance by Pizzolato (2003), showing a correlation between coronal X-ray emission and stellar rotation rate at moderate and high Rossby number $R_o = P_{ro}/\tau_{conv}$ and also that either the surface field or the stellar dynamo, or both, saturates for fast rotation rates. Stellar internal structure also impacts the magnetic field topology and amplitude as shown by Gregory *et al.* (2012) (see also these proceedings). This study show that stars with a massive radiative core, $M_{core} > 0.4M_*$, possess a complex magnetic field, i.e. non axisymmetric and with weak dipole components. Stars with smaller radiative core, $0 < M_{core} < 0.4M_*$, have a less complex and more axisymmetric field. In fully convective stars the behavior of stellar magnetic field might even be bistable with a mixture of different geometries and amplitudes (Morin 2010).

As the stellar radiative core increases as stars age along the PMS, we also wish to know how the magnetic field evolves in the radiative zone as convective dynamo action does not support it anymore. These magnetic fields are observed in massive stars, since their envelop is radiative, where they are often oblique dipole. These fields are highly unstable and Tayler (1973) and Markey & Tayler (1973) showed that only a mixed poloidal-toroidal topology can be stable in a stably stratified radiative zone. Braithwaite (2008) introduced a stability criteria for a stable fossil field in a radiative zone $E_{pol}/E_{tot} < 0.8$. It is interesting to assess if the magnetic field left over by the dynamo process is stable or if it must relax to a different configuration.

2. Model setup

To analyze the evolution of magnetic field during the PMS, we choose to compute 3D global magnetohydrodynamical models of a one solar mass star. However stellar evolution in the PMS lasts for several Myr whereas 3D MHD simulations can only compute, with reasonable resources, stellar evolution for several hundreds of years. Hence, we select specific models that represent the most important stages of the PMS. To characterize these models, we need quantitative values for luminosity, rotation rate, radii ... These physical values are given by stellar evolution models that were computed by the 1D stellar evolution code STAREVOL Amard (2016) as shown in Figure 1. We choose to select five models such that the radiative core radii (in stellar units) are well distributed, almost every 20%, and the ratio between the rotation rate of two consecutive models is smaller than two (see Figure 1). Radial structures provided by STAREVOL simulations enable us to create reference states and thus initialize our 3D ASH simulations. Hence we compute hydrodynamical simulations for each of the fives models. We run the calculations until the hydrodynamical simulations have equilibrated internal flows and coupling between radiative core and the convective envelop.

To study the evolution of the stellar magnetic field through the PMS, we first injected a *seed magnetic field* in the fully convective hydrodynamical model. This weak seed confined dipole represents the field left by the proto-stellar phase. We run the magneto-hydrodynamical simulation of the fully convective model until it reaches an equilibrium state with a dynamo generated field. Then we introduced the magnetic field resulting from this simulation into the 20% hydrodynamical simulation. Hence we can see how the change of internal structure affects the magnetic field. Once this simulation has reached an equilibrium state, in the statistically stationary sense, and the magnetic field has

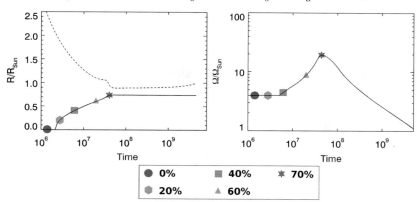

Figure 1. Choice of the 3D ASH models based on stellar evolution STAREVOL code. *Left* : Evolution of stellar radius (solid) and radius of the radiative core (dash) during the PMS and the MS. *Right* : Stellar rotation rate as a function of time. At the beginning of the PMS it is constant since star is still in the disk- locking phase. Then it increases as the star contracts under the effect of gravity until the ZAMS. Stellar rotation rate starts decreasing as the stellar contraction stops and magnetic wind brakes the star.

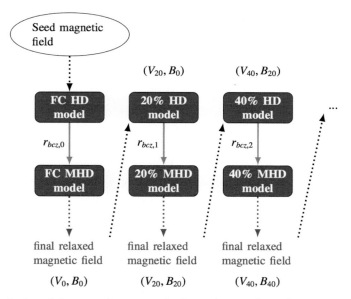

Figure 2. Description of the procedure to study the evolution of a stellar magnetic field through the PMS. In each HD simulation, we introduced the magnetic field resulting from the previous MHD simulation. MHD simulations are computed until they are equilibrated with dynamo process in the convective zone and a relaxation of the magnetic field in the radiative zone, if present. In the first model, i.e. the fully convective one, we choose to inject a weak seed confined dipole representing the field left by the proto-stellar phase.

relaxed in the radiative core, we introduce the resulting magnetic field in the following hydrodynamical model. By reproducing these operations with all HD models, we can analyze the influence on the magnetic field of the changes of internal structure and rotation rate caused by stellar evolution (see details in Emeriau-Viard & Brun 2017).

HD   **MHD**

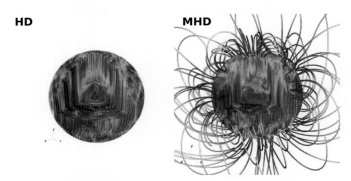

Figure 3. 3D views of the 40% model. *Left*: Radial velocity of the hydrodynamical model. *Right*: Radial velocity of the MHD model with the potential extrapolation of magnetic field outside the star. Upflows in red and downflows in blue. Field lines are color coded with the radial component of the magnetic field.

3. Magnetic field properties and evolution during the PMS phase

The introduction of magnetic field in the HD simulations strongly impacts the internal flows. Differential rotation is flatten due to the presence of Maxwell stresses. Hence the radial extent of the convective motions in the MHD models is larger than in the HD ones, since there is less horizontal shear. The quenching of differential rotation also have an influence on transport of angular momentum. Indeed viscous diffusion becomes negligible for the transport of angular momentum. Since Reynolds stresses is outward, as in the HD models, the inward transport is carried by Maxwell stresses. The 3D topology of radial velocity is illustrated for the 40% model in Figure 3 both for the hydrodynamical and magnetohydrodynamical cases. We notice the cylindrical patterns of convection linked to fast rotation in both cases. In Figure 3, we also see an extrapolation of the magnetic field outside the star. It shows us a potential extrapolation of the magnetic field outside the star which is complex, highly non-axisymmetric and exhibits as well extended transequatorial loops.

As the star evolves along the PMS, we want to study the evolution of the magnetic field both in amplitude and topology. Thus we define the dipole field strength as done by Christensen & Aubert (2006). $f_{\rm dip}$ is time-average ratio at the surface of the mean dipole field strength to the field strength in harmonic degrees $\ell = 112$. As in Schrinner (2012), we also defined the local Rossby number of our simulations

$$R_{o,l} = R_o \frac{\bar{\ell}}{\pi} \quad \text{where} \quad \bar{\ell} = \sum_{\ell} \ell \frac{\langle (\mathbf{v})_{\ell} \cdot (\mathbf{v})_{\ell} \rangle}{\langle \mathbf{v} \cdot \mathbf{v} \rangle} \tag{3.1}$$

with $\mathbf{v}$ the local velocity in spherical coordinates in the frame rotating at constant angular velocity Ω_0 and $\langle \cdot \rangle$ the average on time and radius. As seen in Figure 4, we notice that an increasing radiative core and rotation rate lead to an increasing mean harmonic degree. In our simulations, Rossby number increases as the star ages. This result may seem counterintuitive since faster rotation rate should lead to smaller Rossby number. However, rotation rate is not the only stellar parameter changing during the PMS phase. The thickness of the convective envelop D decreases at the same time. By looking at the product $\Omega_0 D$, we see that it decreases as the star ages, leading to larger Rossby number. As the mean harmonic degree and the Rossby number grow, the local Rossby number increases as the star evolves along the PMS. By plotting $f_{\rm dip}$ as a function of the local Rossby number, Schrinner (2012) noted a transition between the dipolar and multipolar

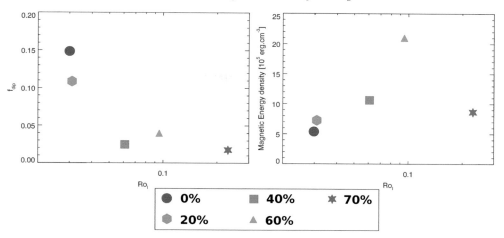

Figure 4. Evolution of the magnetic field topology with respect to the local Rossby number of our simulations. *Left*: Dipole field strength decreases as the local Rossby number increases. *Right*: Magnetic energy density evolution as the star ages along the PMS.

mode at $R_{o,l}^c = 0.1$. The local Rossby number of our simulations are around this transition value. We also observe a transition in the topology evolution of the magnetic field along the PMS. As in Schrinner (2012), dipolar components of the magnetic field are weak when $R_{o,l} > 0.1$. These components are bigger when $R_{o,l} < 0.1$ even if the transition is less strong than the one observed by Schrinner. The amplitude of the magnetic field also changes as the star ages. To take into account the variation in aspect ratio of our simulations, we analyze the evolution of the magnetic energy density (ME). In figure 4, we notice that ME increases until the radiative core reaches 60% of the stellar radius and for the 70% model, ME decreases. The specificity of this simulation is a change in the stellar luminosity evolution. The size of the radiative core and the rotation rate change monotically during the PMS whereas the luminosity first decreases down to the 60% model and then increases until the ZAMS. This can explain a different behavior for the 70% model compared to the four other simulations.

For a better understanding of the dynamics of our MHD simulations, we studied the mean field generation through the analysis of the $\alpha - \Omega$ effect. We found that the influence of the Ω effect mostly increases with respect to the α effect. This can be understood as the Ω effect is strongly link to the differential rotation and we see that the contrast in differential rotation grows as the star evolves along the PMS. Only the 70% model behaves differently. As luminosity increases in this model, we increase both viscous and magnetic diffusivities to keep consistent Reynolds numbers. This increasing magnetic diffusivity possibly explains the behavior of the 70% model.

As the convective envelop get shallower, the radiative core grows. These radiative cores are mostly in solid rotation, i.e. do not possess differential rotation. The lack of ingredients for a dynamo generation leads us to infer, as Braithwaite (2008), that these magnetic fields in radiative zone are *fossil fields*, i.e. stable fields that evolves on diffusive time scales. These fields are very sensitive to instabilities and many of them are unstable and disappears quickly. These instabilities were studied by Tayler (1973) and Markey & Tayler (1973). They showed that both purely poloidal and toroidal magnetic fields cannot be stable. Braithwaite (2008) derived a stability criteria for a stable mixed poloidal-toroidal configuration : the poloidal magnetic energy must be less that 80% of the total magnetic energy. This ratio can be rewritten as $B_{\rm pol}/B_{\rm tor} < 2$. In Figure 5, we

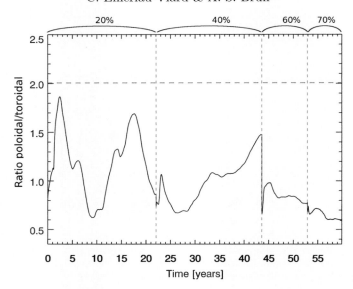

Figure 5. Evolution of the ratio $B_{\mathrm{pol}}/B_{\mathrm{tor}}$ over the PMS.

show this ratio along all our simulations. Hence we notice that the stability criteria is always fulfilled in our MHD simulations. We conclude that interestingly dynamo action tends to generate mixed fields whose properties satisfy stability criteria in stratified radiative core. This result on the stability of the fossil field left over by dynamo action as the convective envelope becomes shallower along the PMS is a direct outcome of our set of simulations and could not be easily anticipated. It has direct consequences on the geometry of fossil field that can be expected in solar-like star's radiative core as stable mixed poloidal-toroidal configurations should be favored.

4. Conclusion

During the PMS, between the protostellar phase and the ZAMS, the stellar radius decreases due to gravitational contraction. This stellar contraction causes an increase of the rotation rate due to angular momentum conservation. Hence we expect the star's dynamical properties to vary significantly and these variations are key to characterize. In order to do so, we have developed a series of 3D MHD simulations of stellar convective dynamo. To make this study more realistic, we used to setup the spherically symmetric background state of our 3D models, radial profiles obtained from 1D stellar evolution model at various stages along the PMS evolutionary track. We choose five different models that represent the star at specific ages of the PMS with different rotation rates and radiative radii. At first, we run hydrodynamical 3D simulations of these models in order to equilibrate internal flows and coupling between the radiative core and the convective envelop. Then we inject a magnetic field into the fully convective model and propagate it through all simulations.

Our five MHD simulations show the mutual influence of the internal magnetic field and internal flows as the star evolves along the PMS. The introduction of the magnetic field in the hydrodynamical models leads to important modifications of the internal mean flows and the convective patterns. As the differential rotation profiles are quenched by the influence of the Maxwell stresses, the radial convective patterns are larger, since they are less sheared. Hence viscous diffusion becomes negligible for the transport of angular momentum and inward transport is carried by Maxwell stresses.

As the star ages along the PMS, the magnetic energy in the convective envelop globally increases with a strong decrease of the mean field energy. This result is coherent with study led by Gregory *et al.* (2012) in which the magnetic field is less axisymmetric and more complex as the radiative core is bigger. Since we follow the evolutionary path of a solar-like star, both rotation rate and aspect ratio of the convective envelop change as the star evolves along the PMS. The specific influence of each parameter is not always easy to disentangle in our study.

The generation of the mean magnetic field shows that as the convective zone becomes shallower and the rotation rate increases, the Ω effect becomes predominant in the generation of the mean toroidal magnetic field. Moreover the α-effect tends to generate more poloidal field than toroidal one. Hence, in each model, we see an $\alpha - \Omega$ dynamo.

As the radiative zone grows in the star, we observe that, in all models, the magnetic field in the core, left over by the convective dynamo, is stable regarding the limit given by Braithwaite (2008): $E_{pol}/E_{tot} < 0.8$. The magnetic field in the radiative core of the star originates from the relaxation of the dynamo field coming from the previous stellar evolution phase in our sequence of models. By looking at this dynamo field, we notice, that in all convection zones, the magnetic field that comes from the dynamo action also fulfill the stability criteria $E_{pol}/E_{tot} < 0.8$ even if this has no obvious consequence in that zone.

The global properties of the magnetic fields we obtain in our study also have direct consequences on the coronae of PMS stars. We have computed the change of the Alfvén radius (e.g. the radius where the stellar wind decouples from the star) that such topological and rotation state implies, following the prescription described in Réville (2015). We find that the Alfvén radius shrink from about $33R_\odot$ to $10 \ R_\odot$. This is in good qualitative agreement with the recent work of Réville (2016) who computed realistic 3D stellar wind along the evolutionary track of a solar mass star using spectro-polarimetric maps from Folsom (2016). We intend in the near future to use the magnetic field coming out of our dynamo simulations to compute similar 3D wind solutions along the PMS. This will allow us to assess the loss of mass and angular momentum, that must also vary significantly given the large change of Alfvén radius we have identified. It is important to notice that in this study, we choose our models to follow an astrophysical path along the PMS. A logical follow-up is therefore to apply this analysis to the evolution of solar-like stars along the following step of stellar evolution, i.e. the main sequence. In that study the main parameter will be the decrease of the rotation rate as the star is braked by the solar wind and the internal stellar structure of the star remains almost unchanged during this evolutionary phase. An additional study would be to study the impact of stellar structure with a fixed rotation rate. We have started doing such studies and their results will be reported in future communications.

References

Amard, L., Palacios, A., Charbonnel, C., Gallet, F. & Bouvier, J. 2016 *A&A*, 587, A105

Braithwaite, J. 2008 *MNRAS*, 386, 1947–1958

Brun, A. S., Miesch, M. S., & Toomre, J. 2004, *ApJ*, 614, 1073

Christensen, U. R. & Aubert, J. 2006, Geophysical Journal International, 166, 97–114

Emeriau-Viard, C. & Brun, A. S. 2017 *ApJ*, submitted

Folsom, C. P., Petit, P. , Bouvier, J., Lèbre, A., Amard, L., Palacios, A., Morin, J., Donati, J.-F., Jeffers, S. V., Marsden, S. C. & Vidotto, A. A. 2016 *MNRAS*, 457, 580–607

Gallet, F. & Bouvier, J. 2013, *A&A*, 556, A36

Gregory, S. G. Donati, J.-F., Morin, J., Hussain, G. A. J., Mayne, N. J., Hillenbrand, L. A. & Jardine, M. 2012 *ApJ*, 755, 97

MacGregor, K. B. & Brenner, M. 1991, *ApJ*, 376, 204–213

Markey, P. & Tayler, R. J. 1973 *MNRAS*, 163, 77

Morin, J., Donati, J.-F., Petit, P., Delfosse, X., Forveille, T. & Jardine, M. M., 2010 *MNRAS*, 407, 2269–2286

Pizzolato, N., Maggio, A., Micela, G., Sciortino, S. & Ventura, 2003 *A&A*, 397, 147–157

Réville, V., Brun, A. S., Matt, S. P., Strugarek, A. & Pinto, R. F. 2015 *ApJ*, 798, 116

Réville, V., Brun, A. S., Strugarek, A., Matt, S. P., Bouvier, J., Folsom, C. P., & Petit, P. 2015, *ApJ*, 814, 99

Réville, V., Folsom, C. P., Strugarek, A.& Brun, A. S. 2016 *ApJ*, 832, 145

Schrinner, M., Petitdemange, L. & Dormy, E. 2012 *ApJ*, 752, 121

Strugarek, A., Beaudoin, P., Charbonneau, P., Brun, A.S. & do Nascimento, J.-D. 2017 *submitted*

Tayler, R. J. 1973 *MNRAS*, 161, 365

Living Around Active Stars
Proceedings IAU Symposium No. 328, 2016
D. Nandy, A. Valio & P. Petit, eds.

© International Astronomical Union 2017
doi:10.1017/S1743921317003830

The Suppression and Promotion of Magnetic Flux Emergence in Fully Convective Stars

Maria A. Weber, Matthew K. Browning, Suzannah Boardman, Joshua Clarke, Samuel Pugsley and Edward Townsend

Department of Physics and Astronomy, University of Exeter,
Stocker Road, EX4 4QL Exeter, UK
email: mweber@astro.ex.ac.uk

Abstract. Evidence of surface magnetism is now observed on an increasing number of cool stars. The detailed manner by which dynamo-generated magnetic fields giving rise to starspots traverse the convection zone still remains unclear. Some insight into this flux emergence mechanism has been gained by assuming bundles of magnetic field can be represented by idealized thin flux tubes (TFTs). Weber & Browning (2016) have recently investigated how individual flux tubes might evolve in a $0.3M_\odot$ M dwarf by effectively embedding TFTs in time-dependent flows representative of a fully convective star. We expand upon this work by initiating flux tubes at various depths in the upper $\sim$50-75% of the star in order to sample the differing convective flow pattern and differential rotation across this region. Specifically, we comment on the role of differential rotation and time-varying flows in both the suppression and promotion of the magnetic flux emergence process.

Keywords. MHD, stars: magnetic fields, stars: spots, stars: interiors, methods: numerical

1. Introduction

M dwarfs are among the most magnetically active and abundant stars in the galaxy. They encompass a broad mass range of $\sim$0.08-0.6$M_\odot$. It is widely thought that the seat of the dynamo in partially convective dwarfs ($\gtrsim$0.35$M_\odot$) resides in the tachocline, a region of shear at the interface between the convection and radiative zones (e.g. Charbonneau 2010). In this interface dynamo, the toroidal magnetic field is amplified in the tachocline and then rises to the surface where it may be observed as starspots and serve as the launching site for strong flares. Yet M dwarfs on the fully convective side of the 'tachocline divide' ($\lesssim$0.35$M_\odot$) still effectively build magnetic fields, with this activity increasing in prevalence toward late M spectral types (e.g. West *et al.* 2015). This magnetism is similar to that observed in solar-like stars; i.e., there is still a rotation-activity correlation that plateaus at rapid rotation (e.g. Wright & Drake 2016).

Many have turned to global-scale dynamo models as a way of self-consistently capturing the strength and morphology of magnetism that may be built in fully convective stars (e.g. Dobler *et al.* 2006; Browning 2008; Yadav *et al.* 2015). Similar simulations are just now beginning to capture some aspects of magnetic flux emergence in rapidly rotating Suns, exhibiting buoyant magnetic loops that emerge naturally from wreaths of magnetism (e.g. Nelson *et al.* 2011). But, these simulations are remarkably expensive to compute. As an alternative and less computationally expensive method, the flux tube model has been invoked to describe the evolution of magnetic field bundles, utilizing either the effectively 1D thin flux tube (TFT) approximation or solving the full 3D magnetohydrodynamic (MHD) equations (see e.g. review by Fan 2009). Applying this model to the Sun has provided a wealth of insight regarding the flux emergence process, replicating many observed features of active regions (e.g. Fan 2009; Weber *et al.* 2013). These simulations assume the dynamo has already built fibril magnetic flux tubes that rise under the combined effects of buoyancy and advection by turbulent flows.

Weber & Browning (2016) (hereafter WB16) recently investigated for the first time how flux tubes in a fully convective M dwarf might rise under the joint effects of buoyancy, differential rotation, and convection. The work presented here expands upon the parameter space explored in WB16 by initiating flux tubes at multiple depths between 0.475-0.75R to sample the varying convective flow field structure across this region. In Section 2, we introduce our model and initial conditions. We present the results of some new TFT simulations and diagnostic metrics in Section 3. We focus on how convection modulates the initially toroidal flux tube (Sec. 3.1), the way differential rotation and time-varying flows may alter the duration and trajectory of the flux tube rise (Sec. 3.2), and discuss how convection can suppress the mean rise of magnetism (Sec. 3.3).

2. Formulating the Problem

The dynamic evolution of isolated, fibril magnetic flux tubes can be modeled by applying the thin flux tube (TFT) approximation (e.g., Roberts & Webb 1978; Spruit 1981). Derived from ideal MHD, the TFT approximation assumes the diameter of the flux tube is small enough that all variables can be represented by their averages over a tube segment, reducing the model to one dimension. To capture the effects of global-scale convection on flux emergence in a $0.3M_\odot$ fully convective star, a convective velocity field computed apart from the TFT models is incorporated through the aerodynamic drag force influencing each flux tube segment. A description of the equations and methods used to solve for the evolution of individual flux tubes is discussed in WB16.

The traditional TFT approach assumes that the magnetic field is generated in the tachocline region of a solar-like star through an interface dynamo. Here we postulate that a distributed dynamo might build toroidal flux tubes as well. We adopt the simplifying assumption that turbulent, large-scale convective motions have built such flux tubes initially co-rotating with the local differential rotation and in thermal equilibrium with the background fluid. The condition of thermal equilibrium renders the flux tube initially buoyant with a density deficit of $\rho_e - \rho = (\rho_e B_0^2)/(8\pi p_e)$.

Flux tubes are initiated at latitudes of 1°-60° in both hemispheres and depths between 0.475-0.75R in order to sample the varying differential rotation and convective pattern with depth. In this investigation, we are interested in studying the evolution of magnetic flux tubes that might produce starspots. To parallel our work in WB16, we perform simulations where B_0=30-200 kG and the cross-sectional radius a of each flux tube is 1.7×10^8 cm. Assuming the constant total flux of the tube is given by $\Phi = B\pi a^2$, the magnetic flux spans from 2.72×10^{21} Mx for 30 kG tubes to 1.82×10^{22} Mx for 200 kG tubes. This range of Φ is typical of solar active regions (e.g. Zwaan 1987). As the flux tube approaches the surface, the cross-section expands quickly due to the rapid decrease of the density and pressure of the external plasma, and the TFT approximation is no longer satisfied. Therefore, we terminate our simulations once the fastest rising portion of the tube reaches $0.95R$, assuming the motion through the remaining $0.05R$ is negligible.

To model the external velocity field, we utilize the Anelastic Spherical Harmonic (ASH) code, which solves the 3D MHD equations under the anelastic approximation. Representative of fluid motions in a low-mass, fully convective star, this hydrodynamic ASH simulation captures differential rotation and giant-cell convection in a rotating spherical domain (Ω_0=2.6$\times10^{-6}$ rad s^{-1}) spanning from 0.10 to 0.97R, with R the stellar radius of 2.013×10^{10} cm (similar to Case C in Browning 2008). In WB16, we examine how the evolution of flux tubes may change when subject to varying differential rotation profiles.

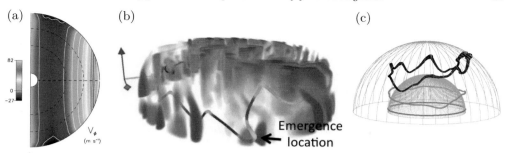

Figure 1. Convective velocity field and representative flux tubes advected by fluid flows in a fully convective star. (a) Meridional plot of the longitudinal velocity $\hat{v}_\phi$ relative to the rotating frame averaged over ~460 days, with contour intervals every 10 m s^{-1}. Dashed lines at 0.5 and 0.75R. (b) Snapshot of a 30 kG flux tube initiated at 5° and 0.75R. Tube emerges in a region of strong upflow (red in online version) next to a strong downflow lane (blue), and is colored according to its magnetic field strength, with darker/lighter tones representing stronger/weaker fields. Only the northern hemisphere from the equator to half the stellar radius in the vertical direction is shown. The tube is rendered with a 3D extent according to the local cross-sectional radius. (c) Snapshots in time of a 30 kG flux tube initiated at 5° and 0.5R. Inner/outer mesh spheres represent surfaces of constant radius at r_0/0.95R. Shown is the initial flux tube (red), 75% in time through the total rise (green), and once reaching 0.95R (black). Evolution of this tube is largely parallel to the rotation axis, unlike the more radial rise of the tube apex in (b).

Here we use the differential rotation profile with an angular velocity contrast at the surface between the equator and 60° of $\Delta\Omega/\Omega_0 \sim 22\%$ (Fig. 1a). An instantaneous view of the radial velocity field is shown in Figure 1b.

To sample the time-varying velocity field in a uniform way, we perform ensemble simulations. Flux tubes in each ensemble are initialized at the same moment, but evolve independently of each other. Therefore, they are advected by the exact same time-varying convective flows. The simulations performed here align with Cases TLf /TLfC in WB16, but only comprise of one ensemble as compared to the three performed in that paper.

3. Results

3.1. *The dynamic evolution of flux tubes and modulation by convective motions*

In the quiescent interior of a 0.3M$_\odot$ star, flux tubes in an axisymmetric configuration in thermal equilibrium with the background plasma rise with a trajectory largely parallel to the rotation axis. Explored in detail in WB16, this motion is governed by the force balance the tube achieves in the direction perpendicular to the rotation axis. The four main forces that contribute to flux tube evolution include: buoyancy, magnetic tension, the Coriolis force, and aerodynamic drag. In the early stages, the flux tube adjusts until there is an equilibrium of the inward and outward-directed forces (toward/away from rotation axis). Once reaching this equilibrium, the tube rises parallel to the rotation axis due in large part to the unbalanced vertical component of the buoyancy force.

Radial convective motions distort the shape of the toroidal ring, promoting buoyantly rising loops (Figs. 1b-c). Strong downflows pin parts of the flux tube in deeper layers, while strong upflows may boost loops toward the surface. By the time the fastest-rising peak (i.e. flux tube apex) reaches the upper boundary, the tube has developed a number of loops with troughs that stretch a large fraction of the star. For flux tubes of similar thickness (i.e. similar a), the modulation by convective flows increases with decreasing magnetic field strength. The 30 kG tubes in Figures 1b-c are strongly advected by convection. In comparison to tubes of stronger magnetic field strength, those of weaker B_0

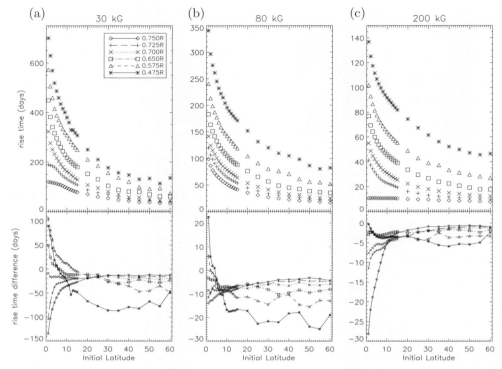

Figure 2. Average apex rise duration (t_{apex}) for flux tubes rising subject to flow fields in a fully convective star (top panels) as a function of $|\theta_0|$. The difference in rise times for tubes with the same initial conditions rising in a quiescent interior (t_q) and those rising through the convective flow field is shown below ($t_{apex} - t_q$). A positive value for this quantity indicates that the tube apex rises slower than the same tube in a quiescent interior. This scenario is more likely to occur for weaker B_0 tubes originating at lower latitudes in deeper layers. Quantities are plotted for tubes originating at depths ranging from 0.475-0.750R. Each plot has a different y-axis range.

have relatively smaller magnetic tension and buoyancy forces, rendering them more susceptible to the aerodynamic drag imparted by turbulent flows. Conversely, tubes of larger B_0 evolve more like the flux rings in a quiescent interior described above.

3.2. *Rise times, emergence latitudes, and the influence of convective flows*

Mean and local flows can significantly alter the duration and trajectory of a buoyantly rising loop's journey to the stellar surface. Figure 2 shows the total rise time of the flux tube apex for those that evolve subject to convection. Also shown is the relative difference in rise times between tubes with the same initial conditions that rise both with and without convective effects. Each symbol in all the plots in this paper represent the average for flux tubes originating with the same $|\theta_0|$. Within the parameter space explored, rise times vary greatly from more than 600 days (30 kG, deeper interior) to as fast as 10 days (200 kG, nearer surface). Generally, flux tubes of a stronger magnetic field strength rise more rapidly because of their greater buoyancy. Tubes initiated at higher latitudes also tend to rise more quickly. This is driven in part by a larger poleward acceleration arising from the magnetic tension force due to a smaller radius of curvature there. Also, from a geometric standpoint, tubes initiated at higher latitudes have a trajectory parallel to the rotation axis that may cover a shorter distance across the convection zone.

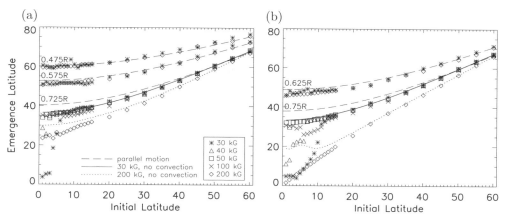

Figure 3. Average emergence latitude of the flux tube apex as a function of $|\theta_0|$. These are shown for flux tubes of various B_0 originating at (a) 0.475, 0.575, and 0.725R and (b) 0.625 and 0.750R. Dashed lines represent the emergence latitude expected if the tube were to rise exactly parallel to the rotation axis. Curves depicted in the legend correspond to tubes rising through a quiescent interior, only shown for those originating at 0.725 and 0.750R. Large deviations from parallel motion for 30-40 kG tubes initiated in shallower regions nearer the equator is largely a result of the strong prograde differential rotation there, suppling angular momentum to these weaker B_0 loops that are strongly advected by convection. The deviation for the 200 kG tubes in shallower regions is related to the initial force imbalance perpendicular to the rotation axis and vigorous time-varying flows closer to the surface.

A few trends emerge in the bottom panels of Figure 2. One is the tendency for the duration of the apex rise to roughly follow the rise of the flux rings that evolve without convection. Yet, weaker 30 kG flux tubes initiated at lower latitudes between 0.475-0.65R rise slower than the same flux tubes rising through a quiescent convection zone. This is indicative of a process akin to magnetic pumping (see Sec. 3.3). Another striking feature is the faster relative rise time of especially the 30 kG and 80 kG flux tubes initiated at latitudes $\gtrsim 10°$ in the deep interior at 0.475R. It is not the case that the whole tube rises faster than the identical tube in a quiescent interior. Rather, the fastest rising loop of the flux tube evolves in such a way that its journey to the surface is boosted from interaction with the mean and time-varying flows. At such depths, the differential rotation is more strongly retrograde (see Fig. 1a), and may be the reason for the faster apex rise time rather than a change in the nature of the giant-cell structure.

Initially low latitude flux tubes of 30 kG and 200 kG originating between 0.725-0.75R show a significant decrease in their rise times relative to the corresponding flux tubes that rise without convection. These same tubes also have apices that rise more radially, in stark contrast to the parallel trajectories exhibited by the majority of flux tubes we study here (see Fig. 3). Explained in detail in WB16, both of these effects are facilitated by the strongly prograde differential rotation in the upper $\sim$30% of the convection zone at lower latitudes. If allowed to continue evolving after the apex reaches 0.95R, the same flux tube might produce multiple starspot regions. For example, the flux tube in Figure 1c initiated at 0.5R could go on to produce three or more high latitude starspots. The tube in Figure 1b initiated at 0.75R could give rise to at least one near-equatorial spot, but may also potentially produce a few higher latitude spots.

3.3. Suppression of flux emergence by convection

The flux tube properties discussed in Section 3.2 are representative of the apex of the fastest rising loop. However, a large fraction of the flux tube may in fact remain in deeper

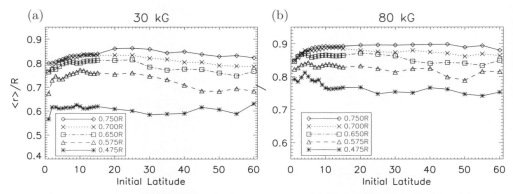

Figure 4. Average magnetic field weighted radial depth $\langle r \rangle$ as of function of $|\theta_0|$ for (a) 30 kG and (b) 80 kG flux tubes originating at various depths. This quantity is calculated once the apex of the tube has reached the simulation upper boundary. At this instant, the majority of the flux tube still resides in deeper layers. For any given depth, tubes of weaker magnetic field strengths are also largely confined to deeper layers. Both plots have a different y-axis range.

layers (see Figs. 1b-c). Through interaction with a sequence of favorable flows, a buoyant loop may reach the surface faster than the commensurate flux tube rising through a quiescent interior. On the other hand, the journey of a rising loop may be prolonged compared to the quiescent case due to buffeting of the flux tube by downflows.

As in WB16, to assess the ability of convection to suppress the mean rise of the flux tube, we compute the average magnetic field weighted radial depth of the flux tube

$$\langle r(t) \rangle = \frac{\int_{u_0}^{u_{N-1}} r(u,t)B(u,t)du}{\int_{u_0}^{u_{N-1}} B(u,t)du}, \tag{3.1}$$

where $u_j = s/L = j/(N-1)$ for $j = 0, .., N-1$ is the fractional arc length along the flux tube, s is the length of the tube up to a mesh point j from the origin, L is the total flux tube length, and N is the number of uniformly spaced mesh points. The quantity $\langle r(t) \rangle$ represents the radial depth where the majority of the flux tube magnetic field resides. It places a reduced weight on the magnetic field in shallower layers, which has diminished in strength as portions of the tube rise and subsequently expand. Similar treatments have been employed for magnetic fields in 3D computational domains (e.g. Tobias *et al.* 2001; Abbett *et al.* 2004).

Figure 4 shows the average magnetic field weighted radial depth $\langle r \rangle$ for 30 kG and 80 kG flux tubes calculated once the apex has reached 0.95R. Near depths of $\sim$0.5R, the critical field strength at which the magnetic buoyancy is roughly equivalent to the drag force from radial convective downflows is $B_c \sim$30 kG. Even at field strength values close to B_c, $\langle r \rangle$ increases with time owing to the unbalanced vertical and poleward force components acting on the entire flux tube. As such, even 30 kG tubes initiated at 0.475R reach values of $\langle r \rangle \sim$0.6R before the tube apex reaches the surface. As one might expect, weaker B_0 flux tubes are effectively pinned to deeper layers by downflows than stronger B_0 tubes originating at the same depth. For any given initial depth, even though rise times may vary greatly over the latitudinal range (see Fig. 2), the quantity $\langle r \rangle$ does not fluctuate by much more than 5% of the stellar radius.

The degree to which $\langle r \rangle$ deviates in time from the radial position r_q of the corresponding flux tube rising in a quiescent interior is an expression of the effectiveness of 'magnetic pumping'. Here we do not refer to magnetic pumping in the traditional sense (e.g. Tobias *et al.* 2001), but rather to the general ability of convective downflows to suppress the rise

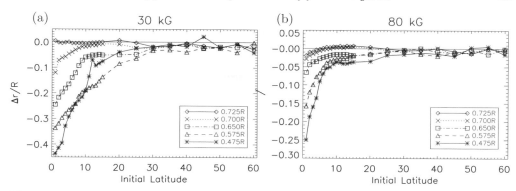

Figure 5. Relative pumping depth Δr as of function of $|\theta_0|$ for (a) 30 kG and (b) 80 kG flux tubes originating at various depths, some of which are different from Fig. 4. Suppression of the global motion of the flux tube by convective motions is more pronounced at lower latitudes and in the deeper interior. Both plots have a different y-axis range.

of buoyant flux tubes. In some sense, many flux tubes subjected to time-varying flows are 'pumped' downwards relative to those that travel through a convection zone void of fluid motions. To capture this process in a single value, we calculate a *relative pumping depth* for each individual flux tube $\Delta r = \langle r(t_{min}) \rangle - r_q(t_{min})$. The value t_{min} is the elapsed rise time corresponding to whichever flux tube reaches the upper boundary at 0.95R first, either the flux tube evolving without convective motions or the tube advected by fluid flows (see also WB16).

Figure 5 shows the relative pumping depth Δr for 30 kG and 80 kG flux tubes originating at various depths. A negative Δr indicates that the overall rise of the tube represented by $\langle r \rangle$ is suppressed compared to the flux tube rising in the quiescent interior at the same instant. Even though the majority of flux tubes launched from the same depth attain a similar $\langle r \rangle$, it is clear from Figure 5 that the mean motion of tubes initiated at lower latitudes and in deeper layers is strongly suppressed by fluid motions. This suppression of flux emergence reduces as the magnetic field strength increases. At higher latitudes, the mean motion of the flux tube evolves similarly to those that rise through a quiescent convection zone (i.e. smaller Δr). This is in part a consequence of the larger poleward acceleration due to a greater magnetic tension force arising from a smaller flux tube radius of curvature. Convective motions are not sufficient enough to reduce this motion as effectively. Finally, for the parameter space we explore here, flux tubes of all B_0 originating in layers >0.7R show very little suppression of the mean motion by convection.

4. Conclusions and Perspectives

We have reported some results from thin flux tube simulations embedded in a rotating spherical domain of mean and time-varying flows depictive of a $0.3M_\odot$ fully convective star. These simulations model how coherent bundles of magnetic fields, assumed to be starspot progenitors, may behave as they travel through the stellar interior. In this work, we focus on studying how differential rotation and time-varying flows can either promote or suppress the rise of magnetism.

The archetype of loop-shaped, buoyantly rising magnetic structures is not achievable in the quiescent interior of a fully convective star, assuming the toroidal flux tube is built in thermal equilibrium with its surroundings. However, when subjected to a time-varying convective velocity, the flux tube develops small undulations that grow into peaks and troughs, promoting rising loops (Sec. 3.1). To a zeroth order approximation, these flux tubes rise parallel to the rotation axis. Yet, flux tubes of a few times B_c (where $B_c \sim$20-30 kG), initiated at lower latitudes and in shallower depths of $\gtrsim$0.725R begin to exhibit near-equatorial emergence and rise times significantly shorter than expected (Sec. 3.2). Such behavior arises because the strong prograde differential rotation in this region supplies angular momentum to the legs of the rising loop. When initiated in the deep interior ($\sim$0.475-0.65R), these same flux tubes exhibit significantly longer rise times than their counterparts rising in a quiescent convection zone (Sec. 3.2). The apices of these tubes are continually pummeled by downflows, which also act to retard the mean motion of the flux tube. A broad trend emerges: the overall suppression of flux emergence by convective motions is more effective for weaker magnetic field strengths initiated at lower latitudes in deeper layers (Sec. 3.3).

These TFT simulations complement 3D dynamo simulations of fully convective stars. They inform us about the processes at work in stellar interiors that may lead to the observed pattern of magnetism on stellar surfaces. They also guide our understanding of the relative timescales over which flux emergence may occur and how the majority of this magnetic field may be effectively retained within a star. We plan to further explore these topics by performing similar TFT simulations in a variety of M dwarfs, some of which may have rapid rotation or small radiative cores.

This work was supported by the ERC under grant agreement No. 337705 (CHASM) and by a Consolidated Grant from the UK STFC (ST/J001627/1). Some calculations were performed on DiRAC Complexity - jointly funded by STFC and the Large Facilities Capital Fund of BIS, and the University of Exeter supercomputer - a DiRAC Facility jointly funded by STFC, the Large Facilities Capital Fund of BIS and the University of Exeter. We thank Isabelle Baraffe for providing the 1D stellar structure model used in this work. Figure 1b was generated by VAPOR (Clyne *et al.* 2007). We acknowledge PRACE for awarding us access to computational resources Mare Nostrum based in Spain at the Barcelona Supercomputing Center, and Fermi and Marconi based in Italy at Cineca. S.B, J.C., S.P, and E.T. are MPhys students in the Dept. of Physics and Astronomy at the University of Exeter.

References

Abbett, W. P., Fisher, G. H., Fan, Y., & Bercik, D. J. 2004, *ApJ*, 612, 557
Browning, M. K. 2008, *ApJ*, 676, 1262
Charbonneau, P. 2010, *Living Reviews in Solar Physics*, 7, 3
Clyne, J., Mininni, P., Norton, A., & Rast, M. 2007, *New Journal of Physics*, 9, 301
Dobler, W., Stix, M., & Brandenburg, A. 2006, *ApJ*, 638, 336
Fan, Y. 2009, Living Reviews in Solar Physics, 6, 4
Nelson, N. J., Brown, B. P., Brun, A. S., Miesch, M. S., & Toomre, J. 2011, *ApJL*, 739, L38
Roberts, B. & Webb, A. R. 1978, *SoPh*, 56, 5
Spruit, H. C. 1981, *A&A*, 98, 155
Tobias, S. M., Brummell, N. H., Clune, T. L., & Toomre, J. 2001, *ApJ*, 549, 1183
Weber, M. A. & Browning, M. K. 2016, *ApJ*, 827, 95
Weber, M. A., Fan, Y., & Miesch, M. S. 2013, *SoPh*, 287, 239
West, A. A., Weisenburger, K. L., Irwin, J., *et al.* 2015, *ApJ*, 812, 3
Wright, N. J. & Drake, J. J. 2016, *Nature*, 535, 526
Yadav, R. K., Christensen, U. R., Morin, J., *et al.* 2015, *ApJL*, 813, L31
Zwaan, C. 1987, *ARA&A*, 25, 83

Living Around Active Stars
Proceedings IAU Symposium No. 328, 2016
D. Nandy, A. Valio & P. Petit, eds.

© International Astronomical Union 2017
doi:10.1017/S1743921317003726

Beyond sunspots: Studies using the McIntosh Archive of global solar magnetic field patterns

Sarah E. Gibson[1], David Webb[2], Ian M. Hewins[2], Robert H. McFadden[2], Barbara A. Emery[2], William Denig[3] and Patrick S. McIntosh[4]

[1] High Altitude Observatory, National Center for Atmospheric Research,
Boulder, CO, 80301, USA
email: sgibson@ucar.edu
[2] Institute for Scientific Research, Boston College,
Chestnut Hill, MA, USA
email: david.webb@bc.edu
[3] National Centers for Environmental Information,
National Oceanic and Atmospheric Administration,
Boulder, CO, 80305, USA
email: William.Denig@noaa.gov
[4] Deceased

Abstract. In 1964 (Solar Cycle 20; SC 20), Patrick McIntosh began creating hand-drawn synoptic maps of solar magnetic features, based on Hα images. These synoptic maps were unique in that they traced magnetic polarity inversion lines, and connected widely separated filaments, fibril patterns, and plage corridors to reveal the large-scale organization of the solar magnetic field. Coronal hole boundaries were later added to the maps, which were produced, more or less continuously, into 2009 (i.e., the start of SC 24). The result was a record of ∼45 years (∼570 Carrington rotations), or nearly four complete solar cycles of synoptic maps. We are currently scanning, digitizing and archiving these maps, with the final, searchable versions publicly available at NOAA's National Centers for Environmental Information. In this paper we present preliminary scientific studies using the archived maps from SC 23. We show the global evolution of closed magnetic structures (e.g., sunspots, plage, and filaments) in relation to open magnetic structures (e.g., coronal holes), and examine how both relate to the shifting patterns of large-scale positive and negative polarity regions.

Keywords. Sun: evolution, Sun: sunspots, Sun: filaments, Sun: solar wind, Sun: magnetic fields

1. Introduction

The solar magnetic field is constantly changing, driven by the dynamo below and driving in turn a field that permeates the heliosphere. Concentrated magnetic flux is generated in the Sun's interior and emerges through its surface, e.g., as sunspots. Ongoing diffusion and transport by solar-surface flows results in a shifting pattern of positive and negative magnetic polarity that is an evolving boundary on the global magnetic field. Because the hot corona results in an expanding solar wind, both "closed" and "open" magnetic fields extend upwards from this boundary. Closed-field regions include sunspots, plage, and, if magnetic shear/twist is concentrated at the polarity inversion line (PIL), filaments. Of particular note on a global scale are polar crown filaments, which may extend nearly 360° around the sun at high latitudes. Open-field regions manifest as

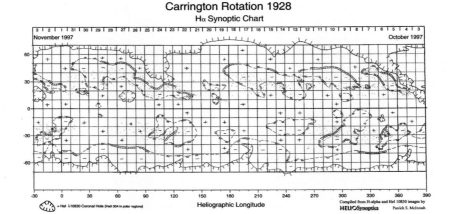

Figure 1. Example of original, hand-drawn McA synoptic solar map. Magnetic polarity is indicated by +/−; PILs are dashed lines with filaments indicated by extensions; coronal hole boundaries are indicated by hashed lines; plage by light dots, and sunspots by darker dots.

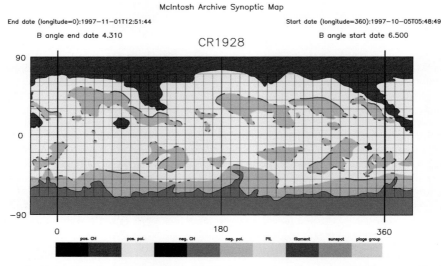

Figure 2. Example of colorized, digitized McA synoptic solar map (for the same CROT as Fig. 1). Magnetic features are identified with a distinct color, as described in the legend.

unipolar coronal holes, which, depending on solar-cycle phase, may appear predominantly at the poles or as isolated structures at lower latitudes.

2. The McIntosh Archive

The McIntosh Archive (McA) synoptic maps synthesize these disparate solar magnetic features, open and closed, into a global representation of the evolving solar magnetic field. Over the four decades of their creation, McIntosh consistently used Hα daily images to compile the maps. By carefully tracing Hα observations of filaments, fibril patterns and plage corridors, McIntosh determined the global structure of the PIL. This Hα representation of global magnetism has been shown to correlate well with large-scale

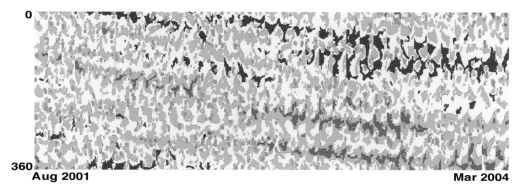

Figure 3. Equatorial-zone stack plot for CROT 1980 - 2014. Vertical axis is longitude; a sequence of equatorial slices ($-20°$ to $+20°$ latitude) is stacked along the horizontal axis with time increasing to the right. Colors are as in Fig. 2, with sunspots (orange), coronal holes (blue=positive, red=negative) and quiet sun (light-blue=positive, grey=negative) indicated.

magnetic field measured with photospheric magnetograms (McIntosh, 1979), and to be particularly useful for tracing the PILs in weak field regions and near the poles of the Sun (Fox *et al.*, 1998, McIntosh, 2003). Starting in 1978, McIntosh added coronal holes to the maps, primarily using ground-based He-I 10830 Å images from NSO-Kitt Peak. Magnetograms were used, when available, to determine the overall dominant polarity of each region. Fig. 1 shows an example of an original McA map.

Some of the original hand-drawn McIntosh maps were published as Upper Atmosphere Geophysics (UAG) reports in McIntosh (1975), McIntosh and Nolte (1975), and McIntosh (1979). Versions of the maps were also routinely published in the Solar-Geophysical Data (SGD) Bulletins. The UAG and SGD reports are all archived in scanned format at the NOAA National Center for Environmental Information (NCEI). However, many maps only existed in hard-copy format in boxes, and none of the scanned maps possessed metadata allowing digital search and analysis.

The intent of the McA project has been first and foremost to preserve the archive in its entirety, by completing the scanning of all of the maps; this has been achieved. Second, we have designed and implemented a procedure to standardize the size and orientation of the digital maps, to remove any unnecessary notes, marks or symbols, and to colorize the maps so that each magnetic feature is uniquely searchable. Fig. 2 shows an example of a McA map processed in this manner. To date, we have completed the full processing of solar cycle 23 (SC 23), and are working our way backwards through the earlier cycles.

3. SC 23 Analysis

We now present preliminary results for SC 23, and describe directions for future study.
Active Longitudes and Periodic Solar-Wind Forcing. Figs. 3 – 5 show "stack plots" in which equatorial slices of maps such as Fig. 2 are placed side by side so that trends over multiple Carrington rotations (CROTs) can be easily examined (see, e.g., McIntosh & Wilson (1985)). Stack plots are useful for illustrating active longitudes, where sunspots and magnetic flux appear to emerge preferentially (e.g., de Toma *et al.* (2000)). In particular, evidence for the persistence of two active longitudes separated by 180° for 100 years has been found in historical records of sunspot locations (e.g., Berdyugina & Usoskin (2003)) and in recently-digitized white-light images from the Kodaikanal Observatory (Mandal *et al.*, 2016).

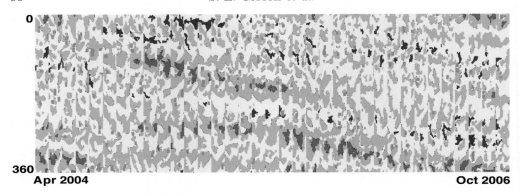

0

360
Apr 2004   **Oct 2006**

Figure 4. Equatorial-zone stack plot as in Fig. 3, CROT 2015 – 2049.

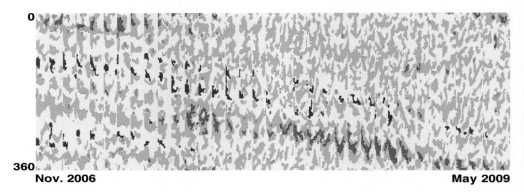

0

360
Nov. 2006   **May 2009**

Figure 5. Equatorial-zone stack plot as in Fig. 3, CROT 2050 – 2084.

Figs. 3 – 5 show not only sunspots (orange) but also near-equatorial coronal holes (red/blue). A 180° longitudinal asymmetry is particularly evident during solar maximum years (Fig. 3), and in general there is a long-lived pattern where coronal holes appear, disappear, and reappear in a preferred (differentially-rotating) set of longitudes. The pattern is also apparent in the quiet sun polarity (light-blue/grey) as noted previously in, e.g., McIntosh & Wilson (1985). (See also Bilenko & Tavastsherna (2016) for a study of coronal hole latitudinal and longitudinal patterns spanning three solar cycles.)

An association of sunspot active longitudes with high-speed solar wind periodicities has been noted for some time (e.g., Balthasar & Schuessler (1983) and references therein). The role of long-lived, low-latitude coronal holes in driving periodic behavior, both in the solar wind and in the Earth's space environment and upper atmosphere, was studied extensively for the SC 23 declining period and the extended solar minimum that followed (see, e.g., Temmer *et al.* (2007), Gibson *et al.* (2009), and Luhmann *et al.* (2009)). As Fig. 5 demonstrates, this otherwise quiet (low sunspot activity) time period was characterized by remarkably sustained longitudinal structure, first with three and then two near-equatorial open-field regions.

Differential Rotation. The downward shift of the active longitude locations with time seen in Figs. 3 - 5 corresponds to an increase in Carrington longitude and so indicates a rotation that is faster than the Carrington rate. This is expected based on solar surface differential rotation, since the Carrington rate (27.2753 days as viewed from the Earth) corresponds to a mid-latitude surface rotation rate. Fig. 6 shows an extended stack plot for all of SC 23, both for the equatorial zone shown in Figs. 3 – 5 and for the northern

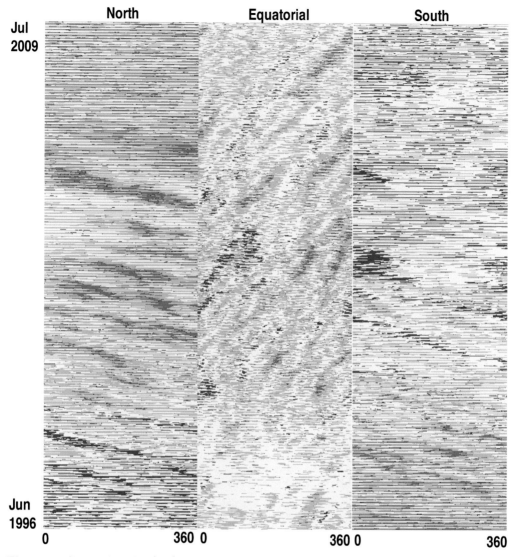

Figure 6. Stack plots for (left) North polar zone (30° to 70°) latitude), (middle) Equatorial zone (−20° to +20° latitude), and (right) South polar zone (−70° to −30° latitude), for all of SC 23. These are rotated 90° in comparison to Figs. 3 – 5, so the horizontal axes = longitude.

and southern polar zones. Near solar minimum (the top and bottom of the plots) the polar zones show unipolar coronal holes at all latitudes, but at solar maximum and in the declining phase the slower polar rotation rate is evident. Note that Fig. 6 is rotated 90° relative to Figs. 3 – 5, so that a feature moving at the Carrington rate would appear vertical (constant Carrington longitude).

Studies of sunspot active-longitude differential rotation (e.g., Usoskin *et al.* (2005), Mandal *et al.* (2016)) find a rotation rate of the active longitude location matching the surface flow at that latitude. This is in contrast to individual sunspots which generally rotate faster than the surface does, indicating they are rooted below the near-surface shear layer (Thompson *et al.* (2003)). Studies of coronal hole rotation indicate that polar

S. E. Gibson *et al.*

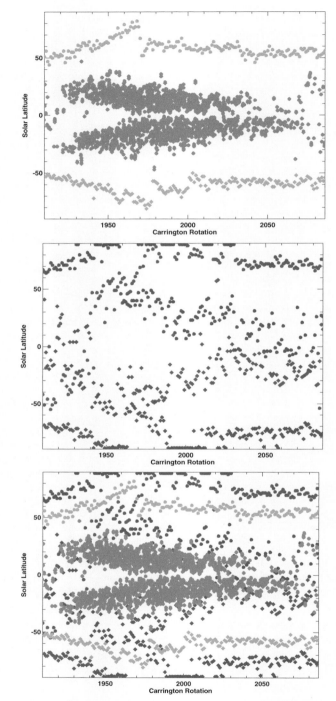

Figure 7. SC 23 patterns of open vs. closed magnetic features. (Top) sunspots (orange) and most poleward filament (green). (Middle) coronal hole boundaries (red=negative, blue=positive; furthest north per CROT=circles, furthest south=diamonds). (Bottom) combined.

coronal holes may rotate rigidly, while low-latitude coronal holes rotate more differentially but with a variability possibly associated with coronal-hole lifetime and solar-cycle phase (see, e.g., Ikhsanov & Ivanov (1999)). Such variability may contribute to the range in slopes of coronal-hole patterns seen in Figs. 3 – 6.

Evolution of Open vs. Closed Fields. Fig. 7 shows the global evolution of coronal holes, sunspots, and filaments over SC 23. The sunspots are plotted as a classic butterfly diagram, appearing after solar minimum first at high latitudes, and emerging progressively closer to the equator as the cycle continues. The green dots show the location of the most poleward filament for each CROT, and show the "rush to the poles" which is a tracer of the solar magnetic polarity reversal process. After this reversal, the old polar crown filament is replaced by a secondary crown filament in both hemispheres (McIntosh (1992)).

The middle panel in Fig. 7 plots the locations of coronal hole boundaries as a function of CROT. In particular, the northmost and southmost boundary position is plotted for both magnetic polarities, capturing the behavior of both polar and low-latitude coronal holes. The result is the "double helix" pattern referred to in S. McIntosh *et al.* (2014) (see also Bilenko & Tavastsherna (2016), Fujiki *et al.* (2016)). This process of polar coronal holes reforming as a consequence of the migration of lower-latitude coronal holes poleward is described in Webb *et al.* (1984) and Harvey & Recely (2002). Finally, the bottom panel in Fig. 7 shows how the positive and negative polarity open fields (blue and red) surround the closed-field filaments (green) and sunspots (orange), on a global scale.

4. Conclusions

The unique power of the McIntosh archive is its capability for simultaneously representing closed and open magnetic structures over a range of time scales. The completion of the full McA digitization will provide the community with a comprehensive resource for addressing key questions including: How do active longitudes vary within and between solar cycles, for both closed and open magnetic features? Where are closed and open magnetic features rooted (as evidenced by rotation rate), and how does this depend on solar cycle phase, feature lifetime, and latitude? How does the evolution of open and closed magnetic features relate to surface flows on solar-cycle time scales (e.g., torsional oscillations, Howe *et al.*, 2013)? Answering any or all of these questions has important implications for our understanding of the solar dynamo, and for our interpretation of periodic variations of Earth's space environment and upper atmosphere.

Acknowledgements

We dedicate this project to Pat McIntosh, who died this year. We are grateful to his daughter Beth Schmidt for granting us permission to use his original data. The work of the authors is supported by NSF RAPID grant number 1540544. NCAR is funded by the NSF. We thank Giuliana de Toma, Larisza Krista, and Scott McIntosh for helpful discussions.

References

Balthasar, H. & Schuessler, M. 1983, *Sol. Phys.*, 87, 23
Berdyugina, S. V. & Usoskin, I. G. 2003, *Astron. Astrophys.*, 405, 1121
Bilenko, I. A., Tavastsherna, K. S. 2016 *Sol. Phys.*, 291, 2329
de Toma, G., White, O. R., & Harvey, K. L. 2000, *Astrophys. J.* 529, 1101

Fox, P., McIntosh, P. S., & Wilson, P. R. 1998, *Sol. Phys.*, 177, 375

Fujiki, K., Tokumaru, M., Hayashi, K., Satonaka, D., & Hakamada, K. 2016, *Astrophys. J. Lett.*, 827, L41

Gibson, S. E., Kozyra, J. U., de Toma, G., Emery, B. A., Onsager, T., & Thompson, B. J. 2009, *J. Geophys. Res.*, 114, A09105

Harvey, K. L. & Recely, F. 2002, *Sol. Phys.*, 211, 31

Howe, R., Christensen-Dalsgaard, J., Hill, F., Komm, R. Larson, T. P., Rempel, M., Schou, J., & Thompson, M. J. 2013, *Astrophys. Journ.*, 767, L20

Ikhsanov, R. N. & Ivanov, V. G. 1999, *Sol. Phys.*, 188, 245

Luhmann, J. G., Lee, C. O., Li, Y., Arge, C. N., Galvin, A. B., Simunac, K., Russell, C. T., Howard, R. A., & Petrie, G. 2009, *Sol. Phys.*, 256, 285

Mandal, S., Chatterjee, S., & Banerjee, D. 2016, *Astron. Astrophys.*, in press

McIntosh, P. S. 1975, *UAG-40* (NOAA Space Environment Laboratory, Boulder, CO)

McIntosh, P. S. 1979, *UAG 70, World Data Center A for Solar-Terrestrial Physics* (NOAA/NGSDC, Boulder, CO)

McIntosh, P. S. 1992, *in The Solar Cycle; Proceedings of the National Solar Observatory Sacramento Peak 12th Summer Workshop, ASP Conference Series* (ASP: San Francisco), 27, 14

McIntosh, P. S. 2003, *in Solar Variability as an Input to the Earth's Environment, A. Wilson (ed.)* (ESA SP-535, ESTEC, Noordwijk, Netherlands), 807

McIntosh, P. S. & Nolte, J. T. 1975, *UAG-41* (NOAA Space Environment Laboratory, Boulder, CO)

McIntosh, P. S. & Wilson, P. R. 1985, *Sol. Phys.*, 97, 59

McIntosh, S. W., Wang, X., Leamon, R. J., Davey, A. R., Howe, R., Krista, L. D., Malanushenko, A. V., Merkel, R. S., Cirtain, J. W., Gurman, J. B., Pesnell, W. D., & Thompson, M. J. 2014, *Astrophys. J.*, 792, 12

Temmer, M., Vrsnak, B., & Veronig, A. M. 2007, *Sol. Phys.*, 241, 371

Thompson, M. J., Christensen-Dalsgaard, J., Miesch, M. S., & Toomre, J. 2003, *Ann. Rev. Astron. Astrophys.*, 41, 599

Usoskin, I. G., Berdyugina, S. V., & Poutanen, J. 2005, *Astron. Astrophys.*, 441, 347

Webb, D. F., Davis, J. M., & McIntosh, P. S. 1984 *Sol. Phys.*, 92, 109

Living Around Active Stars
Proceedings IAU Symposium No. 328, 2016
D. Nandy, A. Valio & P. Petit, eds.

© International Astronomical Union 2017
doi:10.1017/S1743921317004100

Magnetic fields of weak line T-Tauri stars

Colin A. Hill and the MaTYSSE Collaboration

Université de Toulouse / CNRS-INSU, IRAP / UMR 5277, F-31400, Toulouse, France
email: chill@irap.omp.eu

Abstract. T-Tauri stars (TTS) are late-type pre-main-sequence (PMS) stars that are gravitationally contracting towards the MS. Those that possess a massive accretion disc are known as classical T-Tauri stars (cTTSs), and those that have exhausted the gas in their inner discs are known as weak-line T-Tauri stars (wTTSs). Magnetic fields largely dictate the angular momentum evolution of TTS and can affect the formation and migration of planets. Thus, characterizing their magnetic fields is critical for testing and developing stellar dynamo models, and trialling scenarios currently invoked to explain low-mass star and planet formation. The MaTYSSE programme (Magnetic Topologies of Young Stars and the Survival of close-in Exoplanets) aims to determine the magnetic topologies of ~ 30 wTTSs and monitor the long-term topology variability of ~ 5 cTTSs. We present several wTTSs that have been magnetically mapped thus far (using Zeeman Doppler Imaging), where we find a much wider range of field topologies compared to cTTSs and MS dwarfs with similar internal structures.

Keywords. stars: magnetic fields, stars: pre–main-sequence, stars: spots, stars: late-type, stars: activity, stars: imaging, techniques: polarimetric, techniques: radial velocities

1. Introduction

T-Tauri stars (TTS) are young, low mass pre-main-sequence (PMS) stars that are equivalent to young Suns. After emerging from their dust cocoons, newly formed protostars are initially surrounded by a large-scale magnetized accretion disc. These protostars eventually settle as gravitationally contracting PMS stars surrounded by a protoplanetary disc (André *et al.* 2009), and at an age of 0.5–10 Myr, they are known as classical T-Tauri stars (cTTSs) if they are still possess a massive (presumably planet forming) accretion disc, or weak-line T-Tauri stars (wTTSs) if they have exhausted the gas in their inner discs. Both cTTSs and wTTSs are excellent test beds for theories of low-mass star and planet formation, and a such, have been studied intensely over the last few decades.

Magnetic fields in such objects have their largest impact during the star's early evolution, with fields controlling accretion processes and triggering outflows, as well as largely dictating the angular momentum evolution of low-mass PMS stars (e.g. Bouvier *et al.* 2007, Frank 2014). In particular, large-scale fields of cTTSs can evacuate the central regions of accretion discs, funnel the material onto the star, and enforce co-rotation between the star and the inner disc Keplerian flows, causing cTTSs to rotate more slowly than expected from the contraction and accretion of the disc material (e.g. Davies *et al.* 2014). Furthermore, magnetic fields of cTTS and their discs can affect the formation and migration of planets (e.g. Baruteau *et al.* 2014). Moreover, fields of both cTTSs and wTTSs are known to trigger thermally driven winds through heating by accretion shocks and/or Alfvén waves (e.g. Cranmer 2009, Cranmer & Saar 2011), resulting in flares, coronal-mass ejections, and angular momentum loss (e.g. Aarnio *et al.* 2012, Matt *et al.* 2012).

The importance of magnetic fields in both the long and short-term evolution of TTS is clear, and in order to develop and test theoretical models to provide more physical realism and reliable predictions, it is crucial that we characterize the magnetic fields in cTTSs and wTTSs through observations.

2. Observing magnetic fields of T-Tauri stars

Despite first detecting magnetic fields on cTTSs nearly 20 years ago (e.g. John-Skrull *et al.* 1999), their large-scale topologies have only recently been revealed for a dozen stars through the MaPP (Magnetic Protostars and Planets) programme (e.g. Donati *et al.* 2007, 2010, 2013, Hussain *et al.* 2009). The MaPP survey revealed that the large-scale fields of cTTSs remain relatively simple and mainly poloidal when the host star is still fully or largely convective, but become much more complex when the host star turns mostly radiative (Gregory *et al.* 2012, Donati *et al.* 2013). This survey also showed that these fields are likely of dynamo origin, varying over timescales of a few years (Donati *et al.* 2011, 2012, 2013), and resembling those of mature stars with comparable internal structure (Morin *et al.* 2008).

By contrast, only a few wTTS have been magnetically imaged (and published) to date, namely V410 Tau, LkCa 4, V819 Tau, V830 Tau and TAP 26 (Skelly *et al.* 2010, Donati *et al.* 2014, 2015, Yu *et al.* 2017). These few stars show a much wider range of field topologies compared to cTTSs and MS dwarfs with similar internal structures; while V819 Tau and V830 Tau display a mostly poloidal field topology, V410 Tau and LkCa 4 show significant toroidal components despite being fully convective, in surprising contrast to fully convective cTTSs and mature M dwarfs that harbour relatively simple poloidal fields (Morin *et al.* 2008, Donati *et al.* 2013).

Studies of magnetic fields (and associated winds) on wTTSs are of great interest, as these are the initial conditions in which disc-less PMS stars initiate their unleashed spin-up as they contract towards the MS. This is one of the goals of the MaTYSSE (Magnetic Topologies of Young Stars and the Survival of close-in Exoplanets) Large Programme, allocated at the 3.6-m CFHT at Mauna Kea on Hawaii, using the ESPaDOnS spectropolarimeter over semesters 2013a–2016b (510 h), with complementary observations using NARVAL on the 2-m TBL at Pic du Midi in France (420 h), and using the HARPS spectropolarimeter at the 3.6-m ESO Telescope at La Silla in Chile (70 h).

MaTYSSE has several aims; Firstly, to determine the magnetic topologies of ~ 30 wTTSs and compare them to those of cTTSs to discover if their large-scale fields are similar, or significantly different. In particular, as wTTSs are no longer accreting, we may assess the role of accretion and the impact of its interruption, given that accretion or star/disc coupling torques may modify dynamo processes and the corresponding large-scale field topology. Furthermore, we can study magnetic winds of wTTSs and corresponding spin-down rates (e.g. Vidotto *et al.* 2014), painting a picture of the kind of magnetospheres that young Sun-like stars have when they contract and spin-up towards the MS, thus consistently explaining the rotational history of low-mass stars once on the MS (given that magnetic braking is the main cause of their spin down). Moreover, we will monitor the long-term magnetic topology variability of ~ 5 cTTSs, and study the variability of magnetospheric gaps and winds due to non-stationary dynamos, aiming to determine if disc migration is the main process for producing hot Jupiters (hJs), and if magnetospheric gaps and winds key are factors for their survival. If this is the case, we can expect to find at least as many hJs in TTSs as are in mature stars, and significantly more if we account for those absorbed by protostar over the contraction phase.

To map surface features (such as spots and plages) and magnetic fields on our target stars, we use our Zeeman Doppler Imaging (ZDI) code that has been specifically adapted for use with wTTSs. First, we take our Stokes I and V spectra and use Least-Squares

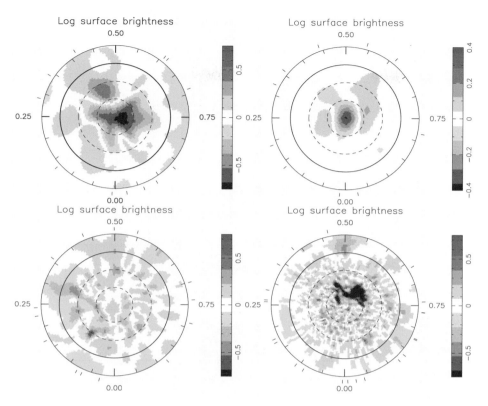

Figure 1. Maps of the logarithmic brightness (relative to the quiet photosphere), at the surfaces of LkCa 4 (top left), Par 1379 (top right), Par 2244 (bottom left) and TWA 6 (bottom left). The stars are shown in flattened polar projection down to latitudes of $-30°$, with the equator depicted as a bold circle, and $30°$ and $60°$ parallels as dashed circles. Radial ticks around each plot indicate phases of observations. This figure is best viewed in colour.

Deconvolution (LSD) to create a 'mean' line profile (with a greatly increased S/N, see Donati *et al.* 1997). Then, we simultaneously invert a time series of these Stokes I and V LSD profiles (using a spherical harmonic decomposition for the magnetic field, and using maximum-entropy regularization) to create our reconstructed maps (see Figures 1 and 2).

As well as determining the magnetic field topology, we can use the brightness maps from ZDI to predict the activity-related radial velocity (RV) jitter due to spots and/or plages. We can then filter the RV measurements to detect potential hot Jupiters (hJs) around wTTSs, and thus verify whether core accretion and migration is the most likely mechanism for forming close-in giant planets (see Chapter ??? by Yu in this volume; Donati *et al.* 2016).

3. Results

Figures 1 and 2 show the brightness and magnetic maps of several wTTSs from the MaTYSSE sample. These few stars (namely, LkCa 4, Par 1379, Par 2244 and Twa 6) demonstrate the wide range of magnetic topologies that can exist in this class of T-Tauri star. In particular, LkCa 4 (0.9 $M_\odot$, 2 Myr, $v \sin i = 28$ kms^{-1}, $P_{\rm rot} = 3.37$ d)

C. A. Hill

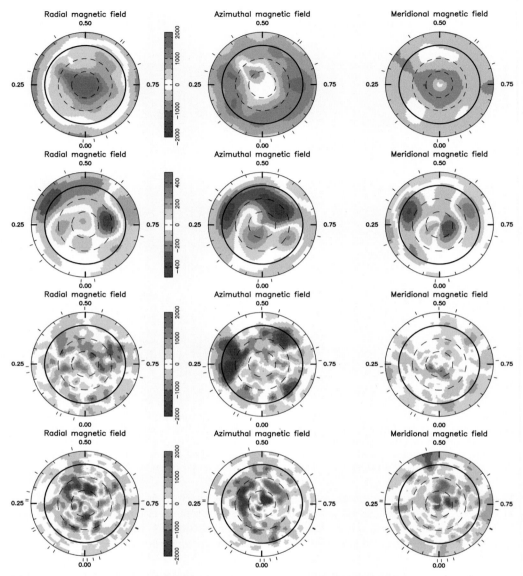

Figure 2. Maps of the radial (left column), azimuthal (middle column) and meridional (right column) components of the magnetic field **B** at the surfaces of (top to bottom) LkCa 4, Par 1379, Par 2244 and TWA 6. Magnetic fluxes in the colourbar are expressed in G. The stars are shown in flattened polar projection as in Figure **??**. This figure is best viewed in colour.

shows a simple, strong and mainly axisymmetric field featuring a 2 kG aligned-poloidal component and a 1 kG toroidal component encircling the star at equatorial latitudes (with the latter feature markedly different from cTTSs of similar mass and age). In contrast, Par 1379 (1.6 $M_\odot$, 1.6 Myr, $v \sin i = 13.7$ kms^{-1}, $P_{\rm rot} = 5.585$ d) shows a much weaker, mostly non-axisymmetric field, with a 75 G aligned-poloidal component and a 400 G toroidal field. The similarly massive wTTS, Par 2244 (1.8 $M_\odot$, 1 Myr, $v \sin i = 57.2$ kms^{-1}, $P_{\rm rot} = 2.8153$ d), features a much more complex field topology, with a mostly non-axisymmetric aligned-poloidal component of 800 G, and a mostly

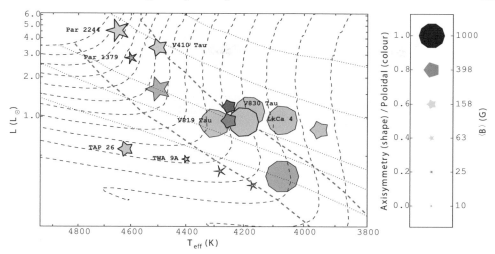

Figure 3. H-R diagram showing the MaTYSSE wTTSs (solid colours) and the MaPP cTTs (translucent/faded colours). The size of the symbols represents the surface-averaged magnetic field strength (with a larger symbol meaning a stronger field), the colour of the symbol represents the fraction of the field that is poloidal (with red being completely poloidal), and the shape of the symbols represents the axisymetry of the poloidal field component (with higher axisymmetry shown as a more circular symbol). Also shown are evolutionary tracks from Siess 2000 (black dashed lines, ranging from 0.5–1.9M$_\odot$), with corresponding isochrones (black dotted lines, for ages of 0.5, 1, 3, 5 & 10 Myr), and lines showing 0% and 50% convective by radius (blue dashed).

axisymmetric toroidal component, with fields in excess of 2 kG. Finally, the much older wTTS, Twa 6 (0.9 M$_\odot$, 20 Myr, $v \sin i$ = 72.6 kms^{-1}, $P_{\rm rot}$ = 0.5409 d), harbours a mostly axisymmetric field with a 900 G aligned-poloidal field and a toroidal component with surface fields in excess of 2 kG.

To put these stars in context, we have plotted in Figure 3 an H-R diagram of the cTTSs from the MaPP programme, as well as the (analysed) MaTYSSE stars. Figure 3 also indicates the fraction of the field that is poloidal, the axisymmetry of the poloidal component, and shows PMS evolutionary tracks from Siess *et al.* 2000. So far, our sample of wTTSs show a much wider range of field topologies compared to cTTSs. We find that large scale fields can be more toroidal and non-axisymmetric than cTTSs, with a significant toroidal component emerging after disc dissipation.

This preliminary result demonstrates the eclectic mix of magnetic field topologies in wTTSs, and once a complete analysis is made of the MaTYSSE sample, we will be able to robustly compare the magnetic topologies of cTTSs and wTTSs, assessing the role of accretion and the presence of a disc on dynamo processes. Furthermore, we will determine the nature of the magnetic winds on wTTSs, and the frequency of hJs, leading to a better understand their formation and migration.

References

Aarnio, A. N., Matt, S. P., & Stassun, K. G., 2012, *ApJ*, 760, 9

André, P., Basu, S., & Inutsuka, S., 2009, *The formation and evolution of pre-stellar cores.* Cambridge University Press, p. 254

Baruteau, C., *et al.*, 2014, *Protostars and Planets VI*, pp 667–689

Bouvier, J., Alencar, S. H. P., Harries, T. J., Johns-Krull, C. M., & Romanova, M. M., 2007, *Protostars and Planets V*, pp 479–494

Cranmer, S. R., 2009, *ApJ*, 706, 824

Cranmer, S. R. & Saar, S. H., 2011, *ApJ*, 741, 54

Davies, C. L., Gregory, S. G., & Greaves, J. S., 2014, *MNRAS*, 444, 1157

Donati, J.-F. *et al.*, 1997, *MNRAS*, 291, 658

Donati, J.-F. *et al.*, 2007, *MNRAS*, 380, 1297

Donati, J.-F. *et al.*, 2010, *MNRAS*, 402, 1426

Donati, J.-F. *et al.*, 2011, *MNRAS*, 412, 2454

Donati, J.-F. *et al.*, 2012, *MNRAS*, 425, 2948

Donati, J.-F. *et al.*, 2013, *MNRAS*, 436, 881

Donati, J.-F. *et al.*, 2014, *MNRAS*, 444, 3220

Donati, J.-F. *et al.*, 2015, *MNRAS*, 453, 3706

Donati, J.-F. *et al.*, 2016, *Nature*, 534, 662

Gregory, S. G., Donati, J.-F., Morin, J., Hussain, G. A. J., Mayne, N. J., Hillenbrand, L. A., & Jardine, M., 2012, *ApJ*, 755, 97

Hussain, G. A. J., *et al.*, 2009, *MNRAS*, 398, 189

Johns-Krull, C. M., Valenti, J. A., & Koresko, C., 1999, *ApJ*, 516, 900

Matt, S P., 2012, *ApJ*, 745, 101

Morin, J., *et al.*, 2008, *MNRAS*, 390, 567

Siess, L., *et al.*, 2000, *A&A*, 358, 593

Skelly, M. B., Unruh, Y. C., Collier Cameron, A., Barnes, J. R., Donati, J.-F., Lawson, W. A., & Carter, B. D., 2008, *MNRAS*, 385, 708

Living Around Active Stars
Proceedings IAU Symposium No. 328, 2016
D. Nandy, A. Valio & P. Petit, eds.

© International Astronomical Union 2017
doi:10.1017/S1743921317004136

Differential rotation of stars with multiple transiting planets

Yuri Netto and Adriana Valio

Center for Radio Astronomy and Astrophysics (CRAAM) Mackenzie Presbyterian University,
Sao Paulo, Brazil
email: dirceuyuri@hotmail.com

Abstract. If a star hosts a planet in an orbit such that it eclipses the star periodically, can be estimated the rotation profile of this star. If planets in multiplanetary system occult different stellar areas, spots in more than one latitude of the stellar disc can be detected. The monitored study of theses starspots in different latitudes allow us to infer the rotation profile of the star. We use the model described in Silva (2003) to characterize the starspots of Kepler-210, an active star with two planets. Kepler-210 is a late K star with an estimated age of 350 ± 50 Myrs, average rotation period of 12.33 days, mass of 0.63 $M_\odot$ and radius of 0.69 $R_\odot$. The planets that eclipses this star have radii of 0.0498 R_s and 0.0635 R_s with orbital periods of 2.4532 ± 0.0007 days and 7.9725 ± 0.0014 days, respectively, where R_s is the star radius.

Keywords. stars: activity, planetary systems, spots

1. Introduction

If a star is eclipsed by an orbiting planet periodically, it is possible to estimate its stellar rotation profile. Small variations in the bottom of the light curve transits may be seen during the transit of the planet. These variations can be interpreted as the planet occulting a spot on the photosphere of the star. Based on these signatures, Silva (2003) developed a method that allows the detection of spots as small as 0.2 planetary radii. Through this method it is also possible to infer properties of individual starspots occulted by the planet, such as size, intensity, and position. This technique was already applied to HD 209458 (Silva 2003), to CoRoT-2 (Silva-Valio & Lanza 2011), and to Kepler-17 (Valio *et al.* 2016). By monitoring spots on later transits or considering the fortuitous geometry of Kepler-63, which presents a planet orbiting in an almost polar orbit, the stellar differential rotation was estimated, since it eclipses different stellar latitudes (Netto & Valio 2016).

In the case of a multiplanetary system, it is possible to infer the stellar rotation profile if the planets occult starspots in different regions of the star, transiting in more than one latitude of the stellar disc. The monitored study of theses starspots at different latitudes allow us to infer the rotation profile of the star.

2. Kepler-210

Kepler-210 is an active star, late K star with an estimated age of 350 ± 50 Myrs. With a radius of 0.69 $R_\odot$, mass of 0.63 $M_\odot$ and an average rotation period of 12.33 days, it hosts two transiting Neptune-like planets (Ioannidis *et al.* 2014).

We have applied the model presented by Silva (2003) to simulate the planetary transits in front of the star Kepler-210. This model assumes a synthesized 2D image of the star with limb-darkening, whereas the planet has a circular orbit.

Star	
M (M$_\odot$)	0.63
R (R$_\odot$)	0.69
P$_{rot}$ (day)	12.33 ± 0.15
Age (Myr)	350 ± 50
T$_{eff}$ (K)	4300
Linear Limb-Darkening coeff., u$_1$	0.8536*
Quadratic Limb-Darkening coeff. u$_2$	0.1990*

Planets		
	Kepler-210b	Kepler-210c
Rp (R$_s$)	0.0490*	0.0590*
a (R$_s$)	7.8706*	22.449*
i (deg)	83.61*	88.87*

Table 1. Parameters of Kepler-210 system. Those marked by an asterisk are the fitted values found in this work.

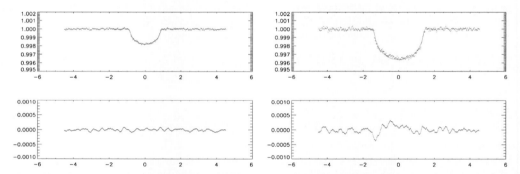

Figure 1. Top: Binned light curve with the fit (red curve) for the transits of planet Kepler-210b (left) and Kepler-210c (right). Bottom: The residuals of the subtraction of the light curve from the modelled spotless transit light curve.

To obtain the spotless light curve for the planet Kepler-210b, we used 302 transits. The data from these transits were binned into a single transit and fitted by a least χ-square minimisation routine (AMOEBA) using the model detailed in Silva (2003) but without spots. The five parameter that were fitted are planet radius, semi-major axis, inclination angle for the planets and limb darkening coefficients for the star. Different values resulted from this fit with respect to the values reported by Ioannidis *et al.* (2014). The same procedure was applied to the 86 transits of the larger planet, Kepler-210c. The values we obtained are listed in Table 1, and the resulting fits for planets Kepler-210b and Kepler-210c are shown as the red line in the top panels of the Figure 1. The residuals of the subtraction of this model from the binned light curve are plotted on the bottom panels of Figure 1.

A simulation of the star Kepler-210 and its planets are presented in the Figure 2. To the values fitted in this work, the planets should eclipse the star at the stellar latitudes of −61° and −26° for planets Kepler-210b and Kepler-210c, respectively.

Unfortunately, no spots signatures in the residuals were found above the detection threshold of 10 times the CDPP (Combined Differential Photometric Precision, Christiansen *et al.* (2012)).

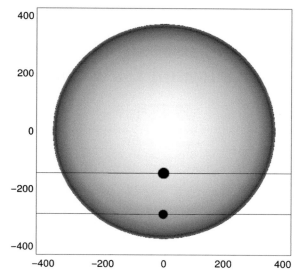

Figure 2. Planetary transit simulation for the two planets orbiting around Kepler-210. Kepler-210c transits closer to the stellar pole, while Kepler-210b is transiting at mid-latitudes.

3. Summary and Conclusions

Using the method described in Silva (2003), we modelled the light curve of the star Kepler-210. A total of 302 transits were used for the first planet, and 86 for the second one. The fitted parameters resulted in an increase in the semi-major axis and in the orbital inclination angle with respect to those reported by Ioannidis *et al.* (2014). On the other hand, the fits resulted in a decrease in the planet radius of 1.6% and 7.0% for the Kepler-210b and Kepler-210c, respectively. Although, Kepler-210 is an active young star, its noise is such that it drowns the signature of any spot in the transit light curve of its planets.

References

Christiansen, J. L., Jenkins, J. M., Caldwell, D. A., *et al.* 2012, *PASP*, 124, 1279

Ioannidis, P., Schmitt, J. H. M. M., Avdellidou, C., von Essen, C., & Agol, E. 2014, *A&A*, 564, A33

Netto, Y.,& Valio, A. 2016, *ApJLett* (Submitted)

Silva, A. V. R. 2003, *ApJLett*, 585, 147

Silva-Valio, A. & Lanza, A. F. 2011, *A&A*, 529, A36

Valio, A., Estrela, R., Netto, Y., Bravo, J. P., & Medeiros, J. R. 2016, *ApJ* (Accepted)

Living Around Active Stars
Proceedings IAU Symposium No. 328, 2016
D. Nandy, A. Valio & P. Petit, eds.

© International Astronomical Union 2017
doi:10.1017/S1743921317003751

Studies of synoptic solar activity using Kodaikanal Ca K data

K. P. Raju

Indian Institute of Astrophysics,
Bangalore 560034, India
email: kpr@iiap.res.in

Abstract. The chromospheric network, the bright emission network seen in the chromospheric lines such as Ca II K and H_α, outline the supergranulation cells. The Ca images are dominated by the chromospheric network and plages which are good indicators of solar activity. Further, the Ca line is a good proxy to the UV irradiance which is particularly useful in the pre-satellite era where UV measurements are not available. The Ca spectroheliograms of the Sun from Kodaikanal have a data span of about 100 years and covers over 9 solar cycles. The archival data is now available in the digitized form. Programs have been developed to obtain the activity indices and the length scales of the chromospheric network from the data. The preliminary results from the analysis are reported here. It is shown that the Ca II K intensity and the network boundary width are dependent on the solar cycle.

Keywords. Sun: activity, Sun: chromosphere, Sun: magnetic fields

1. Introduction

The observations at Kodaikanal Solar Observatory date back to the year 1904 (Bappu 1967). The archival data have a span of about 100 years which covers over 9 solar cycles. The database consists of

- Ca II K spectroheliograms at 3934 Å
- H_α spectroheliograms at 6563 Å
- White light images

The data is now available in the digitized format and can be used for studying the synoptic variations in the Sun (Priyal *et al.* 2014).

The Ca II K line at 3934 Å is a strong chromospheric line. The Ca images are dominated by chromospheric network and plages (Raju & Singh 2014). It is well known that the chromospheric network outlines supergranular cells which play an important role in maintaining solar cycle through magnetic flux dispersal along its boundaries. Therefore the Ca line is a good tracer of solar activity.

Solar irradiance variation is about 0.1 % over solar cycle. However, the variation of UV irradiance is much higher, about 10 % and it plays an important role in Earth's climate. The Ca line is a proxy to UV irradiance which is particularly useful in the pre-satellite era.

The length scales of supergranulation network can be obtained through the autocorrelation method of Ca II K images. The half-width of the autocorrelation function gives the width of the network boundary (Patsourakos *et al.* 1999). The solar cycle variation of the transition region EUV network boundaries has been reported recently by Raju (2016). In the present work, we examine the synoptic variation of chromospheric network in the Ca II K line from the Kodaikanal archival data. The recent IRIS observations have reported

Figure 1. Ca II K Spectroheliogram from Kodaikanal.

the presence of jets and loops in the network boundaries (Tian *et al.* 2014). Hence the variations of network length scales are important in the mass and energy budget of the solar atmosphere.

2. Data and Analysis

The Kodiakanal Ca II K spectroheliograms have a spatial resolution of about 2 arc sec. The data was recently digitized using a 4K x 4K CCD with a pixel resolution of 0.86 arc sec and then calibrated to intensity units (Priyal *et al.* 2013). The spectroheliograms cover a period from 1907 to 2008. An example of the calibrated image is shown in Figure 1.

During the analysis, a strip of width 480 arc sec x 45 arc sec was selected from the solar centre. The strip contains about 12 supergranules which will give the statistical behaviour of the cells. First the mean intensity of the strip was obtained. Any possible large-scale trend was then removed from the strip by fitting a polynomial fit and subtracting it out. The two-dimensional autocorrelation function as a function of lag was obtained through an IDL routine. The width of the network boundary as a function of time was then obtained from the spectroheliograms.

3. Results

The results are plotted in Figure 2. The panels give the sunspot number, Ca II K intensity and the network boundary width against time. It can easily be seen that both the intensity and the boundary width have a rough positive correlation with the sunspot number. Since the network is formed due to the concentration of magnetic flux by solar convection, the correlation is rather expected. As the cycle rises, more and more flux reaches the network boundary and adds to its width. The result agrees well with that from the transition region network (Raju (2016)). The implications of the result to the mass and energy budget of the solar atmosphere need to be examined through modelling. As a future work, we plan to study the variation of these parameters at different solar latitudes which may give insights on surface flows and flux transport.

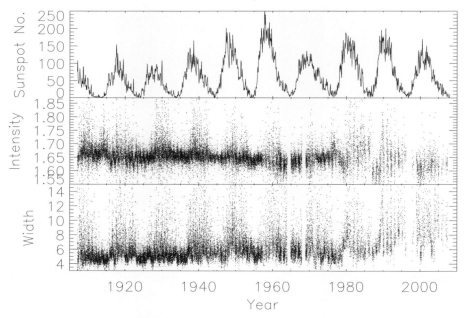

Figure 2. Temporal variation of sunspot number, Ca II K intensity and network boundary width.

References

Bappu, M. K. V. 1967, *Solar Phys.*, 1, 151

Patsourakos, S., Vial, J.-C., Gabriel, A. H., & Bellamine, N. 1999, *ApJ*, 522, 540

Priyal, M., Singh, J., Ravindra, B., Priya, T. G., & Amareswari, K. 2014, *Solar Phys.*, 289, 137

Raju, K. P. 2016, *Solar Phys.*, 291, 3519

Raju, K. P. & Singh, J. 2014, *Research in Astronomy and Astrophysics*, 14, 229

Tian, H., DeLuca, E. E., Cranmer, S. R., De Pontieu, B., Peter, H., Martnez-Sykora, J., *et al.* 2014, *Science*, 346, 1255711

Living Around Active Stars
Proceedings IAU Symposium No. 328, 2016
D. Nandy, A. Valio & P. Petit, eds.

© International Astronomical Union 2017
doi:10.1017/S1743921317004148

Solar Radius at Sub-Terahertz Frequencies

Fabian Menezes and Adriana Valio

Center for Radio Astronomy and Astrophysics (CRAAM)
Mackenzie Presbyterian University, São Paulo, Brazil
email: `fabianme17@gmail.com`
email: `avalio@craam.mackenzie.br`

Abstract. The visible surface of the Sun, or photosphere, is defined as the solar radius in the optical spectrum range located at 696,000 km (Cox *et al.* (Ed. 2015)). However, as the altitude increases, the dominant electromagnetic radiation is emitted at other frequencies. Our aim is to measure the solar radius at frequencies of 212 GHz and 405 GHz through out a solar cycle and, therefore, the altitude where these emissions are generated and that variation along the years. Also we tried to verify the the radius dependence on the solar activity cycle, which can be a good indicator of the changes that occur in the atmosphere structure. For this, we used data obtained by the Submillimetric Solar Telescope (SST) created from daily scans made by SST from 1990 to 2015. From these scans a 2D map of the solar disk was constructed. The solar radius is then determined by adjusting a circumference to the points where the brightness is half of the quiet Sun level, which is set as the most common temperature value in the solar map, *i.e.*, the mode of the temperature distribution. Thus, we determined the solar radius at 212 and 405 GHz and the altitude of the emissions respectively. For 212 GHz, we obtained a radius of 976.5"±8" (707±4 Mm), whereas for 405 GHz, we obtained 975.0"±8" (707±5 Mm). optical spectrum range

Keywords. Sun: activity, Sun: chromosphere, Sun: fundamental parameters, Sun: radio radiation

1. Introduction

The Sun is considered a mildly active star with an 11-year activity cycle. It emits radiation across the electromagnetic spectrum at several wavelengths, from gamma rays to radio waves.

Our aim is to measure the solar radius at frequencies of 212 GHz and 405 GHz and, therefore, the altitude where these emissions are generated. The relevance of this research is the possibility to understand more about the solar atmosphere and what is the radius dependence on the solar activity cycle, which can be a good indicator of the changes that occur in the atmosphere structure.

2. Sub-terahertz observations

We used data obtained by the Submillimetric Solar Telescope (SST), in the Astronomical Complex El Leoncito Observatory (CASLEO) at the Argentinean Andes in partnership with the Center for Radio-Astronomy and Astrophysics Mackenzie (CRAAM). This telescope uses a multibeam system operating at radio frequencies with 4 beams at 212 GHz and 2 beams at 405 GHz (Figure 1a). Maps of the whole Sun were created (Figure 1b) from daily scans made by SST from 1999 to 2015 (Figure 2).

The the quiet Sun level is set as the most common temperature value in the solar map, i.e., the mode of the temperature distribution (Figure 3a). Next, the solar limb is set as the half of the quiet Sun level (Figure 3b). Then, the solar radius is obtained by

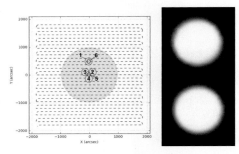

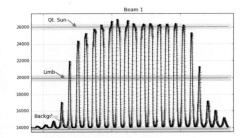

Figure 1. Left: scans over the solar disk made by SST. The colored circles represent the telescope beams. Right: examples of solar maps made from SST's data.

Figure 2. Solar intensity as a function of time during the scan map from SST-beam 1 at 15 UT on 23 February 2012 showing background and quiet Sun values.

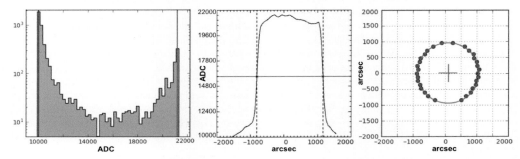

Figure 3. Left: histogram of intensities where we define the value of quiet Sun. Middle: the markup of the limb points. Right: the coordinates of the limb with the circumference fit.

Table 1. Solar radius results for 212GHz and 405GHz frequencies

Frequency	Radius (arcsec)	Radius (Mm)	Altitude (km)
212 GHz	974 ± 5	707 ± 4	11,492
405 GHz	974 ± 7	707 ± 5	11,492
Optical	959.63	695.508 ± 0.026	0

adjusting a circumference to the solar limb points (Figure 3c). Thus, by substracting the sub-THz radius from the visible one, we are able to determine the altitudes where the 212 Ghz and 405 GHz emissions are generated.

3. Preliminary results

For 212 GHz, we obtained a radius of 974"±5" (707±4 Mm), whereas for 405 GHz, we obtained a radius of 974"±7" (707±5 Mm). These results are shown in Table 1 and Figure 4.

In addition to the mean solar radius results at both frequencies, we investigated the relationship of the solar radius variation at submillimetric wavelengths with the solar activity cycle of 11 years. Unfortunately it was not possible to observe this relationship due to high dispersion of data.

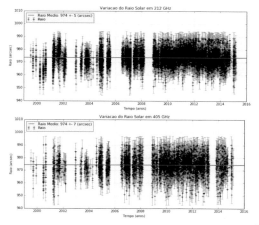

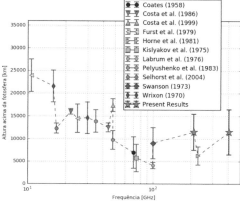

Figure 4. Fitted radius of the solar disk throughout the solar cycle at 212 GHz (above) and 405 GHz (below).

Figure 5. Height of emission above the photosphere as function of the frequency with previous results from other authors.

4. Conclusion

From the radii obtained – from the fit to the SST maps (Figure 4) – we could calculate their respective heights above the photosphere (Table. 1): the 212-GHz emission is produced at an altitude of 11,492±4000 km, whereas 405-GHz one is emitted at an altitude of 11,492±5000 km. In Figure 5 we can also compare our present rusults with previous ones from Coates (1958), Costa *et al.* (1986), Costa *et al.* (1999), Fürst *et al.* (1979), Horne *et al.* (1981), Kisliakov *et al.* (1975), Labrum *et al.* (1978), Pelyushenko & Chernyshev (1983), Selhorst *et al.* (2004), Swanson, P. N. (1973), Wrixon (1970).

The Sun's atmosphere is divided to three structures. According to Cox *et al.* (Ed. 2015), they are photosphere (from 0 to 525 km), chromosphere (from 525 to 2100 km) and the corona (from 2100 km and above). So, we can conclude that both emission frequencies are dominant at the corona layer.

Aknowledgements

Special thanks to MackPesquisa and IAU for the support and aid for the this event (IAUS 328) and to CAPES for the scholarship aimed at the postgraduate program of Geospatial Sciences and Applications.

References

Coates, R. J. 1958, *The Astrophysical Journal*, 128, 83.
Costa, J. E. R., Homor, J. L., & Kaufmann, P. 1986, *In Solar Flares & Coronal Physics Using P/OF as a Research Tool*, 1, 201.
Costa, J. E. R., & Silva, A. V. R., & Makhmutov, V. S., & Rolli, E., & Kaufmann, P., & Magun, A. 1999, *The Astrophysical Journal Letters*, 520, L63.
Cox, A. N. and other authors (Ed.) 2015, *Allen's astrophysical quantities*.
Fürst, E., Hirth, W., & Lantos, P. 1979, *Solar Physics*, 63(2), 257.
Horne, K., Hurford, G. J., Zirin, H., & De Graauw, T. 1981, *The Astrophysical JournaL*, 244, 340.
Kisliakov, A. G., Kulikov, I. I., Fedoseev, L. I., & Chernyshev, V. I. 1975, *Soviet Astronomy Letters*, 1, 79.
Labrum, N. R., Archer, J. W., & Smith, C. J. 1978, *Solar Physics*, 59(2), 331.

Pelyushenko, S. A. & Chernyshev, V. I. 1983, *Soviet Astronomy*, 27, 340.
Selhorst, C. L., & Silva, A. V. R., & Costa, J. E. R. 2004, *Astronomy & Astrophysics*, 420, 1117.
Swanson, P. N. 1973, *Solar Physics*, 32(1), 77.
Wrixon, G. T. 1970, *Nature*, 227, 1231.

Living Around Active Stars
Proceedings IAU Symposium No. 328, 2016
D. Nandy, A. Valio & P. Petit, eds.

© International Astronomical Union 2017
doi:10.1017/S1743921317003921

Exploring shallow sunspot formation by using Implicit Large-eddy simulations

F. J. Camacho[1,*], **G. Guerrero**[1,**], **P. K. Smolarkiewicz**[2], **A. G. Kosovichev**[3] **and N. N. Mansour**[4]

[1] Universidade Federal de Minas Gerais, Av. Pres. Antônio Carlos, 6627 - Pampulha
Belo Horizonte - MG, Brazil
[*] email: `camacho@fisica.ufmg.br`
[**] email: `guerrero@fisica.ufmg.br`
[2] European Centre for Medium-Range Weather Forecasts, Reading RG2 9AX, UK
email: `smolar@ecmwf.int`
[3] New Jersey Institute of Technology, Newark, NJ 07103, USA
email: `alexander.g.kosovichev@njit.edu`,
[4] NASA, Ames Research Center, Moffet Field, Mountain View, CA, USA
email: `nagi.n.mansour@nasa.gov`

Abstract. The mechanism by which sunspots are generated at the surface of the sun remains unclear. In the current literature two types of explanations can be found. The first one is related to the buoyant emergence of toroidal magnetic fields generated at the tachocline. The second one states that active regions are formed, from initially diffused magnetic flux, by MHD instabilities that develop in the near-surface layers of the Sun. Using the anelastic MHD code EULAG we address the problem of sunspot formation by performing implicit large-eddy simulations of stratified magneto-convection in a domain that resembles the near-surface layers of the Sun. The development of magnetic structures is explored as well as their effect on the convection dynamics. By applying a homogeneous magnetic field over an initially stationary hydrodynamic convective state, we investigate the formation of self-organized magnetic structures in the range of the initial magnetic field strength, $0.01 < B_0/B_{eq} < 0.5$, where B_{eq} is the characteristic equipartition field strength.

Keywords. convection, EULAG-MHD, ILES, turbulence, sunspots

1. Anelastic Turbulent Convection with EULAG-MHD

We use the code EULAG-MHD which solves an MHD extension of the Lipps & Hemler (1982) anelastic equations (see Smolarkiewicz & Charbonneau 2013).

$$\frac{d\vec{u}}{dt} = -\nabla\pi' - \vec{g}\frac{\theta'}{\theta_0} + \frac{1}{\mu\rho_0}\vec{B}\cdot\nabla\vec{B}, \tag{1.1}$$

$$\frac{d\theta'}{dt} = -\vec{u}\cdot\nabla\theta_e - \alpha\theta', \tag{1.2}$$

$$\frac{d\vec{B}}{dt} = \vec{B}\cdot\nabla\vec{u} - \vec{B}\nabla\cdot\vec{u}, \tag{1.3}$$

$$\nabla\cdot(\rho_0\vec{u}) = 0, \tag{1.4}$$

$$\nabla\cdot\vec{B} = 0. \tag{1.5}$$

Here, θ is the potential temperature, tantamount of the specific entropy ($s = c_p \ln\theta$). Subscripts "o" denote an isentropic base state in hydrostatic balance. Primes denote perturbations around an ambient state (denoted by subscript "e"), and $\vec{g}$ is a constant gravitational acceleration. π' is a density normalized pressure perturbation encompassing the hydrostatic and magnetic pressure. The domain of the simulations is a rectangular box with dimensions $50 \times 50 \times 20$ Mm^3 in the xyz axes, respectively. In the x and y directions we use periodic boundary conditions. For the z direction, the bottom of the domain is taken where $z = 0$ Mm and the top where $z = 20$ Mm. We use two 3D grids at low and high resolution, 128^3 and 256^3 meshpoints, respectively. For the ambient state we choose a three-layer polytropic atmosphere given by the equations:

$$\frac{dT_e}{dz} = \frac{-g_0}{R_g(m(z)+1)} \tag{1.6}$$

$$, \frac{d\rho_e}{dz} = \frac{\rho_e}{T_e}\left(\frac{-g_0}{R_g} - \frac{dT}{dz}\right), \tag{1.7}$$

where the layers are connected via smooth transitions in the polytropic index profile, $m(z)$. The vertical stratification given by eqs. (1.6) and (1.7) leads to a convectively stable layer at the bottom $0.0 < z < 0.2$ Mm, followed by an adiabatic layer $0.2 < z < 1.8$ Mm, and a convection-unstable layer at the top, $1.8 < z < 2.0$ Mm. The last term on the rhs of (1.2) is a Newtonian cooling that models the divergence of the non-computable Reynolds heat flux. It relaxes θ' toward zero within a time-scale, $\tau = \alpha^{-1}$

$$\frac{d}{dz}\rho_0\langle w'\theta'\rangle = -\rho_0\alpha\theta'. \tag{1.8}$$

For the simulations performed in this work we consider $\tau = 1.1$ days. We impose a uniform vertical magnetic field, $\vec{B} = (0,0,B_0)$, of three different strengths over a HD convective pattern in steady state. The three different strengths for the imposed magnetic field are $B_0/B_{eq} = (0.01, 0.1, 0.5)$, for the low resolution HD simulation, and $B_0/B_{eq} = (0.05, 0.1, 0.5)$, for the high resolution simulation. Here, B_{eq} is the equipartition magnetic field:

$$\frac{1}{2\mu_0}B_{eq}^2 = \frac{1}{2}\bar{\rho}u_{rms}^2. \tag{1.9}$$

2. Results

From the cases considered, only the low resolution simulation with $B_0/B_{eq} = 0.1$, and the high resolution simulations with $B_0/B_{eq} = 0.05$ and $B_0/B_{eq} = 0.1$, show the formation of localized magnetic flux concentrations. The most representative results are obtained for the low resolution simulation with $B_0/B_{eq} = 0.1$, shown in Figures 1 and 2. The simulations with imposed vertical magnetic field reveal that there is an optimal range of field strengths, $0.01 < B_0/B_{eq} < 0.5$, for the formation of localized magnetic flux concentrations. Vertically, these structures have the length scale of the entire adiabatic layer. For the cases with a stronger imposed magnetic field, the flow pattern losses the convective structure.

For the simulations that show the formation of magnetic flux concentrations, an important change in the form of the convective cells is observed. At the upper layers ($z > 18.8$ Mm), the convective cells near the magnetic flux concentrations acquire a elongated shape pointing towards the center of these structures. This is reminiscent of the penumbra around the dark center of sunspots (e.g., Borrero & Ichimoto 2011).

For the moment, it is not clear whether these structures are the results of MHD instabilities or the collapse of the magnetic flux in the downflow lanes. Since the downflows have the horizontal scale of the deep convective motions, the magnetic elements at the surface cover several small-scale convective elements.

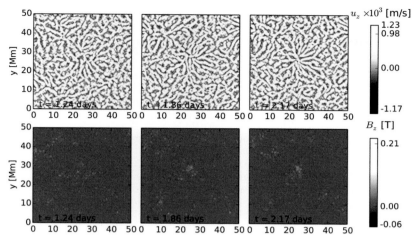

Figure 1. xy snapshots, at 1.2 Mm below the top boundary, of the vertical velocity and magnetic field at different times for the low resolution simulation with $B_0/B_{eq} = 0.1$.

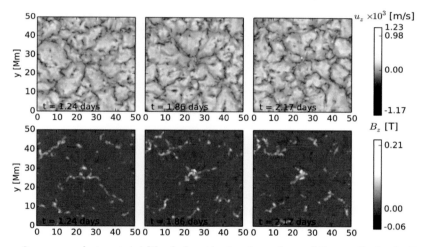

Figure 2. xy snapshots, at 4.4 Mm below the top boundary, of the vertical velocity and magnetic field at different times for the low resolution simulation with $B_0/B_{eq} = 0.1$.

Acknowledgements

Special thanks to the organizer for a wonderful meeting and to the IAU for travel support. This work was partly funded by FAPEMIG grant APQ-01168/14 (FC and GG). The simulations were performed in the NASA cluster Pleiades and Brazilian supercomputer Sdumont of the National Laboratory of Scientific Computation (LNCC).

References

Smolarkiewicz, P. K., & Charbonneau, P. 2013, *J. Comput. Phys.*, 236, 608

Lipps F.B., Hemler R.S., A scale analysis of deep moist convection and some related numerical calculations, *J. Atmos. Sci.*, 39 (1982) 2192-2210

Borrero J. M. & Ichimoto K., Magnetic Structure of Sunspots, *Living Rev. Solar Phys.*, 8, (2011), 4.

Living Around Active Stars
Proceedings IAU Symposium No. 328, 2016
D. Nandy, A. Valio & P. Petit, eds.

© International Astronomical Union 2017
doi:10.1017/S1743921317003829

Contribution of energetic ion secondary particles to solar flare radio spectra

Jordi Tuneu[1], Sérgio Szpigel[1], Guillermo Giménez de Castro[1] and Alexander MacKinnon[2]

[1]Center for Radio Astronomy and Astrophysics, Mackenzie Presbyterian University,
Sao Paulo, Brazil
email: `jordituneu@protonmail.com`
[2]School of Physics and Astronomy, University of Glasgow, Glasgow, UK
email: `alexander.mackinnon@glasgow.ac.uk`

Abstract. Recent observations of solar flares at high frequencies have provided evidence of a new spectral component with flux increasing with frequency in the THz range. Its origin remains unclear. Here, we present preliminary results of simulations of synchrotron emission due to secondary positrons and electrons produced in nuclear reactions during a solar flare. We use the general purpose Monte-Carlo code FLUKA to obtain distributions of secondary particles resulting from accelerated protons interacting in the solar atmosphere. We calculate the synchrotron radiation spectrum and compare our results to observations of the November 4th, 2003 burst event.

Keywords. Solar flares, Secondary positrons and electrons, Synchrotron radiation

1. Introduction

Systematic observations of solar flares at high frequencies (0.2 and 0.4 THz) with the *Solar Submillimeter Telescope* (SST) at the *El Leoncito* observatory in Argentina have provided evidence of a new spectral component with flux increasing with frequency in the THz range (Kaufmann *et al.* 2002, Kaufmann *et al.* 2004). This new component occurs simultaneously but separated from the well-known microwave spectral component which reaches its maximum flux typically at 10's of GHz. Its nature remains unclear (see Krucker *et al.* (2013) for a review on observations and possible radiation mechanisms).

The γ-ray continuum tells us that $\sim GeV$ ions are present in some flares (Vilmer *et al.* 2011). These will produce secondary positrons and electrons with energies in the range from 0.1 to 1 GeV by a variety of mechanisms: pion decay, Compton scattering and pair production of γ-ray photons as well as "knock-on" electrons. These secondary positrons and electrons will radiate in the sub-THz range of frequencies via synchrotron emission. Silva *et al.* (2007) and Trottet *et al.* (2008) both found that the populations of secondaries implied by observed γ-ray fluxes were too low to account for the sub-THz observations. Trottet *et al.* (2008) did not discard this mechanism, however, noting that a more detailed treatment of secondary particle transport was needed.

2. Simulation Methods and Results

We use FLUKA (Ferrari *et al.* 2005), a general-purpose Monte-Carlo code for calculations of particle transport and interactions in matter, to simulate accelerated protons colliding with a thick target with Asplund *et al.* (2009) chemical abundances and producing secondary particles through nuclear reactions (see also MacKinnon *et al.* (2016)).

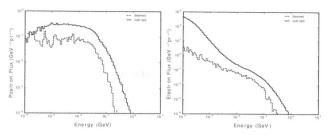

Figure 1. Energy distributions of secondary positrons and electrons.

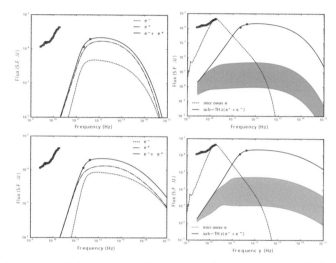

Figure 2. Synchrotron radiation spectra due to secondary positrons and electrons calculated for the downward unidirectional beam (top) and the downward isotropic beam (bottom).

We consider accelerated primary protons with a power-law energy distribution and two different angular distributions: downward unidirectional and downward isotropic. In each case, we monitor the energy distributions of secondary positrons and electrons as they cross from the dense atmosphere to the corona. We assume that these particles are trapped in a magnetic field and use a code based on Ramaty's algorithm (Ramaty *et al.* 1994) to calculate the resulting synchrotron radiation spectrum. We determine the total number of accelerated primary electrons by fitting the observed spectrum in the microwave range of frequencies, and estimate the associated number of primary protons via the correlation between primary electrons and protons found by Shih *et al.* (2009).

In Fig. 1 we show the energy distributions of secondary positrons and electrons calculated for primary protons with a power-law energy distribution in the range from 0.2 to 10 GeV and spectral index $\delta = 2$ in the case of a downward unidirectional beam and in the case of a downward isotropic beam. As one can observe, the energy distributions of secondary positrons and electrons strongly depend on the angular distribution of primary protons. Knock-on electrons are numerically dominant over pion decay electrons below a few MeV, with possible implications for emission at cm wavelengths, particularly so in the case of a downward isotropic beam. In Fig. 2 we show the synchrotron radiation spectra due to secondary positrons and electrons calculated for the downward unidirectional beam (top) and the downward isotropic beam (bottom). In the left panels, we show the results obtained by considering the number of protons as a free parameter, N_p^{free}, which is adjusted to fit the spectrum of the November 4th, 2003 flare at sub-THz frequencies. The values of the other model parameters used in the calculations are kept fixed: magnetic field strength $B = 1000\,G$, viewing angle $\theta = 45°$, source size $\phi_s = 0.3''$, and electronic plasma density $n_p = 1.0 \times 10^8\,cm^{-3}$. In the right panels, we show the results obtained by using the number of primary electrons adjusted to fit the observed spectrum at microwave frequencies to constrain the number of protons, finding a minimum and a maximum number of protons, N_p^{min} and N_p^{max}, based on the spread of the correlation by Shih *et al.* (2009). In Table 1 we compare the values for N_p^{free}, N_p^{min} and N_p^{max} and the corresponding number of secondary positrons and electrons found in each case.

Table 1. Number of protons, positrons and electrons found in the calculations of the synchrotron radiation spectra for the downward unidirectional beam and the downward isotropic beam.

PROTONS	N_p^{free}	N_p^{min}	N_p^{max}
Downward unidirectional	9.0×10^{33}	1.3×10^{30}	6.4×10^{31}
Downward isotropic	2.6×10^{33}	1.3×10^{30}	6.4×10^{31}

POSITRONS	$N_{e^+}^{free}$	$N_{e^+}^{min}$	$N_{e^+}^{max}$
Downward unidirectional	3.6×10^{31}	5.1×10^{27}	2.5×10^{29}
Downward isotropic	3.8×10^{31}	1.9×10^{28}	9.4×10^{29}

ELECTRONS	$N_{e^-}^{free}$	$N_{e^-}^{min}$	$N_{e^-}^{max}$
Downward unidirectional	3.1×10^{31}	4.4×10^{27}	2.2×10^{29}
Downward isotropic	2.4×10^{32}	1.2×10^{29}	6.0×10^{30}

3. Final Remarks

We have used the Monte-Carlo code FLUKA to simulate the nuclear reactions of $\sim GeV$ ions precipitating into the dense atmosphere during a solar flare and obtain the energy distributions of the resulting secondary positrons and electrons escaping from the dense atmosphere to the corona. Our results show that secondary knock-on electrons are not negligible compared to secondary positrons. For a downward unidirectional beam of primary protons we obtain a ratio $N_{e^-}/N_{e^+} \sim 0.9$ and for a downward isotropic beam we obtain $N_{e^-}/N_{e^+} \sim 6.3$. The results obtained in calculations of the synchrotron radiation spectrum due to secondary positrons and electrons show that in the most favorable case (downward isotropic beam) the number of primary protons required to fit the increasing spectrum of the November 4th, 2003 flare at sub-THz frequencies is ~ 40 times more than the maximum number estimated from the correlation by Shih *et al.* (2009).

Acknowledgments

This work was supported by the Royal Society Newton Mobility Grant and CAPES.

References

Asplund, M., Grevesse, N., Sauval, A. J., & Scott, P. 2009, *ARAA*, 47, 481-522
Ferrari, A., Sala, P. A., & Fasso, A., *et al.* 2005, *Technical Report*, CERN-2005-10
Kaufmann, P., Raulin, J.-P., & Melo, A. M., *et al.* 2002, *Astrophys. J.*, 574, 1059-1065
Kaufmann, P., Raulin, J.-P., & de Castro, C. G. G., *et al.* 2004, *Astrophys. J.*, 603, L121-L124
Krucker, S., de Castro, C. G. G., & Hudson, H. S., *et al.* 2013, *Astron. Astrophys. Rev.*, 21, 58
MacKinnon, A., Szpigel, S., de Castro, C. G. G., & and Tuneu, J. 2016, *Sol. Phys., submitted*

Ramaty, R., Schwartz, R. A., Enome, S., & Nakajima, H. 1994, *Astrophys. J.*, 436, 941-949

Shih, A. Y., Lin, R. P., & and Smith, D. M. 2009, *Astrophys. J. Lett.*, 698, L152-L157

Silva, A. V. R. Share, G. H., & Murphy, R. J, *et al.* 2007, *Sol. Phys.*, 245, 311-326

Trottet, G. Krucker, S., Lüthi, T., & Magun, A. 2007, *Astrophys. J.*, 678, 509-514

Vilmer, N., MacKinnon, A. L., & Hurford, G. J. 2011, *Space Sci. Rev.*, 159, 167

Living Around Active Stars
Proceedings IAU Symposium No. 328, 2016
D. Nandy, A. Valio & P. Petit, eds.

© International Astronomical Union 2017
doi:10.1017/S1743921317003659

Low-mass eclipsing binaries in the WFCAM Transit Survey

Patricia Cruz[1], Marcos Diaz[1], David Barrado[2] and Jayne Birkby[3,4]

[1]Instituto de Astronomia, Geofísica e Ciências Atmosféricas, Universidade de São Paulo (IAG/USP), Brazil
email: `patricia.cruz@usp.br`
[2]Departamento de Astrofísica, Centro de Astrobiología (CAB/INTA-CSIC), Spain
[3]Harvard-Smithsonian Center for Astrophysics, USA
[4]NASA Sagan Fellow

Abstract. The characterization of short-period detached low-mass binaries, by the determination of their physical and orbital parameters, reveal the most precise basic parameters of low-mass stars. Particularly, when photometric and spectroscopic data of eclipsing binaries (EBs) are combined. Recently, 16 new low-mass EBs were discovered by the WFCAM Transit Survey (WTS), however, only three of them were fully characterized. Therefore, new spectroscopic data were already acquired with the objective to characterize five new detached low-mass EBs discovered in the WTS, with short periods between 0.59 and 1.72 days. A preliminary analysis of the radial velocity and light curves was performed, where we have derived orbital separations of 2.88 to 6.69 $R_\odot$, and considering both components, we have found stellar radii ranging from 0.40 to 0.80 $R_\odot$, and masses between 0.24 and 0.71 $M_\odot$. In addition to the determination of the orbital parameters of these systems, the relation between mass, radius and orbital period of these objects can be investigated in order to study the mass-radius relationship and the radius anomaly in the low main-sequence.

Keywords. Eclipsing binaries, low-mass stars, spectroscopy

1. Introduction

Combined photometric and spectroscopic phase-resolved data of eclipsing binaries (EBs), and particularly double-lined EBs, provide the most precise ways to measure their fundamental properties without using spectral models.

Presently, only a small number of low-mass eclipsing binaries (LMEBs) was characterized with a good precision. As few as ∼20 low-mass systems have stellar masses and radius measured with uncertainties below 5%. Surprisingly, the data present in the literature when compared to stellar models present significant discrepancy, specially concerning the stellar radius. The M-dwarf stars in LMEBs seem to have radii 5 to 10% greater than the expected radius, derived from spectral models (López-Morales & Ribas 2005, Kraus *et al.* 2011). This is known as the radius anomaly and it is a recurring problem of the determination of physical parameters of low-mass stars.

The discovery and characterization of new LMEBs will help unveiling the radius anomaly problem and set new limits on the mass-radius relationship of low-mass stars.

2. Data analysis

The WFCAM Transit Survey. The WFCAM Transit Survey (WTS) was focused on the discovery and study of exoplanets around low-mass stars. This survey has been awarded 200 nights with the Wide Field Camera (WFCAM) on the 4m UK Infrared

Table 1. System parameters: preliminary results.

	17e-3-02003	17h-4-01429	19c-3-08647	19f-4-05194	19g-2-08064
P$_{orb}$ (*days*)	1.2250074	1.4445895	0.8674656	0.5895297	1.7204092
error	0.0000005	0.0000003	0.0000001	0.0000001	0.0000004
M$_1$ ($M_\odot$)	0.595	0.503	0.392	0.536	0.714
error	0.021	0.016	0.018	0.015	0.027
M$_2$ ($M_\odot$)	0.508	0.409	0.243	0.388	0.642
error	0.017	0.012	0.013	0.011	0.026
R$_1$ ($R_\odot$)	0.643	0.510	0.499	0.608	0.796
error	0.017	0.005	0.017	0.009	0.010
R$_2$ ($R_\odot$)	0.445	0.418	0.401	0.487	0.465
error	0.023	0.005	0.021	0.008	0.007
log g$_1$	4.596	4.724	4.635	4.599	4.490
error	0.022	0.007	0.029	0.011	0.008
log g$_2$	4.847	4.808	4.616	4.652	4.910
error	0.045	0.009	0.045	0.015	0.012
a$_{orb}$ ($R_\odot$)	4.976	5.213	3.288	2.881	6.686
error	0.054	0.052	0.052	0.026	0.083
incl (°)	82.43	89.12	81.47	83.46	83.91
error	0.19	0.14	0.14	0.18	0.05
T$_{eff,SED}$ (K)	3500	3400	3800	4200	4200
T$_{eff,1}$ (K)	3800	3400	3900	4400	4200
T$_{eff,2}$ (K)	3100	3100	3100	3500	3200

Telescope (UKIRT) to search for planets via the transit method at infrared wavelengths. The observations were performed in the WFCAM J band (λ_c at 1.25 μm), near to the maximum of the spectral energy distribution (SED) of low-mass stars. As a secondary objective, the survey also detected several light curves of eclipsing binaries (EBs). For instance, a fine number of light curves of low-mass EBs was discovered with short periods of less than 5 days (Birkby *et al.* 2012). He have then selected light curves of five new detached low-mass EBs discovered in the WTS, with short periods between 0.59 and 1.72 days.

SED fitting. We performed the SED fitting using the Virtual Observatory tool VOSA (Bayo *et al.* 2008). The filters considered were: SDSS u, g, r, i, z; WFCAM Z, Y, J, H, K; 2MASS J, H, Ks; WISE W1, W2. These data were available in the literature and allowed estimates of the combined effective temperatures of the EBs, which are shown in table 1.

Low-resolution spectroscopy. Low-resolution spectroscopic data were acquired with the TWIN spectrograph mounted on the 3.5m-telescope at the Calar Alto Observatory (CAHA) to derive the effective temperature of each component of the binary. We compared the observed spectra with the BT-Settl library of synthetic spectra from Allard *et al.* (2013), by combining two synthetic spectra in order to reproduce the flux-calibrated observed spectra. The obtained individual temperatures (T$_{eff,1}$, T$_{eff,2}$) are also shown in table 1.

Radial velocity measurements. We have gathered intermediate resolution spectra also with TWIN/3.5m-telescope (CAHA) to measure radial velocity shifts (RVs) of the 5

P. Cruz *et al.*

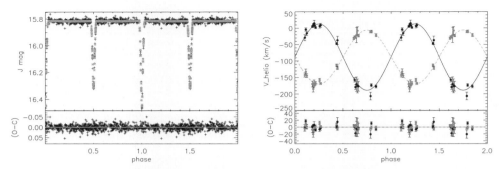

Figure 1. Light curve and RV fittings for the LMEB 17h-4-01429.

double-lined LMEBs. The RVs were computed via the Fourier cross-correlation of Hα emission components using IRAF's FXCOR.

Solving the systems. We have used the jktEBOP code (Southworth *et al.* 2004a,b) to analyze all photometric and spectroscopic data together and obtain an unique solution for each EB system. The orbital and physical parameters obtained from jktEBOP, with respective errors, are presented in table 1. Figure 1 shows, as an example, the best fitting solution for the RV data and light curve of the LMEB 17h-4-01429.

3. Discussion

We have presented the current status of our project on the study of low-mass eclipsing binaries. The complete characterization of five new detached low-mass EBs discovered in the WTS is only possible by combining sets of photometric and spectroscopic data. There are few high precision determination of masses and radii for low-mass stars in short-period binaries. Between the analyzed systems, 3 EBs have M-dwarf stars as primary and secondary components, and 2 EBs with M-dwarfs as secondaries, with late K-dwarfs as primaries. This is an ongoing work, and we have presented preliminary results for all 5 systems. Future analysis steps include placing these objects in the mass-radius diagram and investigating the radius anomaly in low-mass binaries and its relation with their basic properties.

References

Allard, F., Homeier, D., Freytag, B., Schaffenberger, W., & Rajpurohit, A. S. 2013, *MSAIS*, 24, 128

Bayo, A., Rodrigo, C., Barrado, Y., Navascués, D., Solano, E., Gutiérrez, R., Morales-Calderón, M., & Allard, F. 2008, *A&A*, 492, 277

Birkby, J., Nefs, B., Hodgkin, S., Kovács, G., Sipőcz, B., Pinfield, D., Snellen, I., Mislis, D., Murgas, F., Lodieu, N., de Mooij, E., Goulding, N., Cruz, P., Stoev, H., Cappetta, M., Palle, E., Barrado, D., Saglia, R., Martin, E., & Pavlenko, Y. 2012, *MNRAS*, 426, 1507

Kraus, A. L., Tucker, R. A., Thompson, M. I., Craine, E. R., & Hillenbrand, L. A. 2011, *ApJ*, 728, 48

López-Morales, M. & Ribas, I. 2005, *ApJ*, 631, 1120

Southworth, J., Maxted, P. F. L., & Smalley, B. 2004a, *MNRAS*, 351, 1277

Southworth, J., Zucker, S., Maxted, P. F. L., & Smalley, B. 2004b, *MNRAS*, 355, 986

Living Around Active Stars
Proceedings IAU Symposium No. 328, 2016
D. Nandy, A. Valio & P. Petit, eds.

© International Astronomical Union 2017
doi:10.1017/S1743921317003817

Evolution of the Active Region NOAA 12443 based on magnetic field extrapolations: preliminary results

André Chicrala[1], Renato Sergio Dallaqua[1], Luis Eduardo Antunes Vieira[1], Alisson Dal Lago[1], Jenny Marcela Rodríguez Gómez[1], Judith Palacios[2], Tardelli Ronan Coelho Stekel[1], Joaquim Eduardo Rezende Costa[1] and Marlos da Silva Rockenbach[1]

[1]National Institute for Space Research,
12227-010, São José dos Campos, Brazil
email: andre.chicrala@inpe.br

[2]Departamento de Física y Matemáticas,
Universidad de Alcalá
University Campus, Sciences Building, P.O. 28871, Alcalá de Henares, Spain
email: judith.palacios@uah.es

Abstract. The behavior of Active Regions (ARs) is directly related to the occurrence of some remarkable phenomena in the Sun such as solar flares or coronal mass ejections (CME). In this sense, changes in the magnetic field of the region can be used to uncover other relevant features like the evolution of the ARs magnetic structure and the plasma flow related to it. In this work we describe the evolution of the magnetic structure of the active region AR NOAA12443 observed from 2015/10/30 to 2015/11/10, which may be associated with several X-ray flares of classes C and M. The analysis is based on observations of the solar surface and atmosphere provided by HMI and AIA instruments on board of the SDO spacecraft. In order to investigate the magnetic energy buildup and release of the ARs, we shall employ potential and linear force free extrapolations based on the solar surface magnetic field distribution and the photospheric velocity fields.

Keywords. Photosphere, Sunspots, Evolution.

1. Introduction

The Photosphere, commonly referred as the solar surface, is also occasionally populated by Active Regions (ARs) that have a strong magnetic field, which can reach a few thousand Gauss, when compared to its surroundings. The evolution of such regions may be related to energetic events that occur on both small and large scales. The study of the solar magnetism is mainly supported by the Zeeman effect and polarization of light. In this sense, the parameters of Stokes are enough to fully characterize the polarization of light.

The object of this study is the AR NOAA12443 that was observed in the solar disk from 2015/10/30 to 2015/11/10. During the time in which the AR was visible on the disk X-ray flux peaks reached values that characterizes the occurrence of M-class flares. This reveals a complex magnetic structure.

Some physical quantities of the observed region, such as line-of-sight velocity and magnetic field strength, can be directly retrieved from the Stokes parameters measurements or be used as entries in inversion algorithms. For this work the available data from

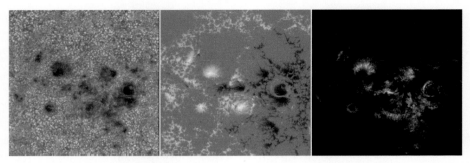

Figure 1. Left panel: it is presented a continuum map of the AR. Mid panel: Weak field approximation was performed over the same data set unveiling features of the AR such as its penumbral structure and magnetic field along the Line Of Sight (LOS). Right panel: is displayed a Linear polarization map of the region highlighting the transverse field structures such as the penumbral field.

Figure 2. On the left-hand side it displays the velocity map produced around the Fe I 6301.5107 Å line using the COG method in Stokes I and on the right-hand side the same calculation was performed using a Gauss fit. White colors indicate downflows and Dark colors represents upflows.

Hinode telescope were used and processed from the instrument Solar Optical Telescope (SOT/SP).

2. Results

Maps of the solar surface, as illustrated in Figure 1, were drawn to identify the AR physical structures. Complex structures, for example a light bridge, are clearly observed.

The solar plasma motion along the LOS can be calculated considering the Doppler shift in the measured wavelengths FeI 6301.5Å and 6302.5Å. In order to map the downflow and upflow velocities, two different methods were applied, Center of Gravity (COG) and fitting a Gaussian, yielding similar results as displayed in Figure 2. Comparing the results, it can be noticed that the map produced with the COG method appears to have more contrast indicating that the velocities obtained were slightly different. This difference arises from the slight asymmetries on Stokes I profiles. Typical values found reach up to 2 km s^{-1} for both downflows and upflows.

Assuming, in a first guess, that flows with a speed greater than 4km s^{-1} are supersonic, the two data sets coordinates of the supersonic flows could be found and then overploted on the continuum map as shown in Figure 3. Even though the results of COG and Gauss fit were different, the supersonic flows obtained were found in the same coordinates.

The integral of the red and blue lobes of Stokes V can be used in order to investigate the presence of velocity and magnetic field gradients being the asymmetry of the line integral

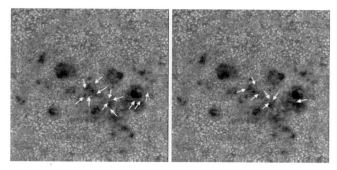

Figure 3. The regions pointed by the white arrows indicated where supersonic velocities were found considering the COG (left) and fitting a Gaussian (right) approaches over the continuum map. No supersonic upflow was found among the calculations performed over all data sets using both methods.

associated with the gradients. The Stokes V profiles were normalized and integrated for each pixel of each map in order that the integral value distribution could be studied.

3. Conclusion and perspectives

The positions where the supersonic downflows were identified are probably regions of penumbral development. Also, considering that the Evershed effect, the downflows might be evidence of regions where there is a net mass flow. It can be seen in Figure 3 and all the other maps that were produced that the supersonic downflows appeared within or at the border of the penumbra region probably due to the penumbra development and the boundary of the Evershed effect. Checking the values of the LOS magnetic field on those points were found values usually between +1kG and -1 kG.

The histogram has a peak at zero, where Stokes V can be considered symmetric, however, there are plenty of values for different levels of asymmetry which are later going to be used to identify regions according to their asymmetry level.

The data analysed so far wield promising results that can be used to further study the AR's behaviour and, eventually, evolution. Advanced algorithms that are able to perform the inversion of Stokes procedure and coronal magnetic field extrapolation are already being employed so that, with the preliminary results, more complex features of this AR can be unveiled.

References

Borrero Santiago, J. M. & IMPRS theses, 2004, *Dissertation.*
del Toro Iniesta, J.C., "Introduction to Spectropolarimetry", Cambridge Ed., 2003
del Toro Iniesta, J. C. & Ruiz Cobo, B., 1992, *ApJ*, 398, 375.
González, N. B. IMPRS theses, 2006, *Dissertation.*
Joshi, J. IMPRS theses, 2014, *Dissertation.*
Riethmüller, T. L. IMPRS theses, 2013, *Dissertation.*
Solanki, S. K. 2003, *A&AR*, 11, 153

Living Around Active Stars
Proceedings IAU Symposium No. 328, 2016
D. Nandy, A. Valio & P. Petit, eds.

© International Astronomical Union 2017
doi:10.1017/S1743921317003763

Deriving the solar activity cycle modulation on cosmic ray intensity observed by Nagoya muon detector from October 1970 until December 2012

Rafael R. S. de Mendonça[1], Carlos. R. Braga[1], Ezequiel Echer[1], Alisson Dal Lago[1], Marlos Rockenbach[1], Nelson J. Schuch[2] and Kazuoki Munakata[3]

[1] Space Geophysics Division, National Institute for Space Research,
São José dos Campos, SP, Brazil,
email: `rafael.mendonca@inpe.br`
[2] Southern Regional Space Research Center (CRS/INPE),
P.O. Box 5021, 97110-970, Santa Maria, RS, Brazil
[3] Physics Department, Shinshu University,
Matsumoto, Nagano 390-8621, Japan

Abstract. It is well known that the cosmic ray intensity observed at the Earth's surface presents an 11 and 22-yr variations associated with the solar activity cycle. However, the observation and analysis of this modulation through ground muon detectors datahave been difficult due to the temperature effect. Furthermore, instrumental changes or temporary problems may difficult the analysis of these variations. In this work, we analyze the cosmic ray intensity observed since October 1970 until December 2012 by the Nagoya muon detector. We show the results obtained after analyzing all discontinuities and gaps present in this data and removing changes not related to natural phenomena. We also show the results found using the mass weighted method for eliminate the influence of atmospheric temperature changes on muon intensity observed at ground. As a preliminary result of our analyses, we show the solar cycle modulation in the muon intensity observed for more than 40 years.

Keywords. Solar Activity Cycle, Cosmic Rays

1. Introduction

Cosmic rays are charged particles (mostly protons) with energy from MeV to ZeV (10^{21} eV) that travel in space and hit the Earth in an almost isotropic flow. They respond to the configuration of the Interplanetary Magnetic Field (IMF) presenting anisotropies related to solar and interplanetary (transient or recurrent) phenomena (Potgieter 2013). Thus, studying the anisotropies in the cosmic ray fluxuseful for understanding the physical aspects of solar and interplanetary phenomena, which can make cosmic rays a useful tool for predicting and monitoring the Space Weather conditions (Bieber & Evenson 1998; Munakata *et al.* 2003; Kudela *et al.* 2000). In this work, we show the procedures performed to making possible to observe clearly the 11 and 22-yrs cosmic ray intensity variations related to the solar activity cycle on the Nagoya muon vertical detector directional channel data recorded between October 1970 and December 2012.

2. Nagoya muon detector and its data analysis

The Nagoya (NGY) muon detector is part of the Global Muon Detector Network (GMDN) and is located in Nagoya - Japan (35.15° N, 136.97° E) at 77 m above sea level.

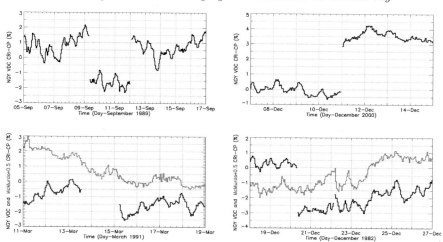

Figure 1. Examples of discontinuities and gaps found on the original hourly Nagoya (NGY) detector Vertical Directional Channel (VDC) Cosmic Ray Intensity Corrected by Pressure effect (CRI-CP).The black curve shows the NGY data, while the red curve represents the McMurdo neutron monitor data multiplied by 0.5.

The Vertical Directional Channel (VDC) of this detector has a detection area of 36 m² and observes secondary cosmic ray (muons) arriving with zenith angle lower than 30° since 1970. However, many discontinuities and gaps are present in the NGY VDC data actually. As shown in the two upper panels of Fig. 1, most of discontinuities can be easily identified as caused by a problem or change in the detector electronics because there are no natural phenomena capable of generating variations similar to them. However, it is necessary to take into account that not every change which occurs together a discontinuity or gap is due to that situation. For example, comparing data from Nagoya detector and McMurdo Neutron monitor, we can see that the decrease associated with the gap occurred on March 1991 (black curve in the lower left panel of Fig. 1) is real because a decrease is also present in McMurdo data (red curve). On the other hand, the decrease associated with the gap observed on December 1982 (black curve in the right panel on the bottom part of Fig. 1) is not present on McMurdo data suggesting that is not caused by a natural phenomenon. It is important to stand out that we used neutron monitor, geomagnectic and interplanetary data only for decide if a adjustment on NGY data is necessary or not. The coefficients used for correcting them are calculated taking into account NGY data only.

The top panel of Fig. 2 shows the hourly mean cosmic ray intensity corrected by pressure observed by NGY VDC from October 1970 until December 2012 without any bad data removal or level adjustment. The vertical light blue lines highlight the periods where non-natural changes are found after analyzing this data considering (when available) the cosmic ray intensity variation observed by other detectors together with parameters related to the interplanetary medium and the geomagnetic field (Dst Index and ACE spacecraft data). We believe that the huge intensity decrease occurred between 1970 and 1974 (highlighted by the slanted yellow line) is related to an inital efficiency detection decrease, since there are no known natural phenomena that can produce this monotonous decrease. After we solved these data problems, we can observe two clear periodic variations with periods around one and eleven years (see red and green curves in the bottom panel of Fig. 2).

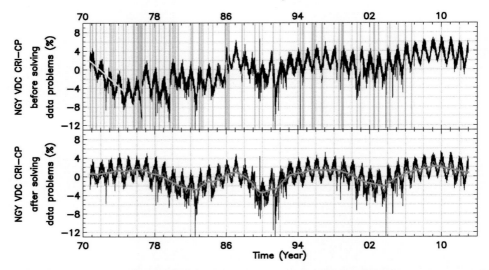

Figure 2. Hourly Nagoya Vertical Directional Channel Cosmic Ray IntensityCorrected by Pressure before and after solving data problems (black curves). The vertical light blue lines in the top panel indicate the periods when we notice non-natural changes in CRI-CP. The yellow slanted line in the top panel highlights the period when we assumed a detection efficiency decrease. The red and green curves in bottom panel show the 6 and 12-month running averages of NGY data after eliminating non-natural variations.

The clear seasonal variation seen in the black and red curves on Fig. 2 is related to the temperature effect that influences the muon creation and disintegration processes in the cosmic ray shower (Sagisaka 1986). For ground muon detectors (like NGY), we expect an anti-correlation between the observed cosmic ray count rate and the atmospheric temperature changes. Because of the atmospheric expansion on summer, most of muons are generated at higher altitude having a longer path to cross before reaching the detector at ground. This allows more of them to decay causing a decrease on its intensity at surface (Sagisaka 1986; Mendonça *et al.* 2016). We removed the temperature effect from the vertical directional channel data of Nagoya muon detector utilizing the Mass Weighted Method and using the methodology shown in Mendonça *et al.* (2016). This method considers the whole profile of the atmospheric temperature weighted by the atmospheric air density profile. In other words, the temperature measured in an atmospheric layer is weighted by the air mass quantity present in this layer. Using atmospheric temperature data from the Global Data Assimilation System (GDAS), we obtain a mass weighted temperature coefficient equal to -0.255 %/K , which is similar to that found by Mendonça *et al.* (2016). Using this coefficient, we eliminated the temperature effect from NGY VDC data. After that, as we can see in the top panel of Fig. 3, NGY VCD data present a good correlation with the cosmic ray intensity observed by McMurdo neutron monitor and show a clear and full solar activity cycle modulation (11-yr variation with a peak-plateau alternation on the intensity maxima). Comparing monthly mean data of Nagoya and WSO sunspot number (bottom panel of Fig. 3), we also notice that the maximum cosmic ray intensity in each solar minimum occurs just before the sunspot number sharp increase.

3. Summary and Final Remarks

After eliminating non-natural changes present in Nagoya muon detector vertical channel data and correcting it for the temperature effect, a clear 11 and 22 years variations

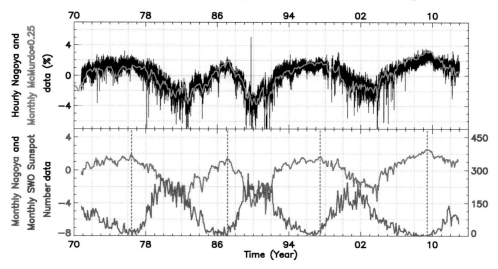

Figure 3. Hourly and Monthly NGY VDC cosmic ray intensity after solving electronic data problems and eliminating the temperature effect (black curve). The green curve shows the monthly average McMurdo neutron monitor data multiplied by 0.25. The red curve shows the monthly mean of NGY data. The purple curve represents the Monthly SWO Sunspot number.

have been clearly observed in a good correlation (R = 0.953) with the cosmic ray intensity observed by the McMurdo Neutron Monitor. This indicates that the instrumental problems, such as data gaps and discontinuities, and the atmospheric temperature effect are successfully removed in the present analysis.. Analyzing data recorded in the last three solar cycles, a clear anti-correlation (R = -0.746) between the solar activity cycle and the cosmic ray intensity observed by Nagoya muon detector corrected was found. These results does not be induced by the data problems correction since it is found in periods with no big problems in the data are found. If we analyze the data (corrected by temperature) between 1986 and 2012 correcting only the huge data problem occurred in 2001, we obtain a satisfactory correlation with NM data and anti-correlation with solar activy for example. Detailed analysis of the anti-correlation with the solar activity will be done in a future work.

Acknowledgements

The authors acknowledges the LOC of the IAUS 328, the Neutron Monitor Database and the institutions responsible to provide Dst Index and ACE data, CNPq for grants 152050/2016-7, 304209/2014-7, 302583/2015-7 and FAPESP for grant 2014/24711-6. The observations with the Nagoya muon detector are supported by Nagoya University. The McMurdo neutron monitor is supported by National Science Foundation award OPP-0739620.

References

Potgieter, M. S. 2013, *Living Rev. Solar Phys.*, 10, 3
Bieber, J. W. & Evenson, P. 1998, *Geophysical Research Letters*, 25, 15, 2955
Munakata, K. *et al.* 2003, *Adv. Sp. Res.*, 36, 12, 2357
Kudela, K., Storini, M., Hofer, M. Y., & Belov, A. 2000, *Space Sci. Revs*, 93, 153
Sagisaka, S. 1986, *Il Nuovo Cimento C*, 9, 4, 809
Mendonça, R. R. S. *et al.* 2016, *ApJ*, 830, 88

Living Around Active Stars
Proceedings IAU Symposium No. 328, 2016
D. Nandy, A. Valio & P. Petit, eds.

© International Astronomical Union 2017
doi:10.1017/S1743921317003787

Coherent Synchrotron Radiation in Laboratory Accelerators and the Double-Spectral Feature in Solar Flares

Wellington Cruz[1], Sérgio Szpigel[1], Pierre Kaufmann[1], Jean-Pierre Raulin[1] and Michael Klopf[2]

[1]CRAAM, Mackenzie University, 01302-907, São Paulo, Brazil
[2]Institute of Radiation Physics, Helmholtz-Zentrum, Dresden, Germany

Abstract. Recent observations of solar flares at high-frequencies have provided evidence of a new spectral component with fluxes increasing with frequency in the sub-THz to THz range. This new component occurs simultaneously but is separated from the well-known microwave spectral component that maximizes at frequencies of a few to tens of GHz. The aim of this work is to study in detail a mechanism recently suggested to describe the double-spectrum feature observed in solar flares based on the physical process known as *microbunching instability*, which occurs with high-energy electron beams in laboratory accelerators.

Keywords. Solar flares, Coherent synchrotron radiation, Microbunching instability

---

1. Introduction

Several events observed at high-frequencies (0.2 and 0.4 THz) using the Solar Sub-millimeter Telescope (SST) at the El Leoncito Observatory in the Argentinean Andes have shown clear evidence of a new spectral burst component with fluxes increasing with frequency in the sub-THz range (Kaufmann *et al.* 2002; Kaufmann *et al.* 2004; Silva *et al.* 2007; Fernandes *et al.* 2016). This new component occurs simultaneously but is distinct from the well-known microwave spectral component that maximizes at frequencies of a few to tens of GHz. More recently, impulsive bursts have been observed in the mid-infrared at 30 THz using small telescopes with a relatively simple optical setup at the El Leoncito Observatory and at Mackenzie Presbyterian University in São Paulo, exhibiting fluxes considerably larger than those measured for the concurrent microwave and sub-THz frequencies (Kaufmann *et al.* 2013; Kaufmann *et al.* 2015; Miteva *et al.* 2016).

A number of interpretations based on different emission mechanisms have been suggested to explain the THz spectral component (see Krucker *et al.* (2013) for a detailed review), but none of them account for the microwave spectral component which is simultaneously observed. An alternative possibility which has been recently investigated (Kaufmann & Raulin 2006; Klopf *et al.* 2010; Klopf *et al.* 2014) is that both spectral components can be produced by a single beam of accelerated electrons undergoing the process known as *microbunching instability*, which occurs in laboratory accelerators (Williams 2002). Such a process is responsible for the production of synchrotron radiation with a double-spectrum structure similar to that observed in several solar flares, showing a broadband coherent synchrotron radiation (CSR) component and a distinct incoherent synchrotron radiation (ISR) component with maximum at higher frequencies.

2. Microbunching Instability and the ISR/CSR Mechanism

Extremely bright photon beams in the form of ISR are produced in laboratory accelerators by highly relativistic electrons moving through a dipole magnet. Several techniques have been developed to further enhance the brightness of the ISR photon beam (Friedman & Herndon 1973), such as the use of periodic magnetic structures known as insertion devices (*wigglers* or *undulators*).

Under certain conditions, instabilities due to inhomogeneities of the magnetic field or wave-particle interactions, which arise from the feedback between the insertion device, the radiation field and the accelerated electron beam, can produce modulations of the density of electrons, generating spatial structures called *microbunches* (Williams 2002). At wavelengths comparable to or longer than the size of the microbunch, the near field of the radiation from each electron overlaps the entire microbunch structure, resulting in a multiparticle coherent interaction which produces the emission of broadband CSR.

The spectrum of synchrotron radiation emitted by a microbunch of highly relativistic electrons accelerated in a dipole magnetic field is derived by generalizing the results obtained from the classical theory of electrodynamics for a single radiating electron to a system with multiple electrons. For a monoenergetic electron beam with discrete microbunches, the spectral intensity (energy radiated per unit of solid angle per unit of frequency) emitted by a microbunch containing N_e electrons is given by (Williams 2002):

$$\frac{d^2W}{d\nu d\Omega} = \left\{ N_e[1 - f(\nu)] + N_e^2 f(\nu) \right\} I_e(\nu) , \qquad (2.1)$$

where $I_e(\nu)$ is the single-electron spectral intensity for synchrotron emission at frequency ν and $f(\nu)$ is a form factor defined from the size and the shape of the microbunch structure, which is given by the square of the Fourier transform of the normalized longitudinal spatial charge distribution $S(z)$ within the microbunch structure,

$$f(\nu) = \left| \int_{-\infty}^{\infty} \exp\left[i\, 2\pi\nu(\hat{n} \cdot z)/c \right] S(z)dz \right|^2 . \qquad (2.2)$$

For a gaussian-shape microbunch structure with a characteristic length scale σ_b, as it is often assumed in a laboratory accelerator scenario, the form factor is given by $f(\nu) = \exp[-4\pi^2 \tau_b^2 \nu^2]$, where $\tau_b = \sigma_b/c$ is a characteristic time-width scale. One should note that in a solar flare accelerator scenario the longitudinal spatial charge distribution $S(z)$ may be far more complex than the gaussian function usually assumed in a laboratory accelerator. Simulations considering different analytical solutions for $S(z)$ were tested (Klopf *et al.* 2010; Klopf *et al.* 2014), showing that particularly good fits can be obtained using hyperbolic secant-shape microbunch structures, for which the form factor is given by $f(\nu) = sech[\pi\tau_b\nu/2]$.

3. Simulations of the ISR/CSR mechanism for THz burst events

In the simulations of the ISR/CSR mechanism presented here, we have considered a beam of accelerated electrons with kinetic energies in the range from E_{min} to E_{max} following a power-law distribution $n(E) = A\, E^{-\delta}$, where δ is the spectral index and A is a normalization constant such that $\int n(e)dE = 1$. We have also considered that only electrons with kinetic energies above a certain threshold E_{th} (set at a few MeV) can form microbunch structures because of Coulomb repulsion (Ingelman & Siegbahn 1998). In this way, we assume that the number of electrons participating in the CSR process is just a fraction x of the number of high-energy electrons, i.e. $N_{CSR} = xN_{high}$.

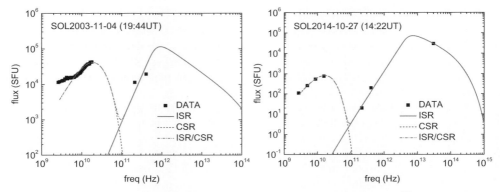

Figure 1. Fits to the spectra of the solar flares SOL2003-11-04 ($N_e = 1.0 \times 10^{35}$; $E_{min} = 50\ keV$; $E_{max} = 100\ MeV$; $\delta = 2.5$; $\phi_s = 0.52''$; $n_p = 10^9\ cm^{-3}$; $B = 1000\ G$; $\theta = 60°$; $N_{CSR}/N_{high} = 5.2 \times 10^{-15}$; $E_{th} = 5.0\ MeV$; $\tau_b = 67\ ps$) and SOL2014-10-27 ($N_e = 1.3 \times 10^{35}$; $E_{min} = 100\ keV$; $E_{max} = 100\ MeV$; $\delta = 2.3$; $\phi_s = 0.031''$; $n_p = 10^9\ cm^{-3}$; $B = 1000\ G$; $\theta = 60°$; $N_{CSR}/N_{high} = 1.2 \times 10^{-17}$; $E_{th} = 5.0\ MeV$; $\tau_b = 80\ ps$). The individual contributions from the ISR and the CSR components are also shown.

In Figure 1 we show the fits to the spectra of the solar flares SOL2003-11-04 (19:44UT) and SOL2014-10-27 (14:22UT), from microwave to THz frequencies, obtained through simulations of the ISR/CSR mechanism considering a hyperbolic secant-shape microbunch structure. The model parameters adjusted to fit the spectra are: the number of accelerated electrons N_e, the spectral index δ, the minimum electron kinetic energy E_{min}, the maximum electron kinetic energy E_{max}, the source angular size ϕ_s, the viewing angle θ, the magnetic field strength B, the plasma density n_p, the fraction of high-energy electrons participating in the CSR process N_{CSR}/N_{high}, the energy threshold for microbunch formation E_{th} and the characteristic time-width scale of the microbunch structure τ_b.

4. Conclusion

We have shown through numerical simulations that the ISR/CSR mechanism can provide a plausible explanation to the double-spectrum observed for several solar flares. Using typical flaring parameters and power-law energy distributions, we have obtained remarkable good fits to the spectra of the solar flares SOL2003-11-04 and SOL2014-10-27.

References

Kaufmann, P. *et al.* 2002, *Astrophys. J.*, v. 574, p. 1059-1065
Kaufmann, P. *et al.* 2004, *Astrophys. J. Lett.*, v. 603, n. 2, p. L121
Silva, A. V. R. *et al.* 2007, *Sol. Phys.*, v. 245, p. 311-326
Fernandes, L. O. T. *et al.* 2016, *ASP Conf. Series*, v. 504, p. 87
Kaufmann, P. *et al.* 2013, *Astrophys. J.*, v. 768, p. 134
Kaufmann, P. *et al.* 2015, *J. Geophys. Res.: Space Phys.*, v. 120, n. 6, p. 4155-4163
Miteva, R. *et al.* 2016, *Astron. Astrophys.*, v. 586, p. A91
Krucker, S. *et al.* 2016, *Astron. Astrophys.*, v. 21:58
Kaufmann, P. & Raulin, J.-P. 2006, *Phys. Plasmas*, v. 13, 070701
Klopf, J. M., Kaufmann, P., & Raulin, J.-P. 2010, *Bul. Am. Astron. Soc.*, v. 42, p. 905
Klopf, J. M., Kaufmann, P., Raulin, J.-P., & Szpigel, S. 2014, *Astrophy. J.*, v. 791, n. 1, p. 31
Friedman, M. & Herndon, M. 1973, *Phys. of Fluids*, v. 16, p. 1982-1995
Williams, G. P. 2002, *Rev. Sci. Instr.*, v. 73, p. 1461-1463
Ingelman, G. & Siegbahn, K. 1998, *Fysik-Aktuellt*, v. 1, p. 3

Living Around Active Stars
Proceedings IAU Symposium No. 328, 2016
D. Nandy, A. Valio & P. Petit, eds.

© International Astronomical Union 2017
doi:10.1017/S1743921317004069

The behavior of the spotless active regions during the solar minimum 23-24

Alexandre José de Oliveira e Silva[1] and Caius Lucius Selhorst[1,2]

[1]IP&D - Universidade do Vale do Paraíba (UNIVAP) - São José dos Campos, SP, Brazil
email: `ajoliveiraesilva@gmail.com`
[2]NAT - Universidade Cruzeiro do Sul - São Paulo, SP, Brazil
email: `caiuslucius@gmail.com`

Abstract. In this work, we analysed the physical parameters of the spotless actives regions observed during solar minimum $23 - 24$ ($2007 - 2010$). The study was based on radio maps at 17 GHz obtained by the Nobeyama Radioheliograph (NoRH) and magnetograms provided by the Michelson Doppler Imager (MDI) on board the Solar and Heliospheric Observatory (SOHO). The results shows that the spotless active regions presents the same radio characteristics of a ordinary one, they can live in the solar surface for long periods (> 10 days), and also can present small flares.

Keywords. Sun: sunspots, actives regions - Sun: radio radiation - Sun: magnetograms

1. Introduction

The study of the solar magnetic field dynamics is very importante to understand the phenomena which may affect the Earth environment. These dynamics have been daily monitored since the years ~ 1600 by the number of sunspot (SSN). Another classical index used to measure the solar activity is the radio flux at 10.7 cm ($F10.7$), which is generated at coronal heights and related to the presence of active regions and the occurrence of flares. The relationship between these indexes, that used to be linear, was destroyed in the last two cycles, which presented a SSN smaller than the expected by the measured $F10.7$ flux (Livingston et al. 2012).

Selhorst *et al.* (2014) studied the number of active regions observed by the NoRH (Nobeyama Radioheliograph) at 17 GHz between the years 1992 and 2013. During the quiet solar period between 2008 and 2009, they reported the presence of active regions during days without sunspots (about 33% of active days in the period). This fact was addressed to regions with magnetic field intensity less than 1500 G (Livingston *et al.* 2012). In this work, we investigated the physical parameters of the spotless actives regions observed during solar minimum period between 2007 and 2010.

2. Data Analyses and Results

In this study, the active regions at 17 GHz were identified, just as Selhorst *et al.* (2014), that is: i) size greater than 150 $pixels^2$ ($\sim 300 MSS$ (millionth solar surface)); ii) latitudes between $\pm 45°$; iii) maximum brightness temperature ($T_{b_{max}}$) at least 40% greater than the quiet Sun value and iv) longitudes $< 70°$. The active region size was calculated considering the pixels with T_b equal to 12000 K or more.

During the last minimum ($2007 - 2010$), our analyses find 92 days without sunspots, that presented active regions at 17 GHz. Since, an active region can live for long periods, those spotless active regions were identified as 48 distinct ones. During their lives, some

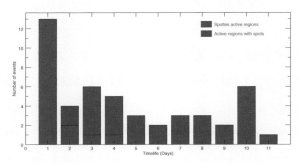

Figure 1. The lifetime of spotless active regions (in days). In red were represented the active regions that presented at least one day with sunspot associated.

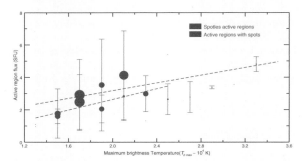

Figure 2. The relation between the active region flux and their maximum brightness temperature ($T_{b_{max}}$). The size of circles are proportional to the percentage of active regions in each group. Moreover, the active regions were separated in with or without spot, respectively, red and blue. The dashed lines are linear adjusts.

of these active regions presented days with sunspot associated, these days were also analysed (128 days). These weak active regions may also present flares, and a small one (class C1.1) was identified.

The lifetimes of the spotless active regions analysed are shown in Figure 1. Almost 50% of them (23) were ephemeral ones and still visible for a maximum of three days. The results show that the spotless active regions present a short lifetime but can also live on the solar surface for long periods (> 10 days). All of these active regions living 5 days or more presented at least one day with a sunspot associated to them.

The active regions presented a minimum brightness temperature of ~ 13700 K and the maximum brightness temperature of ~ 24500 K for the spotless days and 52600 K for thats ones sunspots. Moreover, the the spotless ones showed an average area of $325\ pixels^2$, whereas, those with sunspot were 35% greater ($440\ pixels^2$).

In Figure 2, we compare active region flux (in SFU) with their maximum brightness temperature. The active regions were separated in groups by their $T_{b_{max}}$, every 2000 K, and the result shows the increase of the flux with $T_{b_{max}}$ for both groups the spotless regions (blue circles) and for those with associated sunspot (blue circles). The mean flux difference is 0.36 SFU, however, this difference reaches 0.85 SFU when the $T_{b_{max}}$ is lower, where the largest number of spotless regions are concentrated, and decreases when the spotless active regions reached their maximum values (~ 25000 K).

To analise the active regions magnetic fields intensities, the magnetograms obtained by the MDI (Michelson Doppler Imager) were analysed. Due to sight line, the values of the maximum intensities of magnetic fields ($|B|_{max}$) were corrected by dividing the value obtained in the magnetogram by ($\cos(Lat.) \times \cos(Long.)$) (Schad & Penn 2010). Here, we

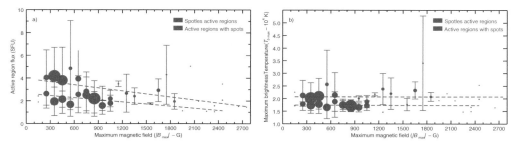

Figure 3. Comparison of the (a) active region flux (SFU) and (b) mean brightness temperature in relation to $|B|_{max}$.

used the absolute maximum magnetic field intensity ($|B|_{max}$) to characterise the active regions.

In the Figure 3, the active regions were grouped by their $|B|_{max}$ each 100 G. The spotless regions are plotted in blue and the active regions with spots in red. For each group, the maximum brightness temperature ($T_{b_{max}}$) and the active region flux were averaged for each group. The panel 3(a) shows a negative trend in the linear adjust for both, active regions with spots and without them, that is, as magnetic fields increase the flux tends to decrease. The averaged flux of the active regions with spots is 0.81 SFU greater than the spotless ones. In panel 3(b), $T_{b_{max}}$ still constant with the increase of the magnetic field. Moreover, the active regions with spots are 3400 K hotter than the spotless ones.

3. Final Remarks

A total of 48 distinct active regions were analysed in the period 2007–2010. About 50% of them were ephemeral living a maximum of three days. On the other hand, those ones living 5 days more presented at least one day with a sunspot. The active regions with sunspots are hotter and presented more flux than the spotless ones. However, the values were significantly smaller than the proposed by Livingston *et al.* (2012) for minimum necessary for the spot formation (1500 G), that could be due to instrumental differences.

Acknowledgements

We would like to thank the Nobeyama Radioheliograph, which is operated by the NAOJ/Nobeyama Solar Radio Observatory. A.J.O.S. acknowledge the scholarship form CAPES. C.L.S. acknowledge financial support from the São Paulo Research Foundation (FAPESP), grant #2014/10489-0.

References

Livingston, W., Penn, M. J., & Svalgaard, L. 2012, *ApJ*, 757, L8
Penn, M. J. & Livingston, W. 2006, *ApJ*, 649, L45
Schad, T. A. & Penn, M. J. 2010, *Sol. Phys.*, 262, 19
Selhorst, C. L., Costa, J. E. R., Giménez de Castro, C. G., *et al.* 2014, *ApJ* , 790, 134

Living Around Active Stars
Proceedings IAU Symposium No. 328, 2016
D. Nandy, A. Valio & P. Petit, eds.

© International Astronomical Union 2017
doi:10.1017/S1743921317004057

Analysis Of Kepler-71 Activity Through Planetary Transit

Eber A. Gusmão[1], Caius L. Selhorst[1,2] and Alexandre S. Oliveira[1]

[1]IP&D - Universidade do Vale do Paraíba - UNIVAP
São José dos Campos, SP, Brazil
email: `eber.gusmao@hotmail.com`
`alexandre@univap.br`
[2]NAT - Núcleo de Astrofísica Teórica - Universidade Cruzeiro do Sul
São Paulo, SP, Brazil
email: `caiuslucius@gmail.com`

Abstract. An exoplanet transiting in front of the disk of its parent star may hide a dark starspot causing a detectable change in the light curve, that allows to infer physical characteristics of the spot such as size and intensity. We have analysed the Kepler Space Telescope observations of the star Kepler-71 in order to search for variabilities in 28 transit light curves. Kepler-71 is a star with 0.923 $M_\odot$ and 0.816 $R_\odot$ orbited by the hot Jupiter planet Kepler-71b with radius of 1.0452 R_J. The physical parameters of the starspots are determined by fitting the data with a model that simulates planetary transits and enables the inclusion of spots on the stellar surface with different sizes, intensities, and positions. The results show that Kepler-71 is a very active star, with several spot detections, with a mean value of 6 spots per transit with size 0.6 R_P and 0.5 I_C, as a function of stellar intensity at disk center (maximum value).

Keywords. Stars: activity - Starspots - Planetary Systems

1. Introduction

More than 2,000 years ago, sunspots had already been reported by the Chinese, but their scientific study began with the advent of the telescope. The spots on the surface of the Sun were first observed with the aid of a telescope by Galileo four centuries ago. The sunspots are colder regions in the photosphere with a strong concentration of magnetic field lines. Furthermore, sunspots are important signatures of the cyclic nature of the star's magnetic field and wealth of information about solar activity. It is considered that other stars also are subject to the same magnetic activity. Nevertheless, nowadays it is not possible to observe or even monitor similar spots on the surface of other stars due to their size and distance.

When a planet moves in front of its parent star and is seen by an observer, the event is called transit. Through the continuous monitoring of these transits, it is possible to study the exoplanet eclipsing its parent star and if a dark stellar spot were occulted, a detectable variation in the light curve (positive variation) can be observed as shown in Fig. 1. From modelling of these transits, it might be possible to infer the physical properties of the spots, such as size, intensity, position, and temperature (e.g., Silva 2003, Silva-Valio *et al.* 2010).

In this work, we have analysed 28 transit light curves of the star Kepler-71 in order to search for the starspot physical parameters and their variabilities.

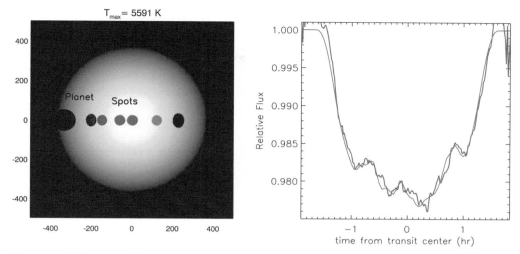

Figure 1. Left panel: Simulated image of the star Kepler-71 with quadratic limb darkening, 6 spots, and its planet, a hot Jupiter planet Kepler-71b with radius of 1.0452 R_J, assumed as a dark disk. Right panel: Observed light curve (blue) and the simulated transit light curve (red).

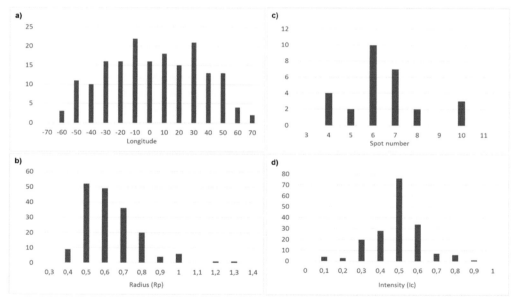

Figure 2. Histograms of the spot parameters obtained from the fits to the light curve transits: a) longitude in the stellar disk, b) radius in units of R_P, c) spot number per transit and d) intensity in units of I_C.

2. Data Analyses and Results

A planet around the star Kepler-71 was detected during one of the long run observations of a field toward the Galactic center performed by the Kepler satellite. A total of 28 transits were detected in the light curve with a high temporal resolution of 58 s, for a total of 144 days. Kepler-71 is a star with 0.923 $M_\odot$ and 0.816 $R_\odot$ orbited by the hot Jupiter (Kepler-71b) with a mean distance of 0.047 AU and a radius of 1.0452 R_J.

The limb darkening of the star as well as the physical characteristics of starspots are obtained by fitting the model described in Silva 2003. The stelar limb darking parameters was simulated with w_1=0.9 and w_2=0.9. The round spots are modeled by three parameters: (i) intensity, as a function of stellar intensity at disk center, I_C (maximum value); (ii) size, or radius, measured in units of planet radius, R_P; and (iii) longitude.

The figure 2 represents the synthesized star with spots of varying intensity (with respect to I_C) and radius (in units of R_P) and longitude. A histogram of the spot longitudes is shown in the Fig. 2 (a). These are topocentric longitudes, that is, they are not the ones located on the rotating frame of the star, but rather are measured with respect to an external reference frame. The number of spots for each transit is shown in Fig. 2 (c) and varied in the range from 4 to 10 with a mean value of 6. For spot radius, the distribution of spot radius obtained from the fits to all transit data is shown in Fig. 2 (b). The results show that the radius of the modelled spots varies from 0.4 to 1.3 R_P with a mean value of 0.6 R_P. Spots with lower intensity values, or higher contrast spots, are spots cooler than those with intensity values close to I_C. The spot intensities obtained from the model are shown in Fig. 2 (d). The figure shows that the spot intensities range from 0.1 to 0.9 I_C with a mean value of 0.5 I_C.

3. Conclusion

The star was evaluated using a model having up to 10 spots at a given time on its surface at certain location (latitude and longitude) during 144 days of observation by Kepler Telescope. During this period a total of 28 transits were detected. Kepler-71 is a very active star, and many intensity variations were identified in each transit (see Fig. 2), implying that there are many spots present on the surface of the star at any given time. The results show spot detections, with a mean value of 6 spots per transit with size 0.6 R_P and 0.5 I_C, as a function of stellar intensity at disk center (maximum value). The spots on Kepler-71 has diameters in a rough order of magnitude of 44.000 km. The mean surface of star area covered by spots within the transit latitudes is in the range of 40%. It was observed that most of spots are smaller than the planet Kepler-71b. The values obtained here can be compared with the star CoRoT-2 (Silva-Valio et al.2010). Both star presents high activity considering the number of spots as well as spot size.

Acknowledgements

E.A.G. acknowledges a CAPES scholarship. C.L.S. acknowledge financial support from the São Paulo Research Foundation (FAPESP), grant number 2014/10489-0.

References

Silva, A. V. R. 2003, *ApJL*, 585, L147
Silva-Valio, A., Lanza, A. F., Alonso, R., & Barge, P. 2010, *A&A*, 510, A25

Living Around Active Stars
Proceedings IAU Symposium No. 328, 2016
D. Nandy, A. Valio & P. Petit, eds.

© International Astronomical Union 2017
doi:10.1017/S1743921317003878

A large catalog of young active RAVE stars in the Solar neighborhood

Maruša Žerjal[1], Tomaž Zwitter[1], Gal Matijevič[2] and RAVE Collaboration[3]

[1]Faculty of Mathematics and Physics, University of Ljubljana,
Jadranska 19, 1000 Ljubljana, Slovenia
email: marusa.zerjal@fmf.uni-lj.si
[2]Leibniz-Institut für Astrophysik Potsdam (AIP),
An der Sternwarte 16, D-14482, Potsdam, Germany
[3]https://www.rave-survey.org/

Abstract. The catalog of 38,000 chromospherically active RAVE dwarfs represents one of the largest samples of young active solar-like and later-type single field stars in the Solar neighbourhood. It was established from the unbiased magnitude limited RAVE Survey using an unsupervised stellar classification algorithm based merely on stellar fluxes (Ca II infrared triplet). Using a newly-calibrated age-activity relation, ∼15,000 active stars are estimated to be younger than 1 Gyr. Almost 2000 stars are presumably younger than ∼100 Myr and possibly still in the pre-main sequence phase, the latter being supported by their significant offset from the main sequence in the $N_{UV} - V$ versus $J - K$ space. 16,000 stars from the sample have positional and velocity vectors available (using TGAS parallaxes and proper motions and radial velocities from RAVE).

Keywords. stars: activity, chromospheres, emission-line, Be, pre–main-sequence, catalogs

1. Automated search for active field stars in large spectroscopic surveys

The latest RAVE data release (DR5, Kunder *et al.* 2016) includes 521,000 spectra of 458,000 southern sky stars with an unbiased magnitude limited selection function ($9 < I < 12$). Due to the large number of spectra, automated, possibly parameter free classification techniques are necessary to discover peculiar objects, e.g., active stars. Their additional flux in the strongest spectral lines (e.g., Ca II IRT in RAVE) makes chromospherically active stars easily recognized by spectral classification techniques, such as locally linear embedding (LLE). Because LLE is a general dimensionality reduction procedure that conserves relations between the neighboring points of the high-dimensional manifold, a selected spectrum in the projected (2D) space is surrounded by its neighbors from the high-dimensional space (for more details on classification of the RAVE spectra see Matijevič *et al.* 2012). In view of a single observation with a moderate signal-to-noise ratio (S/N > 20) and a mid-range resolution being sufficient this approach enables the discovery of a vast number of young, chromospherically active field stars in large spectroscopic surveys.

In RAVE, 38,000 dwarfs were recognized as active candidates (Žerjal *et al.* 2013) with activity levels continuously increasing from marginally active stars to individual cases with emission peaks exceeding the continuum level. Many of the most active stars from the sample show strong X-ray (ROSAT) and ultraviolet excess emission in addition to their position off the main sequence in the $J - K$ (2MASS) versus $N_{UV} - V$ diagram

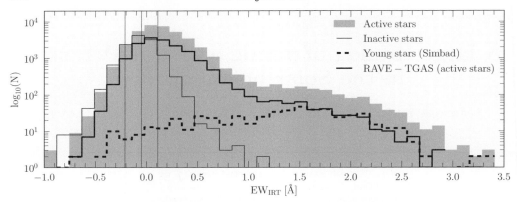

Figure 1. The distribution of activity levels is bimodal: there is 22,000 stars (58%) above 1σ and 7000 stars (18%) above 3σ activity detection level. Stars marked as young in the Simbad database (e.g., T Tau, Young, Pre-main sequence objects) mostly coincide with the more active peak of the distribution. Vertical lines in the plot mark $\pm 1\sigma$ deviation. Image adapted from Žerjal *et al.* 2016.

(using N_{UV} from GALEX and V from APASS) where they overlap with the reference pre-main sequence RAVE stars (according to the Simbad database).

2. Parameter-free data-driven characterization of activity levels in the Ca II IRT

In order to quantitatively characterize emission levels a large database of 12,000 spectra of inactive RAVE dwarfs was used as a template library. After the cores of the calcium lines were removed (± 2.5 Å from the center) from both active and inactive spectra, each normalized active spectrum was compared to the inactive database to find its nearest neighbors. Thanks to the large database of inactive stars covering the entire parameter space (effective temperature, surface gravity, metallicity etc.) it was possible to find inactive counterparts with the same stellar parameters for each active spectrum. The procedure was parameter-free: no atmospheric parameters were used to avoid biases originating from the pipeline designed for inactive dwarfs and giants. Another advantage of the measured inactive template database over synthetic spectra is the absence of the non-LTE problems in the cores of the calcium lines where emission is present.

The sum of the equivalent widths of the disentangled emission flux $\mathrm{EW_{IRT}} = \mathrm{EW_{8498}} + \mathrm{EW_{8542}} + \mathrm{EW_{8662}}$ is used as a proxy for activity levels (Žerjal *et al.* 2013, Figure 1). No photospheric correction is needed and the metallicity term is reduced. The typical uncertainty of the $\mathrm{EW_{IRT}}$ estimation is 0.16 Å.

3. Age–activity relation

It is well known that stellar activity diminishes with age (e.g., Mamajek & Hillenbrand 2008). Age estimates from the literature for 137 active RAVE stars (mostly cluster, moving group and association members) enabled the age-$\mathrm{EW_{IRT}}$ calibration in the range from ~ 1 Gyr down to a few 10 Myr (Žerjal *et al.* 2016, Figure 2). Although the scatter is large due to time variability, the saturation of activity in the youngest stars and age uncertainties, the anti-correlation between the two is clear. The data are divided into three main age–activity regimes: $\geqslant 1$ Gyr ($\mathrm{EW_{IRT}} < 0.25$ Å), between 0.1 and 1 Gyr (0.25 Å $< \mathrm{EW_{IRT}} < 0.75$ Å) and younger than 100 Myr ($\mathrm{EW_{IRT}} \geqslant 0.75$ Å).

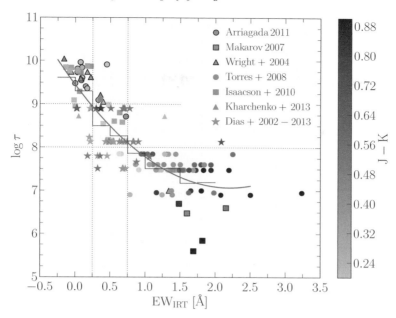

Figure 2. Age–EW_{IRT} activity calibration using 137 RAVE dwarfs with known ages. Most of the stars are cluster, moving group and association members. Although the scatter is large the correlation is clear. Thin black line shows an averaged age for a given activity bin while thick solid (red) line is a parabolic fit to the data. Dotted lines mark the three main activity–age classes. Figure adapted from Žerjal *et al.* 2016.

~15,000 RAVE field dwarf stars are shown to be younger than 1 Gyr and ~2000 younger than 100 Myr.

A combination of an efficient automated stellar classification algorithm and the age–activity relation offers an opportunity to build a young active candidate database and to perform further follow-up studies of dwarfs that possibly harbour exoplanets. The active database will be vastly enhanced with emission catalogs from the ongoing and future surveys, e.g. Galah (De Silva *et al.* 2015), FunnelWeb (Lawson *et al.* 2016) and Gaia (covers Ca II IRT as well).

References

De Silva, G. M., Freeman, K. C., Bland-Hawthorn, J., *et al.* 2015, *MNRAS*, 449, 2604
Kunder, A., Kordopatis, G., Steinmetz, M., *et al.* 2016, arXiv:1609.03210
Lawson, W., Murphy, S., Tinney, C. G., Ireland, M., & Bessell, M. S. 2016, *American Astronomical Society Meeting Abstracts*, 228, 217.08
Mamajek, E. E. & Hillenbrand, L. A. 2008, *ApJ*, 687, 1264-1293
Matijevič, G., Zwitter, T., Bienaymé, O., *et al.* 2012, *ApJS*, 200, 14
Žerjal, M., Zwitter, T., Matijevič, G., *et al.* 2013, *ApJ*, 776, 127
Žerjal, M., Zwitter, T., Matijevič, G., *et al.* 2016, *ApJ*, accepted.

Living Around Active Stars
Proceedings IAU Symposium No. 328, 2016
D. Nandy, A. Valio & P. Petit, eds.

© International Astronomical Union 2017
doi:10.1017/S174392131700391X

Rossby numbers of fully convective and partially convective stars

Natália R. Landin[1] and Luiz T. S. Mendes[2]

[1] Campus UFV Florestal - Universidade Federal de Viçosa,
CEP 35690-000, Florestal-M.G., Brazil
email: nlandin@ufv.br
[2] Dept. de Engenharia Eletrônica, Universidade Federal de Minas Gerais,
CEP 31270-901, Belo Horizonte-M.G., Brazil
email: luiztm@cpdee.ufmg.br

Abstract. In this work, we investigate the stellar magnetic activity in the theoretical point of view, through the use of stellar structure and evolution models. We present theoretical values of convective turnover times and Rossby numbers for low-mass stars, calculated with the ATON stellar structure and evolution code. We concentrate our analysis on fully convective and partially convective stars motivated by recent observations of X-ray emission of slowly rotating fully convective stars, which suggest that the presence of a tachocline is not a central key for magnetic fields generation. We investigate the behavior of the convective turnover time evolution, as well as its radial profile inside the star. A discussion about the location where the convective turnover time is calculated in the stellar interior is also addressed. Our theoretical results are compared to observational data from low-mass stars.

Keywords. stars: rotation, convection, stars: activity, stars: magnetic fields.

1. Introduction

Stars of different spectral types and ages host large-scale magnetic fields, as evidenced by observable phenomena like star spots, flares, chromospheric and coronal emissions, all of which express in some way stellar magnetic activity. Activity in late-type stars is well correlated with rotation, but the activity-rotation correlation is usually better described in terms of the Rossby number R_o, defined as the ratio of the rotational period P_{rot} to the convective turnover time τ_c ($R_o{=}P_{rot}/\tau_c$). Though τ_c can only be assessed theoretically, by using stellar models ($\tau_c{=}\alpha H_p/v_c$, where α is the Mixing Length of Theory's parameter, H_p is the pressure scale height and v_c is the convective velocity), it has been customary in the literature to use its empirical value, obtained as a function of the $B-V$ color index (Noyes *et al.* 1984). Recent observations by Wrigth & Drake, 2016 of X-ray emission of slowly rotating fully convective stars indicate that partially and fully convective stars follow the same rotation-activity relationship and they operate very similar rotation-dependent dynamos. This implies that the presence of a tachoclina is not a central key for magnetic fields generation. In order to contribute to the understanding of the dynamo mechanism existing in fully and partially convective stars, we present theoretical values of τ_c and *Ro* for different stellar masses and ages calculated with the ATON stellar evolution code.

For a given stellar mass and age, the convective turnover time changes significantly depending on the location inside the star in which it is calculated. The location more commonly used in the literature is one half mixing length above the base of the convective zone, but this standard location seems to be not suitable for young very low mass stars. We, then, evaluate τ_c throughout the whole star at different evolutionary stages, in order

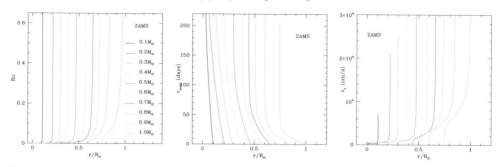

Figure 1. Profiles of Ro (left), τ_c (middle) and v_c (right) for each stellar mass at the ZAMS.

to try to set this location in a alternative way. The convective turnover time profile is also investigated and our values are compared with those available in the literature. Finally, observational data from low-mass stars are used in order to test our theoretical results.

2. Models

In the version of ATON code we use in this work, convection is treated according to the Mixing Length of Theory (with $\alpha = 2$, the parameter that represents the convection efficiency) and non-grey surface boundary conditions is used. Rotating models were generated by assuming differential rotation (Mendes *et al.* 1999). The initial angular momentum of each model is obtained according to the Kawaler, 1987 relation

$$J_{\mathrm{kaw}} = 1.566 \times 10^{50} \left(\frac{M}{M_\odot} \right)^{0.985} \quad \text{cgs.} \tag{2.1}$$

The evolutionary tracks were computed in the mass range of 0.1-1.0 $M_\odot$. We adopted the solar chemical composition with $X = 0.7155$ and $Z = 0.0142$. More details about the physics of the models can be found in Landin *et al.* 2006.

3. Results and Discussions

First, we analyze how Rossby numbers vary inside the stars for all models at the ZAMS.

From the left panel of Fig. 1, we can see that the Rossby number profiles are very steep, mainly for lower mass stars. This behavior depends on the convective turnover time profiles in the stellar interiors at the ZAMS (middle panel of Fig. 1), which in their turn, depend on convective velocities shown in the right panel of Fig. 1. Masses increase from left ($0.1M_\odot$) to right (1.0 $M_\odot$). In the left panel of Fig. 3, is the opposite.

Now, we analyze how Ro and τ_c vary as stars age. Plots of Ro, τ_c and v_c versus age, calculated at the standard location, are shown in Fig. 2†. For main sequence and pre-main sequence models with $M<0.4M_\odot$, the mixing length is so large that the place where τ_c should be calculated is larger than the stellar radius. In these cases, τ_c, v_c and Ro values are not calculated. Our results are in agreement with those by Kim & Demarque, 1996.

Our calculations indicate that the usual location used to determine Ro is not suitable for young fully convective stars. We, then, have taken our τ_c values, computed at the standard location for ZAMS models with $M\geqslant0.4M_\odot$, and have evaluated them in terms of the stellar radius and the local H_p. After that, we have made a linear extrapolation

---

† Masses increase from bottom to top (left and right panel of Fig. 2 and middle and right panel of Fig. 3). In the middle panel of Fig. 2, is the opposite. Color figures in the online version.

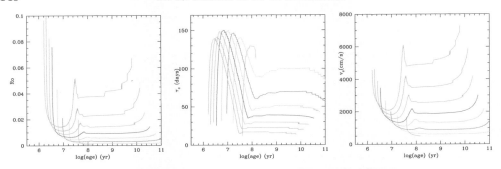

Figure 2. Ro, τ_c and v_c as a function of age and stellar mass.

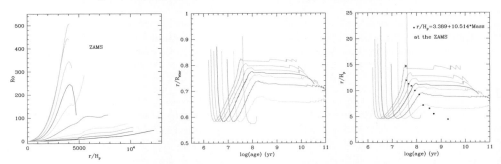

Figure 3. Ro profiles for all models at the ZAMS (left). Location, in terms of the stellar radius (middle) and H_p (right), where the Ro is calculated as a function of mass and age.

of these τ_c values for lower mass stars and have found a suitable location to calculate τ_c as a funcion of mass (see the right panel of Fig. 3). In this way, we keep the equivalence between the standard location and the location derived in this work.

In the left panel of Fig. 3, we show Rossby number profiles at the ZAMS. According to Fig. 3 (middle), this derived location should be among 0.6 and $0.9R_\odot$. Points in the right panel of Fig. 3 show where Ro is calculated in terms of H_p, at the ZAMS, for $M \geqslant 0.4M_\odot$ (standard location) and for $M < 0.4M_\odot$ (extrapolation). As τ_c do not vary significantly during the main sequence (see middle panel of Fig. 2), we can use our new location to determine τ_c values for fully convective stars in this phase of evolution. By using τ_c values of stars with $M < 0.4M_\odot$ at these derived locations, we are able to reproduce the trend that fully convective stars follow in Wrigth & Drake, 2016.

Acknowledgments

The authors thank D'Antona & Mazzitelli for allowing them to use their ATON code. IAU 328 organizing committee is also acknowledged.

References

Kawaler, S. D. 1987, *PASP*, 99, 1322

Kim, Y. C. & Demarque, P. S. 1996, *ApJ*, 457, 340

Landin N. R., Ventura, P. & D'Antona, F., Mendes L. T. S., & Vaz L. P. R.. 2006, *A&A*, 456, 269

Mendes, L. T. S.., D'Antona, F., & Mazzitelli, I. 1999, *A&A*, 341, 174

Noyes, R. W., Hartmann, S.,W., Baliunas, S., Duncan, D. K., & Vaughan, A. 1984, *ApJ*, 279, 763

Wright, N. J. & Drake, J. J. 2016, *Nature*, 535, 526

Living Around Active Stars
Proceedings IAU Symposium No. 328, 2016
D. Nandy, A. Valio & P. Petit, eds.

© International Astronomical Union 2017
doi:10.1017/S1743921317003842

Modelling coronal electron density and temperature profiles of the Active Region NOAA 11855

J. M. Rodríguez Gómez[1], L. E. Antunes Vieira[1], A. Dal Lago[1], J. Palacios[2], L. A. Balmaceda[3] and T. Stekel[1]

[1]National Institute for Space Research (INPE),
Avenida dos Astronautas-12227-010, São José dos Campos-SP, Brazil
email: jenny.gomez@inpe.br

[2]Departamento de Física y Matemáticas, Universidad de Alcalá University Campus, Sciences Building, P.O. 28871, Alcalá de Henares, Spain

[3]Instituto de Ciencias Astronómicas de la Tierra y el Espacio, ICATE-CONICET, Avda. de España Sur 1512, J5402DSP, San Juan, Argentina.

Abstract. The magnetic flux emergence can help understand the physical mechanism responsible for solar atmospheric phenomena. Emerging magnetic flux is frequently related to eruptive events, because when emerging they can reconnected with the ambient field and release magnetic energy. We will use a physic-based model to reconstruct the evolution of the solar emission based on the configuration of the photospheric magnetic field. The structure of the coronal magnetic field is estimated by employing force-free extrapolation NLFFF based on vector magnetic field products (SHARPS) observed by HMI instrument aboard SDO spacecraft from Sept. 29 (2013) to Oct. 07 (2013). The coronal plasma temperature and density are described and the emission is estimated using the CHIANTI atomic database 8.0. The performance of the our model is compared to the integrated emission from the AIA instrument aboard SDO spacecraft in the specific wavelengths 171Å and 304Å.

Keywords. Sun: abundances, atmosphere, magnetic field.

1. Introduction

Flux emergence phenomena play a key role in the dynamics of the solar atmosphere. The physical processes involved in flux emergence contribute to a better understanding of magnetic evolution, transport processes, solar dynamo and emission over the solar cycle. Also, the problem of active region heating requires precise measurements of plasma parameters such as density and temperature (Tripathi *et al.* 2008), but direct measurements are difficult to acquire. Modelling these parameters from photospheric magnetic field are necessary. It is important to understand the relationship of coronal structures, magnetic field, and emission in specific wavelength and their relationship with the Earth's atmosphere. Here we show some preliminary ideas because this is a work is in progress.

2. Data

We have used data from SDO/HMI SHARPs (Bobra *et al.* 2014 and Hoeksema *et al.* 2014), from Sept. 29 (2013) to Oct. 07 (2013) related to the AR NOAA 11855. Images were corrected and aligned with standard procedures through SolarSoftWare (SSW). We used a data cube every six hours, ie. at 05:59 UT, 11:59 UT, 17:59 UT and 23:59 UT. We used the AIA images for some specific wavelengths 171Å and 304Å during

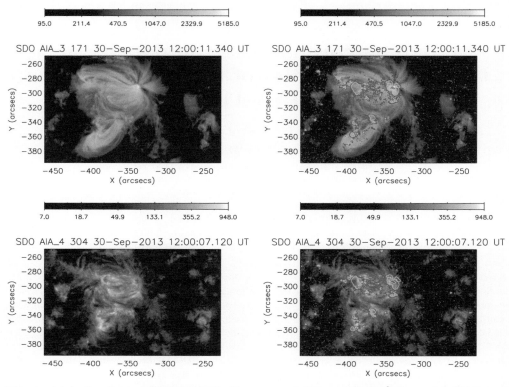

Figure 1. Active Region NOAA 11855. Upper panel: Images at 171Å from AIA/SDO and composite images using photospheric magnetic field from HMI/SDO for the same region Sept. 30 12:00:00 UT. Lower panel: Images at 304Å from AIA/SDO and composite images using photospheric magnetic field from HMI/SDO for the same region Sept. 30 12:00:00 UT.

the period of interest. The region is selected and extracted. The AIA/SDO images are then corrected and aligned with standard procedures through SolarSoftWare (SSW) (left panels in Fig. 1). We used SHARPS magnetic field data overimposed to AIA (right panels in Fig. 1).

3. Photospheric and coronal dynamics

Non-linear force-free-field extrapolation of the solar magnetic field is used through the optimization method available by NLFFF in SSW (Wheatland *et al.* 2000). The magnetic field through the lower corona was obtained from $1R_\odot$ to $1.15R_\odot$. This magnetic field is an important key to modelling plasma parameters through the solar corona. The electron density (N) and temperature (T) profiles are described using the following expressions: $N = N_o \left(B\right)^\gamma$ and $T = T_o \left(B\right)^{b-1} e^{-\left(\frac{B}{a}\right)^{(b-1)}}$. Temperature profile is based on the Weibull distribution function, where a, b and γ are proportional coefficients, N_o and T_o are background density and temperature. From the images of AIA/SDO (left panel in Fig. 1) it is possible to calculate the integrated intensity in different heights of the solar atmosphere. We will use an emission model based on the contribution function and integrated line-of-sight emission. This emission is compared to the emission from AIA/SDO. This procedure allows modelling the electron density and temperature in the emergence flux region until the AR is developed. This diagnosis can be characterised by

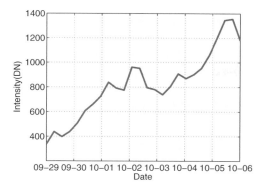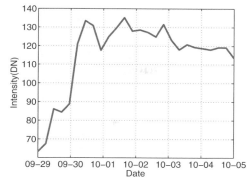

Figure 2. Integrated emission of NOAA 11855 from the images of the AIA instrument on board SDO spacecraft. Left panel: integrated intensity at 171Å AIA/SDO from Sept. 29 18 : 00 UT to Oct.06 18 : 00 UT. Right panel: Integrated intensity at 304Å AIA/SDO from $09 - 29$ 18 : 00UT to $10 - 05$ 18 : 00UT.

these plasma parameters and evaluate the relationship with the evolution of photospheric features (flux emergence and ARs).

4. Discussion

Using these specific wavelengths we will obtain a description of the emission in the solar corona related to the active region NOAA 11855. These wavelengths describe the emission from the hot loops in 171Å and from the active region in the lower transition region in 304Å. The emission is not linked to flares. The intensity increase is steeper in 304Å that in 171Å as we can see from the faster flux emergence emergence in the upper chromosphere, as compared to the corona (Fig. 2). In the next steps, we will include the extrapolated magnetic field in our model. We will obtain the density and temperature profiles and discuss the behavior of the emission in the lower solar corona related to the emergence flux region and Active Regions (ARs).

5. Acknowledgements

This work is partially supported by CNPq/Brazil under the grant agreement no. 140779/2015-9 and no. 304209/2014-7. JP acknowledges UAH travel grant for visiting INPE and MINECO project no. AYA2013-47735-P.

References

Bobra, M. G., Sun, X., Hoeksema, J. T., Turmon, M., Liu, Y., Hayashi, K., Barnes, G. & Leka, K. D. 2014, *Solar Phys.*, 289, 9

Hoeksema, J. T., Liu, Y., Hayashi, K., Sun, X., Schou, J., Couvidat, S., Norton, A., Bobra, M., Centeno, R., Leka, K. D., Barnes, G., & Turmon, M. 2014, *Solar Phys.*, 289, 3483

Tripathi, D., Mason, H. E., Young, P. R., & Del Zanna, G. 2008, *A&A*, 481, 1

Wheatland, M. S., Sturrock, P. A., & Roumeliotis, G. 2000, *ApJ*, 540, 2

Living Around Active Stars
Proceedings IAU Symposium No. 328, 2016
D. Nandy, A. Valio & P. Petit, eds.

Using planetary transits to estimate magnetic cycles lengths in Kepler stars

Raissa Estrela and Adriana Valio

Center for Radio Astronomy and Astrophysics (CRAAM), Mackenzie Presbyterian University,
Sao Paulo, Brazil
Rua da Consolacao 01301-000, 896, Sao Paulo, Brazil
email: `rlf.estrela@gmail.com`

email: `avalio@craam.mackenzie.br`

Abstract. Observations of various solar-type stars along decades showed that they could have magnetic cycles, just like our Sun. These observations yield a relation between the rotation period P_{rot} and the cycle length P_{cycle} of these stars. Two distinct branches for the cycling stars were identified: active and inactive, classified according to stellar activity level and rotation rate. In this work, we determined the magnetic activity cycle for 6 active stars observed by the Kepler telescope. The method adopted here estimates the activity from the excess in the residuals of the transit light curves. This excess is obtained by subtracting a spotless model transit from the light curve, and then integrating over all the residuals during the transit. The presence of long term periodicity is estimated from the analysis of a Lomb-Scargle periodogram of the complete time series. Finally, we investigate the rotation-cycle period relation for the stars analysed here.

Keywords. magnetic, activity, rotation.

1. Introduction

The magnetic activity of the Sun varies throughout the 22 year long magnetic cycle, with the polarity of its magnetic field flipping every 11 years. These cycles are identified by the frequency and number of sunspots in the solar surface, which acts as an indicator of activity. Stellar activity is also present in other stars, that show remarkable lightcurve variations due to starspots and other magnetic phenomena. Skumanich (1972) first suggested that the activity of the star was associated with its rotation rate, and consequently with it age. Therefore, young rapidly rotating stars show higher level of activity and can produce larger spots and energetic flares.

The Mount Wilson Observatory Ca II H K survey was the first to show that hundred of stars could also exhibit long and short periodic cycles between 2.5 and 25 years (Baliunas *et al.*, 1995). Using this data, Saar & Brandenburg (1999) established a relation between the stellar rotation period, P_{orb} and stellar cycle period, P_{cycle} as a function of the Rossby number. This relation divided the stars in two branches: active (A) and inactive (I), according to its activity level and rotation rate. The active sequence is composed by stars that rotates faster than the Sun, while the inactive one has slow rotating stars. Later, it was observed that some of the stars in the active branch also exhibits secondary short cycles that fall in the inactive branch (Böhm-Vitense, 2007).

Recently, the Kepler telescope provided long-term high photometric precision of thousand of stars. These data offer an unique opportunity to increase our understanding about magnetic cycles in other stars. In addition, it could improve stellar dynamo models in different type of stars. Among the studies of magnetic cyles using Kepler data, Vida & Olah (2013) analyzed fast-rotating stars and found activity cycles of 300-900 days for 9

Table 1.
Observational parameters of the stars

	Radius [$R_\odot$]	Age [Gyr]	Effective Temperature [K]	Rotation Period [days]	Reference
Kepler-17	1.05 ± 0.03	< 1.78	5780 ± 80	11.89	1,2
Kepler-63	$0.901^{+0.022}_{-0.027}$	0.2	5580 ± 50	5.40	3
KIC 9705459	$0.951^{+0.159}_{-0.04}$		5900^{+106}_{-125}	2.83	4,5
KIC 5376836	$0.885^{+0.363}_{-0.081}$		5903^{+93}_{-112}	~ 4	4,5
Kepler-96	1.02 ± 0.09	2.34	5690.0 ± 73.0	15.30	5,6
Hat-p-11 (Kepler-3)	0.75 ± 0.02	$6.5^{+4.1}_{-4.1}$	4780.0 ± 50.0	30	7

[1]**References.** (1) Bonomo *et al.* (2012), (2) Désert *et al.* (2011), (3) Sanchis-Ojeda *et al.* (2013), (4) MAST Kepler database, (5) Walkowicz & Basri (2013), (6) Marcy *et al.* (2014) and (7) Sanchis-Ojeda & Winn (2011).

targets, and Marthur *et al.* (2014) found evidences of magnetic cycle in two Kepler solar type star.

2. Spot model

Kepler-17 and Kepler-63 are two active solar-type stars that exhibits rotational modulations in their light curve caused by the presence of starspots, with a peak-to-peak variation of 6% and 4%, respectively. To analyze and characterize the physical parameters of the spots in these stars, we applied the transit model proposed by Silva (2003). This model simulates the passage of a planet in front of its host star. The simulation consider a star with quadratic limb darkening as a 2D image and the planet is assumed to be a dark disk with radius R_p/R_{star}, where R_p is the radius of the planet and R_{star} is the radius of the primary star. The sum of all the pixels in the image (star plus dark planet) yields the transit light curve.

We added round spots to the stellar surface and modelled them by three parameters: intensity (in function of the intensity at disc center I_c), radius (measured in units of planetary radius) and position (longitude and latitude). The latitude remains fixed and equal to the transit latitude. The longitude of the spots is limited to -70 and 70° from the central meridian to avoid any distortions caused by the ingress and the egress of the transit. In the case of Kepler-17, the latitude is -14°.6, while for Kepler-63b the planet occults several latitudes of the star from its equator all the way to the poles due to its high obliquity.

To obtain a better fit for each transit light curve, it was necessary to refine the values for the semi-major axis and planet radius obtained from literature for Kepler-17b (Désert *et al.*, 2011) and for Kepler-63b (Sanchis-Ojeda *et al.*, 2013). The physical parameters of the stars are described in Table 1.

To fit the spots, we subtracted a spotless model from the transit light curve. The result of this subtraction are the residuals, where the spots became more evident, as the "bumps" seen in the residuals. This process is illustrated in Fig. 1c for Kepler-17. We used the CDPP (Combined Differential Photometric Precision, see Christiansen *et al.* (2012) as an estimation of the noise in the Kepler data. Each quarter of the light curve has an associated CDPP value. Here, we considered the uncertainty in the data, σ, as being ten times the average of the CDPP values in all quarters. Only the "bumps" that exceed the detection limit of 10σ are assumed as spots and modelled. Finally, the best fit of the spots parameters is obtained by minimizing χ^2, calculated using the AMOEBA routine (Press *et al.*, 1992).

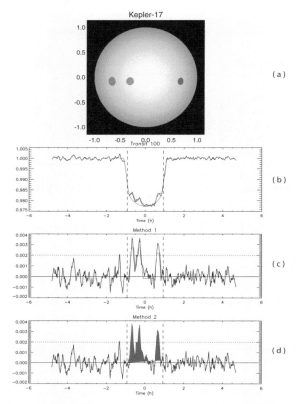

Figure 1. The 100th transit from Kepler-17 illustrates a typical example of the two methods adopted in this work: the spot model developed by Silva (2003) and transit residual excess. (a): Synthetised star with three spots. (b): Transit light curve with the model of a spotless star overplotted (red). (c) Residuals of the transit lightcurve after subtraction of a spotless star model. The red curve shows the fit to the data "bumps". (d): Integration (in red) of the residual excess resulted from the subtraction.

The determination of the stellar cycles was performed by using two approaches (Estrela & Valio (2016)). The first one is the analysis of the variation in the number of spots during the 4 years of observation of the Kepler stars. The latter is the calculation of the flux deficit resulting from the presence of spots on the star surface. The relative flux deficit of a single spot is the product of the spot contrast and its area, thus for each transit the total flux deficit associated with spots was calculated by summing all individuals spots:

$$F \approx \sum (1 - f_i)(R_{\mathrm{spot}})^2 \qquad (2.1)$$

where the spots contrast is taken to be $(1 - f_i)$, and f_i is the relative intensity of the spot with respect to the disk center intensity I_c. A value of $f_i = 1$ means that there is no spot at all.

Possible long duration trends were removed in these time series by applying a quadratic polynomial fit and then subtracting it. Then, a Lomb Scargle periodogram (LS) (Scargle, 1982) was applied on these time series to obtain the period related to the magnetic cycle. In addition, it was applied a significance test to quantify the significance of the peaks from the LS periodogram. The statistical significance associated to each frequency in the periodogram is determined by the p-value (p). The smaller the p-value, the larger the significance of the peak. We adopted the significance level α as being 3σ (p $\pm$ 0.0013).

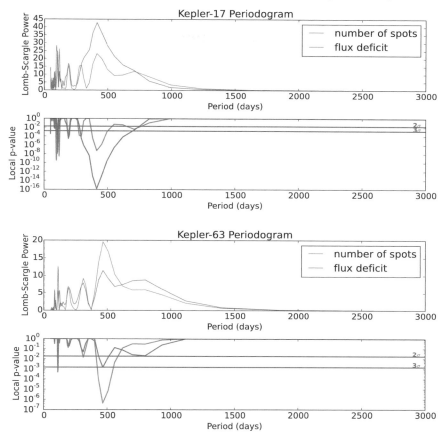

Figure 2. Lomb Scargle periodogram applied to the number of spots and total flux deficit of Kepler-17 (*top*) and Kepler-63 (*bottom*). The highest peak, indicated by a dashed line, corresponds to a periodicity of 410 ± 50 days for Kepler-17 and 460 ± 60 days for Kepler-63.

Each periodogram in Figure 2 has a significance test associated (plotted below). The uncertainty of the peaks in the periodogram is given by the FWHM of the peak power.

The LS periodogram detected a long term periodicity for both stars, as shown in Fig. 2. Kepler-17 shows a prominent peak at 410 ± 60 days (number of spots) and 410 ± 50 days (flux deficit), while Kepler-63 shows a periodicity of 460 ± 60 days for the total number of spots, and 460 ± 50 days for the flux deficit. Detailed analysis of the results using the spot modeling for these two stars are described in Estrela & Valio (2016).

3. Transit residuals excess

In the second method of this work we subtracted a modelled light curve of a star without spots from the transit light curves. The result from this subtraction is the residual that clearly shows the spots signatures. An example of this method is shown in Figure 1 for the 100th transit of Kepler-17. The excess in the residuals (in bold in Fig. 1d) corresponds to the spots signatures. Thus, we integrated the residual excess constrained to ± 70° longitude of the star (delimited by the vertical dashed lines of Fig.1). This

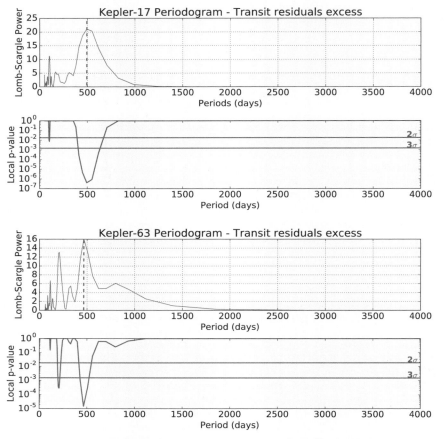

Figure 3. Lomb Scargle periodogram applied to the integrated transit residuals excess of Kepler-17 (*top*) and Kepler-63 (*bottom*). The highest peak, indicated by a dashed line, corresponds to a periodicity of 494 ± 100 days for Kepler-17 and 465 ± 40 days for Kepler-63.

allows us to characterize the magnetic activity level of the star. To remove any possible trends in this time series, we applied a quadratic polynomial fit and subtracted.

We applied the LS periodogram to the time series of integrated transit residuals excess. A peak at 490 ± 100 days was found for Kepler-17 and 460 ± 40 days for Kepler-63. The value obtained for Kepler-63 is similar to that from the first approach, and corresponds to a 1.27 year-cycle. On the other hand, the cycle period estimate for Kepler-17 agree within the uncertainty of the result from the first method.

Therefore, the results of both methods agree with each other. For this reason, we decided to work with more stars applying only the second method (transit residuals excess), which is easier in estimating magnetic cycles.

This method was applied to four more Kepler active stars hosting planets: Hat-p-11 (Kepler-3), Kepler-96, KIC 9705459 and KIC 5376836 (see details in Table 1). However, the planets candidates orbiting KIC 9705459 and KIC 5376836 were later classified as false positive, and the two stars are now known to be a binary eclipsing system. In this case, the transit of a companion star can also occult a spot in the stellar disk and show its signature, which works similar when eclipsed by a planetary transit. For this reason, we decided to keep these stars in our analysis. Then, by applying a Lomb Scargle periodogram, we found a clean peak at 305 ± 60 days for Hat-p-11. Kepler-96 and KIC

Table 2.
Magnetic activity cycle periods

Star	P_{cycle} (days)	P_{cycle} (years)
Kepler-17	490 ± 100	1.35 ± 0.27 yr
Kepler-63	460 ± 60	1.27 ± 0.16 yr
Hat-p-11 (Kepler-3)	305 ± 60	0.83 ± 0.16
Kepler-96	545 ± 128	1.50 ± 0.35
KIC 9705459	100 ± 9	0.27 ± 0.024
KIC 5376836	42 ± 3	0.11 ± 0.007

9705459 also showed a clean peak at 545 ± 128 days and 100 ± 9 days, respectively. Finally, KIC 5376836 showed a significant peak at 42 ± 3 days. These periodicities show a p-value below the 3σ significance level, confirming their significance. Table 2 shows a summary of the results found for the magnetic cycles of all stars analysed in this work.

4. Summary and Conclusions

We have estimated the period of the magnetic cycle, P_{cycle}, for two active solar-type stars, Kepler-17 and Kepler-63, by applying two new methods: spot modelling and transit residuals excess. Since the results of both methods agreed with each other, we used the second method to estimate the magnetic cycle of four more active stars observed by Kepler: Hat-p-11 (Kepler-3), Kepler-96, KIC 9705459 and KIC 5376836. This method is much faster to determine magnetic cycles because it only requires to integrate the area of the residuals due to the activity (spots) in the transit light curve. The first two stars have a transiting planet, while KIC 9705459 and KIC 5376836 had planet candidates, found later to be false positives. These stars were classified as eclipsing binary systems, and we used the primary transit in our analysis to detect spots.

The results found here, with an exception of KIC 5376836 and KIC 9705459, have a P_{cycle} within the same range of 300-900 days found for the 9 Kepler fast-rotators stars analyzed by Vida & Olah (2013). As we are constrained to the duration of observation ($\leqslant 4$ years) of the Kepler telescope, it is not possible to determine longer cycles.

Based on the relation between the stellar rotation period, P_{orb} and stellar cycle period, P_{cycle}, proposed by Saar & Brandenburg (1999), we can verify that Kepler-63, KIC 5376836 and KIC 9705459 fall within the active branch. Kepler-96 and Kepler-17 have their short cycles in the inactive branch, however, as proposed by Böhm-Vitense (2007) stars, the active branch may also shows secondary cycles that fall in the inactive branch. Finally, Hat-p-11 is a special case compared to the other stars in our sample. This star has a long rotation period ($P_{rot} \sim 30$ days) and has an age older than the Sun (6.5 Gyrs), but it shows remarkable activity. Probably, Hat-p-11 is interacting magnetically with its close orbit Hot-Jupiter, and for this reason its activity level is higher. The short cycle found for this star falls close to the inactive branch.

References

Baliunas, S. L., Donahue, R. A., Soon, W. H., *et al.* 1995, *ApJ*, 438, 269
Böhm-Vitense, E. 2007, *ApJ*, 657, 486
Bonomo, A. S., Hébrard, G., Santerne, A., *et al.* 2012, *A&A*, 538, A96
Christiansen, J. L., Jenkins, J. M., Caldwell, D. A., *et al.* 2012, *PASP*, 124, 1279
Désert, J.-M., Charbonneau, D., Demory, B.-O., *et al.* 2011, *ApJS*, 197, 14
Estrela, R. & Valio, A. 2016, *ApJ*, 831, 57

Marcy, G. W., Isaacson, H., Howard, A. W., *et al.* 2014, *ApJS*, 210, 20

Mathur, S., García, R. A., Ballot, J., Ceillier, T., Salabert, D., Metcalfe, T. S., Régulo, C., Jiménez, A., & Bloemen, S. 2014, *A&A*, 562A, 124M

Press, W. H., Teukolsky, S. A., Vetterling, W. T., & Flannery, B. P. 1992, Cambridge: University Press, —c1992, 2nd ed.,

Saar, S. H. & Brandenburg, A. 1999, *ApJ*, 524, 295

Sanchis-Ojeda, R. & Winn, J. N. 2011, *ApJ*, 743, 61

Sanchis-Ojeda, R., Winn, J. N., Marcy, G. W., *et al.* 2013, *ApJ*, 775, 54

Scargle, J. D. 1982, *ApJ*, 263, 835

Silva, A. V. R. 2003, *Ap. Lett.*, 585, L147

Skumanich, A. 1972, *ApJ*, 171, 565S

Vida, K. & Olah, R. 2013, *MNRS*, 441, 2744V

Walkowicz, L. M. & Basri, G. S. 2013, *MNRAS*, 436, 1883

Living Around Active Stars
Proceedings IAU Symposium No. 328, 2016
D. Nandy, A. Valio & P. Petit, eds.

© International Astronomical Union 2017
doi:10.1017/S1743921317003799

Modelling coronal electron density and temperature profiles based on solar magnetic field observations

J. M. Rodríguez Gómez[1], L. E. Antunes Vieira[1], A. Dal Lago[1], J. Palacios[2], L. A. Balmaceda[3] and T. Stekel[1]

[1]National Institute for Space Research (INPE), Avenida dos Astronautas-12227-010, São José dos Campos-SP, Brazil
email: jenny.gomez@inpe.br
[2]Departamento de Física y Matemáticas, Universidad de Alcalá University Campus, Sciences Building, P.O. 28871, Alcalá de Henares, Spain
[3]Instituto de Ciencias Astronómicas de la Tierra y el Espacio, ICATE-CONICET, Avda. de España Sur 1512, J5402DSP, San Juan, Argentina.

Abstract. The density and temperature profiles in the solar corona are complex to describe, the observational diagnostics is not easy. Here we present a physics-based model to reconstruct the evolution of the electron density and temperature in the solar corona based on the configuration of the magnetic field imprinted on the solar surface. The structure of the coronal magnetic field is estimated from Potential Field Source Surface (PFSS) based on magnetic field from both observational synoptic charts and a magnetic flux transport model. We use an emission model based on the ionization equilibrium and coronal abundances from CHIANTI atomic database 8.0. The preliminary results are discussed in details.

Keywords. Sun: magnetic fields, corona.

1. Introduction

The coronal electromagnetic emission provides important information about the dynamics and the characteristics of the solar corona and it is spatial and temporal variability. However, the determination of the plasma parameters, such as the electron density and temperature is difficult because the electromagnetic emission is affected due to the distribution is along the line of sight. Here we present a physical model to reconstruct the electron density and temperature profiles through the solar corona based on the configuration of the magnetic field. In particular, we study the evolution of the electron density and temperature during the two last solar cycles. For this purpose, we use magnetic field from synoptic charts of MDI/SOHO and HMI/SDO instruments.

2. Approach

In order to model electron density and temperature profiles in the solar corona, we used solar surface magnetic field from the surface flux model of Schrijver (2001). The surface flux model was updated each six hours. A diffusion model, the evolution of active regions and the transport process at the solar photosphere were considered. The magnetic field components B_r, B_θ, B_ϕ in each magnetic field line were obtained from PFSS (Schrijver & De Rosa (2003)). They were used to build the density and temperature profiles. The electron density (N) and temperature (T) profiles are described using the

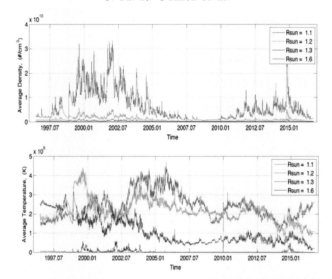

Figure 1. Upper panel: Temperature profile in different layers through the solar corona. Lower panel: Average density profile at different heights in the solar atmosphere.

following expressions:

$$N = N_o \, (B)^{\gamma} \tag{2.1}$$

$$T = T_o \, (B)^{b-1} \, e^{-\left(\frac{B}{a}\right)^{(b-1)}} \tag{2.2}$$

Temperature profile is based on the Weibull distribution function. Where a, b and γ are proportional coefficients, the magnitude of the magnetic field corresponds to $B = \sqrt{B_r^2 + B_\theta^2 + B_\phi^2)^2}$, N_o and T_o are the background density and temperature. We use an emission model based on the solar reference spectra. It is built using ionization equilibrium model Mazzotta *et al.* (1998) and the coronal abundances from Meyer (1985) at the central wavelength. The contribution function $G(\lambda, T)$, from CHIANTI atomic database 8.0 (Del Zanna *et al.* (2000)) is yielded. The emission at a specific wavelength is calculated in each voxel and integrated line-of-sight. This emission is compared to observational data from TIMED/SORCE and it provides the performance of our model.

3. Variability of the electron density and temperature during the solar cycle 23 and 24

The density and temperature profiles were obtained from the solar cycle 23 to 24 in different layers through the solar corona ($Rsun = 1.1$, 1.2, 1.3 and 1.6 $R_\odot$) are shown in Fig. 1. In this case we use the guess parameters: $\gamma = 1.9992$, $a = 2$, $b = 6$, $N_o = 294692000 \; cm^{-3}$, $T_o = 9 \times 10^6$ K, to obtain the density and temperature profiles.

4. Discussion

The temperature profiles show differences during solar cycles 23 and 24 at different heights in the solar corona (lower panel in Fig. 1) due to the relationship with the structure of the coronal magnetic field. The average temperature profile (Fig. 2) is in agreement with the trend of some features in the solar atmosphere shown in Fontenla *et al.* (2011).

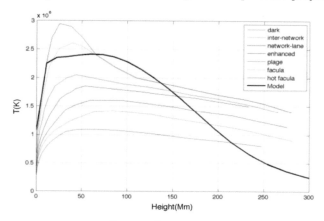

Figure 2. Average temperature profiles in all considered layers from our model (black line). Temperature profiles from Fontenla *et al.* (2011); in coloured lines, average temperature profiles from different structures of the solar photosphere are displayed.

The density profiles are related to magnetic flux variations over the solar cycles (upper panel in Fig. 1). This is the starting point for the study of long-term properties in the solar corona. In the next steps we will use the optimization algorithm to fit the model parameters.

5. Acknowledgements

The Authors want to acknowledge C. Schrijver and M. de Rosa. We would like to thank SOHO/MDI, SDO/HMI Data Science and Teams and the Chianti atomic database 8.0. This work is partially supported by CNPq/Brazil under the grant agreement no. 140779/2015-9 and no. 304209/2014-7. JP acknowledges UAH travel grant for visiting INPE and MINECO project no. AYA2013-47735-P.

References

Charbonneau, P. 1995, *ApJS*, 101, 309
Bobra, M. G., Sun, X., Hoeksema, J. T., Turmon, M., Liu, Y., Hayashi, K., Barnes, G., & Leka, K. D. 2014, *Solar Phys.*, 289, 9
Fontenla, J. M., Harder, J., Livingston, W., Snow, M., & Woods, T. 2011, *JGRD*, 116, D20, D20108
Hoeksema, J. T., Liu, Y., Hayashi, K., Sun, X., Schou, J., Couvidat, S., Norton, A., Bobra, M., Centeno, R., Leka, K. D., Barnes, G., & Turmon, M. 2014, *Solar Phys.*, 289, 3483
Mazzotta, P., Mazzitelli, G., Colafrancesco, S., & Vittorio, N. 1998, *A&A*, 133, 403
Meyer, J. 1985, *ApJS*, 57, 173
Schrijver, C. & De Rosa, M. 2003, *Solar Phys.*, 212, 165
Schrijver, C. 2001, *ApJ*, 547, 475
Tripathi, D., Mason, H. E., Young, P. R., & Del Zanna, G. 2008, *A&A*, 481, 1
Del Zanna,G., Dere, K. P., Young, P. R., Landi, E., & Mason, H. E. 2000, *A&A*, 582, A56
Wheatland, M. S., Sturrock, P. A., & Roumeliotis, G. 2000, *ApJ*, 540, 2

Living Around Active Stars
Proceedings IAU Symposium No. 328, 2016
D. Nandy, A. Valio & P. Petit, eds.

© International Astronomical Union 2017
doi:10.1017/S174392131700374X

Solar and stellar coronae and winds

Moira Jardine

SUPA, School of Physics and Astronomy, University of St Andrews, North Haugh,
St Andrews, KY16 9SS, UK
email: mmj@st-andrews.ac.uk

Abstract. Solar-like stars influence their environments through their coronal emis- sion and winds. These processes are linked through the physics of the stellar magnetic field, whose strength and geometry has now been explored for a large number of stars through spectropolari-metric observations. We have now detected trends with mass and rotation rate in the distribution of magnetic energies in different geometries and on also different length scales. This has impli-cations both for the dynamo processes that generate the fields and also for the dynamics and evolution of the coronae and winds. Modelling of the surface driving processes on stars of various masses and rotation rates has revealed tantalising clues about the dynamics of stellar coronae and their ejecta. These new observations have also prompted a resurgence in the modelling of stellar winds, which is now uncovering the range of different interplanetary conditions that exoplanets might experience as they evolve.

Keywords. stars:magnetic fields, stars:coronae, stars:imaging, stars:spots

---

1. Introduction

In the following I aim to focus mainly on the relationship between stellar coronae and winds, and in particular to explore the ways in which this is governed by the stellar magnetic field geometry. This relationship can be seen very clearly for the Sun when we look at the variation in the coronal X-ray emission and wind speed. At cycle minimum, most of the emission originates close to the equator, where the slow wind originates. The fast wind, in contrast, comes mainly from the polar regions. During the maximum phase of the sunspot cycle, however, a much more complex picture is seen, with fast and slow wind streams interspersed at a range of latitudes. This change in the morphology of both the closed and open field regions is driven by the evolution of the Sun's magnetic field over the course of its cycle. We can analyse these changes by looking at the variation of the dipole and quadrupole modes through the solar cycle. As shown by DeRosa *et al.* (2012), the strength of the quadrupole contribution varies in phase with the sunspot number, while the strength of the dipole mode is in antiphase (see also Vidotto (2016)).

Clearly, if we are to understand the changes in the Sun's corona and wind through its cycle, we need to model the changes in its magnetic field. There are, of course, many approaches to this - in this paper I will focus on one method that is particularly applicable to stars. The Wang-Sheeley-Arge (WSA) model is an empirical approach, that maps the wind speed directly to the magnetic field geometry (Wang & Sheeley 1990; Arge & Pizzo 2000). The magnetic field structure in the corona is extrapolated from surface magnetograms using a potential field source surface method (Altschuler & Newkirk 1969). An inherent assumption of this model is that the magnetic field is forced open by the coronal pressure at some radius r_s, known as the *source surface*. The wind speed is then related directly to the field line expansion factor f as follows:

$$u_i [\mathrm{kms}^{-1}] = 267.5 + \frac{410.0}{f_i^{2/5}} \qquad (1.1)$$

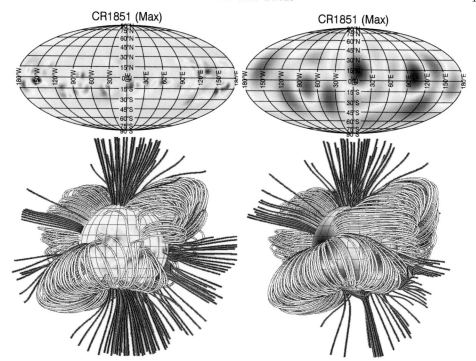

Figure 1. Top: Surface magnetograms for Carrington rotation 1851 (close to cycle maximum). The map is reconstructed for a maximum spherical harmonic degree of (left) $\ell_{\max} = 63$ and (right) $\ell_{\max} = 5$ corresponding to surface spatial scales of 3° and 30° respectively. Colourbars are set to ± 200G (left) and ± 30G (right) . Bottom: The corresponding field extrapolations, with wind-bearing (open) field lines coloured red. The overall structure of the largest fieldlines is very similar.

where the expansion factor f_i of any field line is given by:

$$f_i = \frac{r_\odot^2}{r_{\rm s}^2} \frac{B_i(r_\odot)}{B_i(r_{\rm s})}. \tag{1.2}$$

A large expansion factor therefore leads to a slow wind and conversely a small expansion factor leads to a fast wind. This approach is computationally very efficient and underlies many models of the solar wind. In an attempt to quantify the effect that the observational resolution might have on the output from such a code, Cohen (2015) undertook a detailed comparison of this approach with much more sophisticated MHD models. He showed that for the Sun, an increase in resolution from 2 to 1 in the surface magnetogram leads to only modest changes in the arrival time at Earth of coronal mass ejections. This approach has also been employed in analysing the variation of the solar wind through its cycle that accompanies the changing magnetic field structure (Pinto *et al.* 2011; Gressl *et al.* 2014; Pinto *et al.* 2016).

2. Application to other stars

Maps of the surface magnetic fields of other stars have been available now for some time. These show trends in the field strength and geometry with both mass and rotation rate (Donati and Landstreet 2009). It is clear that many stars have surface magnetic fields that are very non-solar in nature, and this is likely to have direct implications for the structure and dynamics of their coronae and winds.

2.1. *Resolution*

Extending this method to other stars would therefore be very useful, but faces the problem of the limited resolution of stellar magnetograms. Figure 1 shows a typical solar magnetogram obtained from the US National Solar Observatory, Kitt Peak, and the potential field extrapolation in the corona, compared to what would be found for a typical stellar magnetogram. These magnetograms are reconstructed using the Zeeman Doppler Imaging technique, which recovers only the large scale field. Surface resolutions for stellar magnetograms are typically 20° to 30°, although they may be as good as 3°. As can be seen from Figure 1 however, the largest scale field lines in both cases have a very similar geometry. This is simply because only the large scale field extends out into the wind region.

The corresponding wind speeds, derived from the WSA model, are shown in Figure 2 (Jardine, Vidotto, and See 2017). It is clear that a surface resolution of better than about 20° has little influence on the derived wind speed. This suggests that these magnetograms can indeed be used to explore the 3-D structure of stellar winds.

2.2. *Trends with Rossby number*

It has been known for a long time that the behaviour of stellar winds depends on the open flux of magnetic field (Mestel 1968; Mestel & Spruit 1987; Réville et al. 2015b,a). While this is not directly observable, it can be calculated from the surface magnetogram. See (2017) used the WSA method combined with an expression for the torque (Réville et al. 2015b,a) to determine open fluxes for a large number of stars whose surface magnetic fields had been mapped and to derive a scaling between the angular momentum loss rate and the Rossby number. The low mass stars, i.e. those without a tachocline, do not lie on this relation however. These stars could therefore be expected to spin down only very slowly. It is notable that when the results of 3D MHD simulations are overplotted on this relation, the agreement is very good.

The other parameter that varies very closely with Rossby number is the X-ray luminosity. This suggests a close relationship between the mass loss rate and the X-ray luminosity. Mass loss rates from solar like stars are very difficult to measure, as the winds have a very low density. One indirect method of doing this, developed by Brian Wood, uses the enhanced absorption in the hydrogen wall of stellar astrospheres as a measure of the local mass loss rate (see, for example, Wood et al. (2005)). Low activity stars appear to show a clear relation between wind mass loss rate and X-ray flux such that $\dot{M} \propto F_X^{1.34}$ for stars with $F_X \leqslant 10^6 \mathrm{erg\ cm^{-2}s^{-1}}$. The high activity stars, however, with the largest X-ray flux, seem to lie well below this relation. Vidotto et al. (2016) however showed that there was no clear change in the geometry of the magnetic fields of the stars across this "wind dividing line". Using the mass loss rates derived from the surface magnetograms, See (2017) also found no evidence of a wind dividing line, but recovered a scaling relations $\dot{M} \propto F_X^{1.31}$, similar to that found in the observations.

2.3. *Classifying stellar magnetic fields*

The nature of stellar magnetic fields is typically described by their symmetry around the rotation axis, and their geometry. This geometry is expressed by separating the field into its toroidal and poloidal components, as shown in Figure 3. Toroidal fields appear in those stars that have developed a radiative core. See et al. (2015) showed that a plot of the energy in the toroidal field, as a function of the energy in the poloidal field shows

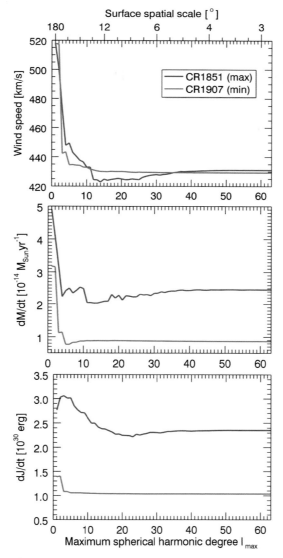

Figure 2. Variation with maximum spherical harmonic degree $\ell_{\max}$ of the average wind speed at the Earth's orbit, the total mass loss rate and the total angular momentum loss rate. Results are shown for two Carrington rotations.

two branches. Stars with radiative cores lie on one branch, while fully-convective stars lie on the other. This suggests that the presence of the tachocline, while not necessary for producing magnetic field, does influence the geometry of a star's magnetic field.

2.4. *Coronal magnetic fields*

The large range in the magnitude of the magnetic energy of these stars suggests that their winds and coronae may be substantially different. In particular the energy release processes may be acting quite differently in more active stars. A comparison between the kinetic energy in solar coronal mass ejections, with the X-ray emission of the associated flare, shows a clear correlation (Drake *et al.* 2013). Extrapolating this correlation,

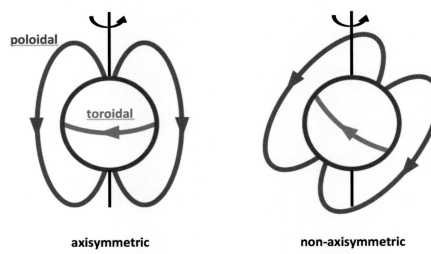

axisymmetric   **non-axisymmetric**

Figure 3. Magnetic fields can be decomposed into poloidal and toroidal components, that may be axisymmetric (left) or non-axisymemetric (right) about the rotation axis.

however, to more active stars, gives an unfeasibly large energy in the coronal mass ejections. So what is happening in the coronae of these very active starts? A pilot study by Gibb, Jardine, and Mackay (2014) suggested that in stars with a large surface shear, the timescales for evolution in the coronae may be surprisingly short. A subsequent large-scale study of the roles of flux emergence and differential rotation, showed that while flux emergence governs the the total energy available, it is the surface shear that determines the topology of the magnetic field (Gibb *et al.* 2016). High shear rates lead to a larger open flux and a higher rate of eruption of flux ropes.

2.5. *Comparing observations and simulations*

Lehmann *et al.* (2016) have recently analysed the surface magnetic fields of these simulations. They sure that the energies of the simulated magnetic fields lie within the same range as the observed poloidal and toroidal energies. By expressing the field as a sum of spherical harmonics, they were able to truncate the series at different spherical harmonic modes. This allows the field structures of different length scales to be analysed separately. This showed that while the highest order modes (i.e. the small-scale field) have a constant relationship between the toroidal and poloidal energies, the dipole mode is particularly poloidal, while the quadruple mode has the largest relative amount of toroidal field.

3. Conclusion

In summary then, it seems that while a tachocline is not necessary for the generation of magnetic field, it may play a role in the geometry of the field that is produced. As a result, it also has a role to play in governing the coronal activity and the efficiency with which the wind can remove angular momentum from the star. The preimpacts both on the evolution of the rotation rate and activity of the star. At any given evolutionary phase for the star, the rate at which flux emerges through the stellar surface governs the total energy available in the corona, but it is the surface shear that controls that geometry of the coronal magnetic field.

Many puzzles remain however. Why does the toroidal field appear so clearly in the quadrupolar component, while the dipole is so poloidal? How is energy distributed on the different lengths scales of the magnetic field? These are puzzles that will absorb the community for quite some time.

References

Altschuler, M. D. & Newkirk, G. 1969, *Solar Phys.*, 9, 131

Arge, C. N. & Pizzo, V. J. 2000, *JGR*, 105, 10465

Cohen, O. 2015, *Solar Phys.*, 290, 2245

DeRosa, M. L., Brun, A. S., & Hoeksema, J. T. 2012, *ApJ*, 757, 96

&onati, J.-F. & Landstreet, J. D. 2009, *Annual Review of Astronomy and Astrophysics* **47**, 333.

Drake, J. J., Cohen, O., Yashiro, S., & Gopalswamy, N. 2013, *Astrophys. J.* **764**, 170.

Gibb, G. P. S.., Jardine, M. M., & Mackay, D. H. 2014, *Monthly Notices of the Royal Astronomical Society* **443**, 3251.

Gibb, G. P. S.., Mackay, D. H., Jardine, M. M., & Yeates, A. R. 2016, *Monthly Notices of the Royal Astronomical Society* **456**, 3624.

Gressl, C., Veronig, A. M., Temmer, M., *et al.* 2014, *Solar Phys.*, 289, 1783

Jardine, M., Vidotto, A. A., & See, V. 2017, *Monthly Notices of the Royal Astronomical Society* **465**, L25.

Lehmann, L. T., Jardine, M. M., Vidotto, A. A., Mackay, D. H., See, V., Donati, J.-F., Folsom, C. P., Jeffers, S. V., Marsden, S. C., Morin, J., & Petit, P. 2016, *ArXiv e-prints*, arXiv:1610.08314.

Mestel, L. 1968, *MNRAS*, 138, 359

Mestel, L. & Spruit, H. C. 1987, *MNRAS*, 226, 57

Pinto, R. F., Brun, A. S., Jouve, L., & Grappin, R. 2011, *apj*, 737, 72

Pinto, R. F., Brun, A. S., & Rouillard, A. P. 2016, ArXiv e-prints

Réville, V., Brun, A. S., Matt, S. P., Strugarek, A., & Pinto, R. F. 2015a, *ApJ*, 798, 116

Réville, V., Brun, A. S., Strugarek, A., *et al.* 2015b, *ApJ*, 814, 99

See, V., Jardine, M., Vidotto, A. A., Donati, J.-F., Folsom, C. P., Boro Saikia, S., Bouvier, J., Fares, R., Gregory, S. G., Hussain, G., Jeffers, S. V., Marsden, S. C., Morin, J., Moutou, C., do Nascimento, J. D., Petit, P., Rosén, L., & Waite, I. A. 2015, *Monthly Notices of the Royal Astronomical Society* **453**, 4301.

See, V. 2017, *MNRAS*, in press

Vidotto, A. A., Donati, J.-F., Jardine, M., See, V., Petit, P., Boisse, I., Boro Saikia, S., Hébrard, E., Jeffers, S. V., Marsden, S. C., & Morin, J. 2016, *Monthly Notices of the Royal Astronomical Society* **455**, L52.

Vidotto, A. A. 2016, *MNRAS*, 459, 1533

Wang, Y.-M. & Sheeley, Jr., N. R. 1990, *ApJ*, 355, 726

Wood, B. E., Müller, H.-R., Zank, G. P., Linsky, J. L., & Redfield, S. 2005, *Astrophys. J.* **628**, L143.

Living Around Active Stars
Proceedings IAU Symposium No. 328, 2016
D. Nandy, A. Valio & P. Petit, eds.

© International Astronomical Union 2017
doi:10.1017/S1743921317003775

The Influences of Stellar Activity on Planetary Atmospheres

Colin P. Johnstone

University of Vienna, Department of Astrophysics, Türkenschanzstrasse 17, 1180 Vienna,
Austria
email: colin.johnstone@univie.ac.at

Abstract. On evolutionary timescales, the atmospheres of planets evolve due to interactions with the planet's surface and with the planet's host star. Stellar X-ray and EUV (='XUV') radiation is absorbed high in the atmosphere, driving photochemistry, heating the gas, and causing atmospheric expansion and mass loss. Atmospheres can interact strongly with the stellar winds, leading to additional mass loss. In this review, I summarise some of the ways in which stellar output can influence the atmospheres of planets. I will discuss the importance of simultaneously understanding the evolution of the star's output and the time dependent properties of the planet's atmosphere.

1. Introduction

Energy released in the cores of stars is transferred into the stellar environment by a multitude of different processes. These include emission of infrared, visible, and ultraviolet light from the star's photosphere, emission of far ultraviolet (FUV), extreme ultraviolet (EUV) and X-rays from the chromosphere and corona, the acceleration of supersonic winds and coronal mass ejections, and the ejection of high energy protons and electrons. Nearby planets are embedded in this environment and are influenced in radically different ways by each of these different processes. In many cases, stellar output causes a variety of atmospheric loss processes, influencing the evolution of the planet's atmosphere. In this review, I will discuss the various manifestations of stellar activity and the resulting losses from planetary atmospheres.

Planets that form to a significant mass within a few Myr of the start of system can gravitationally attract a thick envelope from the circumstellar gas disk; these atmospheres are often called 'primordial atmospheres'. In such cases, the atmospheres are primarily composed of Hydrogen and Helium and can have masses that are a few percent of the planet's own mass, with surface temperatures of several thousand K. Regardless of whether or not they form to large enough masses to gain a primordial atmosphere, bodies of all sizes that form in the disk trap heavier volatile species such as nitrogen, oxygen, and carbon dioxide. After the circumstellar gas disk is gone, terrestrial planets form through the accretion of such bodies. These volatiles can be released when the bodies impact with a forming planet, to form an atmosphere or contribute to an existing atmosphere. Volatile elements that are not released during impacts, or were trapped in the planet as it first formed, can then be released both as the magma ocean that likely exists in the early phases of planet formation solidifies (Elkins-Tanton 2008) and from volcanoes in later phases. Atmospheres formed after the disk phase, i.e. non-primordial atmospheres, can have a range of compositions, with H_2O (e.g. possibly early Venus), CO_2 (e.g. current Venus and Mars), and N_2 (e.g. current Earth) being possible dominant species. In all cases, the later atmospheric evolution can be determined by interactions with the planet's host star.

In this review, I discuss some of the influences that stellar activity related phenomena can have on planetary atmospheres. Given the limited scope of this review, the reader should be aware that there are many interesting and important processes that are not mentioned here. In Section 2, I summarise some of the various phenomena that are part of 'stellar activity' and how they evolve; in Section 3, I discuss the influences of stellar activity on planetary atmospheres; in Section 4, the evolution of primordial atmospheres; finally, in Section 5, I give some final remarks.

2. The various manifestations of stellar activity

In this review, I am interested in the effects of stellar 'activity', i.e. phenomena related to the star's magnetic field. The magnetic field provides a route by which convective energy in the star's photosphere can be transferred upwards and deposited in the overlying layers. The resulting heating leads to gas at EUV and X-ray emitting temperatures, and the acceleration of gas away from the star in the form of a stellar wind.

2.1. *Emission of EUV and X-rays*

Starting in the solar photosphere with a temperature of $\sim$6000 K, the temperature first decreases with increasing height until it reaches a minimum in the chromosphere. At higher altitudes, the temperature first increases in the chromosphere and then shoots up in the transition region to MK temperatures in the corona, as shown in the upper panel of Fig. 1. The plasma in these different layers emit at very different wavelengths. The photosphere dominates in the visible, ultraviolet, and infrared with an approximately blackbody spectrum. The chromosphere mostly emits at far ultraviolet and extreme ultraviolet wavelengths and the corona emits at extreme ultraviolet and X-ray wavelengths. As can be seen in Fig. 10 of Fontenla *et al.* (2009), chromospheric emission dominates at wavelengths of 500 Å and longer and coronal emission dominates at shorter wavelengths. The solar spectrum is shown in Fig. 1.

The chromospheric and coronal heating mechanisms have not yet been unambiguously identified. It is clear, however, that the heating is directly a result of the magnetic field. For example, Pevtsov *et al.* (2003) showed that various features on the solar surface, such as active regions, have a very tight relation between magnetic flux in the photosphere and X-ray luminosity which is valid over many orders of magnitude. Over the course of the solar cycle, the magnetic flux, and therefore the EUV and X-ray emission varies, by large amounts. Pevtsov *et al.* (2003) also showed that their solar relation holds when going to young active stars that have orders of mangitude higher X-ray luminosities.

Observations of other stars in X-rays have shown that similar processes are operating on stars with masses similar to or less than that of the Sun. Unfortunately, interstellar absorption makes it very difficult to observe EUV from stars, and only a few obsevations exist for the nearest stars. The general trend is that the X-ray luminosity, L_X, is higher for stars that rotate faster, up until a certain rotation rate where the L_X-rotation relation saturates. The saturation threshold happens approximately at a constant value of $L_X/L_{bol} \sim -3$ (Pizzolato *et al.* 2003). The rotation rate at which stars saturate in X-rays depends strongly on mass, with a value of $\sim 15\Omega_\odot$ for solar mass stars and values of $\sim 1\Omega_\odot$ for M dwarfs. This means that lower mass stars (e.g. M dwarfs) never become as active as the most active solar mass stars. Observationally, it is known that the coronal temperature depends sensitively on its activity level (e.g. Johnstone & Güdel 2015), which means that the X-ray spectrum is generally shifted to shorter wavelengths (Güdel *et al.* 1997). We also expect signficantly different spectra for lower mass stars, such as M dwarfs, especially in the near ultraviolet (see Fig. 9 of Fontenla *et al.* 2016).

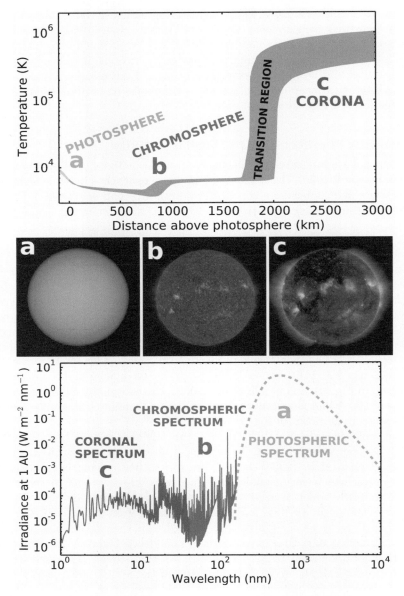

Figure 1. Figure showing the heating and emission of the solar atmosphere. *Upper panel:* the temperature structure of the chromosphere, transition region, and corona. The areas represent a range of temperatures for different features in the solar atmosphere, as shown in Fig. 1 of Fontenla *et al.* (2011). *Middle panel:* images of the Sun from the Solar Dynamics Observatory (SDO) showing the Sun at different wavelengths courtesy of NASA/SDO and the AIA, EVE, and HMI science teams. The three panels show plasmas with different temperatures, and therefore different heights in the solar atmosphere and contribute differently to the solar spectrum; they show (a) the photosphere at ~5800 K, (b) the chromosphere and transition region at $\sim 5 \times 10^4$ K, and (c) the corona at ~ 2 MK. *Lower panel:* the solar spectrum between 1 nm and 10^4 nm. The various parts are from the solar photosphere (*dashed yellow line*), the primarily chromospheric dominated spectrum up to 500 nm (*red line*), and the primarily coronal dominated spectrum at longer wavelengths (*purple line*). The chromospheric and coronal spectra were calculated by Fontenla *et al.* (2011).

2.2. *Winds and coronal mass ejections*

The stellar wind heating and acceleration mechanisms are currently poorly understood. The most likely mechanism is energy and momentum deposition by waves (e.g Alfvén waves) produced in the photosphere that propagate upwards along magnetic field lines. Alternatively, injections of mass and energy into the wind by magnetic reconnection events near the surface might play a role. For a review, see Cranmer (2009).

As a simple approximation, the wind breaks down into three components: these are the slow wind, the fast wind, and coronal mass ejections (CMEs). The slow and fast components are both relatively steady streams of particles, with similar mass fluxes, that are distinguished based primarily on their outflow speeds, with average values of ~ 400 km s^{-1} and ~ 800 km s^{-1} respectively. The spatial distributions of slow and fast winds depend sensitively on the structure of the global magnetic field in the corona, with fast wind originating from the centres of coronal holes (regions where the magnetic field lines are 'open') and slow wind dominating in regions above closed magnetic fields. For the slow wind, the main open question is the exact spatial origin in the corona†. A difficulty for the fast wind is the fact that measurements of the wind temperatures in the corona show that the wind is not hot enough to be accelerated to ~ 800 km s^{-1} by thermal pressure alone, and therefore other acceleration mechanisms must be invoked.

Although the bulk properties of the solar wind are known in a lot of detail, very little is known about the bulk properties of the winds of other low-mass stars, and nothing is known with any certainty. Several studies have attempted to constrain wind properties observationally, using for example radio emission (e.g. Gaidos *et al.* 2000), astrospheric absorption of Lyα radiation (Wood *et al.* 2014), and planetary transits (Kislyakova *et al.* 2014b). Johnstone *et al.* (2015b) estimated a scaling law for wind mass loss rates of $\dot{M}_\star \propto R_\star^2 \Omega_\star^{1.33} M_\star^{-3.36}$ by fitting a rotational evolution model to observational constraints. Stellar spin down on the main-sequence is probably the most unambiguous and well constrained observational signature of stellar winds, though using it to derive wind properties is difficult (see Section 4.3 of Johnstone *et al.* 2015b). Many stellar wind models have been developed and applied: these include theoretical models that consider energy balance in the transition region (Cranmer & Saar 2011), hydrodynamic models of the solar wind scaled to the winds of other stars (e.g. Johnstone *et al.* 2015a), and 3D magnetohydrodynamic (MHD) models based on observed magnetic field geometries (e.g. Vidotto *et al.* 2015). Recently, more physically based MHD models have been applied that heat and accelerate the wind using Alfvén waves (van der Holst *et al.* 2014; Airapetian & Usmanov 2016; Cohen *et al.* 2016), though there are still uncertainties and free parameters in these models. I should also note that for the most rapidly rotating stars, the wind acceleration can be dominated by magneto-centrifugal forces; such winds can have speeds far from the star of several thousand km s^{-1} (Johnstone 2016). For a review of stellar wind observations and models, see Section 2 of Johnstone *et al.* (2015a).

Coronal mass ejections are clouds of coronal plasma that are released by the corona and accelerated at speeds that range from slower than the solar wind up to several thousand km s^{-1}. On the Sun, they are known to be correlated with flares, with the most massive flares almost always being accompanied by a CME (Wang & Zhang (2007)). The CME rate generally varied with the solar cycle, reaching up to a ten per day at cycle maximum

---

† The most natural possibility is that the slow wind comes from the edges of coronal holes and is directed by the magnetic field into the regions above closed field lines. Alternatively, the slow wind could come from closed field regions, and is released into the wind through diffusion of material across field lines or through magnetic reconnection at the closed/open field boundaries and the tops of helmet streamers. For a recent review on the subject, see Abbo *et al.* (2016).

(Robbrecht *et al.* 2009). On other stars, CMEs have not been unambiguously detected, but the relation between flares and CMEs indicate that more active stars release CMEs at higher rates (Aarnio *et al.* 2012). It could be that the winds of active stars are in fact dominated by CMEs, though it is currently not known if this is indeed the case.

2.3. *Evolution of stellar activity*

Given the empirical relation between X-ray emission and rotation, and the likely relation between winds and rotation, it is natural to assume that the evolution of stellar activity is mosty a result of rotational evolution. Stars are born with rotation rates anywhere between a few times to a few tens of times the rotation rate of the Sun ($\Omega_\odot \approx 2.9 \times 10^{-6}$ rad s^{-1}). As they contract on the pre-main-sequence, they spin up (i.e. their rotaton rates increase with time) until just before they reach the zero-age main-sequence (ZAMS). After the ZAMS, the dominant effect is angular momentum removal by stellar winds, causing them to spin down. The initial wide distribution of rotation rates gets wider when spinning up and then converges to a single mass and age independent value early on the main-sequence. The convergence time depends very much on stellar mass, with solar mass stars taking ∼750 Myr to converge fully and lower mass stars taking longer. For M dwarfs, it is possible that a subset of stars do not converge at all and remain rapidly rotating. For a review of stellar rotational evolution, see Bouvier *et al.* (2014).

Spin down results in stellar activity decreasing with age. For solar analogues, this decay in X-rays is approximately $L_X \propto t^{-1.5}$ (Güdel *et al.* 1997). For longer wavelengths, this decay is less steep (Ribas *et al.* 2005). These single time dependent decay laws are however only valid after rotational convergence has taken place. Tu *et al.* (2015) used a rotational evolution model to estimate the different evolutionary tracks for stellar X-ray and EUV emission and showed that these tracks fit very well the distributions of X-ray emission observed in young clusters. Their evolutionary tracks for rotation and XUV emission are shown in Fig. 2. Although it is not known, it is natural to expect that stellar winds also follow different evolutionary tracks for different rotational evolutions. Such tracks were estimated by Johnstone *et al.* (2015b) using $\dot{M}_\star \propto \Omega_\star^{1.33}$, i.e. the relation mentioned in the previous section applied to solar mass and radius stars. I show these evolutionary tracks for the solar wind in time in Fig. 3, as well as a similar estimate calculated assuming $\dot{M}_\star \propto \Omega_\star^{2.43}$ which I estimated for this review based on the wind models of Airapetian & Usmanov (2016). Clearly, our uncertainties for the properties of the early solar wind come from two sources: our lack of knowledge of the relevant wind physics and our lack of knowledge of how rapidly the Sun was rotating. Given that low mass stars often remain rapidly rotating for longer, and the saturation rotation rate is lower (so that even relatively slowly rotating M dwarfs can be highly active), low mass stars remain active for much longer than solar mass stars (West *et al.* 2008).

3. Activity driven atmospheric losses

Atmospheric losses driven by stellar XUV radiation, winds, and high energy particles have been observed both in our own solar system (Lammer *et al.* 2009) and in the transits signatures of planets orbiting other stars (Ehrenreich *et al.* 2008; Kislyakova *et al.* 2014b). To first approximation, the majority of planetary mass loss mechanisms can be broken down into two catagories: 'thermal' and 'non-thermal'. Thermal loss mechanisms are those that are directly a result of the heating of the planetary atmosphere and non-thermal mechanisms are essentially all other processes. In this section, I will instead break down these mechanisms into radiation induced and wind induced losses.

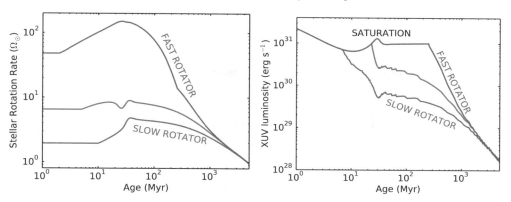

Figure 2. Figure showing the evolution of rotation (*left panel*) and XUV (*right panel*) for solar mass stars with different initial rotation rates, calculated by Tu *et al.* (2015).

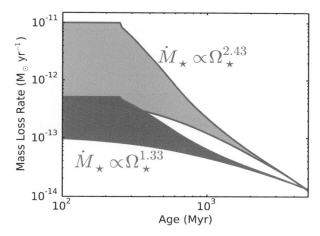

Figure 3. Figure showing the evolution of the mass loss rates of solar mass stars from 100 Myr to 5 Gyr assuming that $\dot{M}_\star \propto \Omega_\star^a$ for two different values of a. The value of $a = 1.33$ was estimated by Johnstone *et al.* (2015b) and the value of $a = 2.43$ is the largest value that can be derived based on the results of the wind models of Airapetian & Usmanov (2016). The upper and lower limits of each shaded areas are the tracks for fast and slow rotating stars respectively.

3.1. *Atmospheric losses due to stellar radiation*

There are several ways in which the various layers in the Earth's atmosphere can be broken down; the most common of these is by temperature gradient. Starting at the surface and going to higher altitudes, the temperature first decreases in the troposphere, then increases in the stratosphere due to absorption of solar UV by ozone, and then decreases again in the mesosphere due to the increasing importance of cooling by CO_2. Above the mesosphere is the thermosphere, where the absorption of solar X-ray and EUV radiation causes the temperature to increase to a few thousand K. Solar radiation also ionises the upper atmosphere, creating the ionosphere, which corresponds to the upper mesosphere and the thermosphere. In general, higher input XUV fluxes lead to a hotter and more expanded thermosphere. The response of the Earth's current atmosphere to increased XUV irradiation was studied by Tian *et al.* (2008) and is summerised in Fig. 4.

The gas in the thermosphere is essentially hydrostatic and as we go to higher altitudes, the particle density decreases rapidly. Eventually, the density becomes so low that

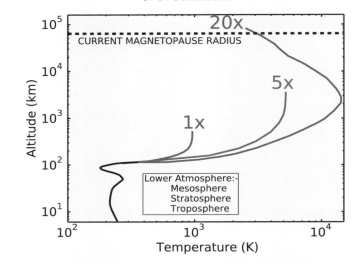

Figure 4. Figure showing how the thermosphere of the Earth reacts to different XUV fluxes (where 5x for example indicated five times what the Earth currently gets on average), adapted from Fig. 6 of Tian *et al.* (2008). Each line stops at the exobase. Adiabatic cooling can be seen in the 20x case due to the wind flowing away from the planet.

particles can travel a large distance without interacting with each other. The thermosphere ends at the point where the gas becomes essentially collisionless; this is the exobase, above which the particles travel on ballistic trajectories. Particles that have upward velocity components that are higher than the escape speed are lost from the atmosphere; this loss mechanism is called Jeans escape. The Jeans escape rate depends sensitively on the altitude and gas temperature of the exobase; another important factor is the molecular mass of the species, with lighter species escaping easier than heavier species.

A useful parameter to calculate at the exobase is the Jeans escape parameter, given by $\lambda_J = GM_p m/k_B T_{exo} R_{exo}$, where m is the molecular mass of the species being considered, and T_{exo} and R_{exo} are the temperature and radius of the exobase respectively. This is simply the ratio of the potential energy to the kinetic energy at the exobase. When λ_J is relatively high, the atmosphere is approximately hydrostatic and the main thermal loss mechanism is Jeans escape, with a mass loss rate that depends strongly on λ_J. When λ_J is close to unity, the atmosphere is not hydrostatic, but has enough energy to flow away from the planet as a planetary wind. In Fig. 5, I show some hydrodynamic simulations from Johnstone *et al.* (2015b) of the planetary wind of an Earth mass planet with a hydrogen dominated envelope under different X-ray and EUV conditions. The winds start out subsonic low in the thermosphere, and as the temperature rises due to XUV heating, the winds accelerate to supersonic speeds. Like any Parker wind, the radius at which the winds become supersonic is the same as the radius at which their outflow speed becomes equal to the escape velocity. This means that in the supersonic part of the wind, the material will be lost from the planet, regardless of other effects such additional heating or stellar wind pick-up. A fundamental requirement for a hydrodynamically outflowing wind is that it becomes supersonic before the exobase, otherwise the particles that move up into the exosphere will mostly not have reached escape velocity.

3.2. *Atmospheric losses due to stellar winds/CMEs*

Stellar winds bring with them a large amount of energy and momentum, which can be given to the atmosphere below the exobase or to individual atmospheric particles in the

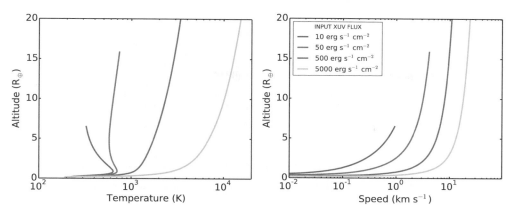

Figure 5. Figure showing the thermospheric temperature and velocity structure a hydrogen dominated atmosphere undergoing hydrodynamic escape with different input XUV fluxes, reproduced from Johnstone *et al.* (2015b).

exosphere. The planetary magnetosphere can play an important role in determining how the wind and the atmosphere interact; they can both protect atmospheres by shielding them from interacting directly with winds, and expose atmospheres more by increasing the effective area of the planet that can collect wind energy and momentum.

Stellar wind/CME protons and electrons can directly enter the atmosphere; for a magnetised planet, this will take place at the magnetic poles. Their effects on the atmosphere include ionisation and heating, which can lead to increased outflow (e.g. Glocer *et al.* 2009). In addition, wind particles that collide with atmospheric particles give them their momentum directly which can lead to the atmospheric particles escaping the planet, or giving this momentum to other atmospheric particles that then escape (Johnson 1994). This effect is called 'sputtering'.

Winds can also pick up atmospheric particles in the exosphere. In order to be picked up, an atmospheric particle should first be ionized so that it can interact with the magnetic field in the wind. Several mechanisms exist that can ionize a particle if it is not already ionized lower in the atmosphere: these include photoionisation from a stellar XUV photon, electron-impact ionisation from a stellar wind electron, and charge exchange. Charge exchange involves a supersonic stellar wind proton taking an electron from a slow atmospheric particle, resulting in an atmospheric ion and a supersonic neutral Hydrogen atom. Heavier ions can also undergo charge exchange, often resulting in the ion emmitting an X-ray photon, which itself can be an additional source of X-ray irradiation of the planet (Kislyakova *et al.* 2015). The atmospheric ions that are produced can either be lost from the planet entirely (e.g. Kislyakova *et al.* 2014a) or be accelerated back into the planet's atmosphere or magnetosphere, potentially increasing the heating and causing additional sputtering (Luhmann & Kozyra 1991).

The high-speed Hydrogen atoms that are produced as a result of charge exchange are called an energetic neutral atoms (ENAs). ENAs can fly unhindered through a planet's magnetic field and deposit their energy in the atmosphere. This was suggested as a significant heating mechanism in the upper atmosphere of early Venus by Chassefière (1997). Recently, Lichtenegger *et al.* (2016) studied this process for early Venus, assuming that the planet had a water vapour atmosphere that was undergoing hydrodynamic escape because of the early Sun's high activity levels. They found that while ENAs bring in a significant energy, the energy is deposited high up in the atmosphere, after the planetary wind had already accelerated to a significant speed, and therefore the heating did not influence the atmospheric loss rates significantly.

There are several other ways in which winds can give energy to planetary atmospheres. For example, wind-magnetospheric interactions can generate electric currents in the ionisphere which then release energy due to Joule heating. This was studied by Cohen *et al.* (2015) for planets in the habitable zones of low mass stars. In this case, planets are expected to be embedded in dense winds because of the close-in habitable zones, and therefore the heating rate from Joule heating is likely much larger.

4. Evolution of primordial atmospheres orbiting active stars

Primordial atmospheres are picked up by planets that form fast enough to have significant masses ($\gtrsim 0.1 M_\oplus$) during the gas disk phases of their systems (i.e. within a few Myr). Lammer *et al.* (2014) and Stökl *et al.* (2016) showed that the mass of the obtained atmosphere depends sensitively on the mass of the planet, with $5.0 M_\oplus$ planets picking up atmosphere that are a factor of 10^5 larger than those picked up by $0.1 M_\oplus$ planets (see Table 2 of Stökl *et al.* 2016 and the lower panels of Fig.6). However, significant uncertainty exists in how much atmosphere is gained, largely due to the its sensitivity on the energy input into the atmosphere and the how quickly this energy can be radiated away†. Of course, we also shouldn't forget that planet's likely undergo significant growth after the disk phase (e.g. the Earth was likely only half its current mass 10 Myr after solar system formation; see Table 6 of Kleine *et al.* 2009), and we won't know for any given planet that we observe what its mass was at the end of the disk phase.

Stökl *et al.* (2015) studied the evolution of these atmospheres after the gas disk dissipated, assuming the planets were located at 1 AU; they found that after the gas disk dissipates, the primordial atmospheres of low mass ($\lesssim 0.5 M_\oplus$) planets would quickly flow away, even without additional XUV heating. Similarly, Owen & Wu (2016) studied this for planets at much smaller orbital distances from their host stars who found that due to the much stronger irradiation from the star's photospheric spectrum, even super-Earths will lose most of their atmospheres after disk dispersal. Lammer *et al.* (2014) combined estimates of the pick up of primordial atmospheres with models for the XUV driven hydrodynamic atmospheric losses and found that planets with masses less than that of the Earth likely lose their primordial atmospheres and planets more massive than the Earth keep them. Similar results were found by Owen & Mohanty (2016), who instead considered terrestrial planets in the habitable zones of M dwarfs. In both cases, the high levels of atmospheric loss are a result of the star's high X-ray and EUV luminosities. These results suggest that we should expect to find many planets with H/He-dominated primordial atmospheres, with the exceptions being planets with masses $\lesssim 1.0 M_\oplus$, and more massive planets that either formed after the dissipation of the gas disk or are on close orbits around their host stars. The observational situation currently appears to support the conclusion that super-Earths mostly have low densities, implying that they have H/He envelopes (Rogers 2015).

Given that atmospheric losses are closely dependent on the star's activity level, it is important that stellar activity evolution is properly taken into account when studying atmospheric evolution (at least when this evolution is a result of losses into space and not

---

† It is useful to picture the pick up of a protoatmosphere in the gas disk by a terrestrial planet as being analogous to the contraction of a star on the pre-main-sequence and not analogous to the planet sucking up the atmosphere around it like a vacuum cleaner. Essentially, the disk gas is contracting around the gravitating body and this contraction rate depends sensitively on the rate at which the thermally-supported atmosphere can cool.

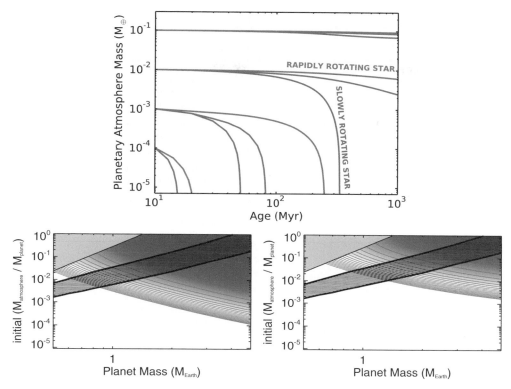

Figure 6. Figure showing the results of Johnstone *et al.* (2015b) for terrestrial planets with H/He dominated primordial atmospheres orbiting at 1 AU around a solar mass star. *Upper-panel:* atmospheric evolution with four different initial starting atmospheric masses; for each case, the red, green, and blue lines correspond to the slow, medium, and fast rotator cases for the host star. *Lower-panels:* remaining atmospheric mass after 1 Gyr for cases with a range of planetary and initial atmospheric masses, assuming the planets are orbiting a slow rotator (*left-panel*) and a fast rotator (*right-panel*). In both panels, dark red and white corresponds to 100% and 0% respectively of the initial atmospheric mass remaining. The grey areas in the upper left corners show areas of parameter space that are unphysical. The shaded black areas show the range of estimates by Stökl *et al.* (2016) for how massive an atmosphere the planets would pick up from the gas disk after 10,000 years (lower boundary) and 1 Myr (upper boundary).

interactions with the surface). As discussed in Section 2.3, the evolution of the activity of solar mass stars can be very different depending on the initial rotation rate of the host star. Tu *et al.* (2015) and Johnstone *et al.* (2015b) studied how these different activity evolution tracks would lead to different evolutions of a terrestrial planet orbiting these stars. Johnstone *et al.* (2015b) calculated a grid of atmospheric evolution models for different planetary and initial atmospheric masses, and their results are summarised here in Fig. 6. For Earth mass planets, they found that when the planets start out (at 10 Myr) with masses of ~ 1% of the Earth mass, as estimated by Stökl *et al.* (2016), the subsequent atmospheric evolution will be very different for different stellar rotation tracks. If the planet orbits a rapid rotator, the planet loses all of its atmosphere in ~ 300 Myr. If the planet orbits a slow rotator, the planet only loses about half of its atmosphere by 1 Gyr; after this, the atmosphere likely remains for the entire lifetime of the planet due to the low activity of the host star. On the other hand, if the planet starts out with a less massive atmosphere of ~ 0.1% of the Earth mass (such as can happen if the planet only grows to a fraction of its final mass during the phase), the atmosphere will be completely eroded in both cases, but the timescale depends sensitively on the star's initial rotation rate. The lower panels of Fig. 6 show the fraction of the atmosphere remaining after 1 Gyr for all models, with the black shaded area showing the range of estimates for the initial atmospheric mass estimated by Stökl *et al.* (2016).

5. Final remarks

The effects I discuss in this review are primarily related to stellar XUV radiation and winds, and I mostly ignore the effects of energetic (i.e. fast moving) particles. Such particles have several sources, including the solar corona, shocks propagating through the inner heliosphere, the planet's own magnetosphere, and from outside the solar system (i.e. galactic cosmic rays). The multitude of ways in which they can influence atmospheres is not discussed here, largely due to lack of space and expertise on the part of the author, and not because they are any less important and interesting than the effects that are discussed. I have also ignored the influences that impacts of larger bodies, ranging from small asteroids up to protoplanets, can have on atmospheric loss, which is expected to be especially important early in a planet's evolution.

6. Acknowledgements

The author acknowledges the support of Austrian Science Fund (FWF) NFN project S11601-N16 "Pathways to Habitability: From Disk to Active Stars, Planets and Life" (`path.univie.ac.at/`) and the related subproject S11604-N16, and also the International Space Science Institute (ISSI) team "The Early Evolution of the Atmospheres of Earth, Venus, and Mars", supported by ISSI.

References

Aarnio, A. N., Matt, S. P., & Stassun, K. G. 2012, *ApJ* 760 9
Abbo, L., Ofman, L., Antiochos, S. K., Hansteen, V. H., Harra, L., Ko, Y.-K., Lapenta, G., Li, B., Riley, P., Strachan, L., von Steiger, R., & Wang, Y.-M. 2016, *SSRv* 201 55
Airapetian, V. S. & Usmanov, A. V. 2016, *ApJ*, 817 24
Bouvier, J., Matt, S. P., Mohanty, S., Scholz, A., Stassun, K. G., & Zanni, C. 2014, *Protostars and Planets VI. Univ. Arizona Press, Tucson, AZ*, p. 433
Chassefière, E. 1997, *Icar*, 126 229
Cohen, O., Ma, Y., Drake, J. J., Glocer, A., Garraffo, C., Bell, J. M., & Gombosi, T. I. 2015, *ApJ*, 806 41
Cohen, O., Yadav, R., Garraffo, C., Saar, S. H., Wolk, S. J., Kashyap, V. L., Drake, J. J., & Pillitteri, I. 2016, *ApJ*, (accepted)
Cranmer, S. R. 2009, *LRSP*, 6 3
Cranmer, S. R. & Saar, S. H. 20092011 *ApJ*, 741 54
Eddy, J. A. & Ise, R. 1979, *Book: A new sun : the solar results from SKYLAB*
Ehrenreich, D., Lecavelier Des Etangs, A., Hébrard, G., Désert, J.-M., Vidal-Madjar, A., Mc-Connell, J. C., Parkinson, C. D., Ballester, G. E., & Ferlet, R. 2008, *A&A*, 483 933
Elkins-Tanton, L. T. 2008, *E&PSL*, 271 181
Fontenla, J. M., Curdt, W., Haberreiter, M., Harder, J., & Tian, H. 2009, *ApJ*, 707 482
Fontenla, J. M., Linsky, J. L., Witbrod, J., France, K., Buccino, A., Mauas, P., Vieytes, M., & Walkowicz, L. M. 2016, *ApJ*, 830 154
Fontenla, J. M., Harder, J., Livingston, W., Snow, M., & Woods, T. 2011, *JGRD*, 11620108F
Gaidos, E. J., Güdel, M., & Blake, G. A. 2000, *GeoRL*, 27 501
Glocer, A., Tóth, G., Ma, Y., Gombosi, T., Zhang, J.-C., & Kistler, L. M. 2009, *JGRA*, 11412203

Güdel, M., Guinan, E. F., & Skinner, S. L. 1997, *ApJ*, 483 947

Gunell, H., Brinkfeldt, K., Holmström, M., Brandt, P. C. :., Barabash, S., Kallio, E., Ekenbäck, A., Futaana, Y., Lundin, R., Andersson, H. *et al.* 2006, *Icar*, 182 431

Johnson, R. E. 1994, *SSRv*, 69 215

Johnstone, C. P., Güdel, M., Lüftinger, T., Toth, G., & Brott, I. 2015a, *A&A*, 577 27

Johnstone, C. P., Güdel, M., Brott, I., & Lüftinger, T. 2015b, *A&A*, 577 28

Johnstone, C. P., Güdel, M., Stökl, A., Lammer, H., Tu, L., Kislyakova, K. G., Lüftinger, T., Odert, P., Erkaev, N. V., & Dorfi, E. A. 2015b, *ApJ*, 815 12

Johnstone, C. P. & Güdel, M. 2015, *A&A*, 578 129

Johnstone, C. P. 2016, *A&A*, (accepted)

Kislyakova, K. G., Johnstone, C. P., Odert, P., Erkaev, N. V., Lammer, H., Lüftinger, T., Holmström, M., Khodachenko, M. L., & Güdel, M. 2014, *A&A*, 562 116

Kislyakova, K. G., Holmström, M., Lammer, H., Odert, P., & Khodachenko, M. L. 2014, *Science*, 346 981

Kislyakova, K. G., Fossati, L., Johnstone, C. P., Holmström, M., Zaitsev, V. V., & Lammer, H. 2015, *ApJ*, 799 15

Kleine, T., Touboul, M., Bourdon, B., Nimmo, F., Mezger, K., Palme, H., Jacobsen, S. B., Yin, Q.-Z., & Halliday, A. N. 2009, *GeCoA*, 73 5150

Lammer, H., Kasting, J. F., Chassefière, E., Johnson, R. E., Kulikov, Y. N., & Tian, F. 2009, *SSRv*, 139 399

Lammer, H., Stökl, A., Erkaev, N. V., Dorfi, E. A., Odert, P., Güdel, M., Kulikov, Y. N., Kislyakova, K. G., & Leitzinger, M. 2014, *MNRAS*, 439 3225

Lichtenegger, H. I. M., Kislyakova, K. G., Odert, P., Erkaev, N. V., Lammer, H., Gröller, H., Johnstone, C. P., Elkins-Tanton, L., Tu, L., Güdel, M., & Holmström, M. 2016, *JGRA*, 121 4718

Luhmann, J. G. & Kozyra, J. U. 1991, *JGR*, 96 5457

Owen, J. E. & Wu, Y. 2016, *ApJ*, 817 1070

Owen, J. E. & Mohanty, S. 2016, *MNRAS*, 459 40880

Pevtsov, A. A., Fisher, G. H., Acton, L. W., Longcope, D. W., Johns-Krull, C. M., Kankelborg, C. C., & Metcalf, T. R. 2003, *ApJ*, 598 1387

Pizzolato, N., Maggio, A., Micela, G., Sciortino, S., & Ventura, P. 2003, *A&A*, 397 147

Ribas, I., Guinan, E. F., Güdel, M., & Audard, M. 2005, *ApJ*, 622 680

Robbrecht, E., Berghmans, D., & Van der Linden, R. A. M. 2009, *ApJ*, 691 1222

Rogers, L. A. 2015, *ApJ*, 801 41

Stökl, A., Dorfi, E., & Lammer, H. 2015, *A&A*, 576 87

Stökl, A., Dorfi, E. A., Johnstone, C. P., & Lammer, H. 2016, *ApJ*, 825 86

Tian, F., Kasting, J. F., Liu, H.-L., & Roble, R. G. 2008, *JGRE*, 113 5008

Tu, L., Johnstone, C. P., & Güdel, M., Lammer 2015, *A&A*, 577 L3

van der Holst, B., Sokolov, I. V., Meng, X., Jin, M., Manchester, IV, W. B., Tóth, G., & Gombosi, T. I. 2014, *ApJ*, 782 81

Vidotto, A. A., Fares, R., Jardine, M., Moutou, C., & Donati, J.-F. 2015, *MNRAS*, 449 4117

Wang, Y. & Zhang, J. 2007, *AJ*, 665 1428

West, A. A., Hawley, S. L., Bochanski, J. J., Covey, K. R., Reid, I. N., Dhital, S., Hilton, E. J., & Masuda, M. 2008, *AJ*, 135 785

Wood, B. E., Müller, H.-R., Redfield, S., Edelman, E. 2014, *ApJ*, 781 33

Living Around Active Stars
Proceedings IAU Symposium No. 328, 2016
D. Nandy, A. Valio & P. Petit, eds.

© International Astronomical Union 2017
doi:10.1017/S1743921317003933

Solar activity forcing of
terrestrial hydrological phenomena

Pablo J. D. Mauas[1], Andrea P. Buccino[1] and Eduardo Flamenco[2]

[1]Instituto de Astronomía y Física del Espacio,
Universidad de Buenos Aires, CONICET
C.C. 67 Suc. 28 - 1428
Buenos Aires, Argentina
email: `pablo@iafe.uba.ar, abuccino@iafe.uba.ar`

[2]Instituto Nacional de Tecnología Agropecuaria,
Rivadavia 1439, 1033,
Buenos Aires, Argentina

Abstract. Recently, the study of the influence of solar activity on the Earth's climate received strong attention, mainly due to the possibility, proposed by several authors, that global warming is not anthropogenic, but is due to an increase in solar activity. Although this possibility has been ruled out, there are strong evidences that solar variability has an influence on Earth's climate, in regional scales.

Here we review some of these evidences, focusing in a particular aspect of climate: atmospheric moisture and related quantities like precipitation. In particular, we studied the influence of activity on South American precipitations during centuries. First, we analyzed the stream flow of the Paraná and other rivers of the region, and found a very strong correlation with Sunspot Number in decadal time scales. We found a similar correlation between Sunspot Number and tree-ring chronologies, which allows us to extend our study to cover the last two centuries.

Keywords. Solar activity, climate

1. Introduction

In the last decades, several authors proposed that global warming is not anthropogenic, but is due instead to an increase in solar activity, a proposition which resulted in a strong interest to study the influence of solar activity on the Earth's climate. This discussion was, of course, of great political interest, and had a strong repercussion in the media. For example, on December 4, 1997, on the Wall Street Journal appeared an article on the subject entitled "Science Has Spoken: Global Warming Is a Myth" (see Fig. 1). This article, together with a copy of a scientific-looking paper (of which there are three versions, e.g. Soon *et al.* 1999), was massively sent to North American scientists, accompanied by a petition to be presented to the Congress of the Unites States opposing the ratification of the Kyoto protocol.

This article was based on the results obtained by Friis-Christensen and Lassen (1991) and Lassen and Friis-Christensen (1995), who found a similarity between the length of the solar cycle (LCS), smoothed with a 1-2-2-2-1 filter, and the 11-yr running mean of the Northern Hemisphere temperature anomalies. However, this studies were seriously objected by Laut and Gunderman (2000) and Laut (2003). In particular, these results were obtained using the actual, non-smoothed, LSCs for the last 4 cycles. Using the right values, already available 10 years later, it can be seen that the solar cycles had

TIDAL INLETS
Hydrodynamics and Morphodynamics

This book describes the latest developments in the hydrodynamics and morphodynamics of tidal inlets, with an emphasis on natural inlets. A review of
morphological features and sand transport pathways is presented, followed by an
overview of empirical relationships between inlet cross-sectional area, ebb delta
volume, flood delta volume and tidal prism. Results of field observations and laboratory experiments are discussed and simple mathematical models are presented
that calculate the inlet current and basin tide. The method to evaluate the cross-
sectional stability of inlets, proposed by Escoffier, is reviewed, and is expanded,
for the first time, to include double inlet systems. This volume is an ideal reference
for coastal scientists, engineers and researchers, in the fields of coastal engineering,
geomorphology, marine geology and oceanography.

J. VAN DE KREEKE is Emeritus Professor at the Rosenstiel School of Marine
and Atmospheric Science, University of Miami, where his research focused on
coastal engineering and estuarine and nearshore hydrodynamics. He has published
extensively on tidal inlets, and is the editor of *Physics of Shallow Estuaries and
Bays* (Springer-Verlag, 1986). In 2004, Professor van de Kreeke received the Bob
Dean Coastal Research Award for world-renowned research on tidal inlets.

R.L. BROUWER, while at Delft University of Technology, The Netherlands,
wrote both his MSc and PhD theses on the subject of cross-sectional stability of
double inlet systems. He continued working on this subject as a postdoctoral fellow
and at the same time did pioneering work in deploying drones for coastal and inlet
research. He has published several papers on double inlet systems in refereed jour-
nals and conference proceedings. Presently, he is employed as a senior researcher
at Flanders Hydraulic Research in Antwerp, Belgium.

TIDAL INLETS

Hydrodynamics and Morphodynamics

J. VAN DE KREEKE
University of Miami, USA

and

R.L. BROUWER
Delft University of Technology, The Netherlands

CAMBRIDGE
UNIVERSITY PRESS

University Printing House, Cambridge CB2 8BS, United Kingdom

One Liberty Plaza, 20th Floor, New York, NY 10006, USA

477 Williamstown Road, Port Melbourne, VIC 3207, Australia

4843/24, 2nd Floor, Ansari Road, Daryaganj, Delhi – 110002, India

79 Anson Road, #06–04/06, Singapore 079906

Cambridge University Press is part of the University of Cambridge.

It furthers the University's mission by disseminating knowledge in the pursuit of education, learning, and research at the highest international levels of excellence.

www.cambridge.org
Information on this title: www.cambridge.org/9781107194410
DOI: 10.1017/9781108157889

First published 2017

A catalogue record for this publication is available from the British Library.

ISBN 978-1-107-19441-0 Hardback

Contents

Preface   *page* xi

1   Introduction   1

2   Geomorphology   6
   2.1   Introduction   6
   2.2   Origin of Tidal Inlets   6
   2.3   Equilibrium Morphology   7
   2.4   Large-Scale Morphological Elements   8
          2.4.1   Inlet   8
          2.4.2   Ebb Delta   9
          2.4.3   Flood Delta   11
   2.5   Back-Barrier Lagoon   11

3   Sand Transport Pathways   13
   3.1   Introduction   13
   3.2   Sediment Budget   14
   3.3   Sand Bypassing   16
          3.3.1   Bypassing Modes   16
          3.3.2   Bypassing Modes and the *P/M* Ratio   17
   3.4   Inlet Closure   19
   3.5   Location Stability   19
   3.6   Effect of Inlets on Adjacent Shoreline   20
          3.6.1   Continuous Bypassing   20
          3.6.2   Intermittent Bypassing   21

4   Sand Transport and Sand Bypassing at Selected Inlets   24
   4.1   Introduction   24
   4.2   Price Inlet   24

4.3	Breach Inlet	25
4.4	Captain Sam's Inlet	26
4.5	Mason Inlet	28
4.6	Wachapreague Inlet	29
4.7	Katikati Inlet	30
4.8	Ameland Inlet	31

5	**Empirical Relationships**		**34**
	5.1	Introduction	34
	5.2	Cross-Sectional Area – Tidal Prism Relationship	34
		5.2.1 Observations	34
		5.2.2 Physical Justification of the $A-P$ Relationship	35
		5.2.3 Examples of $A-P$ Relationships for Natural Inlets	37
		5.2.4 Equilibrium Velocity	40
	5.3	Relationship between Depth and Width of the Cross-Section and Tidal Prism	40
	5.4	Ebb Delta Volume – Tidal Prism Relationship	41
	5.5	Flood Delta Volume – Tidal Prism Relationship	43

6	**Tidal Inlet Hydrodynamics; Excluding Depth Variations with Tidal Stage**		**44**
	6.1	Introduction	44
	6.2	Inlet Schematization	44
	6.3	Governing Equations and Boundary Condition	44
		6.3.1 Dimensional Equations	44
		6.3.2 Non-Dimensional Equations; Lumped Parameter Model	46
	6.4	Analytical Solution (Öszoy–Mehta)	47
		6.4.1 Basin Tide and Inlet Velocity	47
		6.4.2 Nature of the Solution; Resonance	49
	6.5	Semi-Analytical Solution (Keulegan)	50
		6.5.1 Basin Tide and Inlet Velocity	50
		6.5.2 Maximum Basin Level and Maximum Inlet Velocity	53
		6.5.3 Relative Contribution of the Third Harmonic	54
		6.5.4 Multiple Inlets	54
	6.6	Application to a Representative Tidal Inlet	55
		6.6.1 Representative Tidal Inlet	55
		6.6.2 Öszoy–Mehta Solution	55
		6.6.3 Keulegan Solution	57
	6.A	Dynamics of the Flow in the Inlet	58

7 **Tidal Inlet Hydrodynamics; Including Depth Variations with**
    Tidal Stage 61
    7.1   Introduction 61
    7.2   Equations Including Depth Variations with Tidal Stage 61
    7.3   Solution of the Leading-Order Equations 63
    7.4   Solution to the First-Order Equations 65
           7.4.1   First-Order Forcing 65
           7.4.2   Mean Inlet Velocity and Mean Basin Level 66
           7.4.3   First-Order Tide and Velocity 67
    7.5   Tidal Asymmetry 69
    7.6   Application to the Representative Inlet 70
           7.6.1   Leading-Order Solution 70
           7.6.2   First-Order Solution 72
           7.6.3   Tidal Asymmetry 72
    7.A   Reduced System of Equations and Perturbation Analysis 73

8 **Cross-Sectional Stability of a Single Inlet System** 75
    8.1   Introduction 75
    8.2   Equilibrium and Stability 75
           8.2.1   Escoffier Stability Model 75
           8.2.2   Escoffier Diagram 76
           8.2.3   The Shape of the Closure Curve 77
    8.3   Adaptation Timescale 78
    8.4   Cross-Sectional Stability of Pass Cavallo 80
    8.A   Geometric Similarity 82
    8.B   Linear Stability Analysis 84

9 **Cross-Sectional Stability of a Double Inlet System, Assuming**
    a Uniformly Varying Basin Water Level 86
    9.1   Introduction 86
    9.2   Escoffier Stability Model for a Double Inlet System 87
           9.2.1   Schematization 87
           9.2.2   Equilibrium Velocity 88
           9.2.3   Governing Equations 89
           9.2.4   Closure Surface 90
           9.2.5   Equilibrium Velocity Curves 90
           9.2.6   Flow Diagram 91
    9.3   Conditions for a Set of Stable Cross-Sectional Areas 93
    9.4   Basin with Topographic High 95
           9.4.1   Schematization 95

    9.4.2   Governing Equations   96

    9.4.3   Flow Diagrams   97

10 Cross-Sectional Stability of a Double Inlet System, Assuming a Spatially Varying Basin Water Level   100

   10.1   Introduction   100

   10.2   Schematization   100

   10.3   Governing Equations and Boundary Conditions   100

   10.4   Solution Method   103

   10.5   Effect of Spatial Variations in Basin Water Level on Cross-Sectional Stability   104

       10.5.1   Spatial Variations in Basin Water Level   104

       10.5.2   Comparison with Earlier Stability Analysis   105

       10.5.3   Effects of Basin Depth, Coriolis Acceleration, Radiation Damping and Basin Geometry   106

   10.6   Multiple Inlets   109

11 Morphodynamic Modeling of Tidal Inlets Using a Process-Based Simulation Model   110

   11.1   Introduction   110

   11.2   Model Concept and Formulation   110

   11.3   Morphology of a Newly Opened Inlet   112

   11.4   Cross-Sectional Area – Tidal Prism Relationship   115

   11.5   Limitations of Process-Based Morphodynamic Models   118

12 Morphodynamic Modeling of Tidal Inlets Using an Empirical Model   120

   12.1   Introduction   120

   12.2   Modeling Concepts   120

   12.3   Ebb Delta Development at Ocean City Inlet   121

       12.3.1   Ocean City Inlet   121

       12.3.2   Schematization and Model Formulation   121

       12.3.3   Model Results   123

   12.4   Adaptation of the Frisian Inlet after Basin Reduction   124

       12.4.1   Frisian Inlet   124

       12.4.2   Schematization and Model Formulation   125

       12.4.3   Model Results   128

       12.4.4   Analytical Solution; Local and System Timescales   130

       12.4.5   Bumps and Overshoots   134

12.5   Adaptation of an Inlet-Delta System Using a Diffusive
        Transport Formulation   135
12.6   Limitations of Empirical Modeling   138

13   River Flow and Entrance Stability   139
13.1   Introduction   139
13.2   Effect of River Flow on Basin Tide and Inlet Velocity   140
13.3   Effect of River Flow on Cross-Sectional Stability of
        Selected Inlets   143
        13.3.1   Thuan An Inlet: A Permanently Open Inlet   143
        13.3.2   Wilson Inlet: A Seasonally Open Inlet   145
        13.3.3   Lake Conjola Inlet: An Intermittently Open Inlet   145
13.4   A Morphodynamic Model for the Long-Term Evolution of
        an Inlet   147
13.A   Öszoy–Mehta Solution Including River Flow   150

14   Engineering of Tidal Inlets   152
14.1   Introduction   152
14.2   Artificial Opening of a New Inlet   152
14.3   Relocation of an Existing Inlet   155
14.4   Dredging   156
14.5   Sand Bypassing Plants   156
14.6   Jetties; Jetty Length and Orientation   158
14.7   Weir-Jetty Systems   160

References   161
Index   172

Preface

Historically, interest in tidal inlets originates from their importance for commercial shipping and recreational boating. Unfortunately, when in a natural state, most inlets are less than ideal from a navigational point of view and need improvement. They are unstable, i.e., as a result of tide and waves they have a tendency to migrate and shoal. Initially, to stabilize inlets, common sense and practical experience was used as the sole guide. It was not until the nineteen-twenties that research, using field observations, mathematical analysis and laboratory experiments, led to an improved understanding of the complex physical processes that govern the water motion and morphology of tidal inlets. This knowledge could then be used to arrive at science-based improvements.

This book summarizes and synthesizes the scientific advances in inlet research with emphasis on the period 1978 to present. It is a sequel to the earlier books, *Stability of Coastal Inlets* by Per Bruun and Gerritsen (1960) and *Stability of Tidal Inlets: Theory and Engineering* by Per Bruun et al. (1978). The focus is on natural (no man-made modifications) tidal inlets in a sandy environment. The book is intended for anyone who is interested or has dealings with tidal inlets, including coastal engineers, coastal scientists, students and managers. The two authors made an equal contribution to the contents of this book.

Per Bruun and Frans Gerritsen, through their afore mentioned book, were central in introducing Co van de Kreeke to the field of tidal inlets. Discussions with Per Bruun, Robert Dean, Murrough O'Brien and Ashish Mehta have further stimulated this interest. Ronald Brouwer was introduced to the field of tidal inlets by Co van de Kreeke. They worked closely together during his graduate work on tidal inlets at Delft University of Technology, The Netherlands. The support of Henk Schuttelaars and Pieter Roos during that period is acknowledged.

In preparing the manuscript, a number of chapters have benefited greatly from discussions with colleagues. They include Henk Schuttelaars and Pieter Roos on Chapters 9 and 10, Zheng Wang on Chapter 12 and Erroll Mclean and Jon

Hinwood on Chapter 13. Albert Oost was helpful in explaining the geology and sedimentology of the Wadden Sea.

The authors would like to acknowledge Duncan FitzGerald, Todd Walton, Ashish Mehta, Henk Schuttelaars, Marcel Stive, Pieter Roos, Huib de Swart, Zheng Wang and Judith Bosboom for reviewing earlier versions of different chapters.

Co van de Kreeke did most of his research on tidal inlets during the period 1971–2003, while a professor at the Rosenstiel School of Marine and Atmospheric Science of the University of Miami. During that period he cooperated regularly with the National Institute of Coastal and Marine Management of the Dutch Rijkswaterstaat. In that context the many discussions with Job Dronkers should be mentioned. After graduating, Ronald Brouwer worked on the topic of tidal inlets as a Post-Doctoral Fellow at the Delft University of Technology.

In-kind support of the Rosenstiel School of Marine and Atmospheric Science of the University of Miami, by providing work space, computer support and library services during Co van de Kreeke's tenure as an emeritus professor, is acknowledged. Marcel Stive was instrumental in having Ronald Brouwer do his graduate and post-doctoral work on tidal inlets and providing financial support.

We wrote this book out of curiosity, looking for answers to questions such as: why do tidal inlets wander and shift position; why are maximum velocities close to the same for most inlets; how do inlets interact with other inlets; how do they affect the beaches; and many more. We found the answers to some of the questions but certainly not to all of them. There remains room for future research.

Not knowing all the answers made it difficult to decide when and where to stop. We decided that, after having worked on the book for five years, it was enough and could we answer the often-posed question, "are you still working on that book?" with a resounding "No!"

1

Introduction

In the context of this book, tidal inlets are defined as the relatively short and narrow passages between barrier islands. They are sometimes referred to as passes or cuts. Tidal inlets are a common occurrence as barrier island coasts cover some 10 percent of the world's coasts (Glaeser, 1978). According to Hayes (1979), their presence is limited to coasts where the tidal range is less than 4 m.

The earliest interest in tidal inlets originates from their importance to commercial shipping. The relatively protected back-barrier lagoons were a favorite location for harbors. Later, with the increase in recreational boating, small boat basins and marinas were located in back-barrier lagoons. In addition to these commercial and recreational aspects, tidal inlets are ecologically important. Through the exchange of lagoon and ocean water, they contribute to the increase of water quality in the lagoon. Unfortunately, there is also a downside: tidal inlets interrupt the flow of sand along the coast. They not only interrupt but also capture part of the sand, causing erosion of the downdrift coast. For example, in Florida, with some eighty inlets, much of the beach erosion has been attributed to tidal inlets.

Most natural tidal inlets are less than ideal from a navigational point of view. The many shoals, the strong tidal current and the exposure to ocean waves make entering difficult. In addition, on timescales of years to decades, the morphology shows considerable variation, and maintaining sufficient depth and alignment of the channels requires substantial dredging. To minimize dredging and to improve navigation conditions, many inlets have been modified by adding jetties and breakwaters. As a result, tidal currents, waves and sand transport pathways differ from those at inlets without these structures. Nevertheless, in this book, emphasis is on tidal inlets that have not been modified. The reasoning is that understanding the physical processes governing the behavior of tidal inlets in a natural state is a prerequisite for the proper design of engineering measures. This includes the determination of undesirable side effects such as erosion of the adjacent beaches.

The main morphological features of a tidal inlet are the inlet, the ebb delta on the ocean side and the flood delta on the lagoon side. The morphology of the inlet and deltas is shaped by tide and waves. Depending on their relative importance, tidal inlets have been categorized as tide- or wave-dominant (Davis and Hayes, 1984; Hayes, 1994). Tides tend to keep inlets open. In this respect, the tidal prism – the volume of water entering the inlet on the flood and leaving during the ebb – is an important parameter; the larger the tidal prism, the larger the inlet. In turn, waves tend to close the inlet through the wave-driven longshore and cross-shore sand transport.

Scoured in sand, tidal inlets are dynamic features; inlet and channels move, ebb deltas change shape and volume. In discussing these morphological changes, emphasis is on processes with timescales of days (storm timescale), weeks and decades, as opposed to the geological timescale. This excludes the small-scale sand transport processes to which reference is made to Soulsby (1997) and van Rijn (1993). The morphology of the back-barrier lagoon comes into play only as it affects the water motion in the tidal inlet. For the water motion and morphology of the back-barrier lagoon, reference is made to Dronkers (2005).

Examples of barrier island coasts are the East Coast of the US, the Gulf Coast of the US, the Dutch, German and Danish Wadden Sea coast (Fig. 1.1a), the east coast of Vietnam, the northeast coast of the North Island of New Zealand, the Algarve coast of Portugal (Fig. 1.1b) and the Adriatic coast of Italy (Fig. 1.1c). Many inlets along these coasts have been studied extensively. For the origin and morphology of barrier island coasts reference is made to a series of articles in Davis (1994). Rather than one, most barrier island coasts consist of a chain of islands resulting in multiple inlets connected to the same back-barrier lagoon. In fact, it would be difficult to find a tidal inlet that is not affected by a companion inlet. In case of multiple inlets, each inlet competes for part of the tidal prism. This could lead to some inlets closing while others remain open.

Since the 1960s the main tool in studying tidal inlet processes has shifted from laboratory research, including scale models and flume studies, to mathematical models. A distinction is made between process-based models and empirical models (van de Kreeke, 1996). Following Murray (2003), process-based models are divided in exploratory and simulation models. Exploratory models only include the processes that are essential in reproducing the basic behavior. Process-based simulation models start with basic physics and are designed to reproduce the behavior of a natural system, or a schematization thereof, as accurately as possible. Process-based exploratory models are usually simplified to a level allowing analytical or semi-analytical solutions. Although not adequate for predictive purposes, they can be used in a diagnostic mode to help understand phenomena observed in the field

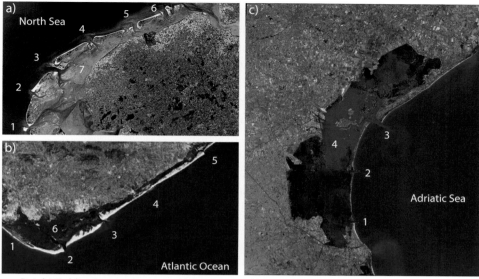

Figure 1.1 a) the Dutch Wadden Sea coast (USGS and ESA, 2011), b) the Algarve coast of Portugal (Esri et al., 2016) and c) the Adriatic coast of Italy (NASA et al., 2003).

and to check the validity of the results of the more complicated simulation models. Solving the equations underlying the process-based simulation models requires a numerical approach. The models provide a far more realistic representation of the physical processes than the exploratory models; however, a drawback is that it is often difficult to pinpoint the interactions that determine the overall behavior.

Empirical models are based on the assumption that after a perturbation the morphology tends towards an equilibrium state. The equilibrium state is defined by empirical relationships between the size or volume of the morphological units and the tidal prism. The return to equilibrium is described by empirical equations. Empirical models are a useful substitute when knowledge of the basic processes is insufficient, as is often the case.

The book summarizes and synthesizes the advances in tidal inlet research over the past 40 years. Emphasis is on natural inlets in a sandy environment with tide and waves as the dominant forcing. The book is organized as follows.

In Chapters 2–4 a description of the morphology and morphological changes of tidal inlets is presented. Chapter 2 describes the origin and major elements: inlet, ebb delta, flood delta and back-barrier lagoon. Chapter 3 focuses on sand transport pathways and sand bypassing. Attention is given to location stability, modes of bypassing and their relationship to the ratio of tidal prism and long-shore sand transport (P/M ratio). Furthermore, the effect of inlets on the adjacent shores is discussed. In Chapter 4, selected inlets are reviewed with emphasis on sand bypassing and location stability.

Chapter 5 deals with the empirical relationships. The empirical relationship between inlet cross-sectional area and tidal prism (A–P relationship) is discussed and a physical explanation for this relationship is given. Additionally, the concept of equilibrium velocity is introduced. An empirical relationship between delta volume, tidal prism and wave energy is also presented.

Chapters 6 and 7 introduce process-based exploratory models that are used to explore the hydrodynamics of tidal inlets. In Chapter 6 a lumped parameter model and the Keulegan and Öszoy–Mehta Solutions are described. The internal generation of the third harmonic is discussed. Solutions are applied to a representative inlet and results are compared to those of a numerical solution. In Chapter 7 the hydrodynamic equations are expanded to include depth variations with tidal stage. The analytical solution to the expanded equations shows the generation of even overtides and the resulting tidal asymmetry, mean inlet velocity and mean basin level.

Chapters 8–10 deal with cross-sectional stability. Cross-sectional stability is determined using the Escoffier Stability Model. The Escoffier Stability Model, including the Escoffier Diagram, is described in Chapter 8. As examples, the model is applied to two single-inlet systems, Pass Cavallo (TX) and a representative inlet. An expression for the adaptation timescale of the inlet cross-sectional area after a storm is presented. Chapters 9 and 10 deal with the cross-sectional stability of double inlet systems, i.e., rather than one inlet the back-barrier lagoon is connected to the ocean by two inlets. In Chapter 9 the water motion in the inlets is described by the lumped parameter model. This model includes the one-dimensional hydro-dynamic equations for the inlet and the assumption of a uniformly fluctuating basin water level. To investigate the effect of this assumption on cross-sectional stability, variations in basin water level are introduced by dividing the basin into two sub-basins connected by an opening, representing a topographic high. As part of the stability analysis a flow diagram is introduced. The flow diagram is the two-dimensional counterpart of the Escoffier Diagram. In Chapter 10, the spatial variations in basin water level are introduced by describing the hydrodynamics of inlets and basin by the shallow water wave equations. A semi-analytical solution is used to solve the governing equations.

Chapters 11 and 12 present applications of a process-based simulation model and an empirical model, respectively. In Chapter 11 a process-based simulation model is used to determine the morphology of a newly opened tidal inlet with emphasis on the inlet and the ebb delta. Using different inlet dimensions, ocean tidal amplitudes and basin surface areas, a series of numerical experiments is carried out to verify the $A-P$ relationship. Chapter 12 describes the use of empirical models to explain the ebb delta development at a newly opened inlet (Ocean City Inlet, MD) and the adaptation of an inlet and ebb delta after basin reduction (Frisian Inlet, NL).

Chapter 13 focuses on the effect of river flow on the entrance stability of tidal inlets. The effect of river flow on the basin tide and the mean basin level is shown by expanding the Öszoy–Mehta Solution to include river flow. The effect of river flow on cross-sectional stability is discussed for a permanently open inlet, a seasonally open inlet and an intermittently open inlet. An exploratory morphodynamic model for the evolution of the depth of an inlet subject to river flow is presented.

Chapter 14 reviews measures to improve navigation and sand bypassing at tidal inlets.

2

Geomorphology

2.1 Introduction

Depending on their origin, tidal inlets are identified as primary or secondary inlets. Regardless of the origin, the morphology is characterized by three major elements: the inlet, the ebb delta and the flood delta. The morphology of each element is determined by tide and waves. In particular, the tidal prism (the volume of water entering on the flood and leaving on the ebb) and the wave-induced longshore sand transport play an important role in determining the cross-sectional area of the inlet and the size and shape of the ebb delta.

2.2 Origin of Tidal Inlets

Following Ehlers (1988), in tracing the origin of tidal inlets a distinction is made between primary and secondary inlets. Primary inlets are those where pre-existing relief, characterized by troughs and adjacent highs, plays a decisive role in the formation of the inlet. During the Holocene, starting some 10,000 BP, onshore sand transport associated with the rapid rise in sea level caused these existing troughs to fill while barrier islands formed on the adjacent highs (Jelgersma, 1983). Examples of primary inlets are the Ameland Inlet (Fig. 4.7) and the Frisian Inlet (Fig. 12.4) along the Dutch Wadden Sea, both relics of drowned river valleys (Beets and van der Spek, 2000).

Apart from these primary tidal inlets, secondary inlets can be identified. Secondary inlets originate from flooding of narrow and shallow parts of barrier islands during a storm. Once the fore-dune ridge is dismantled as a result of storm erosion, a shallow washover channel develops. As the storm passes and winds change direction, return flow forces water against the landward side of the barrier. Often the return flow is funneled across the low portion of the barrier island through the washover channel. Depending on the tidal prism and the longshore sand transport, the washover channel closes or remains open. When remaining open, this channel

Figure 2.1 The breach at Old Inlet, Fire Island (NY) before (left, 2010) and after (right, November 2012) hurricane Sandy (National Park Service, 2012).

is then gradually enlarged by the ensuing tidal currents. Sand from the channel is deposited both offshore and in the basin, forming the onset to, respectively, the ebb delta and the flood delta. All these secondary or washover inlets are located in a sand-rich environment. As discussed in more detail in the following chapters, the ultimate shape and size of newly opened inlets depends on tide and waves.

A description of a washover inlet that ultimately closed on its own can be found in El-Ashry and Wanless (1965). The tidal inlet was located 4.5 miles north of Cape Hatteras (NC). It opened in March 1962 during a severe storm. Large volumes of sand collected at the landward side of the inlet with channels connecting inlet and lagoon. During the first month the inlet width and the volume of the deposits experienced rapid growth, and then the growth slowed down and ultimately the inlet closed on its own. More recently, as a result of hurricane Sandy, a washover inlet was opened at Fire Island (NY), connecting Great South Bay and the ocean (Fig. 2.1). The inlet opened in October 2012 and since then has been extensively monitored. As of September 2016, the inlet is still open (National Park Service, 2012).

In addition to these naturally opened inlets, there are numerous man-made inlets. They are opened or relocated for the purpose of either navigation or water quality. Examples are Government Cut, Lake Worth Inlet, South Lake Worth Inlet and Bakers Haulover Inlet, all located on the southeast coast of Florida (Stauble, 1993). The first two were opened for navigation purposes. Water quality was the main motive for opening South Lake Worth Inlet and Bakers Haulover Inlet.

2.3 Equilibrium Morphology

Newly opened inlets either close or remain open. When remaining open the morphology tends towards equilibrium. In most cases this equilibrium is dynamic rather than static, whereby the morphology oscillates about an equilibrium state.

Oscillations are either the result of variations in the hydrodynamic forcing or are associated with intrinsic instabilities of the morphology.

Examples of variation in forcing are the spring–neap tidal cycle and the seasonal variation in storm severity. Byrne et al. (1974) report fortnightly variations in the cross-sectional area of the inlet channel of Wachapreague Inlet (VA) (Section 4.6), resulting from the spring–neap variation in the tide. Seasonal variations in morphology are reported in Morris et al. (2004). Using video techniques and wave measurements they show that the seasonal behavior of the Barra Nova Inlet (formerly called Ancão Inlet), Rìa Formosa, Portugal, is cyclic in nature. The center and the alignment of the inlet change due to high-energy winter storms, after which the inlet returns to the original morphological state during the remainder of the year. The timing and progression of the morphology through this cycle is thus closely related to the seasonal variation in local wave climate.

Examples of intrinsic instabilities in morphology are spit formation and breaching and the movement of channels on the ebb delta. Both of these lead to cyclic variations in morphology. Examples of tidal inlets where spit formation and breaching plays a role are Captain Sam's Inlet (SC) (Hayes, 1977) and Mason Inlet (NC) (Cleary and FitzGerald, 2003). Channel movement on the ebb delta is observed at Ameland Inlet, The Netherlands (Israel and Dunsbergen, 1999) and Price Inlet (Fitzgerald et al., 1984). In both cases the timescale of the cyclic motion is measured in decades. Detailed information on Captain Sams, Mason, Ameland and Price Inlets is presented in Chapter 4.

In summary, the morphology of tidal inlets is not static but shows variations, with timescales ranging from weeks to decades. In addition, as a result of gradual infilling of the back-barrier lagoon, and depending on the rate of sea level rise, tidal inlets on timescales of centuries can become self-destructive.

2.4 Large-Scale Morphological Elements

In spite of the forcing and sediment conditions being different, the morphology of most tidal inlets is characterized by three large-scale elements: inlet, ebb delta and flood delta (Fig. 2.2). The deltas are an integral part of the tidal inlet as they affect the hydrodynamics and vice versa. In particular, the ebb delta constitutes a bridge that allows bypassing of sand from the updrift to the downdrift coast. In the next sections the main characteristics of each of these elements will be briefly discussed.

2.4.1 Inlet

Here, "inlet" refers to the channel separating two barrier islands. Inlets exhibit a wide range of plan-forms. Some are straight and oriented perpendicular to

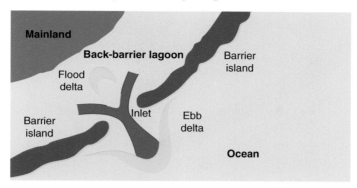

Figure 2.2 Schematized tidal inlet (adapted from Davis and FitzGerald, 2004).

the coast; others exhibit strong curvature. An important role in determining the orientation is played by the longshore sand transport and the orientation of the channels in the back-barrier lagoon. In addition, geological constraints, such as consolidated sediment and rock outcroppings, can play a role. At the seaward end, where the inlet connects to the ebb delta, there is often a tendency for the inlet to branch in separate ebb and flood channels: a result of different flow patterns for ebb and flood. The ebb flow pattern is similar to that of a turbulent jet, whereas the flood flow resembles a streamlined flow into a bell-shaped entrance (Stommel and Farmer, 1952).

Cross-sectional areas of the inlet vary and are assumed to be smallest where the inlet is narrowest and the depth is largest. This part of the inlet is referred to as the gorge or the throat. The cross-sectional area of the throat section is determined by the tidal prism and longshore sand transport (Chapter 5). The shape of the inlet cross-sections is most often asymmetric, being steeper on one side than the other, and the result of the ebb flows having a preference for one of the inlet sides. Dimensions of inlets are: length 500–5,000 m, depth 2–30 m and width 50–2,000 m.

2.4.2 Ebb Delta

An ebb tidal delta, or ebb delta, is the body of sand on the ocean side of a tidal inlet. Typically, ebb deltas consist of a triangular or half-circle-shaped platform: a sandy area of relatively shallow depth containing low-relief bars and shallow channels emanating from the inlet. The shape and volume of the ebb delta are determined by the relative dominance of tidal versus wave energy. In tide-dominated environments, ebb deltas are relatively large, extending far offshore. In wave-dominated environments, ebb deltas are small, hugging the shore (Sha, 1989). In

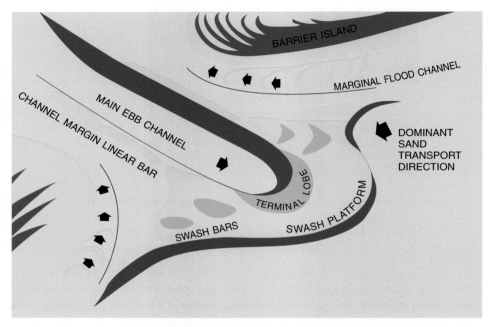

Figure 2.3 Idealized ebb delta (adapted from Hayes, 1980, and Davis and FitzGerald, 2004).

wave-dominated inlets, with a large longshore sand transport, the channels on the platform are forced in a downdrift direction and the delta takes on an asymmetric shape.

Based on field studies in a number of tidal inlets, Hayes (1980) presented a general description of ebb delta morphology. In its simplest form the ebb delta has one main channel conveying the flow from the inlet to the ocean (Fig. 2.3). On both sides of the main ebb channel are channel margin linear bars, levee-like deposits. At the end of the main channel is a relatively steep, seaward-sloping lobe of sand called the terminal lobe. Broad sheets of sand, called swash platforms, flank both sides of the main channel. On the platform are swash bars that form by the swash action of waves. The swash bars migrate across the swash platform by the action of wave-generated currents (King, 1972). The ebb delta is separated from the barrier islands by marginal flood channels. Because this is an idealized representation, some of the features in this simplified model will be more pronounced in some tidal inlets than in others, and some might not be present at all.

The depth of water over the delta affects navigation. In that respect the depth over the shallowest part of the ebb channel is important. It is usually located over the terminal lobe. In general, this depth is much less than the minimum depth in the inlet (Bruun et al., 1978; Dean, 1988).

2.4.3 Flood Delta

Much less is known about flood deltas than ebb deltas. Flood deltas are shaped chiefly by tidal currents as waves play a minor role. At many tidal inlets the flood delta is not even a distinguishable feature as it has merged with the marsh or tidal flats. Flood delta characteristics differ widely from one tidal inlet to another and therefore it is difficult to present their features in a single stylized figure, as was done for the ebb delta. An attempt to identify the major components of flood deltas is described in Hayes (1980). Where flood deltas are a distinct feature, the indications are that their volume increases with increasing values of the tidal prism (Powell et al., 2006).

2.5 Back-Barrier Lagoon

Following Davis and FitzGerald (2004), back-barrier lagoons resemble either an open bay (Fig. 2.4a), an elongated stretch of open water parallel to the mainland (Fig. 2.4b) or a basin with tidal channels marshes and mudflats (Fig. 2.4c). The morphology of the back-barrier lagoon and inlet is shaped by the tide (Dronkers, 2005). In this respect an important role is played by the tidal prism, the volumes of water entering on the flood and leaving on the ebb. In addition to an exchange of water, the tide results in an exchange of sediment. Although the amounts of sediment entering on the flood and leaving on the ebb can be substantial, the net transport over a tidal cycle is usually small, with the direction depending on the tidal velocity asymmetry (Section 7.5). However, on timescales much larger than the tidal period, the net transport can be significant. In case of a net import, sediment is deposited in the relative calm lagoon water resulting in the formation of tidal flats and a decrease in surface area and depth. Where vegetation is established, the originally open lagoon increasingly resembles a marsh area incised with channels. Examples are the tidal inlets along the South and North Carolina coast.

As a result of lagoon infilling, the tidal prism and the velocities decrease. Ultimately, velocities decrease to a level where they are no longer capable of removing

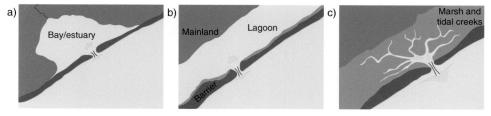

Figure 2.4 Different types of back-barrier lagoons (adapted from Davis and FitzGerald, 2004).

the longshore sand transport and storm deposits out of the inlet, and the inlet closes. Lagoon infilling takes place on timescales of centuries. Therefore, disregarding sea level rise, many inlets on these timescales are self-destructive (van der Spek and Beets, 1992). When accounting for sea level rise, the fate of the tidal inlet depends on the relative rate of sea level rise and the rate of sedimentation in the lagoon. If the sedimentation rate is smaller than the rate of sea level rise, the tidal prism will remain close to the same and the inlet will remain open. When larger than the rate of sea level rise, the lagoon will fill in. As a result, the tidal prism will continue to decrease and the inlet will ultimately close (Louters and Gerritsen, 1994). An example of an ephemeral inlet where the rate of sedimentation was larger than the rate of sea level rise is described in Jelgersma (1983). The Bergen Inlet along the west coast of The Netherlands existed from 5,300 to 3,300 BP and then closed as a result of infilling, in spite of sea level rise.

3

Sand Transport Pathways

3.1 Introduction

The major elements, inlet, ebb delta and flood delta, together with the adjacent coast constitute a sand sharing system (Dean, 1988); sand is transported among these elements by tide- and wave-generated currents. Because at tidal inlets direct measurements are difficult, much of what is known of sediment transport and sediment transport pathways has been inferred from migration and shape of bed forms and swash bars, dredging records, comparison of sequences of bathymetric maps and aerial photographs (Bruun and Gerritsen, 1959; Hanisch, 1981; Hine, 1975).

Sand is transported towards a tidal inlet by longshore currents. Longshore currents and the resulting longshore sand transport result from waves approaching the coast at an oblique angle (Kamphuis, 2006). Some of the longshore sand transport is carried into the inlet by the flood currents and is deposited in the back-barrier lagoon. Another part is jetted to the deeper parts of the ocean and some of it is transported over the ebb delta to the downdrift coast. The sand stored in the lagoon and the deeper parts of the ocean is lost to the littoral zone. As a result, the supply of sand to the downdrift coast is less than the longshore sand transport causing erosion of this part of the coast. The details of the transport of sand from the updrift to the downdrift coasts are discussed in Section 3.3.

An example of sand entering and leaving an inlet is presented in Fig. 3.1. Sand enters through the porous breakwater and is temporarily stored on the updrift side of the inlet in the form of a protruding sand bank. During ebb, sand is carried from the bank in an offshore direction. A similar process was observed in a small inlet in the Bay Islands, Honduras. In that case the clarity of the water and the size of the inlet (width 3 m, depth 0.3 m) made it possible to visually observe the deposition and formation of the sand bank on the updrift side of the inlet and the subsequent removal of some of the sand during ebb.

13

Figure 3.1 St. Lucie Inlet (FL). Sand is carried to the inlet by the longshore current and deposited on the updrift side of the inlet. Sand is removed from the updrift
side of the inlet by the ebb tidal current. Picture taken somewhere in the seventies
(Photo: Paul W. Larsen).

In view of the limited knowledge of sand transport and sand transport pathways, the present conceptual models for the sand transport and the sand transport
pathways at tidal inlets are somewhat speculative. A basic assumption is that each
element (inlet, ebb delta and flood delta), when disturbed, tends towards an equilibrium state. More specifically, when the volume of one element is altered, the whole
system responds to restore the original equilibrium by transporting sand among the
different elements. This is described in some detail in Chapter 12.

3.2 Sediment Budget

Whatever is known about sediment transport, sediment transport pathways and volume changes can be expressed in terms of a sediment budget (Dean-Rosati, 2005).
The sediment budget is a tally of sediment gains and losses, or sources and sinks,
within a specified control area over a given time. In general, the seaward boundary
of the control area corresponds to the seaward limit of the littoral zone, the zone
extending to a depth where sand transport by waves is negligible. This so-called
closure depth usually is between 10 and 20 m (Kamphuis, 2006; Komar, 1998).

The landward boundary of the control area varies: in some budgets it includes the entire back-barrier lagoon and in others only the flood delta. In the longshore direction the boundary extends to an updrift position, where the shoreline is not measurably affected by the inlet, and similarly for the downdrift position.

The control area is divided in interconnected cells representing such tidal inlet elements as ebb delta, flood delta, inlet and parts of the updrift and downdrift coast. For each cell the sediment balance can be written as

$$Tr_i - Tr_o - \Delta V + P - R = \text{residual}, \tag{3.1}$$

where Tr_i is the sand transport into the cell, Tr_o is the sand transport out of the cell, ΔV is the change in volume, P and R are the amounts of material placed in and removed from the cell and residual is the degree to which the cell is balanced. Provided sufficient information on the sand fluxes at the boundary of the control area with the outside world is available and ΔV, P and R are known, the sand transport between cells can be calculated.

To aid in the construction of a sediment budget and to evaluate the different terms in Eq. (3.1), a computer program, Sediment Budget Analysis System (SBAS), was developed (Kraus and Rosati, 1999; Rosati and Kraus, 1999). SBAS also allows the user to record uncertainty for each value entered in the sediment budget. Then, for each cell, SBAS calculates the root mean square uncertainty. The user can apply the root mean square uncertainty to indicate the relative confidence that can be given to each cell and compare alternatives that represent different assumptions about the sediment budget.

Examples of sediment budgets are those at Grays Harbor (Byrnes et al., 2003) and Faro-Olhão Inlet (Pacheco et al., 2008). Grays Harbor is an estuary on the southwest coast of the state of Washington, USA. The entrance to the estuary has two jetties constructed during the period 1898–1916. A sediment budget covering the period 1987–2002 is presented. The aim of the sediment budget was to document the sediment transport pathways and rates during that period. The control area covers the entrance area, including part of the offshore, the beaches on either side of the inlet. Sediment fluxes at the boundaries with the outside world were partly based on regional transport results and numerical modeling estimates. The control area is divided in nine cells, representing the major morphological elements. For each cell the change in sand volume and the dredging and placement volumes were determined. With the known sediment fluxes at the boundaries with the outside world, the sand transport between cells follows from the observed volume changes and the known dredging and placement volumes.

Faro-Olhão is one of the inlets of the Ría Formosa (Portugal) barrier island system, serving the cities of Faro and Olhão. The inlet was opened in 1929 and gradually improved with jetties between 1929 and 1955. Sediment budgets were constructed for the time periods 1962–1978 and 1978–2001. The objective was to explain the observed coastline changes after the opening and the stabilization of the inlet. The control area covered the entire inlet system. The seaward limit of the control area was taken at the transition of the ebb delta slope and sea floor. The landward limit of the control area coincides with the seaward boundary of the salt-marsh in the back-barrier lagoon. The control area was divided into six cells representing the flood delta, ebb delta, inlet channel, ebb delta channel and adjacent updrift and downdrift coast. For each cell the change in volume, including the dredged and placed sediment volumes, was determined. Because the information on sediment fluxes at the boundaries of the control area is insufficient, internal sand transport rates and pathways could not be calculated. In the paper considerable attention is given to the different methods to estimate the volume changes in a cell, including the application of the error analysis presented in (Dean-Rosati, 2005).

3.3 Sand Bypassing

3.3.1 Bypassing Modes

The way sand is transported from the updrift to the downdrift coast of a tidal inlet is commonly referred to as inlet sand bypassing. It is the way that sand, after a short interruption on the ebb delta and in the inlet, is returned to the littoral zone. Bruun and Gerritsen (1959) were the first to address bypassing. They reasoned that the transfer of sand is the result of waves and tidal currents. Based on this observation they distinguished between bar bypassing, where waves are dominant, and tidal flow bypassing, where tide is dominant.

Bar bypassing implies that sand is directly transferred from the updrift coast onto the ebb delta and then to the downdrift coast via the terminal lobe and swash platform (Fig. 2.3). The terminal lobe and swash platform serve as a bridge over which the sand is carried to the downdrift beach. When channels are present on the delta platform, they are small and shallow.

Tidal flow bypassing implies that, during flood, sand is transported into the inlet and main ebb channel. Part of this transport takes place across the channel margin linear bars (Fig. 2.3).This lateral inflow of sand often results in a cyclic variation of the orientation of the ebb channel on a timescale of decades. Sand is transported out of the main ebb channel by the ebb tidal currents and deposited at the distal end (the seaward end) of the channel. Swash bars are formed at the distal end that, as a result of waves and the dominant landward current, move onshore over the swash platform. On their way to the beach they form bar complexes. It takes these bar complexes, any time from several years to decades to reach the beach.

Lately, it has become apparent that there are many bypassing mechanisms that do not fit the mode of bar bypassing or tidal flow bypassing. Some eight of these are categorized in FitzGerald et al. (2000). In addition to bar bypassing and tidal flow bypassing, the ones that stand out are spit formation and ebb delta breaching.

Spit formation implies that the littoral drift, rather than being transferred to the ebb delta, is deposited as a sand-spit in front of the inlet. When the spit grows, it forces the inlet and the ebb delta to move in a downdrift direction. When breached, usually at the original position of the inlet, the spit welds onto the downdrift shore together with the sand from the abandoned ebb delta. At many inlets this process is repeated at decadal timescales, resulting in an episodic transfer of large volumes of sand to the downdrift beaches. In Friedrichs et al. (1993), based on a study at Chatham Inlet (MA), it is suggested that spit breaching is initiated by a washover forced by a storm. For the washover to evolve into a channel requires a certain head difference between the inlet and the ocean at the location of the washover. This head difference between inlet and ocean results from the distortion of the tidal wave in the inlet channel and increases with increasing spit length.

Ebb delta breaching involves transfer of sand over the ebb delta. Sand enters the updrift side of the main ebb channel, forcing it to migrate in a downdrift direction. When the inlet throat position is relatively fixed, this causes an increased curvature in the alignment of the channel. A new channel is formed and the shoal between the old and new channel is transported by waves towards the downdrift coast. At many inlets this process is repeated at annual timescales, resulting in an episodic transfer of large volumes of sand to the downdrift beaches.

Because each inlet has its own peculiarities, it is difficult to arrive at a universal framework for bypassing modes. However, the four bypassing mechanisms, bar bypassing, spit formation and breaching, ebb delta breaching and tidal flow bypassing, are believed to at least capture the gross characteristics of inlet bypassing. Examples of inlets with different bypassing modes are presented in Chapter 4.

3.3.2 Bypassing Modes and the P/M Ratio

Starting with Bruun and Gerritsen (1959), attempts have been made to correlate bypassing mechanisms with the P/M ratio, where P is the tidal prism under spring tide conditions (m^3) and M is the gross longshore sand transport (m^3 year^{-1}). For a large number of both improved and unimproved inlets Bruun et al. (1978) determined the sand bypassing mode (bar bypassing or tidal flow bypassing) together with values of the ratio of tidal prism and gross longshore sand transport; see columns 2 and 3 in Table 3.1. Based on these observations they concluded that for bar bypassing the P/M ratio had to be smaller than approximately 50 and for tidal flow bypassing the P/M ratio had to be larger than approximately 150.

Recently, for a number of inlets more detailed information on bypassing mechanisms and P/M ratio has become available. The results are summarized in

Table 3.1 *Sand bypassing mode,* P/M *ratio and location stability for selected inlets (Bruun et al., 1978).*

Inlet	Sand bypassing mode	P/M	Location stability
Aveiro Inlet (Portugal)	Not known	60	Fair/Poor
Big Pass (FL)	Not known	100	Fair
Brielse Maas (Neth.)	Bar bypassing	30	Poor
Eyerlandse Gat (Neth.)	Tidal flow bypassing	200	Good
Figueira da Foz (Portugal)	Bar bypassing	28	Poor
Texel Inlet (Neth.)	Tidal flow bypassing	1,000	Good
Vlie Inlet (Neth.)	Tidal flow bypassing	1,000	Good
John's Pass (FL)	Tidal flow bypassing	140	Fair
Longboat Pass (FL)	Tidal flow bypassing	200	Good
Oregon Inlet (NC)	Not known	60	Fair/Poor
Ponce de Leon Inlet (FL)	Bar bypassing	30	Poor

P is spring tidal prism (m^3 year^{-1}) and M is gross longshore sand transport (m^3 year^{-1}). Values are of overall character.

Table 3.2 *Sand bypassing mode,* P/M *ratio and location stability for inlets discussed in Chapter 4.*

Inlet	Sand bypassing mode	P/M	Location stability
Captain Sam's Inlet (SC)	Spit formation	12	Poor
Mason Inlet (NC)	Spit formation	17	Poor
Breach Inlet (SC)	Ebb delta breaching	72	Fair
Price Inlet (SC)	Tidal flow bypassing	80	Good
Wachapreague Inlet (VA)	Tidal flow bypassing	182–455	Good
Ameland Inlet (Neth.)	Tidal flow bypassing	960	Good
Katikati Inlet (New Zealand)	Tidal flow bypassing	>190	Good

P is spring tidal prism (m^3 year^{-1}) and M is gross longshore sand transport (m^3 year^{-1}). Values are of overall character.

Table 3.2. For *tidal flow bypassing* the P/M ratios conform to those proposed by Bruun et al. (1978) and are larger than 150. An exception is Price Inlet; the reason for this could be that this inlet is eroded into semi-consolidated Pleistocene deposits (FitzGerald, 1984). The bypassing mode *spit formation and breaching* has the smallest P/M values and the P/M value for *ebb delta breaching* is somewhere in between. No information on inlets with *bar bypassing* is available other than that provided in Bruun et al. (1978). To determine whether bypassing mechanisms are uniquely determined by the P/M ratio, and, if not, what other parameters play a role, additional observations are needed.

3.4 Inlet Closure

Not all sand is necessarily bypassed and some of it is permanently deposited in the inlet, resulting in closure. Following Ranasinghe and Pattiaratchi (2003), a distinction is made between closure by longshore and cross-shore sand transport. Longshore sand transport results from waves approaching the shore at an oblique angle. Onshore transport is due to swell waves approaching the shore more or less perpendicularly. As shown in Section 3.3, inlets with a relatively large longshore transport rate and small tidal prism, and thus a small P/M ratio, are prone to spit formation. If not breached, the spit will continue to accrete and prograde, resulting in a lengthening of the inlet in a shore-parallel direction. The inlet becomes less hydraulically efficient and closes. An example is Midnight Pass on the west coast of Florida (Davis et al., 1987). Assuming the longshore sand transport is small, persistent onshore sand transport by swell waves can become the dominant closure mechanism. An example is Wilson Inlet on the southwest coast of Australia. This inlet is usually closed for a period of 6–7 months every year due to the formation of a sandbar across the inlet (Ranasinghe and Pattiaratchi, 1999). Additional information on Wilson Inlet can be found in Chapter 13.

3.5 Location Stability

As part of the bypassing process, many inlets and channels on the ebb delta migrate. This migration is undesirable from a navigation point of view and has led to the concept of location stability. Location stability refers to the permanence of the location of the inlet and ebb delta channels. Depending on the degree of location stability, Bruun et al. (1978) used the somewhat subjective designations: poor, fair and good. Using these designations, the location stability for the inlets considered in Bruun et al. (1978) and those discussed in Chapter 4, are presented in the last columns of, respectively, Tables 3.1 and 3.2.

From the discussion on sand bypassing in Section 3.3.1, it is evident that location stability is closely related to the way sand bypasses the inlet. In case of *bar bypassing*, there are few navigable channels and, where present, they tend to shift. The location stability is poor. The bypassing mode *spit formation and breaching* involves channel migration but to a lesser extent than for bar bypassing, and the designation poor to fair seems appropriate. When bypassing is by *ebb delta breaching* the throat position is stable but the channels on the ebb delta migrate on a relatively short timescale (years); the designation fair seems applicable. In case of *tidal flow bypassing* the inlet throat position is stable and the channels on the delta show only a slow decadal change in orientation, corresponding to good location stability. The relationship between location stability, mode of bypassing and P/M ratio is reflected in the information presented in Tables 3.1 and 3.2.

3.6 Effect of Inlets on Adjacent Shoreline

From the foregoing, it is obvious that tidal inlets have potential impact on the adjacent shoreline. The nature and severity of the impact, among other factors, depends on the sand bypassing mode. Bar and tidal flow bypassing are more or less continuous processes, whereas spit-formation and ebb delta breaching are discontinuous or intermittent processes. For each of these a simple mathematical model is presented that is helpful in interpreting observed shoreline changes.

3.6.1 Continuous Bypassing

Here, shoreline changes are measured with respect to a $x–y$ coordinate system (Fig. 3.2). The x-axis is in the general direction of the shoreline and $x = 0$ is at the inlet axis. The y-axis is positive in the seaward direction. In case of continuous bypassing the shoreline change is usually in the form of advancement and recession of, respectively, the updrift and downdrift shorelines; or, in unusual cases, both sides can advance or retreat. For a given time, the shoreline change signature includes an even (symmetric) and odd (anti-symmetric) component with the inlet

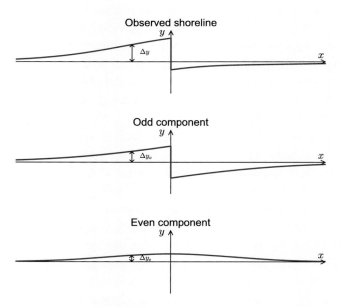

Figure 3.2 Even and odd components of shoreline change.

channel axis as the symmetry axis. The odd component is attributed to the interruption of the longshore sand transport while the even component is attributed to
cross-shore transport.

Given the total shoreline change, Dean and Work (1993) describe a procedure, referred to as *even–odd analysis*, to separate the even and odd components
(Fig. 3.2). The even–odd analysis is demonstrated, assuming a known total shoreline change $\Delta y(x)$. The total shoreline change is composed of an even component,
$\Delta y_e(x)$, and an odd component, $\Delta y_o(x)$, i.e.,

$$\Delta y(x) = \Delta y_e(x) + \Delta y_o(x). \tag{3.2}$$

By definition

$$\Delta y_e(x) = \quad \Delta y_e(-x), \qquad \text{and} \tag{3.3}$$

$$\Delta y_o(x) = -\Delta y_o(-x). \tag{3.4}$$

In terms of the total shoreline change, the even component is

$$\Delta y_e(x) = \frac{\Delta y(x) + \Delta y(-x)}{2}. \tag{3.5}$$

The odd component in terms of the total shoreline change is

$$\Delta y_o(x) = \frac{\Delta y(x) - \Delta y(-x)}{2}. \tag{3.6}$$

Even–odd analysis can be a useful procedure to quantify and interpret the impacts
of tidal inlets on the adjacent shores. Carrying out the even–odd analysis for different time periods in combination with information on waves could potentially
provide information on the processes that are responsible for the shoreline changes.
For a number of improved tidal inlets along the lower east coast of Florida, results
of the even–odd analysis are presented in Dean and Work (1993). Although these
examples pertain to improved tidal inlets, the method should equally well apply to
unimproved tidal inlets. For the jettied tidal inlet at Cape Canaveral Inlet (FL) the
results of the even–odd analysis are presented in Fig. 3.3.

3.6.2 Intermittent Bypassing

An example of intermittent bypassing is the Vlie Inlet in The Netherlands (Bakker,
1968). Similar to the Ameland Inlet described in Section 4.8, waves and flood
currents transport sand from the updrift coast into the channel across the ebb delta.
Ebb tidal currents then are responsible for the transport of the sand to the distal
end of the channel, where it forms a shoal. From here the sand is transported by
waves onto the ebb delta platform, where it forms large bar complexes. The onshore
migration of the bar complexes is intermittent with a timescale of 50–60 years.

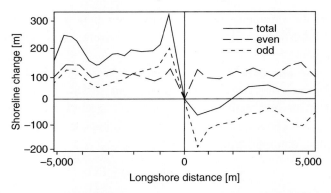

Figure 3.3 Even–odd analysis for Cape Canaveral Inlet (FL) (from Dean and Work, 1993).

The intermittent supply of sand is transported along the downdrift coast in the form of a damped progressive wave. Assuming the transport of sand is by waves, the shoreline position $y(x,t)$ satisfies the diffusion equation (Bakker, 1968, 2013)

$$\frac{\partial y}{\partial t} = \frac{q}{D}\frac{\partial^2 y}{\partial x^2},$$
(3.7)

with y positive in the seaward direction. The direction of the x-axis coincides with the general direction of the coastline and is positive in the downdrift direction. The origin of the x-axis is at the downdrift side of the inlet. The parameter q is a constant depending on wave height, wave period and sand grain characteristics and D is the sum of the closure depth and the berm height. The closure depth is the depth at the seaward limit of the longshore sand transport.

The damped progressive wave, propagating in the positive x-direction, satisfying Eq. (3.7) is

$$y = Ae^{-kx}\cos(\omega t - kx),$$
(3.8)

where ω is the radian frequency of the sand wave, k is the wave number and A is the amplitude of the sand wave to be determined from the boundary conditions. The wave number is related to the radian frequency by

$$k = \sqrt{\frac{\omega D}{2q}}.$$
(3.9)

At the Vlie inlet, the radian frequency corresponding with the intermittent sand supply is $\omega = 0.11$ rad/year. Observations of coastline positions 4,000 m apart show attenuation in coastline position of 75 percent. Using Eq. 3.8, this translates into $e^{-k(x_2-x_1)} = 0.25$. With $(x_2 - x_1) = 4,000$ m, it follows that $k = 3.6 \times 10^{-4}$ m^{-1}. The corresponding wave length is 1,745 m. The celerity of the sand wave

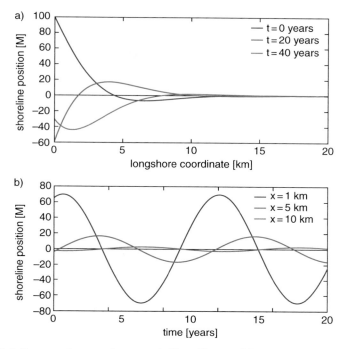

Figure 3.4 Propagating sand wave. a) Shoreline position as a function of long-shore coordinate for $t = 0$ years, $t = 20$ years and $t = 40$ years and b) shoreline position as a function of time at $x = 1$ km, $x = 5$ km and $x = 10$ km. Amplitude at $x = 0$ is $A = 100$ m.

$\omega/k = 310$ m/year. Based on observations extending over a period of 70 years, the amplitude of the sand wave at the downdrift side of the inlet $(x = 0)$ is $\mathcal{O}(100$ m). As an example, for different years the shoreline positions are presented in Fig. 3.4a and for different longshore positions are presented in Fig. 3.4b.

4

Sand Transport and Sand Bypassing at Selected Inlets

4.1 Introduction

This chapter describes sand transport patterns and sand bypassing at seven inlets; five of these are located on the east coast of the USA (Price Inlet, Breach Inlet, Captain Sam's Inlet, Mason Inlet and Wachapreague Inlet), one inlet is located in the Bay of Plenty on the North Island of New Zealand (Katikati Inlet) and another is part of the Dutch Wadden Sea coast (Ameland Inlet). The inlets are selected because they are still in their natural state and have been extensively studied. Emphasis is on the mode of bypassing, location stability and their relationship with the P/M ratio. In judging the results, it should be pointed out that estimates of longshore sand transport have limited accuracy.

4.2 Price Inlet

Price Inlet (Fig. 4.1) is located on the coast of South Carolina. Tides are semi-diurnal with a mean tidal range of 1.5 m and a spring tidal range of 2.1 m. The annual average deep water significant wave height is 0.6 m. The mean tidal prism is 14×10^6 m^3 and the spring tidal prism is 20×10^6 m^3. The throat cross-sectional area is 894 m^2. From this, maximum cross-sectionally averaged velocities are 1.1 m s^{-1} for mean tide conditions and 1.56 m s^{-1} for spring tide conditions.

The ebb delta has a volume of 6×10^6 m^3 and extents approximately 800 m offshore. The gross longshore sand transport is 0.25×10^6 m^3 year^{-1} and is predominantly from the north. Tide- and wave-generated currents carry the sand through marginal flood channels and across the channel margin linear bars to the main ebb channel. The lateral inflow of sand causes the channel to meander on timescales of decades. With ebb currents stronger than flood currents, most sand deposited in the ebb channel is ultimately transported to the seaward portion of the ebb delta. At low tide, waves break on the seaward edge of the delta and transport sand along the periphery of the delta towards the downdrift beaches and onto the

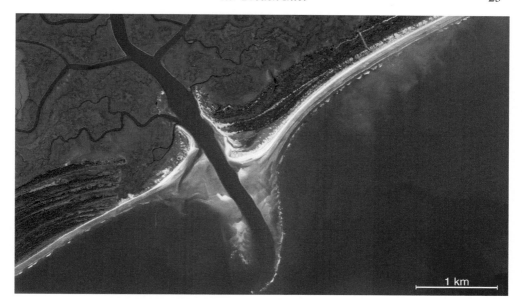

Figure 4.1 Price Inlet (SC) in 2004 (Source: Google Earth).

ebb delta platform. The sand on the ebb delta platform is transported in the form of swash bars. Swash bars travel faster in deeper water than in shallower water and as a result coalesce, forming bar complexes that migrate and attach to the beach. The inlet channel has not migrated, partly because the channel (8.5 m deep) is eroded into semi-consolidated Pleistocene deposits.

Based on these observations it is concluded that the sand bypassing mode is tidal flow bypassing and location stability is good. With a spring tidal prism of 20×10^6 m^3 and a gross longshore sand transport of 0.25×10^6 m^3 year^{-1}, the ratio $P/M = 80$.

Information on Price Inlet is based on Fitzgerald et al. (1984) and Gaudiano and Kana (2001).

4.3 Breach Inlet

Breach Inlet (Fig. 4.2) is located on the South Carolina coast, 19 km south of Price Inlet. The mean tidal range is 1.5 m and the spring tidal range is 2.1 m. The annual average deep water significant wave height is 0.6 m. The tidal prism for mean tide conditions is 13×10^6 m^3 and for spring tide conditions is 18×10^6 m^3. The throat cross-sectional area is 946 m^2. Based on this, maximum cross-sectionally averaged velocities are 0.96 m s^{-1} for mean tide conditions and 1.33 m s^{-1} for spring tide conditions.

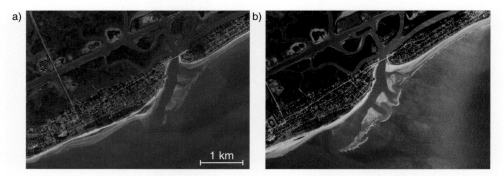

Figure 4.2 Breach Inlet (SC) in a) 2013 (Source: Google Earth) and b) 2015 (Esri et al., 2016).

The ebb delta volume is 7×10^6 m^3. The gross longshore transport is 0.25×10^6 m^3 year^{-1} and is predominantly from the north. The position of the inlet channel has been relatively stable. This is partly attributed to stabilizing structures along the downdrift margin of the inlet. The sequence of bypassing starts with a single channel on the delta (Fig. 4.2a). As a result of the longshore transport the channel and channel-margin linear bar are gradually pushed in a downdrift direction increasing the channel curvature. Because of the increased curvature, the channel becomes hydraulically less efficient and the flow is diverted across the channel-margin linear bar, scouring a new channel (Fig. 4.2b). As a result of wave-generated currents the shoal on the downdrift side of the new channel migrates in a downdrift direction, thereby filling the abandoned channel in the lee of the shoal. The shoal then gradually migrates in a shoreward direction by waves and ultimately welds onto the beach. Shoal attachment is intermittent with an average frequency of once every five years.

Based on the aforementioned, it is concluded that the sand bypassing mode is ebb delta breaching and location stability is fair. With a spring tidal prism of 18×10^6 m^3 and a gross longshore sand transport of 0.25×10^6 m^3 year^{-1}, the ratio $P/M = 72$.

Information on Breach Inlet is based on Fitzgerald et al. (1984) and Gaudiano and Kana (2001).

4.4 Captain Sam's Inlet

Captain Sam's Inlet (Fig. 4.3), also known as Kiawah River Inlet, is located on the central South Carolina coast, approximately 38 km south of Breach Inlet. Tides are semi-diurnal with a mean tidal range of 1.5 m and a spring tidal range of 2.1 m. The annual average deep water significant wave height is 0.6 m. The tidal prism for

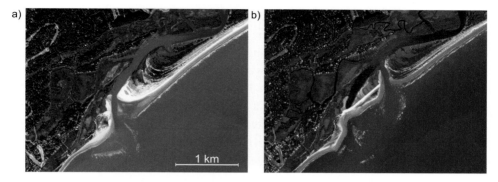

Figure 4.3 Captain Sam's Inlet (SC) in a) 2014 and b) 2015 (Source: Google Earth).

mean tide conditions is 2.3×10^6 m^3 and for spring tide conditions is 3.0×10^6 m^3. The throat cross-sectional area is 210 m^2. From this, it follows that the maximum velocity for mean tide conditions is 0.77 m s^{-1} and for spring tide conditions is 1.0 m s^{-1}.

The ebb delta volume is 4.0×10^6 m^3. The gross longshore sand transport is 0.25×10^6 m^3 year^{-1}. Captain Sam's Inlet has a history of spit formation, elongating the inlet channel and forcing the entrance together with the ebb delta to move in a southerly direction. The average growth rate of the spit is approximately 75 m year^{-1}. As a result of washovers, the spit is periodically breached at the neck. After breaching, the inlet resumes its southerly migration. The abandoned spit and ebb delta move in a southwesterly direction and ultimately weld on to the downdrift beach.

The most recent natural breach at Captain Sams occurred in 1948. The trapping of sand in the spit after this breach caused severe erosion of the downdrift beaches. To alleviate the beach erosion, in 1983 the spit was artificially breached and the abandoned channel was closed. By that time the spit was approximately 1,500 m long and had an average width of 250 m. The shortened inlet was left to adjust naturally by currents and waves. In a period of 3–4 months the initial throat cross-sectional area of 112 m^2 increased to a relatively stable value of 210 m^2. The major impact of the artificial breaching was onshore migration of the abandoned spit and ebb delta. By 1987 most of this sand, estimated to have a volume of 2.0×10^6 m^3 comprising the spit and ebb delta, welded onto the downdrift beaches. After the artificial breaching in 1983, spit extension continued until 1996 when it was artificially breached again. By that time the inlet had moved 1,000 m in a southerly direction.

Based on the foregoing account and inspection of a series of historical photographs, covering the period 1993–2013 (Fig. 4.3), it is concluded that the sand bypassing mode is spit formation and breaching. Smaller volumes of sand are bypassed during breaching events of the main channel of the ebb delta. Location stability is considered poor. With a spring tidal prism of 3.0×10^6 m^3 and a gross longshore sand transport of 0.25×10^6 year^{-1}, the ratio $P / M = 12$.

The artificial breaching of the spit in 1983 and 1996 offered the opportunity to study spit formation and breaching as a mode of sand bypassing in more detail. Morphological changes were documented and from this sediment pathways were inferred. In addition, the experiment provided information on adaptation timescales of inlet cross-section and ebb delta growth.

Information on Captain Sam's Inlet is based on Fitzgerald et al. (1984), Kana and Mason (1988), Kana (1989) and Kana and McKee (2003).

4.5 Mason Inlet

Mason Inlet (Fig. 4.4) is located on the North Carolina coast. Along this part of the coast tides are semi-diurnal with a mean tidal range of 1.0 m and a spring tidal range of 1.2 m. The annual average deep water significant wave height is 0.8 m.

The inlet has a history of migrating to the south, and by 2001 it threatened a resort complex. This condition and the overall degradation of the inlet, including the navigable channels, led to relocation of the inlet and dredging of the back-barrier channels in March 2002. Relocation and dredging resulted in an increase in

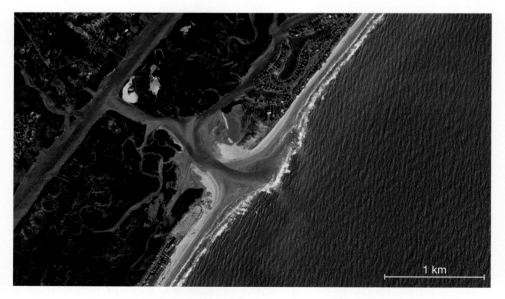

Figure 4.4 Mason Inlet (NC) in 2015 (Source: Google Earth).

mean tidal prism from 0.65×10^6 m^3 prior to 4.2×10^6 m^3 after relocation. The spring tidal prism after relocation is estimated at 5.0×10^6 m^3. The throat cross-sectional area is approximately 300 m^2. From this, it follows that the maximum cross-sectionally averaged velocity for mean tide conditions is 0.97 m s^{-1} and for spring tide conditions is 1.16 m s^{-1}.

The measured ebb delta volume is 0.29×10^6 m^3. The net longshore sand transport is to the south at a rate of approximately 0.3×10^6 m^3 year^{-1}. Because of the persistent southerly direction, the gross longshore sand transport is estimated to be close to the value of the net longshore sand transport.

Based on the foregoing and inspection of a sequence of photographs, covering the period 2002–2013 (Google Earth), it is concluded that the bypassing mode is spit formation and breaching and location stability is poor. With a spring tidal prism of 5.0×10^6 m^3 and a gross longshore sand transport of 0.3×10^6 m^3/year, the ratio $P/M = 17$.

Information on Mason Inlet is based on Cleary and FitzGerald (2003) and Welsh and Cleary (2007).

4.6 Wachapreague Inlet

Wachapreague Inlet (Fig. 4.5) is one of the many inlets along the Atlantic coast of the Delmarva Peninsula (VA). Tides are semi-diurnal with a mean tidal range of 1.16 m and a spring tidal range of 1.37 m. The annual average deep water significant wave height is estimated to be close to the value at Mason Inlet, i.e., 0.8 m. The tidal prism corresponding to the mean tidal range is 77×10^6 m^3 and the tidal prism corresponding to the spring tide is 91×10^6 m^3. The throat cross-sectional area is 4,400 m^2. From this, the maximum cross-sectionally averaged velocity for mean tide condition is 1.22 m s^{-1} and for spring tide condition is 1.44 m s^{-1}.

Wachapreague Inlet connects the coastal ocean to a back-barrier lagoon, consisting largely of tidal flats and marshes. The inlet has a pronounced ebb delta with a single channel. The position of the inlet and the channel has been stable, which in part can be attributed to the presence of cohesive lagoonal mud on the south side of the inlet. Estimates of longshore sand transport, which is primarily from the north, vary between 2×10^5 and 5×10^5 m^3/year (Byrne, personal communication). Based on sequential bathymetric surveys, observed flow patterns and wave information, a qualitative model of sand transport pathways was developed. Longshore sand transport enters the channel from the north. Some of the sand enters the back-barrier lagoon and some of it is transported to the distal end of the channel on the ebb delta. Part of the sand transported to the distal end continues its path along the coast and another part is caught in a sediment loop and returns to the channel.

Figure 4.5 Wachapreague Inlet (VA) in 2006 (Source: Google Earth).

Based on these observations, the sand bypassing is by tidal flow bypassing and location stability is good. With a spring tidal prism of 91×10^6 m^3 and depending on the adopted value of the longshore sand transport, the P/M ratio varies between 182 and 455.

Information on Wachapreague Inlet is based on Byrne et al. (1974, 1975), and DeAlteris and Byrne (1975).

4.7 Katikati Inlet

Katikati Inlet (Fig. 4.6) is located in the Bay of Plenty on the North Island of New Zealand. Tides are semi-diurnal with a neap tidal range of 1.27 m and a spring tidal range of 1.65 m. The annual averaged deep water significant wave height, as determined from measurements, is 0.8 m (Hicks and Hume, 1997). The spring tidal prism is 96×10^6 m^3. The throat cross-sectional area is 4,680 m^2. From this, the maximum cross-sectionally averaged velocity at spring tide is 1.43 m s^{-1}. The mean throat depth is 12 m (Hume and Herdendorf, 1992).

The inlet is bounded by a rocky headland on the north and a barrier island to the south. The ebb delta has a volume of 30×10^6 m^3, extending approximately 3 km offshore. A well-defined channel connects inlet and ocean. The depth over the ebb delta at the distal end of the channel is 2.5 m. Various papers dealing with Katikati Inlet show rather different values for magnitude and direction of the longshore sand transport (Hicks and Hume, 1996, 1997; Hicks et al., 1999; Hume and Herdendorf,

Figure 4.6 Katikati Inlet, New Zealand, in 2015 (Source: Google Earth).

1992). Based on the numbers in these papers, the gross longshore sand transport is estimated to be less than 0.5×10^6 m³ year⁻¹.

Hicks et al. (1999), based on personal communication with T.M. Hume, concluded that the mode of bypassing at Katikati Inlet is tidal flow bypassing. Images of Google Earth, covering the period 2003–2013, show that the position of the main ebb channel has not changed much during this period and thus location stability is good. With a spring tidal prism of 96×10^6 m³ and assuming a maximum gross longshore sand transport not exceeding 0.5×10^6 m³ year⁻¹, the value of P / M is larger than 190.

4.8 Ameland Inlet

Ameland Inlet is one of the tidal inlets on the Dutch Wadden coast (Fig. 4.7). Tides are semi-diurnal with mean and spring tidal ranges of, respectively, 1.96 m and 2.26 m. The annual average deep water significant wave height is 1.10 m with a dominant northwest direction. The mean tidal prism is 434×10^6 m³ and the spring tidal prism is 500×10^6 m³. The inlet cross-sectional area is 27,780 m². Using these values the maximum cross-sectionally averaged velocity for mean tide conditions is 1.10 m s⁻¹ and for spring tide conditions is 1.27 m s⁻¹ (Israel and Dunsbergen, 1999; van de Kreeke, 1998).

The inlet is located between the barrier islands Terschelling to the west and Ameland to the east. The inlet connects the Wadden Sea to the North Sea. It has a

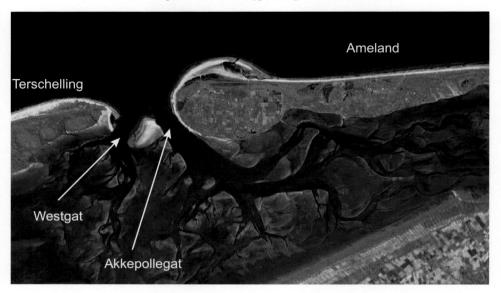

Figure 4.7 Ameland Inlet, The Netherlands, in 2010 (Esri et al., 2016).

stable throat, which can be attributed to channel and bank protection works along
the southwest coast of Ameland. The average depth of the inlet throat is 15 m. In
agreement with the large tidal prism, the Ameland Inlet has a large ebb delta with
a volume of 130×10^6 m^3 (Louters and Gerritsen, 1994). The shape of the ebb
delta is asymmetric which is attributed to the relatively strong, 0.5–1 m s^{-1}, shore-
parallel tidal currents (Sha, 1989; van Veen, 1936). The estimated gross longshore
sand transport is 0.7×10^6 m^3 year^{-1} (Cheung et al., 2007).

The ebb delta has two channels, Westgat to the west and the Akkepollegat to the
east. The size and to a lesser extent also the orientation of these channels show a
cyclic variation with a period of 50–60 years (Israel and Dunsbergen, 1999). This
cycle is associated with the evolution of the inlet throat from a one-channel to a
two-channel system. Throughout the morphological cycle Westgat varies between
ebb dominant and flood dominant. Akkepollegat is always ebb-dominant. The
wave-induced sand transport along the updrift Terschelling coast upon reaching
the inlet is temporarily stored on the Terschelling flat, located at the eastern tip of
that island. Depending on the flood or ebb dominance of the Westgat, sand is carried from the Terschelling flat towards the gorge of the inlet or the distal end of the
Westgat. The flow in Akkepollegat is responsible for the transport of the sand to
the distal end of this channel where it forms a shoal. From here the sand is transported by waves onto the ebb delta platform where it forms large bar complexes.
The onshore migration of the bar complexes is intermittent having a timescale of
50–60 years. Once it reaches the shore, part of the sand is transported back to

the inlet and another part travels along the shore in an eastward direction. The sand traveling along the shore exerts itself in the form of a progressive attenuating sand wave (Cheung et al., 2007). For this type of sand wave reference is made to Section 3.6.2.

Based on the foregoing, it is concluded that the bypassing mode is tidal flow bypassing and location stability is good. With a value of the spring tidal prism of 500×10^6 m^3 and a value of the gross longshore sand transport of 0.7×10^6 m^3 year^{-1}, the ratio $P / M = 714$.

5

Empirical Relationships

5.1 Introduction

For inlets in a sandy environment, relationships exist between parameters characterizing morphology and water motion. The more well known are the relationship between the cross-sectional area and tidal prism and the relationship between the ebb delta volume and the tidal prism. Attempts to correlate width and depth of the inlet cross-section with the tidal prism are only moderately successful.

5.2 Cross-Sectional Area – Tidal Prism Relationship

5.2.1 Observations

Observations suggest a relationship between cross-sectional area (A) and tidal prism (P). The cross-sectional area is measured with respect to MSL at the location of the gorge or throat. Usually, the tidal prism pertains to spring tide conditions. The most common presentation of the A–P relationship is in the form of a power function,

$$A = C P^q, \tag{5.1}$$

with C and q constants to be determined from observations. The relationship implies that the cross-sectional area is in equilibrium with the hydraulic environment. This equilibrium is dynamic rather than static, i.e., the cross-sectional area, and to a lesser extent the tidal prism, oscillates about an annual mean value (Section 2.3). Unfortunately, annual mean values are seldom available as most A and P values pertain to observations on a given day. These daily values can differ considerably from the annual mean. For example, Byrne et al. (1974) reported spring–neap tide variations in the cross-sectional area of Wachapreague Inlet (VA) of 10 percent. Similarly, FitzGerald and Nummedal (1983) reported that Price Inlet

(SC) experienced a change in throat cross-sectional area of 8 percent during a single tidal cycle and 26 percent during a three-year period.

Using Eq. (5.1) as the regression equation, regression analysis has been applied to many data sets. Examples for US inlets include LeConte (1905), O'Brien(1931, 1969), Jarrett (1976), van de Kreeke (1992), and Powell et al. (2006). Results for Japanese inlets are presented in Shigemura (1980). For the Dutch and German Wadden Sea inlets reference is made to van de Kreeke (1998) and (Dieckmann et al., 1988), respectively. Townend (2005) presents results for UK inlets and Heath (1975) and Hume and Herdendorf (1992) for New Zealand inlets. Using the metric system, C-values for the different sets of inlets vary between 10^{-6} and 10^{-3}. Values of the exponent q are close to 1, varying in a narrow band between 0.80 and 1.05. There are several reasons why C and q values differ for different sets of inlets. Among others, they result from differences in longshore sand transport, grain size and density and tide characteristics. This will be further discussed in Section 5.2.2. In addition, indications are that values of C and q depend on the type of least square analysis. The least square analysis is carried out by either solving a nonlinear least square equation or by taking the log on both sides of Eq. (5.1) and applying a linear regression.

Instead of using a power function, O'Brien (1969) showed that for a set of eight natural inlets (no jetties) in the US, the $A–P$ relationship is reasonably represented by the linear equation

$$A = C_l P, \tag{5.2}$$

where $C_l = 6.5 \times 10^{-5}$ m^{-1} and correlation coefficient $r^2 = 0.99$. In addition to being simpler than Eq. (5.1), this equation has the advantage that it is dimensionally correct.

5.2.2 Physical Justification of the A–P Relationship

Physical justifications for the $A–P$ relationship have been presented by a number of investigators (Kraus, 1998; Suprijo and Mano, 2004; van de Kreeke, 1998, 2004). Except for minor differences, the approach by each of these investigators is the same. The basic premise is that when at equilibrium, on an annual average base, a balance exists between the volume of sand entering the inlet on the flood and the volume of sand leaving the inlet on the ebb. Only a negligible volume of sand is assumed to be deposited in the basin. The sand balance is assumed to hold for normal wave conditions, i.e., excluding storms.

Sand is carried towards the inlet by the wave-induced longshore sand transport and enters the inlet on the flood. For normal wave conditions, the volume of sand entering the inlet is taken as a fraction M' of the gross longshore sand transport M (m^3 s^{-1}), i.e.,

$$Tr_{fl} = M', \tag{5.3}$$

where Tr_{fl} is sand transport into the inlet during normal wave conditions. For the sand transport into the inlet during storm conditions reference is made to Section 8.2 and 8.3.

For a physical justification of the A–P relationship given by Eq. (5.2), the sand transport leaving the inlet is taken in proportion to the power n of the velocity amplitude $\hat{u}$.

$$Tr_{ebb} = k\hat{u}^n, \tag{5.4}$$

where Tr_{ebb} is sand transport leaving the inlet and k is a coefficient depending on grain characteristics. Values of n are assumed to be between 3 and 5. Underlying assumptions are that the velocity can be reasonably approximated by a sine curve and velocities are considerably larger than the critical velocity of erosion over most of the ebb cycle.

When the inlet is at equilibrium

$$k\hat{u}^n = M'. \tag{5.5}$$

With the velocity not exactly sinusoidal, the velocity amplitude is defined as

$$\hat{u} = \frac{\pi P}{AT}, \tag{5.6}$$

where P is tidal prism, A is cross-sectional area and T is tidal period (Sorensen, 1977; van de Kreeke, 2004).

Substituting for $\hat{u}$ from Eq. (5.6) in Eq. (5.5) results in Eq. (5.2) with

$$C_l = \sqrt[n]{\frac{k}{M'}\frac{\pi}{T}}. \tag{5.7}$$

With M' being a fraction of M, the constant C_l decreases with increasing values of gross longshore sand transport.

The physical justification for the A–P relationship given by Eq. (5.1) follows along the same lines as the justification for Eq. (5.2), with the exception of the formulation of the ebb transport. Instead of Eq. (5.4), the ebb transport is written as

$$Tr_{ebb} = k\hat{u}^n W, \tag{5.8}$$

where W is the width of the inlet at MSL. Assuming that cross-sections with different cross-sectional area are geometrically similar, the width of the inlet is proportional to the square root of the cross-sectional area, i.e., $W = \beta_1\sqrt{A}$ (Appendix 8B). Substituting for W in Eq. (5.8) results in the expression for the ebb transport

$$Tr_{ebb} = k\hat{u}^n\sqrt{A}, \tag{5.9}$$

where the proportionality coefficient β_1 is incorporated in k. When the inlet is at equilibrium the ebb transport equals the flood transport, i.e.,

$$k\hat{u}^n\sqrt{A} = M'. \tag{5.10}$$

Substituting for $\hat{u}$ from Eq. (5.6) in Eq. (5.10) results in Eq. (5.1) with

$$C = \left(\frac{k\left(\frac{\pi}{T}\right)^n}{M'}\right)^{\frac{1}{n-\frac{1}{2}}} \quad \text{and} \quad q = \frac{n}{n-\frac{1}{2}}. \tag{5.11}$$

With M' being a fraction of M, C decreases with increasing values of the gross longshore sand transport. With $3 \le n \le 5$, values of q range from 1.11 to 1.20. For comparison, observed values of q range from 0.80 to 1.05, suggesting that the present physical justification is an oversimplification of the real physics. More research is needed to resolve this.

Even though some of the assumptions can be questioned, the physical justification for Eqs. (5.1) and (5.2) suggests that $A–P$ relationships should only be expected to hold for sets of inlets that are geologically and hydrodynamically similar, i.e., inlets have the same grain characteristics, longshore sand transport and tidal period. Ocean tidal amplitudes, inlet lengths and basin surface areas can differ. Furthermore, for Eq. (5.1) to hold, cross-sections in the data set have to be geometrically similar. It follows that the coefficients C_l, C and q are not universal constants but are expected to differ from one geologically and hydrodynamically similar set of inlets to another. In this context it is of interest to note that Jarrett (1976) sampled a large number of US inlets and separated them into classes depending on the absence or presence of one or two jetties, reasoning that the jetties would affect the movement of sand into an inlet.

5.2.3 *Examples of* A–P *Relationships for Natural Inlets*

Data for tidal prisms and cross-sectional areas for five inlets on the Dutch Wadden Coast are presented in van de Kreeke (1998) and are reproduced in Table 5.1. The data pertains to mean tide conditions. The inlets are in a natural state (no jetties) and are scoured in fine to medium sand. Tides are semi-diurnal with offshore mean tidal ranges increasing from 1.60 m at Texel Inlet to 2.00 m at the Frisian Inlet (Dillingh, 2013). River flow is insignificant. The gross longshore sand transport is $0.5 - 1 \times 10^6$ m^3 year^{-1} (Spanhoff et al., 1997). Cross-sectional areas vary in a rather narrow range between 16,540 and 63,300 m^2. In view of similar tides, sediment characteristics and gross longshore sand transport, this set of natural inlets can be considered hydrodynamically and geologically similar.

Table 5.1 *Cross-sectional area* A *and tidal prism* P *for inlets of the Dutch Wadden Sea (van de Kreeke, 1998).*

Inlet	$A \ [\mathrm{m}^2]$	$P \ [10^6 \ \mathrm{m}^3]$
Texel Inlet	59,160	957
Eyerlands Gat Inlet	16,540	172
Vlie Inlet	63,300	848
Ameland Inlet	27,780	434
Frisian Inlet (before basin reduction)	24,540	321

P pertains to mean tide

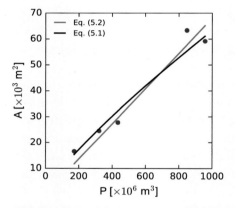

Figure 5.1 Cross-sectional area – tidal prism relationship for five inlets in the Dutch Wadden Sea for different regression equations. Using Eq. (5.1), $A = 3.4 \times 10^{-3} P^{0.81}$ with $r^2 = 0.96$. Using Eq. (5.2), $A = 6.8 \times 10^{-5} P$ with $r^2 = 0.95$.

A regression analysis for the five inlets shows the following results. Using the linear regression equation (5.2), $C_l = 6.8 \times 10^{-5}$ with $r^2 = 0.95$. Using the power function, Eq. (5.1), $C = 3.4 \times 10^{-3}$ and $q = 0.81$ with $r^2 = 0.96$. The trend lines for the two regression equations are presented in Fig. 5.1. Values of the correlation coefficients for the two regression equations are high and differ little. In spite of this, cross-sectional areas of individual inlets still can deviate substantially from the trend line. For example, using the linear regression equation and given the tidal prism, the cross-sectional area of the Eyerlandse Gat Inlet is ($6.8 \times 10^{-5} \times 172 \times 10^6 =$) 11,700 m², as opposed to 16,540 m² from observations. Possible reasons for the observations deviating from the trend line are discussed in Section 5.2.1.

Data for tidal prisms and cross-sectional areas for 16 inlets on the North Island of New Zealand are presented in Hume and Herdendorf (1992) and are reproduced in Table 5.2. The data pertains to spring tide conditions. The inlets are scoured in fine to medium sand. Tides are semi-diurnal with spring tidal ranges varying between 1.15 and 2.34 m. Inlets are located on a coast with a small longshore sand transport (Section 4.7). River flow is insignificant. Given that the wave climate is

Table 5.2 *Cross-sectional area* A *and tidal prism* P *for inlets on the North Island of New Zealand (Hume and Herdendorf, 1992).*

Inlet	A [m^2]	P [10^6 m^3]	Inlet	A [m^2]	P [10^6 m^3]
Whananaki	130	1.46	Whangapoua	980	8.54
Ngunguru	310	3.83	Whitianga	1,300	12.56
Pataua	140	2.24	Tairua	430	5.02
Whangarei	14,610	155	Whangamata	360	3.93
Mangawhai North	100	1.50	Puhoi	130	1.93
Mangawhai South	400	5.05	Maketu	70	0.79
Whangateau	660	10.53	Ohiwa	1,880	28.11
Katikati	4,680	95.82	Tauranga	6,260	130.8

P pertains to spring tide

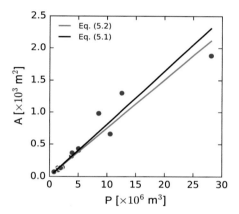

Figure 5.2 Cross-sectional area – tidal prism relationship for 13 inlets on the North Island of New Zealand for different regression equations. Using Eq. (5.1), $A = 6.5 \times 10^{-5} P^{1.01}$ with $r^2 = 0.96$. Using Eq. (5.2), $A = 7.5 \times 10^{-5} P$ with $r^2 = 0.91$.

more or less the same for all the inlets, it is reasonable to assume that this is also the case for the longshore sand transport. To a good approximation, the inlets in the data set are geologically and hydrodynamically similar.

A regression analysis for the 16 inlets shows the following results. Using the linear regression equation, Eq. (5.2), $C_l = 7 \times 10^{-5}$ m^{-1} with $r^2 = 0.875$. Using the power function, Eq. (5.1), $C = 1.68 \times 10^{-4}$ and $q = 0.95$ with $r^2 = 0.98$. Three of the inlets, Whangarei, Katikati and Tauranga, have cross-sectional areas and tidal prisms that are an order of magnitude larger than the other inlets. Omitting these inlets from the data set and using Eq. (5.2), $C_l = 7.5 \times 10^{-5}$ m^{-1} with $r^2 = 0.91$. Using Eq. (5.1), $C = 6.5 \times 10^{-5}$ and $q = 1.01$ with $r^2 = 0.96$. The trend lines for the 13 inlets are presented in Fig. 5.2. Similar to the inlets of the Wadden Sea, the cross-sectional areas of the individual inlets can differ considerably from the trend line.

A direct comparison of the C, q and C_l values for the data sets of the Wadden Sea and New Zealand inlets is difficult because of the difference in the longshore sand transport and because the $A-P$ relationships for the Wadden Sea inlets pertains to mean tide conditions and that for the New Zealand inlets pertains to spring tide conditions.

5.2.4 Equilibrium Velocity

In the equations for the sand balance, Eqs. (5.5) and (5.10), the value of the velocity amplitude $\hat{u}$ pertains to equilibrium conditions and is referred to as the equilibrium velocity $\hat{u}_{eq}$. For $\hat{u} > \hat{u}_{eq}$ the inlet erodes and for $\hat{u} < \hat{u}_{eq}$ the inlet shoals.

A difficulty in determining the value of $\hat{u}_{eq}$ from Eqs. (5.5) and (5.10) is ascertaining the values of k, n and M'. To circumvent this problem, recourse is taken to one of the $A-P$ relationships discussed in Section 5.2.1.

With Eq. (5.1) as the $A-P$ relationship and using Eq. (5.6) as the expression for the velocity, the equilibrium velocity is

$$\hat{u}_{eq} = \frac{\pi A^{\frac{1}{q}-1}}{TC^{\frac{1}{q}}}. \tag{5.12}$$

With values of q close to one, the equilibrium velocity is a weak increasing function of A for $q < 1$ and a weak decreasing function of A for $q > 1$. With Eq. (5.2) as the cross-sectional area–tidal prism relationship and using Eq. (5.6), the expression for the equilibrium velocity is

$$\hat{u}_{eq} = \frac{\pi}{C_l T}, \tag{5.13}$$

In this case the equilibrium velocity is independent of the cross-sectional area. With C and C_l decreasing for increasing values of the gross longshore sand transport, $\hat{u}_{eq}$ increasing with increasing values of the gross longshore sand transport.

Using the results of the linear regression, the equilibrium velocities are calculated for the inlets of the Dutch Wadden Sea and New Zealand. For the Wadden Sea inlets with $C_l = 6.8 \times 10^{-5}$ m^{-1} and $T = 44,712$ s, it follows from Eq. (5.13) that the equilibrium velocity $\hat{u}_{eq} = 1.03$ m s^{-1}. Similarly, for the 13 New Zealand inlets, with $C_l = 7.5 \times 10^{-5}$ m^{-1}, $\hat{u}_{eq} = 0.94$ m s^{-1}.

5.3 Relationship between Depth and Width of the Cross-Section and Tidal Prism

Attempts to correlate the width and depth of the inlet throat cross-section with the tidal prism are reported in Hume and Herdendorf (1992). Using data for a set of inlets on the North Island of New Zealand they conclude that: "While the cross-sectional area shows a strong correlation with the tidal prism, there is a much weaker correlation between depth and tidal prism ($r^2 = 0.82$) and width and tidal prism ($r^2 = 0.63$)." Apparently, the inlet width and depth are free to adjust while the cross-sectional area is controlled by the tidal prism. Other studies, e.g., Bruun (1981) and Marino and Mehta (1988), show similar results with width to mean depth ratios varying widely and no obvious relationship with cross-sectional area or tidal prism.

5.4 Ebb Delta Volume – Tidal Prism Relationship

Similar to the cross-sectional area of the inlet, the ebb delta volume, V, appears to maintain a consistent relationship with the tidal prism P. In terms of a power function,

$$V = aP^b. \tag{5.14}$$

This relationship is empirical and physical justification is based on the idea that, when the ebb jet disperses and its competency decreases, sand is deposited in the nearshore, thereby building up an ebb delta platform. As this platform accretes vertically, wave energy augments the flood tidal currents. The equilibrium volume is reached when the flood tidal and wave generated currents transport the same volume of sand onshore as the ebb currents transports offshore (Fitzgerald et al., 1984).

Based on observations and a regression analysis, values of the coefficients a and b are presented in Walton and Adams (1976). They distinguish between highly, moderately and mildly exposed coasts taking the parameter H^2T^2 as a measure of wave energy. H and T are average values of wave height and period, respectively, determined from wave gages located in the near-shore zone. Restricting attention to the US coast, examples of mildly exposed coasts are the South Carolina, Texas and lower Florida Gulf coasts. The east coast and Panhandle of Florida are in the moderate range and the Pacific coast is in the highly exposed range. For each type of coast the regression analysis showed reasonable correlation with values of $b = 1.08$ for the moderately, $b = 1.23$ for the highly and $b = 1.24$ for the mildly exposed coast. A second regression analysis was then carried out taking $b = 1.23$ for all inlet groupings. Using the metric system, the resulting values of the coefficient a, together with the number of inlets for each type of coast, are listed in Table 5.3. It is not clear in the analysis of Walton and Adams (1976) whether values of the tidal prism refer to spring or mean tide conditions. Information on correlation coefficients is not available. However, the graphs in Walton and Adams (1976), in which ebb delta volume is plotted versus tidal prism, show only

Table 5.3 *Results of the regression analysis performed by Walton and Adams (1976).*

Exposure	a	b	Number of inlets
Mild	10.12×10^{-3}	1.23	16
Moderate	7.70×10^{-3}	1.23	18
High	6.38×10^{-3}	1.23	7

Table 5.4 *Results of the regression analysis performed by Powell et al. (2006).*

Location	a	b	Number of inlets	r^2
Florida Atlantic coast	3.80×10^{-3}	1.26	28	0.57
Florida Gulf coast	1.41×10^{1}	0.73	39	0.70

a limited goodness of fit, as some inlets deviate substantially from the regression equation.

With b positive, the volume of the ebb delta increases with increasing values of the tidal prism. The coefficient a, and therefore the ebb delta volume, decreases, going from mildly to highly exposed coasts; more sand is stored in the ebb delta of a low-energy coast than in the ebb delta of a high-energy coast. This is in agreement with what has been stated in Section 2.4.2 when discussing the geomorphology of the ebb delta.

Powell et al. (2006), using Eq. (5.14), correlated the ebb delta volumes and tidal prisms for 28 inlets on the Florida Atlantic coast and 39 inlets on the Florida Gulf coast. They found values for the coefficients a and b, using the metric system, listed in Table 5.4. Similar to Walton and Adams (1976), with b positive, the ebb delta volume increases with increasing values of the tidal prism. In the same table, the number of inlets and the values of the correlation coefficients are indicated. In their data set they used the spring tidal prism. Values of correlation coefficients are low and, similar to the Walton and Adams (1976) regression analysis, it can be concluded that the goodness of fit is limited.

For the Florida Atlantic coast being moderately exposed, it seems justified to compare the results of Powell et al. (2006) with the results of the moderately exposed coast of Walton and Adams (1976). For Powell et al. (2006), the value of the exponent b is larger and the value of the coefficient a is smaller than for Walton and Adams (1976). As a result, given a value for the tidal prism, the calculated ebb delta volumes do not differ all that much. This is demonstrated for the smallest and largest value of the tidal prism in the Powell et al. (2006) data set, respectively, 2×10^5 m^3 and 3.3×10^8 m^3. For the smallest value, Walton and

Adams (Eq. (5.14), with $a = 7.7 \times 10^{-3}$ and $b = 1.23$) yields an ebb delta volume of 0.26×10^5 m^3 and Powell et al. (Eq. (5.14), with $a = 3.8 \times 10^{-3}$ and $b = 1.26$) yields a value of 0.18×10^5 m^3. Taking the largest tidal prism in the data set, the corresponding ebb delta volumes are 2.31×10^8 m^3 for Walton and Adams and 2.05×10^8 m^3 for Powell et al. In spite of the differences in the coefficients and exponents, the delta volumes for the Walton and Adams and the Powell et al. empirical equations show reasonable agreement.

5.5 Flood Delta Volume – Tidal Prism Relationship

In addition to the ebb delta volume, Powell et al. (2006) looked for a relationship between flood delta volume and tidal prism for the 39 inlets along the Florida Gulf coast. Using Eq. (5.14) as the regression equation and using the metric system, this resulted in $a = 6.95 \times 10^3$ and $b = 0.37$. The scatter in the data is substantial as witnessed by the r^2 value of 0.38. According to FitzGerald (1988, 1996), the limited correlation between flood delta volume and tidal prism is related to the fact that the size and presence of flood deltas depend on the amount of open water space in the back-barrier lagoon. This can vary considerably from one type of back-barrier lagoon to the other; see Fig. 2.4.

6

Tidal Inlet Hydrodynamics; Excluding Depth Variations with Tidal Stage

6.1 Introduction

The dynamics of the flow in the inlet are described by the equation for uniform unsteady open channel flow. Variations in depth with tidal stage are neglected. The dynamic equation is complemented with a continuity condition that assumes a pumping mode for the back-barrier lagoon, i.e., the water level in the back-barrier lagoon fluctuates uniformly. Although these are simplifications, the advantage is that they allow relatively simple analytical solutions that are helpful in identifying mechanisms responsible for phenomena such as resonance, tidal choking and generation of (odd) overtides. As examples, analytical solutions by Keulegan (1951, 1967) and Mehta and Özsoy (1978) are presented. Results of the analytical solutions are applied to a representative inlet and compared with numerical results.

6.2 Inlet Schematization

The tidal inlet system is schematized to an inlet and a back-barrier lagoon (Fig. 6.1). The inlet connects the back-barrier lagoon and the ocean. Its geometry is simplified to a prismatic channel with diverging sections at both ends. The back-barrier lagoon is schematized to a basin with uniform depth. Referring to Chapter 2, in the real world inlets have varying widths and depths and back-barrier lagoons are characterized by tidal flats and marsh areas. Therefore, the schematization presented in Fig. 6.1 is only a rough representation of an actual inlet.

6.3 Governing Equations and Boundary Condition

6.3.1 Dimensional Equations

In deriving the governing equations, the major assumptions are 1) one-dimensional unsteady uniform flow in the inlet, 2) a uniformly fluctuating water level in the basin (pumping or Helmholz mode) and 3) negligible variations in cross-sectional

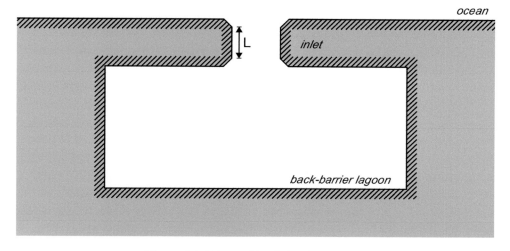

Figure 6.1 Inlet and back-barrier lagoon.

area of the inlet and basin surface area with tidal stage. With these assumptions, the equation for the flow in the inlet is (Appendix 6.A):

$$\frac{L}{g}\frac{du}{dt} + \left(\frac{m}{2g} + \frac{FL}{gR}\right)u|u| = \eta_0 - \eta_b. \qquad (6.1)$$

In this equation u is the cross-sectionally averaged velocity, positive in the flood direction, L is length of the prismatic part of the inlet, g is gravity acceleration, t is time, $F = f/8$ where f is the Darcy–Weisbach friction factor, R is hydraulic radius, m is entrance/exit loss coefficient, η_0 is the ocean tide and η_b is the basin tide.

Eq. (6.1) is complemented by a continuity equation, expressing the equality of the water volume flux in the inlet and the rate of change of the basin volume. Assuming a uniformly fluctuating basin water level, this results in

$$uA = A_b\frac{d\eta_b}{dt}, \qquad (6.2)$$

where A is cross-sectional area of the inlet channel, A_b is basin surface area and t is time. The ocean forcing is a simple harmonic tide

$$\eta_0 = \hat{\eta}_0 \sin \sigma t, \qquad (6.3)$$

where $\hat{\eta}_0$ is the ocean tidal amplitude and σ is the angular frequency of the tide.

To illustrate the relative importance of the terms on the left-hand side of Eq. (6.1), the velocity is assumed to be simple harmonic,

$$u = \hat{u}\cos(\sigma t - \alpha). \qquad (6.4)$$

Here $\hat{u}$ is the velocity amplitude and α is a phase angle. Using Eqs. (6.1) and (6.4), it follows that the order of magnitude values of the inertia, entrance/exit loss and bottom friction terms, respectively, the first, second and third term on the left-hand side of Eq. (6.1), are

$$\frac{\sigma L}{g}\hat{u}, \quad \frac{m}{2g}\hat{u}^2 \quad \text{and} \quad \frac{FL}{gR}\hat{u}^2.$$

Restricting attention to inlets in a sandy environment that have existed for some time, the velocity amplitude $\hat{u}$ is approximately 1 m s^{-1} and is referred to as the equilibrium velocity (see Section 5.2.4). A typical value of σ is 1.4×10^{-4} rad s^{-1}, corresponding with a semi-diurnal tide. From the literature, typical values of the coefficients m and F are, respectively, 1 and 3×10^{-3}. With these values it follows that the ratio of the friction and inertia term is approximately $20/R$. With R between 2 and 30 m (see Section 2.4.1), for the shallower inlets bottom friction is considerably larger than inertia. Similarly, the ratio of the bottom friction term and the entrance/exit loss term is approximately $6 \times 10^{-3}L/R$. Only for short and deep inlets, the value of the entrance/exit loss term approaches that of the bottom friction term.

6.3.2 Non-Dimensional Equations; Lumped Parameter Model

To reduce the number of parameters, Eqs. (6.1) and (6.2) are non-dimensionalized. Introducing the water level scale $\hat{\eta}_0$, velocity scale U and timescale σ^{-1}, the non-dimensional variables are

$$\eta_0^* = \frac{\eta_0}{\hat{\eta}_0}, \quad \eta_b^* = \frac{\eta_b}{\hat{\eta}_0}, \quad u^* = \frac{u}{U}, \quad t^* = \sigma t. \tag{6.5}$$

Using Eq. (6.2) the velocity scale is

$$U = \frac{\sigma \hat{\eta}_0 A_b}{A}. \tag{6.6}$$

Using the non-dimensional variables and the expression for U, the non-dimensional equations, corresponding to Eqs. (6.1) and (6.2), are

$$K_2^2\frac{du^*}{dt^*} + \frac{1}{K_1^2}u^*|u^*| = \eta_0^* - \eta_b^*, \tag{6.7}$$

and

$$u^* = \frac{d\eta_b^*}{dt^*}. \tag{6.8}$$

The dimensionless parameters in Eq. (6.7) are defined as

$$K_1 = \frac{A}{\sigma \hat{\eta}_0 A_b} \sqrt{\frac{\hat{\eta}_0}{\left(\frac{m}{2g} + \frac{FL}{gR}\right)}}, \quad \text{and} \quad K_2 = \frac{\sigma}{\sqrt{\frac{gA}{LA_b}}}. \tag{6.9}$$

In terms of non-dimensional variables the forcing, Eq. (6.3), is

$$\eta_0^* = \sin(t^*). \tag{6.10}$$

The parameter K_1 is the Keulegan repletion factor and is the inverse of a damping factor. K_2 is the ratio of the forcing frequency σ and the natural or Helmholz frequency $\sqrt{gA/LA_b}$. The eight parameters, A_b, A, R, L, F, m, $\hat{\eta}_0$ and σ, in Eqs. (6.1) and (6.2) are lumped into the two non-dimensional parameters K_1 and K_2; hence the name *lumped parameter model* for Eqs. (6.7)–(6.10). For most natural inlets, $K_1 < 2$ (Seabergh, 2002).

6.4 Analytical Solution (Öszoy–Mehta)

6.4.1 Basin Tide and Inlet Velocity

In deriving a solution to Eqs. (6.7) and (6.8), Mehta and Özsoy (1978) assumed Eq. (6.7) to be weakly nonlinear. Together with the boundary condition given by Eq. (6.10), this suggests the following trial solutions for $\hat{\eta}_b^*$ and $\hat{u}^*$:

$$\eta_b^* = \hat{\eta}_b^* \sin(t^* - \alpha), \tag{6.11}$$

and

$$u^* = \hat{u}^* \sin(t^* - \beta). \tag{6.12}$$

Substituting the trial solutions in Eq. (6.8) results in

$$\hat{u}^* \sin(t^* - \beta) = \hat{\eta}_b^* \cos(t^* - \alpha). \tag{6.13}$$

For this equation to hold requires that

$$\beta = \alpha - \frac{\pi}{2}, \tag{6.14}$$

and

$$\hat{u}^* = \hat{\eta}_b^*. \tag{6.15}$$

Substituting for β from Eq. (6.14) in Eq. (6.12) results in

$$u^* = \hat{u}^* \cos(t^* - \alpha). \tag{6.16}$$

In solving Eqs. (6.7) and (6.8), the nonlinear expression $u^*|u^*|$ is linearized by substituting for u^* from Eq. (6.16), i.e.,

$$u^*|u^*| = \hat{u}^{*2}\cos(t^* - \alpha)|\cos(t^* - \alpha)|. \tag{6.17}$$

Developing in a Fourier series, it follows that

$$u^*|u^*| = \frac{8}{3\pi}\hat{u}^{*2}\cos(t^* - \alpha) + \text{odd higher harmonics.} \tag{6.18}$$

Retaining only the first term in the series expansion and substituting in Eq. (6.7) results in

$$K_2^2\frac{du^*}{dt^*} + \frac{1}{K_1^2}\frac{8}{3\pi}\hat{u}^*u^* = \eta_0^* - \eta_b^*. \tag{6.19}$$

Substituting for η_0^*, η_b^* and u^* from, respectively, Eqs. (6.10), (6.11) and (6.16) in Eq. (6.19) and using of Eq. (6.15), it follows that

$$-K_2^2\hat{u}^*\sin(t^* - \alpha) + \frac{1}{K_1^2}\frac{8}{3\pi}\hat{u}^{*2}\cos(t^* - \alpha) = \sin(t^*) - \hat{u}^*\sin(t^* - \alpha). \tag{6.20}$$

Expanding the trigonometric functions and collecting terms proportional to $\sin(t^*)$ results in

$$-K_2^2\hat{u}^*\cos(\alpha) + \frac{1}{K_1^2}\frac{8}{3\pi}\hat{u}^{*2}\sin(\alpha) = 1 - \hat{u}^*\cos(\alpha). \tag{6.21}$$

Collecting terms proportional to $\cos(t^*)$ results in

$$K_2^2\hat{u}^*\sin(\alpha) + \frac{1}{K_1^2}\frac{8}{3\pi}\hat{u}^{*2}\cos(\alpha) = \hat{u}^*\sin(\alpha). \tag{6.22}$$

Multiplying Eq. (6.21) by $\cos(\alpha)$ and Eq. (6.22) by $\sin(\alpha)$ and subtracting, it follows that

$$\cos(\alpha) = (1 - K_2^2)\hat{u}^*. \tag{6.23}$$

Multiplying Eq. (6.21) by $\sin(\alpha)$ and Eq. (6.22) by $\cos(\alpha)$ and adding, it follows that

$$\sin(\alpha) = \frac{1}{K_1^2}\frac{8}{3\pi}\hat{u}^{*2}. \tag{6.24}$$

Solving for α and $\hat{u}^*$ from Eqs. (6.23) and (6.24) gives

$$\hat{u}^* = \sqrt{\frac{\left[(1 - K_2^2)^4 + \frac{4\left(\frac{8}{3\pi}\right)^2}{K_1^4}\right]^{1/2} - (1 - K_2^2)^2}{\frac{2\left(\frac{8}{3\pi}\right)^2}{K_1^4}}}, \tag{6.25}$$

and

$$\alpha = \tan^{-1}\left(\frac{\frac{8}{3\pi}\hat{u}^*}{K_1^2\left(1 - K_2^2\right)}\right). \tag{6.26}$$

It follows from Eqs. (6.23) and (6.24) that for $K_2 < 1$, α is in the first quadrant and for $K_2 > 1$, α is in the second quadrant.

For many inlets the inertia term in the dynamic equation is small compared to the bottom friction term. In that case, neglecting the inertia term in Eq. (6.7), it follows from Eqs. (6.21) and (6.22) that

$$\hat{u}^* = \sqrt{\frac{\left[1 + \frac{4\left(\frac{8}{3\pi}\right)^2}{K_1^4}\right]^{1/2} - 1}{\frac{2\left(\frac{8}{3\pi}\right)^2}{K_1^4}}} \tag{6.27}$$

and

$$\alpha = \tan^{-1}\left(\frac{\frac{8}{3\pi}\hat{u}^*}{K_1^2}\right). \tag{6.28}$$

The same result is obtained by taking $K_2 = 0$ in Eqs. (6.25) and (6.26).

For applications where the bottom friction and entrance/exit loss can be neglected, i.e., $K_1 \to \infty$, it follows from Eqs. (6.21) and (6.22) that

$$\hat{u}^* = \frac{1}{1 - K_2^2} \quad\text{and}\quad \alpha = 0 \quad\text{for}\quad K_2 < 1, \tag{6.29}$$

and

$$\hat{u}^* = \frac{-1}{1 - K_2^2} \quad\text{and}\quad \alpha = \pi \quad\text{for}\quad K_2 > 1. \tag{6.30}$$

Note that the same result is not obtained by taking $K_1 = \infty$ in Eqs. (6.25) and (6.26).

6.4.2 Nature of the Solution; Resonance

As illustrated in Figs. 6.2a and 6.2b, the overall behavior of the Öszoy–Mehta Solution is similar to that of a mass–spring–dashpot system with resonance in the neighborhood of $K_2 = 1$. Depending on the values of K_1 and K_2, the basin tidal amplitude may be smaller or larger than the ocean tidal amplitude. For $K_2 = 0$, the basin tidal amplitude is always smaller than the ocean tidal amplitude. For a

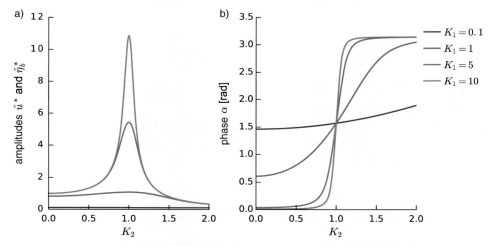

Figure 6.2 a) Amplitudes of non-dimensional basin tide and inlet velocity as a function of K_2 and K_1 and b) Phase of basin tide as a function of K_2 and K_1.

given K_2, the basin tidal amplitude and velocity amplitude increase with increasing values of K_1. For a given K_1, the basin tide and inlet velocity are maximal in the neighborhood of $K_2 = 1$. For values of K_1 smaller than approximately 0.1, the basin tidal amplitude is an order of magnitude smaller than the ocean tidal amplitude. These inlets have been referred to as choked inlets (Kjerfve, 1986).

Analytical solutions similar to the one presented by Mehta and Özsoy (1978) are found in Walton and Escoffier (1981), DiLorenzo (1988), Dean and Dalrymple (2002), Walton (2004b) and de Swart and Zimmerman (2009). Except for DiLorenzo (1988) and Walton (2004b), forcing is by a simple harmonic ocean tide. DiLorenzo (1988) included forcing by higher harmonics (overtides). In Walton (2004b) forcing is by a water level consisting of several astronomical components. Neglecting inertia, analytical solutions are presented in Brown (1928) and van de Kreeke (1967). These solutions agree with Eqs. (6.27) and (6.28).

6.5 Semi-Analytical Solution (Keulegan)

6.5.1 Basin Tide and Inlet Velocity

Keulegan (1951, 1967) derived expressions for the basin tide and the inlet velocity neglecting inertia. With $K_2 = 0$, the simplified set of non-dimensional equations is

$$\frac{1}{K_1^2} u^* |u^*| = \eta_0^* - \eta_b^* \tag{6.31}$$

and

$$u^* = \frac{d\eta_b^*}{dt^*}, \tag{6.32}$$

with the open boundary condition

$$\eta_0^* = \sin(t^*). \tag{6.33}$$

In the following, the major steps in solving for η_b^* and u^* are presented. For details, reference is made to Keulegan (1951, 1967).

It follows from Eqs. (6.31) and (6.32) that the basin and ocean tide curves intersect when the basin tide is maximal and minimal. At those times, the inlet velocity is zero. With this in mind, a qualitative plot of basin tide and ocean tide is presented in Fig. 6.3. In deriving a solution to Eqs. (6.31) and (6.32), Keulegan took the origin of the time axis at the low water intersection of the ocean and basin tide curve, resulting in the expression for the ocean tide

$$\eta_0^* = \sin(t^* - \tau). \tag{6.34}$$

The phase τ is defined in Fig. 6.3. Note that τ is an unknown quantity.

Eliminating the velocity u^* between Eqs. (6.31) and (6.32) results in the following equation for the basin tide:

$$\frac{1}{K_2^2} \frac{d\eta_b^*}{dt^*} \left| \frac{d\eta_b^*}{dt^*} \right| = \eta_0^* - \eta_b^*. \tag{6.35}$$

Rather than the basin tide, Keulegan took the difference between ocean tide and basin tide,

$$z^* = \eta_0^* - \eta_b^*, \tag{6.36}$$

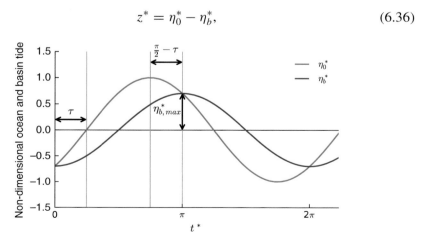

Figure 6.3 Ocean and basin tide curves; definition of time origin.

as the unknown. He assumed, and later verified, that for $0 < t^* < \pi$, z^* is positive and for $\pi < t^* < 2\pi$, z^* is negative. Limiting attention to the range $0 < t^* < \pi$ with $z^* > 0$ and η_0^* given by Eq. (6.34), it follows from Eq. (6.35) that

$$\frac{dz^*}{dt^*} = -K_1\sqrt{z^*} + \cos(t^* - \tau). \tag{6.37}$$

The corresponding equation for the range $\pi < t^* < 2\pi$ with $z^* > 0$ is

$$\frac{dz^*}{dt^*} = -K_1\sqrt{z^*} - \cos(t^* - \tau). \tag{6.38}$$

To solve for z^* from Eqs. (6.37) and (6.38), Keulegan showed that, in the interval $0 < t^* < 2\pi$, z^* the function $z^*(t^*)$ has zero mean and is asymmetric with the property

$$z^*(t^*) = -z^* \left(t^* + \tfrac{\pi}{2}\right). \tag{6.39}$$

This implies that, when developing $z^*(t^*)$ in a Fourier series, only odd harmonics are present. Limiting the Fourier series to a first and a third harmonic results in

$$z^*(t^*) = a_1 \sin(t^*) + a_1 b_3 \cos(t^*) - a_1 b_3 \cos(3t^*) - a_1 a_3 \sin(3t^*). \tag{6.40}$$

Substituting for $z^*(t^*)$ from Eq. (6.40) in Eq. (6.37) results in

$$a_1 \cos(t^*) - a_1 b_3 \sin(t^*) + 3a_1 b_3 \sin(3t^*) - 3a_1 a_3 \cos(3t^*) - \cos(t^* - \tau)$$
$$= -K_1\sqrt{a_1 \sin(t^*) + a_1 b_3 \cos(t^*) - a_1 b_3 \cos(3t^*) - a_1 a_3 \sin(3t^*)}. \tag{6.41}$$

Developing the right-hand side of Eq. (6.41) in a Fourier series results in a series with only odd harmonics. With the right-hand side replaced by the first and third harmonic of the Fourier series, Eq. (6.41) contains terms in $\sin(t^*)$, $\cos(t^*)$, $\sin(3t^*)$ and $\cos(3t^*)$. Collecting terms proportional to each of these harmonics leads to four equations with unknowns a_1, a_3, b_3 and τ. Numerically solving these equations leads to values for a_1, a_3, b_3 and τ as functions of K_1. For K_1 between 0.1 and 5, values are presented in Table 6.1. In particular, the value of τ increases with increasing values of the repletion coefficient K_1 and thus, from Fig. 6.3, the phase between basin and ocean tide decreases for increasing values of K_1.

Referring to Keulegan (1951, 1967), substituting for z^* from Eq. (6.40) in Eq. (6.38) instead of Eq. (6.37) and developing in a Fourier series results in the same values for a_1, a_3, b_3 and τ listed in Table 6.1.

Expressions for the basin tide follow from Eq. (6.36), with $\hat{\eta}_0^*$ given by Eq. (6.34) and z^* given by (6.40), resulting in

$$\eta_b^* = \sin(t^* - \tau) - a_1 \sin(t^*) - a_1 b_3 \cos(t^*) + a_1 b_3 \cos(3t^*) + a_1 a_3 \sin(3t^*). \tag{6.42}$$

Using Eq. (6.32), the corresponding expression for the velocity is

$$u^* = \cos(t^* - \tau) - a_1 \cos(t^*) + a_1 b_3 \sin(t^*) - 3a_1 b_3 \sin(3t^*) + 3a_1 a_3 \cos(3t^*). \tag{6.43}$$

Table 6.1 *Coefficients in the Keulegan Solution.*

K_1	a_1	a_3	b_3	τ (rad)	$\cos \tau$	$\sin \tau$	C
0.1	0.9936	−0.0001	−0.0052	0.1161	0.99327	0.11580	0.8106
0.2	0.9745	−0.0004	−0.0106	0.2314	0.97334	0.22934	0.8116
0.3	0.9435	−0.0009	−0.0164	0.3456	0.94086	0.33874	0.8126
0.4	0.9020	−0.0017	−0.0220	0.4571	0.89735	0.44137	0.8153
0.5	0.8515	−0.0028	−0.0282	0.5656	0.84425	0.53593	0.8184
0.6	0.7942	−0.0043	−0.0347	0.6699	0.78386	0.62091	0.8225
0.7	0.7325	−0.0063	−0.0418	0.7691	0.71856	0.69549	0.8288
0.8	0.6689	−0.0089	−0.0495	0.8620	0.65091	0.75917	0.8344
0.9	0.5997	−0.0123	−0.0579	0.9553	0.57732	0.81649	0.8427
1.0	0.5451	−0.0165	−0.0664	1.0265	0.51783	0.85551	0.8522
1.2	0.4369	−0.0281	−0.0849	1.1599	0.39949	0.91676	0.8751
1.4	0.3489	−0.0448	−0.1038	1.2649	0.30119	0.95357	0.9016
1.6	0.2811	−0.0661	−0.1201	1.3443	0.22449	0.97446	0.9267
1.8	0.2294	−0.0910	−0.1327	1.4041	0.16588	0.98614	0.9484
2.0	0.1893	−0.1177	−0.1401	1.4489	0.12160	0.99258	0.9650
3.0	0.0883	−0.2207	−0.1187	1.5411	0.0295	0.9996	0.9950
4.0	0.0532	−0.2606	−0.0802	1.5608	0.0104	0.9999	0.9993
5.0	0.0323	−0.2740	−0.0532	1.5650	0.0057	1.0000	0.9994

Contrary to the Öszoy–Mehta Solution, the Keulegan Solution includes a third harmonic. This harmonic enters through the quadratic bottom friction.

6.5.2 Maximum Basin Level and Maximum Inlet Velocity

For many practical applications it is the maximum basin water level and maximum inlet velocity that are of interest. Referring to Fig. 6.3, the basin water level is maximum at $t^* = \pi$. From the same figure it follows that the basin half-tidal range is

$$\eta^*_{b_{max}} = \sin(\tau), \qquad (6.44)$$

with τ a function of K_1. Values of $\sin(\tau)$ as a function of K_1 are presented in Table 6.1 for values of K_1 between 0.1 and 5.

It follows from Eq. (6.44) that the maximum basin half-tidal range is never larger than one and thus the maximum basin water level is at best equal to the maximum ocean water level. This differs from the Öszoy–Mehta Solution where, dependent on the ratio of the tidal and natural frequency, water levels in the basin can be larger than those in the ocean.

To arrive at an expression for the maximum velocity, Eq. (6.43) is differentiated with respect to t^*. Taking the derivative equal to zero, solving for t^* and substituting the value in Eq. (6.43), results in an expression for the maximum velocity as a function of the coefficients a_1, a_3, b_3 and τ. These coefficients are functions of K_1 and so is the maximum velocity. Keulegan wrote the relation between the maximum velocity and K_1 as

$$u^*_{max} = C\sin(\tau), \tag{6.45}$$

with both C and $\sin\tau$ functions of K_1. Values of C and $\sin\tau$ are presented in Table 6.1, for values of K_1 between 0.1 and 5.

6.5.3 Relative Contribution of the Third Harmonic

Using Eq. (6.43), the ratios of the velocity amplitudes of the third and first harmonic are calculated for $K_1 < 2$, values typical for most tidal inlets. The values of the coefficients a_1, a_3, b_3 and τ in Eq. (6.43) are listed in Table 6.1. From these calculations it followed that for $K_1 < 2$ the ratios of the velocity amplitudes remained practically the same with a value of approximately 0.12. The relative phase of the third harmonic, γ, as defined by Eq. (7.55) in Chapter 7, varied in a relatively narrow range between $3/4\pi$ and π. Using Fig. 7.2, it then follows that for this phase range the third harmonic lowers the peak velocity.

6.5.4 Multiple Inlets

The Keulegan Solution can be expanded to include multiple inlets, i.e., more than one inlet connecting the same basin to the ocean. For this an overall repletion coefficient,

$$K = \sum K_{1j}, \tag{6.46}$$

is defined, with K_{1j} the repletion coefficient of the jth inlet. The summation is over all the inlets. The overall repletion coefficient may be interpreted as the repletion coefficient of a single equivalent inlet, resulting in the same water level in the basin as the multiple inlets.

Proceeding in the same fashion as for the single inlet, the basin water level, η^*_b, and velocity, u^*, for the equivalent channel are calculated from Eqs. (6.42) and (6.43), respectively. Values of a_1, a_3, b_3 and τ are obtained from the Table 6.1, with K_1 replaced by K. Velocities for the individual inlets follow from

$$u^*_j = u^* \frac{K_{1j}}{K}. \tag{6.47}$$

The basin half-tidal range, $\eta^*_{b_{max}}$, and the maximum velocity for the equivalent channel, u^*_{max}, are given by, respectively, Eqs. (6.44) and (6.45). In these equations

Table 6.2 *Typical parameter values for tidal inlets and parameter values for the representative inlet.*

Parameter	Symbol	Dimensions	Range	Value
Basin surface area	A_b	m^2	$0.5–5 \times 10^8$	1×10^8
Inlet cross-sectional area	A	m^2	$0.01–2 \times 10^4$	6×10^3
Inlet width	b	m	$0.05–2 \times 10^3$	1.5×10^3
Inlet depth	h	m	2–30	4
Inlet length	L	m	$0.5–5 \times 10^3$	3×10^3
Bottom friction factor	F	–	$2–4 \times 10^{-3}$	3.5×10^{-3}
Ocean tidal amplitude	$\hat{\eta}_0$	m	0.25–1.5	0.5
Tidal frequency	σ	rad s^{-1}	$0.7–1.4 \times 10^{-4}$	1.4×10^{-4}
Entrance/exit loss coefficient	m	–	0–2	1

Assumed is a rectangular cross-section with a width to depth ratio of 375, resulting in a shape factor $\beta_2 = 0.051$ (Appendix 8.A).

C and $\sin(\tau)$ are functions of K tabulated in Table 6.1, with K_1 replaced by K. Maximum velocities for the individual inlets follow from

$$u^*_{j\max} = u^*_{\max} \frac{K_{1j}}{K}. \tag{6.48}$$

6.6 Application to a Representative Tidal Inlet

6.6.1 Representative Tidal Inlet

The Öszoy–Mehta and Keulegan Solutions are applied to a representative tidal inlet. The results are compared with a numerical solution. The planform and schematization of the representative tidal inlet are as indicated in Figs. 6.1 and 6.A.1. The selected dimensions and parameter values are typical for real world tidal inlets, and so are the ocean tidal amplitude and frequency. They are presented in the last column of Table 6.2. For comparison, ranges of dimensions and parameter values for real world tidal inlets are presented in column 4 of Table 6.2. A constraint in selecting the dimensions and parameter values of the representative tidal inlet is that the maximum cross-sectionally averaged velocity is approximately 1 m s^{-1}, a value encountered in many inlets located in a sandy environment.

6.6.2 Öszoy–Mehta Solution

The Öszoy–Mehta Solution is an approximate solution to Eqs. (6.7) and (6.8), with the ocean forcing given by Eq. (6.10). The solution is simple harmonic for both the basin tide and inlet velocity. With the parameter values in Table 6.2, the repletion coefficient for the representative inlet is $K_1 = 1.07$ and the natural or Helmholz frequency $K_2 = 0.32$.

The dimensional basin tide and inlet velocity are presented in Fig 6.4a together with the numerical solution. The numerical solution uses the same set of equations as the Öszoy–Mehta Solution. The equations are solved with an explicit time-staggered finite difference scheme (Reid and Bodine, 1969). For both solutions, the basin tide lags and the velocity leads the ocean tide. In contrast to the Öszoy–Mehta Solution, the numerical solution shows higher harmonics that are especially pronounced in the velocity. For the Öszoy–Mehta Solution, using Eq. (6.25), the velocity amplitude is 1.05 m s^{-1} and using Eqs. (6.25) and (6.15), the basin tidal amplitude is 0.44 m. For comparison, values of the numerically calculated maximum basin water level and maximum inlet velocity are 0.46 m and 0.94 m s^{-1}, respectively. It follows that the maximum basin water level for the Öszoy–Mehta Solution is somewhat lower and the maximum inlet velocity is significantly larger than for the numerical solution. Differences are attributed to the approximate nature of the Öszoy–Mehta Solution in which higher harmonics are neglected.

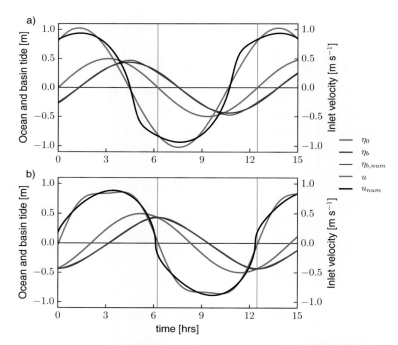

Figure 6.4 Basin tide and inlet velocity for the representative inlet determined from a) the Öszoy–Mehta Solution and b) the Keulegan Solution together with the ocean tide and numerical solutions. For parameter values, reference is made to Table 6.2.

6.6.3 Keulegan Solution

The Keulegan Solution is an approximate solution to Eqs. (6.31) and (6.32), with the ocean forcing given by Eq. (6.33). With the parameter values in Table 6.2, the repletion coefficient for the representative inlet $K_1 = 1.07$.

The dimensional basin tide and inlet velocity for the Keulegan Solution are presented in Fig. 6.4b, together with the numerical solution. The basin tide and inlet velocity are calculated from, respectively, Eqs. (6.42) and (6.43). The values of the coefficients a_1, a_3, b_3 and τ in these equations are determined from Table 6.1. The numerical solution uses the same set of equations as the Keulegan Solution. The equations are solved with an explicit time-staggered finite difference scheme. For both solutions, the basin tide lags and the inlet velocity leads the ocean tide. The ocean and basin tide curves intersect when the basin tide is maximum and minimum. At those times, velocities are zero. Using Table 6.1, it follows that $\sin(\tau) = 0.88$ and $C = 0.86$. Using Eq. (6.44), the maximum basin water level is 0.44 m and using Eq. (6.45), the maximum inlet velocity is 0.88 m s^{-1}. For comparison, values of the numerically calculated maximum basin water level and maximum inlet velocity are, respectively, 0.43 m and 0.89 m s^{-1}. Values of the Keulegan Solution are close to the same as those of the numerical solution.

The Keulegan Solution for the representative inlet consists of a first and third harmonic. The first and third harmonic for the inlet velocity are presented in Fig. 6.5. In agreement with what is stated in Section 6.5.3, the third harmonic lowers the peak velocity. Because there is no third harmonic in the ocean forcing, the third

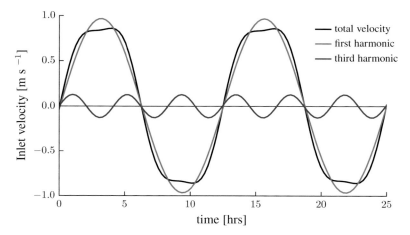

Figure 6.5 Contribution of the first and the third harmonic in the Keulegan Solution to the velocity of the representative inlet. For parameter values, reference is made to Table 6.2.

harmonic in the inlet velocity for the representative inlet is internally generated. The ratio of the velocity amplitudes of the third and first harmonic is 0.12. This seems a reasonable value; for example, the corresponding ratio for the Marsdiep Inlet (Texel Inlet), The Netherlands, as determined from long-term ADCP measurements, is 0.09. The difference can in part be attributed to the presence of the third harmonic in the ocean forcing of the Marsdiep Inlet (Buijsman and Ridderinkhof, 2007).

6.A Dynamics of the Flow in the Inlet

The schematized inlet consists of a prismatic part with diverging sections on both ends; see Fig. 6.A.1. In the prismatic part of the inlet, the flow dynamics is governed by the equations for unsteady uniform open channel flow (Dronkers, 1975), i.e.,

for flood
$$\frac{L}{g}\frac{du}{dt} + \frac{FLu^2}{gR} = \eta_1 - \eta_2, \qquad (6.A.1)$$

for ebb
$$\frac{L}{g}\frac{du}{dt} - \frac{FLu^2}{gR} = \eta_1 - \eta_2. \qquad (6.A.2)$$

Where u is the cross-sectionally averaged velocity, assumed to be independent of x, L is length of the prismatic part of the inlet, R is hydraulic radius, g is gravity acceleration, F is a friction factor, t is time and η_1 and η_2 are water levels at, respectively, x_1 and x_2; see Fig. 6.A.1. In the equations, variation in depth with tidal stage and hypsometric effects, the dependence of the surface area of the inlet on the water level, are neglected. For the equations including variation in depth with tidal stage reference is made to Chapter 7. Equations including hypsometric effects are presented in de Swart and Volp (2012).

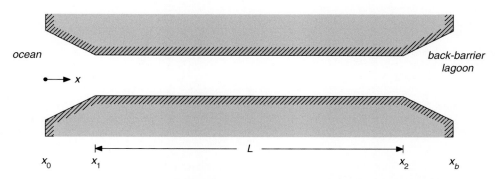

Figure 6.A.1 Schematization of the inlet.

The flow in the diverging sections constitutes a balance between advective accelerations and pressure gradients (Serrano et al., 2013; Vennell, 2006). Assuming u is independent of the cross-channel direction, this results in

$$u\frac{\partial u}{\partial x} + g\frac{\partial \eta}{\partial x} = 0. \tag{6.A.3}$$

For flood, integrating with respect to x between x_0 and x_1 and multiplying by the density ρ gives

$$\frac{1}{2}\rho u^2 + \rho g\eta_1 = \rho g\eta_0, \tag{6.A.4}$$

with η_0 is the ocean tide. At x_0, the cross-sectional area is relatively large, justifying the assumption of zero velocity. The left-hand side of Eq. (6.A.4) is the total energy (kinetic and potential) per unit volume at x_1. The right-hand side represents the potential energy per unit volume at x_0. When accounting for a turbulent kinetic energy loss, $\beta\rho u^2/2$ with $0 \le \beta \le 1$, Eq. (6.A.4) becomes

$$(1+\beta)\frac{1}{2}\rho u^2 + \rho g\eta_1 = \rho g\eta_0. \tag{6.A.5}$$

For flood, integrating with respect to x between x_2 and x_b and multiplying by the density ρ gives

$$\frac{1}{2}\rho u^2 + \rho g\eta_2 = \rho g\eta_b, \tag{6.A.6}$$

where η_b is the basin tide. At x_b the cross-sectional area is relatively large, justifying the assumption of zero velocity. The left-hand side of Eq. (6.A.6) represents the total energy per unit volume at x_2. The right-hand side of the equation is the potential energy per unit volume at x_b. When accounting for a turbulent kinetic energy loss, $\gamma\rho u^2/2$ with $0 \le \gamma \le 1$, Eq. (6.A.6) becomes

$$(1-\gamma)\frac{1}{2}\rho u^2 + \rho g\eta_2 = \rho g\eta_b. \tag{6.A.7}$$

Similar equations as (6.A.5) and (6.A.7) hold for the ebb. For the ocean side and basin side diverging section this results in, respectively,

$$(1-\gamma)\frac{1}{2}\rho u^2 + \rho g\eta_1 = \rho g\eta_0, \tag{6.A.8}$$

and

$$(1+\beta)\frac{1}{2}\rho u^2 + \rho g\eta_2 = \rho g\eta_b. \tag{6.A.9}$$

Solving for η_1 and η_2 from, respectively, Eqs. (6.A.5) and (6.A.7) and substituting in Eq. (6.A.1) results in

$$\frac{L}{g}\frac{du}{dt} + \left(\frac{m}{2g} + \frac{FL}{gR}\right)u^2 = \eta_0 - \eta_b, \tag{6.A.10}$$

with $m = \beta + \gamma$.

Solving for η_1 and η_2 from, respectively, Eqs. (6.A.8) and (6.A.9) and substituting in Eq. (6.A.2) results in

$$\frac{L}{g}\frac{du}{dt} - \left(\frac{m}{2g} + \frac{FL}{gR}\right)u^2 = \eta_0 - \eta_b. \tag{6.A.11}$$

With u positive in the flood direction and negative in the ebb direction, Eqs. (6.A.10) and (6.A.11) can be replaced by a single equation

$$\frac{L}{g}\frac{du}{dt} + \left(\frac{m}{2g} + \frac{FL}{gR}\right)u|u| = \eta_0 - \eta_b. \tag{6.A.12}$$

This equation corresponds to Eq. (6.1) in the main text.

7

Tidal Inlet Hydrodynamics; Including Depth Variations with Tidal Stage

7.1 Introduction

A difference with the previous chapter is that the hydrodynamic equations include depth variations with tidal stage. When forcing the tidal inlet system with a simple harmonic ocean tide, this results in a mean inlet velocity, a mean basin level different from that in the ocean and overtides (Boon and Byrne, 1981; LeProvost, 1991). A distinction is made between odd and even overtides. Odd overtides have frequencies that are odd multiples and even overtides have frequencies that are even multiples of the frequency of the ocean tide. Both odd and even overtides result in asymmetries in the inlet velocity and the basin tide (van de Kreeke and Robaczewska, 1993). It is the asymmetry in the inlet velocity introduced by even overtides that contributes to a long-term net sand transport (Pingree and Griffiths, 1979). For this reason, emphasis in this chapter is on even overtides, together with mean inlet velocity and mean basin level. The generation of the mean inlet velocity, the mean basin level and the even overtides is demonstrated by applying a perturbation analysis to the governing equations.

In addition to depth variations with tidal stage, variations in inlet width and basin surface area with tidal stage (sometimes referred to as hypsometric effects) will result in a mean inlet velocity, a mean basin level different from zero and even overtides. For this, reference is made to de Swart and Volp (2012) and Ridderinkhof et al. (2014).

7.2 Equations Including Depth Variations with Tidal Stage

Expanding Eqs. (6.1) and (6.2) to include depth variations with tidal stage results in

$$\frac{L}{g}\frac{du}{dt} + \left(\frac{m}{2g} + \frac{FL}{g(h+\eta_m)}\right)u|u| = \eta_0 - \eta_b, \tag{7.1}$$

and

$$b(h + \eta_m)u = A_b \frac{d\eta_b}{dt}. \tag{7.2}$$

Assumed is a wide rectangular cross-section with mean depth h and width b. L is the length of the prismatic part of the inlet (Fig. 6.A.1), g is gravity acceleration, u is cross-sectional averaged velocity, positive in the flood direction, t is time, F is a nonlinear friction factor, m is an entrance/exit loss coefficient, η_0 is ocean tide, η_b is basin tide and A_b is basin surface area. The representative water level for the inlet, η_m, is defined as

$$\eta_m = \frac{\eta_0 + \eta_b}{2}. \tag{7.3}$$

To explain the generation of even overtides, mean inlet velocity and mean basin level, an analytical solution is sought to Eqs. (7.1) and (7.2). For this, the product $u|u|$ is linearized (Lorentz, 1926). Because the focus is on even rather than odd overtides, this is an acceptable simplification. After linearization, the term in Eq. (7.1) involving $u|u|$ is

$$\left(\frac{m}{2g} + \frac{FL}{g(h + \eta_m)} \right) u|u| = \left(\frac{m_l}{2g} + \frac{F_l L}{g(h + \eta_m)} \right) u, \tag{7.4}$$

with

$$m_l = \frac{8}{3\pi} \hat{u}_0 m \qquad \text{and} \qquad F_l = \frac{8}{3\pi} \hat{u}_0 F. \tag{7.5}$$

In Eq. (7.5), $\hat{u}_0$ is the leading-order velocity amplitude to be defined in Section 7.3, F_l is a linear friction factor and m_l is a linear entrance/exit loss coefficient. Using Eq. (7.4), Eq. (7.1) is written as

$$\frac{L}{g} \frac{du}{dt} + \left(\frac{m_l}{2g} + \frac{F_l L}{g(h + \eta_m)} \right) u = \eta_0 - \eta_b. \tag{7.6}$$

The ocean forcing is assumed to be a simple harmonic ocean tide

$$\eta_0 = \hat{\eta}_0 \sin \sigma t, \tag{7.7}$$

where $\hat{\eta}_0$ is the amplitude and σ is the angular frequency.

The nonlinear Eqs. (7.2) and (7.6) with the boundary condition, Eq. (7.7), are solved using a perturbation technique. The details of this method are described in Appendix 7.A. As shown in the appendix, the solution for u and η_b can be approximated by the sum of the solutions of a set of leading-order equations and a set of first-order equations.

The set of leading-order equations is

$$\frac{L}{g} \frac{du_0}{dt} + F' u_0 = \eta_0 - \eta_{b_0}, \tag{7.8}$$

$$A_b \frac{d\eta_{b0}}{dt} - bhu_0 = 0,$$ (7.9)

with

$$F' = \frac{m_l}{2g} + \frac{F_l L}{gh}.$$ (7.10)

In these equations u_0 is the leading-order velocity and η_{b0} is the leading-order basin water level. The set of first-order equations is

$$\frac{h^2}{F_l} \frac{du_1}{dt} + \left(\frac{m_l h^2}{2F_l L} + h \right) u_1 + \frac{gh^2}{F_l L} \eta_{b_1} = u_0 \eta_{m_0},$$ (7.11)

$$\frac{A_b}{b} \frac{d\eta_{b_1}}{dt} - hu_1 = u_0 \eta_{m_0},$$ (7.12)

with

$$\eta_{m_0} = \frac{\eta_0 + \eta_{b_0}}{2}.$$ (7.13)

In these equations, u_1 is the first-order velocity and η_{b_1} is the first-order basin water level. The leading-order term on the right-hand side of Eqs. (7.11) and (7.12) forces the first-order water motion and is the result of including variations of depth with tidal stage in the hydrodynamic equations. The complete solution consists of the sum of the leading- and first-order solutions.

7.3 Solution of the Leading-Order Equations

With the boundary condition given by Eq. (7.7), the following trial solutions are used to solve for η_{b0} and u_0 from Eqs. (7.8) and (7.9),

$$\eta_{b0} = \hat{\eta}_{b0} \sin(\sigma t - \alpha),$$ (7.14)

and

$$u_0 = \hat{u}_0 \sin(\sigma t - \beta).$$ (7.15)

Substituting the trial solutions in Eq. (7.9) results in

$$A_b \sigma \hat{\eta}_{b0} \cos(\sigma t - \alpha) = bh\hat{u}_0 \sin(\sigma t - \beta).$$ (7.16)

This equation can only be satisfied for

$$\beta = \alpha - \frac{\pi}{2},$$ (7.17)

and

$$\hat{\eta}_{b0} = \frac{bh}{\sigma A_b} \hat{u}_0.$$ (7.18)

Substituting for β from Eq. (7.17) in Eq. (7.15) results in

$$u_0 = \hat{u}_0 \cos(\sigma t - \alpha). \tag{7.19}$$

It follows from Eqs. (7.14) and (7.19) that the leading-order velocity and the leading-order basin tide are $\pi/2$ radians out of phase, with the velocity leading the basin tide.

Substituting the expression for η_{b_0} and u_0, respectively, Eqs. (7.14) and (7.19), in Eq. (7.8) and collecting terms proportional to $\sin \sigma t$ and $\cos \sigma t$ results in

$$\left(\hat{\eta}_{b_0} - \frac{\sigma L}{g} \hat{u}_0 \right) \cos \alpha + F' \hat{u}_0 \sin \alpha = \hat{\eta}_0, \tag{7.20}$$

and

$$-\left(\hat{\eta}_{b_0} - \frac{\sigma L}{g} \hat{u}_0 \right) \sin \alpha + F' \hat{u}_0 \cos \alpha = 0. \tag{7.21}$$

Eliminating $\hat{\eta}_{b_0}$ between Eqs. (7.18) and (7.21), and assuming $\hat{u}_0 \neq 0$, results in an equation for α,

$$-\left(1 - K_2^2\right) \sin \alpha + K' \cos \alpha = 0, \tag{7.22}$$

with

$$K' = F' \frac{\sigma A_b}{bh} \qquad \text{and} \qquad K_2 = \frac{\sigma}{\sqrt{\frac{gbh}{LA_b}}}. \tag{7.23}$$

Here, K' is a damping coefficient and K_2 is the ratio of the forcing and natural or Helmholtz frequency. It follows from Eq. (7.22) that

$$\tan \alpha = \frac{K'}{1 - K_2^2}. \tag{7.24}$$

For $K_2 < 1$, $\tan \alpha$ is positive and α is in the first or third quadrant and for $K_2 > 1$, $\tan \alpha$ is negative and α is in the second or fourth quadrant.

Substituting for $\hat{\eta}_{b_0}$ from Eq. (7.18) in Eq. (7.20) results in an equation for $\hat{u}_0$,

$$\left[\left(1 - K_2^2\right) \cos \alpha + K' \sin \alpha\right] \hat{u}_0 = \frac{\sigma A_b}{bh} \hat{\eta}_0. \tag{7.25}$$

For $K_2 < 1$ and α in the first or third quadrant, it follows from Eq. (7.25) that for $\hat{u}_0$ to be positive, α has to be in the first quadrant. Similarly, for $K_2 > 1$ with α in the second or fourth quadrant, it follows from Eq. (7.25) that for $\hat{u}_0$ to be positive, α has to be in the second quadrant.

Using Eq. (7.24), with $K_2 < 1$ and α in the first quadrant, the expressions for $\cos \alpha$ and $\sin \alpha$ are, respectively,

$$\cos \alpha = \frac{1 - K_2^2}{\sqrt{K'^2 + (1 - K_2^2)^2}}, \qquad (7.26)$$

and

$$\sin \alpha = \frac{K'}{\sqrt{K'^2 + (1 - K_2^2)^2}}. \qquad (7.27)$$

Using Eq. (7.24), the same expressions are found for $K_2 > 1$ and α in the second quadrant.

Substituting for $\cos \alpha$ and $\sin \alpha$ in Eq. (7.25) and solving for $\hat{u}_0$ results in

$$\hat{u}_0 = \frac{\sigma A_b}{bh} \frac{\hat{\eta}_0}{\sqrt{K'^2 + (1 - K_2^2)^2}}. \qquad (7.28)$$

Substituting for $\hat{u}_0$ from Eq. (7.28) in Eq. (7.18) gives the expression for the leading-order basin tidal amplitude

$$\hat{\eta}_{b_0} = \frac{\hat{\eta}_0}{\sqrt{K'^2 + (1 - K_2^2)^2}}. \qquad (7.29)$$

Because K', through Eqs. (7.5) and (7.10), depends on $\hat{u}_0$, evaluating $\hat{u}_0$ and $\hat{\eta}_{b_0}$ requires iterating. A first estimate of the value of $\hat{u}_0$ can be obtained using the Öszoy–Mehta Solution presented in Section 6.4.

7.4 Solution to the First-Order Equations

7.4.1 First-Order Forcing

The term on the right-hand side of Eqs. (7.11) and (7.12) constitutes the first-order forcing. By substituting the leading-order expressions η_0, η_{b_0} and u_0, given by Eqs. (7.7), (7.14) and (7.19), respectively, the first-order forcing term reads

$$u_0 \eta_{m_0} = \tfrac{1}{4} \hat{\eta}_0 \hat{u}_0 \left[\sin(2\sigma t - \alpha) + \sin \alpha \right] + \tfrac{1}{4} \hat{\eta}_{b_0} \hat{u}_0 \sin(2\sigma t - 2\alpha). \qquad (7.30)$$

In this equation α, $\hat{u}_0$ and $\hat{\eta}_{b_0}$ are given by Eqs. (7.24), (7.28) and (7.29), respectively.

The forcing consists of harmonics with frequency 2σ and a time-independent part. As a result, the solution to the first-order equations, Eqs. (7.11) and (7.12), consists of a harmonic with frequency 2σ and a constant. For the velocity, this is written as

$$u_1 = \tilde{u}_1 + \langle u_1 \rangle, \qquad (7.31)$$

and for the basin tide,

$$\eta_{b_1} = \tilde{\eta}_{b_1} + \langle \eta_{b_1} \rangle. \tag{7.32}$$

The tilde implies a harmonic with frequency 2σ, and thus $\tilde{u}_1$ and $\tilde{\eta}_{b_1}$ represent the first even overtides. The angle brackets denote tidal averaging, where $\langle u_1 \rangle$ is the mean velocity and $\langle \eta_{b_1} \rangle$ is the mean basin water level.

7.4.2 Mean Inlet Velocity and Mean Basin Level

Equations for the mean inlet velocity and mean basin level follow by tidally averaging the first-order equations, Eqs. (7.11) and (7.12). The resulting equations are

$$\left(\frac{m_l h^2}{2 F_l L} + h \right) \langle u_1 \rangle + \frac{g h^2}{F_l L} \langle \eta_{b_1} \rangle = \langle u_0 \eta_{m0} \rangle, \tag{7.33}$$

and

$$\langle u_0 \eta_{m0} \rangle + h \langle u_1 \rangle = 0. \tag{7.34}$$

Using Eq. (7.30), the tidal average of the first-order forcing is

$$\langle u_0 \eta_{m0} \rangle = \tfrac{1}{4} \hat{\eta}_0 \hat{u}_0 \sin \alpha. \tag{7.35}$$

Eliminating $\langle u_0 \eta_{m0} \rangle$ between Eqs. (7.34) and (7.35) and using Eq. (7.27) results in the expression for the mean inlet velocity

$$\langle u_1 \rangle = -\frac{1}{4} \frac{A_b \sigma}{b h^2} \frac{\hat{\eta}_0^2 K'}{\left(K'^2 + (1 - K_2^2)^2 \right)}. \tag{7.36}$$

The mean inlet velocity is negative, i.e., in the ebb direction.

Eq. (7.34) expresses the balance between a tidally averaged water transport in the ebb direction and in the flood direction. The first term on the left represents a tidally averaged transport associated with the phase difference between the leading-order inlet tide and inlet velocity. It follows from Eq. (7.35) that, with $\sin \alpha$ positive (Section 7.3), this transport is in the flood direction. The second term on the left represents a tidally averaged transport in the ebb direction resulting from the mean inlet velocity.

Eliminating $\langle u_0 \eta_{m0} \rangle$ between Eqs. (7.33) and (7.34) results in the expression for the mean basin level in terms of the mean inlet velocity

$$\langle \eta_{b_1} \rangle = -\left(\frac{m_l}{2g} + 2 \frac{F_l L}{gh} \right) \langle u_1 \rangle. \tag{7.37}$$

Substituting for $\langle u_1 \rangle$ from Eq. (7.36), the mean basin level is

$$\langle \eta_{b_1} \rangle = \left(\frac{m_l}{8g} + \frac{F_l L}{2gh} \right) \frac{A_b \sigma}{bh^2} \frac{\hat{\eta}_0^2 K'}{\left(K'^2 + (1 - K_2^2)^2 \right)}. \tag{7.38}$$

The mean basin level is positive and thus the basin level encounters a set-up.

For most tidal inlets the tidally averaged inlet velocity and basin tide are an order of magnitude smaller than the corresponding leading-order amplitudes. For example, using long-term measurements in the Frisian Inlet, van de Kreeke and Hibma (2005) report a leading-order velocity amplitude of 0.77 m s^{-1} and a tidally averaged velocity in the ebb direction of 0.05 m s^{-1}. The Frisian Inlet is a tidal inlet on the Dutch Wadden Sea coast; see Fig. 1.1a.

7.4.3 First-Order Tide and Velocity

Equations for the first-order tide follow by subtracting Eq. (7.33) from Eq. (7.11) and subtracting Eq. (7.34) from Eq. (7.12) resulting in, respectively,

$$\frac{h^2}{F_l} \frac{d\tilde{u}_1}{dt} + \left(\frac{m_l h^2}{2 F_l L} + h \right) \tilde{u}_1 + \frac{gh^2}{F_l L} \tilde{\eta}_{b_1} = \tfrac{1}{4} \hat{\eta}_0 \hat{u}_0 \sin(2\sigma t - \alpha)$$

$$+ \tfrac{1}{4} \hat{\eta}_{b_0} \hat{u}_0 \sin(2\sigma t - 2\alpha), \tag{7.39}$$

and

$$\frac{A_b}{b} \frac{d\tilde{\eta}_{b_1}}{dt} - h\tilde{u}_1 = \tfrac{1}{4} \hat{\eta}_0 \hat{u}_0 \sin(2\sigma t - \alpha) + \tfrac{1}{4} \hat{\eta}_{b_0} \hat{u}_0 \sin(2\sigma t - 2\alpha). \tag{7.40}$$

The following trial solutions for the unknowns $\tilde{\eta}_{b_1}$ and $\tilde{u}_1$ are proposed:

$$\tilde{\eta}_{b_1} = \hat{\tilde{\eta}}_{b_1} \sin(2\sigma t - \varphi), \tag{7.41}$$

and

$$\tilde{u}_1 = \hat{\tilde{u}}_1 \sin(2\sigma t - \psi). \tag{7.42}$$

Eq. (7.41) is written as

$$\tilde{\eta}_{b_1} = \hat{\tilde{\eta}}_{b_{1s}} \sin 2\sigma t - \hat{\tilde{\eta}}_{b_{1c}} \cos 2\sigma t, \tag{7.43}$$

with

$$\hat{\tilde{\eta}}_{b_1} = \sqrt{\hat{\tilde{\eta}}_{b_{1s}}^2 + \hat{\tilde{\eta}}_{b_{1c}}^2}, \tag{7.44}$$

and

$$\tan \varphi = \frac{\hat{\tilde{\eta}}_{b_{1c}}}{\hat{\tilde{\eta}}_{b_{1s}}}. \tag{7.45}$$

Similarly, Eq. (7.42) is written as

$$\tilde{u}_1 = \hat{\tilde{u}}_{1s} \sin 2\sigma t - \hat{\tilde{u}}_{1c} \cos 2\sigma t, \tag{7.46}$$

with

$$\hat{\tilde{u}}_1 = \sqrt{\hat{\tilde{u}}_{1s}^2 + \hat{\tilde{u}}_{1c}^2}, \tag{7.47}$$

and

$$\tan \psi = \frac{\hat{\tilde{u}}_{1c}}{\hat{\tilde{u}}_{1s}}. \tag{7.48}$$

Substituting for $\tilde{\eta}_{b_1}$ and $\tilde{u}_1$ from, respectively, Eqs. (7.43) and (7.46) in Eqs. (7.39) and (7.40), and collecting terms in $\sin 2\sigma t$ and $\cos 2\sigma t$, results in four algebraic equations with unknowns $\hat{\tilde{\eta}}_{b_{1s}}$, $\hat{\tilde{\eta}}_{b_{1c}}$, $\hat{\tilde{u}}_{1s}$ and $\hat{\tilde{u}}_{1c}$. In matrix form:

$$\begin{bmatrix} 0 & \frac{gh^2}{F_l L} & \frac{2\sigma h^2}{F_l} & \left(\frac{m_l h^2}{2F_l L} + h\right) \\ -\frac{gh^2}{F_l L} & 0 & -\left(\frac{m_l h^2}{2F_l L} + h\right) & \frac{2\sigma h^2}{F_l} \\ \frac{2\sigma A_b}{b} & 0 & 0 & -h \\ 0 & \frac{2\sigma A_b}{b} & h & 0 \end{bmatrix} \begin{bmatrix} \hat{\tilde{\eta}}_{b_{1c}} \\ \hat{\tilde{\eta}}_{b_{1s}} \\ \hat{\tilde{u}}_{1c} \\ \hat{\tilde{u}}_{1s} \end{bmatrix} = \begin{bmatrix} F_c \\ F_s \\ F_c \\ F_s \end{bmatrix}, \tag{7.49}$$

with

$$F_c = -\tfrac{1}{4}(\hat{\eta}_0 \hat{u}_0 \cos \alpha + \hat{\eta}_{b_0} \hat{u}_0 \cos 2\alpha), \tag{7.50}$$

and

$$F_s = \tfrac{1}{4}(\hat{\eta}_0 \hat{u}_0 \sin \alpha + \hat{\eta}_{b_0} \hat{u}_0 \sin 2\alpha). \tag{7.51}$$

Solving for $\hat{\tilde{\eta}}_{b_{1s}}$, $\hat{\tilde{\eta}}_{b_{1c}}$, $\hat{\tilde{u}}_{1s}$ and $\hat{\tilde{u}}_{1c}$, the amplitude of the first-order basin tide and inlet velocity follow from Eqs. (7.44) and (7.47), respectively. The phase of the first-order basin tide and inlet velocity follows from Eqs. (7.45) and (7.48), respectively.

Solving the matrix equations (7.49) analytically is a cumbersome exercise. Instead, the system of equations is solved numerically using the parameter values of the representative inlet listed in Table 6.2. The results are presented in Section 7.6.

At most tidal inlets the amplitudes of the first-order velocity and basin tide are an order of magnitude smaller than the corresponding leading-order amplitudes. For example using long-term observations, for the Texel Inlet the ratio of the first and leading-order velocity amplitudes is 0.17 (Buijsman and Ridderinkhof, 2007) and for the Frisian Inlet is 0.08 (van de Kreeke and Hibma, 2005). Both inlets are part of the Dutch Wadden Sea coast; see Fig. 1.1a.

7.5 Tidal Asymmetry

Tidal asymmetry is associated with the combination of a fundamental tidal harmonic and its overtides. A distinction is made between water level asymmetry and
velocity asymmetry. Here, emphasis is on velocity asymmetry because it results in
a net sand transport (Pingree and Griffiths, 1979; van de Kreeke and Robaczewska,
1993).

To define tidal asymmetry, consider a rectilinear current velocity $u(t)$ that is
periodic with zero mean. Defining the time origin at slack water, $u(t)$ is symmetric
when

$$|u(t)| = |u(-t)|, \tag{7.52}$$

and is asymmetric when

$$|u(t)| \neq |u(-t)|. \tag{7.53}$$

Assuming the velocity consists of a fundamental harmonic and its first even
overtide

$$u(t) = \hat{u}_0 \cos(\sigma t) + \hat{u}_1 \cos(2\sigma t - \beta). \tag{7.54}$$

In this equation, β is the phase of the even overtide with respect to the phase of the
fundamental harmonic. It follows from Eqs. (7.52) and (7.53) that $u(t)$ is asymmetric except for $\beta = \pi/2$ and $\beta = 3\pi/2$. This is further demonstrated in Fig. 7.1.
The asymmetry implies that for $-\pi/2 < \beta < \pi/2$ the maximum flood velocity
(defined as positive) is larger than the maximum ebb velocity and the flood duration is shorter than the ebb duration. Assuming sand transport is proportional to
$|u|^n \text{sign}(u)$, this leads to a long-term net sand transport in the flood direction. For
$\pi/2 < \beta < 3\pi/2$, the maximum ebb velocity is larger than the maximum flood
velocity, resulting in a net sand transport in the ebb direction (van de Kreeke and
Robaczewska, 1993).

For a fundamental harmonic and its first odd overtide the expression for the
velocity is

$$u(t) = \hat{u}_0 \cos(\sigma t) + \hat{u}_2 \cos(3\sigma t - \gamma). \tag{7.55}$$

In this equation, γ is the phase of the odd overtide relative to the phase of the
first harmonic. Using Eqs. (7.52) and (7.53), it follows that the velocity $u(t)$ is
asymmetric except for $\gamma = 0$ and $\gamma = \pi$. This is further demonstrated in Fig. 7.2.
Contrary to the velocity asymmetry for the fundamental harmonic and its first even
overtide, the velocity asymmetry for the fundamental harmonic and its first odd
overtide does not lead to a net sand transport. The reason is that for the fundamental
tidal harmonic and its first odd overtide holds that

$$u(t) = -u\left(t + \tfrac{T}{2}\right), \tag{7.56}$$

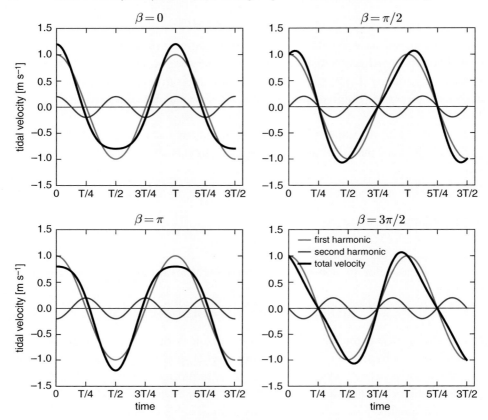

Figure 7.1 Leading-order tide, first even overtide and their sum for different values of the relative phase β. For definition of the symbols reference is made to Eq. (7.54) (adapted from van de Kreeke and Robaczewska, 1993).

with T is the period of the fundamental harmonic. Eq. (7.56) implies that regardless of the value of the relative phase γ, maximum flood velocities equal maximum ebb velocities.

For additional information on tidal asymmetry, reference is made to Boon and Byrne (1981), Dronkers (1986), Friedrichs and Aubrey (1988), Fry and Aubrey (1990), van de Kreeke and Dunsbergen (2000) and Dronkers (2005).

7.6 Application to the Representative Inlet

7.6.1 Leading-Order Solution

Parameter values for the representative inlet are presented in Table 6.2. Using this table, the value of K_2 defined by Eq. (7.23) is 0.32.

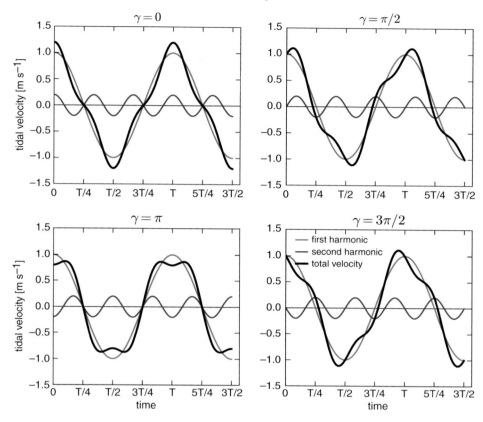

Figure 7.2 Leading-order tide, first odd overtide and their sum for different values of the relative phase γ. For definition of the symbols reference is made to Eq. (7.55) (adapted from van de Kreeke and Robaczewska, 1993).

Calculating the amplitudes and phases of the leading-order inlet velocity and basin tide requires an iterative approach. Calculations start with assuming a value for the velocity amplitude. The values of m_l and F_l are then calculated from Eq. (7.5). These values are used to calculate F' from Eq. (7.10). With the known value of F', K' is calculated from Eq. (7.23). With the known value of K', a new value of the velocity amplitude is calculated using Eq. (7.28). Calculations are carried out until the new and old velocity amplitudes are close to the same. A value of 1.05 m s^{-1} was taken as a first estimate of the amplitude of the leading-order velocity. This velocity amplitude was calculated using the Öszoy–Mehta Solution (Section 6.6.2). After iterating, the amplitude of the leading-order inlet velocity $\hat{u}_0 = 1.05$ m s^{-1}. The amplitude of the leading-order basin tide, calculated from Eq. (7.29), is $\hat{\eta}_{b_0} = 0.45$ m and using Eq. (7.24) the phase is $\alpha = 0.63$ rad. The values of the various coefficients are $m_l = 0.89$, $F_l = 3.12 \times 10^{-3}$, $F' = 0.28$ and $K' = 0.66$.

7.6.2 First-Order Solution

The first-order solution consists of the first even overtide of the velocity and basin tide and the tidally averaged inlet velocity and basin level. The amplitudes and phases of the overtides of the inlet velocity and basin level are numerically calculated using the matrix equation, Eq. (7.49). The results are $\hat{\tilde{u}}_1 = 0.06$ m s^{-1}, $\psi = -0.23$ rad, $\hat{\tilde{\eta}}_{b_1} = 0.02$ m and $\varphi = 1.91$ rad. The first-order amplitudes are an order of magnitude smaller than the corresponding leading-order amplitudes. Using Eq. (7.36), the mean velocity $\langle u_1 \rangle = -0.02$ m s^{-1} and from Eq. (7.37), the mean basin level $\langle \eta_{b_1} \rangle = 0.01$ m.

7.6.3 Tidal Asymmetry

Using Eqs. (7.19) and (7.42), the sum of the leading-order and first-order velocity is

$$u = \hat{u}_0 \cos(\sigma t - \alpha) + \hat{\tilde{u}}_1 \sin(2\sigma t - \psi). \tag{7.57}$$

To determine the tidal asymmetry, i.e., ebb or flood dominance, this equation is written in the form of Eq. (7.54) by introducing the transformation of the time axis

$$\sigma t' = \sigma t - \alpha, \tag{7.58}$$

resulting in

$$u = \hat{u}_0 \cos(\sigma t') + \hat{\tilde{u}}_1 \cos\left(2\sigma t' - \left((\psi + \tfrac{\pi}{2}) - 2\alpha\right)\right). \tag{7.59}$$

With the values of $\psi = -0.23$ rad and $\alpha = 0.63$ rad, the relative phase $\left((\psi + \tfrac{\pi}{2}) - 2\alpha\right)$ is 0.08 rad. With the relative phase smaller than $\pi/2$ and larger than $-\pi/2$, it follows from what has been stated in Section 7.5 that the velocity is flood dominant. This agrees with Fig. 7.3, in which the leading- and first-order velocity and the sum of the two are presented.

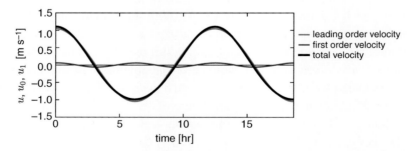

Figure 7.3 Leading-order velocity, u_0, first-order velocity, u_1, and their sum u for the representative inlet.

7.A Reduced System of Equations and Perturbation Analysis

This appendix is an extension of Section 7.2 and shows the derivation of the leading- and first-order equations. As a first step, the variables in the system of equations (7.2) and (7.6) are scaled. Similar to the scaling analysis in Section 6.3.2, the water level scale is $\hat{\eta}_0$, the velocity scale is $U = \sigma \hat{\eta}_0 A_b / bh$ and the timescale is σ^{-1}, leading to the following non-dimensional variables:

$$\eta_0^* = \frac{\eta_0}{\hat{\eta}_0}, \quad \eta_b^* = \frac{\eta_b}{\hat{\eta}_0}, \quad \eta_m^* = \frac{\eta_m}{\hat{\eta}_0}, \quad u^* = \frac{bh}{\sigma \hat{\eta}_0 A_b} u, \quad t^* = \sigma t.$$

Substituting these scaled variables into Eqs. (7.2) and (7.6) leads to the non-dimensional equations for the water motion

$$\frac{\sigma A_b}{gbh}\left[\sigma L \frac{du^*}{dt^*} + \left(\frac{m_l}{2} + \frac{F_l L}{h}\frac{1}{(1 + \epsilon \eta_m^*)}\right)u^*\right] = \eta_0^* - \eta_b^*, \tag{7.A.1}$$

and

$$\left(1 + \epsilon \eta_m^*\right)u^* = \frac{d\eta_b^*}{dt^*}. \tag{7.A.2}$$

In Eqs. (7.A.1) and (7.A.2) the dimensionless parameter ϵ denotes the ratio of the ocean tidal amplitude $\hat{\eta}_0$ and the water depth h. With $\epsilon = \hat{\eta}_0 / h \ll 1$, the following approximation can be made: $1/(1 + \epsilon \eta_m^*) = 1 - \epsilon \eta_m^* + \mathcal{O}(\epsilon^2)$. Neglecting terms of $\mathcal{O}(\epsilon^2)$ and higher, Eq. (7.A.1) reduces to

$$\frac{\sigma A_b}{gbh}\left[\sigma L \frac{du^*}{dt^*} + \left(\frac{m_l}{2} + \frac{F_l L}{h}\right)u^* - \frac{F_l L}{h}\epsilon \eta_m^* u^*\right] = \eta_0^* - \eta_b^*. \tag{7.A.3}$$

The nonlinear product $\eta_m^* u^*$ in Eqs. (7.A.2) and (7.A.3) is the result of including variation in depth with tidal stage.

The solution to Eqs. (7.A.2) and (7.A.3) is written in terms of a power series of the small parameter ϵ, i.e.

$$u^* = \epsilon^0 u_0^* + \epsilon^1 u_1^* + \mathcal{O}(\epsilon^2), \tag{7.A.4}$$
$$\eta_b^* = \epsilon^0 \eta_{b_0}^* + \epsilon^1 \eta_{b_1}^* + \mathcal{O}(\epsilon^2), \tag{7.A.5}$$
$$\eta_m^* = \epsilon^0 \eta_{m_0}^* + \epsilon^1 \eta_{m_1}^* + \mathcal{O}(\epsilon^2). \tag{7.A.6}$$

Terms with ϵ^0 are of leading-order and terms with ϵ^1 are of first-order. The scale of a first-order non-dimensional variable is ϵ times the scale of the corresponding leading-order non-dimensional variable. As a result, all non-dimensional variables in the equations are of $\mathcal{O}(1)$. As an example, the scale of the leading-order velocity, u_0, is $bh/\sigma \hat{\eta}_0 A_b$ and the scale of the first-order velocity, u_1, is $\epsilon bh/\sigma \hat{\eta}_0 A_b$. Substituting the expansions of u^*, η_b^* and η_m^*, Eqs. (7.A.4), (7.A.5) and (7.A.6) respectively, in Eqs. (7.A.2) and (7.A.3) and collecting terms of equal powers of ϵ, results in a set of equations with non-dimensional terms of $\mathcal{O}(\epsilon^0)$ and one with non-dimensional terms of $\mathcal{O}(\epsilon^1)$. Using the appropriate scales, the two sets of equations are written in terms of dimensional variables.

The resulting the system of equations at leading-order, $\mathcal{O}(\epsilon^0)$, is

$$\frac{L}{g}\frac{du_0}{dt} + F'u_0 = \eta_0 - \eta_{b_0}, \tag{7.A.7}$$

$$A_b\frac{d\eta_{b_0}}{dt} - bhu_0 = 0, \tag{7.A.8}$$

where F' is given by Eq. (7.10). The system of equations at first-order, $\mathcal{O}(\epsilon^1)$, is

$$\frac{h^2}{F_l}\frac{du_1}{dt} + \left(\frac{m_l h^2}{2F_l L} + h\right)u_1 + \frac{gh^2}{F_l L}\eta_{b_1} = u_0\eta_{m_0}, \tag{7.A.9}$$

$$\frac{A_b}{b}\frac{d\eta_{b_1}}{dt} - hu_1 = u_0\eta_{m_0}. \tag{7.A.10}$$

8

Cross-Sectional Stability of a Single Inlet System

8.1 Introduction

This chapter deals with the equilibrium and stability of the cross-sectional area of the inlet. The cross-sectional area is in equilibrium with the hydrodynamic environment when the volume of sand entering equals the volume of sand leaving the inlet. The equilibrium is dynamic rather than static; the inlet cross-sectional area oscillates about a mean value. The mean value is referred to as the equilibrium cross-sectional area. The equilibrium is stable when, after a perturbation, the cross-sectional area returns to its equilibrium value.

Predicting the equilibrium cross-sectional area and its stability requires coupling of hydrodynamics and sand transport. Although progress has been made, the morphodynamic models containing this element are still in a developmental state, as discussed in Chapter 11. In the meantime, recourse is taken to a simple semi-empirical approach in which the hydrodynamics is described by Eqs. (6.1) and (6.2) and the sediment dynamics is introduced empirically. This approach was first proposed by Escoffier (1940).

8.2 Equilibrium and Stability

8.2.1 Escoffier Stability Model

The Escoffier Stability Model is described in Escoffier (1940). Escoffier assumed a sinusoidal inlet velocity and reasoned that when the inlet is in equilibrium with the hydrodynamic environment, the velocity amplitude $\hat{u}$ equals the equilibrium velocity $\hat{u}_{eq}$,

$$\hat{u} = \hat{u}_{eq}. \tag{8.1}$$

In general, both $\hat{u}$ and $\hat{u}_{eq}$ are functions of the cross-sectional area, A. When the amplitude of the inlet velocity is larger than the equilibrium velocity, the inlet cross-sectional area increases and when the amplitude is smaller, the inlet cross-sectional area decreases. Escoffier took $\hat{u}_{eq} \sim 1 \text{ m s}^{-1}$.

Partly based on earlier work by Sorensen (1977), the Escoffier Stability Model is expanded in van de Kreeke (2004). Because the inlet velocity for most applications is not sinusoidal, the velocity amplitude is defined in terms of the tidal prism by Eq. (5.6). The tidal prism in this equation is calculated by solving for $u(A)$ from Eqs. (6.1) and (6.2) and integrating over the ebb cycle. A rational expression for the equilibrium velocity is introduced by substituting for P/A in Eq. (5.6) from one of the $A-P$ relationship discussed in Section 5.2. Using Eq. (5.1) as the $A-P$ relationship the equilibrium velocity follows from Eq. (5.12). Using Eq. (5.2) as the $A-P$ relationship, the equilibrium velocity follows from Eq. (5.13).

To calculate $u(A)$ from Eqs. (6.1) and (6.2), the hydraulic radius, R, in Eq. (6.1) is expressed in terms of the cross-sectional area, A, by using the assumption of geometric similarity (O'Brien and Dean, 1972). Referring to Appendix 8.A, geometric similarity implies that the hydraulic radius is proportional to the square root of the cross-sectional area, i.e., $R = \beta_2 \sqrt{A}$, with β_2 a shape factor. As a note of caution, although there are indications that cross-sections are geometrically similar (Winton and Mehta, 1981), it seems wise to heed the observation of (Walton, 2004a) that the assumption of geometric similarity is not trivial and to also calculate the closure curve for hydraulic radius cross-sectional area scenarios other than geometric similarity.

8.2.2 Escoffier Diagram

The Escoffier Diagram consists of a closure curve, $\hat{u}(A)$ and an equilibrium velocity curve, $\hat{u}_{eq}(A)$. The closure curve represents the relationship between velocity amplitude and cross-sectional area. The equilibrium velocity curve represents the relationship between equilibrium velocity and cross-sectional area. The diagram is used to calculate the equilibrium cross-sectional area(s) and their stability. As an example, the Escoffier Diagram for the representative inlet is presented in Fig. 8.1. The parameter values for the representative inlet are given in Table 6.2. The closure curve is calculated using the Öszoy–Mehta Solution (Section 6.4). Assuming the $A-P$ relationship for the representative is of the form given by Eq. (5.2), the equilibrium velocity curve is calculated using Eq. (5.13). Taking $C_l = 6.8 \times 10^{-5}$ m^{-1}, the value for the Dutch Wadden Sea inlets, and the tidal period $T = 44,712$ s, it follows from Eq. (5.13) that $\hat{u}_{eq} = 1.03 \text{ m s}^{-1}$. The equilibrium velocity is independent of the cross-sectional area. Selecting Eq. (5.1) rather than Eq. (5.2) as the $A-P$ relationship, the equilibrium velocity is given by Eq. (5.12). This makes $\hat{u}_{eq}$ dependent on the cross-sectional area. The intersections of the closure curve and equilibrium velocity curve correspond to equilibrium cross-sectional areas. For equilibrium cross-sectional areas to exist, the two curves must intersect. In

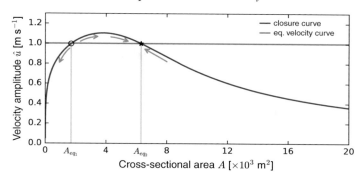

Figure 8.1 Escoffier Diagram for the representative inlet; for parameter values reference is made to Table 6.2.

that case there are always two equilibrium cross-sectional areas, $A_{1_{eq}}$ and $A_{2_{eq}}$. In the example for the representative inlet, $A_{1_{eq}} = 1,700 \text{ m}^2$ and $A_{2_{eq}} = 6,300 \text{ m}^2$.

In calculating the closure curve and equilibrium velocity curve, the velocity amplitude, $\hat{u}(A)$, and the A–P relationship used to calculate the equilibrium velocity, $\hat{u}_{eq}(A)$, must pertain to the same (mean or spring) tide conditions.

Using the Escoffier Diagram, the stability of the equilibriums is evaluated realizing that for $\hat{u} > \hat{u}_{eq}$ the inlet erodes and for $\hat{u} < \hat{u}_{eq}$ the inlet shoals (Section 5.2.2). It then follows that the equilibrium corresponding to A_{eq_1} is unstable; a decrease in cross-sectional area leads to closure and an increase causes the cross-sectional area to increase until it reaches the value of A_{eq_2}. The equilibrium corresponding to A_{eq_2} is stable; after a decrease in cross-sectional area, and provided its value remains larger than A_{eq_1}, the cross-sectional area returns to the equilibrium value, when increasing the cross-sectional area it will always return to the equilibrium value. The arrows in Fig. 8.1 show the direction in which the cross-sectional areas evolve after a perturbation. It follows from the Escoffier Diagram that the smaller the equilibrium velocity, the larger the stable equilibrium cross-sectional area. The difference between the stable and unstable equilibrium cross-sectional areas may be looked upon as a measure for the stability of the cross-sectional area A_{eq_2}. As shown in Appendix 8.B, the stability of the two equilibrium cross-sectional areas can be more formally investigated using a linear stability analysis.

The Escoffier Diagram is not only a valuable tool in determining the equilibrium cross-sectional areas and their stability, but it also provides insight into the response of the tidal inlet when the equilibrium is disturbed.

8.2.3 The Shape of the Closure Curve

The shape of the closure curve in Fig. 8.1 is typical for most tidal inlets. This is explained by using approximate expressions for the velocity amplitude for small and large values of the cross-sectional area. For small values of the cross-sectional area, the friction term in Eq. (6.1) is large compared to the inertia term. Neglecting the inertia term, the expression for the dimensionless velocity amplitude $\hat{u}^*$ is given by Eq. (6.27). Furthermore, for small values of A the repletion coefficient K_1, given by Eq. (6.9), approaches zero. Neglecting entrance and exit losses, replacing the hydraulic radius by $R = \beta_2\sqrt{A}$, it then follows that the dimensional velocity amplitude is

$$\hat{u} \cong \sqrt{\frac{3\pi g \beta_2 \sqrt{A}}{8FL}}\,\hat{\eta}_0. \tag{8.2}$$

The velocity amplitude $\hat{u}$ approaches zero when A goes to zero and increases as $A^{1/4}$ for increasing values of A.

For large values of A, the basin water level approaches the ocean water level and the tidal prism attains a value

$$P = 2A_b\hat{\eta}_0. \tag{8.3}$$

Using Eqs. (5.6) and (8.3), the velocity amplitude for large values of the cross-sectional area is

$$\hat{u} = \frac{2\pi A_b \hat{\eta}_0}{AT}. \tag{8.4}$$

The velocity amplitude decreases as A^{-1} for increasing values of A.

8.3 Adaptation Timescale

The adaptation timescale is a measure for the time it takes the cross-sectional area to return to its original equilibrium after a perturbation. To determine the adaptation timescale, it is assumed that, as a result of a storm, an excess volume of sand is deposited in the entrance section of the inlet, thereby reducing the cross-sectional area. The entrance section has a length L_e. Following the storm, the excess volume of sand and the fraction M' of the gross longshore sand transport that continues to enter the inlet are gradually removed by the ebb tidal current until the cross-sectional area again attains its equilibrium value. Taking the sand transport by the ebb tidal current proportional to the velocity amplitude to the power n, the conservation of sand equation for the entrance section is

$$L_e\frac{dA}{dt} = k\hat{u}^n - M'. \tag{8.5}$$

When at equilibrium, this equation reduces to

$$k\hat{u}^n_{eq} = M'. \tag{8.6}$$

Eliminating k between Eqs. (8.5) and (8.6), it follows that

$$\frac{dA}{dt} = \frac{M'}{L_e}\left[\left(\frac{\hat{u}}{\hat{u}_{eq}}\right)^n - 1\right]. \tag{8.7}$$

In this equation the velocity amplitude $\hat{u}$ is a function of the cross-sectional area A. This makes solving Eq. (8.7) a somewhat cumbersome (numerical) exercise.

To demonstrate the nature of the solution, the simplifying assumption is made that during the adaptation period the tidal prism remains constant. Using Eq. (5.6), this implies

$$\frac{\hat{u}}{\hat{u}_{eq}} = \frac{A_{eq}}{A}. \tag{8.8}$$

Substituting for $\hat{u}/\hat{u}_{eq}$ in Eq. (8.7) results in

$$\frac{dA}{dt} = \frac{M'}{L_e}\left[\left(\frac{A_{eq}}{A}\right)^n - 1\right]. \tag{8.9}$$

The quotient $1/A^n$ is linearized by introducing

$$A = A_{eq} + \Delta A, \tag{8.10}$$

and thus

$$\frac{1}{A^n} = \frac{1}{A_{eq}^n\left(1 + \frac{\Delta A}{A_{eq}}\right)^n}. \tag{8.11}$$

With

$$\frac{\Delta A}{A_{eq}} \ll 1, \tag{8.12}$$

Eq. (8.11) is approximated by

$$\frac{1}{A^n} \cong \frac{1}{A_{eq}^n}\left(1 - n\left(\frac{\Delta A}{A_{eq}}\right)\right). \tag{8.13}$$

Using Eq. (8.10), it follows from Eq. (8.13) that

$$\frac{A_{eq}^n}{A^n} \cong \left(1 + n\left(\frac{A_{eq} - A}{A_{eq}}\right)\right). \tag{8.14}$$

Substituting for A_{eq}^n/A^n from Eq. (8.14) in Eq. (8.9) results in

$$\frac{d(A - A_{eq})}{dt} + \frac{nM'}{L_e A_{eq}}(A - A_{eq}) = 0. \tag{8.15}$$

From Eq. (8.15) it follows that the cross-sectional area adjusts exponentially. The time for the initial deviation to reduce by a factor e (the Naperian timescale) is

$$\tau = \frac{L_e A_{eq}}{n M'}. \tag{8.16}$$

For the same transport of sand, M', entering the inlet, the adaptation timescale decreases with decreasing values of cross-sectional area; smaller inlets adapt faster than larger inlets.

As an example, for the Frisian Inlet, one of the inlets of the Dutch Wadden Sea, with $A_{eq} = 15,300$ m^2 (van de Kreeke, 2004), a storm deposit of 50,000 m^3 extending over a distance $L_e = 50$ m reduces the cross-sectional area by about 6.5 percent. With $M' = 500,000$ m^3 year^{-1} and n between 3 and 5, it then follows from Eq. (8.16) that the timescale for the cross-sectional area to return to its equilibrium value is 0.3–0.5 year. Similar values are reported in Kraus (1998).

In the foregoing, it is assumed that, as a result of a storm, a volume of sand is deposited in the entrance section of an inlet. The same reasoning and results hold when assuming that, as a result of a storm, a volume of sand is removed from the entrance section of the inlet.

8.4 Cross-Sectional Stability of Pass Cavallo

An aerial view of Pass Cavallo is presented in Fig. 8.2. Until 1963 Pass Cavallo was the sole inlet connecting Matagorda Bay (TX) and the Gulf of Mexico. In 1963 a second inlet, the Matagorda shipping channel, was artificially opened. This affected the stability of Pass Cavallo. Here the equilibrium and stability of Pass Cavallo prior to opening of the second inlet is evaluated using the Escoffier Stability Model described in Section 8.2.1. The analysis is in part based on an earlier study by van de Kreeke (1985), with additional data derived from later studies by Kraus et al. (2006) and Batten et al. (2007).

Prior to the opening of the second inlet, Pass Cavallo was in stable equilibrium with the hydrodynamic environment. It had been open for at least 200 years. Using observations from 1959, the cross-sectional area was 8,000 m^2. The cross-section had a parabolic shape with a width to maximum depth ratio of 333. Offshore tides are dominantly diurnal with a great diurnal amplitude of 0.27 m. The great diurnal amplitude is half the distance between the tidal datum planes of Mean Higher High (MHH) and Mean Lower Low (MLL). It is used to characterize the tide in areas where it has a mixed character (Jarrett, 1976). The observed tidal prism corresponding to the great diurnal tide is $P = 2.1 \times 10^8$ m^3. Compared to the tidal prism, freshwater inflow is negligible. The observed maximum velocity corresponding to the great diurnal tide is 1 m s^{-1}.

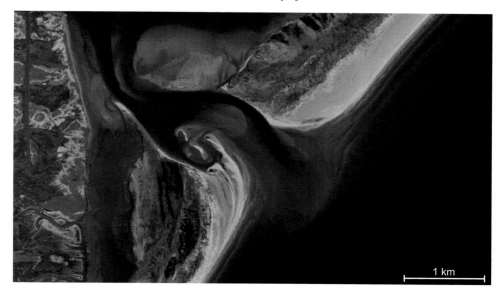

Figure 8.2 Pass Cavallo (TX) in 2015 (Esri et al., 2016).

Lacking an A–P relationship, the equilibrium velocity for Pass Cavallo is determined from the observed values of the tidal prism $P = 2.1 \times 10^8$ m³ and the cross-sectional area $A = 8,000$ m² at the time the inlet was at equilibrium. Using Eq. (5.6) with $T = 86,400$ s results in a value of the equilibrium velocity $\hat{u}_{eq} = 0.96$ m s⁻¹. The closure curve $\hat{u}(A)$ is calculated using Eq. (5.6) with the parameter values given in Table 8.1. The value of P in Eq. (5.6) is calculated from the Keulegan Solution (Section 6.5), realizing that $P = 2\hat{\eta}_0 A_b \sin \tau$, with $\sin \tau$ a function of the repletion coefficient K_1 tabulated in Table 6.1. In calculating the values of the repletion coefficient for different values of the cross-sectional areas, cross-sections are assumed to be geometrically similar with the value of the shape factor given in Table 8.1 (Appendix 8.A). With a relatively large basin surface area of 9.4×10^8 m² and shallow depth, the assumption of a pumping mode for Matagorda Bay, as used in the Keulegan equations, is questionable. To account for this and to match the observed and calculated tidal prism, the basin surface area is reduced to a value of 4.0×10^8 m². In this respect the basin surface area serves as a calibration parameter.

The Escoffier Diagram for Pass Cavallo is presented in Fig. 8.3. The value of the stable equilibrium cross-sectional area agrees with the observed cross-sectional area of 8,000 m². The value of the unstable equilibrium is approximately 300 m². The large difference between the values of the stable and unstable equilibrium cross-sectional areas explains why the inlet has remained open for at least 200 years despite several tropical storms and hurricanes.

Table 8.1 *Parameter values for Pass Cavallo prior to opening of the Matagorda shipping channel.*

Parameter	Symbol	Dimension	Value
Inlet length	L	m	1,600
Friction factor	F	–	2.5×10^{-3}
Entrance/exit loss coefficient	m	–	0.5
Shape factor	β_2	–	0.045
Basin surface area	A_b	m^2	4×10^8
Tidal amplitude	$\hat{\eta}_0$	m	0.27
Tidal frequency	σ	rad s^{-1}	7.3×10^{-5}

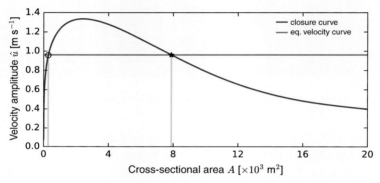

Figure 8.3 Escoffier Diagram for Pass Cavallo. For parameter values reference is made to Table 8.1.

As for many inlets, parameter values used in constructing the Escoffier Diagram for Pass Cavallo are at best estimates. In spite of this, at a minimum the diagram explains the observed equilibrium cross-sectional area and its stability.

8.A Geometric Similarity

Objects that are geometrically similar have the same shape, and the ratio of any two linear dimensions is the same. An example is the triangles in Fig. 8.A.1. A property of geometrically similar objects is that any linear dimension is proportional to the square root of the area. This relationship is demonstrated for the widths and hydraulic radii of the triangular cross-sections.

The two triangles in the figure have the same shape. The shape is determined by the two ratios

$$\frac{L_1}{h} = \alpha_1, \quad \text{and} \quad \frac{L_2}{h} = \alpha_2. \tag{8.A.1}$$

The ratios α_1 and α_2 for the larger cross-section are the same as the corresponding ratios for the smaller cross-section.

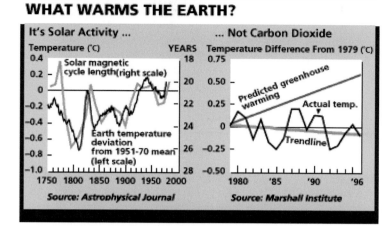

Figure 1. Article on the Wall Street Journal (12/4/1997)

approximately the same length than in the 1970s, while the temperature continued to increase (e.g. see Damon and Peristykh 2005).

Several years later, Friis-Christensen and Svensmark (1997) and Svensmark (1998) found that total cloud cover changed in phase with the flux of galactic cosmic rays (GCR), which are modulated by the interplanetary magnetic field associated with the solar wind and, therefore, with solar activity. They proposed a mechanism for the influence of solar activity on climate, in which GCR would affect cloud formation on Earth, through ionization of the atmosphere. Therefore, during periods of higher solar activity, when the interplanetary magnetic field is larger, and therefore less GCR hit the Earth, the cloud cover would be smaller.

Later on the observed agreement was lost, although Marsh and Svensmark (2000) proposed that it was still visible with low clouds. This theory was criticized by different reasons (e.g. Laut 2003), in particular because GCR should affect more strongly high clouds than low ones. Furthermore, Udelhofen and Cess (2001) studied observations from the ground obtained at 90 meteorological stations in the US during more than 90 years, and found the opposite correlation. At present, the correlation found by Svensmark and collaborators cannot be seen in the data. In fact, Lockwood and Fröhlich (2007) found that all possible solar forcings of climate had trends opposite to those needed to account for the rise in temperatures measured in the last century.

Moreover, the idea that the Sun has played a significant role in modern climate warming was mainly based on a general consensus that solar activity has been increasing during the last 300 years, after the Maunder Minimum, with a maximum in the late 20th century, which some researchers called the Modern Grand Maximum. However, this increase in solar activity has been identified as an error in the calibration of the Group Sunspot Number. When this error is corrected, solar activity appears to have been relatively stable since the end of the Maunder Minimum (see e.g. Svalgaard 2012, and the official IAU release†).

However, even if global warming cannot be attributed to an increase in solar activity, there is strong evidence that activity can influence terrestrial climate, in local scales.

† https://www.iau.org/news/pressreleases/detail/iau1508/

In what follows we will review some of that evidence, in particular the one referred to hydrological phenomena, and review our recent work on the subject.

2. Solar activity and hydrological phenomena

Usually, studies focusing on the influence of solar activity on climate have concentrated on Northern Hemisphere temperature or sea surface temperature. However, climate is a very complex system, involving many other important variables. Recently, several studies have focused in a different aspect of climate: atmospheric moisture and related quantities like, for example, precipitation.

Perhaps the most studied case is the Asian monsoon, where correlations between precipitations and solar activity have been found in several time scales. For example, Neff et al. (2001) studied the monsoon in Oman between 9 and 6 kyr ago, and found strong coherence with solar variability. Agnihotri et al. (2002) found that the monsoon intensity in India followed the variations of the solar irradiance on centennial time scales during the last millennium. Fleitmann et al. (2003) studied the Indian monsoon during the Holocene, and found that intervals of weak solar activity correlate with periods of low monsoon precipitation, and viceversa. On shorter time scales, Mehta and Lau (1997), found that, at multidecadal time scales, when solar irradiance is above normal there is a stronger correlation between the El Niño 3 index and the monsoon rainfall, and viceversa. Bhattacharyya and Narasimha (2005) and Kodera (2004), among others, also found correlations between solar activity and Indian monsoon in decadal time scales.

The monsoon in southern China over the past 9000 years was studied by Wang et al. (2005) who found that higher solar irradiance corresponds to stronger monsoon. They proposed that the monsoon responds almost immediately to the solar forcing by rapid atmospheric responses to solar changes.

Tiwari and Rajesh (2014) studied groundwater recharge rates in the Chinese region of Mongolia. Groundwater recharge is the hydrologic process where water moves downward from surface water to an aquifer. They found a strong stationary power at 200-220 years, significant at more than 95% confidence level, with wet periods coincident with strong solar activity periods.

All these studies found a positive correlation, with periods of higher solar activity corresponding to periods of larger precipitation. In contrast, Hong et al. (2001) studied a 6000-year record of precipitation and drought in northeastern China, and found that most of the dry periods agree with stronger solar activity and viceversa. In the American continent, droughts in the Yucatan Peninsula have been associated with periods of strong solar activity and have even been proposed to cause the decline of the Mayan civilization (Hodell et al. 2001).

In the same sense, studies of the water level of the East African Lakes Naivasha (Verschuren et al. 2000) and Victoria (Stager et al. 2005), found that severe droughts were coincident with phases of high solar activity and that rains increased during periods of low solar irradiation. To explain these differences it has been proposed that in equatorial regions enhanced solar irradiation causes more evaporation increasing the net transport of moisture flux to the Indian region via monsoon winds (Agnihotri et al. 2002).

However, these relationships seem to have changed sign around 200 years ago, when strong droughts took place over much of tropical Africa during the Dalton minimum, around 1800-1820 (Stager et al. 2005). Furthermore, recent water levels in Lake Victoria were studied by Stager et al. (2007), who found that during the 20th century, maxima of the ~11-year sunspot cycle were coincident with water level peaks caused by positive rainfall anomalies ~1 years before solar maxima. These same patterns were also observed

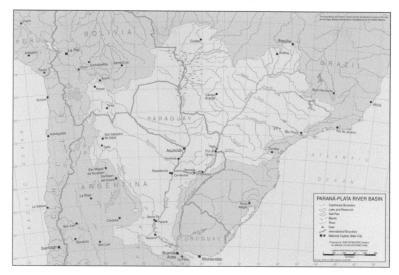

Figure 2. The Paraná basin

in at least five other East African lakes, hinting that these relationships between sunspot number and rainfall were regional in scale.

In Mauas and Flamenco (2005) we took a different approach, and we proposed to use the stream flow of a large river, the Paraná in southern South America, to study precipitations in a large area (see below). In this direction, Ruzmaikin *et al.* (2006) found signals of solar activity in the river Nile using spectral analysis techniques. They reported an 88-year variation present both in solar variability and in the Nile data. Zanchettin *et al.* (2008) studied the stream flow of the Po river, and found a correlation with variations in solar activity, on decadal time scales.

3. Stream flow of the Paraná River

River stream flows are excellent climatic indicators, and those with continental scale basins smooth out local variations, and can be particularly useful to study global forcing mechanisms. In particular, the Paraná River originates in the southernmost part of the Amazon forest, and it flows south collecting water from the countries of Brazil, Paraguay, Bolivia, Uruguay, and Argentina (see Fig. 2). It has a basin area of over 3.100.000 km^2 and a mean stream flow of 20.600 m^3/s, which makes it the fifth river of the world according to drainage area and the fourth according to stream flow.

Understanding the different factors that have an impact on the flow of these rivers it is fundamental for different social and economic reasons, from planning of agricultural or hydroenergetic conditions to the prediction of floods and droughts. In particular, floods of the Paraná can occupy very large areas, as can be seen in Fig. 3. During the last flood, in 1997, 180 000 km^2 of land were covered with water, 125 000 people had to be evacuated, and 25 people died. Together, the three largest floods of the Paraná during the 20th century caused economic losses of five billion dollars.

In Mauas *et al.* (2008) we studied the stream flow data measured at a gauging station located in the city of Corrientes, 900 km north of the outlet of the Paraná. It is measured continuously from 1904, on a daily basis. The yearly data are shown in Fig. 4 together with the yearly sunspot number (SN), which we use as a solar-activity indicator. Also

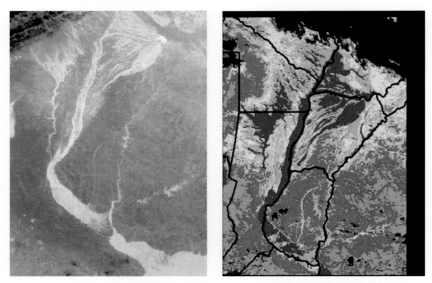

Figure 3. Left: Image taken with the AVHRR instrument on board a NOAA satellite, showing an area of 1200 km x 500 km, during the flood of the Paraná of 1997-1998. Right: An image in false colors, with political divisions over-imposed. The main area is Argentina. To the East, Uruguay and Brazil. To the North, Paraguay.

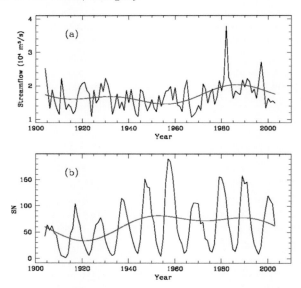

Figure 4. a) Paraná's annual stream flow at the Corrientes gauging station. (b) Yearly international sunspot number (SN). The secular trends, obtained with a low-pass Fourier filter with a 50-yr cutoff, are shown as thick lines.

shown in the figure are the trends, obtained with a low-pass Fourier filter with a 50 years cutoff.

In Fig. 5 we show the stream flow and the SN together. In both cases we have subtracted the secular trend shown in Fig. 4 from the annual data, and we have performed an 11 yr running-mean to smooth out the solar cycle. We have also normalized both quantities by subtracting the mean and dividing by the standard deviation of each series. These lasts

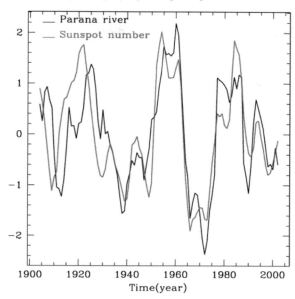

Figure 5. The detrended time series for the Paraná's stream flow and the Sunspot Number. The detrended series were obtained by subtracting from each data series the corresponding secular trend and were smoothed by an 11-yr-running mean to eliminate the solar cycle. Both series were standardized by subtracting the mean and dividing by the standard deviation, to avoid introducing arbitrary free parameters. The Pearsons correlation coefficient is R=0.78.

steps have been done to avoid introducing two free parameters, the relative scales and the offset between both quantities.

It can be seen that there is a remarkable visual agreement between the Paraná's stream flow and the sunspot number. In fact, the Pearson's correlation coefficient is $r = 0.78$, with a significance level, obtained through a t-student test, higher than 99.99%. It can also be noted that in this area wetter conditions coincide with periods of higher solar activity.

A few years later, in Mauas *et al.* (2011) we found that the correlation still held when more years of data were added. In particular, between 1995 and 2003 the Paraná's stream flow and the mean Sunspot Number have both decreased by similar proportions. This is of particular interest, since Solar Cycle 23 was the weakest since the 1970s: SN for the years 2008 (2.9) and 2009 (3.1) have been the lowest since 1913, and the beginning of Solar Cycle 24 was delayed by a minimum with the largest number of spotless days since the 1910s. At the same time, the mean levels of the Paraná discharge were also the lowest since the 1970s (see Fig. 5).

4. Other South-American Rivers

In Mauas *et al.* (2011) we followed up on the study of the influence of solar activity on the flow of South American rivers. In that paper we studied the stream flow of the Colorado river, and two of its tributaries, the San Juan and the Atuel rivers. We also analyzed snow levels, measured near the sources of the Colorado (see Fig. 6).

The Colorado river marks the north boundary of the Argentine Patagonia, separating it from the Pampas, to the northeast, and the Andean region of Cuyo, to the Northwest. Its origin is on the eastern slopes of the Andes Mountains, from where it flows southeast

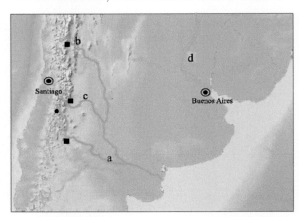

Figure 6. Colorado hydrologic system. The rivers we studied are a. Colorado, b. San Juan, c. Atuel and d. the lower part of the Paraná river. The stream flow (squares) and snow (dot) measuring stations are also indicated.

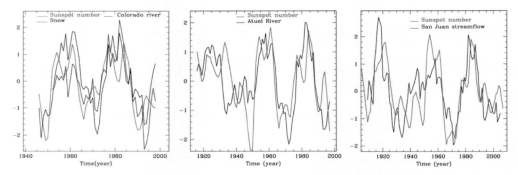

Figure 7. The detrended and renormalized stream flows compared with Sunspot Number. In the left panel the snow level is also shown.

until it discharges in the Atlantic Ocean. The Atuel, which originates in the glacial Atuel Lake, at 3250 m above sea level in the Andes range, and the 500 km long San Juan river, join the Colorado downstream of its gauging station. Therefore, the data given by the three series are not directly related.

Unlike the Paraná, whose stream flow is directly related to precipitation, the regime of all these rivers is dominated by snow melting, and their stream flows reflect precipitation accumulated during the winter, and melted during spring and summer. To directly study the snow precipitation, we complete our data with measurements of the height of snow accumulated in the Andes at 2250 m above Sea level, close to the origin of the Colorado (see Fig. 6), which were measured in situ at the end of the winter since 1952. In fact, the correlation between the stream flow of the Colorado and the snow height is very good, with a correlation coefficient $r = 0.87$, significant to a 99% level.

In Fig. 7 we compare the multidecadal component of the stream flows with the corresponding series for the sunspot number. In all cases we proceed as with the data in Fig. 4: we smoothed out the solar cycle with an 11-year running mean, we detrended the series by subtracting the long term component, and we standardized the data by

Figure 8. NorthWestern Argentina. Marked in red are the locations from where we obtained the tree-ring chronologies. Image acquired by the LANDSAT satellite.

subtracting the mean and dividing by the standard deviation. In the panel corresponding to the Colorado, we also include the snow height.

It can be seen that in all cases the agreement is remarkable. The correlation coefficients are 0.59, 0.47, 0.67 and 0.69 for the Colorado, the snow level, the San Juan and the Atuel, respectively, all significant to the 96-97% level. Although all these rivers have maximum stream flow during Summer, there is a big difference, however, between the regimes of the Paraná and the remaining rivers: for these ones, the important factor is the intensity of the precipitation occurring as snow during the winter months, from June to August. For the Paraná, what is most important is the level of the precipitation during the summer months. It should also be noticed that, here again, stronger activity coincides with larger precipitation.

5. Tree rings

Tree rings are the most numerous and widely distributed high-resolution climate archives in South America. During the last decades, variations of temperature, stream flow, rainfall and snow were reconstructed using tree-ring chronologies from subtropical and temperate forests, which are based on ring width, density and stable isotopes (see Boninsegna *et al.* 2009 and references therein).

Villalba *et al.* (1992) studied the spatial patterns of climate and tree-growth anomalies in the forests of northwestern Argentina. The tree-ring data set consisted of seven chronologies developed from Juglans and Cedrela (see Fig. 8). They show that tree-ring widths in subtropical Argentina are affected by weather conditions from late winter to early summer. Tree-ring patterns mainly reflect the direct effects of the principal types of rainfall-patterns observed. One of these patterns is related to precipitation anomalies concentrated in the northeastern part of the region.

To extend back in time and to a larger geographical area the results obtained previously, we study the relation between the Sunspot Number and the tree-ring chronologies studied

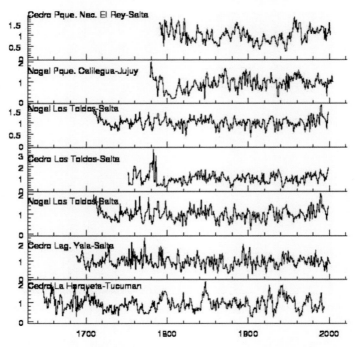

Figure 9. Tree-rings chronologies from Cedrela and Junglans from NorthWestern Argentina. Marked in red are the data used in this work.

by Villalba *et al.* (1992). These data-sets are shown in Fig. 9. It can be seen that the shortest series starts in 1797, while the longest one goes back to the XVI century. Here we study only the data from 1797, where all the series overlap.

The individual sets respond to local conditions, in the particular location of the studied tree. To obtain an indication of global conditions in the region, we built an index in the following way. First, we shifted in time each tree-ring series to obtain the best correlation with the Paraná's stream flow. In particular, in 1982 and 1997 there were two very large annual discharges of the Paraná that are associated with two exceptional El Niño events (see Fig. 4). These two events, although weaker, can be seen in the individual tree-ring series, with a small delay, different in each case. We therefore built a composite series as the average between each individual chronology, shifted to match the Paraná's discharge. Finally, we took the 11-years running-mean, and normalized the series as in the previous cases.

The resulting index is shown in Fig. 10, together with the Sunspot Number. It can be seen that also here the agreement is quite good. The Pearson's correlation coefficient between both series is R=0.69.

6. Discussion

Although the theory that Global Warming is caused by an increase in Solar activity has been dismissed, particularly because activity and temperature do not have similar trends anymore, it gave a strong impulse to the studies on the relation between climate and activity. In particular, there is strong evidence that the Sun could have an influence in different climatic variables, in different regions of the globe, and not always in the same sense. In particular, we reviewed different studies which concentrate on different aspects of

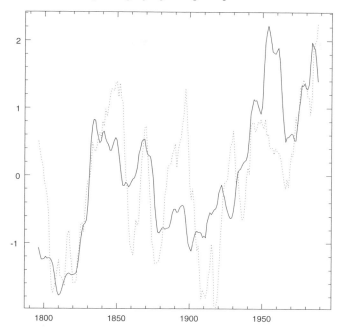

Figure 10. Composite of the tree-rings chronologies (dashed-line) and Sunspot Number (full-line) smoothed by an 11-yr running mean to smooth out the solar cycle. Both series were normalized by subtracting the mean and dividing by the standard deviation. The correlation coefficient is r=0.69.

atmospheric moisture, which in some regions reported positive correlations, with stronger activity related to stronger precipitations, and in others the opposite correlation, with strong droughts coincident with solar activity maxima. There are also regions of the world where this relation changed signs over time.

In particular, we studied different climatic indicators in southern South America. First, we concentrated in the stream flow of one of the largest rivers of the world, the Paraná. We found a strong correlation on decadal time scales between the river's discharge and Sunspot Number. We later found that this correlation was still present during the large solar minimum between Cycles 23 and 24, which corresponded to a period of very low flows in the Paraná.

We can also find in historical records this coincidence between periods with smaller solar activity and low Paraná's discharge. In particular, during the period known as the Little Ice Age there are different reports pointing out to low discharges. For example, a traveler of that period mentions in his diary that in 1752 the level of the river was so small that the small ships of that time could not navigate it. At present, the river can be navigated as far north as Asunción in Paraguay by ships 4 times larger (Iriondo 1999). There other climatic records which point out to reduced precipitations in this region during the Little Ice Age (see Piovano *et al.* 2009 and references therein). It is well known that the Little Ice Age was coincident with the Maunder Minimum, and was perhaps caused by low solar activity (e.g. Eddy 1976).

To check if the solar influence is also present in other areas of South America, we studied the flow of three other rivers of the region, and the snow level from a mountain-high station of the same area. Also in this cases we found a strong correlation between the

Sunspot Number and the stream flows, after removing the secular trends and smoothing out the solar cycle.

Finally, to extended both the area coverage and the temporal baseline, we studied a composite of seven tree-ring chronologies affected by precipitations, starting at the end of the XVIII century. Also in this case we found the same correlation with Sunspot Number.

We point out that, in all cases, we found a correlation in the intermediate time scale. We removed the secular trends when present (e.g, for the Paraná and the Sunspot Number), which are not correlated. We also smoothed out the solar cycle, since on the yearly timescale, the dominant factor influencing precipitations is El Niño. The results we found show that *decades* of larger precipitations correspond to *decades* of higher activity, with these variations overimposed on the corresponding secular trends.

In all cases, the correlation we found is positive, i.e., higher precipitations correspond to larger solar activity, in a very large area.

Since another mechanism that has been proposed to explain the Sun-Earth connection involve the modulation of Galactic Cosmic Rays, we also studied the correlation between the Paraná's discharge and two other solar-activity indexes: the neutron count at Climax, Colorado, available since 1953, and the aa index, which is an indication of the disturbance level of the magnetic field of the Earth based on magnetometer observations of two stations in England and Australia, which is available since 1868. Both indexes can be used to test the GCR hypothesis.

We found that the Paraná's stream flow is correlated with both neutron count and the aa index. This was expected, since all activity indexes are correlated among them. However, the correlation with Sunspot Number was strongest, suggesting a direct link between solar irradiance and precipitations.

It has been shown that variations in solar insolation affect the position of the Inter Tropical Convergence Zone (ITCZ) (Poore *et al.* 2004, Haug *et al.* 2001). Newton *et al.* (2006) proposed that a displacement southwards of the ITCZ would enhance precipitations in the tropical regions of southern South America. We found that the increase in precipitations are seen both in the Southern Hemisphere's summer when the ITCZ is over the equator, close to where the Paraná has its origin, and during winter, when the ITCZ moves north, and precipitations increase further South.

References

Agnihotri, R., K. Dutta, R. Bhushan, & B. L. K. Somayajulu 2002, *Earth and Planetary Science Letters* **198**, 521.

Berri, G. J., & E. A. Flamenco 1999. *Water Resources Research* **35**, 3803.

Bhattacharyya, S., & R. Narasimha 2005. *Geophysical Review Letters* **32**, 5813.

Boninsegna, J. A., Argollo, J., Aravena, J. C. *et al.*, 2009, Palaeogeography, Palaeoclimatology, Palaeoecology 281, 210.

Damon, P. E. & Peristykh, A. N. 2005, Clim.Change 68, 101

Eddy, J. A. 1976. *Science* **192**, 1189.

Fleitmann, D., S. J. Burns, M. Mudelsee, U. Neff, J. Kramers, A. Mangini, & A. Matter 2003. *Science* **300**, 1737.

Friis-Christensen E. & Lassen K. 1991, Science 254, 698700.

Friis-Christensen, E. & Svensmark, H. 1997, Ad.Spa. Res 20, 913

Haug, G. H., K. A. Hughen, D. M. Sigman, L. C. Peterson, & U. Röhl 2001. *Science* **293**, 1304.

Hodell, D. A., C. D. Charles, & F. J. Sierro 2001. *Earth and Planetary Science Letters* **192**, 109.

Hong, Y. T., Z. G. Wang, H. B. Jiang, Q. H. Lin, B. Hong, Y. X. Zhu, Y. Wang, L. S. Xu, X. T. Leng, & H. D. Li 2001. *Earth and Planetary Science Letters* **185**, 111.

Iriondo, M. 1999. *Quat. Int.* **57-58**, 112.

Kodera, K. 2004. *Geophysical Review Letters* **31**, 24209.

Lassen K. & Friis-Christensen E. 1995 JATP 57, 835

Laut, P. 2003, JASTP 65, 801

Laut P. & Gunderman J. 2000 SOLSPA I, p. 189

Lockwood, M. & Fröhlich, C., 2007, Proc. R. Soc. A 463, 2447

Marsh, N. D. & Svensmark, H. 2000, Phys. Rev. Lett. 85, 5004

Mauas, P., & E. Flamenco 2005. *Memorie della Societa Astronomica Italiana* **76**, 1002.

Mauas, P. J. D.., Flamenco, E., & Buccino, A. P. 2008, *Phys. Rev. Let.* 101, 168501

Mauas, P. J. D.., Flamenco, E., & Buccino, A. P. 2011, *JASTP* 73, 377

Mehta, V. M. & K.-M. Lau 1997. *Geophysical Review Letters* **24**, 159.

Neff, U., S. J. Burns, A. Mangini, M. Mudelsee, D. Fleitmann, & A. Matter 2001. *Nature* **411**, 290.

Newton, A., R. Thunell & L. Stott 2006. *Geophysical Review Letters* **33**, 19710.

Piovano, E., D. Ariztegui, F. Córdoba, M. Cioccale, & F. Sylvestre 2009. *Past Climate Variability in South America and Surrounding Regions From the Last Glacial Maximum to the Holocene*, Chapter 14. Hydrological Variability in South America Below the Tropic of Capricorn (Pampas and Patagonia, Argentina) During the Last 13.0 Ka, pp. 323–351. Springer Netherlands.

Poore, R. Z., T. M. Quinn, & S. Verardo 2004. *Geophysical Review Letters* **31**, 12214.

Ruzmaikin, A., J. Feynman, & Y. L. Yung 2006. *Journal of Geophysical Research (Atmospheres)* **111**, 21114.

W. Soon, W. Baliunas, S. L., Robinson, A. B., & Robinson, Z. W. 1999, Climate Research. 13, 149.

Stager, J. C., A. Ruzmaikin, D. Conway, P. Verburg, & P. J. Mason 2007. *Journal of Geophysical Research (Atmospheres)* **112**, 15106.

Stager, J. C., D. Ryves, B. Cumming, L. Meeker, & J. Beer 2005. *J. Paleolimnol.* **33**, 243.

Svalgaard, L, 2012, in *Proc. Iau Symp.* 286, 27

Svensmark, H. 1998, *Phys. Rev. Let.* 81, 5027

Udelhofen, P., & Cess, R. 2001, 28, 2617

Tiwari, R. K. & Rajesh, R. 2014, *Geophys. Res. Lett.*, 41, 3103

Verschuren, D., K. R. Laird & B. F. Cumming 2000. *Nature* **403**, 410.

Villalba, R. Holmes, R. L., & Boninsegna, J. A. 1992, *Journal of Biogeography* 19, 631

Wang, Y., H. Cheng, R. L. Edwards, Y. He, X. Kong, Z. An, J. Wu, M. J. Kelly, C. A. Dykoski, & X. Li 2005. *Science* **308**, 854.

Wang, Y.-M., J. L. Lean, & N. R. Sheeley, Jr. 2005. *ApJ* **625**, 522.

Zanchettin, D., A. Rubino, P. Traverso, & M. Tomasino 2008. *Journal of Geophysical Research (Atmospheres)* **113**, 12102.

Living Around Active Stars
Proceedings IAU Symposium No. 328, 2016
D. Nandy, A. Valio & P. Petit, eds.

© International Astronomical Union 2017
doi:10.1017/S174392131700388X

On the influence of magnetic fields in neutral planetary wakes

C. Villarreal D'Angelo[1], M. Schneiter[1] and A. Esquivel[2]

[1]Instituto de Astronomía Teórica y Experimental, Conicet-UNC,
Laprida 854, X5000BGR, Córdoba, Argentina
email: `carolina.villarreal@unc.edu.ar`
[2]Instituto de Ciencias Nucleares, UNAM,
C. Univ., AP 70-543 México D.F., México

Abstract. We present a 3D magnetohydrodynamic study of the effect that stellar and planetary magnetic fields have on the calculated Lyα absorption during the planetary transit, employing parameters that resemble the exoplanet HD209458b. We assume a dipolar magnetic field for both the star and the planet, and use the Parker solution to initialize the stellar wind. We also consider the radiative processes and the radiation pressure.

We use the numerical MHD code GUACHO to run several models varying the values of the planetary and stellar magnetic moments within the range reported in the literature.

We found that the presence of magnetic fields influences the escaping neutral planetary material spreading the absorption Lyα line for large stellar magnetic fields.

Keywords. line: profiles, methods: numerical, magnetic fields, stars: winds, stars: individual (HD209458).

1. Introduction

Despite the large number of Hot Jupiter discovered, only a few of them have observations that indicates that they are losing neutral material from their atmospheres. The first one of this cases was HD 209458b (Vidal-Madjar *et al.* 2003). The Lyα observations analysed by Vidal-Madjar *et al.* (2003) showed an extra absorption during the planetary transit that, together with the detection of heavier elements (Vidal-Madjar *et al.* (2004, 2013); Ben-Jaffel & Sona Hosseini (2010)) in the upper atmosphere, confirmed the "blow off" state of the atmosphere. Later, similar observation were made for HD 189733b (Lecavelier Des Etangs *et al.* (2010); Bourrier *et al.* (2013)), 55 Cnc b (Ehrenreich *et al.* (2012)) and GJ 436b (Kulow *et al.* (2014); Ehrenreich *et al.* (2015)).

In transit, the Lyα profile of HD 209458b show an absorption of approximately 10 per cent in the line wings, at ± 100 km s^{-1}, with absorptions at higher velocities in the blue side. To explain this observational behaviour several models have been proposed. Tremblin & Chiang (2013) studied the charge exchange mechanism together with the interaction region between the winds with 2D hydrodynamic models. Schneiter *et al.* (2016) studied the influence of the photoionization process in the interaction of the winds using a hydrodynamic code. All the models in this work where able to reproduce the absorption in the blue part of the Lyα line, failing in reproducing the absorption in the red part.

A logic step towards a self consistent model is to include the magnetic fields of both the star and the planet. For the star, the most straight forward assumption is to consider it as a solar analogue. For the planet, several works agree that the value of magnetic dipole moment is of the order of $0.1\mu_J$, with $\mu_J = 1.56 \times 10^{26}$ A m^2 the magnetic moment of Jupiter (Durand-Manterola 2009; Sánchez-Lavega 2004; Khodachenko *et al.* 2012, 2015;

Stellar parameters	Value
Radius	1.2 $R_\odot$
Mass	1.1 $M_\odot$
r_{ws}	R_*
$T_*(r_w)$	1.56×10^6 K [a]
$\rho(r_w)$	1.544×10^{-16} g cm^{-3} [a]
log L_{EUV}	< 27.74 erg s^{-1} [b]
Ionizing photon rate	2.4×10^{38} s^{-1}

Planetary parameters	Value
Radius	1.3 R_{Jup}
Mass	0.6 M_{Jup}
a	0.047 AU
Orbital period	3.52 day
r_{wp}	$3R_p$
$T_p(r_w)$	1×10^4 K [c]
$v_p(r_w)$	10 km s^{-1} [c]
$\dot{M}_p$	1×10^{10} g s^{-1} [c]

Table 1. Stellar and planetary wind parameters used in the simulations. [a] Vidotto *et al.* (2009), [b] Sanz-Forcada *et al.* (2011), [c] Murray-Clay *et al.* (2009).

Kislyakova *et al.* 2014). But, higher values have been used (Owen & Adams 2014; Vidotto *et al.* 2015). The lack of observational constraints made this an open debate. In this work, we assume that μ_p is in the range of [0.2-1.2] μ_J).

The current work is an effort to study the effect that different magnetic field moments (stellar and planetary) might have on the Lyα absorption profile.

2. Numerical models

The interaction between the planet and stellar winds was modelled with the 3D/MHD-radiative code GUACHO, in a Cartesian mesh of 256x128x256 cells, with a resolution of 5×10^{-4} *au*.

The code evolves the ideal MHD equations together with an extra equation that tracks the neutral density in each cell, taking into account the process of recombination, collisional ionization and photoionization of the gas.

The model includes the gravitational forces of the planet and the star, together with the radiation pressure implemented as a reduction of the stellar gravity on the neutral planetary material. A full explanation of the model is given in Schneiter *et al.* (2016).

In the numerical models, the star is at the center of the computational domain with the planet orbiting around it in a circular orbit. The ionized stellar wind is launched at the stellar radius, and it is initialized with the parameters used in the work of Pneuman & Kopp (1971) for an isothermal wind. The value of the polytropic index is set to $\gamma = 1.01$ for the stellar wind base but is allow to varies as a function of the neutral fraction.

The star has a magnetic dipole oriented perpendicular to the orbital plane. The values of the stellar magnetic dipole, B_*, ([1-5] Gauss) are taken from Vidotto *et al.* (2009). To take into account the photoionization process we divide the EUV stellar flux into 10^7 photon packages. This packages are launched from the surface of the star at random position in random directions. This is the same approach used in the work of Schneiter *et al.* (2016).

Model	B_* [G]	B_p [G]
A1	1	1
A2	1	5
B1	5	1
B2	5	5

Table 2. Model parameters.

We use the planetary parameters of HD 209458b. The planetary wind is launched at $3\ R_p$ with a density, temperature and velocity values taken from the work of Murray-Clay *et al.* (2009). As in the previous work (Schneiter *et al.* 2016), we assume that the escaping atmosphere of the planet is composed of Hydrogen and at the base of the wind it remains 20% neutral. The mass loss rate of the planet is assumed to be $\dot{M}_p = 1 \times 10^{10}$ g s^{-1} based on the results of Schneiter *et al.* (2007); Villarreal D'Angelo *et al.* (2014); Schneiter *et al.* (2016). This value of $\dot{M}_p$, together with the value of stellar F_{EUV} used, gives a heating efficiency (Yelle 2004) at the wind base of 0.2.

The planetary magnetic field, B_p is assumed dipolar with it's axis oriented perpendicular to the orbital plane and in the same direction as the magnetic field of the star. The values used for the planetary magnetic field are [1,5] Gauss at the poles, extrapolated at the radius where the wind is launched.

The initial parameters in the simulations are shown in the Table 1.

To simplify the analysis we decided to run varying only B_* & B_p. The models are introduce in Table 2.

3. Results & Discussion

The global behaviour of the numerical models show that a cometary tail is formed trailing the planet. As depicted on the left panel of figure (1), this tail can be divided in two parts, one of completely ionized material (farther away from the planet), as a result of the interaction with both the stellar wind and the EUV flux, and another with partially neutral material (closer to the planet). Around to the planet, these partially neutral cloud is composed of atmospheric material and stellar wind particles that became neutralized when interact with the planetary wind. This neutral material is the one responsible for the absorption in the Lyα line.

The right panel of figure (1) show the temperature contour in the orbital plane. The variation of the polytropic index with the neutral fraction is only important around the planet. Hence, the temperature of the stellar wind shows no significant variation except for the interaction region, where the planetary material gets heated and the stellar wind cooled at $T \sim 10^5$ K.

In order to reproduce the Lyα observations we computed the absorption profile in the same manner as in Schneiter *et al.* (2007); Villarreal D'Angelo *et al.* (2014) and Schneiter *et al.* (2016). In figure (2) we can see that the resulting line profile is asymmetric for both models, i.e there is more absorption in the blue side than in the red side, as in the observations. The models show a 10% of absorption in the blue wing, corresponding to velocities close to -100 km s^{-1} for models A1 & A2, reaching -200 km s^{-1} for the models B1 & B2. As mentioned earlier, four different models were tested, two planetary magnetic field values ([1,5] G) for each stellar magnetic field ([1,5] G). As observed in the Lyα profiles of figure (2), the change in planetary magnetic field has almost none influence on the absorption. On the other hand, increasing the stellar wind magnetic field magnitude widens the line profile. In the wings of the profile, the models with higher

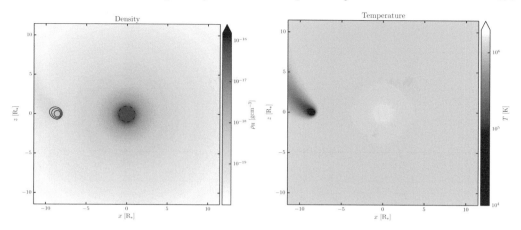

Figure 1. Contour of density and temperature in the orbital plane for model A1. In the left panel, the black contour lines around the planet correspond to the value [0.9,0.99,0.999] of the ionization fraction and the dashed black line is the base of the stellar wind at the stellar radius.

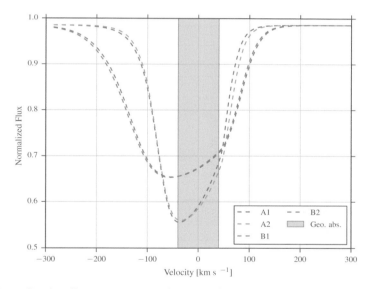

Figure 2. Normalized stellar emission as a function of the line of sight velocity in the Lyα line. The yellow stripe correspond to part of the line contaminated with the geocoronal glow, omitted in the total absorption calculations.

stellar magnetic field (5 G) produce more absorption than the models with 1 G. Given the stellar magnetic field structure, the higher the magnetic pressure ($P_{mag} = B^2/8\pi$), the higher the compression of the planetary material on the poles. This compression produces an enhanced acceleration mainly towards the tail, in the direction of the magnetic field lines (see right panel of figure 3), but also, less pronounced in the direction towards the host star. B_* is also responsible for the enhancement of the neutral density for the more compressed case.

For models A1 & A2, the partially neutral cloud surrounding the planet has a more prolate shape with neutrals farther away from the star. This is depicted in figure (4),

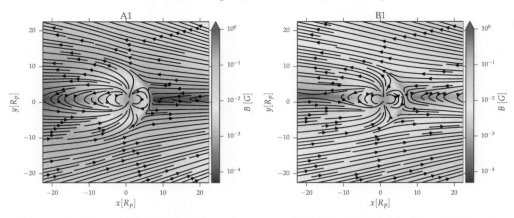

Figure 3. Magnetic field lines and total magnetic field value for models A1 and B1 in the vicinity of the planet position in a plane perpendicular to the orbital plane.

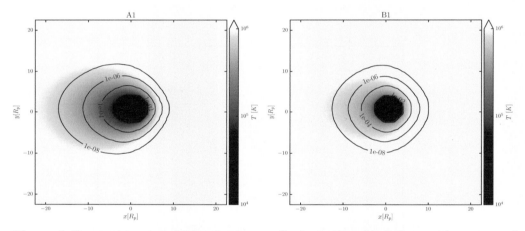

Figure 4. Temperature map, in the plane perpendicular to the orbital plane, with contours of the neutral material normalized to the total density of the planet ($\rho_p = 9.2 \times 10^{-18}$ g cm^{-3}) for models A1 and B1.

where contours of the neutral density, normalized to the planetary density, are show around the planet.

As stated before, the shape of this cloud is modelled by the total pressure of the stellar wind. When the magnetic stellar pressure at the poles compresses the planetary wind (models B1 & B2), it enhances his neutral density and temperature, allowing the planetary wind to balance the total pressure of the stellar wind farther away from the planet towards the star. The neutral material that escapes to the tail is ionized much faster in this case due to the higher temperature.

For cases A1 & A2, the magnetic stellar compression is lower, hence, the partially neutral cloud can expand a bit more away from the planet. However, the balance between the planetary and the stellar pressures take place closer to the planet since this neutral cloud is now less denser and cooler (see figure (4)).

4. Summary

A study of the influence of both the stellar and planetary magnetic fields on the Lyα absorption profile, based on a full 3D magnetohydrodynamic + photoionization model was introduced. From this study we were able to establish, within the range of parameters proposed, that the planetary magnetic field has no important imprint on the Lyα line, whereas the stellar magnetic field has a great influence on the blue part of the line, and a more subtle influence on the red part. Moreover, we were able to produce absorption at velocities below -100 km s^{-1} with models with a magnetic field of 5 G. This absorption are the result of neutrals originated from recombination of stellar wind ions. As stated in Khodachenko *et al.* (2015), the higher the planetary magnetic field the lower the planetary mass loss rate. According to this work, the mass loss rate corresponding to $B_p = 5$ G is an order of magnitude lower than the one used here. A lower value of $\dot{M}_p$ would help to distinguish the absorption profile between the models.

References

Ben-Jaffel, L. & Sona Hosseini, S. 2010, *ApJ*, 709, 1284
Bourrier, V., Lecavelier des Etangs, A., Dupuy, H., *et al.* 2013, *A&A*, 551, A63
Durand-Manterola, H. J. 2009, Plan. & Space Sci., 57, 1405
Ehrenreich, D., Bourrier, V., Bonfils, X., *et al.* 2012, *A&A*, 547, A18
Ehrenreich, D., Bourrier, V., Wheatley, P. J., *et al.* 2015, *Nature*, 522, 459
Khodachenko, M. L., Alexeev, I., Belenkaya, E., *et al.* 2012, *ApJ*, 744, 70
Khodachenko, M. L., Shaikhislamov, I. F., Lammer, H., & Prokopov, P. A. 2015, *ApJ*, 813, 50
Kislyakova, K. G., Holmström, M., Lammer, H., Odert, P., & Khodachenko, M. L. 2014, Science, 346, 981
Kulow, J. R., France, K., Linsky, J., & Loyd, R. O. P. 2014, *ApJ*, 786, 132
Lecavelier Des Etangs, A., Ehrenreich, D., Vidal-Madjar, A., *et al.* 2010, *A&A*, 514, A72
Murray-Clay, R. A., Chiang, E. I., & Murray, N. 2009, *ApJ*, 693, 23
Owen, J. E. & Adams, F. C. 2014, *MNRAS*, 444, 3761
Pneuman, G. W. & Kopp, R. A. 1971, *Sol. Phys.*, 18, 258
Sanz-Forcada, J., Micela, G., Ribas, I., *et al.* 2011, *A&A*, 532, A6
Sánchez-Lavega, A. 2004, *ApJL*, 609, L87
Schneiter, E. M., Velázquez, P. F., Esquivel, A., Raga, A. C., & Blanco-Cano, X. 2007, *ApJ*, 671, L57
Schneiter, E. M., Esquivel, A., D'Angelo, C. S. V., *et al.* 2016, *MNRAS*, 457, 1666
Tremblin, P. & Chiang, E. 2013, *MNRAS*, 428, 2565
Vidotto, A. A., Opher, M., Jatenco-Pereira, V., & Gombosi, T. I. 2009, *ApJ*, 699, 441
Vidotto, A. A., Fares, R., Jardine, M., Moutou, C., & Donati, J.-F. 2015, *MNRAS*, 449, 4117
Vidal-Madjar, A., Lecavelier des Etangs, A., Désert, J.-M., *et al.* 2003, *Nature*, 422, 143
Vidal-Madjar, A., Désert, J.-M., Lecavelier des Etangs, A., *et al.* 2004, *ApJ*, 604, L69
Vidal-Madjar, A., Huitson, C. M., Bourrier, V., *et al.* 2013, *A&A*, 560, A54
Villarreal D'Angelo, C., Schneiter, M., Costa, A., *et al.* 2014, *MNRAS*, 438, 1654
Yelle, R. V. 2004, *Icarus*, 170, 167

Living Around Active Stars
Proceedings IAU Symposium No. 328, 2016
D. Nandy, A. Valio & P. Petit, eds.

© International Astronomical Union 2017
doi:10.1017/S1743921317003969

Hunting for Stellar Coronal Mass Ejections

Heidi Korhonen[1], Krisztián Vida[2], Martin Leitzinger[3], Petra Odert[4,3] and Orsolya Eszter Kovács[2,5]

[1] Dark Cosmology Centre, Niels Bohr Institute, University of Copenhagen, Juliane Maries Vej 30, DK-2100 Copenhagen Ø, Denmark
e-mail: heidi.korhonen@nbi.ku.dk
[2] Konkoly Observatory, MTA CSFK, Konkoly Thege M. út 15-17, 1121, Budapest, Hungary
[3] University of Graz, Institute of Physics, Department for Geophysics, Astrophysics and Meteorology, NAWI Graz, Universitätsplatz 5, 8010, Graz, Austria
[4] Space Research Institute, Austrian Academy of Sciences, Schmiedlstrasse 6, 8042 Graz, Austria
[5] Department of Astronomy, Eötvös Loránd University, Pázmány Péter sétány 1/A, 1117, Budapest, Hungary

Abstract. Coronal mass ejections (CMEs) are explosive events that occur basically daily on the Sun. It is thought that these events play a crucial role in the angular momentum and mass loss of late-type stars, and also shape the environment in which planets form and live. Stellar CMEs can be detected in optical spectra in the Balmer lines, especially in Hα, as blue-shifted extra emission/absorption. To increase the detection probability one can monitor young open clusters, in which the stars are due to their youth still rapid rotators, and thus magnetically active and likely to exhibit a large number of CMEs. Using ESO facilities and the Nordic Optical Telescope we have obtained time series of multi-object spectroscopic observations of late-type stars in six open clusters with ages ranging from 15 Myrs to 300 Myrs. Additionally, we have studied archival data of numerous active stars. These observations will allow us to obtain information on the occurrence rate of CMEs in late-type stars with different ages and spectral types. Here we report on the preliminary outcome of our studies.

Keywords. stars: activity, corona, late-type

1. Introduction

Due to the strong radiation coming from hot stars, the cooler stars having later spectral types (F–M) are thought to be more benign hosts for habitable planets. Still, also the cool stars show strong flaring activity and probably numerous coronal mass ejections (CMEs). Magnetic activity of the planet host star, especially the strong magnetic activity exhibited by many rapidly rotating young stars, has profound effects on the planets and their habitability; strong flares and CMEs are thought to even be able to strip a close-by planet of its atmosphere (e.g., Lammer *et al.* 2007). Additionally, these processes are also relevant for stellar evolution, as stars lose mass and angular momentum via stellar winds and CMEs.

In the Sun CMEs are seen regularly, the average daily occurrence rate during the activity minimum being 0.5 and during the maximum 6, with one solar CME containing on average 10^{11} kg of material (see, e.g., Gopalswamy *et al.* 2009). On the Sun, flares and CMEs are often closely correlated, their association increasing with flare energy (e.g. Yashiro *et al.* 2006). A few studies have therefore aimed to estimate stellar CME rates from flares, and relations between flare and CME parameters known from the Sun (Aarnio *et al.* 2012; Drake *et al.* 2013; Leitzinger *et al.* 2014; Osten & Wolk 2015). Many studies have shown that young stars have very energetic Ultra Violet flares (see, e.g., Osten &

Wolk 2009), but very few systematic studies of the stellar coronal mass ejections exists. There is a handful of detections of stellar CMEs, but they are very rare (e.g., Houdebine *et al.* 1990; Fuhrmeister & Schmitt 2004; Günther & Emerson 1997; Leitzinger *et al.* 2011). The observations indicate velocities ranging from about twice the value for very fast solar CMEs (5800km/s seen in AD Leo; Houdebine *et al.* 1990) to slightly slower than the values seen in the slowest solar CMEs (84 km/s also seen in AD Leo; Leitzinger *et al.* 2011). The slow velocities observed at times can be explained by the CME being seen in projection.

For their large impact on the environments of planets, and also on the stellar angular momentum evolution, it is crucial to decipher the CME occurrence rate with stellar age and spectral type.

2. Detecting CMEs using optical spectra

The Doppler signal of a mass ejection is always seen in projection. It is strongest when plasma is moving towards the observer and weaker for other directions. Doppler shifts and line asymmetries with enhanced blue wings of stellar emission lines can be interpreted as plasma ejected from the star. The material released in CMEs is mainly hydrogen as the CME core is often built by a filament, and they can be detected in the Hα line.

The blue enhancement caused by a CME is an absorption feature when seen against the stellar disc, and an emission feature when seen outside the disc. The time that the ejecta can be seen against the stellar disc is quite short for many projections, meaning that emission features would be more common than absorption features. On the other hand, the material would need to be quite dense to be seen in emission.

The velocities of CMEs span from few hundreds of km/s to some thousands of km/s. The exact value naturally depends on the energetics of the event itself, but also on the projection effects. The maximum velocity is only achieved if the material is ejected directly towards the observer. These high speeds also mean that detecting CMEs does not require very high spectral resolution.

3. Results on single stars

We have gone through the archival data of more than 40 single active stars observed with Narval and ESPaDOnS, and also obtained new observations for a handful of targets. Even with a careful analysis no CMEs were found, except on V374 Peg.

V374 Peg is a young, active M4 dwarf which has been extensively studied (e.g., Donati *et al.* 2006; Morin *et al.* 2008). We investigated the long-term photometric and spectroscopic variability of this star, and also studied archival ESPaDOnS data of the Hα line-profiles (Vida *et al.* 2016). The spectra obtained in 2005 showed a CME event: one real CME that was preceded by two failed events (see Fig. 1). We estimated that the minimum CME mass for the real event was 10^{16}g. In addition, we predicted the CME rate using the formalism by Leitzinger *et al.* (2014), and found out that the star should have 15-60 CMEs per day. We detected only one in 10 hours of observations – clearly less than predicted.

4. Monitoring of open clusters

An efficient way for getting a better handle on the CME rate in young stars is to monitor open clusters with different ages using multi-object spectroscopy. In this way

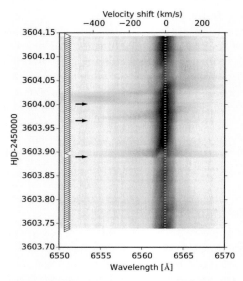

Figure 1. Dynamic spectrum of Hα line of V374 Peg. The arrows show the complex CME event with two failed eruptions and finally one real CME. (from Vida *et al.* 2016)

Table 1. Overview of the clusters observed for this project. Table gives the name of the cluster, age, number of observed targets, instrument used, spectral resolution, typical signal-to-noise ratio (S/N), and the observing time allocated to the cluster.

Cluster	Age [Myrs]	Targets	Instrument	Resolution	Cadence	S/N	Obs time
Blanco 1	65	28	VIMOS, ESO	2500	3 min	40	11h
IC 2391	40	7	EFOSC, ESO	2500	10 min	40	1n
NGC 2516	110	8	EFOSC, ESO	2500	6 min	40	1n
NGC 3532	300	9	EFOSC, ESO	2500	6 min	40	1n
h Per	15	17	ALFOSC, NOT	2500	10 min	30	3n
IC 348	45	29	ALFOSC, NOT	2500	10 min	30	3n

many cool stars of known age can be simultaneously observed, increasing the chances of detecting CMEs.

Our first attempt was done at ESO VLT using VIMOS spectrograph of a young open cluster Blanco 1 (PI Leitzinger). The results from these observations have been published, but unfortunately no CMEs were detected (Leitzinger *et al.* 2014). Still, we estimated an upper limit of four CMEs per day per star, and that we should have detected at least one CME per star with a mass of $1-15 \times 10^{16}$ g. After the Blanco 1 campaign we have observed also other young open clusters. IC 2391, NGC 2516, and NGC 3532 were observed using EFOSC2 at ESO's New Technology Telescope (PI Leitzinger). Three further clusters h Per, IC 348, and NGC 1662 were targeted with the ALFOSC instrument on the Nordic Telescope (PI Korhonen). Unfortunately, many of the NOT observing campaigns were hampered by bad weather, and therefore we decided not to observe the oldest cluster, NGC 1662, which anyway only had few observable cluster members. In total the ALFOSC observations consist of some 2.5 nights worth of good quality data on h Per and IC 348 (out of 9 allocated nights). More details on the observed clusters is given in Table 1. In this table the observing time is the total time allocated for that cluster.

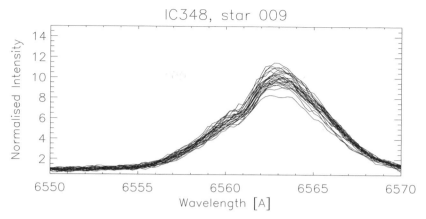

Figure 2. Examples of Hα line profiles of a star in IC 348.

From the EFOSC observations of IC 2391, NGC 2516, and HGC 3532 we only detected flares. No CME signatures were seen. Also the h Per observations obtained with ALFOSC only show some flare events, but IC 348 looks more promising. Our targets in this cluster include stars ranging spectral types G–M. Most of the targets with earlier spectral types have Hα in absorption, and the later ones in emission.

An example of Hα line-profiles for star number 9 in our multi-object spectroscopy of IC 348 is given in Fig. 2. This star is of spectral type late K – early M. As can be seen, the line-profile shows clear variability. Also, note that the persistent extra emission bump in the blue part of the profile is not from a CME event, but is due to fringing in the CCD at these wavelengths. The dynamic Hα spectra for the same target in November–December 2015 are shown in Fig. 3. Both the dynamical spectra with original line-profiles (left side) and with the average line-profile subtracted (right side) are shown. Clear variability in the Hα region is seen, and several events with blue shifted extra emission are discovered. More thorough analysis of these data are needed for firmly confirming the nature of the events, and calculating their occurrence frequency.

5. Conclusions and future prospects

We are detecting stellar CMEs, but less than expected. The detection is limited both by the intrinsic properties of the star and observing constraints. The intrinsic properties include the CME occurrence rate, which depends on the activity level/age of the star, and the typical CME parameters (velocity, mass, etc.). The observing constraints are on one hand related to the timing of the observations (time coverage and cadence), and on the other to the spectral properties (resolution and S/N). One has to also keep in mind that a fraction of the CMEs are also lost due to the projection effects.

Typical S/N of our observations is around 30–40, which is not high for studying small features in the spectral line-profiles. With these observations we can detect only the massive CMEs, comparable to the most massive solar events. These events would be in the range of $10^{16} - 10^{17}$ g. If we increase the S/N then we should also see less massive CMEs. To test this we have been granted UVES bad weather programme on Xi Boo A and B, a 200Myrs old G and K dwarf system. The observations have a resolution of 40000, S/N 600, and time resolution of less than a minute. With these observations we will be able to test the detectability of smaller CMEs.

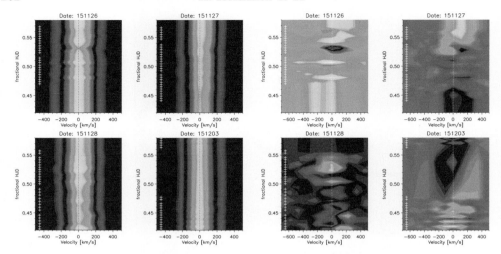

Figure 3. Dynamic spectrum of Hα line of a star in IC 348 (the same star as in Fig. 2). The observations are from November–December 2015, and the title of the plot gives the exact observing date. The crosses show the times from which the observations are (in the HJD minus HJD of that date at noon). The four leftmost plots show the dynamic spectra from the original profiles, and the four rightmost plots the residual dynamic spectra after the average profile has been subtracted.

The so-far detected stellar CMEs have all been on M dwarfs, and the ejecta were found to be in emission. Possibly the CMEs on K and G stars are not dense enough to be detected in emission, and we are not quick enough to detect them in absorption. Another intriguing question is, whether the very active stars really have numerous CMEs. Maybe we are not detecting many CMEs, because there actually are only few of them? It has been hypothesised that the strong magnetic fields on young stars could actually prevent a filament from erupting in analogy to solar failed eruptions (see Drake *et al.* 2016).

Acknowledgement

H.K. acknowledges the support from the *Fonden Dr. N.P. Wieth-Knudsens Observatorium* for the travel grant that made it possible for her to attend the IAU Symposium 328. K.V. was supported by the János Bolyai Research Scholarship of the Hungarian Academy of Sciences. M.L. and P.O. acknowledge the support from the FWF project P22950-N16. P.O. acknowledges also support from the Austrian Science Fund (FWF): P27256-N27. The authors acknowledge support from the Hungarian Research Grants OTKA K-109276, OTKA K-113117, the Lendület-2009 and Lendület-2012 Program (LP2012-31) of the Hungarian Academy of Sciences, and the ESA PECS Contract No. 4000110889/14/NL/NDe.

References

Aarnio, A. N., Matt, S. P., & Stassun, K. G. 2012, *ApJ*, 760, 9
Donati, J.-F., Forveille, T., Collier Cameron, A., *et al.* 2006, *Science*, 311, 633
Drake, J. J., Cohen, O., Yashiro, S., & Gopalswamy, N. 2013, *ApJ*, 764, 170
Drake, J. J., Cohen, O., Garraffo, C., & Kashyap, V. 2016, *arXiv:1610.05185*
Fuhrmeister B., Schmitt J. H. M. M. 2004, *A&A*, 420, 1079
Gopalswamy, N., Akiyama, S., Yashiro, S., & Mäkelä, P. 2010, in: S.S. Hasan & R.J. Rutten
    (eds) *Astrophysics and Space Science Proceedings* (Springer Berlin Heidelberg) p. 289
Günther, E. W. & Emerson, J. P. 1997, *A&A*, 321, 803;

Houdebine, E. R., Foing, B. H., & Rodono, M. 1990, *A&A*, 238, 249;

Lammer, H., Lichtenegger, H. I. M., Kulikov, Y. N., *et al.* 2007, *Astrobiology*, p. 185

Leitzinger, M., Odert, P., Ribas, I., *et al.* 2011, *A&A*, 536, A62;

Leitzinger, M., Odert, P., Greimel, R., *et al.* 2014, *MNRAS*, 443, 898

Morin, J., Donati, J.-F., Forveille, T., *et al.* 2008, *MNRAS*, 384, 77

Osten, R. A. & Wolk, S. J. 2009, *ApJ*, 691, 1128

Osten, R. A. & Wolk, S. J. 2015, *ApJ*, 809, 79

Vida, K., Kriskovics, L., Oláh, K., *et al.* 2016, *A&A*, 590, A11

Yashiro, S., Akiyama, S., Gopalswamy, N., & Howard, R. A. 2006, *ApJ (Letters)*, 650, L143

Living Around Active Stars
Proceedings IAU Symposium No. 328, 2016
D. Nandy, A. Valio & P. Petit, eds.

© International Astronomical Union 2017
doi:10.1017/S1743921317004598

Carrington Class Solar Events and How to Recognize Them

C. T. Russell[1], J. G. Luhmann[2], P. Riley[3]

[1]University of California, Los Angeles, CA, USA 90095
e-mail: dr_russell@igpp.ucla.edu; ctrussell@igpp.ucla.edu
[2]University of California, Berkeley, CA, USA 94720
e-mail: jgluhman@ssl.berkeley.edu
[3]Predictive Science, San Diego, CA, USA 92121
e-mail: pete@predsci.com

Abstract. The so-called Carrington Event on September 1, 1859, is clearly the solar outburst that brought the realization to the inhabitants of Earth that weather existed in space, and that space weather was important to the rapidly developing technological infrastructure on Earth. It is important to understand not only how space weather affects our technological systems, but like the case of atmospheric weather, the possible intensity of such weather, the frequency of extreme events, and how to predict them. This paper reviews what we know about one class of extreme space weather events, the superfast arrival events, how best to compare them given our limited diagnostics in past events and even at the current time, and suggests a direction for progress in this field.

1. Introduction

The work featured at this conference largely focused on other stars and planetary systems, but we know much more about our local conditions—at least in the present epoch—from relatively close-up, comprehensive observations over decades. In living around our own active star, the Sun, we have learned much about how the Sun works and its many and varied impacts on the planets of this solar system. Here we take a brief look at a topic of growing interest in light of the observations of superflares on sun-like stars, and of ongoing efforts to predict extreme solar events and their effects on the Earth.

By the middle of the nineteenth century, the solar cycle had been studied for approximately 250 years, and its average properties were well determined. However, we had no realization that there were connections between the Sun and the Earth by any mechanisms than through light and heat. Furthermore, during these times, it was relatively unimportant that there were sunspot cycles, solar storms and space weather, because civilizations were not sensitive to these variations. The industrial revolution brought new technology with new requirements to humankind, and they became dependent upon this technology with its attendant risks. The Carrington Event on September 1, 1859 (Carrington, 1859), brought about the realizations that the Sun was violently unstable at times and that this violence could be sensed on Earth with its new electrical distribution systems. Since that time, the reputation of the Carrington event has reached mythic proportions as possibly the greatest solar event of recorded history, far outstripping present-day activity. If we are to both evaluate the magnitude of the Carrington event compared to major events of the space age, as well as the likelihood of such events occurring in the future, we need to use a method of classifying these rare, large events by some measure of size available at the time. However, first we quickly review the

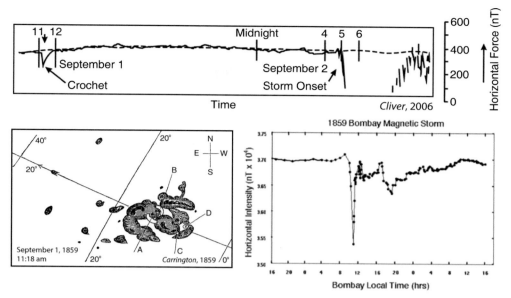

Figure 1. Top panel. Horizontal components of the magnetic field measured in London, England, on September 1 and 2, 1859 (Carrington, 1859). Bottom left panel. The solar flare seen on the Sun at the beginning of the September 1, 1989 event. Bottom right panel. The temporal behavior of the horizontal component of the magnetic field near the equatorial electrojet during the period of the Carrington event. (Tsurutani *et al.*, 2003)

Carrington Event itself to separate fact from fiction regarding what occurred at that time, based on what we know about the physics of the Sun-Earth connection.

2. The Carrington Event of September 1, 1859

The Earth's magnetosphere is a very good monitor of the solar wind, and the ionosphere is quite sensitive to changes in solar x-rays and extreme ultraviolet emissions. The pressure of the outflowing solar wind pushes on the Earth's magnetic field, shaping it into its characteristic teardrop shape with a long geomagnetic tail. When the outward (from the Sun) dynamic pressure of the solar wind increases, the magnetic field in the magnetosphere is compressed. This can be measured on the surface of the Earth. The magnetic field expelled by the Sun in such an outburst can interact with the Earth's magnetic field through a process known as reconnection in which the momentum in the solar wind is transferred to the magnetosphere pulling on it. For the nearly dipolar field of Earth, the reconnection efficiency is highest for interplanetary fields with a large southward component. It is this momentum transfer that produces the most dramatic (for those on the surface of the Earth) effects of the space weather storms: the auroras, the ring current or energetic plasma content of the magnetosphere, and the radiation belts.

The effect of the Carrington Event at the Earth's surface was recorded in the ground based magnetic field measurements shown in Figure 1 -, including the magnetic record obtained in London, England, on September 1, 1859, by Carrington. The lower panel shows a sketch of the magnetic complex observed on the surface of the Sun that was either illuminated by the solar flare or erupted as part of the flare. We know from the magnetic record when the flare occurred as it left its mark in the ionosphere almost immediately as the energetic solar photons hit the Earth's atmosphere and ionized it. The ionosphere is constantly in motion, and this motion creates an electric field. When the photons hit the ionosphere, it became more electrically conducting, creating a larger current than what previously existed which then decayed as the solar photon flux declined. Later another electric current appeared. We now interpret this second event as the arrival of the solar plasma that erupted with the flare, but traveled far more slowly (even at an inferred 1000s of km/s) than the speed of light. While this impact initially produced a brief increase in the magnetic field on the surface of the Earth via compression, its longer term effect was to weaken the Earth's field by inflating it with energetic plasma.

Since the solar observatory where the flare was observed was near the magnetic observatory, the scientists involved could communicate easily and quickly confirm that the two events were associated, and the field of space weather was born.

Clearly this was the first widely recognized space weather event, but was it the greatest of all time (or even since 1859)? There are several ways to characterize the Carrington event in retrospect. The first would be the initial compression of the magnetosphere, the sudden impulse or sudden compression as it is called. This is not possible from the available information. Another way would be to look at the depression of the Earth's magnetic field after the arrival of the plasma. However, the depth of that depression is not well resolved, and its the duration is also shorter than one would expect. Perhaps changes in the solar wind counteracted the effect of the plasma injection. The only measure we can reliably trust is the timing. We know when the event began and when it reached Earth, so we can deduce the average speed of the erupted material. Simultaneous records also exist near the Earth's magnetic equator. This region is sensitive to changes in the circulation of the ionospheric plasma, and due to an effect called Cowling conductivity causes a narrow but strong current along the magnetic equator known as the equatorial electrojet. Also recorded at Bombay (now Mumbai) was a very short but intense drop in the magnetic field strength near the equator. Its suddenness indicates that it was not an increase in the ring current caused by an injection of hot plasma deep into the magnetosphere, so this does not represent an extreme value of the Dst index, rather a healthy equatorial electrojet, responding to increased electric fields associated with the arrival of the plasma from the Sun.

One important additional aspect of these events that we have not mentioned until now is the accompanying generation of enhanced fluxes of solar energetic particles, accelerated either at the flare site, or by the shock wave that precedes the erupted plasma as it travels outward, or both. We now know that solar flares can produce intense bursts of 100s of keV to 100s of MeV protons and electrons that are dangerous to living organisms. We only recently have been able to monitor these properly and our records are still spotty. However, as discussed below, this energetic particle contribution to the solar wind seems to play a special role in the coronal and interplanetary propagation and evolution of extreme events like the Carrington event.

3. The Speed of Solar Initiated Disturbances

It is instructive to compare the speeds of modern events with historical ones. Cliver and Svalgaard (2004) provide a list of the fastest events between 1938 and 2003. We add two more: October 19, 1989, and July 23, 2012, as well as the 1859 Carrington event. In calculating an event speed, we must recognize that the speed will vary with angle away from the flare site. The strongest force will be upward along the line joining the center of the Sun to the flare site. Much weaker forces will extend horizontally from the flare site, tangential to the solar surface.

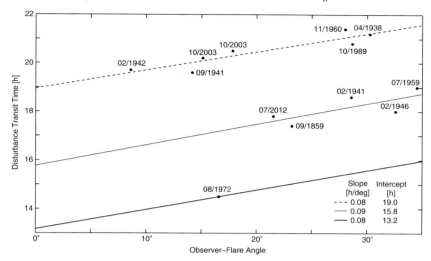

Figure 2. Travel time of the solar disturbance to one AU from the flare site for all reported fast events since 1938, plus the Carrington event. Plotted versus the angle between the flare site and the observer (Freed and Russell, 2014).

Because results from recent missions have shown that large events appear to expand outward self-similarly from the flare site, we can empirically determine the velocity angular dependence by simply plotting the transit times versus angle from the flare site, as shown in Figure 2. These can be viewed as defining three distinct sets of events. The slowest consist of the seven events near the top of this figure. This set includes the two "Halloween events" in 2003 and another modern event for which we have in situ data in 10/1989. These events would have caused disturbances at Earth in 19 hours if the Earth were directly over the flare site.

The middle line shows five faster events lying on a line with the same slope as the slower seven above. This includes the Carrington event and an extreme modern event in July 2012, which we will call the STEREO event. The twin STEREO spacecraft are in solar orbit at 1 AU. The STEREO event erupted plasma appears to have made a direct hit on the STEREO Ahead (-A) spacecraft, providing an extraordinary data set from which to reconstruct solar cause and 1 AU effects.However, although they arewell instrumented for interplanetary and solar studies, these observations cannot tell us what would have happened had this hit the Earth's magnetosphere.

The third event is the fastest of all. It of course cannot definea slope by itself, so we have used the slope that was appropriate for the other 12 events. This event occurred between the Apollo 16 and 17 moon landings. In this case the related energetic particle flux was so intense that some have speculated that the astronauts would have been harmed by this event had it occurred while they were on the surface of the Moon. It is not evident how these events are connected to the magnetic solar cycle as the STEREO event occurred during the current lull in solar activity (Russell *et al.*, 2013) that appears to be as deep as the lull during the Dalton minimum 200 years ago. Assuming symmetry around the radial axis through the flare site, fitting the arrival time versus angle from that axis gives the shapes seen in Figure 3.

4. The Ejected Particles and Plasma

With these rare events, the data are not usually sufficient to perform convincing statistical studies of their properties but what data we do have is very interesting. Figure

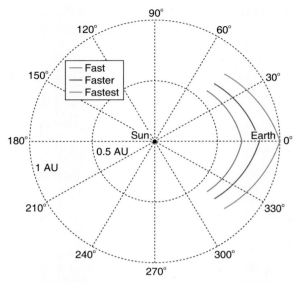

Figure 3. The inferred shapes of the expanding blast wave from the solar flare site (Freed and Russell, 2014).

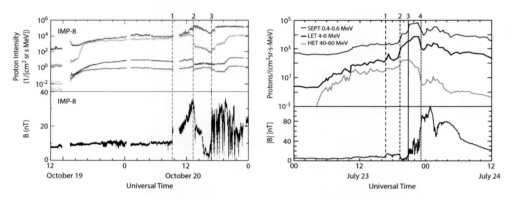

Figure 4. (Left) Measurements of the magnetic field strength and the energetic proton flux from the IMPS spacecraft on October 19–20, 1989, and (right) the STEREO A spacecraft during the July 23–24, 2012 event (Freed and Russell, 2014).

4 shows the in-situ energetic particle and magnetic field data from two of the extreme events. The solar wind becomes filled with highly-energetic, relativistic particles immediately, and they determine the temperature of the solar wind plasma. The solar wind continues to flow outward, but the high thermal speed of the particles in the event causes phenomena to occur at speeds less than that of the fast magnetosonic mode. In other words, these superfast disturbances from the Sun are traveling at subsonic speeds. So when the magnetic flux rope (ICME) arrives, it compresses the plasma in front of it, not as a fast-mode wave, but as a slow magnetosonic wave. Thus, the density of the solar wind plasma increases as it usually does across the compression front, but the magnetic field decreases as shown by the data between vertical line 2 and 3 on the panels of Figure 4. The 2012 STEREO event is accompanied by a very large magnetic cloud, while a strong magnetic cloud is not as evident in the weaker 1989 event. Figure 5 compares the plasma and magnetic pressures in the STEREO event directly. Until the magnetic cloud

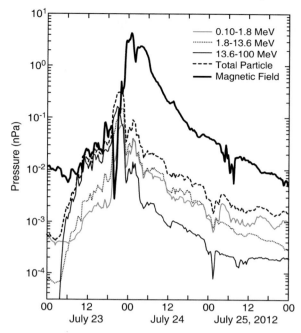

Figure 5. The pressure in the proton fluxes (gray lines) and in the magnetic field (black line) during the July 23–25, 2012 event (Russell *et al.*, 2013).

arrives, there is a steady increase in the thermal pressure of the energetic particles and the magnetic field maintains a beta equals unity plasma. The arrival of the magnetic cloud puts magnetic stresses firmly in control.

There is still much to learn from the STEREO data of the July 2012 event, and with the paucity of these events (cf. Riley, 2012), we should not expect to observe further such ejections during STEREO's lifetime. However, these events are so important that it makes sense from a space weather monitoring perspective to station a network of particle monitors around the Sun at 1 AU, perhaps with ion engines to position them at fixed locations relative to the Earth, and small but powerful energetic particle analyzers to characterize the relativistic plasma.

5. Conclusions

Our analysis of these 13 fast events has revealed that groups of large events become rarer as their intensity increases. However, the statistics of such large events does not allow us to be certain that there are three separate classes, as proposed here. We have used their arrival times as a proxy for their intensity. This inverse "correlation" of occurrence rate with intensity is consistent with the idea that these eventsresult from coronal magnetic configurations that store energy in at least three different ways, with one able to store the greatest magnetic potential. Regardless of whether the number is three or not, it is important to focus our efforts on the question of what allows so much energy energy to be stored for such rapid release? What magnetic configurations can lead to a greater storage capacity for magnetic and particle flux but only rarely occur? Thus we encourage renewed examination of magnetic data for these largest space weather events, and in particular the study of magnetic configurations on the

Sun with high meta-stability with potential for rapid and strong outbursts when they destabilize.

Acknowledgements

This research was supported by the National Aeronautics and Space Administration under a research grant awarded to Predictive Science Inc. (PSI).

References

Carrington, R. C. 1859, *Monthly Not. R. Astron. Soc.*, XX, 13

Cliver, E. W., & Svalgaard, L. 2005, *Solar Physics*, 224, 407-422

Freed, A. J., & Russell, C. T. 2014, *Geophys. Res. Lett.*, 41, 19, 6590-6594, doi:10.1002/2014GL061353

Russell, C. T., Mewaldt, R. A., Luhmann, J. G. *et al.* 2013, *Astrophys. J.*, 770, 1, 38, doi:10.1088/0004-637x/770/1/38

Riley, P. 2012, *Space Weather,* 10(2), 1

Russell, C. T., Luhmann, J. G., & Jian, L. K. 2010, *Rev. Geophys.*, 48, RG2004, doi:10.1029/2009RG000316

Tsurutani, B. T., Gonzalez, W. D., Lakhina, G. S. *et al.* 2003, *J. Geophys. Res.*, 108, 1268, doi:10.1029/2002JA009504

Living Around Active Stars
Proceedings IAU Symposium No. 328, 2016
D. Nandy, A. Valio & P. Petit, eds.

© International Astronomical Union 2017
doi:10.1017/S1743921317003702

Space Weather Storm Responses at Mars: Lessons from A Weakly Magnetized Terrestrial Planet

J. G. Luhmann[1], C. F. Dong[2], Y. J. Ma[2], S. M. Curry[1], Yan Li[1],
C. O. Lee[1], T. Hara[1], R. Lillis[1], J. Halekas[4], J. E. Connerney[5],
J. Espley[5], D. A. Brain[6], Y. Dong[6], B. M. Jakosky[6], E. Thiemann[6],
F. Eparvier[6], F. Leblanc[7], P. Withers[8] and C. T. Russell[3]

[1] Space Sciences Laboratory, University of California, Berkeley, CA, USA,
email: `jgluhman@ssl.berkeley.edu`
[2] Princeton University, Princeton, NJ, USA
[3] Institute of Geophysics and Planetary Physics, UCLA, Los Angeles, CA, USA
[4] University of Iowa, Iowa City, IA, USA
[5] NASA Goddard Space Flight Center, Greenbelt, MD, USA
[6] LASP, University of Colorado, Boulder, CO, USA
[7] LATMOS/IPSL, UPMC University Paris, Paris, France
[8] Boston University, Boston, MA, USA

Abstract. Much can be learned from terrestrial planets that appear to have had the potential to be habitable, but failed to realize that potential. Mars shows evidence of a once hospitable surface environment. The reasons for its current state, and in particular its thin atmosphere and dry surface, are of great interest for what they can tell us about habitable zone planet outcomes. A main goal of the MAVEN mission is to observe Mars' atmosphere responses to solar and space weather influences, and in particular atmosphere escape related to space weather 'storms' caused by interplanetary coronal mass ejections (ICMEs). Numerical experiments with a data-validated MHD model suggest how the effects of an observed moderately strong ICME compare to what happens during a more extreme event. The results suggest the kinds of solar and space weather conditions that can have evolutionary importance at a planet like Mars.

Keywords. Mars, Magnetosphere, Magnetic Storm

1. Introduction

Exploration of the terrestrial planets over the last ~5 decades has shown how diverse the outcome of their evolutionary paths can be. The many influences include both internal processes and interactions with their external environments, including the Sun and solar wind, over time. Of the other planets in or near our star's currently habitable zone, Mars has the particular distinction of having been observed since telescopes came into existence. It moreover appears to have been a possible abode of life early in its history, and is a natural destination for human exploration. As a consequence a host of space missions, including orbiters, landers and rovers, have been sent to carry out robotic investigations of the state of Mars' surface and atmosphere, with the Viking landers in the 70s having searched for evidence of extant life on or near the surface. Today we know a great deal about the solid planet, whose radius is roughly half that of Earth, from extensive topographical, mineralogical and gravitational mapping by orbiters. We know that Mars has a highly multipolar crustal magnetic field that, while strong compared to Earth's field in some locations, falls of relatively rapidly in strength with altitude. We also know from the landed platforms and remote sensing that the atmosphere is thin

($\sim$10 mb surface pressure compared to Earth's $\sim$1 bar), and composed mainly of carbon dioxide. Both the atmosphere and surface are relatively dry as well, though there is a strong seasonal variation in both carbon dioxide and water content as the polar ice caps periodically sublime over a Mars year when exposed to greater solar illumination and warming. Huge dust storms on this dry planet are also observed to produce significant heating effects in the upper atmosphere and ionosphere (e.g., Jakosky and Phillips (2001).

Interest has steadily grown in how Mars came to its present state, with surface features and mineralogy both suggesting it once had (or has) at least occasional flowing water episodes with enough volume to create the many channels observed from the ground and space- and possibly once a northern ocean. The isotopic evidence moreover supports the interpretation that a significant part of the original (post-impact-period) atmosphere of Mars, including its water, either has been lost through mineralogical alteration and sequestration of carbon, oxygen and other constituents, or has escaped. Much work continues on evaluating the still present reservoirs and the processes of surface and subsurface uptake. But comparably intensive work on the atmosphere escape side only started in earnest with the Mars Express mission (e.g. Lundin *et al.* (2013) and references therein), stimulated in part by the discovery by Mars Global Surveyor of small-scale remanent magnetic fields- presumably left from an early era of planetary dynamo operation (Acuna *et al.* (1999)). A popular, though still widely-debated idea holds that planetary magnetospheres can 'shield' a planetary atmosphere from processes that lead to massive escape on evolutionary timescales. Mars Express (MEX) mainly measured the escaping ionized atmospheric species, which constitute only part of the overall escaping population. Theory and modelling expectations, bolstered by related observations at Venus, have long suggested a (currently) more important contribution at Mars from escaping suprathermal or 'hot' oxygen atoms produced by dissociative recombination in its mainly O2+ ionosphere, with a possible contribution from sputtering by precipitating planetary ions (e.g. Lillis *et al.* (2015)). But the historical extrapolation of integrated loss that includes these and other possible escape processes has been compromised by the lack of a more complete complement of related measurements toward evaluating them.

The MAVEN (Mars Atmosphere and Volatile EvolutioN) mission has been in Mars orbit for over a Mars year with a broad instrument complement designed to investigate atmosphere escape processes and consequences (Jakosky *et al.* (2016)). In addition to solar wind and planetary plasma spectrometers and a magnetometer, it includes a thermal ion and neutral mass spectrometer, a UV spectrograph and EUV photometer designed to detect the rarefied exospheric gases and solar activity indicators respectively, and solar energetic particle telescopes (Jakosky *et al.* (2015a)). Over the course of is prime mission period (late 2014 through 2015), it has observed the variability of the upper atmosphere and ionosphere, and measured atmospheric ion escape rates under a range of conditions. Of particular interest here is that MAVEN finds the average global escape rates of both ions and neutrals, derived with the aid of global extrapolations of data-validated models, are too low by several orders of magnitude to explain the loss of the one or more bar atmosphere inferred to have been present $\sim$3.5 Gyr ago. This age is an often adopted benchmark in time because the evidence from cratering records and surface magnetic field measurements indicate that the remanent field has been present since that time, with the implication that any global dynamo was terminated or decayed away by then.

The reasons why loss to space by currently active processes has not been abandoned as a still feasible explanation for the apparent loss of the atmosphere are twofold. One strong argument is the well-established isotopic evidence, for which there is little alternative explanation (although see Guedel *et al.* (2010), for a discussion of escape processes prior to the 3.5 Gyr mark). The other involves the growing inferences from observations of

Sun-like stars of the early Sun's properties, and in particular its probable higher level of magnetic activity. Flare responses in the form of upper atmosphere heating, with implied increased thermal escape, have been observed on MAVEN (Thiemann *et al.* (2015)), though the involved time intervals are brief. But MEX and MAVEN observations have both indicated order-of-magnitude increases in escaping atmospheric ion fluxes when an interplanetary coronal mass ejection (ICME), with its enhanced solar wind plasma and field parameters lasting a day or more, passes Mars (e.g. Edberg *et al.* (2010); Jakosky *et al.* (2015b)). A potentially answerable question relevant to Mars, and more general planetary atmosphere evolution, concerns how large escape rates due to current processes can become, and whether these can explain the evident loss of a potentially habitable surface environment.

In this paper we focus on the specific process of atmospheric ion escape, neglecting the related sputtering and independent photochemical losses to obtain a lower limit for this cometary ion tail-like erosion. Global ion escape rates for both quiet times and times of enhanced solar wind parameters have been determined from MAVEN observations with the aid of MHD simulations of the solar wind interaction (Jakosky *et al.* (2015b); Dong *et al.* (2015)). We first revisit the case when enhanced solar wind conditions occurred when an ICME passed Mars in March 2015. As previously described, inferred global planetary ion outflow enhancements occur at different stages during that ∼1.5 day-long event, which involves an initial shock/sheath phase with compressed solar wind plasma and field, followed by an ejecta phase dominated by the strong field of the coronal erupted material or driver. The model gives a sense of the complex topology of the Mars magnetosphere reconfiguration and ionosphere energization that leads to the increased ion escape. We then ask how much larger such event-related ion escape rates might become, at least by present day standards, by using radially extrapolated solar wind parameters from a particularly extreme ICME event observed at 1 AU by the STEREO-A spacecraft in July 2012. By modelling the Mars atmospheric ion escape response to this example we obtain a better idea of how proposed stronger solar activity conditions on the young Sun might have regularly affected the early Mars atmosphere, at least through this single process of ion escape.

2. Description of the disturbed external conditions and model results

The results of the simulations shown here are described in more detail in a paper submitted to the MAVEN special issue of the Journal of Geophysical Research (Luhmann *et al.* (2016)). Basic CO_2 atmosphere photochemistry in the models produces an ionosphere composed of O_2+, $O+$ and CO_2+ major ion species affected by the solar wind interaction in a self-consistent manner. In the single fluid MHD approximation used, details of the ion escape relating to the heavy atmospheric ion gyroradii (e.g. Dong *et al.* (2015)) are not included. However, it has been found that in spite of this simplification the single fluid treatment can describe spacecraft observations of the magnetic fields, thermal ions, and solar wind around Mars extremely well (Ma *et al.* (2015)). We therefore assume that even above the exobase, physical process that act like collisions determine much of the basic solar wind interaction character, including the bulk planetary plasma energization by the interaction. Within the MHD framework there are several mechanisms that energize the atmospheric ions, including solar wind convection electric fields mapped along open field lines into the ionosphere in a manner similar to what happens in Earth's high latitude open field regions, magnetic field and thermal pressure gradient forces, and magnetic tension forces. Details not explicitly included are the energy deposition and sputtering effects of those accelerated planetary ions whose trajectories carry

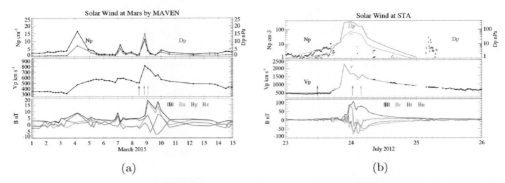

(a)   (b)

Figure 1. In-situ time series showing the interplanetary conditions (solar wind plasma density and velocity and magnetic field) observed by MAVEN at Mars in March 2015, including the passage of an ICME on March 8-10 (left panel), and a much larger event observed at 1 AU on STEREO-A in July 2012 (right panel). The latter can be used to simulate the effects of a realistic, more extreme event at Mars (see text)

them back into the atmosphere (see Curry *et al.* (2015)), related ion outflow reductions, wave-particle interactions, and fluid instabilities at the solar wind-ionosphere boundary. However, because of the apparent data-model agreement, these effects can be considered second-order in the current context.

As discussed above, we are interested in the more extreme interplanetary conditions experienced today, which occur when major interplanetary coronal mass ejection disturbances pass. The in-situ solar wind conditions for the March 2015 and July 2012 ICMEs at Mars and at STEREO-A respectively are shown in Figures 1a and 1b. Even accounting for the 0.5 AU heliocentric distance difference, it can be seen that the notable ICME observed at Mars on March 9-10, 2015 pales in comparison to the STEREO-A event. This provides important perspective because of the likelihood of more frequent strong activity on the early Sun (e.g. Ribas *et al.* (2005)).

The times indicated by arrows in Figures 1a,b were used to define the external parameters during each event for BATS-R-US solar wind interaction runs, assuming the same, nominal Mars atmosphere and solar EUV fluxes appropriate for each period. These represent the pre-ICME conditions in the solar wind, the piled up solar wind or ICME sheath phase when the incident dynamic pressure is largest, and the driver or ejecta phase when the magnetic field dominates the plasma pressure. As STEREO-A is located at 1 AU, in the July 2012 event case, the magnetic fields and densities were scaled to Mars heliocentric distance. The basic parameters used in the simulations for the pre-event and peak incident pressures for both events are summarized in Table 1 together with the related ion escape rates.

The enhancement of the planetary ion fluxes around Mars obtained with the solar wind interaction simulation for the more extreme July 2012 external conditions is illustrated in Figure 2 which shows (log) flux contours in a noon-midnight meridian cross section (where x is the Mars-Sun axis and z points north from the Mars orbital plane). These represent the Mars response to the external conditions in Figure 1b before the ICME hit (black arrow) and during the peak of the ICME's incident dynamic pressure (red arrow). The superposed black lines are projections of the models' 3D magnetic field lines traced from a circle of points located at 150 km altitude in the plane of the contour plot. These show how the planetary ion flux structure is influenced by the combination of the interplanetary magnetic field and Mars' crustal magnetic field. At the passage of the ICME, the magnetic fields of Mars undergo enhanced reconnection with the strong

Table 1.

Case	Nsw (cm^{-3})	Vsw (km/s)	B (nT)	Ion Escape (#/s)
PreMar2015 ICME	1.8	505	4.8	$1.5 \cdot 10^{24}$
PmaxMar2015	11.0	817	7.7	$1.1 \cdot 10^{25}$
Pre STA ICME	0.8	480	2.6	$3.9 \cdot 10^{24}$
Pmax STA ICME	24.8	1700	54.9	$1.4 \cdot 10^{27}$

Notes:values for March 2015 are from Curry et al. (2015), and for the STA event from Luhmann et al. (2016)

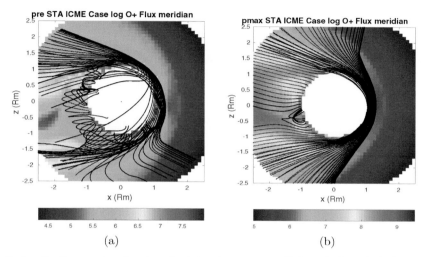

(a)   (b)

Figure 2. In-situ time series showing the interplanetary conditions (solar wind plasma density and velocity and magnetic field) observed by MAVEN at Mars in March 2015, including the passage of an ICME on March 8-10 (left panel), and a much larger event observed at 1 AU on STEREO-A in July 2012 (right panel). The latter can be used to simulate the effects of a realistic, more extreme event at Mars (see text)

external fields. As a result, the solar wind interaction looks more Venus-like on the dayside because the ionosphere induced fields are too weak to exclude deep magnetosheath field penetration and pile-up. However, away from the subsolar region reconnection produces 'open' Martian magnetosphere fields that nearly fill the solar wind wake. The consequence is an ionosphere that is much more widely exposed to solar wind convection electric fields, and thus widely subject to outward thermal pressure gradient forces. Magnetic reconnection also enhances the ion outflows.

As discussed in Luhmann *et al.* (2016), the several order of magnitude escape rate enhancement seen in the extreme case can have a significant impact on Mars atmosphere escape on evolutionary time scales. However, this depends on whether similar interplanetary conditions were sufficiently long-lived or extreme events were more common.

3. Implications from Sun-as-a-Star Research for Mars and Exoplanet Research

Several papers presented at this conference have addressed important issues relevant to using present-day observations of Mars ion escape to infer how much of its atmosphere could have been lost as a result of this ion escape process acting over the past $\sim$3-4 billion years. Guedel *et al.* (2010) points out the recent results from Tu *et al.* (2015) suggesting order-of-magnitude uncertainties in the solar ionizing fluxes in the early period prior to ∼3.5 Gyr ago. As the solar EUV flux evolution figures prominently in historical escape estimates, this represents an important uncertainty that must be taken into account in backward extrapolations of present escape processes including not only ion escape, but also neutral escape mechanisms (photochemical production and escape of hot atoms, in particular) not included here. The ion loss models discussed here all use a nominal present day EUV flux at Mars.

Another important uncertainty is the extent to which the planets experienced enhanced solar wind parameters simply from a stronger early solar wind. The implication of Airapetian and Usmanov's (2016) evolving solar wind models is that between ∼700 Myr and a few Gyr of age, the Sun produced a solar wind with nominal speeds of up to ∼800-1400 km/s, velocities that today are associated with major ICMEs. Moreover, the much higher mass flux of this early solar wind (by up to several orders of magnitude) would have produced regularly high plasma dynamic pressures on the same order as those during the ICMEs described here. The magnetic fields at the solar surface, assumed to be ∼3-4x higher than the present Sun's based on stellar analogs, combined with the higher younger Sun rotation rates (periods 5-10 days in their models) would have increased the interplanetary magnetic field by at least as much. Moreover, these already higher fields and dynamic pressures would be regularly enhanced by the formation of stream interaction regions, which would have corotated by at the higher solar rotation rates. With these relatively intense ambient conditions prevailing over a billion years, loss rates on the order of the ICME loss rates described here should have routinely occurred even without coronal eruptions. The fact that the Airapetian and Usmanov early solar wind model mass fluxes compare favorably with those inferred by Wood (2006) from astrosphere observations lends credence to this possible early ion escape scenario.

At the same time, estimates of increased flaring activity on the young Sun from the stellar analogs suggest those were also part of its history (e.g. Maehara *et al.* (2012)). Moreover, the flare energies may have been larger by orders of magnitude. The implications for the related early interplanetary medium are likely to remain unclear. As suggested by other speakers at this conference (e.g. Osten *et al.* (2017)), it has been difficult to establish the existence of the stellar counterparts of coronal mass ejections, although work is ongoing. Some efforts to imagine how a superflare would affect the planets have been made but they are necessarily speculative. Our broad challenge is to find ways to constrain scenarios for star-planet interactions over their lifetimes. Solar system observations like those from the MAVEN mission at Mars can help provide some 'ground truth' for reality checks, as described here.

4. Acknowledgments

This work has been supported by NASA through contracts to the Laboratory for Atmospheric and Space Physics at the University of Colorado, Boulder, in support of the MAVEN Mission. The authors are grateful to all who made the MAVEN mission possible, and also to the developers of the BATS-R-US MHD model at the University of Michigan, Ann Arbor and the NASA High End Computing Program.

References

Acuna, M. H. *et al.*, Global distribution of crustal magnetization discovered by the Mars Global Surveyor 1999, *Science*, 284, 790-793

Airapetian, V. S. & Usmanov, A. V. Reconstructing the solar wind from its early history to current epoch 2016, *ApJ*, 817, L24-L30

Curry, S.M., *et al.*, Response of Mars O+ pickup ions to the 8 March 2015 ICME: Inferences from MAVEN data-based models 2015, *Geophys. Res. Lett.*, 42, 9095-9102

Dong, C. F., *et al.*, 2015, Multifluid MHD study of the solar wind interaction with Mars' upper atmosphere during the 2015 March 8 th ICME event 2015, *Geophys. Res. Lett.*, 42, 9103-9112

Edberg, N. J. T. , Nilsson, H., A. O. Williams, M. Lester, S. E. Milan, S. W. H.. Cowley, M. Fraenz, S. Barabash, & Y. Futaana, Pumping out the atmosphere of Mars through solar wind pressure pulses 2010, *Geophys. Res. Lett.*, 37,

Guedel *et al.*, *this volume*

Jakosky, B. & Phillips, R., Mars' volatile and climate history 2001, *Nature*, 412, 237-244

Jakosky, B. M. *et al.*, 2015a, The Mars atmosphere and volatile evolution (MAVEN) mission 2015a, *Space Sci. Rev.*, 195,

Jakosky, B. M. *et al.*, 2015b, MAVEN observations of the response of Mars to an interplanetary coronal mass ejection 2015b, *Science*, 350,

&Jakosky, B. M., J. M. Grebowsky, J. G. Luhmann, & D. A. Brain, The MAVEN mission to Mars at the end of one Mars year of science observations 2016, *J. Geophys. Res.*, in press

Lillis, R. *et al.*, Characterizing atmospheric escape from Mars today and through time 2015, *Space Sci. Rev.*, 195, 357-422

Luhmann, J. G., C. F. Dong, Y. J. Ma, S. M. Curry, S. Xu, C. O. Lee, T. Hara, J. Halekas, & Yan Li, J. R. Gruesbeck, J. Espley, D. A. Brain, & C. T. Russell, 2016, Martian magnetic storms 2016, submitted to *J. Geophys. Res.*

&Lundin, R., S. Barabash, M. Homstrom, H. Nillson, Y. Futaana, R. Ramstad, M. Yamauchi, E. Dubinin, & M. Fraenz, Solar cycle effects on the ion escape from Mars 2013, *Geophys. Res. Lett.*, 40, 6028-6032

Ma, Y. J., *et al.*, MHD model results of solar wind interaction with Mars and comparison with MAVEN plasma observations 2015, *Geophys. Res. Lett.*, 42, 9113-9120

&Maehara, , H., T. Shibayama, S. Notsu, Y. Notsu, T. Nagao, S. Kusaba, S. Honda, D. Nogami, & K. Shibata, Superflares on solar-type stars 2012, *Nature*, 485, 478-481

Osten *et al.*, *this volume*

Ribas, L., E. F. Guinan, M. Guedel, & M. Audard, Evolution of the Solar Activity over Time and Effects on Planetary Atmospheres. I. High-Energy Irradiances (1-1700 +) 2005, *ApJ*, 622, 680-694

Terada, N., Y. N. Kulikov, H. Lammer, H. Lichtenegger, T. Tanaka, H. Shinagawa, & T. Zhang, Atmosphere and water loss from early Mars under extreme solar wind and extreme ultraviolet conditions 2009, *Astrobiology*, 9, 55-70

Thiemann, E. M., *et al.*, Neutral density response to solar flares at Mars 2015, *Geophys. Res. Lett.*, 42, 8986-8992

&Tu, L., C. P. Johnstone, M. Guedel, & H. Lammer, The Extreme Ultraviolet and X-Ray Sun in Time: High-Energy Evolutionary Tracks of a Solar-Like Star 2015, *Astronomy and Astrophysics*, 577

Wood, B. E., The Solar Wind and the Sun in the Past 2006, *Space Sci. Rev.*, 126, 3-14

Living Around Active Stars
Proceedings IAU Symposium No. 328, 2016
D. Nandy, A. Valio & P. Petit, eds.

© International Astronomical Union 2017
doi:10.1017/S1743921317003714

Coronal Mass Ejections travel time

Carlos Roberto Braga, Rafael Rodrigues Souza de Mendonça, Alisson Dal Lago and Ezequiel Echer

National Institute for Space Research,
Av. Dos Astronautas 1758, Jd. Granja, São José dos Campos, SP, Brazil
email: carlos.braga@inpe.br

Abstract. Coronal mass ejections (CMEs) are the main source of intense geomagnetic storms when they are earthward directed. Studying their travel time is a key-point to understand when the disturbance will be observed at Earth. In this work, we study the CME that originated the interplanetary disturbance observed on 2013/10/02. According to the observations, the CME that caused the interplanetary disturbance was ejected on 2013/09/29. We obtained the CME speed and estimate of the time of arrival at the Lagrangian Point L1 using the concept of expansion speed. We found that observed and estimated times of arrival of the shock differ between 2 and 23 hours depending on method used to estimate the radial speed.

Keywords. coronal mass ejections (CMEs), halo CMEs, interplanetary disturbances, magnetic clouds

1. Introduction

Coronal mass ejections (CMEs) and their corresponding structures in the interplanetary medium (known as ICMEs) are of great importance because they cause most of the intense geomagnetic storms (Gosling *et al.* 1990; Gonzalez *et al.* 1990). Those events are frequently called geoeffective CMEs when they have direction of propagation close to the Sun-Earth line (Möstl *et al.* 2014).

In this work we identify the CME that caused the ICME observed on 2013/10/03 and then estimate its travel time using only the speed extracted from coronagraph observations.

2. Identifying the CME and estimating its travel time

In order to identify the CME associated to the interplanetary disturbance, we used simultaneous observation from three white light coronagraphs onboard at three different spacecraft: the twin STEREO (Solar TErrestrial RElations Observatory, Kaiser *et al.* 2008) spacecraft and SOHO (Solar and Heliospheric Observatory, Domingo *et al.* 1995). While the later is located in the Lagrangian point L1, the STEREO probes were located approximately 1 AU away from the Sun, one ahead of the Earth orbit by 147° and the other behind by 139° during the ICME observation period. Their separation angle was approximately 73°.

First, we inspected white-light coronagraph observations from the LASCO -C3 (Brueckner *et al.* 1995) onboard SOHO looking for identification of any CME possibly directed toward the Earth. Those CMEs are seen as halos or partial halos surrounding the occulting disk. From previous studies, we know that the travel time of the CME ranges basically from 1 to 5 days (Gopalswamy *et al.* 2001, Schween *et al.* 2005). From our analysis, the unique CME observed as a partial halo on the period was first seen on the LASCO coronagraph C2 at 2013/09/29 22:12. No full halo event was identified.

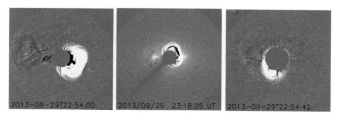

Figure 1. The observation of the CME on 2013/09/29 on both COR2 A (left), COR2B (right) and LASCO -C3 FOVs.

Secondly, we identified the corresponding CME observed on the COR2 coronagraphs onboard the STEREO spacecraft. Due to its viewpoint, a CME directed toward the Earth is observed as a eastward limb on COR2A and as a westward limb on COR2B. From this analysis we confirm that the CME observed on 2013/09/29 22:12 is directed toward the Earth and must be the cause of the ICME (Fig. 1).

According to the in situ observation from the instruments MAG and SWEPAM onboard the ACE spacecraft (located at the Lagrangian Point L1), an interplanetary shock was observed on 2013/10/02 01:18, more than 50 hours after the first observation of the CME on the solar corona. The solar wind increased to approximately 600 km/s and the interplanetary magnetic field reached values of 32 nT. On the first half of 2013/10/03, a structure with magnetic cloud (MC) signatures was observed, in accordance with the criteria from Lepping *et al.* (2005)

The definition of the CME time of arrival (ToA) at L1 will depend on the specific signature observed in situ interplanetary data and there is no consensus about the parameter to be used (see, e.g., Gopalswamy *et al.* 2003). Beyond the time of the shock observation, other parameters used are: (a) the peak magnetic field intensity associated to the shock (2013/10/02 04:24); (b) the beginning time of the MC (2013/10/02 22:35) and (c) peak density (2013/10/02 03:58).

The ToA can be estimated using the expansion speed, a speed perpendicular to the radial direction of the CME (Dal Lago *et al.* 2003). The CME travel time T_{travel} can be estimated by an optimum fit function (from Schwenn *et al.* 2005): $T_{travel} = 203 - 20.77 \ln(v_{exp})$ where T_{travel} is given in hours and v_{exp} is the expansion speed (given in km/s). Using the pseudo-automatic segmentation of the CME from CORSET methodology, we derived $v_{exp} = 837 km/s$ (the reader is referred to Braga *et al.* 2013 for details). In this case the travel time estimated is 63 hours and the ToA is 2013/10/02 13:25. When we do not have the expansion speed, it can be derived from from the fit between the v_{exp} and v_{rad} obtained by Dal Lago *et al.* (2004): $v_{exp} = v_{rad}/0.88$.

We took the radial speed derived by CME catalogs such as the CDAW CME catalog (Yashiro *et al.* 2004), CACTus (Computed Aided CME Track, Robbrecht & Berghmans 2004) and SEEDS (Solar Eruptive Event Detection System, Olmedo *et al.* 2008). By estimating the expansion speed from the radial speed, we calculated the ToA. In the specific case of CORSET, the ToA was calculated in two ways: (i) using the expansion speed provided from the method ignoring the radial speed and (ii) ignoring the expansion speed and using the radial speed to estimate it.

Among all methods cited in the upper paragraph, the ToA estimated differs from the shock observation time the least when using the radial speed from CDAW catalog. In this case, the estimated ToA is 2013/10/02 03:39, approximately 2 hours later than the observed shock. When using the results from CORSET, the difference increases to approximately half a day: 2013/10/02 13:25 using the expansion speed, and 2013/10/02

12:29 using the radial speed to estimate the expansion speed. The speeds from fully automatic catalogs (CACTus and SEEDS) resulted in estimated ToA 20 hours later than the shock (2013/10/03 04:17 and 2013/10/03 00:04, respectively). Notice that these two ToAs lie in the first hours of the MC-like observation. One possible explanation for the best ToA estimation obtained when using the speed derived by CDAW catalog is the similarity between the methodology used in this catalog and the one adopted by Schween *et al.* (2005) and Dal Lago *et al.* (2004): both methodologies identify the CME by eye. On the other methodologies, the CME speed is derived using automatic or pseudo-automatic identification.

3. Final remarks

The ToA of the CME observed on 2013/09/29 was estimated using the methodology from Schween *et al.* (2005) and taking plane-of-sky projected speed measurements from LASCO-C3 derived using several methods. For all radial speeds considered here, the estimated ToA was found to be latter than the actual shock observation and the difference ranges from 2 to 23 hours, depending on the method used.

4. Acknowledgments

C. R. Braga acknowledges grants #2014/24711-6, #2013/02712-8 and #2012/05436-9 from São Paulo Research Foundation (FAPESP). A. Dal Lago acknowledges CNPq for grant 304209/2014-7. E. Echer acknowledges CNPq for grant 302583/2015-7. R. R. S. de Mendonça acknowledges and CNPq for grant 152050/2016-7. LASCO is one of a complement of instruments on the Solar Heliospheric Observatory satellite (SOHO) built in an international collaboration between ESA and NASA. The Sun Earth Connection Coronal and Heliospheric Investigation (SECCHI) was produced by an international consortium and it is part of the STEREO spacecraft.

References

Braga, C. R., Dal Lago, A., & Stenborg, G. 2013, *ASR*, 51,1949
Brueckner *et al.* 1995, *Solar Phys.*, 162, 357
A. Dal Lago, Schwenn, R. & Gonzalez, W. D. 2003, *ASR*, 32, 2637
Domingo, V., Fleck, B., & Poland, A. I. 1995, *Solar Phys.*, 162, 1
Gonzalez, W. D., Tsurutani, B. T., & Clúa de Gonzalez, A. L. C. 1999, *SSRv*, 88, 529.
Gopalswamy, N., Lara, A., Yashiro, S. *et al.* 2001, *JGRA*, 106, 12, 29207
Gopalswamy, N., Manoharan, P. K., & Yashiro, S. 2003, *GRL*, 30, 2232
Gosling, J. T. 1990, in: Russell, C. T. , Priest, E. R. & Lee,L. C. *Physics of Magnetic Flux Ropes*, Geophysics Monograph, 58 (Washington, DC: American Geophysical Union), 343
Kaiser, M. L., Kucera, T. A., Davila, J. M. *et al.* 2008, *SSRv*, 136, 5
Lepping, R. P., Wu, C. C., & Berdichevsky, D. B. 2005, *AnGeo*, 23, 2687
Liu, Y. D., Luhmann, J. G., Möstl, C. *et al.* 2012 *ApJ* (Letters), 746, L15
Möstl, C., Amla, K., Hall, J. R. *et al.* 2014, *ApJ*, 787, 119
Robbrecht, E. & Berghmans, D. 2004, *A&A*, 425, 1097
Schwenn, R., Dal Lago, A., Huttenen, E., & Gonzalez, W. D. 2005, *AnGeo*, 23, 1033
Yashiro, S., Gopalswamy, N., Michalek, G. *et al.* 2004, *JGRA*, 109, A07105

Living Around Active Stars
Proceedings IAU Symposium No. 328, 2016
D. Nandy, A. Valio & P. Petit, eds.

© International Astronomical Union 2017
doi:10.1017/S1743921317003805

Nickel-Phosphorous Development for Total Solar Irradiance Measurement

F. Carlesso[1], L. A. Berni[1], L. E. A. Vieira[1], G. S. Savonov[1], M. Nishimori[1], A. Dal Lago[1] and E. Miranda[1]

[1]National Institute for Space Research, INPE
Av. dos Astronautas 1758, 12227-010 , São José dos Campos, SP, Brazil
email: `fccarlesso@gmail.com`

Abstract. The development of an absolute radiometer instrument is currently a effort at INPE for TSI measurements. In this work, we describe the development of black Ni-P coatings for TSI radiometers absorptive cavities. We present a study of the surface blackening process and the relationships between morphological structure, chemical composition and coating absorption. Ni-P deposits with different phosphorous content were obtained by electroless techniques on aluminum substrates with a thin zincate layer. Appropriate phosphorus composition and etching parameters process produce low reflectance black coatings.

Keywords. Black Nickel-Phosphorous, Active Cavity Radiometers, Low Reflectance Coating

1. Introduction

Variations in total solar irradiance (TSI) influence the Earths climate (Steinhiber *et al.* 2009). The global energy balance and climate can be impacted by changes in TSI, making accurate solar measurements important to discern the long-term effects of this influence (Haigh 2007). TSI measurements have been made continually by different space-borne instruments since 1978. These measurements rely on active cavity radiometers, which use black interior surfaces to absorb incident sunlight. The conical cavity is constructed mainly of electrodeposited silver due to the high thermal conductivity and use black film interior surfaces to absorb incident sunlight. Cavities have a wire resistor in the outer wall encapsulated with epoxy, which provides heat close to the region heated by solar radiation. On the external wall of the cavity is deposited chromium and copper to be a base and gold to reduce the radiative losses to the environment. Ni-P was first applied in active cavity radiometers for measuring solar irradiance in TIM (Total Irradiance Monitor) instrument aboard SORCE spacecraft and has advantages over traditionally used black paints providing robustness to solar exposure and having good thermal conductivity (Kopp 2014). Ni-P alloy coating has been found important applications in different fields of science and finishing industries due high hardness, corrosion and wear resistance (Pillai *et al.* 2012). Although etched NiP by the oxidizing acids produces high absorptivity coating across the entire solar spectrum for terrestrial and space-born optical applications (Saxena *et al.* 2006). This work presents development black Ni-P coating to be used in a TSI radiometer absorptive cavity.

2. Experimental

The Ni-P coatings were deposited on aluminum circular discs with an area of $7.07 cm^2$. The aluminum substrates were mechanically polished, degreased in an ultrasonic acetone bath and after in an alkaline solution cleaner. Before the electroless plating the substrates

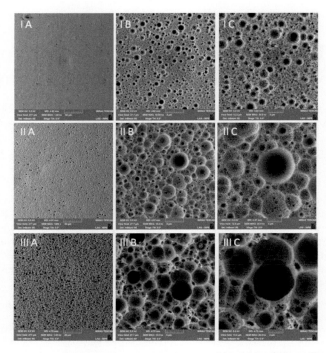

Figure 1. SEM image obtained in three magnification (A) 1 kX, (B) 2 kX and (C) 3 kX of black Ni-P coating with (I) 7.6 %P, (II) 7.0 %P and (III) 5.9 %P.

were deoxidized in acid cleaning solution and then a double zincate process aluminum was realized. Ni-P electroless bath was composed of Nickel sulfate, Sodium hypophosphite, Sodium citrate, Sodium acetate and Thiourea. Electroless conditions of the solution were were: ph $4.80 - 4.00$, bath temperature 363.15 K, time deposition $180 - 420$ min to get different thickness. After electroless deposition Ni-P layers were etched by oxidant acid (HNO_3) for $30 - 60$ s at 313.15 K to produce Black Ni-P. Coatings morphology were observed by Field Emission Gun - Scanning Electron Microscopy (FEG-SEM, Tescan Mira 3 FEG) using accelerating voltage of 5 kV. The chemical microanalysis were realized with an energy dispersive spectroscopy (EDS) at an accelerating voltage of 15 kV and working distance of 15 mm. Spectral reflectance of black NIP were measured using a spectrophotometer (HITACHI U-3501) equipped with an integrating sphere, in 250 nm to 850 nm wavelength region. The solar absorptance were measured by solar reflectometer (410-Solar Reflectometer - Surface Optics Corporation). This instrument provide solar-weighted absorptance and operate from 335 nm to 2500 nm wavelength region.

3. Results and Discussion

The composition of the Ni-P as deposited were around $5.9 - 7.5$ Wt%P and $25 - 50\mu$m of thickness. The observed difference on the P percentage depends on the control of the bath operations conditions. Fig. 1 presents the morphological study of etched Ni-P. The blackned surface is characterized by high conical pore density with disordered distribution and different sizes. Black Ni-P coating with 5.9 %P and 50μm of thickness has larger and more pronounced crater which vary between $0.03 - 6\mu$m of diameter and $2 - 7\mu$m of depth. Black Ni-P exhibited lowest reflectance when greater pore quantity was formed

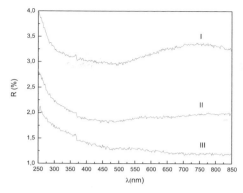

Figure 2. The total reflectance were measured as a function of wavelength of black Ni-P coating with different phosphorus content: (I) 7.6 %P, (II) 7.0 %P and (III) 5.9 %P.

during blackening process. The lower pre-etch P content studied in this work and relation with lower reflectance are accordingly with results obtained by Brown *et al.* (2002).

The spectral reflectance of electroless black coatings is shown in Fig. 2. Total reflectance change from 1.79 to 3.98 % for pre-etch coatings with higher phosphorous content while the coating with 5.9 %P is between 1.16 and 2.21 % in the range of $250-850$nm. The solar-weighted absorptance measured by solar reflectometer is obtained integrating the spectral reflectance over the standard spectral irradiance distribution. Solar absorption (α_{SOL}) measured for sample III is 0.997 ± 0.004. The low reflectance is due conical pores that work as light trap resulting in better values of absorption.

4. Summary

The low reflectance of the black coating is associated with surface morphology obtained during blackening process and the chemical composition has influence in the size and depth of the pores. These results suggest that Black Ni-P can be used in high absorptivity cavities for a future Brazilian TSI radiometer project.

References

Steinhilber, F., Beer, J., & Fröhlich, C. 2009, *Geophys. Res. Lett.*, 36, 19

Haigh, J. D. 2007, *Living Rev. Solar Phys*, 4, 1

Kopp. G 2014, *J. Space Weather Space Clim.*, 4, A14

Pillai,A. M.; Rajendra, A. & Sharma, A. K. 2012, *J Coat Technol Res*, 9, 6

Saxena, V.; Uma Rani, R., & Sharma, A. K. 2006, *Surf. Coat. Technol.*, 201, 3-4

Brown, R. J. C.; Brewer, P. J., & Milton, M. J. T. 2002, *J. Mater. Chem.*, 12, 9

Living Around Active Stars
Proceedings IAU Symposium No. 328, 2016
D. Nandy, A. Valio & P. Petit, eds.

© International Astronomical Union 2017
doi:10.1017/S1743921317003866

Preliminary Design of the Brazilian's National Institute for Space Research Broadband Radiometer for Solar Observations

L. A. Berni[1], L. E. A. Vieira[1], G.S. Savonov[1], A. Dal Lago[1], O. Mendes[1], M. R. Silva[1], F. Guarnieri[1], M. Sampaio[1], M, J, Barbosa[1], J. V. Vilas Boas[1], R. H. F. Branco[1], M. Nishimori[1], L. A. Silva[1], F. Carlesso[1], J. M. Rodríguez Gómez [1], L. R. Alves[1], B. Vaz Castilho[2], J. Santos[2], A. Silva Paula [2] and F. Cardoso[3]

[1]National Institute for Space Research, INPE
Av. dos Astronautas 1758, 12227-010, São José dos Campos, SP, Brazil
email: fccarlesso@gmail.com
[2]Laboratrio Nacional de Astrofsica, LNA
R. dos Estados Unidos 154 - Nações, Itajubá, MG, Brazil, 37530-000
[3]University of São Paulo, USP
Butantã, São Paulo, SP, 03178-200, Brazil

Abstract. The Total Solar Irradiance (TSI), which is the total radiation arriving at Earth's atmosphere from the Sun, is one of the most important forcing of the Earths climate. Measurements of the TSI have been made employing instruments on board several space-based platforms during the last four solar cycles. However, combining these measurements is still challenging due to the degradation of the sensor elements and the long-term stability of the electronics. Here we describe the preliminary efforts to design an absolute radiometer based on the principle of electrical substitution that is under development at Brazilian's National Institute for Space Research (INPE).

Keywords. Total solar irradiance, absolute radiometer, radiometry, absorptive cavities, black Ni-P

1. Introduction

The measurement of the total solar irradiance (TSI) and its variations is very important to understand the influence of the radiation output of the Sun on the Earth's climate. For a better understanding of the importance of these influences upon the Earth's climate, long-term TSI measurements are required. Several instruments aboard space-based platforms have obtained the most reliable observations since around 1978. The records of these instruments overlap in some time periods and differences are observed in the measured values. The Total Irradiance Monitor (TIM) show different values around $4W/m^2$ in relation to others experiments as Variability of Irradiance and Gravity Oscillations Sun PhotoMeter (VIRGO) and Active Cavity Radiometer Irradiance Monitor (ACRIM III) (Kopp *et al.* 2012). The reconstruction of the irradiance based on these observations is challenging due to in-flight degradation of the sensors as well as the different designs and calibration of them. Here we describe the development of an active cavity radiometer by a team at INPE. We describe the preliminary efforts at INPE to build a bench prototype for initial tests and to assist in the characterization of the mechanical, electronic

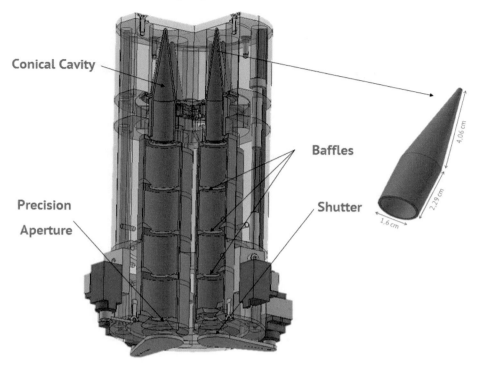

Figure 1. The design of the absolute radiometer broadband to measure the Total Solar Irradiance (TSI).

and optical components. In a second stage, some parts must be tested in an available small satellite. Our main goal is to understand the constraints to design a broadband radiometer to operate on a space-based platform. Additionally, we are aiming to identify incremental improvements in the current design.

2. Overview

The Brazilian Broadband Radiometer is an electrical substitution radiometer. The conceptual design of the instrument is based on existing radiometers, such as TIM instrument on board SORCE spacecraft (Kopp & Lawrence 2005). The device will contain four conical absorptive cavities. The cavities will be made mainly of silver due to the high thermal conductivity and the interior will be coated with a Ni-P film that is highly absorptive and presents excellent thermal conductivity of the substrate of the cone. Three cavities will work in an alternately regime to increase the devices lifetime while the fourth cavity will be kept as a reference. The conical cavities will be maintained at an equilibrium temperature through an electrical current. When the shutter is opened the temperature of cavity will increase due to exposure to solar radiation that depends on parameters system such as thermal link and heat sink (Datla & Parr 2005). Whereas some corrections are needed for final TSI measurement is possible to correlate radiation power and ohmic heating.

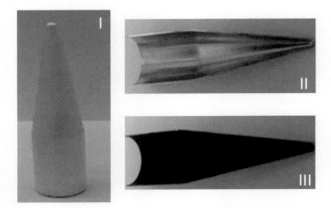

Figure 2. Silver cavity to trap incident sunlight (I) and cavity cross section with Ni-P as deposited (II) and Black Ni-P (III).

3. Concluding Remarks

We are following a formal and structured systems engineering procedure for the development of the broadband radiometer. This approach allows us to define objectively the requirements for the instrument to be developed. Specifically, the systems engineering is being used to develop and manage of the following: definition of the goals, the scope, and the products of the project; identification of constraints, assumptions, risks and dependencies; concept development; technical requirements; definition of the budgets; definition of technical standards; preparation of the documentation to acquire components; performance modeling; verification and validation; and, assembly, integration, testing, and commissioning. As we are developing the system from scratch, we decided to employ a model for the system development lifecycle that describes the activities to be performed and the results that have to be produced during the development. In this initial phase of the project, we are building a bench prototype for initial tests and to assist in the characterization of the mechanical structure, electronics and optics. Although we are designing a proof of concept instrument for laboratorial operation, we are defining the main requirements close to the requirements for space-based operation. The schematic drawing of the absolute radiometer is illustrated in Fig. 1. The precision aperture and the conical cavity are the key elements of the radiometer that determine the performance of the instrument. At this point, we have developed the process to build and characterize the sensor element, including the deposition of the thin film designed to absorb the solar radiation (see Fig. 2).

References

Kopp, G., Fehlmann, A., Finsterle, W., Harber, D., Heuerman, K., & Willson, R. 2012, *Metrologia*, 49, 2
Kopp, G. & Lawrence, G. 2005, *Sol Phys*, 230, 1
Datla, R. U. & Parr, A. C. 2005, *Geochim. Cosmochim. Acta*, 41

Living Around Active Stars
Proceedings IAU Symposium No. 328, 2016
D. Nandy, A. Valio & P. Petit, eds.

© International Astronomical Union 2017
doi:10.1017/S1743921317003738

Ground-based observations of the [SII] 6731 Å emission lines of the Io plasma torus

Fabíola P. Magalhães[1], Walter Gonzalez[1], Ezequiel Echer[1], Mariza P. Souza-Echer[1], Rosaly Lopes[2], Jeffrey P. Morgenthaler[3] and Julie Rathbun[3]

[1]National Institute of Space Research, Av. dos Astronautas, 1758, SP, Brazil
email: `fabiola.magalhaes@inpe.br`
[2]Jet Propulsion Laboratory, California Institute of Technology, Pasadena, CA, USA
email: `rosaly.m.lopes-gautier@jpl.nasa.gov`
[3]Planetary Science Institute, Tucson, AZ, USA
email: `jpmorgen@psi.edu`

Abstract. The Io Plasma Torus (IPT) is a doughnut-shaped structure of charged particles, composed mainly of sulfur and oxygen ions. The main source of the IPT is the moon Io, the most volcanically active object in the Solar System. Io is the innermost of the Galilean moons of Jupiter, the main source of the magnetospheric plasma and responsible for injecting nearly 1 ton/s of ions into Jupiter's magnetosphere. In this work ground-based observations of the [SII] 6731 Å emission lines are observed, obtained at the MacMath-Pierce Solar Telescope. The results shown here were obtained in late 1997 and occurred shortly after a period of important eruptions observed by the Galileo mission (1996-2003). Several outbursts were observed and periods of intense volcanic activity are important to correlate with periods of brightness enhancements observed at the IPT. The time of response between an eruption and enhancement at IPT is still not well understood.

Keywords. Io; Io Plasma Torus; Instrumentation; Volcanism; Jupiter magnetosphere

1. Introduction

Jupiter is a complex system and has an unique interaction with its innermost moon, Io. Jupiter has the largest magnetosphere in the Solar System and Io is immersed within it. Io has an intense volcanic activity and its volcanism is due to tidal heating produced by its forced orbital eccentricity. It is mainly generated by Jupiter's gravitational pull but also, in a smaller scale, by the other Galilean moons (Europa and Ganymedes) (Lopes & Williams, 2005). Io's volcanic plumes expel a considerable amount of material to the atmosphere. The release of lava flows and plumes produces Io's patchy atmosphere, which is mainly composed by SO_2, SO, S, O (Mendillo *et al.*, 2004). A significant fraction of the material lost to the atmosphere escapes as neutral atoms and molecules, mainly oxygen and sulfur atoms. The material which escapes from the gravitational pull of Io forms a neutral cloud that extends for several Jupiter's radii. The neutrals follow Io in its orbit about Jupiter until they are ionized through electron impact and charge exchange (Wilson *et al.*, 2002). Once ionized, the ions are accelerated to the nearly co-rotational flow of the ambient plasma, forming a torus of ions surrounding Jupiter, the so called Io Plasma Torus (IPT).

2. Overview

The Io Plasma Torus (IPT) is a ring-shaped cloud of S^+, S^{++}, S^{+++}, O^+, O^{++}, and a small concentration of other ions. Material escaping from Io creates an environment in constant change. Its physics can only be understood by measuring and correlating the variations in its structure and properties such as density, temperature and state of motion (Herbert *et al.*, 2003). The conditions in the plasma torus are highly variable both spatially and temporally due to the dynamic environment at Io and in the magnetosphere. The plasma torus can be observed in the extreme ultraviolet (EUV) emission and in the optical wavelengths. Optical emissions come from interactions between thermal electrons and sulfur ions. The optical emission traces the densest part of the torus, the EUV traces the hottest part of the torus. It is most dense around Io's orbit ($\sim 5.9\ R_J$). Therefore, from observations it is possible to obtain the temporal and spatial variability of the Io plasma torus (Lopes & Williams, 2005).

Kupo *et al.*, 1976 were the first to identify the individual forbidden emission lines of sulfur ([SII], λ 6731 Å, 6716 Å). Since then a number of ground-based observations were obtained. The plasma torus variability is still not well understood, this is partially because to the lack of continuous observations. From the IPT is possible to observe both sides of the torus simultaneously and monitor the positions of maximum brightness relative to the center of Jupiter. Another effect possible to obtain during observations is the brightness asymmetry produced by the modified electron density and temperature. The IPT variability is believed to be related to Io's volcanic activity as it does not occur steadily. Jupiter's magnetic field confines the plasma torus toward the equator, not by the strength of the magnetic mirror, but by centrifugal forces (Bagenal & Sullivan, 1981), causing the plasma torus to tilt, mainly due to the inclination of the planet magnetic field (10° inclination from Jupiter's rotation axis), but also by the planet rapid rotation ($\sim$ 10 hours). The torus extends along the field lines to $\pm$ 1 R_{Io} from the centrifugal equator (Bagenal, 1989). At the torus, it is estimated that the ions remain there up to a time relative to 100 revolutions of Jupiter, which means that every rotation of the torus through the neutral cloud adds only 1 % of new plasma (Connerney *et al.*, 1993).

3. Data & Preliminary results

Ground-based observations imply that the spatial distribution of SII inside Io's orbit is generally wedge-shaped. The apex of the ansa as observed from the ground corresponds to the maximum SII density (Pilcher 1980, Bagenal & Sullivan, 1981). The results here presented are from the [SII] 6731 Å emission lines. They were observed at the McMath-Pierce Solar Telescope (MMPST) during September and October of 1997. A total of 276 science (Jupiter) images were obtained. In Figure 1 is possible to observe the dawn (east) and dusk (west) ansae of the IPT. From the dawn side 154 images were analyzed, while for the dusk side 122 images were analyzed. The method to obtain the ansae brightness (dawn and dusk) was a photometric box around it (Figure 1). A method for processing [SII] torus data was developed during this work, as well as the routines used for basic image calibrations (bias and flat field) and the corrections necessary due to observations obtained from Earth's surface.

Once the [SII] images were fully reduced and analyzed, it was obtained a time sequence for 1997. The results in Figure 2 presents the average intensity the dawn and dusk ansa sides per day of observation. The abscissa shows the day of the year (DOY), from 240 to 320, versus intensity in Rayleighs (R). It is possible to observe a higher intensity on the dusk side when comparing the results between September and October. The highest

Figure 1. Image observed in October 11, 1997 at the McMath-Pierce Solar Telescope. Jupiter is at the center of the image, attenuated by a neutral density filter. The dawn and dusk ansae are at the borders of the image, delimited by a box. The box area was the region used to obtain each ansa brightness.

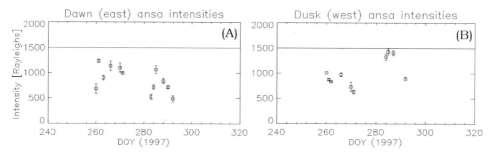

Figure 2. Variations of the ansae intensity of the 6731 Å emissions at dusk side (A) and dawn side (B). A total of 154 frames from the east ansa and 122 frames from the west ansa were used. Average [SII] intensity at dusk side shows slightly increase in October when compared with September observations and also for the same period when compared with the dawn side.

intensities observed are 1332 R at day 284 (October 11, 1997), 1434 R at day 285 (12 October 1997) and 1403 R at day 287 (October 14 1997). The highest intensity observed at the dawn ansa was 1238 R at 261 (18 September 1997).

Future Works: The next steps are to process the rest of the ground-based observations (1998-2003) and apply the Lomb-Scargle periodogram to obtain Jupiter's periodicities to study System's III and IV. Once the [SII] brightness time series is complete, the next step is to correlate it with the thermal emissions of Io's surface obtained with the Near Infrared Mapping Spectrometer (NIMS) during the Galileo mission (1996-2003).

Acknowledgement

This work was supported by CNPq under Grants $N°$. 302583/2015-7, 232274/2014-2 and 152713/2016-6; and CAPES.

References

Bagenal, F. & Sullivan, J. D., 1981, *Journal of Geophysical Research*, 86, 8447-8466

Bagenal, F., 1989, *NASA Special Publication*, 494

Connerney, J. E. P., Baron, R., Satoh, T., & Owen, T., 1993, *Science*, 262, 1035-1038

Herbert, F., Schneider, N. M., Hendrix, A. R., & Bagenal, F., 2003, *JGR*, 108, 1167

Hill, T. W. and Dessler, A. J., & Michel, F. C., 1974, *Geophysical Research Letters*, 1, 3-6

Kupo, I. and Mekler, Y. & Eviatar, A., 1976, *Astrophysical Journal*, 205, L51-L53

Lopes, R. M. C. & Williams, D. A., 2005, *Reports on Progress in Physics*, 68, 303-340

Mendillo, M., Wilson, J., Spencer, J., & Stansberry, J., 2004, *Icarus*, 170, 430-442

Pilcher, C. B., 1980. *Science*, 207, 181-183

Wilson, J. K. *et al.*, 2002, *Icarus*, 157, 476-489

Living Around Active Stars
Proceedings IAU Symposium No. 328, 2016
D. Nandy, A. Valio & P. Petit, eds.

© International Astronomical Union 2017
doi:10.1017/S1743921317003672

A study on Electron Oscillations in the Magnetosheath of Mars with Mars Express observations

Adriane M. de Souza[1]**, Ezequiel Echer**[1]**, Mauricio J. A. Bolzam**[2]
and Markus Fränz [3]

[1] National Institute For Space Research
Sao Jose dos Campos, Brazil
[2] Federal University of Goias, Jatai, Brazil

[3] Max Planck Institute for Solar System Research, Göttingen, Germany,
email: `adriane.souza@inpe.br`

Abstract. Wavelet analysis was employed to identify the major frequencies of low-frequency waves present in the Martian magnetosheath. The Morlet wavelet transform was selected and applied to the electron density data, obtained from the Analyzer of Space Plasmas and Energetic Atoms experiment (ASPERA-3), onboard the Mars Express (MEX) spacecraft. We have selected magnetosheath crossings and analyzed electron density data. From a preliminary study of 502 magnetosheath crossings (observed during the year of 2005), we have found 1409 periods between 0.005 and $0.06 Hz$. The major frequencies observed were in the range 0.005-0.02 Hz with 58.5 % of the 1409 frequencies identified.

Keywords. Mars Magnetosheath, Electron Oscillations, Wavelet Transform

1. Introduction

In planets lacking an internal magnetic field, the principal form of interaction between the solar wind and the body is through electro-magnetic induction. This induction can occur in conductive layers inside the planet or in an ionosphere, if the planet has an atmosphere. The induced electric currents flow through the planet or through the ionosphere and create forces that causes deceleration and deflection of the incident flow. Thereby, the deflected solar wind stream flows around a region similar to a magnetosphere created by an intrinsic magnetic field. Magnetospheres created due to this type of interaction are called induced magnetospheres (Cloutier & Daniell (1973), Cloutier& Daniell (1979), Luhman *et al.* (2004), Kivelson *et al.* (2007),Echer (2010) and Dieval *et al.* (2012)).

The magnetosheath of such unmagnetized planets has typically an addition of a new population of planetary ions to the hydrogen magnetoplasma, which drastically alters the dispersion of hydromagnetic waves and can produce new types of MHD (Magneto-hydrodynamics) discontinuities in the transition region (Sauer *et al.* (1998)). The study of wave propagation is very important due to the fact that they have an important role in the energy and momentum transfer between the solar wind and the magnetosphere for example at Mars (Espley *et al.* (2004)). Consequently, these waves are related to the processes of atmospheric loss at Mars via interaction with the solar wind. Considering the importance to study waves in the Martian magnetosheath, the aim of this work is to determine the major frequency of electron oscillations in that region.

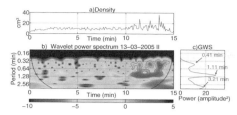

Figure 1. a- ASPERA n_e time serie - 13 March 2005. b- Spectrum Wavelet. c- GWS.

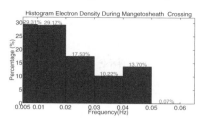

Figure 2. Histogram with the main frequencies identified in the GWS for all 2005 magnetosheath crossings.

2. Data and methodology

In order to perform this work, we have used electron density data (n_e) from the electron spectrometer (ELS) within the Analyzer of Space Plasma and Energetic Atoms Experiment (ASPERA3) on board the spacecraft Mars Express (MEX) (Barabash *et al.* (2004)). A total of 502 magnetosheath crossings during the year 2005 have been analyzed.

The Wavelet Transform (WT) was applied on the data set studied in this work. The wavelet functions are generated by expansion, $\psi(t) \rightarrow \psi(2t)$ and translations, $\psi(t) \rightarrow \psi(t+1)$ from a simple generating function, the wavelet-mother, given by $\psi_{a,b}(t) = (1/\sqrt{a})\psi((t-b)/a)$. In this work we used the Morlet wavelet given by Torrence & Compo (1998) in equation $\psi(t) = e^{i\zeta_0 t}e^{-t^2/2}$, where ζ is the dimensionless frequency. The WT applied on f(t) time is defined as $TW(a,b) = \int f(t)\psi_{a,b}(t)^* dt$, where $\psi_{a,b}(t)$ is the wavelet function and $\psi_{a,b}$ represents the complex conjugate thereof. The Global Wavelet Spectrum (GWS) was used to identify the most energetic periods in each magnetosheath crossing. The GWS is given by $GWS = \int |TW(a,b)|^2 \, db$.

3. Results

In order to obtain the main periods of low frequency electron oscillations in the Mars magnetoseath , the WT was applied to 502 magnetosheath crossings for data of n_e. Figure 1 shows an example of the WT results applied to the n_e for the interval wherein the MEX crossed the Mars magnetosheath, between $10:56$ UT and $11:11$ UT on March 13, 2005. Figure 1-a presents n_e time seires data and Figure 1-b the wavelet power spectrum. In the GWS of the Figure 1-c, we note the presence of three main periods: 0.41 (0.04Hz), 1.11(0.02Hz) and 3.21 minutes (0.005 Hz), approximately.

After applying the WT to n_e during the 502 magnetosheath crossings, 1409 frequencies were identified. These were divided into ranges to do a statistical analysis. Figure 2 shows the histogram with the results of the electron oscillation frequencies for these 502 magnetosheath crossings.

From the histogram we may note that the principal frequencies observed in the n_e were in the range $0.005 - 0.01$ Hz with 29.3% and in the range $0.01 - 0.02$ with 29.1% of the 1409 frequencies identified.Considering those two ranges, we have 58.5% of the frequencies in the interval of $0.005 - 0.02$ Hz.

These results agree with the findings by Winningham *et al.* (2006), where they used the integrated electron energy flux to study the electron oscillations in the induced magnetosphere of Mars. Those authors also observed a peak between 0.01 and 0.02 Hz in the magnetosheath. They have interpreted it as corresponding to the oxygen ion gyrofrequency in the Mars magnetosheath, calculated using data from the magnetometer from Mars Global Surveyor (Espley *et al.* (2004)). It is believed that for large scales (greater than the Debye Length λ_D), the plasma should be electrically neutral ($N_e \cong N_{i+}$). When ion wave modes, of large scale develop, they carry with them, the electrons which are associated with the oscillating ions. By this the electrons take their wavelike behavior. Electrons oscillate at higher frequency, but then also respond to lower frequencies to maintain quasi-equilibrium with positive ions. Therefore, when we observe in the ultra low frequency domain, the electrons trace the motion of the ions and the wave modes thereof (Winningham *et al.* (2006)). These authors also suggested that the bow shock may be the source of the oscillations.

4. Summary

In this paper we have determined the main frequencies of n_e in the Martian magnetosheath. Given the obtained results it is emphasized that:

• The principal frequencies observed in n_e were in the range $0.005 - 0.02$ Hz with 58.5% of the 1409 frequencies identified;

• Those frequencies are near the local oxygen gyrofrequency in the Mars magnetosheath. Electrons trace the low frequency waves that affect ions;

• The bow shock may be the source of the oscillations;

As future work the authors intend to study all the Mars magnetosheath crossings during the whole MEX interval (2004-2015). We also will add electron temperature and pressure data and separate the analyzed data by position in the sheath, by solar cycle phases and solar wind conditions.

5. Acknowledgments

The authors would like to thank FAPESP for the financial support to AMS. The MJAB was supported by FAPEG (grant n. 201210267000905) and CNPq (grants n. 303103/2012-4). EE would like to thank to the Brazilian CNPq/PQ (302583/2015-7) agency for financial support.

References

Barabash *et al.* 2004, in: A. Wilson (ed.), *Mars Express: the scientific payload.* (Noordwijk, Netherlands: Esa publication), p. 121

Cloutier, P. A. & Daniell, R. E. Jr. 1973, *Planet. Space Sci.*, 21, 463

Cloutier, P. A. & Daniell, R. E. Jr. 1979, *Planet. Space Sci.*, 27, 1111

Dieval *et al.* 2012, *Journal of Geophysical Research*, 117, A06222

Echer, E. 2010, *Revista Brasileira de Ensino de Fsica*, 32, 230

Espley *et al.* 2004, *Journal of Geophysical Research*, 109,A4

Kivelson, M. G. & Bagenal, F. 2007, in: L. A. MacFadden, P. R. Weissman, & T. V. (eds.), *Encyclpedia of the Solar System.* (San Diego, CA: Academic), p. 519

Luhmann, J. G., Ledvina, S. A., & Russel, C. T. 2004, *Advances in Space Research*, 33, 1905

Sauer, K., Dubinin, E., & Baumgartel, K. 1998, *Earth Planets Space*, 50, 793

Torrence, C. & Compo, G. P. 1998, *Journal of Geophysical Research*, 118, 1

Winningham *et al.* 2006, *Icarus* 182, 360.

Living Around Active Stars
Proceedings IAU Symposium No. 328, 2016
D. Nandy, A. Valio & P. Petit, eds.

© International Astronomical Union 2017
doi:10.1017/S1743921317004185

Extreme solar-terrestrial events

A. Dal Lago[1], L. E. Antunes Vieira[1], E. Echer[1], L. A. Balmaceda[2], M. Rockenbach[1] and W. D. Gonzalez[1]

[1] National Institute for Space Research (INPE),
Avenida dos Astronautas-12227-010, São José dos Campos-SP, Brazil
email: `alisson.dallago@inpe.br`

[2] Instituto de Ciencias Astronómicas de la Tierra y el Espacio, ICATE-CONICET, Avda. de
España Sur 1512, J5402DSP, San Juan, Argentina

Abstract. Extreme solar-terrestrial events are those in which very energetic solar ejections hit the earth?s magnetosphere, causing intense energization of the earth?s ring current. Statistically, their occurrence is approximately once per Gleissberg solar cycle (70-100yrs). The solar transient occurred on July, 23rd (2012) was potentially one of such extreme events. The associated coronal mass ejection (CME), however, was not ejected towards the earth. Instead, it hit the STEREO A spacecraft, located 120 degrees away from the Sun-Earth line. Estimates of the geoeffectiveness of such a CME point to a scenario of extreme Space Weather conditions. In terms of the ring current energization, as measured by the Disturbance Storm-Time index (Dst), had this CME hit the Earth, it would have caused the strongest geomagnetic storm in space era.

Keywords. geomagnetic storms, solar-terrestrial events

1. Introduction

It is well established that the interplanetary origins of geomagnetic storms are related to the southward component of the interplanetary magnetic field combined with the solar wind velocity (Gonzalez & Tsurutani (1987) , Gonzalez *et al.* (1994)). These out-of-the-ecliptic magnetic fields have several sources, the most important being the interplanetary counterparts of coronal mass ejections (ICMEs) and their related shock-sheath fields. As the intensity of the storm increases, the more predominant these ICMEs and shocks become as drivers. Current estimates show that 95% of very intense storms (Dst$\leqslant$-250nT) are cause by ICMEs or by their associated shock (Gonzalez *et al.* (2007); Echer *et al.* (2008) and Gonzalez *et al.* (2011)). Szajko *et al.* (2013) has found that all Dst$\leqslant$-200nT geomagnetic storms of solar cycle 23 have ICME-sheath origins. If we consider the level of storms as stronger than Dst$\leqslant$-400nT, we find a frequency of occurrence of once every 11 year solar cycle. Only 5 of such ?extreme solar-terrestrial events? occurred during the space era (Gonzalez *et al.* (2011)). A complete coverage of geomagnetic and interplanetary observations was available for few of these events. We highlight the November 2003 extreme storm, in which the peak Dst reached -422 nT. In July 2012, a very fast coronal mass ejection (CME) was observed at the sun by LASCO/SOHO (Brueckner *et al.* (1995)) and SECCHI/STEREO (Howard *et al.* (2008)) instruments. From the combined observations, it was possible to conclude that the event was ejected towards the STEREO A satellite, which was 120 degrees away from the Sun-Earth line. STEREO A in situ observations of this event from IMPACT (Luhmann *et al.* (2008)) and PLASTIC (Galvin *et al.* (2008)) instruments at 0.96 AU indicate an extreme scenario in which IMF peak surpassed 100nT, with considerable southward component (Russell *et al.* (2013)). The aim of this work is to estimate the intensity of the hypothetic geomagnetic storm had this

ICME hit the earth. In order to achieve this goal, we use the simple model from Burton *et al.* (1975) for the Dst estimate, which requires only interplanetary observations as input. The Dst index measures the average horizontal component of the geomagnetic field measured by mid-latitude and equatorial stations around the world. Model parameters for the Dst index were tunned using as a reference the solar-terrestrial event occurred in November 2003.

2. Model description

The approach of this work was to estimate the quantitative geomagnetic response of an interplanetary structure using Burton *et al.* (1975) model for the Dst index. This model requires as input the southward interplanetary magnetic field and solar wind velocity observations. Details on the model can be found in Fenrich & Luhmann (1988) and Dal Lago *et al.* (2015). Burton *et al.* (1975) model assumes a ring current decay time of the order of 7.7 hours. Fenrich & Luhmann (1988) found that for intense and very intense storms (-80$\geqslant$Dst$\geqslant$-300 nT) decay times ranging from 3 to 5 hours are more adequate for correctly estimate the observed Dst index. We estimate this decay time for extreme events (Dst$\leqslant$-400nT) using the November 2003 geomagnetic storm. This event originated a peak Dst of the order of -420 nT, making this geomagnetic storm the most intense of the entire solar cycle 23. Burton *et al.* (1975) formula was used to reproduce the Dst index of this event. A ring current decay time of 4.5 hours was used, in order to correctly reproduce the peak intensity of the storm. More details on this event can be found in Dal Lago *et al.* (2015).

3. The July 2012 event

The UVI instrument, aboard the STEREO A satellite, observed a solar eruption on the 23rd of July (2012), around 02h30 UT. From the earth perspective, it was a behind the limb event. A CME was observed by SOHO/LASCO C2 instrument on July 23rd (2012) at 02:36. SECCHI/COR1 A also observed this CME starting at 02:30UT. Latter on, SECCHI/COR 2 A observed a full halo CME, starting at 02:54UT. On the 23rd of July, the in situ plasma instrument aboard the STEREO A spacecraft detected an extreme event, with magnetic field intensity higher than 100 nT, 20 times stronger than the average solar wind magnetic field. Russell *et al.* (2013) presented a detailed description of this interplanetary event, classified as a magnetic cloud. They reported very high solar wind velocities, above 2000 km/s. Figure 1 shows, from top to bottom, the interplanetary magnetic field, its Bz component, plasma velocity, interplanetary electric field. The bottom panel shows the hypothetical Dst index estimated using Burton *et al.* (1975) model, had this event hit the earth. In the model estimate, a ring current decay time of 4.5 hours was used, following the estimate from the November 2003 event. The estimate peak Dst value was -1113.2 nT, which is nearly twice the strongest Dst value ever measured.

4. Summary and conclusions

In this work, we investigate the hypotetic geomagnetic response to an extreme solar-terrestrial event using the model from Burton *et al.* (1975), using as input the interplanetary observations from STEREO A on the 23rd of July 2012 CME. Results indicate that had had this CME been ejected towards the earth, it would have given rise to strongest geomagnetic storm in space era, with peak Dst values of ? 1113.20 nT. These results are

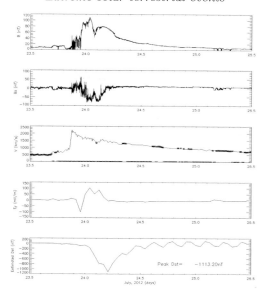

Figure 1. From top to bottom, the interplanetary magnetic field, its Bz component, plasma velocity, interplanetary electric field and the hypothetical geomagnetic Dst index estimated using Burton *et al.* (1975) model, considering a ring current decay time of 4.5 hours.

in good agreement with Baker *et al.* (2013) study of the same event, using other model approaches, in which their estimate peak Dst was -1182nT. Furthermore, the methodology presented by Gonzalez *et al.* (2011) for extreme events also would place this event in the same disturbance level.

5. Acknowledgements

The Authors would like to acknowledge SOHO/LASCO, STEREO/SECCHI, ACE and Dst Kyoto Data used in this work. This work is partially supported by FAPESP and CNPq/Brazil under the grants no. 304209/2014-7 and 302583/2015-7.

References

Baker, D. N., Li, X., Pulkkinen, A., Ngwira, C. M, Mays, M. L., Galvin, A. B., & Simunac, K. D. C. 2013, *SPACE WEATHER*, 11, 585
Brueckner, G. E., Howard, R. A., Koomen, M. J., *et al.* 1995, *SoPh*, 162, 357
Burlaga, L. F., E. Sittler, F. Mariani, & R. Schwenn 1981, *J. Geophys. Res.*, 86, 6673
Burton, R. K., McPherron, R. L., & Russell, C. T. 1975, *J. Geophys. Res.*, 80, 4204
Dal Lago, A., L. E. A. Vieira, E. Echer, L. A. Balmaceda, R. Rawat, & Gonzalez, W. D. 2015, *15thSBGf*, 1500
Echer, E., Gonzalez, W. D., & Tsurutani, B. T. 2008, *GRL*, 35
Fenrich, F. R. & Luhmann, J. 1998, *GRL*, 25, 15
Galvin, A. B., Kistler, L. M., Popecki, M. A., *et al.* 2008, *SSR*, 136, 431
Gonzalez, W. D. & Tsurutani, B. T., 1987, *Planetary and Space Science*, 35, 1101
Gonzalez, W. D., Joselyn, J. A., Kamide, Y., Kroehl, H. W., Rostoker, G., Tsurutani, B. T., & Vasyliunas, V. M., 1994, *JGR*, 99, A4
Gonzalez, W. D., Echer, E., Tsurutani, B. T., & Gonzalez, A. L. C. 2007, *GRL*, 34

Gonzalez, W. D., Echer, E., Tsurutani, B. T., Clua de Gonzalez, A. L., & Dal Lago, A. 2011, *SSR*, 158, 69

Howard, R. A., Moses, J. D., Vourlidas, A., *et al.* 2008, *SSR*, 136, 67

Luhmann, J. G., *et al.* 2008, *SSR*, 136, 117

Russell, C. T. *et al.* 2013, *ApJ*, 770, 38

Szajko, N. S., Cristiani, G., Mandrini, C. H., & Dal Lago, A. 2013, *ASR*, 51, 10

Living Around Active Stars
Proceedings IAU Symposium No. 328, 2016
D. Nandy, A. Valio & P. Petit, eds.

© International Astronomical Union 2017
doi:10.1017/S1743921317003908

How to make the Sun look less like the Sun and more like a star?

A. A. Vidotto[1]

[1] School of Physics, Trinity College Dublin, the University of Dublin, Dublin-2, Ireland
email: `Aline.Vidotto@tcd.ie`

Abstract. Synoptic maps of the vector magnetic field have routinely been made available from stellar observations and recently have started to be obtained for the solar photospheric field. Although solar magnetic maps show a multitude of details, stellar maps are limited to imaging large-scale fields only. In spite of their lower resolution, magnetic field imaging of solar-type stars allow us to put the Sun in a much more general context. However, direct comparison between stellar and solar magnetic maps are hampered by their dramatic differences in resolution. Here, I present the results of a method to filter out the small-scale component of vector fields, in such a way that comparison between solar and stellar (large-scale) magnetic field vector maps can be directly made. This approach extends the technique widely used to decompose the radial component of the solar magnetic field to the azimuthal and meridional components as well, and is entirely consistent with the description adopted in several stellar studies. This method can also be used to confront synoptic maps synthesised in numerical simulations of dynamo and magnetic flux transport studies to those derived from stellar observations.

Keywords. stars: magnetic fields

1. Introduction

Direct comparison between stellar and solar magnetic maps are hampered by their dramatic differences in resolution. Figure 1a illustrates the vector synoptic map of the Sun (Gosain *et al.* 2013), which can be directly compared to the vector synoptic map of a star (Figure 1b). The difference between both maps can be immediate recognised: the solar magnetic field has a "salt-and-pepper" structure, with high intensity magnetic fields localised in very small regions, while the stellar map shows a more smooth distribution of magnetic fields. This happens because the stellar map cannot reach the spectacular resolution of solar observations and are limited to only the large scale surface magnetic fields. More quantitatively, if we use spherical harmonics (see next section) to decompose the magnetic field, a solar magnetic field map can be decomposed out to quite high maximum spherical harmonics degree $l_{max} > 190$ (DeRosa *et al.* 2012), while stellar maps such as the one shown in Figure 1b, would typically have $l_{max} \lesssim 10$ (i.e., l_{max} is used as proxy for spatial resolution).

The goal of this work is to devise a way to compare high-resolution magnetic maps of the Sun with low-resolution maps of stars. This method was published in Vidotto (2016). In this article, I present a summary of the method and how it can be applied to compare solar and stellar magnetic field maps.

2. The method and a case study

Stellar magnetic maps can be observationally derived using the Zeeman Doppler Imaging (ZDI) technique (Donati & Brown 1997). ZDI can reconstruct the three components of the magnetic fields (radial r, meridional θ and azimuthal φ), but due to flux cancelation

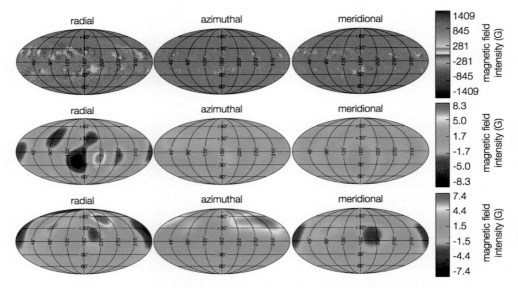

Figure 1. (a) Solar vector magnetic map (data from Gosain *et al.* 2013). (b) Vector magnetic map of a solar-type star where $l_{\max} = 8$ (Fares *et al.* 2009). (c) The solar magnetic field as seen from a "stellar" perspective $l_{\max} = 5$ (Vidotto 2016).

of small-scale fields within an element of resolution, ZDI is restricted to mapping only the large-scale magnetic field of the stellar surface (e.g. Lang *et al.* 2014). ZDI studies decompose the magnetic field using spherical harmonics as follows (Donati *et al.* 2006):

$$B_r(\theta, \varphi) = \sum_{lm} \alpha_{lm} P_{lm} e^{im\varphi} \,, \tag{2.1}$$

$$B_\theta(\theta, \varphi) = \sum_{lm} \frac{\beta_{lm}}{l+1} \frac{dP_{lm}}{d\theta} e^{im\varphi} + \gamma_{lm} \frac{im P_{lm} e^{im\varphi}}{(l+1)\sin\theta} \,, \tag{2.2}$$

$$B_\varphi(\theta, \varphi) = -\sum_{lm} \beta_{lm} \frac{im P_{lm} e^{im\varphi}}{(l+1)\sin\theta} - \frac{\gamma_{lm}}{l+1} \frac{dP_{lm}}{d\theta} e^{im\varphi} \,, \tag{2.3}$$

where $P_{lm} \equiv P_{lm}(\cos\theta)$ is the associated Legendre polynomial of degree l and order m. $\alpha_{lm}, \beta_{lm}, \gamma_{lm}$ are the coefficients that provide the best fit to the spectropolarimetric data, i.e., they are derived from ZDI studies. Therefore, once we have $\alpha_{l,m}$, $\beta_{l,m}$ and $\gamma_{l,m}$, we use equations (2.1) to (2.3) to derive the three components of the magnetic field.

The solar data, on the other hand, provide the three components of the magnetic field. If we want to compute the coefficients, following the formalism used in ZDI studies, we first need to invert equations (1) to (3) to obtain $\alpha_{l,m}$, $\beta_{l,m}$ and $\gamma_{l,m}$. The mathematical description to invert these equations is fully presented in Vidotto (2016). Given arrays of $\{B_r, B_\theta, B_\varphi\}$ spaced in n_θ latitudinal grid points and n_φ longitudinal grid points, the real and imaginary parts of $\alpha_{l,m}$, $\beta_{l,m}$ and $\gamma_{l,m}$ are given by

$$\Re(\alpha_{lm}) = \sum_{i=1}^{n_\varphi} \sum_{j=1}^{n_\theta} B_r^{ji} P_{lm}(\theta_j) \cos(m\varphi_i) \sin\theta_j \delta \,, \tag{2.4}$$

$$\Im(\alpha_{\mathrm{lm}}) = -\sum_{i=1}^{n_\varphi}\sum_{j=1}^{n_\theta} B_r^{ji} P_{\mathrm{lm}}(\theta_j)\sin(m\varphi_i)\sin\theta_j\delta\,, \qquad (2.5)$$

$$\Re(\beta_{\mathrm{lm}}) = \sum_{i=1}^{n_\varphi}\sum_{j=1}^{n_\theta}\left\{ B_\theta^{ji}\cos(m\varphi_i)\sin\theta_j\frac{\mathrm{d}P_{\mathrm{lm}}(\theta_j)}{\mathrm{d}\theta} + B_\varphi^{ji}m\sin(m\varphi_i)P_{\mathrm{lm}}(\theta_j)\right\}\frac{\delta}{l}\,, \qquad (2.6)$$

$$\Im(\beta_{\mathrm{lm}}) = -\sum_{i=1}^{n_\varphi}\sum_{j=1}^{n_\theta}\left\{ B_\theta^{ji}\sin(m\varphi_i)\sin\theta_j\frac{\mathrm{d}P_{\mathrm{lm}}(\theta_j)}{\mathrm{d}\theta} - B_\varphi^{ji}m\cos(m\varphi_i)P_{\mathrm{lm}}(\theta_j)\right\}\frac{\delta}{l}\,, \qquad (2.7)$$

$$\Re(\gamma_{\mathrm{lm}}) = -\sum_{i=1}^{n_\varphi}\sum_{j=1}^{n_\theta}\left\{ B_\theta^{ji}m\sin(m\varphi_i)P_{\mathrm{lm}}(\theta_j) - B_\varphi^{ji}\cos(m\varphi_i)\sin\theta_j\frac{\mathrm{d}P_{\mathrm{lm}}(\theta_j)}{\mathrm{d}\theta}\right\}\frac{\delta}{l}\,, \qquad (2.8)$$

$$\Im(\gamma_{\mathrm{lm}}) = -\sum_{i=1}^{n_\varphi}\sum_{j=1}^{n_\theta}\left\{ B_\theta^{ji}m\cos(m\varphi_i)P_{\mathrm{lm}}(\theta_j) + B_\varphi^{ji}\sin(m\varphi_i)\sin\theta_j\frac{\mathrm{d}P_{\mathrm{lm}}(\theta_j)}{\mathrm{d}\theta}\right\}\frac{\delta}{l}\,, \qquad (2.9)$$

where $\delta = \Delta\theta\Delta\varphi$, $\Delta\theta = \pi/n_\theta$ and $\Delta\varphi = 2\pi/n_\varphi$.

Therefore, from solar data (i.e., B_r, B_θ and B_φ), we can derive $\alpha_{l,m}$, $\beta_{l,m}$ and $\gamma_{l,m}$ for very large values of l-degrees using Equations (2.4) – (2.9). We then use these coefficients in Equations (1) to (3), but we limit the maximum value of l-degree in the sums to values similar to the ones reached in stellar ZDI studies. In other words, by restricting the sums to low-l degrees, we can filter out the small-scale field, which is not assessable in ZDI stellar studies, and the solar "large-scale" map can then be directly compared to stellar maps. A filtered solar magnetic field map is shown in Figure 1c.

3. Conclusions

Here, I presented a method to filter out the small-scale component of vector fields, so that comparison between solar and stellar large-scale magnetic fields can be directly made in all three components. This approach is entirely consistent with the description adopted in several stellar studies.

The method can be used to confront synoptic maps synthesised in numerical simulations of dynamo and magnetic flux transport studies to those derived from stellar observations. A recent application of the method was presented in Lehmann *et al.* (2016).

Acknowledgements

The solar synoptic map in the top panel of Figure 1 was acquired by SOLIS instruments operated by NISP/NSO/AURA/NSF.

References

DeRosa, M. L., Brun, A. S., & Hoeksema, J. T. 2012, *ApJ*, 757, 96
Donati, J., Howarth, I. D., Jardine, M. M., *et al.* 2006, *MNRAS*, 370, 629
Donati, J.-F. & Brown, S. F. 1997, *A&A*, 326, 1135
Fares, R., Donati, J., Moutou, C., *et al.* 2009, *MNRAS*, 398, 1383
Gosain, S., Pevtsov, A. A., Rudenko, G. V., & Anfinogentov, S. A. 2013, *ApJ*, 772, 52
Lang, P., Jardine, M., Morin, J., *et al.* 2014, *MNRAS*, 439, 2122
Lehmann, L. T., Jardine, M. M., Vidotto, A. A., *et al.* 2016, *MNRAS*, in press. ArXiv: 1610.08314
Vidotto, A. A. 2016, *MNRAS*, 459, 1533

Living Around Active Stars
Proceedings IAU Symposium No. 328, 2016
D. Nandy, A. Valio & P. Petit, eds.

© International Astronomical Union 2017
doi:10.1017/S1743921317004318

The X-ray Light-Curves and CME onset of a M2.5 flare of July 6, 2006

J. E. Mendoza-Torres[1] and J. E. Pérez-León[2]

[1]Instituto Nacional de Astrofísica Optica y Electrónica,
C. Luis Enrique Erro No 1, Tonantzintla, Pue., México
email: mend@inaoep.mx
[2]UANL, Monterrey, Nuevo León, México
email: enrique584@hotmail.com

Abstract. A M2.5 solar flare observed by RHESSI in the 6-100 keV range on July 6, 2006 led to a Coronal Mass Ejection (CME). Two compact sources at 12-100 keV are seen at the beginning of the flare, whose further evolution fits well in a loop. Also, time-profiles of the flare at radio wavelengths are compared. The X-ray light-curves at different bands in the 6-100 keV range and radio time profiles show some peaks superimposed on smooth variations. The aim of this work is to compare the X-ray light-curves, of fluxes integrated over the whole source, with the physical parameters of the sources of the flare. Yashiro and Gopalswamy (2009) have found that the fraction of flares that produce CME increases with the flare energy. Here, we look for the characteristics of an M2.5 flare that could make it a generator of a CME. The idea is, in future works, to look in the light-curves of similar flares at other stars for these features. It is found that the CME onset takes place around the time when an X-ray source at 12-25 keV of Chromospheric evaporation stagnates at the loop apex, before the main peak at the light-curve at 25-50 keV and at the radio emission curves. Probably, the amount of evaporated plasma could play some role in triggering the CME.

Keywords. Sun: X-rays, Sun: flares, Sun: coronal mass ejections (CMEs)

1. Introduction

It is widely accepted that flares are caused by magnetic reconnection somewhere in the Corona, where particles are accelerated and part of them precipitate to lower layers. Radio emission is expected to arise due to gyrosynchrotron emission near the loop feet, while X-ray emission arises when the downstream reaches the Chromosphere. For this reason, it is frequently observed that the time variations of the flux at microwaves are similar to X-ray light-curves, in particular the peaks, although with time delays (Liu *et al.* 2015). However, the electron populations that generate peaks and smooth variations could be different from each other (Lin & Johns 1993). The precipitating electrons lead to Chromospheric Evaporation, filling the magnetic loop with heated plasma. Then, the X-ray light-curves exhibit peaks and smooth time-variations.

It has been observed (Yashiro and Gopalswamy 2009) that the fraction of flares accompanied by CMEs increases with flare energy and practically all the X-class events produce CMEs. Then, it is expected that the larger the flare energy, the higher possibilities that a CME will take place. This result could have important implications in the estimation of the mass loss of the Sun in early phases and also in stars (Drake *et al.* 2016). However, it is not known what characteristics of strong flares are responsible of this relation, characteristics that some weak flares may noy display.

In the last years, it has been possible to image the sources of solar X-ray flares with a 4-second cadence. With this good time resolution, it is possible to identify details of

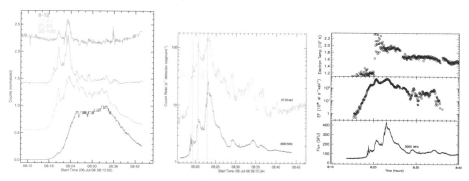

Figure 1. Left. Light-curves of the X-ray emission during the flare, from top to bottom: 50-100 keV, 25-50 keV, 12-25 keV and 6-12 keV. **Middle**. The time profiles observed at Ondrejov at 3 GHz and at RHESSI at 25-50 keV. **Right**. Electron Temperature (upper panel) and Electron Flux (middle panel) obtained from the spectra of the integral field. Time profile observed at Ondrejov at 3 GHz (lower panel).

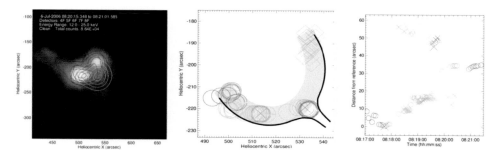

Figure 2. Left. X-ray sources observed by RHESSI at 12-25 keV, the image corresponds to 08:20 UT, black contours to 08:21 UT, blue to 08:22: UT and white to 08:29 UT. **Middle**. The 12-25 keV locations of the sources at different times of the flare, circles are for the South source and triangles for the North one. **Right**. Distance from the Southern loop foot to the 12-25 keV sources, as seen at different times.

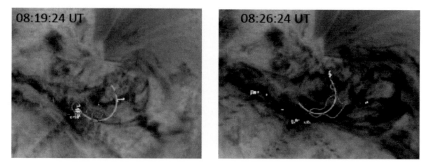

Figure 3. TRACE 171 A images with the times given in each panel. The green arc denotes the loop fitted to the 12-25 keV sources and the red contours the sources. **Left**. The sources at different locations of the loop. **Right**. The source at the loop apex when the CME is triggered.

the time evolution of the flare sources and their spectra and light-curves. We estimate physical parameters, at different times of the flare, from the X-ray spectra of the emission integrated over the whole field and for the spectra of three different Regions of Interest (ROIs) which are located at the loop feet and at the loop apex.

The X-ray light-curves, integrated over the RHESSI field, are compared with those obtained for the ROIs. Also, the time profiles of the Electron Temperature and Emission Measure (EM) obtained from the integrated spectra are compared with the corresponding time-curves obtained for the ROIs.

2. Data and Results

The M2.5 flare began at about 08:18 UT, as observed by RHESSI at X-ray and at radio wavelengths at San Vito and Ondrejov. In the right-panel of Figure 1, the light-curves at different energy bands in the 6-100 keV range are shown. In the middle panel the 25-50 keV light-curve and 3 GHz time profile are shown. In the right panel, the electron temperature (ET) and the electron flux (EF), for the whole image, are shown and the flux at some radio wavelengths. In Figure 2 (left) the 12-25 keV image at 18:20 UT is shown. In the middle-panel the locations of the source on the loop at different times are shown, purple corresponds to the initial times and red to the last ones. At the right-panel, the distances to the Southern source are plotted. At $\sim$ 08:20-08:22 UT the source stagnates at the loop top. The evaporating source also stagnates at a mid location between the feet and the apex, at this time a peak is seen in the light-curve.

According to the values obtained from the ROIs at the loop feet, the ET grows from $\sim$ 10 MK at 08:18 UT to $\sim$ 50 MK at 08:19 UT and decreases to $\sim$ 20 MK at 08:20 UT, which is the time when the evaporating source stagnates at the loop apex. The Emission Measure (EM) grows during the two more intense peaks. For the ROI at the apex of the loop, the ET varies similarly to that at the feet but the EM almost linearly grows from 08:18 UT to 08:25 UT. In the integrated EM, TE and Electron Flux (EF) the time-curves show both, peaks and smooth variations (right panel of Figure 1).

Lets recall that at the apex the EM continuously grows. Then, a possible characteristic of the flares that could play a role in triggering an CME, could be the amount of plasma reaching the loop apex due to Chromospheric evaporation. We have to point out that, the estimation of the source location seems to indicate that, the source underwent various back and forth small displacements while being at the apex. However, the displacements are short compared with the size source. Nevertheless, we cannot rule out other possible causes, such as oscillations (Stepanov et al. 2010). However, in this flare, they should have to be of small amplitude.

3. Conclusions

The more intense peaks at the X-ray and radio time-curves coincide with the times when the loop sources are most intense and when an evaporating source stagnates. The more intense peak occurs when the source stagnates at the loop apex. The onset of the CME takes place after this.

References

Brown, J. C., MacKinnon, A. L., Zodi, A. M., & Kaufmann, P. 1983, A&A, 123, 10

Drake, J. J., Cohen, O., Garraffo, C. O., & Kashya, V. 2016, Solar and Stellar Flares and Their Effects on Planets, Proceedings IAU Symposium No. 320

Lin, R. P. & Johns, C. M. 1993, ApJ, 417, 53L

Liu, Z-Y., Li, Y-P., Gan, W-Q., & Firoz K. A. 2015, Research in Astronomy and Astropnysics, 15, 64

Stepanov A. V., Yuri T. Tsap, Y. T., & Kopylova, Y. G. 2010, Solar and Stellar Variability: Impact on Earth and Planets, Proceedings IAU Symposium No. 264, 288

Yashiro, S. & Gopalswamy, N. 2009, IAU Symposium, Vol. 257, IAU Symposium, ed. N. Gopalswamy & D. F. Webb, 233

Living Around Active Stars
Proceedings IAU Symposium No. 328, 2016
D. Nandy, A. Valio & P. Petit, eds.

© International Astronomical Union 2017
doi:10.1017/S1743921317004252

A Framework for Finding and Interpreting Stellar CMEs

Rachel A. Osten[1,2] † and Scott J. Wolk[3]

[1] Space Telescope Science Institute
3700 San Martin Drive
Baltimore, MD 21218
email: osten@stsci.edu
[2] Dept. of Physics & Astronomy, Johns Hopkins University,
3400 N. Charles St.
Baltimore, MD 21218
[3] Center for Astrophysics, 60 Garden Street, Cambridge, MA 02138
email: swolk@cfa.harvard.edu

Abstract. The astrophysical study of mass loss, both steady-state and transient, on the cool half of the HR diagram has implications both for the star itself and the conditions created around the star that can be hospitable or inimical to supporting life. Stellar coronal mass ejections (CMEs) have not been conclusively detected, despite the ubiquity with which their radiative counterparts in an eruptive event (flares) have been. I will review some of the different observational methods which have been used and possibly could be used in the future in the stellar case, emphasizing some of the difficulties inherent in such attempts. I will provide a framework for interpreting potential transient stellar mass loss in light of the properties of flares known to occur on magnetically active stars. This uses a physically motivated way to connect the properties of flares and coronal mass ejections and provides a testable hypothesis for observing or constraining transient stellar mass loss. Finally I will describe recent results using observations at low radio frequencies to detect stellar coronal mass ejections, and give updates on prospects using future facilities to make headway in this important area.

Keywords. stars:flare, stars:activity, stars: late-type, radio continuum: stars

1. Overview

Stellar magnetic activity includes a whole gamut of observational signatures related to the presence and action of magnetic fields above the visible portion of the star's atmosphere. Stellar magnetic eruptions are the most dramatic releases of energy that a star on the cool half of the main sequence will undergo during its time on the main sequence. For a recent review of stellar flaring activity throughout the age of the solar system see Osten (2016). Solar physicists now recognize that a solar flare is only part of the eruptive event: a triad occurs in which the flare takes place low in the atmosphere, the coronal mass ejection takes place at much larger physical scales, and the energetic particles are the third component (Emslie *et al.* 2012). These all result from magnetic field reconfigurations and resultant reconnection and liberation of energy which goes into particle acceleration, plasma heating, mass motions, and shocks.

Astronomers have long been able to study stellar flares, starting from Hertzsprung's "peculiar nova of short duration" (Hertzsprung 1924). With advent of increasingly sensitive astronomical observatories at different wavelengths, we now see observational signatures of short-lived magnetic activity enhancements on stars across a large swath of

† Thanks to collaborator Scott Wolk for giving this talk in my stead.

the electromagnetic spectrum, from long wavelength radio waves (Konovalenko *et al.* 2012) to high energy gamma rays (Osten *et al.* 2007). Seeing multi-wavelength interrelationships between flare observables on stars as on the Sun (such as the Neupert effect; Hawley *et al.* 1995; Güdel *et al.* 1996)) firms up the conclusion that stellar flares are the same basic physical process as in solar flares, bolsters the comparison and extrapolation to stellar events, which can be up to 10^6 times more energetic (Osten *et al.* 2007) than the largest solar flare energies at about 10^{32} erg. Also, scaling relations established for solar flares which appear to extend to the stellar regime, as in the scaling between flare temperature and volume emission measure (Aschwanden *et al.* 2008), lend credence to the connection between solar and stellar flares.

2. Motivation

Current efforts to find Earth-like planets in the habitable zones of nearby stars are centering on nearby M dwarfs (Shields *et al.* 2016). The proximity to the star, with a habitable zone distance of less than 0.1 AU, renders these planets much more susceptible to the radiative and particle environment that the star creates. M dwarf stars in particular can produce frequent and extreme flares, and these have been the subject of astrobiological studies to examine the impact on an Earth-like planet in an M dwarf habitable zone. Khodachenko *et al.* (2007) examined the consequence of a high CME rate associated with that high flare rate. A high rate of coronal mass ejections could act like a dense, fast stellar wind, compressing the magnetosphere and exposing the planetary atmosphere to ionizing radiation.

CMEs are "geoeffective", and energetic particles have the potential to influence planetary atmospheres. There are currently little to no observational constraints on these from a stellar perspective, and astrobiological studies take scaling relations from detailed solar studies and extrapolate, often by orders of magnitude, to the stellar cases (Segura *et al.* 2010; Venot *et al.* 2016). These extrapolations need to be empirically tested to confirm the validity of extrapolating to the more energetic regime of stellar events.

3. Solar Physicists' Toolbox

Since their discovery in 1971, solar physicists have developed several methods of studying coronal mass ejections. The following is a brief review, to provide perspective on what tools might be applicable to the stellar case:

Direct Imaging via Thomson scattering
The coronagraph is the workhorse of solar CME observations (Howard 2011). Observations enable identification of the CME event in difference images, as well as determination of height-time progression. Measurements of velocity, mass, kinetic energy, potential energy, and acceleration can be made as well. See Yashiro *et al.* (2004) for a catalog of solar CMEs and derived properties.

Direct Imaging via Synchrotron Emission
Another somewhat novel solar observing technique is to directly image the synchrotron emission from a CME Bastian *et al.* (2001), resulting from the interaction of energetic particles (energies 0.5-5 MeV) with a magnetic field of 0.1 Gauss to a few Gauss.

Type II Bursts
Type II bursts have a strong association with CMEs (Gopalswamy 2006a; Gopalswamy *et al.* 2008a). A type II burst is coherent emission that is produced as the CME travels

outward through the outer solar atmosphere. When the CME is super-Alfvenic, a shock is produced, and Langmuir waves produced as the result of the disturbance propagating through the solar atmosphere result in a drifting radio burst. The drifting radio burst reveals the location of the shock; as the density in the atmosphere changes, the characteristic frequency of emission changes. Often, two curves are seen in dynamic spectra of the variation of intensity as a function of time and frequency: these correspond to fundamental and harmonic emission from the MHD shock produced by the CME. Given knowledge of the run of electron density with height in the solar corona, the observed frequency drift can be used to constrain the exciter speed, which will be a lower limit to the CME velocity.

Scintillation of Background Radio Sources

Another technique utilizes the CME producing scintillation of background radio sources, as a means to study the ambient density at large distances from the Sun. Manoharan (2010) describes how the plasma disturbances change the large-scale structure of the heliosphere, and interplanetary scintillation remote-sensing observations can be used to study the CMEs and structures within them.

Coronal Dimming

Thompson *et al.* (2000) noted the association between coronal dimmings and CMEs. Dimmings are large-scale changes in the coronal intensity, and can occur during passage of a CME. These coronal dimming can also be seen in spectro-temporal observations (Harra *et al.* 2016); they primarily manifest at cooler solar coronal plasma (typical of the quiet Sun) than is seen during solar flares.

4. Detecting CMEs on Stars

The following is a non-exhaustive list of different methods which have been used to attempt to observe CMEs on stars. Studies often use the "unusual flare" approach, in which some departure of a given flare from the standard flare scenario is invoked as an explanation for seeing transient mass loss associated with a flare. Another issue is that many of these signatures are flare signatures, and it is necessary to find a signature that is associated uniquely with the CME and not the flare. But of course this is difficult due to the high degree of correlation between flares and CMEs seen (at least on the Sun).

High Velocity Outflows Seen in Emission Lines

High velocity outflows in principal are signatures of the escaping mass, which from solar CMEs can have maximum velocities near 3000 km s^{-1}. Houdebine *et al.* (1990) reported evidence for a ~5800 km/s blueshifted outflow during a moderate flare on the M dwarf AD Leo; the outflow is a low level enhancement which is difficult to see above the quiescent emission. This requires the right orientation to see the maximum effect, and velocities of a few hundred km s^{-1} can be confused as originating from stellar flares. There have been no additional claims of high velocity outflows since Houdebine *et al.* (1990), despite signifiant spectro-temporal monitoring of optical and ultraviolet flare emissions (e.g., Kowalski 2012).

Pre-flare "dips"

Optical observations of stellar flares often see pre-flare diminutions in the light of the star prior to the start of the impulsive phase of the flare. Giampapa *et al.* (1982) reported perhaps the largest of these, with a decrease of about 25% of the stellar light prior to a large flare on the M dwarf EQ Peg. Much smaller dips (of order 1%) have recently been noted by Leitzinger *et al.* (2014) on flares in the Blanco 1 cluster with age of 30-145 MY. Such diminutions could result from the destabilitization of an off-limb filament, which deposits material into the disk line of sight. The concommitant increase in line and continuum opacity temporarily decreases chromospheric line emission, and Balmer continuum emission. This explanation requires a favorable geometry to work. Apart from this not being a unique CME signature, the decrease can also be explained as an increase in H$^-$ opacity resulting from chromospheric heating during the flare.

Increase in X-ray Absorption During Flare

Spectral fitting of stellar coronal X-ray spectra must take into account absorption by intervening material. In low-resolution X-ray spectra, the low energy end of the spectrum must be attenuated by the amount of hydrogen column density N_H that the X-ray emission travels through. For the case of stellar coronal emission the intervening material is usually absorption by the ISM. Occasionally, such as the flare noted by Franciosini *et al.* (2001) in BeppoSAX observations of a large long duration flare on UX Ari, there is an excess amount of absorption seen early in the flare, which requires a higher N_H value than that attributable to the intervening ISM. In this particular event N_H increased by about a factor of five. Since it is variable, one explanation is that the extra absorption arises from an increase in circumstellar material, such as might happen from ejection of matter during a CME. These are incredibly rare, though, and questions about instrumental calibration effects at low energy have plagued this interpretation. There is now a sufficient databse of stellar flares seen with the Chandra and XMM-Newton satellite (thousands of flares reported in McCleary & Wolk 2011) that such an investigation could be launched from a statistical perspective.

Effect of CMEs on stellar environment

Melis *et al.* (2012) showed the rapid disappearance of a debris disk on timescales of ~1 year, which was surprising and unexpected. The star, TYC 8241 2652 1, is a K2 dwarf with an age about 10 MY. As Melis *et al.* (2012) reported, grains present during observations in 2009 and before could be modelled as a characteristic size of ~0.3 μm, temperature 450 K, 0.4 AU separation from the star, and total mass of 5×10^{21} g; they could not come up with plausible mechanism for removing this material. X-rays would be ineffective at heating grains of this size, requiring an X-ray luminosity L_x of 10^2-10^3 $L_\odot$, and flare energy E$\approx10^{39}$ erg. Osten *et al.* (2013) noted that grains smaller than 0.2μm would be radiatively ejected from the system, and investigated whether a CME would be sufficient to remove the grains. A flare could charge the dust grains in the debris disk, and with sufficiently small gyroradii they would be swept up by the magnetic plasma in the CME and removed. For the case noted by Melis *et al.* (2012), removing ~10^{21}g in grains requires a CME with mass ~10^{20}g, with a timescale for removal on the order of a few days. The sudden disappearance of emission from the debris disk material then would stem from a CME occuring along with a large flare; the flare could charge the dust grains, and then the CME and its magnetic field would sweep up the material and remove it from the disk. The timescale for the entire sequence would only take a few days. While other smaller scale bulk changes in debris disk emission have been noted, this has not been used as a diagnostic of CMEs, largely due to the paucity of young active stars with close-in warm debris disks which would be the most likely targets for such a phenomenon.

Observing stellar CMEs through scintillation of background radio sources

This is the astronomical equivalent to interplanetary scintillation used to study solar CMEs. Applying this to the stellar case requires a sufficient number of background sources at different distances from the star to have potential probes of scintillation as well as test particles farther out. As the source density of flat-spectrum radio sources is low, this would require a favorable target and patch of the sky to work, and would favor nearby stellar sources for a larger angular extent of the astrosphere.

Type II-like bursts associated with stellar flares

This technique has the best likelihood of yielding a detection or robust constraints, due to existence of new generation of low frequency telescopes and receivers which can enable wide bandwidth simultaneous observations to detect and diagnose events in a dynamic spectrum. The expected behavior of a radio source changing with frequency and time due to intrinsic motion rather than propagation effects can be written simply as

$$\frac{d\nu}{dt} = \frac{\partial\nu}{\partial n_e}\frac{\partial n_e}{\partial h}\frac{\partial h}{\partial s}\frac{\partial s}{\partial t} \qquad (4.1)$$

where ν is the frequency of the burst at time t, n_e is the electron density, h is the radial height in the atmosphere and s is the path length traveled in the atmosphere. For a barometric atmosphere this reduces to $\dot{\nu} = \nu\cos\theta v_B/(2H_n)$, where v_B is the exciter speed whose motions cause the plasma radiation, H_n is the density scale height, and θ the angle between the propagation direction and the radial direction from the surface. For a source associated with plasma emission, these partial derivatives simplify, for a barometric atmosphere, to the second equation. Thus, given an observed frequency drift rate, frequency, and scale height (which can be estimated from X-ray observations), a constraint on the exciter speed is obtained.

Low frequency radio bursts have been seen from active stars: Konovalenko *et al.* (2012); Kundu & Shevgaonkar (1988); van den Oord & de Bruyn (1994) Drifting radio bursts in particular have been observed from nearby dMe flare stars, with a range of drift rates, as in Osten & Bastian (2006). The expected drift rate for coronal parameters of an active star, scale height $H_n=2\times10^{10}$ cm for a T$\sim10^7$ K, weakly super Alvenic shock gives drift rates of a few MHz s^{-1}, dependent on magnetic field strength and electron density.

Type II radio bursts are one particular example of a drifting solar radio burst, and based on solar observations, are the current best candidates to search for signatures of stellar transient mass loss. Large flares should have a large CMEs occurring along with them. While the overall rate of type II bursts is low, the association rate increases with CME speed. Therefore the fastest, most energetic CMEs should have type IIs associated with them. Yashiro *et al.* (2006) showed a 100% association between large solar flares and large CMEs; while only $\sim$10% of solar CMEs overall show decimetric type II bursts (Gopalswamy 2006b), the Type II burst association with CME increases with increasing CME speed (Gopalswamy *et al.* 2008b).

The LOw Frequency ARray, LOFAR, and the Jansky Very Large Array, are two new radio facilities that can be used to try to detect stellar CMEs. LOFAR expands the frequency range of sensitive interferometric capabilities to frequencies between 10 MHz and 200 MHz. The upgrade of the JVLA's low frequency receivers expanded the bandwidth by about a factor of 10.

Initial searches, reported by Crosley *et al.* (2016), did not see any evidence for bursting behavior. Figure 1 shows the range of parameter space that we have constrained where we would have seen emission – the X axis is the source size, and the y axis is the brightness temperature. Figure 1 also shows the expected burst shape in frequency and time, based on a couple of different assumptions bout the magnetic field strength and density. We are just barely sensitive with these observations to a very large solar-like type II burst. This was based on 15 hours of observations of a highly active dMe flare star, during which time we expected to see about 5 flares and associated CMEs.

Connecting Flares and CMEs

Using statistical associations between flares and CMEs is the next most promising approach to understanding stellar transient mass loss, as stellar flares are commonly observed in many wavelength regions. Numerous observations of solar and

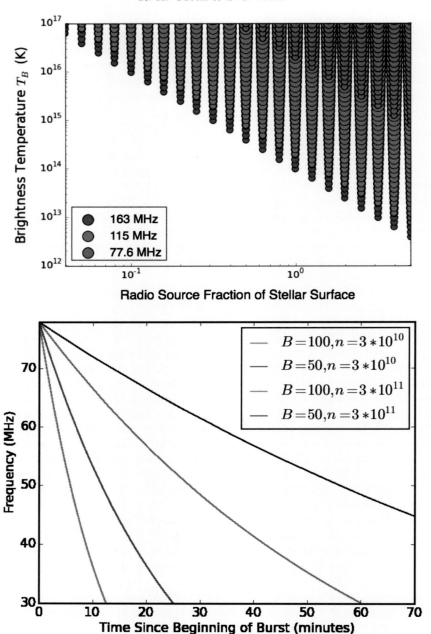

Figure 1. *(top panel)* Figure 5 from Crosley *et al.* (2016), showing region of parameter space constrained by the lack of detection of type II stellar bursts at low frequencies in 15 hours of observation. Typical brightness temperatures of solar type II bursts are ∼10^{14}K. The current sensitivities in the stellar case permit ruling out very large and high brightness temperature events. *(bottom panel)* Figure 6 from Crosley *et al.* (2016), showing expected path in the frequency-time space for a type II burst. The magnetic field strength and electron densities shown are valid from constraints on the plasma properties of active M dwarfs, and set the speed expected for a slightly super-Alfvenic (1.2 M_a) shock.

stellar flares show commonalities, which suggests a similar origin Solar eruptive events show scalings between solar flares and CMEs, which if applied to stellar flares & CMEs could be a powerful (& relatively easy) way to probe stellar eruptive events Emslie *et al.* (2012) studied 38 solar eruptive events and determined the energetics of various components. They found that the bolometric radiated flare energy showed a good correlation with CME energy (kinetic + potential). Additionally, empirical solar CME mass - flare energy scalings have been established, from Aarnio *et al.* (2011, 2012) and Drake *et al.* (2013) Both papers find a relation of the form $M_{\text{CME}} \propto E_{\text{GOES}}^{\beta}$, with $\beta \sim 0.6$. Drake *et al.* (2013) additionally determined an empirical relationship between CME kinetic energy and flare radiated energy in the GOES 1-8 Å bandpass for solar flares. These relationships can in principle be extended to the stellar flare energy regime, and predict quite massive CMEs, but potentially requiring about an order of magnitude more energy in the coronal energy budget.

Osten & Wolk (2015) used the approximate equipartition between CME kinetic energy and flare bolometric radiated energy in the Emslie *et al.* (2012) study to explore the implications for transient mass loss. Applying this technique to stellar flare frequency distributions requires having a way to correct the flare radiative losses in a particular band to determine the total bolometric radiated energy losses, which Osten & Wolk (2015) did. The relatively simple equation is then

$$1/2M_{\text{CME}}v^2 = E_{\text{rad}}/(\epsilon f_{rad}) \tag{4.2}$$

where M_{CME} is the CME mass, v is the CME velocity, E_{rad} is the radiated flare energy in a particular bandpass, ϵ is the equipartition factor, and f_{rad} is the fraction of the bolometric flare radiated energy that appears in the bandpass under consideration. Then the observed flare frequency distributions can be related to an inferred rate of mass loss associated with the flares, and this method can be applied to any wavelength range where the fraction of total flare energy in that bandpass can be estimated. Osten & Wolk (2015) applied this to published flare frequency distributions of several different types of magnetically active stars. For the case of two nearby well-studied M dwarf flare stars, AD Leo and EV Lac, which had flare frequency distributions in the optical and at coronal wavelengths, consideration of the partition fraction f_{rad} for optical and coronal wavelengths, respectively, leads to a similar value of estimated stellar mass loss. The overall rates of stellar mass loss are quite high, 10^{-11} $\dot{M}_{\odot}$ yr^{-1}, for these M dwarfs. The same issue as found by Drake *et al.* (2013) is present here, and the resolution awaits definitive constraint on the detection of stellar coronal mass ejections.

5. Do Stars Produce Coronal Mass Ejections? Can we observe them?

The solar perspective is useful to establish scaling relations which can potentially be extended to the much enhanced energetics of active stars. However, it can also be the source of some cautionary tales for such interpretations. the solar active region 12191 in the fall of 2014 produced many X-class flares but few CMEs. Sun *et al.* (2015) suggested that large overlying fields above the active region may have prevented breakout or eruption. Active stars are known to have large magnetic field strengths on their surfaces, and it would not be out of the realm of possibility that this might lead to arcades of strong magnetic fields which would prevent material from lower in the stellar atmosphere from escaping. Proving a negative is difficult, however, but observationally based methods to constrain the rate of stellar coronal mass ejections is a needed next step.

6. Conclusions

There are many ways to try to observe stellar CMEs; all have some kind of bias associated with them. The solar perspective is needed especially to interpret potential observational diagnostics of stellar CMEs. The stellar perspective is also needed to reveal what is feasible as a method without spatial resolution and exquisite sensitivity. In addition, stellar studies continue to reveal the differences between our well-studied Sun and the panoply of stars, which needs to be folded in to this study as well.

References

Aarnio, A. N., Matt, S. P., & Stassun, K. G. 2012, *Astrophysical Journal*, 760, 9

Aarnio, A. N., Stassun, K. G., Hughes, W. J., & McGregor, S. L. 2011, *Solar Physics*, 268, 195

Aschwanden, M. J., Stern, R. A., & Güdel, M. 2008, *Astrophysical Journal*, 672, 659

Bastian, T. S., Pick, M., Kerdraon, A., Maia, D., & Vourlidas, A. 2001, *Astrophysical Journal Letters*, 558, L65

Crosley, M. K., Osten, R. A., Broderick, J. W., *et al.* 2016, *Astrophysical Journal*, 830, 24

Drake, J. J., Cohen, O., Yashiro, S., & Gopalswamy, N. 2013, *Astrophysical Journal*, 764, 170

Emslie, A. G., Dennis, B. R., Shih, A. Y., *et al.* 2012, *Astrophysical Journal*, 759, 71

Franciosini, E., Pallavicini, R., & Tagliaferri, G. 2001, *Astronomy & Astrophysics*, 375, 196

Giampapa, M. S., Africano, J. L., Klimke, A., *et al.* 1982, *Astrophysical Journal Letters*, 252, L39

Gopalswamy, N. 2006a, Washington DC American Geophysical Union Geophysical Monograph Series, 165, 207

—. 2006b, Washington DC American Geophysical Union Geophysical Monograph Series, 165, 207

Gopalswamy, N., Yashiro, S., Akiyama, S., *et al.* 2008a, Annales Geophysicae, 26, 3033

—. 2008b, Annales Geophysicae, 26, 3033

Güdel, M., Benz, A. O., Schmitt, J. H. M. M., & Skinner, S. L. 1996, *Astrophysical Journal*, 471, 1002

Harra, L. K., Schrijver, C. J., Janvier, M., *et al.* 2016, *Solar Physics*, 291, 1761

Hawley, S. L., Fisher, G. H., Simon, T., *et al.* 1995, *Astrophysical Journal*, 453, 464

Hertzsprung, E. 1924, Bulletin of the Astronomical Institutes of the Netherlands, 2, 87

Houdebine, E. R., Foing, B. H., & Rodono, M. 1990, *Astronomy & Astrophysics*, 238, 249

Howard, T., ed. 2011, Astrophysics and Space Science Library, Vol. 376, Coronal Mass Ejections%

Khodachenko, M. L., Ribas, I., Lammer, H., *et al.* 2007, Astrobiology, 7, 167

Konovalenko, A. A., Koliadin, V. L., Boiko, A. I., *et al.* 2012, in European Planetary Science Congress 2012, EPSC2012–902

Kowalski, A. F. 2012, PhD thesis, University of Washington

Kundu, M. R., & Shevgaonkar, R. K. 1988, *Astrophysical Journal*, 334, 1001

Leitzinger, M., Odert, P., Greimel, R., *et al.* 2014, *Monthly Notices of the Royal Astronomical Society*, 443, 898

Manoharan, P. K. 2010, *Solar Physics*, 265, 137

McCleary, J. E., & Wolk, S. J. 2011, Astronomical Journal, 141, 201

Melis, C., Zuckerman, B., Rhee, J. H., *et al.* 2012, *Nature*, 487, 74

Osten, R., Livio, M., Lubow, S., *et al.* 2013, *Astrophysical Journal Letters*, 765, L44

Osten, R. A. 2016, in Heliophysics Active Stars, their Astrospheres, and Impacts on Planetary Environments, ed. C. J. Schrijver, F. Bagenal, & J. J. Sojka, 406

Osten, R. A., & Bastian, T. S. 2006, *Astrophysical Journal*, 637, 1016

Osten, R. A., Drake, S., Tueller, J., *et al.* 2007, *Astrophysical Journal*, 654, 1052

Osten, R. A., & Wolk, S. J. 2015, *Astrophysical Journal*, 809, 79

Segura, A., Walkowicz, L. M., Meadows, V., Kasting, J., & Hawley, S. 2010, Astrobiology, 10, 751

Shields, A. L., Ballard, S., & Johnson, J. A. 2016, ArXiv e-prints, arXiv:1610.05765

Sun, X., Bobra, M. G., Hoeksema, J. T., *et al.* 2015, *Astrophysical Journal Letters*, 804, L28

Thompson, B. J., Cliver, E. W., Nitta, N., Delannée, C., & Delaboudinière, J.-P. 2000, Geophysical Research Letter, 27, 1431

van den Oord, G. H. J., & de Bruyn, A. G. 1994, *Astronomy & Astrophysics*, 286, 181

Venot, O., Rocchetto, M., Carl, S., Hashim, A., & Decin, L. 2016, ArXiv e-prints, arXiv:1607.08147

Yashiro, S., Akiyama, S., Gopalswamy, N., & Howard, R. A. 2006, *Astrophysical Journal Letters*, 650, L143

Yashiro, S., Gopalswamy, N., Michalek, G., *et al.* 2004, Journal of Geophysical Research (Space Physics), 109, A07105

Discussion

Living Around Active Stars
Proceedings IAU Symposium No. 328, 2016
D. Nandy, A. Valio & P. Petit, eds.

© International Astronomical Union 2017
doi:10.1017/S1743921317003891

The long-term evolution of stellar activity

Scott G. Gregory

School of Physics & Astronomy, University of St Andrews, St Andrews, KY16 9SS, U. K.
email: sg64@andrews.ac.uk

Abstract. I review the evolution of low-mass stars with outer convective zones over timescales of millions-to-billions of years, from the pre-main sequence to solar-age, $\sim$4.6 Gyr (Bahcall *et al.* 1995; Amelin *et al.* 2010), and beyond. I discuss the evolution of high-energy coronal and chromospheric emission, the links with stellar rotation and magnetism, and the emergence of the rotation-activity relation for stars within young clusters.

Keywords. stars: activity, stars: chromospheres, stars: coronae, stars: evolution, stars: interiors, stars: late-type, stars: magnetic fields, stars: pre-main sequence, stars: rotation, X-rays: stars.

1. Introduction

The study of stellar activity encompasses observations at multiple wavelengths over a variety of timescales. Analysis of stars of different age yields information about the long-term evolution of stellar activity, their magnetism, and about how the various layers of their atmospheres and their surface features change over time. In a short review I can only scratch the surface of this vast field of several, complementary, research areas. I focus of a select few topics which are currently at the forefront of astrophysical research: the long-term evolution of stellar magnetism, and of coronal / chromospheric activity.

In §2, I discuss the timescales for the main phases of stellar evolution, the pre-main sequence (PMS), the main sequence (MS), and the post-main sequence (post-MS), using a solar mass star as an example. Stellar activity and rotation are intimately linked, as I discuss in §3. In §4, I discuss the evolution of the X-ray emission from PMS stars and the age at which we start to observe a rotation-activity relation. §5 deals with the long-term evolution, over Gyr, of the coronal and chromospheric activity, and the use of the latter as a proxy for stellar age. Stellar activity diagnostics are related to the dynamo magnetic field generation process and the evolution of stellar magnetism. In §6, I discuss the long-term evolution of stellar magnetic fields, and summarise in §7.

2. The evolution of low-mass stars

The protostellar phase of embedded star formation is complete in $0.1 - 0.4$ Myr (e.g. Dunham & Vorobyov 2012; Offner & McKee 2011). Low-mass stars ($\sim 0.1 - 2$ M$_\odot$) then enter the PMS phase as they contract under gravity, initially interacting magnetospherically with circumstellar disks. A median disk lifetime is $\sim 2 - 3$ Myr (e.g. Haisch *et al.* 2001), which is a small fraction of the total PMS contraction time.

2.1. *Timescale for the PMS, MS, and post-MS phase*

Pre-main sequence stars are luminous because of the release of the gravitational potential energy as they contract. Many PMS stars are more luminous than the Sun due to their large surface area ($L_* \propto R_*^2$). We can estimate the total PMS lifetime of a star by taking the ratio of available energy to the rate at which that energy is being used: in other words, the ratio of the gravitational energy to the stellar luminosity,

$$\tau_{\text{PMS}} = \frac{GM_*^2}{R_* L_*} \approx 3 \times 10^7 \left(\frac{M_*}{M_\odot}\right)^2 \left(\frac{R_*}{R_\odot}\right)^{-1} \left(\frac{L_*}{L_\odot}\right)^{-1} \quad \text{years}, \quad (2.1)$$

where the symbols have obvious meaning. τ_{PMS} is the Kelvin-Helmholtz timescale. A solar mass stars takes $\sim 30\,\text{Myr}$ to complete its PMS contraction and reach the ZAMS.

Once a star has reached the MS we can estimate how long it will take to exhaust its supply of hydrogen fuel - this is the nuclear timescale. Similar to the timescale for PMS contraction, the MS lifetime is obtained by taking the ratio of energy available to the luminosity, although in this case reduced by the fraction of the stellar interior over which the nuclear reactions occur. In the fusion of H into He about 0.7% of the mass is released as energy (heat), with reactions occurring within the inner 10% only for a solar mass star (Maeder 2009). Therefore, the amount of energy available from hydrogen fusion is roughly $7 \times 10^{-4} M_* c^2$, giving an approximate main sequence lifetime of,

$$\tau_{\text{MS}} \approx 7 \times 10^{-4} \frac{M_* c^2}{L_*} \approx 10^{10} \left(\frac{M_*}{M_\odot}\right) \left(\frac{L_*}{L_\odot}\right)^{-1} \quad \text{years}. \quad (2.2)$$

A solar mass star spends approximately $10\,\text{Gyr}$ on the main sequence and $\tau_{\text{MS}} \gg \tau_{\text{PMS}}$.

Post-main sequence evolution is highly dependent on the stellar mass and metallicity. A solar mass star evolves through the subgiant phase as it turns off the MS and onto the red giant branch in the Hertzsprung-Russell (H-R) diagram. H burning continues in a shell surrounding the extinct He core as the star expands. The He flash occurs as the star reaches the tip of the red giant branch and He fusion begins. A solar mass star then moves onto the horizontal branch (or to a region of the H-R diagram known as the red clump if of solar metallicity). Next is the asymptotic giant branch phase as He continues burning in a shell around the core, and eventually the planetary nebula phase which leaves behind a white dwarf. The total timescale for the post-MS evolution of a $1\,M_\odot$ and solar metallicity star is roughly $\tau_{\text{postMS}} \approx 3.3\,\text{Gyr}$ (Pols *et al.* 1998) with $\tau_{\text{postMS}} < \tau_{\text{MS}}$ and $\tau_{\text{postMS}} \gg \tau_{\text{PMS}}$.

2.2. *The evolution of low-mass stars in human terms*

As astronomers we are used to working with stellar ages of order millions, and billions, of years. However, it is close to impossible for us to fully comprehend the immensity of such numbers, unless we rescale them to something commensurate with our life experiences.

In the United Kingdom the life expectancy of a newborn girl (boy) is 82.8 (79.1) years, assuming mortality rates remain the same as they were between 2013 and 2015 throughout their lives.† Let's assume that a person will live to be 80 years of age. From formation to ending up as a white dwarf takes $\sim 13.33\,\text{Gyr}$ for a solar mass star, see §2. Of this total lifetime $\sim 30\,\text{Myr}$ ($\sim 0.2\%$) is the duration of the PMS phase, $\sim 10\,\text{Gyr}$ ($\sim 75\%$) the duration of the MS phase, and $\sim 3.3\,\text{Gyr}$ ($\sim 24.8\%$) the duration of the post-MS evolution. This means, that for an 80 year life expectancy, the PMS phase is complete in only ~ 9 weeks. At 9 weeks of age babies cannot bring their hands to their mouths (Gerber *et al.* 2010). The MS phase, in human terms, is complete by an age of ~ 60 years. This means that the remaining ~ 20 years of the life of our average person, mainly the retirement years, correspond to the post-MS evolution of a solar mass star.

When framed in human terms, the brevity of the PMS phase is readily apparent. However, the PMS evolution sets the initial conditions for the subsequent evolution of the star. As one example, Figure 1, from Gallet & Bouvier (2015), illustrates the

† Office for National Statistics licensed under the Open Government Licence v.3.0.

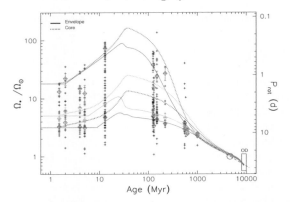

Figure 1. The rotation rate of solar mass stars normalised to the solar rotation rate, $\Omega_*/\Omega_\odot$, versus age, from Gallet & Bouvier (2015). Points are rotation rates of stars in clusters of different age. The red/green/blue points represent the lower quartile/median/90th percentile of the rotation rate distribution in each cluster. The solid/dotted lines are models showing the rotation rate variation of the convective envelope/radiative core. Ω_* is assumed constant during the disk-locked phase, with stars then spinning up as they complete their PMS contraction. Stars then spin down, and rotation rates converge, through angular momentum lose in magnetised winds. Credit: Gallet & Bouvier, A&A, 577, A98, 2015, reproduced with permission ©ESO.

theorised behaviour of the stellar rotation rate Ω_* as a function of stellar age. Following a period of disk-locking, where it is assumed that PMS stars are losing enough angular momentum that they accrete from their disks and contract while maintaining a constant rotation period, they are free to spin-up as they conserve angular momentum for their remaining PMS evolution (e.g. Davies *et al.* 2014). Within young clusters there remains a considerable range of rotation rates until an age of $\sim 0.5\,\mathrm{Gyr}$. Stars spin down via angular momentum lose from magnetised winds, and as the spin down rate is greatest for the faster rotating stars, there is convergence in the spread of rotation rates. Beyond approximately $0.5\,\mathrm{Gyr}$ stars are spinning down with age as $\Omega_* \propto t^{-1/2}$, the Skumanich spin-down law (Skumanich 1972). This convergence in rotation rates corresponds to a reduction in the scatter of activity levels with age, as I discuss in the following sections.

3. Stellar activity and rotation

It has long been known that rotation and activity are linked. Skumanich (1972) demonstrated that the decay in chromospheric CaII H & K emission with increasing age was coupled with a decrease in stellar rotation rate, $\Omega_* \propto t^{-1/2}$. Pallavicini *et al.* (1981) reported that the coronal activity of main sequence stars scales with rotation, with X-ray luminosities $L_X \propto (v_* \sin i)^2$. If we take from dynamo theory that dynamos are linear with the magnetic field $B \propto \Omega_*$ [roughly what is found from magnetic mapping studies, see Vidotto *et al.* (2014) and §6], then this implies that $L_X \propto B^2$; that is, that the X-ray luminosity scales with the magnetic energy density. We may have expected this given that X-ray emission is the ultimate consequence of the magnetic activity and reconnection.

X-ray emission from MS stars does not increase ad infinitum with increasing rotation rate. For rapidly rotating stars the fractional X-ray luminosity, L_X/L_*, saturates at about 10^{-3}. Figure 2, from Wright *et al.* (2016), shows the X-ray rotation-activity relation for MS stars, with the unsaturated (where L_X/L_* decreases with decreasing rotation rate) and saturated (where L_X/L_* is roughly constant) regimes highlighted.†

---

† A third regime, super-saturation, exists where L_X/L_* decays for the most rapid rotators, likely as a result of coronal stripping (Jardine & Unruh 1999; James *et al.* 2000; Jardine 2004).

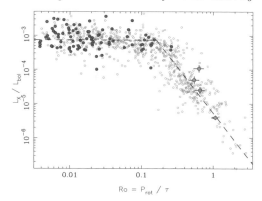

Figure 2. The rotation-activity relation for MS stars - the fractional X-ray luminosity versus Rossby number, from Wright & Drake (2016). Grey points are stars with partially convective interiors. Red points are fully convective stars, four of which (with error bars on their positions) lie within the unsaturated regime. Reprinted by permission from Macmillan Publishers Ltd: Nature (Wright & Drake 2016, Nature, 535, 526-528), copyright 2016.

Rotation-activity relations, first quantified by Noyes *et al.* (1984) using chromospheric CaII H & K emission, are also found for other activity diagnostics, such as Hα emission (e.g. Soderblom *et al.* 1993; Douglas *et al.* 2014; Newton *et al.* 2016), flare activity (Audard *et al.* 2000; Davenport 2016), the magnetic flux Bf, where f is the magnetic filling factor (Reiners *et al.* 2009), and surface averaged magnetic field strengths (Vidotto *et al.* 2014).

Rotation-activity relations are usually plotted with the Rossby number, Ro, as abscissae values, where $Ro = P_{\rm rot}/\tau_{\rm c}$ is the ratio of rotation period to the convective turnover time of gas cells within the stellar interior. Plotting against Rossby number does tighten rotation-activity relations compared to plotting against rotation period alone. However, this is often by design, with (model dependent) convective turnover times calibrated to minimise scatter for stars that fall in the unsaturated regime (e.g. Pizzolato *et al.* 2003). It is not certain if Ro is the best quantity to use for rotation-activity studies, or if the rotation period itself is sufficient (Reiners *et al.* 2014).

The cause of the saturation of activity diagnostics is still debated. Multiple ideas have been proffered. It may be due to the stellar surface becoming completely filled with active regions (Vilhu 1984). This is unlikely, however, as rotationally modulated X-ray emission has been detected from saturated regime stars [from young PMS stars at least, which all lie in the saturated regime e.g. Flaccomio *et al.* (2005) and §4], suggesting that there are X-ray dark regions, analogous to solar coronal holes, the location of stellar wind-bearing open field lines. Other causes may be the saturation of the underlying dynamo itself, or centrifugal stripping of the corona (Jardine & Unruh 1999).

A star's position in the rotation-activity plane depends on multiple parameters. Age, rotation rate, mass, spectral type, stellar internal structure, and magnetic field topology are all interlinked factors. In particular, the rotation-activity relation plotted in Figure 2 contains a mixture of fully convective stars (red points) and partially convective stars with radiative zones and outer convective envelopes (grey points). Saturation occurs at $L_{\rm X}/L_* \approx 10^{-3}$, regardless of spectral type (e.g. Vilhu 1984), and for $Ro \lesssim 0.1$. Therefore, the rotation period for which saturation occurs is longer for later spectral type stars as they have larger convective turnover times (e.g. Landin *et al.* 2010; Wright *et al.* 2011).

MS stars of spectral type $\sim$M3.5 and later have fully convective interiors and are expected and found to show saturated levels of activity. Slowly rotating earlier spectral

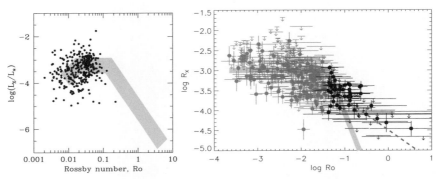

Figure 3. X-ray rotation-activity plots for PMS stars. (left) Stars in the young Orion Nebula Cluster (ONC) and IC 348 star forming regions (black points). The MS rotation-activity relation, with (left-to-right) the super-saturated, saturated, and unsaturated regimes is represented by the grey box. PMS stars show saturated levels of X-ray emission with far more scatter in L_X/L_* compared to saturated MS stars. Figure from Gregory *et al.* (in prep.) using X-ray data from Stelzer *et al.* (2012) and Broos *et al.* (2013). (right) PMS stars (points) in the ~ 13 Myr old cluster h Per ($R_X \equiv L_X/L_*$) from Argiroffi *et al.* (2016). The solid blue line represents the saturated and unsaturated regime for MS stars, from Wright *et al.* (2011). The dashed red line is the fit to h Per members expected to be in the unsaturated regime (black points). Credit: Argiroffi *et al.*, A&A, 589, A113, 2016, reproduced with permission ©ESO.

type stars have partially convective interiors and follow the unsaturated regime with their activity level depending on their rotation rate. However, Wright *et al.* (2016) have recently demonstrated that fully convective M-dwarfs, at least those which have sufficiently spun down, also follow the unsaturated regime, see Figure 2. This behaviour was hinted at previously in data published by Kiraga & Stepien (2007), and Jeffries *et al.* (2011) who split the rotation activity-relation by stellar mass. In their lowest mass sample, stars of mass $< 0.35\,\mathrm{M}_\odot$ (which is the mass below which MS stars have fully convective interiors; Chabrier & Baraffe 1997), it can be seen that 4 stars display unsaturated rotation-activity behaviour. However, the stars considered by Jeffries *et al.* (2011) lie close to the fully convective boundary, and 3 may be partially convective or saturated. The fully convective stars studied by Wright *et al.* (2016) that fall in the unsaturated regime are of sufficiently late spectral type that they are almost certainly fully convective. Therefore, the existence of a tachocline between a radiative core and outer convective envelope is not required in order to generate stellar magnetic fields that link the activity level with rotation rate.

4. The emergence of the rotation-activity relation on the PMS

PMS stars in the youngest star forming regions do not follow the rotation-activity relation. PMS stars show saturated levels of X-ray emission regardless of their rotation rates, but with orders of magnitude more scatter in L_X/L_* compared to saturated regime MS stars (Preibisch *et al.* 2005; Alexander & Preibisch 2012; Figure 3 left panel). The origin of this scatter is intrinsic differences between PMS stars as it is greater than what can be attributed to observational uncertainties in the parameters (Preibisch *et al.* 2005).

The scatter in the rotation-activity plane for stars in young PMS clusters must eventually evolve to form the MS rotation-activity relation. Recently, Argiroffi *et al.* (2016) have reported that PMS stars (at least those of mass $> 1\,\mathrm{M}_\odot$) in the ~ 13 Myr cluster h Per have begun to display activity regimes like MS stars. This can be seen in Figure 3 (right panel). For h Per the gradient of the fit to the unsaturated regime PMS stars is noticeably less when compared to the equivalent MS stars. By ~ 13 Myr PMS coronal activity has evolved from the scatter evident for younger clusters (see Figure 3, left

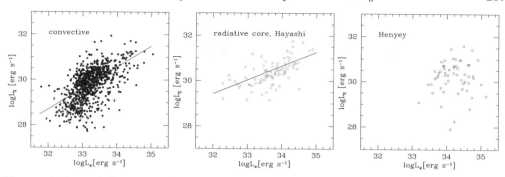

Figure 4. L_X and L_* are correlated for Hayashi track PMS stars (left / middle) but there is no correlation for Henyey track PMS stars (right), from Gregory *et al.* (2016). With $L_X \propto L_*^a$, $a = 0.93 \pm 0.04$ for fully convective PMS stars with the exponent reducing to $a = 0.61 \pm 0.08$ for partially convective Hayashi track PMS stars (those with small radiative cores).

panel) to beginning to show a link with rotation, at least for stars more massive than a solar mass. This may be due to the establishment of "solar-type" $\alpha\Omega$-dynamos, with the development of a shear layer, the tachocline, between an inner radiative core and an outer convective envelope within stellar interiors.

The age of h Per, $\sim 13\,\mathrm{Myr}$, is (in rough terms) the age at which a solar-mass star makes the transition to the Henyey track in the Hertzsprung-Russell diagram (e.g. Siess *et al.* 2000). By the time it does so, the stellar interior is already mostly radiative with the core containing the majority of the stellar mass (Gregory *et al.* 2016). The stellar internal structure transition, from fully to partially convective, and particularly when PMS stars evolve from the Hayashi track to the Henyey track, corresponds to an increase in the complexity of their large-scale magnetic field topology (Gregory *et al.* 2012, 2016). Hayashi track PMS stars (at least those of mass $\gtrsim 0.5\,\mathrm{M_\odot}$) have simple axisymmetric large-scale magnetic fields which evolve into complex, multipolar, and dominantly non-axisymmetric magnetic fields once they develop large radiative cores and evolve onto Henyey tracks. This evolution from the Hayashi to Henyey track also corresponds to a decay of the coronal X-ray emission (Rebull *et al.* 2006; Gregory *et al.* 2016).

Pre-main sequence stars that have evolved onto Henyey tracks have lower L_X/L_* compared to those on Hayashi tracks (Rebull *et al.* 2006; Gregory *et al.* 2016). Henyey track stars are also, on average, more luminous than Hayashi track stars. Therefore, is the reduction in L_X/L_* driven mainly by the increase in L_* or a decay of L_X when PMS stars evolve onto Henyey tracks? It can be seen from Figure 4 that the coronal X-ray emission decays with radiative core development, and especially once PMS stars have evolved onto Henyey tracks. Considering a sample of almost 1000 PMS stars in 5 young star forming regions (Gregory *et al.* 2016), there is an almost linear relationship with $L_X \propto L_*^{0.93 \pm 0.04}$ for fully convective PMS stars. This is not surprising given that they all fall in the saturated regime of the rotation-activity plane where $L_X/L_* \approx$ const. (see Figure 3). For stars that have begun to develop radiative cores, but which are still on Hayashi tracks (where L_* is decreasing with increasing age) the exponent is less with $L_X \propto L_*^{0.61 \pm 0.08}$. In both cases, the probability of there not being a correlation is $< 5e\text{-}5$ from a generalised Kendall's τ test. For Henyey track PMS stars, those with substantial radiative cores, there is no correlation between L_X and L_*, with many having a X-ray luminosity well below what would be expected for their bolometric luminosity.

Coronal X-ray emission decays with substantial radiative core development on the PMS (Gregory *et al.* 2016). This is also apparent when considering the anti-correlation

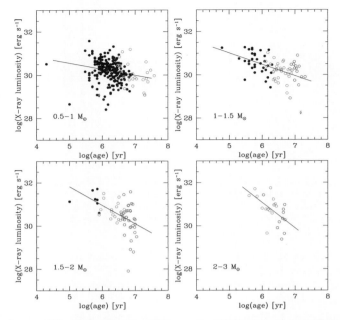

Figure 5. The decay of X-ray luminosity with age for PMS stars in specified mass bins (from Gregory *et al.* 2016). Black / blue / red points are fully convective / partially convective stars on Hayashi tracks / partially convective on Henyey tracks. L_X decays faster with age for stars in the higher mass bins, which have a greater proportion of partially convective stars.

between L_X and stellar age, t, for PMS stars (see Figure 5). Preibisch & Feigelson (2005) reported that $L_X \propto t^{-1/3}$ using a sample of mostly fully convective PMS stars from the ONC. With a larger sample, and a substantial fraction of partially convective PMS stars, it is clear that the coronal X-ray emission decays faster with age for higher mass PMS stars (Figure 5; Gregory *et al.* 2016).

The Henyey track PMS stars considered in Figures 4 & 5 are currently early K-type and G-type stars (with a few F-types also). These stars will evolve into MS A-type stars which lack outer convective zones (e.g. Siess *et al.* 2000). Some 85-90% of MS A-types are undetected in X-rays, with almost all of those that are known / suspected close binaries (Schröder & Schmitt 2007).† The decay of the coronal X-ray emission when PMS stars become substantially radiative is consistent with the lack of X-ray emission from MS A-type stars. Higher mass PMS stars that evolve onto Henyey tracks appear to lose their coronae as they develop large radiative cores, while lower mass PMS stars retain high levels of X-ray emission for millions of years of stellar evolution, as I discuss in the below.

5. The long-term evolution coronal and chromospheric acitivty

PMS stars are 10-to-10^4 times more X-ray luminous than the Sun is today, and stars maintain high levels of X-ray emission for several 100 Myr of evolution before their activity begins to decay at a faster rate (see Figure 6, left panel). For ages beyond ~ 0.5 Gyr the X-ray luminosity decays with age as $L_X \propto t^{-3/2}$ for both solar-type and lower mass stars (e.g. Güdel *et al.* 1997; Guinan *et al.* 2016). This corresponds to roughly the same age at which we observe convergence of the stellar rotation rates (see §3 and Figure 1).

† A small number of single late A-types do show very weak levels of X-ray emission (e.g. Robrade & Schmitt 2010; Günther *et al.* 2012).

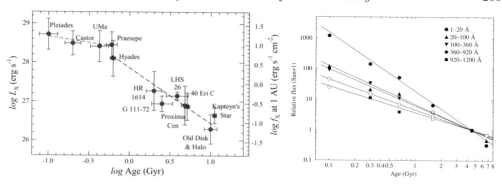

Figure 6. (left) The decay of X-ray luminosity versus stellar age for M0-M5 stars. Figure from Guinan *et al.* (2016). The right hand vertical axis shows the X-ray irradiance f_X at a distance of 1 au. The dashed lines are least squares fits: for age > 0.5 Gyr, that is $\log(\text{age[Gyr]}) > -0.3$, the fit is $\log L_X = 27.857 - 1.424 \log(\text{age[Gyr]})$. X-ray activity decreases sharply with increasing age and as stars spin down (see §3). ©AAS. Reproduced with permission. (right) The decay of the (solar normalised) UV flux versus age. Figure from Ribas *et al.* (2005). Harder (shorter wavelength) emission reduces faster with age. ©AAS. Reproduced with permission.

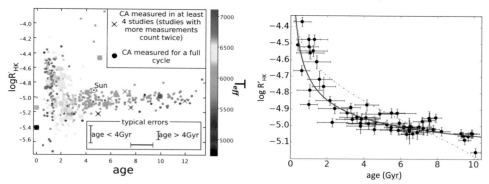

Figure 7. The chromospheric CaII activity index R'_{HK} versus stellar age. (left) Figure from Pace (2013) for stars with effective temperatures T_{eff} as indicated. Credit: Pace, A&A, 551, L8, 2013, reproduced with permission ©ESO. (right) Figure modified from Freitas *et al.* (submitted), see also Freitas *et al.*, these proceedings, with the fit to the points shown as the solid blue line. The dashed line is the relationship derived by Mamajek & Hillenbrand (2008).

The UV emission also decays over Gyr timescales, although more slowly with age than the higher energy X-ray emission. At FUV wavelengths, Guinan *et al.* (2016) report that the Lyα flux (a good proxy for FUV emission as a whole) decays with stellar age as roughly as $t^{-2/3}$ for early-to-mid spectral type M-dwarfs, more slowly than the $t^{-3/2}$ behaviour of the X-ray emission. As can be seen from Figure 6 (right panel), the softer (longer wavelength) the emission the slower the decay with stellar age (e.g. Ribas *et al.* 2005; Güdel 2007; Shkolnik & Barman 2014). Quantifying this further for solar analogues, Güdel (2007) provides a list of enhancement factors of the high energy emission at various wavebands moving backwards in time from the present day solar values to those expected for the young Sun (see his table 6). There is, however, considerable uncertainty in the activity level of the young Sun (Tu *et al.* 2015). Due to the large scatter in rotation rates at young cluster ages (see Figure 3) with the associated large scatter in activity levels, the rotation history of the Sun, and therefore of its activity, is unknown.

Chromospheric activity as measured by R'_{HK} also decays over Gyr timescales. The chromospheric CaII activity index R'_{HK}, see Noyes *et al.* (1984) for its introduction and definition, versus age is shown in Figure 7. R'_{HK} has been advocated as a proxy for stellar age (e.g. Mamajek & Hillenbrand 2008), or at least for young ages (Pace 2013; Figure 7 left panel) due to the flattening of the relation for older ages.

Recently, Freitas *et al.* (submitted) have re-examined the relationship between R'_{HK} and age, see Figure 7 right panel and Freitas *et al.*, these proceedings. Studying CaII H & K activity with HARPS spectra they conclude that Pace (2013) was right to argue that R'_{HK} is only a useful proxy for stellar age for younger clusters, with sensitivity to ages of $\lesssim 3\,\text{Gyr}$. The Freitas *et al.* (submitted) data suggests that the continuing decrease in R'_{HK} with age reported by older studies (e.g. Mamajek & Hillenbrand 2008 - their relation is also plotted in Figure 7 right panel for comparison) is incorrect, and that the chromospheric activity levels flatten off after $\sim 3\,\text{Gyr}$. The flattening of the chromospheric activity at ages approaching and beyond solar age likely reflects the convergence of rotation rates during stellar MS evolution (see §3 and Figure 3). However, the flattening of the chromospheric activity may also be related to a recently reported flattening of rotation rates beyond $\sim 2-3\,\text{Gyr}$, which has been interpreted as being due to a break-down of the Skumanich spin-down law (e.g. dos Santos *et al.* 2016).

6. The long-term evolution of stellar magnetism

The indicators of stellar activity are driven by the interior dynamo and via the external magnetic field that it generates. The past decade has seen major advances in the observational study of stellar magnetism, with magnetic field information now available for stars of all spectral types and evolutionary phases (e.g. Donati *et al.* 2006, 2008, 2011; Morin *et al.* 2008; Petit *et al.* 2010; Fares *et al.* 2013; Folsom *et al.* 2016).

On the PMS, stars of mass $\gtrsim 0.5\,\text{M}_\odot$ appear to be born with simple, axisymmetric, often dominantly octupolar large-scale magnetic fields, that transition to complex and non-axisymmetric with significant radiative core development (Gregory *et al.* 2012; Folsom *et al.* 2016). PMS stars of lower mass have been predicted to host a variety of magnetic field topologies based on the similarities of PMS magnetism and the field topologies found for MS M-dwarfs with similar internal structures (Gregory *et al.* 2012). The lowest mass MS M-dwarfs host a mixture of simple and complex magnetic fields and exist in a bistable dynamo regime (Morin *et al.* 2010). The lowest mass PMS stars may follow similar behaviour, but nIR spectropolarimetry (with, for example, the SPIRou instrument; Moutou *et al.* 2015) is required to properly survey their magnetic fields.

For MS stars, snapshot surveys have confirmed that stars with deeper convective zones (K-types) have stronger average longitudinal magnetic field components, $\langle|B_\ell|\rangle$, than those with smaller convective envelopes (F and G-types; Marsden *et al.* 2014). This, however, may be due to the selection of active K-type stars for the survey, or a flux cancellation effect if the F and G-types have more numerous and smaller magnetic spots than K-types. $\langle|B_\ell|\rangle$ increases with rotation rate, decreases with age, and is well correlated with the chromospheric activity (Marsden *et al.* 2014).

Zeeman-Doppler imaging studies, which allow magnetic maps to be constructed from a time series of circularly polarised spectra, have revealed how the magnetic topologies of stars are linked to the stellar parameters - see the reviews by Donati (2013) and Morin (2012). Using almost all of the stellar magnetic maps of low-mass stars available at the time, Vidotto *et al.* (2014) examined trends in the large-scale stellar magnetism with age. Multiple correlations between age, activity, and rotation parameters were reported including that shown in Figure 8. Over Gyr timescales the average unsigned magnetic field strength of FGKM-type stars, as measured from magnetic maps, decays with age

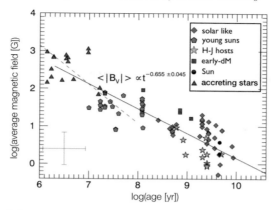

Figure 8. The decay of the mean magnetic field strength measured from magnetic maps derived from Zeeman-Doppler imaging for stars of various age / evolutionary stage. Combined with the Skumanich spin-down law, $\Omega_* \propto t^{-1/2}$, this is roughly consistent with the stellar dynamos being linear ($\langle B \rangle \propto \Omega_*$). Figure modified from Vidotto *et al.* (2014).

as $\langle |B| \rangle \propto t^{-0.66 \pm 0.05}$. This, combined with the Skumanich spin-down law, confirms that stellar dynamos are close to being linear.

7. Summary: key points

• There is a large distribution of rotation rates of stars in young clusters. Rotation rates converge for ages $\gtrsim 0.5$ Gyr (e.g. Gallet & Bouvier 2015). Following the PMS evolution, main sequence stars spin-down via angular momentum loss in magnetised winds with the faster rotators spinning down at a faster rate than the slower rotators.

• Over Gyr timescales, low-mass stars spin down as $\Omega_* \propto t^{-1/2}$, the Skumanich spin-down law (Skumanich 1972). However, there are recent suggestions this relation breaks down, with the rotation rate decay flattening for ages $\gtrsim 3$ Gyr (dos Santos *et al.* 2016).

• Many stellar activity diagnostics, e.g. X-ray emission, flare activity, Hα activity, chromospheric CaII H & K emission, and others, follow a rotation-activity relation, whereby activity increases with increasing rotation rate, equivalently with decreasing Rossby number, Ro. Activity saturates for the fastest rotators. For example, X-ray activity saturates at $L_X/L_* \approx 10^{-3}$ for Ro $\lesssim 0.1$ (e.g. Wright *et al.* 2011).

• X-ray activity saturates at Ro $= P_{\rm rot}/\tau_c \lesssim 0.1$ regardless of spectral type (e.g. Vilhu 1984). Later spectral types have longer turnover times, τ_c, than for earlier spectral types. Therefore, fully convective M-dwarfs ($\sim$M3.5 and later) display saturated emission. However, Wright *et al.* (2016), have recently shown that M-dwarfs that have sufficiently spun down can fall in the unsaturated regime. This means that a tachocline is not a requirement to generate magnetic fields that link the activity level with rotation rate.

• Stars in young PMS clusters (e.g. the ONC) do not follow the MS rotation-activity relation. All young PMS stars show saturated levels of X-ray activity but with orders of magnitude more scatter in L_X/L_* (Preibisch *et al.* 2005).

• The youngest region where stars (at least those of mass $1 < M_*/M_\odot < 1.4$) have begun to follow MS-like rotation-activity behaviour is the ~ 13 Myr old cluster h Per (Argiroffi *et al.* 2016). However, the gradient of the unsaturated regime in h Per is less than found for stars in MS clusters. The emergence of unsaturated behaviour may be driven by stars developing substantial radiative cores.

- PMS stars (at least those of mass $\gtrsim 0.5\,M_\odot$) are born with simple axisymmetric magnetic fields that evolve to dominantly non-axisymmetric and highly multipolar with substantial radiative core development (Gregory *et al.* 2012). Lower mass PMS stars may exist in a bistable dynamo regime and host a variety of large-scale magnetic field topologies (Gregory *et al.* 2012). Additional observations are required for confirmation.

- PMS stars that have developed substantial radiative cores, those which have evolved onto Henyey tracks in the H-R diagram, have lower fractional X-ray luminosities (L_X/L_*) than those on Hayashi tracks (Rebull *et al.* 2006; Gregory *et al.* 2016). Many such PMS stars are the progenitors of MS A-type stars, and the loss of their coronal X-ray emission is consistent with the lack of X-ray detections of MS A-types (Gregory *et al.* 2016).

- Over Gyr timescales, surface average magnetic field strengths reduce with age as $\langle |B| \rangle \propto t^{-2/3}$ (Vidotto *et al.* 2014). Snapshot surveys of MS stars have shown that the average longitudinal magnetic field component, $\langle |B_\ell| \rangle$, is well correlated with chromospheric activity (Marsden *et al.* 2014).

- Over Gyr timescales, harder emission decays faster with age than softer emission (Ribas *et al.* 2005; Güdel 2007). For example, for ages $\gtrsim 0.5\,$Gyr the X-ray emission decays faster with age, $\propto t^{-3/2}$, than the FUV emission, $\propto t^{-2/3}$ (Guinan *et al.* 2016).

- The chromospheric CaII activity index $R'_{\rm HK}$ decays rapidly with, and is a useful proxy for, stellar age for $t \lesssim 3\,$Gyr (Pace 2013; Freitas *et al.* submitted). The relationship between R'_{HK} and t is almost flat for older ages.

Acknowledgements

I acknowledge support from the Science & Technology Facilities Council (STFC) via an Ernest Rutherford Fellowship [ST/J003255/1] and the meeting organisers for the invitation to speak. I thank Nick Wright & Stephen Marsden for comments on their work and Fabrício Freitas for kindly providing and modifying Figure 7 (right panel).

References

Alexander, F. & Preibisch, T. 2012, *A&A*, 539, A64
Amelin, Y., Kaltenbach, A., Iizuka, T., *et al.* 2010, *Earth and Planetary Science Letters*, 300, 343
Argiroffi, C., Caramazza, M., Micela, G., *et al.* 2016, *A&A*, 589, A113
Audard, M., Güdel, M., Drake, J. J., & Kashyap, V. L. 2000, *ApJ*, 541, 396
Bahcall, J. N., Pinsonneault, M. H., & Wasserburg, G. J. 1995, *Reviews of Modern Physics*, 67, 781
Broos, P. S., Getman, K. V., Povich, M. S., *et al.* 2013, *ApJS*, 209, 32
Chabrier, G. & Baraffe, I. 1997, *A&A*, 327, 1039
Davenport, J. R. A.. 2016, *ApJ*, 829, 23
Davies, C. L., Gregory, S. G., & Greaves, J. S. 2014, *MNRAS*, 444, 1157
Donati, J.-F. 2013, EAS Publications Series, 62, 289
Donati, J.-F., Gregory, S. G., Alencar, S. H. P., *et al.* 2011, *MNRAS*, 417, 472
Donati, J.-F., Morin, J., Petit, P., *et al.* 2008, *MNRAS*, 390, 545
Donati, J.-F., Howarth, I. D., & Jardine, M. M., *et al.* 2006, *MNRAS*, 370, 629
Douglas, S. T., Agüeros, M. A., Covey, K. R., *et al.* 2014, *ApJ*, 795, 161
dos Santos, L. A., Meléndez, J., do Nascimento, J.-D., *et al.* 2016, *A&A*, 592, A156
Dunham, M. M. & Vorobyov, E. I. 2012, *ApJ*, 747, 52
Fares, R., Moutou, C., Donati, J.-F., *et al.* 2013, *MNRAS*, 435, 1451
Folsom, C. P., Petit, P., Bouvier, J., *et al.* 2016, *MNRAS*, 457, 580
Flaccomio, E., Micela, G., Sciortino, S., *et al.* 2005, *ApJS*, 160, 450
Freitas, F. C., Meléndez, J., Bedell, M., *et al.* 2017, *A&A*, submitted

Gallet, F. & Bouvier, J. 2015, *A&A*, 577, A98

Gerber, R. J., Wilks T. & Erdie-Lalena, C. 2010, *Pediatrics in Review*, 31, 267

Gregory, S. G., Adams, F. C., & Davies, C. L. 2016, *MNRAS*, 457, 3836

Gregory, S. G., Donati, J.-F., Morin, J., *et al.* 2012, *ApJ*, 755, 97

Güdel, M. 2007, *Living Reviews in Solar Physics*, 4, 3

Güdel, M., Guinan, E. F., & Skinner, S. L. 1997, *ApJ*, 483, 947

Guinan, E. F., Engle, S. G., & Durbin, A. 2016, *ApJ*, 821, 81

Günther, H. M., Wolk, S. J., Drake, J. J., *et al.* 2012, *ApJ*, 750, 78

Haisch, K. E., Jr., Lada, E. A., & Lada, C. J. 2001, *ApJL*, 553, L153

James, D. J., Jardine, M. M., Jeffries, R. D., *et al.* 2000, *MNRAS*, 318, 1217

Jardine, M. & Unruh, Y. C. 1999, *A&A*, 346, 883

Jardine, M. 2004, *A&A*, 414, L5

Jeffries, R. D., Jackson, R. J., Briggs, K. R., Evans, P. A., & Pye, J. P. 2011, *MNRAS*, 411, 2099

Kiraga, M. & Stepien, K. 2007, *Acta Astronomica*, 57, 149

Landin, N. R., Mendes, L. T. S.., & Vaz, L. P. R.. 2010, *A&A*, 510, A46

Maeder, A. 2009, *Physics, Formation and Evolution of Rotating Stars*, Astronomy and Astrophysics Library, Springer Berlin Heidelberg

Mamajek, E. E. & Hillenbrand, L. A. 2008, *ApJ*, 687, 1264

Marsden, S. C., Petit, P., Jeffers, S. V., *et al.* 2014, *MNRAS*, 444, 3517

Morin, J. 2012, EAS Publications Series, 57, 165

Morin, J., Donati, J.-F., Petit, P., *et al.* 2010, *MNRAS*, 407, 2269

Morin, J., Donati, J.-F., Petit, P., *et al.* 2008, *MNRAS*, 390, 567

Moutou, C., Boisse, I., & Hébrard, G., *et al.* 2015, SF2A-2015: Proceedings of the Annual meeting of the French Society of Astronomy and Astrophysics, 205

Newton, E. R., Irwin, J., Charbonneau, D., *et al.* 2016, *ApJ*, in press [astro-ph/1611.03509]

Noyes, R. W., Hartmann, L. W., Baliunas, S. L., Duncan, D. K., & Vaughan, A. H. 1984, *ApJ*, 279, 763

Offner, S. S. R.. & McKee, C. F. 2011, *ApJ*, 736, 53

Pace, G. 2013, *A&A*, 551, L8

Pallavicini, R., Golub, L., Rosner, R., *et al.* 1981, *ApJ*, 248, 279

Petit, P., Lignières, F., Wade, G. A., *et al.* 2010, *A&A*, 523, A41

Pizzolato, N., Maggio, A., Micela, G., Sciortino, S., & Ventura, P. 2003, *A&A*, 397, 147

Pols, O. R., Schröder, K.-P., Hurley, J. R., Tout, C. A., & Eggleton, P. P. 1998, *MNRAS*, 298, 525

Preibisch, T., Kim, Y.-C., Favata, F., *et al.* 2005, *ApJS*, 160, 401

Preibisch, T. & Feigelson, E. D. 2005, *ApJS*, 160, 390

Rebull, L. M., Stauffer, J. R., Ramirez, S. V., *et al.* 2006, *AJ*, 131, 2934

Reiners, A., Schüssler, M., & Passegger, V. M. 2014, *ApJ*, 794, 144

Reiners, A., Basri, G., & Browning, M. 2009, *ApJ*, 692, 538

Ribas, I., Guinan, E. F., Güdel, M., & Audard, M. 2005, *ApJ*, 622, 680

Robrade, J. & Schmitt, J. H. M.. M. 2010, *A&A*, 516, A38

Schröder, C. & Schmitt, J. H. M.. M. 2007, *A&A*, 475, 677

Shkolnik, E. L. & Barman, T. S. 2014, *AJ*, 148, 64

Siess, L., Dufour, E., & Forestini, M. 2000, *A&A*, 358, 593

Skumanich, A. 1972, *ApJ*, 171, 565

Soderblom, D. R., Stauffer, J. R., Hudon, J. D., & Jones, B. F. 1993, *ApJS*, 85, 315

Stelzer, B., Preibisch, T., Alexander, F., *et al.* 2012, *A&A*, 537, A135

Tu, L., Johnstone, C. P., Güdel, M., & Lammer, H. 2015, *A&A*, 577, L3

Vidotto, A. A., Gregory, S. G., Jardine, M., *et al.* 2014, *MNRAS*, 441, 2361

Vilhu, O. 1984, *A&A*, 133, 117

Wright, N. J. & Drake, J. J. 2016, *Nature*, 535, 526

Wright, N. J., Drake, J. J., Mamajek, E. E., & Henry, G. W. 2011, *ApJ*, 743, 48

Living Around Active Stars
Proceedings IAU Symposium No. 328, 2016
D. Nandy, A. Valio & P. Petit, eds.

© International Astronomical Union 2017
doi:10.1017/S1743921317004422

Stellar Midlife Crises: Challenges and Advances in Simulating Convection and Differential Rotation in Sun-like Stars

Nicholas J. Nelson, Charles Payne and Cameron Michael Sorensen

Dept. of Physics, California State University, Chico
Chico, CA, USA 95929-0202
email: `njnelson@csuchico.edu`

Abstract. Low mass, main sequence stars like our Sun exhibit a wide variety of rotational and magnetic states. Observational and theoretical advances have led to a renewed emphasis on understanding the rotational and magnetic evolution of sun-like stars has become a pressing problem in stellar physics. We use global 3D convection and convective dynamo simulations in rotating spherical shells and with realistic stellar stratification to explore the behavior of "middle-aged" stars. We show that for stars with slightly less rotational influence than our Sun a transition occurs from solar-like (fast equator, slow poles) to anti-solar (slow equator, fast poles) differential rotation. We investigate this transition using two different treatments for the upper boundary of our simulations and we hypothesize that this transition from solar-like to anti-solar differential rotation may be responsible for observations of anomalously rapid rotation for stars older than our Sun.

Keywords. hydrodynamics, MHD, Sun: interior, stars: activity, stars: rotation

---

1. The Challenges of Stellar Rotation and Convection

One of the most pressing and exciting challenges of modern astrophysics is the search for habitable exoplanets. With recent discoveries of potentially habitable Earth-like planets around nearby stars including Proxima Centauri (Anglada-Escudé *et al.* 2016) and TRAPPIST-1 (Gillon *et al.* 2016), there is considerable interest to understand if these planets could support life. With advances in exoplanet detection on the horizon there with doubtless be many more such systems discovered in the near future. The potential for extensive observational follow-up of these systems, however, is largely limited for the coming decade to refinements in orbital characteristics, stellar activity, and possibly atmospheric absorptions spectra for transiting planets. There is still likely many years between the present and the ability to directly image most potential Earth-analogues. This begs for a general theoretical understanding of habitability that can be applied to all planets around any star.

A comprehensive "theory of habitability" is an enormous theoretical challenge. To date, the question of habitability has largely focused simply on the mean surface temperature of the planet. Taking the recent discovery of Proxima Centauri b as an example, we see that in this case a more detailed and systematic examination by the community have raised serious questions on issues such as orbital dynamics (Kane *et al.* 2017; Méndez & Rivera-Valentín 2017), planetary mass and composition (Bixel & Apai 2017), space weather (Garrafo *et al.* 2016; Airapetian *et al.* 2017), formation scenarios Coleman *et al.* (2017), magnetospheric dynamics (Luger *et al.* 2016), climate (Boutle *et al.* 2017), and convective dynamo action (Yadav *et al.* 2016). This volume attempts to at least begin to bring a coherent focus on questions of habitability across a wide range of fields. Here

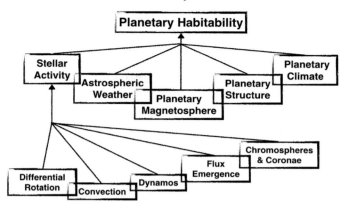

Figure 1. A schematic flow chart of how studies of stellar convection and differential rotation relate to questions of planetary habitability.

we will focus on the issue of stellar activity and even more specifically, on the modeling of convection and differential rotation in sun-like stars.

Figure 1 attempts to illustrate the broad range of inputs needed to theoretically predict the habitability of a particular planet around a particular star. While clearly incomplete, Fig. 1 highlights some of the topics from stellar interiors to climate that play key roles in habitability. Not shown but still of great importance are how these processes vary and evolve over geological and astrophysical timescales. In this paper focus on a specific contribution to the quest for a theory of habitability – the generation and emergence of stellar magnetism. We present simulations that explore recent advances in modeling convection and differential rotation in solar-like stars.

1.1. *Modeling Stellar Convection and Differential Rotation*

With the rapid increase in observational capabilities for both solar and stellar convection, rotation, differential rotation, and magnetic activity, there has been a corresponding advance in 3D numerical modeling which seeks to investigate the physical workings of these difficult-to-observe processes. As an example, after decades of searching, observers are possibly only now achieving detection of convective giant cells (e.g., Hanasoge *et al.* 2010, 2012; Hathaway *et al.* 2013; McIntosh *et al.* 2014; Greer *et al.* 2014, 2015). For other stars the observational guidance is even less clear and becomes progressively harder to interpret moving away from solar parameters such as age, mass, and rotation rate (see Reiners 2012).

Beginning with the work of Gilman (1983) and Glatzmaier (1985), 3D global convection and convective dynamo models have made key advances in improving understanding of the physical mechanisms which drive differential rotation, dynamo action, and magnetic activity in low-mass stars. Modeling stellar convection and dynamo action in 3D requires the use of advanced numerical techniques to utilize modern massively parallel computational resources. The simulations presented here use the ASH (Clune *et al.* 1999; Brun *et al.* 2004; Nelson *et al.* 2013b) and *Rayleigh* (Featherstone & Hindman 2016a) codes. Both ASH and *Rayleigh* solve the equations of anelastic magnetohydrodynamics in rotating spherical shells with solar-like stratification. ASH simulations have been used to model solar-like differential rotation and meridional circulation (Brun & Toomre 2002; Miesch *et al.* 2006; Featherstone & Miesch 2015), convection and differential rotation in a variety of low-mass stars (Brown *et al.* 2008; Brun *et al.* 2017), and dynamo action in stars ranging from solar-like (Brown *et al.* 2010, 2011; Nelson *et al.* 2011, 2013b,a; Nelson & Miesch 2014; Augustson *et al.* 2015) to core dynamos in high-mass stars (Featherstone *et al.* 2009; Augustson *et al.* 2016) to some very low-mass stars (Browning 2008). *Rayleigh* has recently been developed to optimize the use of modern massively parallel computational architectures, enabling some of the highest-resolution simulations of stellar convection to date (Featherstone & Hindman 2016b,a).

In this work we will not attempt to detail the variety of numerical methods used, but will refer the reader to published descriptions elsewhere in the literature in order to focus on presenting some recent results. Specifically, we will outline how the use of a stochastic plume boundary condition may permit an improved understanding of giant cell convection, how the degree of rotational influence yields sharp changes in differential rotation for solar-like stars, and how some of our least diffusive models achieve buoyant magnetic loops which rise through the convection zone and provide a window into the flux emergence process.

2. Modeling Convection with a Plume Boundary Condition

Even with continued exponential growth, computational resources provide strict bounds on the level of resolution a global-scale 3D convection simulation can achieve. This limit makes the inclusion of granular scales on the order of 1 Mm beyond the reach of global-scale models for the foreseeable future. Even super-granular scales of 30 Mm will require years of continued growth in computing resources. This means that for simulations with solar-like stratification the region above $0.98R_\odot$ is inaccessible simply due to a lack of resolution. Thus modelers generally limit themselves to motions below that depth and place an impenetrable boundary near that point. One interesting alternative is to adjust the stratification to provide a stable layer on top of the convection zone (Warnecke *et al.* 2016).

2.1. *A Plume Boundary Condition*

An alternative to an impenetrable boundary condition is a semi-open formulation which permits flows to enter and exit the domain subject to some constraints such as mass conservation, a constant pressure surface, or a specification of some of the flows. These formulations are inherently challenging as they attempt to admit dynamics into the simulation that come from or end up beyond the simulated domain. Semi-open boundaries are successfully used in a variety of codes designed for near-surface solar convection, including MuRAM (Rempel *et al.* 2009; Cheung *et al.* 2010), *Stagger* (Trampedach & Stein 2011), and CSS (Augustson *et al.* 2011), though it should be noted that all of these codes are fully compressible and use finite-difference algorithms. ASH uses an anelastic formulation and is pseudo-spectral, which presents additional limitations.

In general, semi-open boundaries tend to generate pressure perturbations which are the result of the flows being specified rather than resulting from solutions of the differential equations being solved. In compressible codes this generates acoustic modes which are often described as box modes, which can often be damped either volumetrically or by use of an absorbing boundary somewhere else in the domain. In ASH, which does not permit acoustic waves, these perturbations can only be dampened by coupling to viscous or thermal diffusion, which restricts the physical and temporal scales over which flows can vary.

In ASH we have chosen to implement what we term a plume boundary condition in which radial velocity and entropy are specified on the top boundary and horizontal velocities and pressure are determined by the interior of the simulation. For full details of

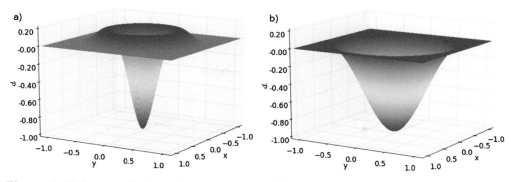

Figure 2. Volume rendering of plume profiles. (a) A so-called "Mexican hat" profile which integrates to zero total contribution. (b) A simpler cold-only plume shape which is offset by a global adjustment to either velocity or thermal fields.

the implementation we refer to Nelson (2013). We choose to populate our boundary with large numbers of small-scale plumes, which are shown in Figure 2. These plumes are local functions with compact support which smoothly match to zero outside of their footprint. The two options explored here are a fifth-order polynomial that gives a "Mexican hat" plume which integrates to zero over a spherical surface, and a third-order polynomial that has a non-zero integral, which we term a cold-only plume. For quantities such as the mass flux or entropy perturbation which we require to be zero the Mexican hat plumes are truly local, while the cold-only plumes require a net outflow or high entropy somewhere else to provide the zero average value we require. We have experimented with a variety of combinations, but the simulation shown here uses Mexican hat shapes for both the radial velocity and entropy profiles of the plumes.

One of the consequences of the plume boundary condition is that it does not locally conserve angular momentum. This allows the global angular momentum of the simulation to vary. To counteract this effect we apply a volumetric torque fitted to counter the time-averaged angular momentum flux through the plume boundary. This assures global angular momentum conservation over long timescales, however it also introduces an effective non-local transport of angular momentum as losses through the boundary are not immediately replaced at the same location. Thus this formulation of the plume boundary condition makes it difficult to study the resulting differential rotation in detail.

Figure 3 shows a snapshot of the plume boundary condition applied to a convective model using four-fold symmetry to save computational expense. The radial velocity and entropy perturbations are specified, while the pressure field is influenced implicitly by these fields as well as the interior of the simulation domain. Generally, however we see that the downflow, low-entropy cores of the plumes correspond to low pressure regions, ensuring that plumes converge as is appropriate in a spherical geometry.

Each individual plume is given its own size, duration, amplitude in both velocity and entropy, and location which are randomly chosen from a specified range. The choices for the simulation presented here are shown in Table 1. The choice of radial velocity and entropy amplitudes are constrained such that a solar luminosity of enthalpy flux is transported through the boundary. Correlations between some of these parameters might reasonably be expected. For example, one might expect larger plumes to be faster, or smaller plumes to have shorter lifetimes. Here we chose to correlate the velocity amplitude, entropy amplitude, plume width, and plume lifetime. Thus each plume is randomly assigned each of the six parameters with the given correlations from the specified range. Finally, we choose to advect longitudinal position of the plume centers ϕ_p with the mean

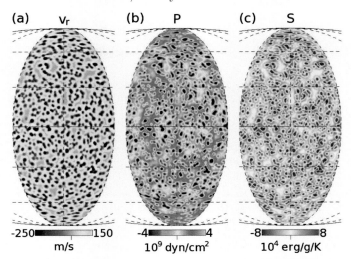

Figure 3. Snapshot of the plume boundary condition applied on the upper boundary of an ASH simulation. The plume boundary condition applies (a) explicit radial velocity structures with a "Mexican-hat" profile, (b) implicit pressure perturbations, and (c) explicit entropy structures, with either a Mexican-hat profile (shown here) or a cold-only profile.

Table 1. Values for Plume Parameters

Parameter	Value Range	Units	Correlation
$\mathcal{V}$	$[240, 360]$	m s^{-1}	–
$\mathcal{E}$	$[1.28, 1.56]$	10^5 erg K^{-1} g^{-1}	0.5
δ	$[0.08, 0.12]$	rad.	0.5
$\delta R_{\rm top}$	$[54.6, 82.0]$	Mm	0.5
τ	$[12.0, 18.0]$	days	0.5
θ_p	$[0, \pi]$	rad.	0
ϕ_p	$[0, \pi/2]$	rad.	0

Notes: Plume boundary parameters used where $\mathcal{V}$ is the peak downflow velocity, $\mathcal{E}$ is the peak entropy perturbation, δ is the plume's angular radius, τ is the plume lifetime, and θ_p and ϕ_p give the coordinate location of the center of the plume on the outer boundary. Also given is the plume width given by $\delta R_{\rm top}$ in Mm for ease of comparison. Correlations are expressed with respect to $\mathcal{V}$.

differential rotation. Plumes are initiated with a linear ramp-up phase corresponding to 10% of their total lifetimes, followed by a constant phase for 80% of their lifetimes, and a ramp-down phase for the final 10%. After ending their lives they are given new random parameters and restarted at a new location.

2.2. *Coalescence of Plumes into Giant Cells*

The use of the plume boundary condition provides a means to test a long-standing theoretical proposition put forward by Spruit (1997) which argues that giant cell convection is formed not by distributed driving due to a super-adiabatic gradient throughout the convection zone, but instead driving occurs almost exclusively at the photosphere (see also Brandenburg 2016). Convection below the photospheric layers are then a self-organization of plumes. This "entropy rain" model of convection is essentially what is achieved with the plume boundary condition reported here. Because our simulation transports energy through the upper boundary by enthalpy rather than diffusion, it does not create the strong superadiabatic gradients seen in other ASH simulations. In fact compare to a comparable simulation with a closed boundary, this plume boundary simulation's mean entropy gradient is reduced by a factor 10 at the upper boundary, reduced by a

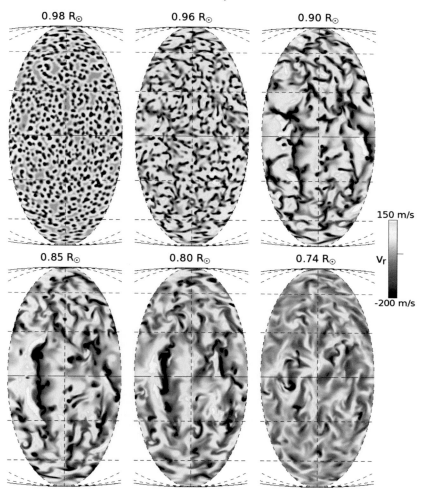

Figure 4. Snapshot of radial velocity at six depths in a simulation using the plume boundary condition with depths indicated. Small-scale, isotropic, randomly-placed plumes are applied on the outer boundary, but quickly coalesce first into sheets with rotational alignment and then into large banana cells with clear rotational alignment.

factor of 2 at mid-convection zone, and is even slightly subadiabatic between $0.93R_\odot$ and $0.97R_\odot$. This essentially removes the majority of the convective driving due to the bulk stratification.

Figure 4 shows a snapshot of the radial velocity at six depths in a well-equilibrated ASH simulation using the plume boundary model with the parameters shown in Table 1. At the boundary plumes are isotropic and very small in scale, but as the plumes descend they rapidly self-organize first into elongated structures resembling sheets at $0.96R_\odot$. By $0.90R_\odot$ these sheets have become rotationally aligned, particularly at low latitudes. Individual downflow plumes can still be seen, but most have merged into much larger scale structures which resemble the "banana cells" seen in closed-boundary models with solar-like differential rotation (e.g., Miesch *et al.* 2006; Brown *et al.* 2010; Guerrero & Käpylä 2011; Gastine & Wicht 2012; Featherstone & Miesch 2015). These giant cells are much larger in scale than the plumes and their scale roughly goes at the local pressure

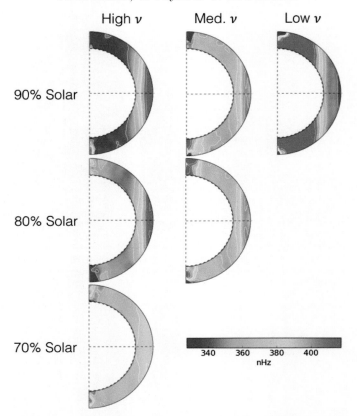

Figure 5. Time-averaged differential rotation profiles for six simulations of solar-like stars varying the bulk rotation rate of the simulation as well as the level of diffusion. Simulations show a systematic trend towards lower differential rotation with decreasing rotation rate, and a transition from solar-like to anti-solar differential rotation for decreasing viscosity ν at constant Prandtl number.

scale height. Finally near the base of the convection zone the convection slows and is dominated by return flows from the lower boundary.

We appear to have discovered a hybrid between the traditional giant cell model and the entropy-rain model with this plume boundary treatment. In our model we achieve both a strong reduction in the super-adiabatic gradient and yet still achieve convective patterns that morphologically fit the giant cells expected by the traditional model. More study is needed to investigate the properties of these solutions further, but this presents an intriguing possible unification of these two models for convective giant cells in sun-like stars.

3. The Solar/Anti-Solar Transition in Differential Rotation

Stellar spin down with age has long been modeled using the Skumanich relationship $\Omega \propto t^{-1/2}$ (Skumanich 1972). Recent results using asteroseismic ages have shown that older stars systematically deviate from this trend and show excessively fast rotation rates (van Saders *et al.* 2016). Specifically, stars that show less rotational constraint on their giant cell convection than the Sun seem to be spinning down more slowly than predicted. Based on these results Metcalfe *et al.* (2016) have proposed that a transition may occur near the solar Rossby number which causes a shift in the dynamo mechanism and thus a change in the rate of angular momentum loss.

Motivated by this observational trend, we have begun to investigate the behavior of stellar differential rotation at rotation rates slower than the current Sun. It has been shown by Gastine *et al.* (2013) and Featherstone & Miesch (2015) that lower levels of rotational constraint than our Sun's leads to anti-solar differential rotation profiles with slowly rotating equatorial regions and rapidly rotating mid- to high-latitudes. This anti-solar differential rotation can be understood as a loss of correlation in the Reynold stress transport seen in solar-like differential rotation (see Brun & Toomre 2002), leading to a mixing of specific angular momentum throughout the convection zone. Previous efforts have focused on changing the Rossby number (the ratio of convective to planetary vorticities) by altering the convective driving by changing the diffusion coefficients of the models or by increasing the rotation rate above the solar value. Here we present some initial results of simulations below the solar rotation rate at a variety of levels of diffusion.

We are conducting an on-going sweep of parameter space of rotation rate and turbulent viscosity with initial results shown here. All simulations are similar to case AB2 of Brun & Toomre (2002) but extend to $0.98R_\odot$ and do not use different treatments for the spherically-symmetric diffusive transport compared to the diffusion felt by other modes. They are also similar to the 4 scale-height models of Featherstone & Miesch (2015), but use a solar stratification based on a stellar structure model rather than a polytropic reference state. All models shown here use a closed boundary condition. Here we examine the results of six of these models.

Figure 5 shows initial results form a suite of simulations exploring the relative effects of decreasing diffusion (represented by the viscosity ν) and decreasing rotation rate Ω. We see clear trends in differential rotation with both decreasing diffusion and decreasing rotation rate. For our most turbulent case at $0.9\Omega_\odot$ we achieve an anti-solar differential rotation profile. For our slowest rotator at $0.7\Omega_\odot$ we see essentially solid body rotation even over ~ 200 years of evolution. This clearly shows that for 3D global convection models both the bulk rotation rate and the level of convective driving can cause transitions from solar-like to anti-solar differential rotation.

4. Future Directions

In this paper we have presented the development and initial use of a plume boundary model for global 3D solar-like convection models. As we continue to gain confidence in the effects of our plume boundary model this treatment may provide some important paths toward better models of stellar dynamo action. To list a few, a plume boundary condition may provide paths toward:

- Improved convective driving in highly turbulent convection simulations, permitting solar-like differential rotation for solar values of the rotation rate and luminosity at levels of turbulence beyond those currently achievable.
- Eventual coupling between global solar convection simulations and local, near-surface models (e.g., Stein & Nordlund 2012; Leake *et al.* 2013). Early forms of one-way coupling can take the form of using near-surface models to generate plume statistics for our plume boundary model or applying ASH convective patterns to the bottom of near-surface models. More advanced dynamic coupling schemes may involve tiled near-surface models and global simulations running interactively.
- Buoyant magnetic loops generated by global convective dynamo models (e.g., Nelson *et al.* 2011; Warnecke *et al.* 2012; Nelson *et al.* 2013b) could be simulated much closer to the photosphere than currently permitted. In current closed-boundary models such

loops are limited to heights below the upper boundary layer. This could be particularly interesting when combined with a coupling to a near-surface simulation, which could then in principle take a flux rope all the way from generation to emergence as an active region.

We have also presented initial results from a suite of models seeking to explore the transition from solar-like to anti-solar differential rotation as a function of both convective driving and bulk rotation rate. Continued exploration at still lower levels of diffusion is needed. Moving forward we will explore the magnetic fields generated by simulations in this parameter space. These varied states of differential rotation should clearly host a wide variety of dynamos. Additionally it has been shown in some cases that the presence of a dynamo can sustain solar-like differential rotation beyond the transition seen in hydrodynamic simulations (Fan & Fang 2016).

While only one of many issues facing the question of habitability, the ability to predict magnetic activity across a wide range of stellar ages would provide a key step towards a coherent theoretical understanding of life around an active star. In order to have confidence in a dynamo model, it is imperative to resolve questions of the nature of convective driving of giant cells and of the evolution of differential rotation over evolutionary timescales.

Acknowledgements

The authors would like to thank Juri Toomre, Mark Miesch, and Nicholas Featherstone for many useful discussions related to the ideas presented here. Special additional thanks to Nicholas Featherstone for his development of *Rayleigh* and allowing us to be early users. Some of this work was conducted while N.J.N. was supported by a Nicholas C. Metropolis Post-Doctoral Fellowship as part of the Advanced Simulation Capabilities Initiative at Los Alamos National Laboratory. C.P. and C.M.S. were supported by the Physics Summer Research Institute at California State University, Chico.

References

Airapetian, V. S., Glocer, A., Khazanov, G. V., *et al.* 2017, The *Astrophysical Journal*, 836, L3

Anglada-Escudé, G., Amado, P. J., Barnes, J., *et al.* 2016, *Nature*, 536, 437

Augustson, K., Brun, A. S., Miesch, M., & Toomre, J. 2015, *The Astrophysical Journal*, 809, 149

Augustson, K. C., Brun, A. S., & Toomre, J. 2016, *The Astrophysical Journal*, 829, 1

Augustson, K. C., Rast, M., Trampedach, R., & Toomre, J. 2011, *Journal of Physics: Conference Series*, 271, 012070

Bixel, A., & Apai, D. 2017, *The Astrophysical Journal*, 836, L31

Boutle, I. A., Mayne, N. J., Drummond, B., *et al.* 2017, arXiv preprint, 1

Brandenburg, A. 2016, *Astrophysical Journal*, 832, 1

Brown, B. P., Browning, M. K., Brun, A. S., Miesch, M. S., & Toomre, J. 2008, *The Astrophysical Journal*, 689, 1354

—. 2010, *The Astrophysical Journal*, 711, 424

Brown, B. P., Miesch, M. S., Browning, M. K., Brun, A. S., & Toomre, J. 2011, *The Astrophysical Journal*, 731, 69

Browning, M. K. 2008, *The Astrophysical Journal*, 676, 1262

Brun, A. S., Miesch, M. S., & Toomre, J. 2004, *The Astrophysical Journal*, 614, 1073

Brun, A. S., & Toomre, J. 2002, *The Astrophysical Journal*, 570, 865

Brun, A. S., Strugarek, A., Varela, J., *et al.* 2017, *The Astrophysical Journal*, 836, 1

Cheung, M. C. M., Rempel, M., Title, A. M., & Schüssler, M. 2010, *The Astrophysical Journal*, 720, 233

Clune, T., Elliott, J., Miesch, M. S., Toomre, J., & Glatzmaier, G. A. 1999, *Parallel Computing*, 25, 361

Coleman, G. A. L., Nelson, R. P., Paardekooper, S. J., *et al.* 2017, *Monthly Notices of the Royal Astronomical Society*, arXiv:1608.06908

Fan, Y., & Fang, F. 2016, *Advances in Space Research*, 58, 1497

Featherstone, N. A., Browning, M. K., Brun, A. S., & Toomre, J. 2009, *The Astrophysical Journal*, 705, 1000

Featherstone, N. A., & Hindman, B. W. 2016a, *The Astrophysical Journal*, 830, L15

—. 2016b, *The Astrophysical Journal*, 818, 32

Featherstone, N. A., & Miesch, M. S. 2015, *The Astrophysical Journal*, 804, 1

Garrafo, C., Drake, J. J., & Cohen, O. 2016, The Astrophysical Journal Letters, 833, 1

Gastine, T., & Wicht, J. 2012, Icarus, 219, 428

Gastine, T., Yadav, R. K., Morin, J., Reiners, A., & Wicht, J. 2013, Monthly Notices of the Royal Astronomical Society: Letters, 438, L76

Gillon, M., Jehin, E., Lederer, S. M., *et al.* 2016, Nature, 533, 221

Gilman, P. 1983, The Astrophysical Journal Supplement Series, 53, 243

Glatzmaier, G. A. 1985, *The Astrophysical Journal*, 291, 300

Greer, B. J., Hindman, B. W., Featherstone, N. a., & Toomre, J. 2015, *The Astrophysical Journal*, 803, L17

Greer, B. J., Hindman, B. W., & Toomre, J. 2014, Solar Physics

Guerrero, G., & Käpylä, P. J. 2011, Astronomy & Astrophysics, 533, A40

Hanasoge, S. M., Duvall, T. L., & DeRosa, M. L. 2010, *The Astrophysical Journal*, 712, L98

Hanasoge, S. M., Duvall, T. L., & Sreenivasan, K. R. 2012, Proceedings of the National Academy of Sciences, 109, 11928

Hathaway, D. H., Upton, L., & Colegrove, O. 2013, Science (New York, N.Y.), 342, 1217

Kane, S. R., Gelino, D. M., & Turnbull, M. C. 2017, The Astronomical Journal, 153, 52

Leake, J. E., Linton, M. G., & Torok, T. 2013, *The Astrophysical Journal*, 778, 12

Luger, R., Lustig-Yaeger, J., Fleming, D. P., *et al.* 2016, eprint arXiv:1609.09075, 837, 1

McIntosh, S. W., Wang, X., Leamon, R. J., & Scherrer, P. H. 2014, *The Astrophysical Journal*, 784, L32

Méndez, A., & Rivera-Valentín, E. G. 2017, *The Astrophysical Journal* Letters, 837, 1

Metcalfe, T. S., Egeland, R., & van Saders, J. L. 2016, *The Astrophysical Journal*, 826, L2

Miesch, M. S., Brun, A. S., & Toomre, J. 2006, *The Astrophysical Journal*, 641, 618

Nelson, N. J. 2013, Proquest dissertations & theses: 3607343, University of Colorado

Nelson, N. J., Brown, B. P., Brun, A. S., Miesch, M. S., & Toomre, J. 2011, *The Astrophysical Journal*, 739, L38

—. 2013a, *The Astrophysical Journal*, 762, 73

Nelson, N. J., Brown, B. P., Sacha Brun, a., Miesch, M. S., & Toomre, J. 2013b, *Solar Physics*

Nelson, N. J., & Miesch, M. S. 2014, *Plasma Physics and Controlled Fusion*, 56, 064004

Reiners, A. 2012, *Living Reviews in Solar Physics*, 8, 1

Rempel, M., Schüssler, M., & Knölker, M. 2009, *The Astrophysical Journal*, 691, 640

Skumanich, A. 1972, *The Astrophysical Journal*, 171, 565

Spruit, H. 1997, *Mem. Soci. Astron. Ital.*, 68, 397

Stein, R. F., & Nordlund, A. k. 2012, *The Astrophysical Journal*, 753, L13

Trampedach, R., & Stein, R. F. 2011, *The Astrophysical Journal*, 731, 78

van Saders, J. L., Ceillier, T., Metcalfe, T. S., *et al.* 2016, *Nature*, 529, 181

Warnecke, J., Käpylä, P. J., Käpylä, M. J., & Brandenburg, A. 2016, *Astronomy & Astrophysics*, 596, A115

Warnecke, J., Käpylä, P. J., Mantere, M. J., & Brandenburg, A. 2012, *Solar Physics*, 280, 299

Yadav, R. K., Christensen, U. R., Wolk, S. J., & Poppenhaeger, K. 2016, *The Astrophysical Journal*, 833, L28

Living Around Active Stars
Proceedings IAU Symposium No. 328, 2016
D. Nandy, A. Valio & P. Petit, eds.

© International Astronomical Union 2017
doi:10.1017/S1743921317004203

Improved rotation-activity-age relations in Sun-like stars

Jorge Meléndez, Leonardo A. dos Santos and Fabrício C. Freitas

Universidade de São Paulo, IAG, Departamento de Astronomia
Rua do Matão 1226, 05508-090 São Paulo, SP, Brazil
email: jorge.melendez@iag.usp.br

Abstract. The evolution of rotational velocity and magnetic activity with age follows approximately a $t^{-1/2}$ relation, the famous Skumanich law. Using a large sample of about 80 solar twins with precise ages, we show departures from this law. We found a steep drop in rotational velocity and activity in the first 2-3 Gyr and afterwards there seems to be a shallow decrease. Our inferred rotational periods suggest that the Sun will continue to slow down, validating thus the use of gyrochronology beyond solar age. The Sun displays normal rotational velocity and activity when compared to solar twins of solar age. We also show that stars with exceedingly high stellar activity for their age are spectroscopic binaries that also exhibit enhanced rotational velocities and chemical signatures of mass transfer.

Keywords. Sun: activity, Sun: rotation, stars: activity, stars: rotation, techniques: spectroscopic

1. Introduction

Solar twins are defined as stars with effective temperatures within 100 K, and $\log g$ and [Fe/H] within 0.1 dex around the Sun's values (Ramírez *et al.* 2014). The first solar twin, 18 Sco, was discovered in 1997 by Porto de Mello & da Silva (1997, see also Soubiran & Triaud 2004). Only in the mid-2000s new solar twins were identified and in the last decade their number has grown to about a hundred (Mahdi *et al.* 2016; Yana Galarza *et al.* 2016b; Porto de Mello *et al.* 2014; Datson *et al.* 2014, 2012; Ramírez *et al.* 2014; Meléndez *et al.* 2014b, 2009, 2006; do Nascimento *et al.* 2013; Takeda & Tajitsu 2009; Pasquini *et al.* 2008; Takeda *et al.* 2007; Meléndez & Ramírez 2007; King *et al.* 2005).

An important application of solar twins is to set the zero-point of fundamental calibrations in astrophysics (Datson *et al.* 2015, 2012; Casagrande *et al.* 2014, 2010; Ramírez *et al.* 2012; Meléndez *et al.* 2010), but many other studies have been possible thanks to the precise stellar parameters ($\sigma(\mathrm{T_{eff}}) \leqslant 10$ K, $\sigma(\log g) \leqslant 0.02$ dex) and chemical abundances ($\sigma \sim 0.01$ dex) that can be achieved in these stars (e.g., Nissen 2015; Bedell *et al.* 2014; Monroe *et al.* 2013; Kiselman *et al.* 2011). One landmark result of the first high precision studies showed that the Sun has a distinct abundance pattern (Meléndez *et al.* 2009; Ramírez *et al.* 2009, 2010). Strikingly, this peculiar pattern may be related to the formation of rocky planets in our solar system (Chambers 2010). In a related investigation, Tucci Maia *et al.* (2014) showed that the binary system of solar twins 16 Cyg displays chemical abundance differences likely due to planets, and other planet-hosting wide binary systems composed of solar-type stars also show distinct abundances probably due to planets (Teske *et al.* 2016a,b; Adibekyan *et al.* 2016; Saffe *et al.* 2016; Ramírez *et al.* 2015; Biazzo *et al.* 2015).

Solar twins have also been crucial to obtain reliable isochrone ages (Tucci Maia *et al.* 2016; Nissen 2015), determine improved mass and radius of benchmark planets (Bedell *et al.* 2016), study the possible effects of radioactive elements on rocky planet interior

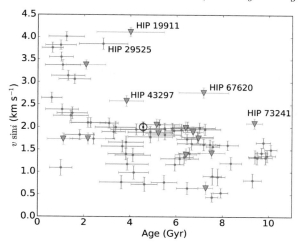

Figure 1. Projected rotational velocities in solar twins vs. age. The Sun is shown with its usual symbol and the solar twins by blue points, except the spectroscopic binaries that are shown by inverted orange triangles (outliers are labeled). Figure from dos Santos *et al.* (2016).

dynamics and habitability (Unterborn *et al.* 2015), infer the effects of mass transfer by the progenitors of white dwarfs (Desidera *et al.* 2016; Schirbel *et al.* 2015), observe stellar oscillations to constrain their masses and ages (Metcalfe *et al.* 2012; Li *et al.* 2012; Bazot *et al.* 2011), inspire new stellar models including episodic accretion (Baraffe & Chabrier 2010), observe depletion of the light elements Li (Carlos *et al.* 2016; Monroe *et al.* 2013; Baumann *et al.* 2010) and Be (Tucci Maia *et al.* 2015) and model their decrease through non-standard stellar models (Andrássy & Spruit 2015; Castro *et al.* 2011; Denissenkov 2010; do Nascimento *et al.* 2009), develop chemical clocks through the [Y/Mg] and [Y/Al] ratios (Tucci Maia *et al.* 2016; Spina *et al.* 2016; Nissen 2016, 2015), and to uncover the nucleosynthetic history of our Milky Way empirically through cosmic phylogeny (Jofré *et al.* 2017) or through the detailed study of abundance ratios (e.g., Nissen 2016, 2015; Spina *et al.* 2016; Yana Galarza *et al.* 2016a; Meléndez *et al.* 2014a).

Since 2011 we have been monitoring about 70 solar twins for planets at the ESO La Silla Observatory (Meléndez *et al.* 2015; Ramírez *et al.* 2014), and have found a Jupiter-twin (Bedell *et al.* 2015) and a super-Neptune and super-Earth (Meléndez *et al.* 2017). Using the individual and combined HARPS spectra obtained for our sample, and additional spectra of 9 solar twins gathered by other ESO programs, we studied the evolution of rotational velocity and magnetic activity using about 80 solar twins with precise stellar ages measured by a differential comparison of our precise stellar parameters with isochrones (Tucci Maia *et al.* 2016). We also assess how common is the Sun by a comparison with solar twins.

2. The Skumanich law and the Sun among stars

More than four decades ago, Skumanich (1972) showed that there is a decline of rotational velocity (v) and activity with age (t), following approximately a $t^{-1/2}$ relation, the famous "Skumanich law". Impressively, this widely used relation was obtained using only a few data points.

Under some assumptions, the $v \propto t^{-1/2}$ relation can be derived from the angular momentum loss due to stellar winds (e.g., Soderblom 1983; Kawaler 1988). However, there are evidences for departures from the Skumanich law. The work by Pace & Pasquini

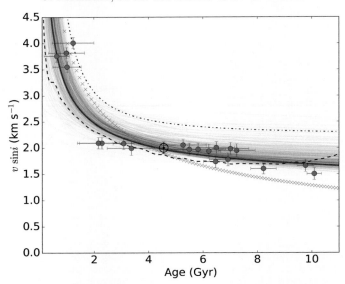

Figure 2. Projected rotational velocities in solar twins of the selected sample vs. age. The Sun is shown with its usual symbol. Our best fit (solid line) and the Skumanich law (red symbols) are shown. Figure from dos Santos *et al.* (2016).

(2004) suggests a steep decline of rotational velocities and stellar activity in the first 1-2 Gyr and then a flat trend with age. Our large sample of solar twins is ideal to verify whether the decay of rotational velocity and activity is smooth, or if they behave as suggested by Pace & Pasquini (2004).

Many previous works sought to evaluate the place of the Sun among other stars (e.g., Gustafsson 2008, 1998; Shapiro *et al.* 2013). This is of fundamental importance because the Sun is used to validate stellar models, so we must know whether the Sun is a regular star or an oddball. Unfortunately, conclusions by previous studies could be hampered by biased samples, and the Sun has not been compared to strictly Sun-like stars (e.g., Robles *et al.* 2008). And, even when the Sun is compared to similar stars, there is the issue of systematics due to the observations of the Sun and the stars being carried out with different instruments. This is clearly shown in a recent study by Egeland *et al.* (2016), where the Sun's magnetic activity S-index is shown to be lower than previously estimated.

In our work the Sun and the comparison stellar sample are analysed using spectra obtained by the HARPS spectrograph at the ESO La Silla Observatory, all gathered with the same high resolution ($R = 115000$) and at high signal-to-noise ratio (S/N) for our planet search around solar twins (Meléndez *et al.* 2015). Furthermore, our reference sample of solar twins have approximately solar mass and metallicity, therefore they follow evolutionary paths similar to the Sun. Thus, our sample is ideal to verify if the Sun is normal in its activity level and rotational velocity.

3. Rotational velocities

In order to determine the projected rotational velocities in our sample, we fit the line profiles considering instrumental, rotational and macroturbulent broadening, as described in dos Santos *et al.* (2016).

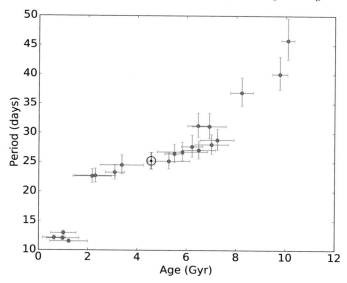

Figure 3. Evolution of inferred rotational period vs. stellar ages in solar twins, using the rotational velocities from dos Santos *et al.* (2016) and taking into account the variation of stellar radius with age predicted by the solar model of do Nascimento *et al.* (2009).

We removed all spectroscopic binaries (Fig. 1), as some of them have anomalously high $v$sin$i$ for their age, likely due to interactions with their companions.

We divided the sample in bins of 2 Gyr and in each bin we only kept stars with $v \sin i$ in the upper 30 %, because these stars have the highest chance of having $\sin i$ above 0.9 (dos Santos *et al.* 2016). The selected sample is shown in Fig. 2. It is clear that the decay in rotational velocity in the first 2-3 Gyr is steeper than the Skumanich law, and that the further decline of rotational velocity with age is flatter than the Skumanich relation.

Using rotational periods from solar-type stars obtained from *Kepler* observations, it has been suggested that there is a weakened magnetic braking after the Sun's age (van Saders *et al.* 2016). We use our selected sample of solar twins with $\sin i \sim 1$, to evaluate the variation of rotational period (P) with age. P is obtained from our rotational velocities and the predicted variation of the solar radius with age using the model by do Nascimento *et al.* (2009). Our results (Fig. 3) indicate that the Sun will keep slowing down, unlike the suggestion by van Saders *et al.* (2016) of weakened braking for old ages. If we obtain the stellar radius using the observed $\log g$ and inferred mass, we obtain similar results but with a larger scatter. Thus, our analysis suggests that gyrochronology can be used to determine stellar ages based on observed rotational periods even for stars older than the Sun.

4. Magnetic activity

We study the variation of magnetic activity with age using the Ca II H and K lines. We measured activity indexes using triangular filters in the core of the lines and flat filters in the continuum, as in Wright *et al.* (2004). Our indexes were calibrated to the S-index of the Mount Wilson system (e.g., Baliunas *et al.* 1995).

We used archival data of asteroid observations from other HARPS programs to obtain the reflected solar spectra covering different epochs of the solar cycle. This guarantees that the Sun and the stars are in the same system, eliminating thus potential systematic

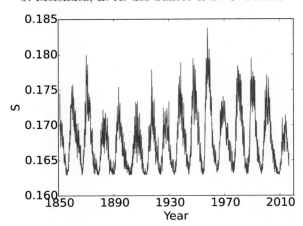

Figure 4. Evolution of the S-index since 1850, inferred through a calibration of the S-index and International Sunspot Number. Figure from Freitas *et al.* (2017).

problems. In order to increase the time baseline of the solar data, we calibrated the S-index to the International Sunspot Number, estimating the S-index since 1850 (Fig. 4). This allowed us to infer an accurate average solar S-index, considering cycles of different amplitudes.

The S-index is converted to the log R'$_{HK}$ activity index following standard procedures (e.g., Wright *et al.* 2004). The average of this index as a function of stellar age is shown in Fig. 5. Our data departs from the Skumanich law, suggesting a steep decrease in activity in the first 3 Gyr and then a moderate decrease of stellar activity for increasing ages.

Some spectroscopic binaries present high stellar activity for their age (Fig. 5). These stars also show high rotational velocity (Fig. 1) and chemical signatures of mass transfer, as revealed by their high yttrium abundances (Tucci Maia *et al.* 2016).

We obtained improved log R'$_{HK}$ values by using new calibrations based on T$_{eff}$ rather than (B-V), as described in detail in Freitas *et al.* (2017). These improved values are plotted in Fig. 6. Qualitatively we obtain similar results for the decay of activity with age, but the activity-age correlation is somewhat improved, with a steep decline of activity in the first 3 Gyrs, and a shallower decrease for increasing ages. For comparison we show the relations by Donahue (1993) and Mamajek & Hillenbrand (2008), which do not reproduce well the pattern observed in solar twins.

5. Conclusions

Our consistent analysis of solar twins and the Sun suggests that the evolution of rotational velocity and magnetic activity in the Sun depart from the widely used Skumanich law. The decay is steeper in the first 3 Gyr, with a slight decrease for older ages.

Our inferred longer rotation periods for increasing ages suggest that gyrochronology (e.g. Barnes 2007) is valid even beyond solar age. This result obtained with solar twins is in contrast to recent results based on solar-type stars from *Kepler* observations (van Saders *et al.* 2016). This shows the importance of having a proper comparison sample of solar twins, rather than solar-type stars that do not necessarily have the same properties as the Sun.

Furthermore, we show that the Sun is normal in rotational velocity and magnetic activity when compared to solar twins of the same age, validating thus the use of the Sun as a cornerstone calibrator in Astrophysics.

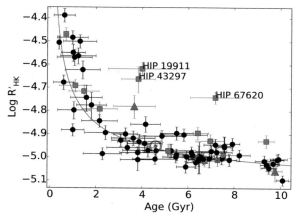

Figure 5. Log R'$_{HK}$ activity index vs. age in solar twins (black circles). Spectroscopic binaries are shown by red squares (outliers are labeled). The Sun is shown with its usual symbol. Our best fit is shown. Figure from Freitas *et al.* (2017).

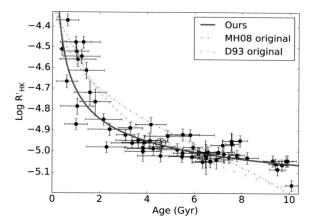

Figure 6. Log R'$_{HK}$ activity index vs. age in solar twins (black circles) using our improved calibrations based on T$_{eff}$ (see Freitas *et al.* 2017). Spectroscopic binaries are excluded. Our fit is shown by the solid blue line. The other lines correspond to the relations by Mamajek & Hillenbrand (2008) and Donahue (1993). Figure adapted from Freitas *et al.* (2017).

References

Adibekyan, V., Delgado-Mena, E., Figueira, P., *et al.* 2016, *A&A*, 591, A34
Andrássy, R. & Spruit, H. C. 2015, *A&A*, 579, A122
Baliunas, S. L., *et al.* 1995, *ApJ*, 438, 269
Baraffe, I. & Chabrier, G. 2010, *A&A*, 521, A44
Barnes, S. A. 2007, *ApJ*, 669, 1167
Baumann, P., Ramírez, I., Meléndez, J., Asplund, M., & Lind, K. 2010, *A&A*, 519, A87
Bazot, M., *et al.* 2011, *A&A*, 526, L4
Bedell, M., Meléndez, J., Bean, J. L., *et al.* 2014, *ApJ*, 795, 23
Bedell, M., Meléndez, J., Bean, J. L., *et al.* 2015, *A&A*, 581, A34
Bedell, M., Bean, J. L., Melendez, J., *et al.* 2016, arXiv:1611.06239
Biazzo, K., Gratton, R., Desidera, S., *et al.* 2015, *A&A*, 583, A135
Carlos, M., Nissen, P. E., & Meléndez, J. 2016, *A&A*, 587, A100
Casagrande, L., Ramírez, I., Meléndez, J., Bessell, M., & Asplund, M. 2010, *A&A*, 512, A54
Casagrande, L., Portinari, L., Glass, I. S., *et al.* 2014, *MNRAS*, 439, 2060

Castro, M., Do Nascimento, J. D., Jr., Biazzo, K., Meléndez, J., & de Medeiros, J. R. 2011, *A&A*, 526, A17

Chambers, J. E. 2010, *ApJ*, 724, 92

Datson, J., Flynn, C., & Portinari, L. 2012, *MNRAS*, 426, 484

Datson, J., Flynn, C., & Portinari, L. 2014, *MNRAS*, 439, 1028

Datson, J., Flynn, C., & Portinari, L. 2015, *A&A*, 574, A124

Denissenkov, P. A. 2010, *ApJ*, 719, 28

Desidera, S., D'Orazi, V., & Lugaro, M. 2016, *A&A*, 587, A46

Donahue, R. A. 1993, Ph.D. Thesis,

Do Nascimento, J. D., Jr., Castro, M., Meléndez, J., Bazot, M., Théado, S., Porto de Mello, G. F., & de Medeiros, J. R. 2009, *A&A*, 501, 687

do Nascimento, J.-D., Jr., Takeda, Y., Meléndez, J., *et al.* 2013, *ApJ*, 771, L31

dos Santos, L. A., Meléndez, J., do Nascimento, J.-D., Jr., *et al.* 2016, *A&A*, 592, A156

Egeland, R., Soon, W., Baliunas, S., *et al.* 2017, *ApJ*, 835, 25

Freitas, F. C., Meléndez, J., Bedell, M., *et al.* 2017, submitted

Gustafsson, B. 1998, *Space Sci. Rev.*, 85, 419

Gustafsson, B. 2008, *Physica Scripta*, Volume T 130, pp. 014036

Jofré, P., Das, P., Bertranpetit, J., & Foley, R. 2017, *MNRAS*, 467, 1140

Kawaler, S. D. 1988, *ApJ*, 333, 236

King, J. R., Boesgaard, A. M., & Schuler, S. C. 2005, *AJ*, 130, 2318

Kiselman, D., Pereira, T. M. D., Gustafsson, B., *et al.* 2011, *A&A*, 535, A14

Li, T. D., Bi, S. L., Liu, K., Tian, Z. J., & Shuai, G. Z. 2012, *A&A*, 546, A83

Mahdi, D., Soubiran, C., Blanco-Cuaresma, S., & Chemin, L. 2016, *A&A*, 587, A131

Mamajek, E. E. & Hillenbrand, L. A. 2008, *ApJ*, 687, 1264

Meléndez, J., Dodds-Eden, K., & Robles, J. A. 2006, *ApJ*, 641, L133

Meléndez, J. & Ramírez, I. 2007, *ApJ*, 669, L89

Meléndez, J., Asplund, M., Gustafsson, B., & Yong, D. 2009, *ApJ*, 704, L66

Meléndez, J., Schuster, W. J., Silva, J. S., Ramírez, I., Casagrande, L., & Coelho, P. 2010, *A&A*, 522, A98

Meléndez, J., Ramírez, I., Karakas, A. I., *et al.* 2014a, *ApJ*, 791, 14

Meléndez, J., Schirbel, L., Monroe, T. R., *et al.* 2014b, *A&A*, 567, L3

Meléndez, J., Bean, J. L., Bedell, M., *et al.* 2015, The Messenger, 161, 28

Meléndez, J., Bedell, M., Bean, J. L., *et al.* 2017, *A&A*, 597, A34

Metcalfe, T. S., Chaplin, W. J., Appourchaux, T., *et al.* 2012, *ApJ*, 748, L10

Monroe, T. R., Meléndez, J., Ramírez, I., *et al.* 2013, *ApJ*, 774, L32

Nissen, P. E. 2015, *A&A*, 579, A52

Nissen, P. E. 2016, *A&A*, 593, A65

Pace, G. & Pasquini, L. 2004, *A&A*, 426, 1021

Pasquini, L., Biazzo, K., Bonifacio, P., Randich, S., & Bedin, L. R. 2008, *A&A*, 489, 677

Porto de Mello, G. F. & da Silva, L. 1997, *ApJ*, 482, L89

Porto de Mello, G. F., da Silva, R., da Silva, L., & de Nader, R. V. 2014, *A&A*, 563, A52

Ramírez, I., Meléndez, J., & Asplund, M. 2009, *A&A*, 508, L17

Ramírez, I., Asplund, M., Baumann, P., Meléndez, J., & Bensby, T. 2010, *A&A*, 521, A33

Ramírez, I., Michel, R., Sefako, R., *et al.* 2012, *ApJ*, 752, 5

Ramírez, I., Meléndez, J., Bean, J., *et al.* 2014, *A&A*, 572, A48

Ramírez, I., Khanal, S., Aleo, P., *et al.* 2015, *ApJ*, 808, 13

Robles, J. A., Lineweaver, C. H., Grether, D., *et al.* 2008, *ApJ*, 684, 691-706

Saffe, C., Flores, M., Jaque Arancibia, M., Buccino, A., & Jofré, E. 2016, *A&A*, 588, A81

Schirbel, L., Meléndez, J., Karakas, A. I., *et al.* 2015, *A&A*, 584, A116

Shapiro, A. I., Schmutz, W., Cessateur, G., & Rozanov, E. 2013, *A&A*, 552, A114

Skumanich, A. 1972, *ApJ*, 171, 565

Soderblom, D. R. 1983, *ApJS*, 53, 1

Soubiran, C. & Triaud, A. 2004, *A&A*, 418, 1089

Spina, L., Meléndez, J., Karakas, A. I., *et al.* 2016, *A&A*, 593, A125

Takeda, Y. & Tajitsu, A. 2009, *PASJ*, 61, 471
Takeda, Y., Kawanomoto, S., Honda, S., Ando, H., & Sakurai, T. 2007, *A&A*, 468, 663
Teske, J. K., Khanal, S., & Ramírez, I. 2016a, *ApJ*, 819, 19
Teske, J. K., Shectman, S. A., Vogt, S. S., *et al.* 2016b, *AJ*, 152, 167
Tucci Maia, M., Meléndez, J., & Ramírez, I. 2014, *ApJ*, 790, LL25
Tucci Maia, M., Meléndez, J., Castro, M., *et al.* 2015, *A&A*, 576, L10
Tucci Maia, M., Ramírez, I., Meléndez, J., *et al.* 2016, *A&A*, 590, A32
Unterborn, C. T., Johnson, J. A., & Panero, W. R. 2015, *ApJ*, 806, 139
van Saders, J. L., Ceillier, T., Metcalfe, T. S., *et al.* 2016, *Nature*, 529, 181
Wright, J. T., Marcy, G. W., Butler, R. P., & Vogt, S. S. 2004, *ApJS*, 152, 261
Yana Galarza, J., Meléndez, J., Ramírez, I., *et al.* 2016a, *A&A*, 589, A17
Yana Galarza, J., Meléndez, J., & Cohen, J. G. 2016b, *A&A*, 589, A65

Living Around Active Stars
Proceedings IAU Symposium No. 328, 2016
D. Nandy, A. Valio & P. Petit, eds.

© International Astronomical Union 2017
doi:10.1017/S1743921317004264

Hunting for hot Jupiters around young stars

Louise Yu[1,2] and the MaTYSSE collaboration

[1]Université de Toulouse, UPS-OMP, IRAP,
14 avenue E. Belin, Toulouse, F–31400 France
[2]CNRS, IRAP / UMR 5277,
Toulouse, 14 avenue E. Belin, F–31400 France
email: `louise.yu@irap.omp.eu`

Abstract. This conference paper reports the recent discoveries of two hot Jupiters (hJs) around weak-line T Tauri stars (wTTS) V830 Tau and TAP 26, through the analysis of spectropolarimetric data gathered within the Magnetic Topologies of Young Stars and the Survival of massive close-in Exoplanets (MaTYSSE) observation programme. HJs are thought to form in the outskirts of protoplanetary discs, then migrate inwards close to their host stars as a result of either planet-disc type II migration or planet-planet scattering. Looking for hJs around young forming stars provides key information on the nature and time scale of such migration processes, as well as how their migration impacts the subsequent architecture of their planetary system. Young stars are however extremely active, to the point that their radial velocity (RV) jitter is around an order of magnitude larger than the potential signatures of close-in gas giants, making them difficult to detect with velocimetry. Three techniques to filter out this activity jitter are presented here, two using Zeeman Doppler Imaging (ZDI) and one using Gaussian Process Regression (GPR).

Keywords. methods: statistical, stars: activity, stars: evolution, stars: imaging, stars: individual (V830 Tau, TAP 26), stars: magnetic fields, (stars:) planetary systems: formation, stars: pre–main-sequence, stars: rotation, stars: spots

1. Introduction

Studying young forming stars stands as our best chance to progress in our understanding of the formation and early evolution of planetary systems. For instance, detecting hot Jupiters (hJs) around young stars (1-10 Myrs) and determining their orbital properties can enable us to clarify how they form and migrate, and to better characterize the physical processes (e.g. planet-disc interaction, planet-planet scattering, Baruteau *et al.* 2014, in-situ formation, Batygin *et al.* 2016) responsible for generating such planets.

However, young stars are enormously active, rendering planet signatures in their spectra and / or light-curves extremely difficult to detect in practice. Until very recently, most planets found so far around stars younger than 20 Myr were distant planets detected with imaging techniques (e.g. β Pic b, Lagrange *et al.* 2010, and LkCa 15, Sallum *et al.* 2015). Early claims of hJs orbiting around T Tauri stars (TTSs)(e.g. TW Hya, Setiawan *et al.* 2008) finally proved to be activity signatures mistakenly interpreted as radial velocity (RV) signals from close-in giant planets (Huélamo *et al.* 2008).

Systematic exploration of hJs around TTSs, and in particular the so called weak-line T Tauri stars (wTTSs), whose accretion disc has just dissipated, is one of the main goals of the MaTYSSE (Magnetic Topologies of Young Stars and the Survival of close-in massive Exoplanets) large programme allocated on the 3.6m Canada-France-Hawaii Telescope (CFHT). This paper focuses on the study of wTTSs V830 Tau (Donati *et al.* 2016, Donati *et al.* 2017) and TAP 26 (Yu *et al.* 2017).

	V830 Tau[1]	TAP 26[2]
Age [Myr]	≃2.2	≃17
$M_\star$ [M$_\odot$]	1.00±0.05	1.04±0.10
P_{rot} [d]	2.741	0.7135
$R_\star$ [R$_\odot$]	2.0±0.2	1.17±0.17
$v \sin i$ [km.s^{-1}]	30.5±0.5	68.2±0.5
i [°]	55±10	55±10
P_{orb} [d]	4.927±0.008	10.79±0.14
$M \sin i$ [M$_{Jup}$]	0.57±0.10	1.66±0.31
a [au]	0.057±0.001	0.0968±0.0032
a [$R_\star$]	6.1±0.6	17.8±1.3

Table 1. Summary of the properties on both hot Jupiters found within the MaTYSSE programme. From top to bottom: age, stellar mass in terms of solar masses, stellar rotation period expressed, stellar radius in terms of solar radii, line-of-sight-projected equatorial rotation velocity, inclination of stellar rotation axis to the line of sight, orbital period, minimal mass in terms of jovian masses, semimajor axis and semimajor axis in terms of stellar radii.

References:
[1] Donati *et al.* (2017)
[2] Yu *et al.* (2017)

2. Modelling the stellar activity

Targets. V830 Tau and TAP 26 are wTTSs in the Taurus stellar formation region. Their characteristics are shown in table 1.

Observations. Unpolarized (Stokes I parameter) and circularly polarized (Stokes V parameter) spectra were collected between 2015 November and 2016 February for both stars, from spectropolarimeters ESPaDOnS (Echelle SpectroPolarimetric Device for the Observation of Stars) at CFHT and Narval at TBL (Télescope Bernard Lyot). Complete information about the observations is found in Donati *et al.* (2016) and Donati *et al.* (2017) for V830 Tau and in Yu *et al.* (2017) for TAP 26.

LSD (Least Squares Deconvolution). The high value of $v \sin i$ creates a strong correlation between brightness features at the surface of the stars and distorsions within the line profiles of the unpolarized spectra. As a star rotates, these distorsions move across line profiles creating a modulation within the subsequent computed RV, following a temporal period equal to the stellar rotation period. In the presence of a hJ, the planet-induced reflex motion causes the spectrum to be blue-shifted or red-shifted as a whole, following a temporal period equal to the hJ's orbital period. LSD, a multiline technique, is applied to each spectrum in order to concentrate the information repeated in more than 7000 spectral lines into a unique profile, describing power as a function of local RV, with an SNR (signal to noise ratio) around one order of magnitude bigger than in the spectrum (see Donati *et al.* 1997b)(see Fig. 1). The global RV for each spectrum is computed as the first-order moment of the continuum-subtracted Stokes I LSD profile. Lunar pollution was corrected from some profiles.

Brightness tomography. In order to model the activity jitter of TAP 26, we applied ZDI (Semel 1989, Donati *et al.* 2014) to our data. ZDI is a tomography technique that enables to infer the bidimensional brightness distribution at the surface of the star from a time series of monodimensional LSD profiles, given the values of the rotation axis inclination to the line-of-sight, of the line-of-sight-projected equatorial rotation velocity and of differential rotation parameters. Minimizing the χ^2 and the spot coverage leads to the values of these stellar parameters (see Table 1) as well as the brightness maps which fit our data best. In the case of TAP 26, the data set was subdivided into two subsets which were fitted separately, because ZDI cannot fit the whole data set down to noise level with only one model. Figure 1 shows the time-series of TAP 26 Stokes I LSD profiles

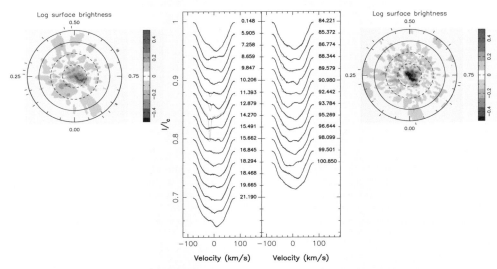

Figure 1. ZDI applied on TAP 26 data. *Center*: maximum entropy fit (thin red lines) to the observed (thick black lines) Stokes I LSD profiles. The 2015 Nov dataset is represented in the 1st column and the 2016 Jan dataset in the 2nd column. The Stokes I LSD profiles before the removal of lunar pollution are coloured in cyan. The rotational cycles are written beside their corresponding profiles, in concordance with Table 1. *Sides*: flattened polar view of the surface brightness maps for the 2015 Nov dataset (left panel) and 2016 Jan dataset (right panel). The equator and the 60°, 30° and -30° latitude parallels are depicted as solid and dashed black lines respectively. The colour scale indicates the logarithm of the relative brightness, with brown/blue areas representing cool spots/bright plages, the outer ticks mark the phases of observation. *Source*: Yu *et al.* (2017).

and brightness maps reconstructed from them. The brightness maps reconstructed for V830 Tau and the time-series of LSD profiles used for the fit are found in Donati *et al.* (2016) and Donati *et al.* (2017).

3. Looking for a planetary signature

After using ZDI to derive the surface brightness of both stars, revealing the presence of cool spots and warm plages, we applied three different methods to search for a planetary signature in the observed spectra. The first method studies the radial velocities filtered out from the activity jitter predicted by ZDI. The second method looks for the planet parameters that enable ZDI to fit the LSD profiles best, after having corrected them from the tested planet-induced reflex motion. The third method uses Gaussian-Process Regression (GPR) to fit the activity jitter in the raw RVs, and like the second method, searches for the orbital parameters which enable GPR to fit the raw RVs corrected from the reflex motion best.

ZDI #1. The first technique consists of using the previously reconstructed maps to predict the pollution to the RV curve caused by activity (called activity jitter in the following) and subtract it from the raw RVs. From the observed Stokes I LSD profiles, we compute the raw RVs RV_{raw}, as the first-order moment of the continuum-subtracted corresponding profiles. Likewise, the synthesised Stokes I LSD profiles derived from the brightness maps yield the synthesised activity jitter of the star (RV signal caused by the brightness distribution and stellar rotation). By subtracting the activity jitter from the raw RVs, we obtain filtered RVs RV_{filt} (see Fig. 2). Looking for a planet signature, we

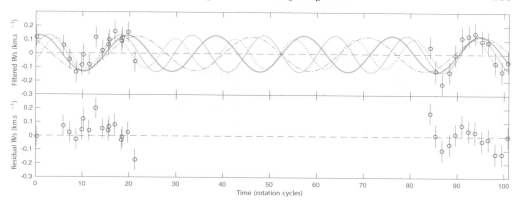

Figure 2. Top: filtered RVs of TAP 26 and four sine curves representing the best fits. The thick green curve represents the case $P_{\rm orb}/P_{\rm rot}=18.80$, the thin magenta one $P_{\rm orb}/P_{\rm rot}=15.27$, the dash-and-dotted blue one $P_{\rm orb}/P_{\rm rot}=24.56$ and the dotted black one $P_{\rm orb}/P_{\rm rot}=12.76$. Bottom: residual RVs resulting from the subtraction of the best fit (green curve) from the filtered RVs. The residual RVs feature a rms value of 51 m.s^{-1}.

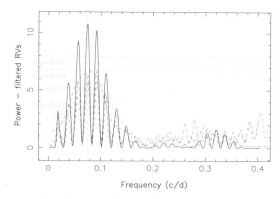

Figure 3. Periodogram of the TAP 26 filtered RV curve over the whole dataset (black line). The red line represents the 2015 Nov dataset, the green line the 2016 Jan dataset, the blue line the odd data points and the magenta line the even data points. False-alarm probability (FAP) levels of 0.33, 0.10, 0.03 and 0.01 are displayed as horizontal dashed cyan lines. The frequency that has the smallest FAP (0.06% at 0.075 cycles per day, corresponding to $P_{\rm orb}$=13.41 d) is marked by a cyan dashed line. Bottom: Zoom in the periodogram of filtered RVs.

want to fit a sine curve (of semi-amplitude K, period $P_{\rm orb}$, phase of inferior conjunction ϕ, and offset RV$_0$) to the RV$_{\rm filt}$, which corresponds to a circular orbit. The fitted curves are shown in Fig. 2 for TAP 26, and are found in Donati *et al.* (2016) and Donati *et al.* (2017) for V830 Tau. Plotting a Lomb-Scargle periodogram for the RV$_{\rm filt}$ further demonstrates the presence of a periodic signal (Fig. 3 for TAP 26, see Donati *et al.* 2016, Donati *et al.* 2017 for V830 Tau). By fitting the filtered RVs with a Keplerian orbit rather than a circular orbit, we obtain an eccentricity of 0.21±0.15 for V830 Tau and 0.16±0.15 for TAP 26, indicating that there is no evidence for an eccentric orbit following the precepts of Lucy & Sweeney (1971). We can thus conclude that the orbits of V830 Tau b and TAP 26 b are likely close to circular, or no more than moderately eccentric.

ZDI #2. In the previous method, we were trying to model the activity jitter by fitting data that were potentially the sum of both activity jitter and planet signature, leading to a risk of modelling part of the planet signature by brightness features on the reconstructed maps. To counter this, a next method is to add the presence of a planet into the physical

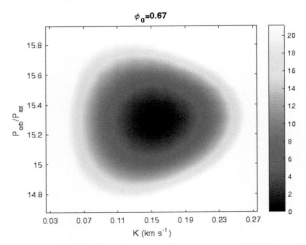

Figure 4. $\Delta\chi^2$ map as a function of K and $P_{\rm orb}/P_{\rm rot}$, derived with ZDI from corrected Stokes I LSD profiles at constant information content. Here the phase is fixed at 0.67, i.e., the value of ϕ at which the 3D paraboloid is minimum. The outer colour delimits the 99.99% confidence level area (corresponding to a χ^2 increase of 21.10 for 2581 data points in our Stokes I LSD profiles). The minimum value of $\chi_{\rm r}^2$ is 0.96824.

model and try to fit simultaneously the surface brightness and the planetary parameters to the LSD profiles. Technically, we try different values for the planetary parameters (amplitude K, period $P_{\rm orb}$ and phase ϕ of the RV signature, under the assumption of a circular orbit), and for each set of values, we subtract the planet-induced reflex motion from the spectra before applying ZDI to them. We look for the planet parameters that result in the best ZDI fit of the remaining activity jitter (see Petit *et al.* 2015, Yu *et al.* 2017). For both stars, we find consistent values of the planet parameters between methods #1 and #2 (Donati *et al.* 2017, Yu *et al.* 2017). A slice of the $\Delta\chi^2$ 3D-map, as a function of K, $P_{\rm orb}$ and ϕ, for TAP 26, is given on Fig. 4.

GPR. The third method we used works directly with the raw RVs and aims at modelling the activity jitter and its temporal evolution with GPR, assuming it obeys an a priori covariance function (Haywood *et al.* 2014, Rajpaul *et al.* 2015). Similarly to the previous method, we fit both the orbit model and the jitter model simultaneously. For a planet with given parameters, we first remove the planet-induced reflex motion from the RVs, then we fit the corrected RVs with a Gaussian process (GP) of pseudo-periodic covariance function:

$$c(t,t') = \theta_1^2 \cdot \exp\left[-\frac{(t-t')^2}{\theta_3^2} - \frac{\sin^2\left(\frac{\pi(t-t')}{\theta_2}\right)}{\theta_4^2}\right] \tag{3.1}$$

where t and t' are two dates, θ_1 is the amplitude (in km.s^{-1}) of the GP, θ_2 the recurrence timescale (in units of $P_{\rm rot}$), θ_3 the decay timescale (i.e., the typical spot lifetime in the present case, in units of $P_{\rm rot}$) and θ_4 a smoothing parameter (within [0,1]) setting the amount of high frequency structures that we allow the fit to include. From a given set of orbital parameters (K, $P_{\rm orb}$, ϕ) and of covariance function parameters (θ_1 to θ_4, called hyperparameters), we can derive the GP that best fits the corrected RVs (noted y below) as well as the log likelihood $\log\mathcal{L}$ of the corresponding set of parameters from:

$$2\log\mathcal{L} = -n\log(2\pi) - \log|C+\Sigma| - y^T(C+\Sigma)^{-1}y \tag{3.2}$$

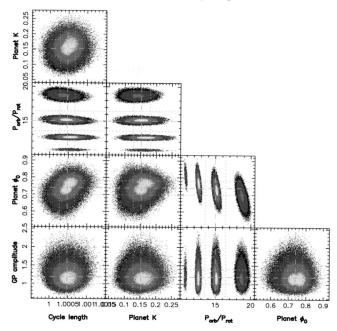

Figure 5. Phase plots of our 5-parameter MCMC run with yellow, red and blue points marking respectively the 1σ, 2σ and 3σ confidence regions. The optimal values found for each parameters are: $\theta_1 = 1.19 \pm 0.21$ km.s^{-1}, $\theta_2 = 1.0005 \pm 0.0002$ $P_{\rm rot}$, $K = 0.152 \pm 0.029$ km.s^{-1}. Several optima are detected for $P_{\rm orb}$: 12.61 ± 0.13 $P_{\rm rot}$, 15.12 ± 0.20 $P_{\rm rot}$ and 18.74 ± 0.34 $P_{\rm rot}$, ordered by decreasing likelihood. The corresponding phases ϕ are: 0.766 ± 0.030, 0.728 ± 0.033 and 0.694 ± 0.042 respectively. *Source*: Yu *et al.* (2017).

where n is the number of data points, C is the covariance matrix of all the observing epochs and Σ is the diagonal variance matrix of the raw RVs. Coupled with a Markov Chain Monte Carlo (MCMC) simulation to explore the parameter domain, this method generates samples from the posterior probability distributions for the hyperparameters of the noise model and the orbital parameters. From these we can determine the maximum-likelihood values of these parameters and their uncertainty ranges. Fig. 5 shows the phase plot for TAP 26, where, after an initial run where all the parameters are free to vary, we fixed θ_4 and θ_3 to their respective best values (0.50 ± 0.09 and 180 ± 60 $P_{\rm rot} = 128 \pm 43$ d) before carrying out the main MCMC run and looking for the best estimates of the 5 remaining parameters.

For both stars, GPR finds results that are consistent with those found using the ZDI methods.

4. Summary and discussion

All three methods demonstrate the clear presence of a planet signature in the data, although in the case of TAP 26 the gap between both data sets generates aliasing problems, causing multiple nearby peaks to stand out in the periodogram (of the dominant periods, the 10.8 d one emerges strongly for all three methods). Allowing ZDI to model temporal evolution of spot distributions and magnetic topologies should bring all methods on an equal footing; this upgrade is planned for a forthcoming study.

The hJs in nearly circular orbits that we have discovered in the young systems V830 Tau and TAP 26 are better explained by type II disc migration than by planet-planet scattering coupled to tidal circularization. When compared to V830 Tau, a 2 Myr wTTS of $\simeq 1.0 M_\odot$, TAP 26, at $\simeq 17$ Myr and of similar mass, appears as an evolved version, rotating 4x faster than its younger sister, likely as a direct consequence of its 4x smaller moment of inertia (according to the evolutionary models of Siess *et al.* 2000).

Regarding the hJs we detected around TAP 26 and V830 Tau and despite their differences (in mass in particular), it would be tempting to claim that, like its host star, TAP 26 b is an evolved version of V830 Tau b. This would actually imply that TAP 26 b migrated outwards under tidal forces from a distance of $\simeq 0.057$ au (where V830 Tau b is located) to its current orbital distance of 0.094 au, as a result of the spin period of TAP 26 being $\simeq 15$x shorter than the orbital period of TAP 26 b. This option seems however unlikely given the latest predictions of tidal interactions between a young T Tauri star and its close-in hot Jupiter (Bolmont & Mathis 2016), indicating that tidal forces can only have a significant impact on a hJ within 0.06 au of a solar-mass host star (for a typical TTS with a radius of $\simeq 2$ $R_\odot$). The most likely explanation we see is thus that TAP 26 b:

• ended up its type-II migration in the accretion disc at the current orbital distance, when TAP 26 was still young, fully convective and hosting a large-scale dipole field of a few kG similar to that of AA Tau (Donati *et al.* 2010), i.e., strong enough to disrupt the disc up to a distance of 0.09 au,

• was left over once the disc has dissipated at an age significantly smaller than 2 Myr, i.e., before the large-scale field had time to evolve into a weaker and more complex topology, and the inner accretion disc to creep in as a result (e.g., Blinova *et al.* 2016).

Admittedly, this scenario requires favorable conditions to operate; in particular, it needs the accretion disc to vanish in less than 2 Myr, which happens to occur in no more than 10% of single T Tauri stars in Taurus (Kraus *et al.* 2012). In fact, since both TAP 26 and V830 Tau have the same angular momentum content, it is quite likely that TAP 26 indeed dissipated its disc very early. Quantitatively speaking, assuming (i) that the hJ we detected tracks the location of the inner disc when the disc dissipated, (ii) that the spin period at this time was locked on the Keplerian period of the inner disc (equal to the orbital period of the detected hJ) and (iii) that stellar angular momentum was conserved since then, we derive that the disc must have dissipated when TAP 26 was about three times larger in radius, at an age of less than 1 Myr (according to Siess *et al.* 2000). Generating a magnetospheric cavity of the adequate size would have required TAP 26 to host at this time a large scale dipole field of 0.3-1.0 kG for mass accretion rates in the range 10^{-9} to 10^{-8} $M_\odot$ /yr, compatible with the large-scale fields found in cTTSs of similar masses (e.g., GQ Lup, Donati *et al.* 2012).

Along with other recent reports of close-in giant planets (or planet candidates) detected (or claimed) around young stars (van Eyken *et al.* 2012, Mann *et al.* 2016, Johns-Krull *et al.* 2016, David *et al.* 2016), our results may suggest a surprisingly high frequency of hJs around young solar-type stars, with respect to that around more evolved stars ($\simeq 1\%$, Wright *et al.* 2012). However, this may actually reflect no more than a selection bias in the observation samples (as for their mature equivalents in the early times of velocimetric planet detections). Planets are obviously much easier to detect around non-accreting TTSs as a result of their lower level of intrinsic variability; observation samples (like that of MaTYSSE) are thus naturally driven towards young TTSs whose accretion discs vanished early, i.e., at a time when their large-scale fields were still strong and their magnetospheric gaps large, and thus for which hJs had more chances to survive type-II migration. A more definite conclusion must wait for a complete analysis of the full MaTYSSE sample.

More observations are currently being planned to better determine the characteristics of the newborn hJs we detected. Furthermore, analyzing thoroughly the full MaTYSSE data set to pin down the frequency of newborn hJs within the sample observed so far will bring a clearer view on how the formation and migration of young giant planets is occurring. Ultimately, only a full-scale planet survey of young TTSs such as that to be carried out with SPIRou, the new generation spectropolarimeter currently being built for CFHT and scheduled for first light in 2018, will be able to bring a consistent picture of how young close-in planets form and migrate, how their population relates to that of mature hJs, and more generally how young hJs impact the formation and early architecture of planetary systems like our Solar System.

References

Baruteau, C. *et al.* 2014, *Protostars and Planets VI*, pp. 667–689
Batygin, K., Bodenheimer, P. H., & Laughlin, G. P. 2016, *ApJ*, 829, 114
Blinova, A. A., Romanova, M. M., & Lovelace, R. V. E. 2016, *MNRAS*, 459, 2354
Bolmont, E. & Mathis, S. 2016, *Celestial Mechanics and Dynamical Astronomy*, 126, 275
Brown, S. F., Donati, J.-F., Rees, D. E., & Semel, M. 1991, *A&A*, 250, 463
David, T. J. *et al.* 2016, *Nature*, 534, 658
Donati, J.-F., Semel, M., Carter, B. D., Rees, D. E., & Collier Cameron, A. 1997, *MNRAS*, 291, 658
Donati, J.-F. & Brown, S. F. 1997, *A&A*, 326, 1135
Donati, J.-F. *et al.* 2006, *MNRAS*, 370, 629
Donati, J.-F. *et al.* 2008, *MNRAS*, 385, 1179
Donati, J. *et al.* 2010, *MNRAS*, 409, 1347
Donati, J.-F. *et al.* 2012, *MNRAS*, 425, 2948
Donati J.-F. *et al.* 2014, *MNRAS*, 444, 3220
Donati, J.-F. *et al.* 2015, *MNRAS*, 453, 3706
Donati, J.-F. *et al.* 2016, *Nature*, 534, 662
Donati, J.-F. *et al.* 2017, *MNRAS*, 465, 3343
van Eyken, J. C. *et al.* 2012, *ApJ*, 755, 42
Haywood, R. D. *et al.* 2014, *MNRAS*, 443, 2517
Huélamo, N. *et al.* 2008, *A&A*, 489, L9
Johns-Krull, C. M. *et al.* 2016, *ApJ*, 826, 206
Kraus, A. L., Ireland, M. J., Hillenbrand, L. A., & Martinache, F. 2012, *ApJ*, 745, 19
Lagrange, A.-M. *et al.* 2010, *Science*, 329, 57
Lucy, L. B. & Sweeney, M. A. 1971, *AJ*, 76, 544
Mann, A. W. *et al.* 2016, *AJ*, 152, 61
Moutou, C. *et al.* 2007, *A&A*, 473, 651
Rajpaul, V., Aigrain, S., Osborne, M. A., Reece, S., & Roberts, S. 2015, *MNRAS*, 452, 2269
Sallum, S. *et al.* 2015, *Nature*, 527, 342
Semel, M. 1989, *A&A*, 225, 456
Setiawan, J., Henning, T., Launhardt, R., Müller, A., Weise, P., & Kürster, M. 2008, *Nature*, 451, 38
Siess, L., Dufour, E., & Forestini, M. 2000, *A&A*, 358, 593
Vogt, S. S., Penrod, G. D., & Hatzes, A. P. 1987, *ApJ*, 321, 496
Wright, J. T., Marcy, G. W., Howard, A. W., Johnson, J. A., Morton, T. D., & Fischer, D. A. 2012, *ApJ*, 753, 160
Yu, L. *et al.* 2017, *MNRAS*, 467, 1342

Living Around Active Stars
Proceedings IAU Symposium No. 328, 2016
D. Nandy, A. Valio & P. Petit, eds.

© International Astronomical Union 2017
doi:10.1017/S1743921317004161

Observable Impacts of Exoplanets on Stellar Hosts – An X-Ray Perspective

Scott J. Wolk[1], Ignazio Pillitteri[1,2] and Katja Poppenhaeger[1,3]

[1]Harvard Smithsonian Center for Astrophysics, 60 Garden Street,
Cambridge MA 02138, USA
email: swolk@cfa.harvard.edu
[2]Osservatorio Astronomico di Palermo, piazza del Parlamento 1, I-90134 Palermo, Italy
email: pilli@astropa.unipa.it
[3]Queen's University Belfast, University Road, Belfast, BT7 1NN, Northern Ireland, UK
email: K.Poppenhaeger@qub.ac.uk

Abstract. Soon after the discovery of hot Jupiters, it was suspected that interaction of these massive bodies with their host stars could give rise to observable signals. We discuss the observational evidence for star-planet interactions (SPI) of tidal and magnetic origin observed in X-rays. Hot Jupiters can significantly impact the activity of their host stars through tidal and magnetic interaction, leading to either increased or decreased stellar activity – depending on the internal structure of the host star and the properties of the hosted planet. We provide several examples of these interactions. In HD 189733, the strongest X-ray flares are preferentially seen in a very restricted range of planetary phases. Hot Jupiters, can also obscure the X-ray signal during planetary transits. Observations of this phenomena have led to the discovery of a thin upper atmospheres in HD 189733A. On the other hand, WASP-18 – an F6 star with a massive hot Jupiter, shows no signs of activity in X-rays or UV. Several age indicators (isochrone fitting, Li abundance) point to a young age ($\sim 0.5 - -1.0$ Gyr) and thus significant activity was expected. In this system, tidal SPI between the star and the very close-in and massive planet appears to disrupt the surface shear layer and thus nullify the stellar activity.

Keywords. X-rays: stars, magnetic fields, exoplanets, interactions

1. Introduction

Over the two decades, the discovery of exoplanets has fundamentally changed our perception of the universe and humanity's place within it. The role of X-rays in the study of exoplanets is subtle, but recent work indicates exoplanets, especially hot Jupiter systems, are unique X-ray environments and the impact of X-rays may be significant for the evolution of the system. The effects can work several ways; the intense high energy flux alters the thermal budget of the upper atmosphere of planet, the angular momentum and magnetic field of the planet can induce more activity on the star and the enhanced X-rays are absorbed by the transiting planet, which, in turn, act as a probe of the planetary upper atmosphere. In addition, an overall enhancement of the stellar host activity can significantly influence the chemistry of any additional planet in the habitable zone of the same star and thus the evolution of life in the system.

X-rays play a significant role in the evolution of close-in exoplanets. X-rays can modify the chemistryof exoplanet upper atmospheres and which can become over-inflated and evaporate becauseof the strong UV/X-ray flux to which they are exposed (see also presentations in these proceedings by Airapetian and Vidotto). As first noted by Lammer *et al.* (2003), inclusion of stellar X-ray and EUV flux in irradiance calculations leads to energy-limited escape and atmospheric expansion not found in models incorporating stellar UV/optical/IR insulation alone. The increased mass loss rates are of order 10^{12} g s^{-1}, implying hydrogen-rich exoplanets may evaporate and shrink to levels at which heavier atmospheric constituents may prevent hydrodynamic escape. The generation of an exosphere due to local X-ray luminosity has been directly detected in the case of the planets HD 209458b and HD 189733b. Absorption by atmospheric gas has been used to probe the layer where the gas escapes in the upper atmosphere (e.g. Vidal-Madjar 2003, Ballester *et al.* 2007 and Poppenhaeger *et al.* 2013).

Modeling by Penz *et al.* (2008) shows evolution of close-in exoplanets strongly depends on the detailed X-ray luminosity history of their host stars. Stars located at the high end of the X-ray luminosity distribution evaporate most of their planets' atmospheres within 0.05 AU, while a significant fraction of planets can survive if exposed to a moderate X-ray luminosity. At lower X-ray luminosities, they find that the mass loss is negligible for hydrogen-rich Jupiter-mass planets at orbits >0.02 AU, while Neptune-mass planets are influenced up to 0.05 AU (see also Murray-Clay *et al.* 2009).

Analytic models and MHD simulations show that the X-ray environment can bemodified by Star-Planet Interaction (SPI) when Jupiter mass planets are very close to their parentstars (< 0.1 AU; c.f. Lanza 2008, 2009, Cohen *et al.* 2009) SPI arises through two mechanisms,either gravitational tides, or interactions between magnetic fields. Both processes shouldincrease in strength as a^{-3} (where a is the separation distance; Cuntz *et al.* 2000).

SPI not only proceeds from star to the planet, but theoretical arguments demonstrate that hot Jupiters (HJs; planets with masses approximately equal to or greater than Jupiter with an orbital semi-major axis of less than 0.1 AU) can also influence their hosts. Cuntz *et al.* (2000) showed that energy generation due to tidal perturbations is also proportional to a^{-3}, More detailed work, demonstrates HJs induce tidal bulges on the host star. Cool stars can dissipate the energy contained in the bulges much more effectively than hot stars due to turbulent eddies in the convective envelopes (Zahn 2008). Similarly, Saar *et al.* (2004) estimated The energy released via reconnection during an interaction of the planetary magnetosphere with the stellar magnetic field is estimated as $F_{int} \propto B_* \times B_P v_{rel} a_P^n$ (with $n \sim -3$).

The manifestation of SPI in X-ray band is matter of debate. Kashyap *et al.* (2008)show that stars with hot Jupiters are statistically brighter in X-rays than stars without hotjupiters. On the other hand, Poppenhaeger *et al.* (2010) have found no statistical evidence ofX-ray SPI. Sample selection seems to be at the heart of the issue. Local solar analogs selectedfor inactivity (perhaps introducing a fundamental bias in the method) showed no significant correlationbetween common proxies for interaction strength (M_p/a^2 or M_p/a) versus coronal activity(L_x or L_x/L_{bol}), but a full sample of $\sim$ 200 FGK stars does (Miller *et al.* 2015). Miller *et al.* argue that there is a threshold for SPI and the effect is driven by hot Jupiters ($M_p/a^2 > 450 M_{jup}/AU^2$). In both analysis, a bias was noted in the data that stars hosting hot Jupiters were more active, but in both cases the effect was related to a limited number of extreme outliers in the sample, not a general trend.

The focus of this presentation is to discuss signs of feedback between the stars and planets in some of these extreme examples. In some cases there is a detectable enhancement, other times the high energy signal appear nullified. These cases may prove crucial for evolution of planets as well as our estimation of their habitability. For our purposes, star-planet interaction (SPI) is driven by magnetic interaction between the stellar and planetary magnetic fields, or by tidal interaction (Cuntz *et al.*, 2000). Both effects strongly depend on the planet-star separation, which is directly measurable. But the observation result is also a function of the intensity and topology of the magnetic fields, and the internal structure of the star – which are less apparent.

2. Observations

For the purposes of this contribution we will focus on observations of three systems with HJs. The second and third cases are edge cases representing a highly eccentric system and a high mass planet respectively. The first case, on the other hand, is the gold standard of HJs.

2.1. *HD 189733*

HD 189733 is one of the best studied systems with a transiting HJ. It is in a binary system with a quite inactive and old ($\tau > 5$ Gyr) M4 secondary star. This system was originally thought to be about 600 Myr, based on a relatively high activity level of the primary which led to the assumption it was a Hyades member. Pillitteri *et al.* (2010) noted they did not detect the M4 companion in the XMM image although it should have been bright enough if it were indeed 600 Myr. They speculated that the activity in the primary was enhanced by the interaction with the exoplanet. Indeed, when Poppenhaeger, Schmitt & Wolk (2013) detected the secondary star with the *Chandra X-ray Observatory*, they found an activity level consistent with an age of about 5 Gyr. This latter finding has led to the conclusion that the primary has been spun up by the hot Jupiter (Poppenhaeger & Wolk 2014).

2.1.1. *In Eclipse*

In addition to the age anomaly, Pillitteri *et al.* (2014) discuss three eclipse observations of HD 189733b (Fig. 3). Each time they noted a significant flare within hours after the eclipse. They speculated this was due to a hot spot forward phased from the sub planetary point by about 90°, consistent with analytic predictions by Lanza (2008). Using HST, Pillitteri *et al.* (2015) acquired high quality COS spectra in the wavelength range 1150-1450 Å. Again, flares were observed just after the eclipse. They found two episodes of strong variability of the line fluxes of ions of Si, C, and N that had not been observed in planetary transits. The details of the flares were consistent with an MHD model (Matsakos *et al.* 2015). The flow morphology in the model provides a natural explanation of the FUV line and X-ray variability of HD 189733. Specifically, the plasma is liberated from the upper atmosphere of the planet and funneled by the magnetized stellar wind in an almost radial trajectory close to the star. The flow forms a knee structure that consists of hot and dense plasma which then accretes in a region of the star fixed with the synodic phase (Fig. 2).

On the other hand, the X-ray flare observed in 2012, had aspects of reverberation. During the decay of the flare, three successively smaller peaks are observed separated by about 4 ks. This appears to have been a damped magneto acoustic oscillation in a flaring loop (e.g. Mitra-Kraev *et al.* 2005). In such loop the change of the intensity of the successive peaks can be described as:

$$\frac{\Delta I}{I} \sim \frac{4\pi n k_B T}{B^2}$$

Since fits to the XMM spectra can be used to determine the temperature and the density, this formulation can be used to measure the magnetic field – if magneto acoustic oscillation is the cause of the flare structure. In this case, the derived B field (~ 40 G) is consistent with results of spectropolarimetry (Fares *et al.* 2010). With the reverberation hypothesis thus supported, the length of the loop can be calculated by simple arguments about the sound speed, a function of the density, and the travel time (the time between the successive peaks). The result is a loop length of about 4 R_* indicating a flare covering half the distance to the planet. While there is no evidence that the flare is actually

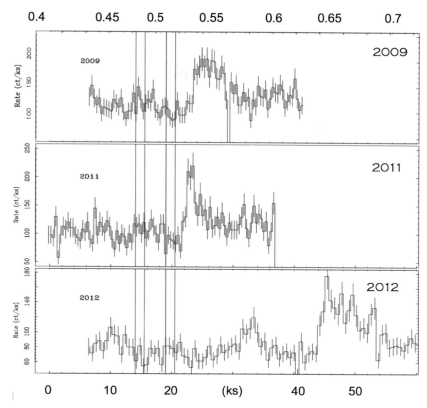

Figure 1. Light curves for the XMM-Newton EPIC/PN observations of three eclipses of HD 189733b. The top X-axis indicates phase. The bottom X-axis indicates time (normalized to the phase start of the 2011 observation). First through fourth contacts are indicated by the vertical lines. In all cases, the quiescent count rates are within 20% of 100 cnt/ks.

directed to the planet, the result is reminiscent of Favata *et al.* (2005) and McCleary & Wolk (2011). Both groups found that for flares in PMS stars, long length flares only occurred in cases where the star is surrounded by a disk. Both groups concluded that the flares stretched from the star to the disk in these cases. By analogy the magnetic field of the planet could be acting as a footpoint for occasional, massive, flares. Such flares would have significant impacts on the planet, ionizing material in its upper atmosphere. Since it appears to be tidal interaction between the star and the planet is the *prima facia* cause of the enhanced stellar field, planetary orbital energy is the ultimate source of the flare that scorches the planet's atmosphere.

2.1.2. *In Transit*

Transit observations of HD 189733b across HD 189733A have been performed with both XMM and *Chandra*. Popenhaeger *et al.* (2013) reported a detection of the planetary transit in soft X-rays using several coadded observations mostly relying on five clean (flare–free) *Chandra* observations. They noted a significantly deeper X-ray transit depth than observed in the optical. The X-ray data favor a transit depth of 6% – 8%, versus a broadband optical transit depth of 2.41%. Because several observation were co-added they were able to exclude transits of active regions and other possible stellar origins for this deep transit. They interpret the deep X-ray transit to be caused by a thin

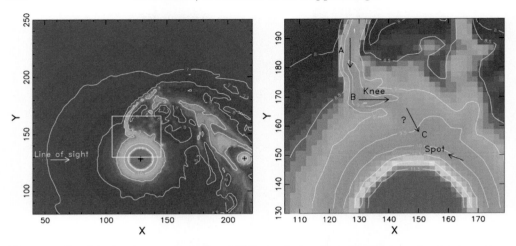

Figure 2. Particle density contours of an MHD simulation that models star–planet interactions between a HJ and its host (polar view). The star rotates counterclockwise, and the planet orbits the star along the same direction. The two "+" symbols shown on the left panel indicate the location of the star (red disk) and the planet (green disk). Right panel: a close-up of the impact region, where the motion of the accreting plasma is marked with arrows. Specifically, the shocked plasma is funneled by the magnetized stellar wind in an almost radial trajectory close to the star (A), forms a "knee" structure that consists of hot and dense plasma (B), and then accretes in a spot ahead of the orbital phase (C). The precise details of accretion are not investigated by the simulation (zone marked with ?). The knee (B) of the stream and the active spot upon impact on the surface (C) are the main sites of production of the enhanced flux observed in the FUV and X-ray bands and phased with the orbital motion (from Pillitteri *et al.* 2015).

outer planetary atmosphere extended to about 1.75 times the nominal radius which is transparent at optical wavelengths, but dense enough to be opaque to X-rays. The X-ray radius appears to be larger than the radius observed at far-UV wavelengths, most likely due to high temperatures in the outer atmosphere at which hydrogen is mostly ionized.

2.2. *HD 17156*

In addition to HJs close to their host stars, eccentric systems – in which the Jupiter is "hot" only a fraction of the time – represent another type of extreme system. These *should* be a good test bed for the existence of SPI. The prediction is that SPI impacts only occur when the star and planet are close. One candidate for such a test is HD 17156.

HD 17156 is a G0 star with a HJ in a 21 day orbit (Barbieri *et al.* 2007). The eccentricity of the orbit is 0.68. The planet reaches a minimum separation of ~ 15 R$_*$. XMM-Newton observed HD 17156 when the planet was at the periastron and a second time, when the planet was more distant. HD 17156 was not detected by XMM when the two were separated. However, just after periastron passage, there was a marked rise in the X-ray luminosity with a corresponding rise in the chromospheric activity. Maggio *et al.* (2015) suggest that this could have been either due a magnetic reconnection and or flaring activity when the planet was at its minimum separation or due to material stripped from the planet and falling onto the star. The excess of X-rays has a soft spectrum and could favor the tidal stripping hypothesis.

2.3. *WASP-18*

In a sense, WASP-18 (F6V) is the most extreme exoplanet system. It hosts one of the fastest orbiting and most massive HJs. The planet has a mass of over 10M$_{Jup}$ and the

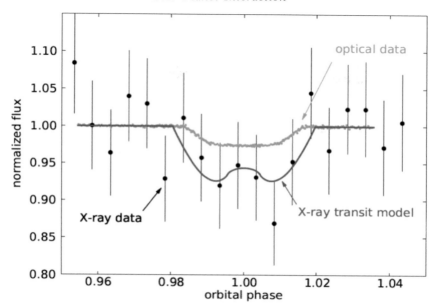

Figure 3. The co-added lightcurve of 5 Chandra X-ray Observatory observations (without flares) of a transit of HD 189733A by HD 189733b. The X-axis indicates phase. The orange line indicates typical optical data and an eclipse depth of 2-3 %. The dots with the error bars are the co-added and normalized X-ray data. The purple line is a best fit to the X-ray data using a limb-brightened model appropriate of an optically thin stellar corona. The best fit X-ray eclipse depth is about 7%.

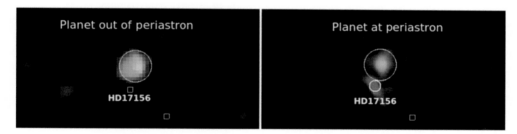

Figure 4. X-ray images of HD 17156 taken far from the planetary periastron (left panel) and near the periastron (right panel). The image is colored by photon energy (red = 0.3-1.0 keV, green = 1.0-2.5 keV, blue=2.5-5.0 keV). Smoothing is applied to the images, with a Gaussian $\sigma = 4''$ kernal. Positions of the only two objects in the SIMBAD catalog are shown with small squares. Circle sizes indicate the wavelet detection scales of HD 17156 and of an unrelated background object with a harder spectrum (From Maggio *et al.* 2015).

period is about 22 hours. The estimate of the age for this system is about 600 Myrs. This is based on both isochrone fitting and the strong Li absorption. Indeed, the observed Li strength is a near perfect match for Hyades and Beehive cluster stars of similar temperature. The expected X-ray luminosity from such a star should be at least $\log L_X \sim 28.5$. This is nearly the same as observed from HD 189733 and *a priori* might indicate strong magnetic re-connection events.

In fact, exactly the opposite is seen. No X-rays were detected with a flux limit of log $L_X \sim 26.5$. This result is part of a trend. Recently, Staab *et al.* (2016) showed that a quarter of known short period planet hosts exhibit anomalously *low* activity levels. Fossati *et al.* (2014) asserted that the lack of chromospheric activity detected from WASP-18 (and WASP-12) may have been due to local absorption of UV by atmospheric material stripped off of the planet by tidal forces. We find this unlikely for several reasons: First, given the high mass of WASP-18b, atmospheric stripping should be 100 times less effective in this case than HD 189733. Second, the calcium absorption line is observed. It is only the emission reversal which is not observed. As this should occur above the Ca absorption feature it is hard to understand how you can observed the lower lying feature while absorbing the higher altitude feature. Third, no X-rays are detected at all, not even high energy X-rays which might be expected from the Planck tail of a 1 keV thermal distribution. Pillitteri et al. (2014) conclude the the star is X-ray dark and suggest tidal interaction with the planet must have a role in destroying the dynamo efficiency and the overall activity of the star. Based on the formulae given in Cuntz et al. (2000), tides on WASP-18 are the largest of any known exoplanet host, of order of 500 km because of planet proximity and its mass. However, a complete break down in convection is not possible as this is key to energy transport in the star. On the other hand, the tidal wave on the surface may reach heights of $\sim$ 500 km every $\sim$ 10 hours. This could disrupt the shear layer at the top of the convection zone. This would prevent the build-up and concentration of magnetic energy close to the surface. Meanwhile the convective thermal transport from the core to near the surface is free to occur. Near the surface, radiative cooling processes dominate.

3. Conclusions

Over the past five years there has been a great deal of case specific evidence gathered that stars can interact with their HJ companions via tides and magnetic fields. The level of confidence in each kind of evidence varies.

One example that has not been touched on here comes from binary stars. Poppen-haeger & Wolk (2014) discussed using binaries to test for activity induced by HJs. They presented at least 3 examples of binaries in which the HJ host had an activity level incompatible with the age indicated by the activity level of the non-hot Jupiter host. They concluded that **very close in HJs spin-up stars when the HJ host has a large convective zone**. All the observed data are consistent with this hypothesis and it is strongest direct evidence of SPI. The corollary to this is an uncertainty when using the activity–age relation to date stars with close in planets.

Prior to eclipse observations, analytical calculations predicted the existence of an active region on HD 189733 forward phased by about 90° from the sub-planetary point. This would imply enhanced activity between phases 0.5 and 1.0. The propensity for the star to flare between phase 0.55 and 0.65 has been taken as evidence of the active region, perhaps being observed over the limb. In total there have been 5 high energy flares observed in four observations (including the FUV observations). The significance of this is unclear and fraught with over–interpretation. So while the result is suspicious, it is not yet compelling.

This could be an observational bias, certainly some (albiet smaller) flares have been noted in other phases and to be fair any flare observed between phase 0.45 and 0.6 would probably have been counted by Pillitteri et al. as evidence of this effect. Still, the odds of all the strong flares being observed to happen on one half of the star and not the other half is about 10% and this is taken as support of the hot spot hypothesis.

The two remaining pieces of evidence, the long length flare in HD 189733 and enhanced activity and periapses for HD 17156 are tantalizing. On the other hand, each of these has only been seen to occur once. Until additional flares are seen in HD 17156 or another eccentric system in periapses passage this will stand as a one-time occurrence.

The expected frequency for such long length flares is unknown, but it appears long. In 850 ks observing the ONC only 25 such flares were observed among about 400 disked stars (Favata *et al.* 2005). The situation is similar for the long flare seen on HD 189733. Overall the System has been monitored for several hundred kiloseconds in X-rays. While this is a long time, we expect such flares to be exceptionally rare. Simply based on the COUP, Favata *et al.* (2005) reported 25 loops longer than 10^{11} cm. In that project there was a total of of 850 ks observing time put in on about 400 stars known to possess disks. From this, the derived "long flare" rate is about 1 per 13 megaseconds or about 2 per year among active young stars with disks. The occurrence rate is probably somewhat higher than this. The Favata study focused on luminous flares and the one observed on HD 189733 would not have met this criterion. Still, it appears such long flares are not the most common.

While the result of the extreme example appear compelling, statistical evidence is a different story. Several groups (Kayshap *et al.* 2008, Poppenhaeger *et al.* 2011 and Miller *et al.* 2015) have looked for statistical evidence of SPI, with mixed results. The problem is that many parameters effect the eventual outcome of the star-planet interaction. Mass and distance ratios are the obvious parameters, but depth of convection and relative field orientation and strength are clearly others. The cautionary tale of WASP-18 (and others such as WASP-12) indicates that tidal forces can be constructive or destructive when it comes to modifying stellar activity. Any statistical test needs to account properly for outliers. Both the high and low outliers may be the result of SPI while the median and mean may not be very affected. The coming few years should prove very interesting as we continue to gather evidence.

References

Barbieri, M., *et al.* 2007, *A&A*, 476, L13

Favata, F., Flaccomio, E., Reale, F., *et al.* 2005, *ApJS*, 160, 469

Fares, R., *et al.* 2010, *MNRAS*, 406, 409

Fossati, L., Ayres, T. R., Haswell, *et al.* 2014, *APSS*, 354, 21

Kashyap, V. L., Drake, J. J., & Saar, S. H. 2008, *ApJ*, 687, 1339

Lanza, A. F. 2008, *A&A*, 487, 1163

Maggio, A., *et al.* 2015, *ApJL*, 811, L2

McCleary, J. E. & Wolk, S. J. 2011, *AJ*, 141, 201

Miller, B. P., Gallo, E., Wright, J. T., & Pearson, E. G. 2015, *ApJ*, 799, 163

Mitra-Kraev, U., Harra, L. K., Williams, D. R., & Kraev, E. 2005, *A&A*, 436, 1041

Murray-Clay, R. A., Chiang, E. I., & Murray, N. 2009, *ApJ*, 693, 23

Pillitteri, I., Wolk, S. J., Cohen, O., Kashyap, V., *et al.* 2010, *ApJ*, 722, 1216

Pillitteri, I., Wolk, S. J., Lopez-Santiago, J., Günther, H. M., *et al.* 2014, *ApJ*, 785, 145

Pillitteri, I., May A., Micela, G., Sciortino, S., Wolk, S. J., & Matsakos, T. 2015 *ApJ*, 805, 52

Poppenhaeger, K. & Wolk, S. J. 2014, *A&A*, 565, L1

Poppenhaeger, K., Schmitt, J. H. M. M., & Wolk, S. J. 2013, *ApJ*, 773, 62

Poppenhaeger, K. & Schmitt, J. H. M. M. 2011, *ApJ*, 735, 59

Staab, D., Haswell, C. A., Smith, G. D., Fossati, L., Barnes, J. R., Busuttil, R., & Jenkins, J. S. 2016, arXiv:1612.01739

Vidal-Madjar, A., Lecavelier des Etangs, A., Désert, *et al.* 2003, Nature, 422, 143

Discussion

Living Around Active Stars
Proceedings IAU Symposium No. 328, 2016
D. Nandy, A. Valio & P. Petit, eds.

© International Astronomical Union 2017
doi:10.1017/S1743921317003647

Possible effects on Earth's climate due to reduced atmospheric ionization by GCR during Forbush Decreases

Williamary Portugal[1], **Ezequiel Echer**[1], **Mariza Pereira de Souza Echer**[1], **and Alessandra Abe Pacini**[2]

[1] INPE - National Institute for Space Research
email: `williamary.portugal@inpe.br`
[2] Johns Hopkins University/Applied Physics Lab.

Abstract. This work presents the first results of a study about possible effects on the surface temperature during short periods of lower fluxes of Galactic Cosmic Rays at Earth, called Forbush Decreases. There is a hypothesis that the Galactic Cosmic Ray flux decreases cause changes on the physical-chemical properties of the atmosphere. We have conducted a study to investigate these possible effects on several latitudinal regions, around the ten strongest FDs occurred from 1987 to 2015. We have found a possible increase on the surface temperature at middle and high latitudes during the occurence of these events.

Keywords. Cosmic rays, solar activity, Forbush Decreases, solar-terrestrial relations, surface temperature.

1. Introduction

The possible effects of Galactic Cosmic Rays (GCR) on the physical - chemical atmospheric processes with influence on the surface temperature and consequently on the Earth's climate, has been widely studied recently (Svensmark & Friis - Christensen 1997, Marsh & Svensmark 2000, Marsh & Svensmark 2000a, Usoskin *et al.* 2004, Svensmark, Bondo & Svensmark 2009, Dragic *et al.* 2011, Marsh & Svensmark 2011 and Svensmark, Enghoff & Svensmark 2012). The hypothesis is that the ions present on the atmosphere, resulted from the induced ionization by GCR, can be enhanced by ion-ion or by ion-molecule combination, increasing the total number of aerossols (Usoskin & Kovaltsov 2008). Since some ions / aerossols can behave like cloud condensation nuclei, it is known that the presence of ions in the troposphere, that vary with the GCR flux variation, may change the water vapor condensation patterns (Usoskin & Kovaltsov 2008).

Some works have investigated the atmospheric parameter variation during the sporadic periods of lower GCR flux, called Forbush Decreases (FD) (Svensmark, Bondo & Svensmark 2009, Dragic *et al.* 2011 and Svensmark, Enghoff & Svensmark 2012). FDs are events characterized by a lower flux of GCR particles reaching the atmosphere, that occur mainly due to the passage of interplanetary remnants of Coronal Mass Ejections (CMEs) on the Earth (Cane 2000). So, if more energetic particles are impinging on the Earth's atmosphere (lower solar magnetic activity) leading to more ionization, it is possible that the presence of a lower GCR flux, will lead to atmospheric changes too. Then, we have investigated the possible effects of solar variability related to surface temperature on several global regions, including high and middle latitude stations of the two hemispheres (Northern and Southern), around the ten strongest FDs occurred from 1987 to 2015.

Table 1. The strongest FD events from 1987 to 2015, the 0 day, the periods analised, the decrease intensity (%) considering Oulu Neutron Monitor data, solar cycle and the solar polarity of the periods investiged.

FD	Day 0 and Period Analised	Intensity (%)	Solar Cycle and Polarity	FD	Day 0 and Period Analised	Intensity (%)	Solar Cycle and Polarity
1	Oct. 31, 2003 (10/26 to 11/10)	33.76	Cycle #23 A>0	6	Jul. 27, 2004 (07/22 to 08/06)	15.27	Cycle #23 A>0
2	March 24, 1991 (03/19 to 04/03)	22.00	Cycle #22 A<0	7	Nov. 30, 1989 (11/25 to 12/10)	14.73	Cycle #22 A<0
3	Oct. 29, 1991 (10/24 to 11/08)	20.20	Cycle #22 A<0	8	Sep. 13, 2005 (09/08 to 09/23)	14.35	Cycle #23 A>0
4	Mar. 13, 1989 (03/08 to 03/23)	17.44	Cycle #22 A<0	9	Nov. 10, 2004 (11/05 to 11/20)	13.00	Cycle #23 A>0
5	Mar. 9, 2012 (03/04 to 03/19)	15.60	Cycle #24 A<0	10	Jun. 25, 2015 (06/20 to 07/05)	11.00	Cycle #24 A<0

2. Methodology

Using GCR data collected from Oulu neutron monitor (Lat: 65.01^0; Long: 25.51^0) where the local vertical geomagnetic cutoff rigidity (P_c) is about 0.8 GV and surface temperature data obtained from NOAA - National Oceanic Atmospheric Administration / GSOD - Global Surface Summary of the Day, we constructed time series of surface temperature daily mean from ten meteorological stations of high (60^0 - 70^0) and of middle (40^0 - 50^0) latitudes, of Northern and Southern Hemispheres, for a period of fifteen days around the ten strongest FD that occurred from 1987 to 2015 (Table 1). We have used five days before and ten days after the FD day - day 0, characterized by the day with the lowest flux of GCR, measured by a neutron monitor station. A superposed epoch analysis was performed considering the surface temperature daily mean of the meteorological stations of each latitude range of each hemisphere. Here we have investigated only the FD events not related to the occurrence of Ground Level Enhancement events, or GLEs.

3. Preliminary Results and Discussion

In this work we are showing the preliminary results on the temperatute variation during FD events at high and middle latitudes of the two hemispheres (Northern and Southern). Figure 1 suggests that there was an increase of the surface temperature during the oc-curence of FD events, reaching its maximum value some days after the 0 day. This delay of 3 to 6 days, between the FD events and the response in an atmospheric parameter, is found also by other works (e.g: Svensmark, Bondo & Svensmark 2009 and Svensmark, Enghoff & Svensmark 2012). This might indicate that there is a connection between GCR decrease and some changes on the atmosphere parameters, such as: aerosols, cloud cover, cloud microphysics and temperature (e.g.: Svensmark, Bondo & Svensmark 2009, Dragic *et al.* 2011 2011 and Svensmark, Enghoff & Svensmark 2012). A possible mechanism could be by effects of lower ionization on the chemical-physical atmospheric conditions due to lower GCR flux, that could lead to less aerosols, less cloud cover and conse-quently to the increase of solar irradiance reaching the surface, leading to warming / surface temperature increase. In future works we will investigate these effects on the low latitude region and the difference between the surface temperature on the same regions during the occurrence of FD and during normal periods, represented by 1987, 1996 and

W. Portugal *et al.*

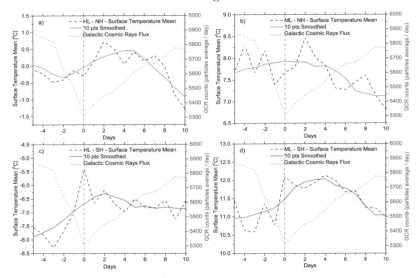

Figure 1. Superposed epoch analysis of surface temperature daily means of ten meteorological stations of diferent regions (black dashed line) during the ten strongest FD events: a) high latitude of Northern Hemisphere; b) middle latitude of Northern Hemisphere; c) high latitude of Southern Hemisphere; d) middle latitude of Southern Hemisphere. The continuous lines (pink and blue) represent the surface temperature daily mean data smoothed by 10 points. The red dot lines are representing the superposed epochs of GCR Flux from Oulu Neutron Monitor data and the grey vertical dashed line indicates the 0 day of FD. HL: High Latitude; ML: Middle Latitude; NH: North Hemisphere; SH: South Hemisphere.

2008 years (minimum solar periods). The final goal of this study is to understand the effects of the GCR on the surface temperature and the statistical significance of these effects in several global regions, in short and long time scales.

4. Acknowledgements

The authors would like to thanks for CAPES, INPE and Ezequiel Echer - CNPQ/PQ (302583/2015-7).

References

Cane, H. V. 2000, *Space Sci. Revs*, 93, 55

Dragić, A., Aničin, I., Banjanać, R., Udovičić, V., Joković, D., Maletić, D., & Puzović, J. 2011, *Ap&SS*, 7, 315

Marsh, N. & Svensmark, H. 2000, *Space Sci. Revs*, 94, 215

Marsh, N. & Svensmark, H. 2000a, *Phys. Rev. Lett.*, 85, 5004

Svensmark, H. & Friis - Christensen, E. 1997, *Journal of Atmospheric and Solar - Terrestrial Physics*, 59, 1225

Svensmark, H. & Bondo, T, Svensmark, J. 2009, *Geophysical Research Letters*, 36, L 15101, DOI 10.1029/2009GL038429

Svensmark, J., Enghoff, M. B., & Svensmark, H. 2012, *Atmos. Chem. Phys. Discuss.*, 12, 3595

Usoskin, I. G., Marsh, N., Kovaltsov, G. A., Mursula, K., & Gladysheva, O. G. 2004, *Geophysical Research Letters*, 31, L16109

Usoskin, l. G. & Kovaltsov, G. A. 2008, *C. R. Geoscience*, 340, 441

Living Around Active Stars
Proceedings IAU Symposium No. 328, 2016
D. Nandy, A. Valio & P. Petit, eds.

© International Astronomical Union 2017
doi:10.1017/S1743921317004501

Interaction of extra solar planets with their host star

Douglas Silva and Adriana Valio

Center for Radio Astronomy
and Astrophysics Mackenzie (CRAAM),
São Paulo, Brazil
email: **douglas93f@gmail.com**

Abstract. Transit is the passage of the planet in front of its star. During one of these transits, the planet may occult a spot on the photosphere of the star, causing small variations in its light curve. By detecting the same spot in a later transit, it is possible to estimate the stellar rotation period. The comparison between the rotation period of star at the equator and the planets orbital period showed the existence of resonances between these periods. Two types of mechanisms are proposed in the literature: electromagnetic interaction between the stellar and planetary fields and gravitational interaction. Our results have shown that for planets CoRoT-2b, CoRoT-5b and CoRoT-8b, tidal effects seem to dominate, whereas for planets CoRoT-4b and CoRoT-6b electromagnetic interaction dominates over tidal effects. A distinct characteristic of these last two systems is that the orbital period is larger than the rotation period of the star.

Keywords. resonances: rotation period, orbital period, tidal effects, electromagnetic interac

1. Introduction

According to the website exoplanet.eu there are already more than 3,700 exoplanets discovered orbiting other stars, with more than 2,700 transiting their host star. During the passage of the planet in front of its host star, there is a decrease in the intensity of light from the star. When the planet eclipses the star, it can occult spots causing slight variations perceptible in the star transit light curve.

These small changes detected during planetary transit interpreted as due to the spots were fit using the model developed by Silva (2003). In this way it was possible to obtain the physical characteristics of the spots and by applying this technique to many transist, estimate the rotation profile of the star. Five stars detected by CoRoT satellite have been analyzed and their rotation periods and differential rotation determined. Using the rotation profile, we estimated the period at the stellar equator.

The results showed a resonance between the planetary orbital period and the rotational period of the star, as shown on Table 1. This observed resonance may indicate an interaction between the planet and its host star. Two interaction mechanisms are proposed in the literature. The first type of interaction is the gravitational type, or tidal effect, and the second type of interaction is the magnetic interaction that occurs between the magnetic fields of the host star and planet (Cuntz *et al.* (2000)).

To study the gravitational interaction, we adapted the model of Jackson *et al.* (2008), which studies the evolution of the planet's orbital semi-major axis and eccentricity. The application of this model shows the evolution caused by the tidal effects.

For the magnetic interaction, the models of Lanza (2009) and Lanza (2012) were used, to calculate the energy dissipated by magnetic reconnection between the lines of magnetic fields of both the planet and the star. In Lanza (2012), the authors emphasize the energy release processes in the stellar corona due to changes in helicity.

Table 1. Differential Rotation and Rotation of CoRoT stars

Star	CoRoT -2a	CoRoT -4a	CoRoT -5a	CoRoT -6a	CoRoT -8a	Sol
$P_{rot}(days)$	4.54	8.87	26.63	6.35	21.7	27.6
$P_{rot}(\alpha)(days)$	4.48	8.71	26.49	6.08	21.42	
Diff. Rot.(rd/day)	0.042	0.026	0.103	0.101	0.014	0.050
Relative Diff. Rot. (%)	3.0	3.6	42.9	10.2	4.9	22.1
$P_{equat}(days)$	4.47	8.53	20.52	5.98	20.33	24.7
$P_{orb}(days)$	17.43	9.203	4.038	8.886	6.212	
$P_{equat} : P_{orb}$	2 : 5	1 : 1	1 : 5	3 : 2	1 : 3	

The objective of this study is to better understand planetary systems of solar-type stars orbited by hot Jupiters, and their interaction with one another.

Star-planet interactions

To study the gravitational interaction we used the system of equations described in Jackson et al. (2008) to determine the orbital evolution the eccentricity and semi-major axis of the planets:

$$\frac{1}{e}\frac{de}{dt} = - \left[\frac{63}{4}(GM_s^3)^{1/2} \frac{R_p^5}{Q_p' M_p} \pm \frac{171}{16} \left(\frac{G}{M_s} \right)^{1/2} \frac{R_s^5 M_p}{Q_s'} \right] a^{\frac{-13}{2}} \tag{1.1}$$

$$\frac{1}{a}\frac{da}{dt} = - \left[\frac{63}{2}(GM_s^3)^{1/2} \frac{R_p^5}{Q_p' M_p} e^2 \pm \frac{9}{2} \left(\frac{G}{M_s} \right)^{1/2} \frac{R_s^5 M_p}{Q_s'} \right] a^{\frac{-13}{2}} \tag{1.2}$$

where e is the eccentricity, a the semi-major axis, G the constant gravitational, M_S the mass of the star, M_p the planet's mass, R_s the radius of the star, R_p the radius of the planet, Q_p' is the planetary dissipation parameter and Q_s' is the stellar dissipation parameter. The (+) sign applies for systems where the orbital period is smaller than the stellar rotation period, whereas the (−).

To obtain the energy dissipated by the system, we used the following equation of Jackson et al. (2008)

$$h = \left(\frac{63}{16\pi} \right) \frac{(GM_s)^{3/2} M_s R_p^3}{Q_i^p} a^{-15/2} e^2 \tag{1.3}$$

where h is the internal heating per unit surface area of the planet. As the planetary orbit evolves through the coupling of tide, the orbital energy can result in substantial internal heating of the planet caused by tidal effects. The equations were integrated over the estimated age of each planet studied using the Runge-Kutta algorithm. The input data for the model for our planetary systems were taken from Table 1. For the stellar dissipation parameter, we assume $Q_s = 10^{5.5}$ and for terrestrial planets, we used $Q_p = 10^{6.5}$ (Jackson et al. (2008)).

For the power dissipated by the magnetic reconnection we used the equation Lanza (2009) described as follows:

$$P_{rec} = \gamma_{rec} \frac{\pi}{\mu} R_p^2 B_{rp}^{4/3} B_{p0}^{2/3} V_{rel} \tag{1.4}$$

where P_{rec} is the power dissipated, $0 < \gamma < 1$ is a factor that depends on the angle between the interacting magnetic field lines, R_p is the radius of the planet, B_{rp} is the coronal field of the star on the stellar equatorial plane at $r = a$ and B_{p0} is the magnetic field strength at the poles of the planets and V_{rel} is the relative velocity between the planet and the stellar coronal field.

Table 2. Dissipated tidal energy of the CoRoT planets

System	a	e	year (10^9)	*Power (W)*
CoRoT 2	0.028	$0,023$	$0,13 - 0,5$	$1,53 \times 10^{20}$
CoRoT 4	0.09	$0,09$	$0,7 - 2,0$	$1,15 \times 10^{18}$
CoRoT 5	0.04947	$0,09$	$5,5 - 8,3$	$5,08 \times 10^{19}$
CoRoT 6	0.0855	$0,11$	$1,0 - 3,3$	$2,06 \times 10^{18}$
CoRoT 8	0.063	$0,09$	$2,0 - 3,0$	$5,47 \times 10^{18}$

Table 3. Power dissipated magnetic by reconnection and Helicity magnetic

System	a (R_{star})	R_p (m)	V_{rel} $10^5_{km/s}$	$P_{rec}(W)$	$P_{Hel}(W)$
CoRoT 2	6.7	1.2×10^7	1.6	1.52×10^{17}	7.69×10^{17}
CoRoT 4	17.47	8.3×10^7	1.1	4.66×10^{18}	5.97×10^{17}
CoRoT 5	9.877	9.2×10^7	1.4	7.99×10^{18}	9.11×10^{18}
CoRoT 6	17.95	8.1×10^7	0.96	4.094×10^{18}	4.72×10^{17}
CoRoT 8	17..61	1.5×10^7	1.1	1.639×10^{17}	2.035×10^{16}

The magnetic models developed by Lanza (2012) estimated the energy released by changes in helicity of magnetic fields. The model was developed for linear force-free field, asymmetric forces free fields and non-linear fields. In this work, we only adopted the approach of non-linear field as shown below:

$$P_{hel} = \frac{27\pi}{16\mu} \frac{n+1}{n} \lambda^2 B_0^{4/3} B_{p0}^{2/3} R_p^2 \left(\lambda_2 + n^2\right)^{\frac{-1}{3}} V_{rel} \left(\frac{r}{R}\right)^{-(n+11)/3} \tag{1.5}$$

where n is a positive constant, λ^2 is the eigenvalue, r is radius of the star and R radius of the planet. The values used for B_{p0} and B_0 are respectively 100 Gauss and 10 Gauss, $n = 5$ and $\lambda^2 = 0.82343$, the same values adopted by Lanza (2012). These values are intended to maximize power dissipation. The relative velocity was taken as the difference between the orbital velocity of the planet and the rotational velocity of the star

Results

The gravitational and magnetic interactions were modeled for CoRoT 2, CoRoT 4, CoRoT 5, CoRoT 6 and CoRoT 8. To calculated the energy dissipated by tidal effects we used the variation of the orbital parameters obtained by Solving the Equations (1.1) and (1.2).

The values listed on Table 2 were obtained by substituting the present values of the semi-major axis and the eccentricity in Equation (1.3) of this work.

To obtain the values of the power dissipated by magnetic reconnection, listed in Table 3, we used Equation 1.4 of this work, and for the values of power due to the helicity variation, we used Equation 1.5.

Conclusions

Comparison of the results listed on Tables 2 and 3 shows that the power dissipated by gravitational effects is greater than the power lost by magnetic reconnection or helicity for the CoRoT 2, CoRoT 5 and CoRoT 8 systems. Lanza(2012) also concluded that the power dissipated by magnetic interaction was not sufficient to explain the observations of the stars studied.

However, the power dissipated by magnetic reconnection was greater than the energy dissipated by tidal effect for the CoRoT 4 and CoRoT 6 systems, these systems have planets with an orbital period greater than the period of rotation of the stars, unlike the others.

Therefore the mechanism that dominates, tidal or magnetic effects, depends on the type of star and the individual parameters of the planet and its host star.

References

Silva, A. V. R. 2003, *Bulletin of the Astronomical Society of Brazil*, 23, 15
Lanza, A. F. 2012, *A & A*, 544, A23
Lanza, A. F. 2009, *A & A*, 505, 339
Jackson, B., Greenberg, R., & Barnes, R. 2008, *ApJ*, 681, 1631-1638
Jackson, B., Barnes, R., & Greenberg, R. 2008, *MNRAS*, 391, 237
Cuntz, M., Saar, S. H., & Musielak, Z. E. 2000, *ApJL*, 533, L151

Living Around Active Stars
Proceedings IAU Symposium No. 328, 2016
D. Nandy, A. Valio & P. Petit, eds.

© International Astronomical Union 2017
doi:10.1017/S1743921317003854

The influence of eclipses in the stellar radio emission

Caius Lucius Selhorst[1,2] and Adriana Valio[3]

[1] NAT - Núcleo de Astrofísica Teórica - Universidade Cruzeiro do Sul
São Paulo, SP, Brazil
email: **caiuslucius@gmail.com**

[2] IP&D - Universidade do Vale do Paraíba - UNIVAP
São José dos Campos, SP, Brazil

[3] CRAAM - Universidade Presbiteriana Mackenzie, São Paulo, SP, Brazil

Abstract. Here we simulate the shape of a planetary transit observed at radio wavelengths. The simulations use a light curve of the K4 star HAT-P-11 and its hot Jupiter companion as proxy. From the HAT-P-11 optical light curve, a prominent spot was identified (1.10 R_P and 0.6 I_C). On the radio regime, the limb brighting of 30% was simulated by a quadratic function, and the active region was assumed to have the same size of the optical spot. Considering that the planet size is 6.35% of the the stellar radius, for the quiet star regions the transit depth is smaller than 0.5%, however, this value can increase to $\sim 2\%$ when covering an active region with 5.0 times the quiet star brightness temperature.

Keywords. Eclipses - Stars: activity - Radio Continnum: stars

1. Introduction

Since the discovery of the first exoplanets in the nineties, the number of confirmed ones exceed 3500 at this time (http://exoplanet.eu). Part of this success can be addressed to dedicated projects like HARPS and Kepler.

Some of these exoplanets can observed by the dimming of the light from the parent star during the planetary transit (e.g., Alonso *et al.* 2004). Besides planet detection, the transit can be used to detect spots on the stellar surface (Silva 2003) and estimate the stelar activity (Silva-Valio *et al.* 2010).

Despite the great number of exoplanets, the observations are still restricted to the optical wavelengths range. Although recently, a transit observation was reported in X-ray (Poppenhaeger *et al.* 2013), with interesting results indicating that the hot Jupiters atmosphere can be broader in X-ray than the observed in the optical.

In Selhorst *et al.* 2013, the authors considered the physical contributions of the planetary transits observations at radio frequencies. However, the attempts to detect the exoplanet radio emission were restricted to trying to observe the emission from the planet atmosphere, but, without success (e.g., Hallinan *et al.* 2013).

Here, we present a simple model to estimate the influence of eclipses in the stellar radio emission based on the observed optical light curves.

2. Simulations

To model the optical limb darkening observed in the stellar light curves, Silva 2003 used the following quadratic function: $I(\mu)/I(1) = 1 - w_1(1 - \mu) - w_2(1 - \mu)^2$, where μ is the cosine of the angle between the line of sight and the normal to the local stellar surface.

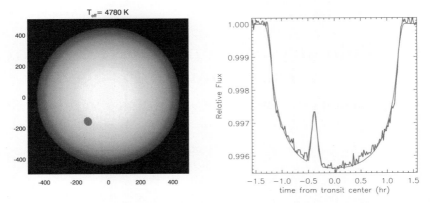

Figure 1. Left panel: Synthesised model of HAT-P-11 with a dark spot $(0.6\ T_{eff})$ aiming to reproduce the observed transit light curve. Right panel: The simulated transit light curve (red) in comparison with the observed one (blue).

In order to test the model, we choose a light curve of the star HAT-P-11 and its planet HAT-P-11b (Béky *et al.* 2014), which present a single dark spot . The HAT-P-11 is a K4 star, with a temperature of 4780.0 ($\pm$ 50.0) K and a radius of 0.75 ($\pm$0.02) $R_{\odot}$. Its planet is a hot Jupiter with a semi-major axis of $\sim 0.05\ AU$ and a radius of $R_P \sim 0.42\ R_J$ or 6.35% of the the stellar radius. The left panel of Figure 1 shows the synthesised model (Silva 2003) of HAT-P-11 with a dark spot aiming to reproduce the observed transit light curve, whereas the right panel compares the simulated transit light curve (red) with the observed one (blue). The stelar limb darkening was simulated with $w_1 = 0.3$ and $w_2 = 1.0$, whereas, the spot had 1.10 R_P and 0.6 I_C, where R_P is the planet radius and I_C is the stellar central intensity.

While the stellar optical observations show characteristic limb darkening, solar maps at radio wavelengths present a limb brightening (see Selhorst *et al.* 2003 and references therein). A simple change in the equation operators can reproduce the observed limb brightening, i.e., $I(\mu)/I(1) = 1+w_1(1-\mu)+w_2(1-\mu)^2$. Setting $w_1 = 0.1$ and $w_2 = 0.2$ the synthesised model present a limb brightening 30% greater than the central temperature, that is compatible with the solar radio observations (Selhorst *et al.* 2003).

For the radio simulations, we assume an opaque planet with the same radius of the optical observations, however, the X-ray observations (Poppenhaeger2013) suggested that the atmosphere of hot Jupiters can be optically thick at radio. We also adopt the active region with the same size of the spot simulation above and its temperature was defined as uniform.

The left panel of Figure 2 shows a synthesised image of the star with limb brightening and the active region of the same size and position as simulated in the optical. The active region brightness temperature was chosen based on the solar observations, where active regions with maximum brightness due to free-free emission varies from 0.2 up to 5.0 times the quiet values depending on the wavelength (e.g., Silva 2005, Selhorst *et al.* 2008). The effects in the light curves caused by active regions with temperatures of 1.5, 2.0 and 5.0 times the quiet star temperature are plotted on the right panel Figure 2.

3. Discussion and Conclusions

Despite the great number of planetary systems discovery in the last two decades, no detection yet has been reported at radio wavelengths. However, new radio interferometers,

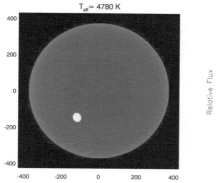
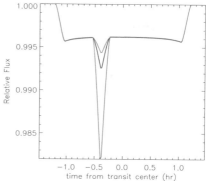

Figure 2. Left panel: Synthesised model of HAT-P-11 free-free radio emission. Right panel: light curves considering active regions temperatures of 1.5, 2.0 and 5.0 times the quiet star temperature, which can be observed at the Sun at mm and cm wavelengths.

like ALMA and the future SKA, may be able to fill this gap. In this work we simulated the use of planetary transits to investigate the radio emission coming from active regions.

To test the simulations, the Silva (2003) spot model was applied to determine parameters of the spot observed in the light curve of HAT-P-11. The simulations suggested that the starspot is bigger than the planet size ($1.10 \, R_P$) and cold with only $0.6 \, I_C$. This low temperature could be due to much more intense magnetic field than those observed in sunspots.

The radio simulations were performed with a quadratic function to estimate the limb brightening, while the active region was assumed to have an uniform temperature. The reduction in the light curve during the transit was smaller than 0.5% in the quiet star region, however the depth increased to 2% when the planet crossed the active region with a temperature 5.0 times the quiet star temperature. These values are consistent with the values suggested in the previous work (Selhorst *et al.* 2013) that used 17 GHz solar maps as a proxy for stellar emission.

Acknowledgements

C.L.S. acknowledge financial support from the São Paulo Research Foundation (FAPESP), grant #2014/10489-0.

References

Alonso, R., Brown, T. M., & Torres, G., *et al.* 2004, *ApJL*, 613, L153
Béky, B., Holman, M. J., Kipping, D. M., & Noyes, R. W. 2014, *ApJ*, 788, 1
Hallinan, G., Sirothia, S. K., Antonova, A., *et al.* 2013, *ApJ*, 762, 34
Poppenhaeger, K., Schmitt, J. H. M. M., & Wolk, S. J. 2013, *ApJ*, 773, 62
Selhorst, C. L., Barbosa, C. L., & Valio, A. 2013, *ApJL*, 777, L34
Selhorst, C. L., Silva, A. V. R., Costa, J. E. R., & Shibasaki, K. 2003, *A&A*, 401, 1143
Selhorst, C. L., Silva-Valio, A. V. R., & Costa, J. E. R. 2008, *A&A*, 488, 1079
Silva, A. V. R. 2003, *ApJL*, 585, L147
Silva, A. V. R., Laganá, T. F., Gimenez Castro, C. G., *et al.* 2005, *SoPh*, 227, 265
Silva-Valio, A., Lanza, A. F., Alonso, R., & Barge, P. 2010, *A&A*, 510, A25

Living Around Active Stars
Proceedings IAU Symposium No. 328, 2016
D. Nandy, A. Valio & P. Petit, eds.

© International Astronomical Union 2017
doi:10.1017/S1743921317004045

Tidal effects on stellar activity

K. Poppenhaeger[1,2]

[1] Astrophysics Research Centre, Queen's University Belfast, BT7 1NN Belfast, United Kingdom
email: k.poppenhaeger@qub.ac.uk
[2] Harvard-Smithsonian Center for Astrophysics, 60 Garden Street, Cambridge, 02138 MA, USA

Abstract. The architecture of many exoplanetary systems is different from the solar system, with exoplanets being in close orbits around their host stars and having orbital periods of only a few days. We can expect interactions between the star and the exoplanet for such systems that are similar to the tidal interactions observed in close stellar binary systems. For the exoplanet, tidal interaction can lead to circularization of its orbit and the synchronization of its rotational and orbital period. For the host star, it has long been speculated if significant angular momentum transfer can take place between the planetary orbit and the stellar rotation. In the case of the Earth-Moon system, such tidal interaction has led to an increasing distance between Earth and Moon. For stars with Hot Jupiters, where the orbital period of the exoplanet is typically shorter than the stellar rotation period, one expects a decreasing semimajor axis for the planet and enhanced stellar rotation, leading to increased stellar activity. Also excess turbulence in the stellar convective zone due to rising and subsiding tidal bulges may change the magnetic activity we observe for the host star. I will review recent observational results on stellar activity and tidal interaction in the presence of close-in exoplanets, and discuss the effects of enhanced stellar activity on the exoplanets in such systems.

Keywords. stars: activity, stars: evolution, (stars:) planetary systems, (stars:) binaries (including multiple): close, stars: late-type

1. Stellar activity

Stellar activity is a collective term for a variety of magnetic phenomena observed in cool stars, i.e. stars with outer convective envelopes (spectral types mid-F to mid-M) or stars that are fully convective (mid-M and later). Manifestations of magnetic activity include the presence of flares, coronal mass ejections, chromospheres and coronae, starspots, and faculae.

All of these phenomena are ultimately driven by the stellar rotation through the magnetic dynamo (Parker 1955). The differential rotation of a star, both latitudinally and radially, causes not only a long-term inversion of the global magnetic field polarity, but also the localized phenomena in the stellar atmosphere which make up the individual facets of magnetic activity. Understanding the evolution of stellar rotation, from the formation of stars through the gigayears of their lifetime, is therefore fundamental to our understanding of stellar magnetic activity.

One important factor of the rotational evolution of cool stars the the spin-down that occurs due to magnetic braking (Schatzman 1962). This happens because cool stars shed an ionized stellar wind, which moves away from the star along the stellar magnetic field lines. Finally, it decouples from the magnetic field, and at this moment the angular momentum is carried out of the system by the stellar wind. This continuous loss of angular momentum causes a spin-down of the star over time, which can be studied observationally (Barnes 2003, 2010; Meibom *et al.* 2015; van Saders *et al.* 2016). Typically, stars set out on the main sequence with short rotation periods of half a day to a few days, and spin down to long periods of ca. 30 days over a few gigayears in the case of the Sun, and much longer rotation periods of the order of 100 days for low-mass stars on the main-sequence at old ages (Irwin *et al.* 2011). Studying slow rotation of stars is observationally challenging, because the main observables of rotation (rotational broadening of spectral lines and photometric variability due to star spots on the stellar surface) provide only weak signatures in the slow-rotation regime.

As rotation slows down, the stellar magnetic activity decreases. While this qualitatively makes sense, since (differential) rotation is the motor for stellar activity, the physical details of this are not fully understood. For example, it is not clear how the rotational period of a star and the presence and duration of activity cycles (i.e. the 11-year activity cycle of the Sun) are related. Still, the overall effects of activity, such as coronal X-ray emission (Güdel *et al.* 1997; Preibisch & Feigelson 2005; Telleschi *et al.* 2005), chromospheric line emission such as the Ca II H and K lines and the H alpha line (Skumanich 1972; Noyes *et al.* 1984; Mamajek & Hillenbrand 2008; Reiners *et al.* 2012), and photospheric variability (Bastien *et al.* 2014; Stelzer *et al.* 2016) can be related to stellar rotation and also directly to stellar age.

However, magnetic braking is not the only physical effect that influences stellar rotation and activity over time, which brings us to tidal interaction.

2. Tidal interaction

Whenever we have two astronomical objects in close proximity to each other, tides start to play a role. For our topic of interest, examples for relevant systems are: a close binary system consisting of two stars, a star with a massive planet in a close orbit, or a planet-moon system. Tides cause a deformation of the involved bodies due to the gravitational force acting on them, and due to their rotation around their common center of mass. There are three main observational effects of tides in such systems: alignment of the spin axes perpendicular to the orbital plane; synchronization, meaning that over time the rotational periods of the bodies and the orbital period become equal; and circularization, meaning that bodies in an eccentric orbit slowly lose their eccentricity and adopt a circular orbit (Zahn 2008; Mathis & Le Poncin-Lafitte 2009).

For our topic of stellar activity, the synchronization effect is the most relevant one. Therefore we take a more detailed look at how this effect plays out in different combinations of orbital and rotational periods. Let us assume a system of two bodies A and B in a close-in orbit, as depicted in Fig. 1. Both objects get deformed by tides, and here we focus on what happens to the central body (A) due to the tides.

Assume that A has a longer rotational period than the orbital period of B. B raises a tidal bulge on A, and because B moves faster on its orbit than A rotates (in terms of angular velocity), the tidal bulge on A will lag behind (see Fig. 1 left side). The gravitational pull of B on this bulge will therefore induce a tidal torque, and pull A into a somewhat faster rotation. Angular momentum is transferred from the orbit of B to the spin of A. Since the total angular momentum of the system is conserved, the semi-major axis of B decreases as the angular momentum of its orbit decreases, meaning B spirals slowly closer to A.

In the opposite configuration, where A has a shorter rotational period than the orbit of B, angular momentum is transferred into the other direction. The tidal bulge on A runs ahead, and gets "pulled back" by B as it does so (see Fig. 1 right side). A therefore slows down, and the angular momentum of the orbit of B increases. This causes B to move to a larger semi-major axis. This is actually what happens in the Earth-Moon system, where Earth's rotation period of one day is shorter than the Moon's orbital period of

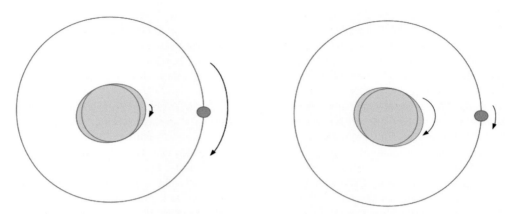

Figure 1. Example of tidal interaction in a two-body system. In the example to the left, the rotation period of the central object (A) is longer than the orbital period of the smaller body (B), in the example to the right the rotation period of the central object is shorter. Typically, for star-planet systems with Hot Jupiters and relatively old host stars, the left example is representative of the orbital and rotational periods observed.

ca. 27.3 days: the Earth's rotation slows down, and the Moon moves outwards over time. This effect has a magnitude of ca. 38 mm increase of the Moon's semi-major axis per year, measured through laser reflectors left on the Moon during the Apollo program (Chapront *et al.* 2002).

3. Activity in stellar binaries

For binary stars in close orbits, the tidal synchronization has strong observable effects on their stellar activity. After the stellar rotation of both stars has synchronized with the orbital period of the binary, it stays locked to this period. Even though the stars still lose some angular momentum due to the stellar wind they expel, they are kept at a high rotation rate due to this tidal locking, i.e. over long time scales the orbital distance of the stellar binary decreases as angular momentum is lost (Stepien 1995). For the activity, this means that even though the stars in a binary may have a relatively old age, they are still rotating at a period of a few days and are similarly magnetically active as a single star of that rotation period systems.

In addition to the magnetic activity from the rotation of the individual stars, there can be interactions of the stellar magnetic fields with each other, such as magnetic loops connecting the two stars. This can lead to further magnetic activity effects (Siarkowski *et al.* 1996; Peterson *et al.* 2010).

The activity of stars in close binaries has been investigated thoroughly in various activity observables. Flare rates of M dwarfs in close pairs with white dwarfs have been found to be higher than for single M dwarfs (Morgan *et al.* 2016), and also their ambient activity, i.e. the overall activity level outside of time-resolved flares, which is thought to be a superposition of smaller-scale activity events, is high (Morgan *et al.* 2012). An interesting observation is that in systems with somewhat larger orbital distances of a few AU, where tidal synchronization should not be relevant, there is still an elevated activity observed (Meibom *et al.* 2007); this may be due to differences in the stellar formation and a different rotation period with which these moderate-distance binaries set out on the main sequence. Large-distance binaries (semi-major axes of several hundred AU) do not show this effect. But generally, activity indicators are found to be high for tidally interacting close binaries with periods of a few days (Schrijver & Zwaan 1991).

4. Activity in planet-hosting stars

Thinking of a star-planet system as a scaled-down version of stellar binaries, with a very small mass ratio of the components, lets one expect that there may be relevant tidal effects as well. One active line of research in the exoplanetary field is the study of how exoplanetary orbits evolve over time, and how quickly exoplanets may spiral into their host stars (Penev & Sasselov 2011; Jackson *et al.* 2010). The tidal quality factor of stars, which specifies how quickly the kinetic energy of tidal deformations and waves is dissipated, is not well constrained yet by current theories and observations, and requires further study (Zahn 2008; Ogilvie & Lin 2007; Penev & Sasselov 2011). A "smoking gun" of an inspiraling exoplanet in the form of an actual measurement of a decreasing orbital period is yet to be found.

Concerning the tidal effects on the activity of the host star, initial theoretical studies were performed early-on. The two main interaction scenarios were identified as tidal interaction and magnetic interaction (Cuntz *et al.* 2000). Magnetic interaction is expected to follow scenarios either similar to loop interactions in close binaries, or similar to the Jupiter-Io unipolar inductor interaction. For tidal interaction, both the general tidal spin-up (or spin-down) of a star and activity effects due to a tidally induced increased turbulence in the outer convection layer of the star were proposed.

The observational search for star-planet interactions has been challenging. Initial detections of magnetic star-planet interaction were reported for two out of 13 stars with Hot Jupiters, where the chromospheric emission in the Ca II lines was observed to modulate with the planetary orbital period, not the stellar rotation period (Shkolnik *et al.* 2005). Later observational campaigns of those targets, however, showed that during those later epochs a modulation with the stellar rotation period was present (Shkolnik *et al.* 2008; Poppenhaeger *et al.* 2011). Other magnetic effects like flare triggering or hot spots in the stellar chromosphere and corona were expected from theoretical investigations (Lanza 2008; Cohen *et al.* 2009). For the Hot Jupiter host HD 189733, several small flares in X-rays and the UV were observed during the time shortly after the secondary transit (Pillitteri *et al.* 2011). As the orbit of that planet is circular and not eccentric, it is not obvious why a certain phase of the orbit should show preferential flaring (as opposed to consistent flaring during one full half of the orbit when a stellar hot spot would be visible). Later observations and modelling suggested that a plasma trail of infalling material from the planet onto the star may be the source of the high-energy emission, with the largest viewing cross-section shortly after the secondary transit (Pillitteri *et al.* 2014a). Another possible magnetic interaction effects has been reported for the system HD 17156, which hosts a Jupiter in a strongly eccentric orbit. The system showed elevated X-ray emission during two periastron passages of the planet, and low X-ray emission during two apoastron passages (Maggio *et al.* 2015). This may be analogous to colliding magnetospheres observed for some young binary stars in eccentric orbits (Getman *et al.* 2011, 2016).

Systematic investigations of stellar activity in larger samples of planet-hosting stars have been performed. While initial studies observed a trend of stars with close-in and massive planets to be more active than stars with small and far-away planets (Kashyap *et al.* 2008), these trends have simultaneously a large scatter over the whole sample (Poppenhaeger *et al.* 2010) and can in part be traced back to selection effects from the efficiency of planet-detection methods for active and inactive stars (Poppenhaeger & Schmitt 2011). Some effects, especially for extremely close and massive planets, seem

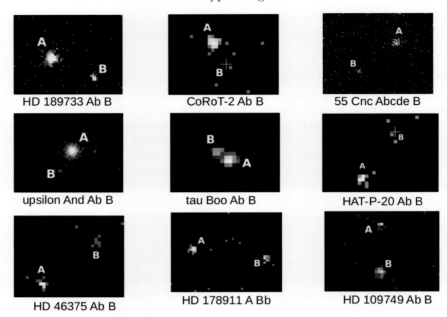

Figure 2. Examples of several wide binary systems where one star hosts a known planet, observed in the X-ray band with *Chandra* and *XMM-Newton*.

to be still present when strictly controlling for the spectral type of the sample stars (Miller *et al.* 2015); however, not all Hot Jupiters necessarily have an active host star (Poppenhäger *et al.* 2009; Miller *et al.* 2012; Pillitteri *et al.* 2014b).

One important point is that even if planets may increase the stellar activity through some form of star-planet interaction, the magnetic braking of the star due to its stellar wind is going on at the same time (Penev *et al.* 2012). Taking into acocunt the age of a star-planet system is therefore crucial in order to distinguish whether a star is active because it is relatively young (and would therefore be active no matter if there was a planet or not), or if the star is actually old and is only active because it has been influenced by its planet. Unfortunately, ages for single field stars with ages over a gigayear are hard to estimate (Soderblom 2010). One way around this problem is using wide stellar binaries in which one of the stars hosts a known planet. In a wide stellar binary, the two stars will have the same age, and their activity levels should be similar (after adjusting for differences due to stellar mass). If the planet-hosting star has a much higher activity level than the companion star, one can deduce that the high activity level is not due to youngness of the system, but due to a planetary influence. In a sample of 18 such systems (see some examples in Fig. 2), the stars for which a strong tidal influence is expected from their planet preferentially display higher activity levels than their companion stars (Poppenhaeger & Wolk (2014), Poppenhaeger *et al.* submitted). This effect is absent for stars with planets that are not expected to have a strong tidal influence on their host stars.

Systematic effects on stellar activity can have important consequences for exoplanets: since the atmospheric mass loss of exoplanets is thought to be driven by X-ray and extreme UV irradiation (Lecavelier des Etangs *et al.* 2004; Sanz-Forcada *et al.* 2010), an elevated stellar activity level can lead to higher evaporation rates for planets. Indeed, extended planetary atmospheres and/or active atmospheric escape have been observed for several exoplanets (Vidal-Madjar *et al.* 2003; Lecavelier Des Etangs *et al.* 2010; Poppenhaeger *et al.* 2013; Bourrier *et al.* 2013; Kulow *et al.* 2014; Ehrenreich *et al.* 2015). For small exoplanets, such evaporation may lead to the total loss of their atmosphere (Lopez & Fortney 2013; Poppenhaeger *et al.* 2012). Especially for habitability considerations, such as for planets in the habitable zones around M dwarfs, this is an important concern (Segura *et al.* 2010). M dwarfs can produce frequent flares even at older ages (Güdel *et al.* 2004; Robrade *et al.* 2010; Davenport *et al.* 2016). From a stellar perspective this is particularly interesting in the fully convective M dwarf regime, where a different dynamo than in the solar case needs to be present due to the lack of a stellar radiative core, and different models have been developed to investigate the possible magnetic field structures for these stars (Browning 2008; Yadav *et al.* 2015). Especially since a habitable-zone exoplanet has been detected for the nearest neighbor of the Sun (Anglada-Escudé *et al.* 2016), the fully convective M dwarf Proxima Centauri, investigations of stellar activity and its impact on exoplanet habitability will continue to be a prime concern for studying near-by exoplanets.

5. Conclusion

Magnetic activity is not only an interesting stellar phenomenon, but also an important topic for exoplanets. Tidal influences on stellar activity are well-known in stellar binaries, and there is some observational evidence accumulating that also massive planets in close-in orbits can influence the stellar activity. Further investigations into the observational magnitudes of tidal effects as well as into stellar tidal quality factors will be necessary for understanding of these systems.

References

Anglada-Escudé, G., Amado, P. J., Barnes, J., *et al.* 2016, *Nature*, 536, 437
Barnes, S. A. 2003, *ApJ*, 586, 464
Barnes, S. A. 2010, *ApJ*, 722, 222
Bastien, F. A., Stassun, K. G., Pepper, J., *et al.* 2014, *AJ*, 147, 29
Bourrier, V., Lecavelier des Etangs, A., Dupuy, H., *et al.* 2013, *A&A*, 551, A63
Browning, M. K. 2008, *ApJ*, 676, 1262
Chapront, J., Chapront-Touzé, M., & Francou, G. 2002, *A&A*, 387, 700
Cohen, O., Drake, J. J., Kashyap, V. L., *et al.* 2009, *ApJL*, 704, L85
Cuntz, M., Saar, S. H., & Musielak, Z. E. 2000, *ApJL*, 533, L151
Davenport, J. R. A., Kipping, D. M., Sasselov, D., Matthews, J. M., & Cameron, C. 2016, *ApJL*, 829, L31
Ehrenreich, D., Bourrier, V., Wheatley, P. J., *et al.* 2015, *Nature*, 522, 459
Getman, K. V., Broos, P. S., Kóspál, Á., Salter, D. M., & Garmire, G. P. 2016, *AJ*, 152, 188
Getman, K. V., Broos, P. S., Salter, D. M., Garmire, G. P., & Hogerheijde, M. R. 2011, *ApJ*, 730, 6
Güdel, M., Audard, M., Reale, F., Skinner, S. L., & Linsky, J. L. 2004, *A&A*, 416, 713
Güdel, M., Guinan, E. F., & Skinner, S. L. 1997, *ApJ*, 483, 947
Irwin, J., Berta, Z. K., Burke, C. J., *et al.* 2011, *ApJ*, 727, 56
Jackson, B., Miller, N., Barnes, R., *et al.* 2010, *MNRAS*, 407, 910
Kashyap, V. L., Drake, J. J., & Saar, S. H. 2008, *ApJ*, 687, 1339
Kulow, J. R., France, K., Linsky, J., & Loyd, R. O. P. 2014, *ApJ*, 786, 132
Lanza, A. F. 2008, *A&A*, 487, 1163
Lecavelier Des Etangs, A., Ehrenreich, D., Vidal-Madjar, A., *et al.* 2010, *A&A*, 514, A72
Lecavelier des Etangs, A., Vidal-Madjar, A., McConnell, J. C., & Hébrard, G. 2004, *A&A*, 418, L1

Lopez, E. D. & Fortney, J. J. 2013, *ApJ*, 776, 2

Maggio, A., Pillitteri, I., Scandariato, G., *et al.* 2015, *ApJL*, 811, L2

Mamajek, E. E. & Hillenbrand, L. A. 2008, *ApJ*, 687, 1264

Mathis, S. & Le Poncin-Lafitte, C. 2009, *A&A*, 497, 889

Meibom, S., Barnes, S. A., Platais, I., *et al.* 2015, *Nature*, 517, 589

Meibom, S., Mathieu, R. D., & Stassun, K. G. 2007, *ApJL*, 665, L155

Miller, B. P., Gallo, E., Wright, J. T., & Dupree, A. K. 2012, *ApJ*, 754, 137

Miller, B. P., Gallo, E., Wright, J. T., & Pearson, E. G. 2015, *ApJ*, 799, 163

Morgan, D. P., West, A. A., & Becker, A. C. 2016, *AJ*, 151, 114

Morgan, D. P., West, A. A., Garcés, A., *et al.* 2012, *AJ*, 144, 93

Noyes, R. W., Hartmann, L. W., Baliunas, S. L., Duncan, D. K., & Vaughan, A. H. 1984, *ApJ*, 279, 763

Ogilvie, G. I. & Lin, D. N. C. 2007, *ApJ*, 661, 1180

Parker, E. N. 1955, *ApJ*, 122, 293

Penev, K., Jackson, B., Spada, F., & Thom, N. 2012, *ApJ*, 751, 96

Penev, K. & Sasselov, D. 2011, *ApJ*, 731, 67

Peterson, W. M., Mutel, R. L., Güdel, M., & Goss, W. M. 2010, *Nature*, 463, 207

Pillitteri, I., Günther, H. M., Wolk, S. J., Kashyap, V. L., & Cohen, O. 2011, *ApJL*, 741, L18+

Pillitteri, I., Wolk, S. J., Lopez-Santiago, J., *et al.* 2014a, *ApJ*, 785, 145+

Pillitteri, I., Wolk, S. J., Sciortino, S., & Antoci, V. 2014b, *A&A*, 567, A128

Poppenhaeger, K., Czesla, S., Schröter, S., *et al.* 2012, *A&A*, 541, A26

Poppenhaeger, K., Lenz, L. F., Reiners, A., Schmitt, J. H. M. M., & Shkolnik, E. 2011, *A&A*, 528, A58+

Poppenhaeger, K., Robrade, J., & Schmitt, J. H. M. M. 2010, *A&A*, 515, A98+

Poppenhaeger, K. & Schmitt, J. H. M. M. 2011, *ApJ*, 735, 59

Poppenhaeger, K., Schmitt, J. H. M. M., & Wolk, S. J. 2013, *ApJ*, 773, 62

Poppenhaeger, K. & Wolk, S. J. 2014, *A&A*, 565, L1

Poppenhäger, K., Robrade, J., Schmitt, J. H. M. M., & Hall, J. C. 2009, *A&A*, 508, 1417

Preibisch, T. & Feigelson, E. D. 2005, *ApJS*, 160, 390

Reiners, A., Joshi, N., & Goldman, B. 2012, *AJ*, 143, 93

Robrade, J., Poppenhaeger, K., & Schmitt, J. H. M. M. 2010, *A&A*, 513, A12

Sanz-Forcada, J., Ribas, I., Micela, G., *et al.* 2010, *A&A*, 511, L8

Schatzman, E. 1962, Annales d'Astrophysique, 25, 18

Schrijver, C. J. & Zwaan, C. 1991, *A&A*, 251, 183

Segura, A., Walkowicz, L. M., Meadows, V., Kasting, J., & Hawley, S. 2010, Astrobiology, 10, 751

Shkolnik, E., Bohlender, D. A., Walker, G. A. H., & Collier Cameron, A. 2008, *ApJ*, 676, 628

Shkolnik, E., Walker, G. A. H., Bohlender, D. A., Gu, P., & Kürster, M. 2005, *ApJ*, 622, 1075

Siarkowski, M., Pres, P., Drake, S. A., White, N. E., & Singh, K. P. 1996, *ApJ*, 473, 470

Skumanich, A. 1972, *ApJ*, 171, 565

Soderblom, D. R. 2010, *Ann. Rev. Astron. Ast.*, 48, 581

Stelzer, B., Damasso, M., Scholz, A., & Matt, S. P. 2016, *MNRAS*, 463, 1844

Stepien, K. 1995, *MNRAS*, 274, 1019

Telleschi, A., Güdel, M., Briggs, K., *et al.* 2005, *ApJ*, 622, 653

van Saders, J. L., Ceillier, T., Metcalfe, T. S., *et al.* 2016, *Nature*, 529, 181

Vidal-Madjar, A., Lecavelier des Etangs, A., Désert, J., *et al.* 2003, *Nature*, 422, 143

Yadav, R. K., Christensen, U. R., Morin, J., *et al.* 2015, *ApJL*, 813, L31

Zahn, J.-P. 2008, in *EAS Publications Series*, Vol. 29, EAS Publications Series, ed. M.-J. Goupil & J.-P. Zahn, 67–90

Living Around Active Stars
Proceedings IAU Symposium No. 328, 2016
D. Nandy, A. Valio & P. Petit, eds.

© International Astronomical Union 2017
doi:10.1017/S1743921317004288

The Environment of the Young Earth in the Perspective of An Young Sun

Vladimir S. Airapetian

NASA GSFC, Code 671, Greenbelt, MD, USA
email: vladimir.airapetian@nasa.gov

Abstract. Our Sun, a magnetically mild star, exhibits space weather in the form of magnetically driven solar explosive events (SEE) including solar flares, coronal mass ejections and energetic particle events. We use Kepler data and reconstruction of X-ray and UV emission from young solar-like stars to recover the frequency and energy fluxes from extreme events from active stars including the young Sun. Extreme SEEs from a magnetically active young Sun could significantly perturb the young Earth's magnetosphere, cause strong geomagnetic storms, initiate escape and introduce chemical changes in its lower atmosphere. I present our recent simulations results based on multi-dimensional multi-fluid hydrodynamic and magnetohydrodynamic models of interactions of extreme CME and SEP events with magnetospheres and lower atmospheres of early Earth and exoplanets around active stars. We also discuss the implications of the impact of these effects on evolving habitability conditions of the early Earth and prebiotic chemistry introduced by space weather events at the early phase of evolution of our Sun.

Keywords. The Sun, Earth, space weather, CME, SEP, atmosphere, chemistry, exoplanets, active stars

1. Introduction

The early Earth in the late Hadean period was a highly energetic and dynamic planet. Despite of hellish conditions introduced by intensive volcanic and tectonic activity, frequent impacts by Late Heavy Bombardment events and high fluxes of X-ray and UV radiation from the early Sun as well as frequent eruptive ejections from young Sun, liquid water was an essential ingredient of our planet in its earliest history (Wilde *et al.* 2001; Gomes *et al.* 2005; Abramov & Mojzsis 2009; Airapetian *et al.* 2016). Recent biogenic carbon data suggest that our young planet in the first 0.7 Gyr managed to support the initiation and development of life (Bell *et al.* 2015). The conditions controlling life include of the appropriate surface temperature and pressure to support liquid water and biochemical processes on the early Earth along with X-ray & UV fluxes and solar wind energy fluxes from the young Sun. Geophysical factors include internal dynamics between the inner and outer Earth core driving generation of geomagnetic field and the magnetosphere and plate tectonics and volcanic processes that contributed to mantle degassing and atmosphere refueling. The geodynamo has been operating over last 4 billion years and provided the magnetic shield, which was strong enough to withstand the pressures from the young Sun wind (Tarduno *et al.* 2015).

What combination of geophysical and astrophysical conditions driven by the internal dynamics of the Earth and activity of the young Sun create habitability factors favorable for initiation of life? This is one of the most fundamental questions of the modern science because the answer to this question will provide a unique opportunity to understand how life might form on other planets and how to search for their observational signatures.

In this review, I describe our recent progress in understanding space weather from the current and the young Sun at the time when life started on Earth and discuss physical processes that drive their interactions with out planet. Here, I emphasize the negative and positive factors affecting habitability conditions on the young Earth as an important test case for understanding the habitability conditions on terrestrial type exoplanets around main-sequence stars of G, K and M spectral classes recently discovered by Kepler Space Telescope, HST and ground-based telescopes. Finally, I discuss the role of space weather processes in initiation of prebiotic chemistry that set favorable conditions for the origin of life on the young Earth.

2. The Activity of the Young Sun

Recent X-ray and UV missions including CHANDRA, XMM-NEWTON, the Hubble Space Telescope and recent Kepler Space telescope opened new windows in studying the lives of stars resembling our Sun at various phases of evolution. These observations provided a unique opportunity to trace the properties of activity of the young Sun by observing other young solar-type stars. This may provide crucial information in our understanding of habitability of the early Earth, Mars and Venus and factors controlling the origin of life on our planet.

In order to reconstruct the properties of explosive events from the young Sun, we need to examine observations of young solar-like stars resembling our Sun in its infancy. Observations of young (a few hundred Myr old) solar-like stars show that our Sun had about 30% less luminous (at the time when life arose on our planet) due to less dense core driven by thermonuclear fusion of hydrogen into helium (Gough 1981).

As Sun evolves, its luminosity increases roughly 10% per every billion years. Despite its lower bolometric luminosity, the young Sun represented a very magnetically active and rapidly rotating star. Rapid rotation in combination with deep convection zones of these stars produce strong surface magnetic field that emerge to the surface and form compact, dense and hot corona (10 MK) of the young Sun. Recent direct measurements of surface magnetic fields from young suns (Bcool project) shows that the surface magnetic flux of an 0.5 Myr old star is by a factor of 30 greater than that measured from the current Sun as a star (Vidotto *et al.* 2015). Such strong magnetic fields serve as the major energy source to produce frequent and energetic flares in their coronae contributing to plasma heating and production of large X-ray luminosities that by 3-4 orders of magnitude greater than that observed today (Pevtsov *et al.* 2003; Gudel *et al.* 1997; Tu *et al.* 2015).

The photospheric convection motions excite Alfvén waves that propagate upward and contribute to the initiation of the fast solar and dense solar wind. The winds from young active sun can play a crucial role in removal of angular momentum from them and resulting in spin-down in times as stars age (Sterenborg *et al.* 2011; Garraffo *et al.* 2016). We have recently used a three-fluid three-dimensional magnetohydrodynamic Alfven wave driven solar wind model, ALF3D, to study the evolution of the young solar wind. Our model treats the wind thermal electrons, protons and pickup protons as separate fluids and incorporates turbulence transport, eddy viscosity, turbulent resistivity, and turbulent heating to properly describe proton and electron temperatures of the solar wind. We used three input model parameters, the plasma density, Alfven wave amplitude and the strength of the magnetic dipole field at the wind base for each of three solar wind evolution models. We concluded that the terminal velocity of the young solar wind was twice faster, 100 times denser and 5 times hotter at 1 AU in its early history (at 0.7 Gyr) (Airapetian & Usmanov 2016).

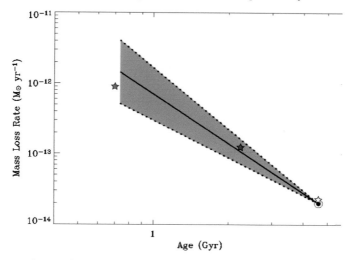

Figure 1. The total mass-loss rates from the solar wind at 0.7 Gyr (red star), 2.2 Gyr (blue star) and 4.6 Gyr (yellow star) superimposed on the empirically derived values of mass loss rates (grey area) from a sample of solar-type stars of various ages.

Figure 1 shows the total mass-loss rates from Sun at at the three evolutionary phases of the Sun, 0.7 Gyr, 2 and 4.65 Gyr (the current epoch) superimposed on the range of empirically derived mass-loss rates for solar-like stars at various phases of evolution (Wood *et al.* 2005). The evolution of the solar wind was driven mostly by the coronal magnetic field, the plasma density at the wind base and the amplitude of Alfven waves. Our models suggest that the dynamic pressure from the young solar wind at 0.7 Gyr is expected to be up to 170 times greater than the wind pressure from the current Sun.

Frequent flares on the young Sun were the source of fast and dense CMEs forming as a result of the global restructuring of the solar coronal magnetic field. The frequency of CMEs from young Sun and other active stars can be estimated from their association with solar/stellar flares. Recent SOHO/LASCO and STEREO observations of energetic and fast (> 1000 km/s) CMEs show strong association with powerful solar flares (Yashiro & Gopalswamy 2009; Aarnio *et al.* 2011; Tsurutani & Lakhina 2014). This empirical correlation established for the events from the current Sun provides an estimate for CME occurrence frequencies. Kepler data suggest that stellar superflare events with energy of 3 x 10^{33} ergs (referred to as superflares) occur on young and active K-G type main-sequence stars at the rate of $\sim$ 250 events/day (Maehara *et al.* 2012; Shibayama *et al.* 2013; Airapetian *et al.* 2016). This suggests that CME events associated with such superflares (referred to as super-CME events) should have the kinetic energy by a factor of 10 greater than the energy of associated flare events. The frequency of such events directed toward the young Earth is estimated to be at least a few events per day.

Another evidence for the presence of high frequency energetic flares from the young Sun comes from our direct comparison of the reconstructed X-ray to UV flux (XUV) for the young Sun using k1 Cet as a proxy for it (Airapetian *et al.* 2017). In Figure 2 we present the reconstructed spectral energy distribution (SED) of the current Sun at the average level of activity (between solar minimum and maximum with the total flux, F0 (5 - 1216Å) = 5.6 erg/cm^2/s; yellow dotted line), the X5.5 solar flare occurred on March 7, 2012 (blue line), the young Sun at 0.7 Gyr (yellow solid line) and an inactive M1.5 red dwarf, GJ 832 (red line). The spectra for the current Sun and the solar X5.5 flare in the XUV band (0.5-10Å) are constructed from the Solar Dynamic Observatory (SDO)/EVE instrument data.

V. S. Airapetian

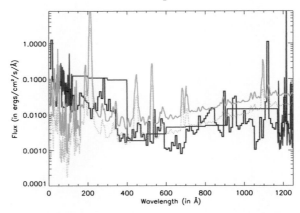

Figure 2. The spectral energy distribution (SED): reconstructed for the solar X5.4 flare (blue curve), the young Sun's SED (orange curve) and the quite Sun at the average (intermediate between solar minimum and maxium) magnetic activity (dotted orange curve) scaled to 1 AU and GJ 832 SED (red curve) scaled to 0.16 AU

.

The Vacuum Ultraviolet (VUV) contribution of the total radiative output is obtained by implementing the Flare Irradiance Spectral Model (FISM, Chamberlin *et al.* 2008), which represents an empirical model developed for space weather applications that estimates the solar irradiance at wavelengths from 1 to 1900Å at 10Å resolution with a time cadence of 60 s. We also reconstructed the XEUV spectrum of a moderately-old and inactive M1.5 dwarf, GJ 832, that hosts a super-Earth planet at 0.16 AU using the Measurements of the Ultraviolet Spectral Characteristics of Low-mass Exoplanetary Systems (MUSCLES) Treasury Survey data (Loyd *et al.* 2016). Finally, to approximate the spectrum of the young Sun at 0.7 Gyr, we used the data obtained from the parameterization of the two young solar analogs of the Sun at around 0.7 Gyr, k1 Cet and EK Dra (Claire *et al.* 2012). The total XEUV flux from the young Sun and the red dwarf are 8.3 F_0 (at 1 AU) and 7.7 F_0 (at 0.16 AU) respectively. The XEUV flux from the young Sun, and GJ 832 are comparable in magnitude and shape at wavelengths shorter (and including) Ly-alpha emission line. This suggests contribution of X-type flare activity flux is dominant in the "quiescent" fluxes from the young Sun and inactive M dwarfs.

3. Astrophysics of Habitability of the Early Earth

Recent paleomagnetic observations suggest that the geodynamo has been active over at least the past 3.5 Ga (Biggin *et al.* 2011) and possibly even as early as 4.2 Ga (Tarduno *et al.* 2015). However, the early Earth's magnetic field is expected to be weaker by an up to 50% during the Archean, while other researchers suggested a magnetic field of a quarter of the present-day intensity (Miki *et al.* 2009). As discussed in the previous sectons, fast and dense winds and energetic CMEs associated with superflares from the young magnetically active Sun should have exerted larger dynamic pressures on weak Earth's magnetosphere and generate energy flux at the magnetopause that may cause the atmospheric erosion. The XUV fluxes from the young Sun in the first 0.5 Gyr should have been at least 10 times of the present day solar flux (see Figure 3). Such large dynamic pressures and XUV fluxes could have ignited significant atmospheric escape from the early Earth (Airapetian *et al.* 2017). Indeed, recent fossilized raindrop imprint data suggest that atmospheric pressure of the Earth 2.7 Gyr ago was at least the half of the current pressure (Som *et al.* 2016). Such low atmospheric pressure creates problems in explaining the existence of oceans on Earth as early as 4.4 Gyr ago as supported by recent zircon data (Wilde *et al.* 2001). This problem is further challenged by the faint (30% less bright) young Sun at that time that makes is difficult to find an efficient warming mechanism to support liquid surface water on the early Earth. In this Section, we discuss physical processes responsible for atmospheric escape from the early Earth and ways to resolve the FYS paradox.

3.1. *Effects of super CMEs on the early Earth's magnetosphere*

As Interplanetary CMEs (ICMEs) propagate toward the Earth, they interact with Earth magnetosphere compressing its dayside and night sides. If the interplanetary magnetic field (IMF) is directed southward (or oppositely directed to the Earth's dipole field), then CMEs trigger geomagnetic storms due to the combined effects of magnetic reconnection on the dayside (as recently directly observed by MMS mission observations, Birch *et al.* 2016) and dynamic pressure effects (Birch *et al.* 2016). Also, CMEs perturb the nightside geomagnetic field producing magnetic reconnection in the Earth's magnetotail (Zhao *et al.* 2016).

One of the strongest CME events characteristic of the young Sun's conditions had occurred on July 23-24 2012. This event was observed as a series of two successive CME events. The first CME on July 23 had the peak speed over 2500 km/s with the peak southward magnetic field Bz = - 201 nT (Riley *et al.* 2016). This catastrophic event was comparable in its energy to the kinetic energy of the Carrington event of Sep 1-2, 1859. This rare type energetic event missed the Earth. The modeling by Ngwira *et al.* 2013 using SWMF at CCMC/GSFC suggests that if these events would have hit the Earth magnetosphere, the stand-off distance would have been as low as 2RE. The height integrated Joule heating rate deposited in the Earth thermosphere widened polar regions would have been as high as 2.5 W/m^2 or by a factor of 50 greater than in St. Patrick event. This would suggest that the temperature increase and the thermospheric expansion to at least 100,000K at 150 km. Our estimates of the frequency of such events suggest that they would have hit the magnetosphere of the young Earth 4 billion years ago at a rate of few events per day (Airapetian *et al.* 2015).

Here we present our results of the SWMF/CCMC based simulations of an extreme CME event referred to as a super-Carrington event interacting with magnetosphere of the young Earth. To characterize this event, we utilized the Space Weather Modeling Framework (SWMF) available at Community Coordinated Modeling Center (CCMC) at NASA Goddard Space Flight Center (see at http://ccmc.gsfc.nasa.gov). A single-fluid, time dependent fully non-linear 3D magnetohydrodynamic (MHD) code BATS-R-US (Block-Adaptive-Tree Solar-wind Roe-type Upwind Scheme) is a part of SWMF and was developed at the University of Michigan Center of Space Environment Modeling (CSEM). The spine of the SWMF is the BATS-R-US code (Powell *et al.* 1999). The global magnetosphere model is coupled to the inner magnetosphere through the Rice Convection Model (De Zeeuw *et al.* 2004). Field-aligned currents, Jpar, calculated at the lower boundary are mapped to the ionospheric height of 110 km under the assumption of a dipole magnetic field. From the electric currents, J, mapped at the lower ionospheric boundary the conductance obtained in the inner magnetosphere model, we calculate the Joule heating (JH) at 110 km.

For a super-Carrington event, we implemented the model of the young solar wind discussed in Airapetian and Usmanov (2016). The simulations were carried out using a block adaptive high resolution grid with the minimum cell size of 1/16 R$_E$. The inner boundary is set at 1.25 R$_E$. The young solar wind conditions are set at the upstream

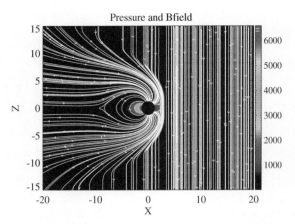

Figure 3. Magnetospheric storm at t=10 h. Plasma pressure and the magnetic field lines. X
and Y axis are given in the units of Earth radius.

boundary and some period of local time stepping is used to get an initial steady state
solution.

We assume the solar wind input parameters including the three components of interplanetary magnetic field, Bx , By and Bz, the plasma density and the wind velocity, V_x
using the physical conditions associated with a Carrington-type event as discussed by
Tsurutani *et al.* (2003) and Ngwira *et al.* (2014) and Airapetian *et al.* (2016). Figure 3
presents a 2D map of the steady-state plasma density superimposed by magnetic field
lines for the magnetospheric configuration in the Y=0 plane corresponding to the initial 30 minutes of the simulations, when the Earth's magnetosphere was driven only by
dynamic pressure from the solar wind. At t=30 min, we introduce a super-Carrington
CME event characterized by the time profile of Vx as the CME approaches the Earth at
the maximum velocity of 1800 km/s. The CME magnetic field is directed southward or
is sheared by 180 degrees with respect to the dipole field with the $B_z = -212$ nT.

As the CME propagates inward, its large dynamic pressure compresses and convects the
magnetospheric field inducing the convective electric field. It also compresses the nightside magnetosphere and ignites magnetic reconnection at the nigh-side of the Earth's
magnetosphere causing the magnetospheric storm as particles penetrate the polar regions
of Earth. Another effect appears to be crucial in our simulations. The strong sheared
magnetic field on the dayside (sub-solar point) of Earth is also subject to reconnection,
which dissipates the outer regions of the Earth's dipole field up until 1.5 R_E above the
surface. The boundary of the open-closed field shifts to 36 degrees in latitude.

The CME drives large field aligned electric currents that provide a Joule ionospheric
heating at 110 km reaching 4 W/m2. The more extreme geomagnetic events introduced
by stellar winds and CMEs around magnetically actiove M dwarfs can introduce much
stronger currents in Earth-like exoplanetary ionospheres with the Joule heating rates as
high as 10 W/m^2 (Cohen *et al.* 2014).

The thermospheric temperature of Earth in the quiet state at 150 km is 900K. From
the first law of thermodynamics we an obtain the high estimate on the temperature
change

$$JH = \frac{7}{2} m_H k_B \frac{dT}{dt} \tag{3.1}$$

where JH is the volumetric heating rate. Assuming the height of the thermosphere
of 150 km, we can derive the temperature rate change at 13,600 K/day at the heating

rate of 0.04 W/m^2. However, the part of the energy we be spent to drive the thermospheric expansion that will increase the thermospheric density. Such energy deposition will induce the NO (at 5.3 μm) and CO2 (at 15 μm) mediated radiative cooling of the thermosphere, which is scaled linearly with the temperature increase (Weimer *et al.* 2015). Recent observations suggest that strongest storms ignite thermospheric overcooling due to such effects (Knipp *et al.* 2017). One can see that if the dissipation rate becomes 100 times greater, this will heat the thermosphere to huge temperatures and ignite adiabatic expansion of ionospheric plasma driving atmospheric evaporation and radiative cooling by NO and CO2. These processes will be simulated in the near future using our ionosphere-thermosphere code (Smithro and Sojka 2005). Ionospheric heating develops large gradients of plasma pressure in addition to forces which drive mass outflow at velocities greater than 20 km/s. This is greater than the Earth's escape velocity, and thus, this bulk flow contributes to the mass loss during the storm. Ionospheric cross cap potential drives large energy flux of non-thermal precipitating electrons   24 erg/cm^2/s with the mean energy of 5 keV.These processes are crucial factors that could contribute to habitability conditions on early Earth and Mars and exoplanets around active stars.

3.2. *XUV Driven Atmospheric Escape From the Young Earth*

We have recently modeled the effects of X-ray and UV (XUV) radiation from the young Sun on atmospheric escape from the 0.7 Gyr young Earth, when XUV fluxes were by a factor of 10 greater than that at the current epoch (Airapetian *et al.* 2017a). XUV radiation induces non-thermal heating via photo-absorption and photoionization raising the temperature of the exosphere, and therefore, its pressure scale height. At high XUV fluxes, this process initiates hydrodynamic atmospheric escape of neutral atmospheric species, with the loss rate dependent on the molecular mass of atmospheric species. Hydrogen, as the lightest component, escapes more readily than any other species by this mechanism (Lammer *et al.* 2008; Tian *et al.* 2008). For the environments of active solar-type stars, much of the hydrogen likely escapes from a planet's atmosphere during the system's early evolution, leaving behind an atmosphere enriched in heavier elements such as N and O. Thus, processes of atmospheric ionization and loss via non-thermal mechanisms are crucial for modeling the evolution of oxygen and nitrogen-rich atmospheres as well the efficiency of atmospheric loss of water as a critical factor of habitability of the young Earth. In the region above an Earth-size planet's exobase, the layer where collisions are negligible, the incident XUV flux ionizes atmospheric atoms and molecules and produces photoelectrons. The upward propagating photoelectrons outrun ions in the absence of a radially directed polarization electric field and forms the charge separation between electrons and atmospheric ions. Thus, a radially directed polarization electric field is established that enforces the quasi-neutrality and zero radial current. For ionospheric ions with energies over 10 eV, the polarization electric field cancels a substantial part of the Earth's gravitational potential barrier, greatly enhancing the flux of escaping ions and forming an ionospheric outflow.

We modeled the effects of XUV flux on the ionosphere by coupling the ion hydrodynamics of the Polar Wind Outflow Model (PWOM) to the latest version of the SuperThermal Electron Transport (STET) code (Glocer *et al.* 2009; Glocer *et al.* 2012; Khazanov 2011; Khazanov *et al.* 2015). Full details of the model coupling will appear in a separate publication (Glocer *et al.* 2016). The XUV fluxes from the evolving Sun are expressed in terms of the total XUV flux, F_0, of the Sun at the average level of magnetic cycle. We find that the photoelectron flux increases approximately linearly with the input XUV flux. We then used PWOM to calculate the ionized atmospheric escape rates along an open single magnetic field line of the polar region at heights between 200 km and 6000 km.

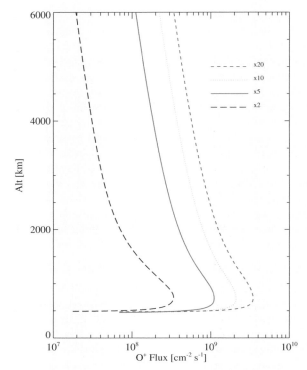

Figure 4. The mass loss rate of oxygen ions from the Earth atmosphere due to XUV irradiation from the young Sun at F=2 (long dash), 5 (dash-dot), 10 (dot), 20 (short dash).

The XUV flare flux at 10F0 corresponds to the associated super-Carrington type CME event. We then calculated the steady state outflow rate of O+ ions driven by the input XUV flux and the value of the neutral temperature specified at the exospheric base at 200 km. In order to evaluate the effect of the base temperature on the O^+ outflow rate, we calculated two escape models for the XUV flux of $10F_0$ for these two exobase temperatures. We find that as we increase the base temperature by a factor of 2 (from 1000K to 2000K), the resulting O+ outflow rates increase by a factor of 10. The total loss rate of O+ at h=1000 km is found from the integration of this value over the whole area. Figure 4 shows that the mass loss of oxygen ions increases roughly linearly with the solar flux and reaches 400 kg/s for $F=10F_0$ (Airapetian *et al.* 2017a). This estimate does not account for a number of effects typically contributing to the ion escape during space weather events associated with large solar flares. This mass loss rate can also be affected by precipitated energetic electrons from the day and night sides of the Earth magnetosphere. This input efficiently produces secondary superthermal electrons due to collisional ionization of species in the ambient ionosphere (Strangeway *et al.* 2005) and needs further study.

Our simulations of the atmospheric escape suggests that if we account for the escape for nitrogen ions along with oxygen ions, then the upper limit of escape rate at 10 F_0, characteristic of the Sun's flux 3.8 billion years ago, is ∼ 400 kg/s. This suggests that Earth could have lost half of its 1-bar atmosphere in 300 million years after the secondary atmosphere was formed on the early Earth. Given that the Earth had an intensive volcanic and tectonic activity, this suggests that the XUV impact on the Earth's habitability was

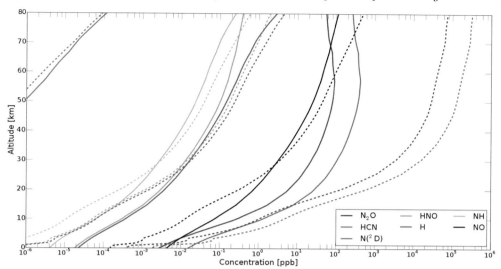

Figure 5. Aeroplanets model predictions of mixing ratios of species under chemical equilibrium driven by frequent energetic SEPs on the early Earth (Airapetian *et al.* 2016a).

pretty mild. This is consistent with the recent data suggesting that the atmosphere of the early Earth 2.7 Gyr ago was at least 0.5 bars (Som *et al.* 2016).

4. Space Weather as a Factor of Life

Our global magnetospheric simulations suggest that a disturbance of the early Earth magnetosphere by a super-Carrington CME event should shift the boundary of the open-closed field to 36^{deg} in latitude, producing a polar cap opening 70% of the planet′s dipole magnetic field. Thus, extended polar caps may provide a pathway for energetic electrons and protons accelerated in CME-driven shocks to penetrate the Earth atmosphere along the open field lines (Airapetian *et al.* 2016a).

The secondary atmosphere of the early Earth at 0.5 Gyr was nitrogen rich (80-90%) and CO_2 rich (10-20%) with traces of methane, CH_4, and water vapor, H_2O. Molecular nitrogen was mostly supplied by tectonic activity from the highly oxidized mantle wedges driven by subduction processes, while carbon dioxide, methane and water vapor were released by intensive volcanic activity (Mikhail & Sverjensky 2014). We have recently applied our Aeroplanet model (Gronoff *et al.* 2014) to simulate the atmospheric chemistry of such highly reduced nitrogen-dominated (79% N_2, 20% CO_2, 0.4% CH_4, 1% H_2O) prebiotic Earth atmosphere at a surface pressure of 1 bar with the photochemistry controlled by the EUV-XUV flux fro the young Sun and a proton energy flux of 50 times that of the Jan 20, 2005 SEP event (Airapetian *et al.* 2016).

The Aeroplanets model calculates photoabsorption of the EUV-XUV flux from the early Sun and electron and proton fluxes to compute the corresponding energetic fluxes at all altitudes between 200 km to the surface (Gronoff *et al.* 2014). These fluxes are then used to calculate the photo and particle impact ionization/dissociation rates of the atmospheric species producing secondary electrons due to ionization processes. Then, using the XUV flux and the photoionization-excitation- dissociation cross-sections, we calculated the production of ionized and excited state species and resulted photoelectrons. In this steady-state model of the early Earth atmosphere, energetic protons from an SEP event precipitate into the middle and lower atmosphere (stratosphere, mesosphere and troposphere) and produce ionization, dissociation, dissociative ionization, and excitation of atmospheric species. The destruction of N_2 into reactive nitrogen, N(2D) and N(4S) and the subsequent destruction of CO_2 and CH4 produces NOx, CO and NH in the polar regions of the atmosphere as shown in Figure 5. NO_x then converts in the stratosphere to NO_3, HNO_2 and HNO_3.

One of the major predictions of our atmospheric model is efficient production of nitrous oxide, N_2O, which is a potent greenhouse gas (Airapetian *et al.* 2016). This could represent a pathway to the resolution of the faint Young Sun's (FYS) paradox that suggest that the energy from faint young Sun would be insufficient to support liquid water on the early Earth contrary to geological evidence of its presence that time (Sagan and Mullen 1972; Ramirez 2016). The proposed models of the atmospheric warming due to large atmospheric concentration of CO_2, H_2O and/or CH_4 cannot resolve the FYS paradox (Kasting 2010; Rosing *et al.* 2010). This problem becomes even worse for the Martian atmosphere that would require up to 4 bars of the atmospheric abundance of CO (Ramirez *et al.* 2014). Our model proposes a resolution of the FYS paradox due to collisional dissociation of the atmospheric N_2, CO_2, CH_4 and NH_3 producing abundant NOx and NH molecules and efficient formation of N_2O through NO + NH $\rightarrow$ N_2O + H (Airapetian *et al.* 2016b). Atmospheric N_2O density reaches a concentration with the mixing ratio of 0.3 to 1 ppmv in the lower atmosphere depending of availability of gases shielding nitrous oxide from photodestruction. The sources and sinks for N_2O depend strongly on the chemical composition of the initial atmosphere and the energy flux in accelerated protons. Specifically, our simulations show that N_2O's abundance increases with increasing CO_2/CH_4 ratio in the initial atmosphere. Moreover, the derived value should be considered as a lower bound, because our model does not account for a number of factors including eddy diffusion and convection effects, concentration of hazes, inclusion of SO_2 and H_2S volcanic outgassing sources and Rayleigh scattering of solar EUV radiation that significantly reduces photodestruction of N_2O, and therefore increases its production. Indeed, Earth's atmospheric data suggest that stratospheric-tropospheric exchange provide flat vertical profiles of [7]Be and [10]Be from 30km to 2-3 km above the ground at higher lattitudes (60-90 deg) (Land & Feichter 2003). Thus, we expect the same profiles for N^2O and HCN vertical profiles. This will provide the mixing ratio of N_2O at least 1 ppm in the lower troposphere required to provide efficient greenhouse warming. Also, energetic protons associated with SEP events significantly enhance atmospheric ion production rates, which in turn that drive increased rate of formation/nucleation of newly formed and/or existing production of stratospheric aerosol particles by up to one order of magnitude in the polar regions at 10-25 km, which provides an efficient shield from UV emission around 240 nm (Mironova & Usoskin 2014).

Greater typical energies of SEP events from the young Sun could be another factor that contribute to the increased production rate by a factor of 5-10 greater than that conservatively assumed in our model. This is due to the fact that young Sun's corona that represents the source of CMEs was at least by a factor of 10 denser as compared to the current Sun. Thus, denser corona provided correspondingly larger concentration of seed particles that participated in acceleration processes driven by CME driven shock waves closer to the solar surface. Also, our recent simulations suggest that the particle acceleration via diffusive shock acceleration mechanisms on a quasi-parallel shocks produce mostly SEPs with harder spectrum (Airapetian *et al.* 2017b) similar to the Feb 1956 SEP event (see Figure 6). Also, recent statistical study of These spectra suggests the particle flux at 1 GeV by 2 orders of magnitude greater than that assumed in the current model by Airapetian *et al.* 2016. Because the production of reactants is proportional to the number of incoming particles, it should be linearly scaled for the incident

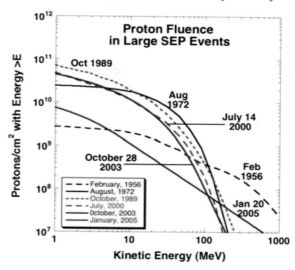

Figure 6. Protons fluence spectra from the largest SEPs observed over the last 50 years (Mewaldt *et al.* 2005).

flux. Thus, the concentration of produced N_2O and HCN should be boosted by a factor of 100 at 10 - 15 km.

Recent statistics of CMEs associated with SEP events shows that SEPs with hardest particle fluence spectra are correlated with CMEs with large initial acceleration when they were observed close to the Sun (Gopalswamy *et al.* 2016). Large initial acceleration of CMEs suggests that they were originated from structures with stronger magnetic field strength that rapidly inflate producing strong shock waves in the outer solar corona. These shocks then provide a fertile ground for particle acceleration in SEP events. This empirical picture is consistent with our recent simulations of SEP accelerated events from the young Sun formed from the strong shocks produced by magnetized CMEs via diffusive shock acceleration mechanisms on a quasi-parallel shocks produce mostly SEPs with harder fluence spectra (Airapetian *et al.* 2017a) similar to the Feb 1956 SEP event.

Such hard particle spectra suggest the particle flux at 1 GeV which is by a factor of 20 at 600 MeV greater than the Jan 20 2005 SEP (see Figure 6) event taken in our recent prebiotic chemistry model of the early Earth (Airapetian *et al.* 2016a). Because the production rates of N_2O and HCN is proportional scaled with the number of incoming protons at 1 GeV, we should therefore scale the produced concentration of these molecules in the lower stratosphere (at 20 km) by a factor of 30 larger than discussed in our paper. Thus, the concentration of produced N2O and HCN should be boosted by a factor of 100 in the Earth's lower atmosphere. Thus, the combination of discussed factors would boost the concentration of N_2O to the level of 100 ppmv. Climate models by Roberson *et al.* 2011 of the for the Proterozoic Earth with the concentration of methane at 1.6 ppm at the fixed concentration of carbon dioxide at 320 ppm yield surface warming above $0^0 C$.. This suggest that our model presents an opportunity to obtain temperatures above the freezing point in our ongoing 3D GCM simulations, and thus resolve the longstanding Faint Young Sun paradox for the early Earth and Mars.

5. Conclusions

As we discussed in this paper, space weather effects from the young Sun can contribute to the habitability conditions on the early Earth in a variety of ways. First, we have shown that the high magnitude southward IMF and large dynamic pressures from super Carrington type CME events can restructure the Earth's magnetosphere due to reconnection events and widen its polar caps. Then, the dissipation of large induced geomagnetic currents can heat the thermospheric plasma to high temperature that can support its escape from the planet in its earliest phase of evolution when CME events were frequent and energetic. Second, high XUV fluxes from associated superflare flare events can support the escape of oxygen and possibly nitrogen ions due to production of photoelectrons. However, in our models, where the escape process was uncoupled with the thermal effects of inflated exosphere due, and therefore, they provide the lower bound of escape rates. In these ways, CMEs can provide negative conditions for habitability on the early Earth and Mars especially in the first 0.5 billion years of the Sun's magnetically active phase of evolution. The conditions on the early Mars were more severe because of much lower surface gravity and the efficiency of photochemical escape via dissociative recombination of O2 and possibly N2 producing hot atomic oxygen and nitrogen.

However, frequent SEP events from the young Sun probably played a positive role in setting the conditions for formation of hydrogen cyanide and nitrous oxide in the lower stratosphere and upper troposphere of the early Earth and Mars. Organic molecules may subsequently rain out into surface reservoirs and ignited higher order chemistry producing more complex organics. For example, the hydrolysis of HCN produces formamide, $HCONH_2$. When irradiated with energetic protons, formamide can serve as a precursor of complex biomolecules that are capable of producing amino acids, the building blocks of proteins and nucleobases, sugars and nucleotides, the constituents of RNA and DNA molecules (Saladino *et al.* 2015). In our recent experiments, irradiation of gas mixture resembling the young Earth atmosphere with high-energy protons (2.5 MeV) produced amino acids including glycine and alanine (Kobayashi *et al.* 2001; 2017). The irradiation of the same mixture by the spark discharge (accelerated electrons) or UV irradiation (2500/AA), produced no amino acids at CH_4 mixing ratio (less than 15%). Thus, considering fluxes of various energies on the primitive Earth, energetic protons appear to be an efficient factor to produce N-containing organics than any other conventional energy sources like thundering or solar UV emission irradiated the early Earth atmosphere. Also, abiotic production of nitrous oxide in the lower troposphere at 10 - 20 ppm driven by energetic protons can provide an efficient way to resolve a long standing Faint Young Sun paradox to explain the warming of our young planet to keep water in the liquid state in its early history (Airapetian *et al.* 2016; Airapetian *et al.* 2017b).

In our future work, we plan to develop a comprehensive 1D and 2D photo-collisional models of the early Earth that will describe the production of greenhouse gases and biological molecules in the troposphere with implementation newly develop models of solar energetic particle events from the young Sun as inputs along with more realistic representation of volcanic gasses and aerosols in the early Earth atmosphere. These models will provide insights in understanding the challenging problem of warming of early Mars and also expand the definition of habitability zones around main-sequence G and K stars with volcanically active planets.

References

Abramov, O. & Mojzsis, S. J. 2009, *Nature*, 459, 419

Airapetian, V. S., Glocer, A., Khazanov, G. V., Loyd, R. O. P., France, Kevin *et al.* 2017a, *ApJ Letters*, 836, L3

Airapetian, V. S., Zank, G., Verkhodlyadova, O. Li, G., Gronoff, G. 2017b, in *preparation*.

Airapetian, V. S. , Glocer, A. , Gronoff, G., E. H/ebrard, E., Danchi, W. 2016, *Nature Geoscience*, DOI:10.1038/NGEO2719

Airapetian, V. & Usmanov, A. 2016, *ApJ Let.*, 817, L24

Airapetian, V., Glocer, A., & Danchi, W. 2015, *Proc. of the 18th Cambridge Workshop on Cool Stars*, Stellar Systems, and the Sun, Lowell Observatory (9-13 June 2014) Eds: G. van Belle & H. Harris (eprint arXiv:1409.3833)

Aarnio, A. N., Stassun, K. G., Hughes, W. J., & McGregor, S. L. 2011, *Sol.Phys.*, 268, 195

Thakur, N., Gopalswamy, N., Xie, H., Makela, P., Akijama, P., Davila, J. M., & Rickard, L. J. 2014, *ApJLet*, 790, L13

Birch, J. L. *et al.* 2016, *Science*, 352, id.aaf2939

Biggin, A. J., Strik, G. H. M. A., & Langereis, C. G., 2008, *Nature Geoscience 1*, 395

Bell, E. A., Boehnke, P., Harrison, T.M. & Mao, W. L. 2015, *PNAS*, 112, 14,518

Claire, M. W. *et al.* 2012, *ApJ.* 757, 95

Cohen, O., Drake, J. J., Glocer, A., Garraffo, C., Poppenhaeger, K., Bell, J. M., Ridley, A. J., Gombosi, T. I. 2014, *ApJ*, 790, 57

Garraffo, C., Drake, J. J., Cohen, O. 2016, *A&A*, 595, id.A110

Gomes, R., Levison, H. F., Tsiganis, K., & Morbidelli, A. 2005, *Nature*, 435, 466

de Zeeuw, D. L., Sazykin, S., Wolf, R. A., Gombosi, T. I., Ridley, A. J., Toth, G. 2004, *JGR*, 109, A12

Emslie, A. G. *et al.* 2012, *ApJ*, 759, 71

Glocer, A., G. Toth, T. Gombosi, & D. Welling, 2009, *JGR.*, 114

Glocer, A., N. Kitamura, G. Tóth, & T. Gombosi, 2012, *JGR.*, 117, A04318, doi: 10.1029/2011JA017136

Gopalswamy, N. 2011, *First Asia-Pacific Solar Physics Meeting ASI Conference Series*, 2011, Vol. 2, pp 2, Edited by Arnab Rai Choudhuri & Dipankar Banerjee

Gopalswamy, N., Yashiro, S., Thakur, N., Mäkelä, P., Xie, H., & Akiyama, S. 2016, *ApJ*, 833, Issue 2, article id. 216

Gough, D. O. 1981, *Solar Phys.* 74, 21

Gronoff, G. *et al.* 2014, *Geophys. Res. Lett.* 41, 4844

Güdel, M, Guinan, E. F., & Skinner, S. L. 1997, *ApJ*, 483, 947

Kasting, J. F. 2010, *Nature* 464, 687

Knipp, D. J., Pette, D. V., Lilcommons, L. M. *et al.* 2017, *Space Weather*, 15, doi:10.1002/20165W001567

Khazanov, G. 2011, *Kinetic Theory of the Inner Magnetospheric Plasma, Astrophysics and Space Science Library*, Vol. 372. ISBN 978-1-4419-6796-1. Springer

Khazanov, G. V., Tripathi, A. K., Sibeck, D., Himwich, E., Glocer, A., & Singhal, R. P. 2015, *JGR*, 20, Issue 11, 9891

Kobayashi, K., Masuda, H., Ushi, K., Ohashi, A. Yamanashi, H. *et al.* 2001, *Adv. Space Res.* 27, No. 2, 207

Kobayashi, K,. Aoki, R., Abe, H., Kebukawa, Y., Shibata, H., Yoshida, S., Fukuda, H., Kondo, K., Oguri, Y., & Airapetian, V. S. 2017, *Astrobiology Science Conference 2017*, Abstract #3 259

Kramar, M., Airapetian, V., & Lin, H. 2016, *Frontiers in Astronomy and Space Sciences*, Volume 3, id.25

Lammer, H., Kasting, J. F., Chassefiere, Johnson, R. E., Kulikov, Y. N., & Fian, F. 2008, *Space Sci. Rev.*, 139, 399

Land, C. & Feichter, J. 2003, *JGR*, 108, Issue D12, 10.1029/2002JD002543

Le *et al.* 2016, *GRL*, 43, 2396

Love, J. J. 2012, *GRL.*, 39, L10301

Loyd, R. O. P., France, Kevin, Youngblood, A., Schneider, C. Brown, A., Hu, R., Linsky, K., Froning, C. S., Redfield, S., Rugheimer, S., & Tian, F. 2016, *ApJ*, 824, 102

Maehara, H. *et al.* 2012, *Nature* 485, 478

Miyake, F., Nagaya, K., Masuda, K., & Nakamura, T. 2012, *Nature*, 486, 240

Ngwira, C. M., Pulkkinen, A., Kuznetsova, M. M. & Glocer, A. 2014, *JGR*, 119, 4456

Nitta, N. V., Aschwanden, M. J., Freeland, S. L., Lemen, J. R., Wülser, J.-P., & Zarro, D. M. 2014, *Sol. Phys.*, 289, 1257

Mikhail & Sverjensky 2014, *Nature Geoscience* 7, 816

Mewaldt, R. A., Cohen, C. M. S., Labrador, A. W., Leske, R. A., Mason, G. M., Desai, M. I., Looper, M. D., Mazur, J. E., Selesnick, R. S., & Haggerty, D. K. 2005, *JGR*, 110, A09S18

Mironova, I. A. & Usoskin, I. G. 2014, *Environ. Res. Lett.*, 9

Powell, K. G., Roe, P. L., Linde, T. J., Gombosi, T. I., & De Zeeuw, D. L. 1999, *JCP*, 154, 284

Pevtsov, A. A., Fisher, G. H., Acton, L. W., Longcope, D. W., Johns-Krull, C. M., Kankelborg, C. C., & Metcalf, T. R. 2003, *ApJ*, 589, 1387

Ramirez, R. M., Kopparapu, R., Zugger, M. E., Robinson, T. D., Freedman, & R. Kasting, J. 2014, *Nature Geoscience*, 7, Issue 1, 59

Ramirez, R. M. 2016, *Nature Geoscience*, 9, Issue 6, 413

Ridley, A. J., Hansen, K. C., Tóth, G., de Zeeuw, D. L., Gombosi, T. I., & Powell, K. G. 2002, *JGR*, 107, 1290

Riley, P., Caplan, R. M., Giacalone, J., Lario, D., & Liu, Y. 2016, *ApJ*, 819, 57

Roberson, A. L., Roadt, J., Halevy, I., & Kasting, J. F. 2011, *Geobiology*, 9, 313

Rosing, M. T., Bird, D. K., Sleep, N. H. & Bjerrum, C. J. 2010, *Nature* 464, 744

Sagan, C. & Mullen, G. 1972, *Science*, 177, 52

Saladino, R., Carota, E., Botta, G., Kapralov, M., Timoshenko, G. N., Rozanov, A. Y., Krasavin, E., & Ernesto Di Mauro, E. 2015, *Publ. Nat. Acad. Sci.*, 112 (21), E2746

Schrijver, C. J., Kauristie, Kirst A., Alan D., Denardini, C. M., Gibson, S. E., Glover, A., Gopalswamy, N., Grande, M., Hapgood, M., Heynderickx, D., & 16 coauthors, 2015, *Advances in Space Res.*, 55, 2745

Shibayama, T., Maehara, H., Notsu, S., Notsu, Y., Nagao, T., Honda, S., Ishii, T. T., Nogami, D., & Shibata, K. 2013, *ApJSS*, 209, 5

Schrijver, C. J. & Title, A. M. 2013, *ApJ*, 619, 1077

Som, S. M., Buick, R., Hagadorn, J. W., Blake, T. S., Perreault, J. M., Harnmeijer, J. P., & Catling D. C. 2016 *Nature Geoscience*, 9,448

Sterenborg, M. G., Cohen, O. , Drake, J. J., & Gombosi, T. I. 2011, 116, A01217

Strangeway, R. J., Ergun, R. E., Su, Y.-J., Carlson, C. W., & Elphic, R. C. 2005, *JGR*, 110, Issue A3,CiteID A03221

Tarduno, J. A., Blackman, E. G., & Mamajek, E. E. 2014, *Physics of the Earth and Planetary Interiors*, 233, 68

Tarduno, J. A., Cottrell, R. D., Watkeys, M. K., Hofmann, A., Doubrovine, P. V., Mamajek, E. E., Liu, D., Sibeck, D. G., Neukirch, L. P., & Usui, Y. 2010, *Science*, 327, Issue 5970, 1238

Tian, F., Kasting, J. F. Liu, H.-Li, & Roble, R. G. 2008, *JGR*, 113, Issue E5, CiteID E05008

Thomas, B. C., Melott, A. L., Arkenberg, K. R., & Snyder, B. R. 2013, *Geoph. Res. Let*, 40, 1237

Tsurutani, B. T., Gonzales, W. D., Lakhina, G. S. & Alex, S. 2003, *J. Geophys. Res.* 108, 1268

Tsurutani, B. T. & Lakhina, G. S. 2014, *Geoph. Res. Let*, 41, 287

Tu, L., Johnstone, C., Gudel, M., & Lammer, H.2016, *A&A*, 577, L3

Vidotto, A. A., Gregory, S. G., Jardine, M., Donati, J. F., Petit, P. *et al.* 2014, *NMRAS*, 441, 2361

Yashiro, S. & Gopalswamy, N. 2009, in *IAU Symp.257, Universal Geophysical Processes*, ed. N. Gopalswamy & D. F. Webb (Cambridge: Cambridge Univ. Press), 233

Weimer, D. R., Mlynczak, M. G., Hunt, L. A., & Tobiska, W. K. 2015, *JGR*, 120, 5998

Wilde, S., Valley, J. W., Peck, W. H. & Graham, C. M. 2001, *Nature*, 409, 175

Wood, B. E., Müller, H-R, Zank, G. P., Linsky, J. L.& Redfield, S. 2005, *ApJ*, 628, L143

Zhao, Y., Wang, R., Lu, Q., Du, A., Yao, Z., & Wu, M. 2016, *JGR*, 121, Issue 11, 10,898

Living Around Active Stars
Proceedings IAU Symposium No. 328, 2016
D. Nandy, A. Valio & P. Petit, eds.

© International Astronomical Union 2017
doi:10.1017/S1743921317004173

Evolution of Long Term Variability in Solar Analogs

Ricky Egeland[1,2], Willie Soon[3], Sallie Baliunas[4], Jeffrey C. Hall[5] and Gregory W. Henry[6]

[1]High Altitude Observatory/NCAR, 3080 Center Green Dr, Boulder CO, 80301, USA
email: egeland@ucar.edu
[2]Dept. of Physics, Montana State University, P.O. Box 173840, Bozeman MT 59717, USA
[3]Harvard-Smithsonian Center for Astrophysics, Cambridge, MA 02138, USA
[4]No affiliation
[5]Lowell Observatory, 1400 West Mars Hill Road, Flagstaff, AZ 86001, USA
[6]Center of Excellence in Information Systems, Tennessee State University, 3500 John A. Merritt Blvd., Box 9501, Nashville, TN 37209, USA

Abstract. Earth is the only planet known to harbor life, therefore we may speculate on how the nature of the Sun-Earth interaction is relevant to life on Earth, and how the behavior of other stars may influence the development of life on their planetary systems. We study the long-term variability of a sample of five solar analog stars using composite chromospheric activity records up to 50 years in length and synoptic visible-band photometry about 20 years long. This sample covers a large range of stellar ages which we use to represent the evolution in activity for solar mass stars. We find that young, fast rotators have an amplitude of variability many times that of the solar cycle, while old, slow rotators have very little variability. We discuss the possible impacts of this variability on young Earth and exoplanet climates.

Keywords. Sun Activity Chromosphere Variability Climate Habitability

1. Introduction

The cyclic variability of the solar sunspot count was noted by Schwabe (1844), but observations of surface *activity* in Sun-like stars came more than a century later. Wilson (1968) introduced the Mount Wilson Observatory (MWO) HK Project, which began synoptic monitoring of the emission in the cores of Fraunhofer H (3968.47 Å) and K (3933.66 Å) K lines for a sample of Sun-like (F, G, and K-type) stars. Formed by singly-ionized calcium, these lines have a reversal feature for which the central emission has long been known to correlate with regions of strong magnetic field on the Sun (Leighton 1959; Linsky and Avrett 1970). Using Ca II H & K as a proxy for magnetic activity on stars, Wilson (1978) presented observations of 91 main-sequence stars, showing that they do in fact vary, and that several of the stars appeared to have completed a cycle in HK flux variations. Baliunas *et al.* (1995) summarized ~25 years of synoptic observations for 112 stars and conclusively showed the existence of cyclic variability, as well as other patterns of variability. More than half the sample showed either erratic variability, long-term trends, or flat activity that may be analogous to the solar Maunder Minimum, a long period of subdued solar activity from about 1650–1700 (Eddy 1976).

Wilson (1968) discussed the difficulty of detecting variability in broad-band visible observations, estimating a 0.001 magnitude (1 millimagnitude (mmag); ≈0.1%) change in solar luminosity due to the passage of spots covering about 1400 millionths of the solar surface. This is comparable to the later measurements of the average cyclic variation in the total solar irradiance (TSI) from the Solar Maximum Mission (Willson and Hudson

1991). The challenge of measuring visible-band variability in Sun like (FGK-type) stars was taken up by researchers at Lowell Observatory, who used differential photometry of the Strömgren b and y bands to achieve the required precision (Lockwood *et al.* 1997). They found short-term (inter-year) and long-term (year-to-year) rms amplitudes ranging from 0.002 mag (0.2%) to 0.07 mag (7%) for about 41 program stars. Overlap with MWO targets allowed the comparison of photometric variability in the combined bandpass $((b + y)/2)$ to Ca II H & K activity expressed with the R'_{HK} index, the ratio of HK flux to the bolometric luminosity. Lockwood *et al.* (1997) generally found that active stars (high R'_{HK}) have larger rms photometric variability, and Radick *et al.* (1998) found a power law relationship between the two quantities. Furthermore, Radick *et al.* (1998) found that stars were either *faculae-dominated*, with a positive correlation between brightness and H & K activity, or *spot-dominated*, with a negative correlation. The terminology here refers to the dominant features contributing to visible-band brightness variations. The faculae are the photospheric counterpart to the *plage* in the chromosphere, which are bright features in Ca II H & K, while spots are dark features in both H & K emission and visible bandpasses.

Stars like the Sun emit most of their flux in the visible spectrum, and for a planet with an atmosphere like the Earth's, the majority of the radiant energy reaching the surface likewise comes in the visible. The ∼0.1% variability in TSI from the present day Sun is thought to be of little consequence to the globally averaged Earth temperature (Stocker *et al.* 2013), however this may not have always been the case. The climate impact of the Maunder Minimum period, and its coincidence with the Medieval Little Ice Age are actively debated, however interpretations are crucially dependent on the use of proxy records to extrapolate the present TSI into the Maunder Minimum period (e.g. Kopp 2014; Solanki *et al.* 2013). The stellar studies of Radick *et al.* (1998) and Lockwood *et al.* (2007) show a clear relationship between visible band variability and Ca II H & K activity, and furthermore it has long been known that stellar activity decreases with age as a star loses angular momentum (Skumanich 1972; Noyes *et al.* 1984; Barnes 2007). We therefore ask the question, "how has solar variability impacted Earth's climate on stellar evolution (billion year) timescales?", and the related question, "how might stellar variability affect exoplanet climates?"

Clearly the most important impact of stellar evolution on planetary climate is the total flux reaching the top of the atmosphere. According to standard solar evolution models, the luminosity of the Sun has been steadily increasing from an initial value of ∼70% the present-day luminosity when hydrogen burning began ∼4.6 billion years ago (e.g. Gough 1981). Because of the lower luminosity, from first-order calculations we would expect the mean temperature of the Earth to be below the freezing point of water, which is in contradiction to geological evidence for wet conditions and the development of life on Earth 3–4 billion years ago (Sagan and Mullen 1972). This problem is known as the "Faint Young Sun Paradox," which was discussed by Dr. Martens at this symposium. In this contribution, we shall ignore the mean luminosity and consider the climate impact of decadal scale variability from younger, more active stars.

2. Long-term Variability of Solar Analogs

To begin to address the questions of the relationship between stellar variability and planetary climate, we look at a sample of five solar-analog stars that may represent the behavior of the Sun at various points in the history of the solar system, as shown in Table 1. This sample is drawn from a larger sample of 26 solar-analog stars with Ca II H & K observational records up to 50 years in length. These long records are obtained

Table 1. Stellar Properties & Variability

Quantity	HD 20630	HD 30495	HD 76151	HD 146233	Sun	HD 9562
M_V	5.04	4.87	4.81	4.79	4.82	3.41
$T/\mathcal{T}_\odot^N$	0.99	1.00	0.98	1.00	1	1.01
$R/\mathcal{R}_\odot^N$	0.93	0.97	1.05	1.02	1	1.85
$L/\mathcal{L}_\odot^N$	0.83	0.95	1.03	1.03	1	3.62
[Fe/H]	0.00	-0.08	-0.04	-0.02	0	$+0.13$
$P_{\rm rot}$ [d]	9.2	11.36	15.0	22.7	26.09	29.0
Age [Gyr]	0.5 ± 0.1	1.0 ± 0.1	1.4 ± 0.2	$3.66^{+0.44}_{-0.50}$	4.57	$3.4^{+1.7}_{-0.2}$
$\widehat{S}$	0.3606	0.3020	0.2363	0.1703	0.1686	0.1369
$A_{S,98\%}$	0.1169	0.0708	0.0679	0.0414	0.0275	0.0226
$A_{S,s}$	0.0902	0.0502	0.0576	0.0313	0.0203	0.0159
$A_{by,s}$ [mmag]	30.1	21.5	7.9	1.3	1.5^*	1.8

Notes: Stellar M_V, $T_{\rm eff}$, and [Fe/H] are from the Geneva-Copenhagen survey (Holmberg *et al.* 2009). Stellar luminosities are derived using the empirical bolometric correction of (Flower 1996), and radii follow from the Stephan-Boltzmann Law. T, R, and L are given in solar units using the IAU 2015 resolution B2 nominal values (Prša *et al.* 2016) and have an uncertainty of 1–2%. Rotation periods are from (in order) Gaidos *et al.* (2000), Egeland *et al.* (2015) (E15), Donahue *et al.* (1997), Petit *et al.* (2008), Donahue *et al.* (1996), Baliunas *et al.* (1996). Ages are from (in order) Barnes (2007) (B07), E15, B07, Li *et al.* (2012), Bouvier (2008), Holmberg *et al.* (2009). The solar $(b+y)/2$ amplitude is estimated in Lockwood *et al.* (2007) by applying a scaling factor to the TSI record.

by combining observations from the MWO HK Project (1966–2003) and the Lowell Observatory SSS (1994–present). Some initial results from this study were presented in Egeland *et al.* (2016) and Egeland *et al.* (2016). A similarly long Sun-as-a-star Ca II H & K record was developed in Egeland *et al.* (2017), which accurately placed the long NSO Sacramento Peak Ca II K-line record on the S-index scale using coincident observations of the Moon from the MWO HKP-2 instrument. Figure 1 shows the solar S-index record and three other stars from our sample on the same scale, illustrating the range of mean activity levels and amplitudes. The youngest, most active star in our sample is HD 20630 (κ^1 Ceti), which was discussed at this symposium by Dr. Dias do Nascimento, Jr. Not shown are HD 30495 and HD 146233 (18 Sco), the former which is studied in detail in Egeland *et al.* (2015), and the latter which is a solar twin (Porto de Mello and da Silva 1997; Meléndez *et al.* 2014) and has a mean activity and amplitude very similar to the Sun (Hall *et al.* 2007; Egeland *et al.* 2017).

Table 1 shows the properties of the sample. All the stars are within 2% of the solar effective temperature. All but HD 9562 lie very near to the 1 $M_\odot$ evolutionary track, and therefore approximate the Sun at different points in its lifetime from an age of 0.5 Gyr to the present Sun. HD 9562 is a subgiant which has cooled into the temperature range of our "solar analog" definition, and is more massive than the Sun and the other stars in the sample. Using $\log g = 3.99 \pm 0.01$ from Lee *et al.* (2011) and the radius from Table 1 we obtain $M/\mathcal{M}_\odot^N = 1.24 \pm 0.05$. Its slow rotation and increased radius are representative of a future Sun, however the Sun is expected to have a lower surface temperature when it similarly expands (see Bressan *et al.* 2012).

We have computed two estimates of the amplitude of variability in the S-index records of this sample. The first is the inter-98% range, $A_{S,98\%}$, taken as the difference between the top and bottom 1% of the $\sim$50 year time series. The thin bars in Figure 1 demonstrate this estimate of amplitude. We also computed full range of the timeseries of seasonal median activity, $A_{S,s}$. Both of these measures of amplitude increase monotonically with

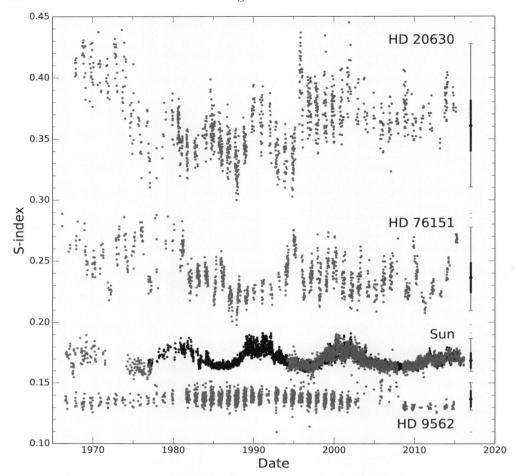

Figure 1. Calibrated composite MWO (red) + SSS (blue) time series for the Sun and three solar analogs. The relative offsets of each time series are real. Data from the Sun are those described in Egeland *et al.* (2017). The black bar symbol on the right of each time series indicates four quantities: (1) the middle diamond is at the median S for the complete time series, (2) the thin capped bar indicates the location of the 1st and 99th percentile of the data (3) the small dashes indicate the minimum and maximum points and (4) the thick bar is the median seasonal inner-98% amplitude.

median activity, $\widehat{S}$, but decrease with rotation period, P_{rot}. In fact, from the larger sample of 26 stars we find good linear relationships between median activity and the amplitude, while the relationship with rotation period has significant scatter (Egeland *et al.* 2017, in preparation). On the Sun, the S-index is a proxy for surface magnetic flux (e.g. Harvey and White 1999; Pevtsov *et al.* 2016). From Table 1 we conclude that the younger Sun had not only higher mean levels of surface flux, but also significantly larger variation in surface flux over decadal timescales. The most active star in our sample, HD 20630 (κ^1 Ceti), has an amplitude of S-index variability over four times the solar amplitude. The variability is quite erratic, as can be seen in Figure 1, but a period of reduced activity persists for about two decades from 1975 to 1995. HD 30495 varies by 2.5 times the solar amplitude, though it appears to have a semi-regular cycle with a period of ∼12 yr (Egeland *et al.* 2015). HD 76151 is also varying with about 2.5 times the solar

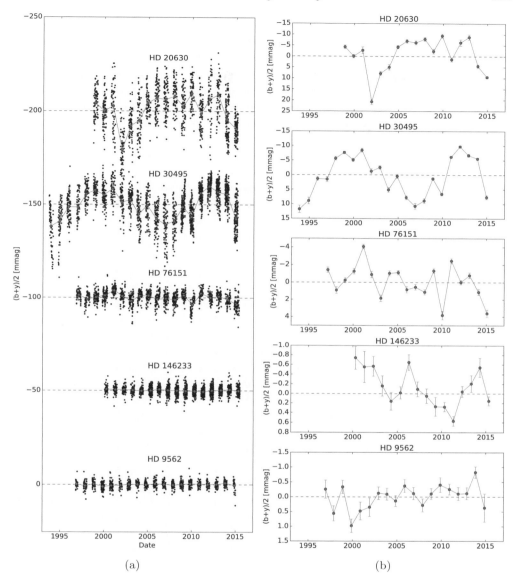

Figure 2. Variability in the combined Strömgren $(b + y)/2$ bandpass from the Fairborn Observatory APT differential photometry. Note that the y-axis is reversed so that higher points represent higher brightness. Panel (a) shows the nightly measurements for each star on the same scale, with the arbitrary mean value shifted in increments of 50 mmag. Seasonal means are shown with red points, and are shown again with a smaller scale in panel (b). Error bars indicate the uncertainty of the seasonal mean.

amplitude, but quite erratically. HD 146233 (18 Sco) varies with a 50% larger amplitude than the Sun in a cyclic fashion (Hall *et al.* 2007), while the subgiant HD 9562 has an amplitude about 20% smaller.

Figure 2 shows the variability of our sample in Strömgren $(b+y)/2$ from the Fairborn Observatory Automated Photometric Telescopes (APT; Henry *et al.* 1995). The photometric brightness is measured in millimagnitudes (mmag). In these visible bandpasses the range of variability across the sample is even more pronounced. The full range of the seasonal means, $A_{by,s}$, is shown in Table 1. The Sun's variability in $b + y$ is not well known, but Lockwood *et al.* (2007) estimate it from the TSI variations and a blackbody approximation of spectral irradiance.† Note that the solar twin 18 Sco has a similar amplitude of variability to the solar estimate. We find that for HD 20630 the amplitude of variability in the visible varies by about *twenty times* the estimated solar value. HD 30495 varies by about 14 times the solar amplitude, and HD 76151 about 5 times. The flat-activity subgiant 9562 varies slightly more in the visible than the Sun and 18 Sco.

3. Consequences for Planetary Climate

What would be the impact on Earth climate if Sun were to vary by *twenty times* its present value in the visible, as does the young solar analog HD 20630 (κ^1 Ceti)? If HD 20630 represents the Sun at an age of ~500 Myr, then this greatly enhanced variability took place ~4.1 Gya, at a time when life may have been forming on Earth (Bell *et al.* 2015). Did the enhanced variability play a role in the development of life on Earth? Is such stellar variability a significant factor in determining the habitability of exoplanets?

To begin to address these questions, we consider the Earth climate study of Meehl *et al.* (2013), who asked whether a future Maunder Minimum-type event might significantly slow global warming. Meehl *et al.* (2013) used the Whole Atmosphere Community Climate Model (WACCM) and modified the solar TSI input to include a step-function 0.25% decrease lasting 50 years. The model produced an immediate response in globally averaged temperature to this small decrease in TSI compared to the baseline case with no prolonged TSI decrease, reducing global temperature by several tenths of a degree centigrade. However, following the period of decreased TSI the warming trend resumed and caught up with the baseline case. Thus, Meehl *et al.* (2013) concludes that a future Maunder Minimum-like event could slow down, but not stop the global warming trend.

For our purposes, the significant point is that the global temperature registered an immediate response to the small 0.25% decrease in TSI. When the Sun was like HD 20630, it may have produced a much larger variations of the order 1-2%. What would solar variability such as this entail for the primitive Earth surface atmosphere and oceans, which were significantly different not only in composition (much less oxygen), but in structure (continental shifts)? More detailed theoretical work is required to determine the importance of such enhanced stellar variability on ancient Earth and exoplanet climate.

R.E. thanks the organizers for the invitation to this symposium and the travel funding provided by the AAS-SPD Thomas Metcalf award. R.E. is supported by the Newkirk Fellowship at the NCAR High Altitude Observatory.

References

Baliunas, S. L., Donahue, R. A., Soon, W. H., Horne, J. H., Frazer, J., Woodard-Eklund, L., Bradford, M., Rao, L. M., Wilson, O. C., Zhang, Q., Bennett, W., Briggs, J., Carroll, S. M., Duncan, D. K., Figueroa, D., Lanning, H. H., Misch, T., Mueller, J., Noyes, R. W., Poppe,

† The SORCE SIM instrument measured a *negative* correlation between solar activity and irradiance from 400–691 nm, which covers the b and y bands (Harder *et al.* 2009). This surprising result remains controversial (e.g. Yeo *et al.* 2014), and new observations will be required to settle the matter. Judge and Egeland (2015) proposes to place well-characterized reflector into geosynchronous orbit from which a very long, stable timeseries of spectral irradiance using differential photometry could be obtained.

D., Porter, A. C., Robinson, C. R., Russell, J., Shelton, J. C., Soyumer, T., Vaughan, A. H., & Whitney, J. H. Chromospheric variations in main-sequence stars. *ApJ*, 438: 269–287, Jan. 1995. doi:10.1086/175072.

Baliunas, S. L., Nesme-Ribes, E. Sokoloff, D., & Soon, W. H. A Dynamo Interpretation of Stellar Activity Cycles. *ApJ*, 460: 848, Apr. 1996. doi:10.1086/177014.

Barnes, S. A. Ages for Illustrative Field Stars Using Gyrochronology: Viability, Limitations, and Errors. *ApJ*, 669: 1167–1189, Nov. 2007. doi:10.1086/519295.

Bell, E. A., Boehnke, P., Harrison, T. M., & Mao, W. L. Potentially biogenic carbon preserved in a 4.1 billion-year-old zircon. *Proceedings of the National Academy of Sciences*, 112 (47): 14518–14521, 2015. doi:10.1073/pnas.1517557112.

Bouvier, J. Lithium depletion and the rotational history of exoplanet host stars. *A&A*, 489: L53–L56, Oct. 2008. doi:10.1051/0004-6361:200810574.

Bressan, A., Marigo, P., Girardi, L., Salasnich, B., Dal Cero, C., Rubele, S., & Nanni, A. PARSEC: stellar tracks and isochrones with the PAdova and TRieste Stellar Evolution Code. *MNRAS*, 427: 127–145, Nov. 2012. doi:10.1111/j.1365-2966.2012.21948.x.

Donahue, R. A., Saar, S. H., & Baliunas, S. L. A Relationship between Mean Rotation Period in Lower Main-Sequence Stars and Its Observed Range. *ApJ*, 466: 384, July 1996. doi:10.1086/177517.

Donahue, R. A., Dobson, A. K., & Baliunas, S. L. Stellar Active Region Evolution - II. Identification and Evolution of Variance Morphologies in CA II H+K Time Series. *Sol. Phys.*, 171: 211–220, Mar. 1997. doi:10.1023/A:1004922323928.

Eddy, J. A. The Maunder Minimum. *Science*, 192: 1189–1202, June 1976. doi:10.1126/science.192.4245.1189.

Egeland, R., Metcalfe, T. S., Hall, J. C., & Henry, G. W. Sun-like Magnetic Cycles in the Rapidly-rotating Young Solar Analog HD 30495. *ApJ*, 812: 12, Oct. 2015. doi:10.1088/0004-637X/812/1/12.

Egeland, R., Soon, W., Baliunas, S., Hall, J. C., Pevtsov, A. A., & Henry, G. W. Dynamo Sensitivity in Solar Analogs with 50 Years of Ca II H & K Activity. In G. A. Feiden, editor, *Proceedings of the 19th Cambridge Workshop on Cool Stars, Stellar Systems, and the Sun*. Zenodo, Sept. 2016. doi:10.5281/zenodo.154118.

Egeland, R., Soon, W., Baliunas, S., Hall, J. C., Pevtsov, A. A., & Henry, G. W. The solar dynamo zoo. In *The 19th Cambridge Workshop on Cool Stars, Stellar Systems, and the Sun*. Zenodo, 2016. doi:10.5281/zenodo.57920. URL https://doi.org/10.5281/zenodo.57920.

Egeland, R., Soon, W., Baliunas, S., Hall, J. C., Pevtsov, A. A., & Bertello, L. The Mount Wilson Observatory S-index of the Sun. *ApJ*, 835 (1), January 2017. doi:10.3847/1538-4357/835/1/25.

Flower, P. J. Transformations from Theoretical Hertzsprung-Russell Diagrams to Color-Magnitude Diagrams: Effective Temperatures, B-V Colors, and Bolometric Corrections. *ApJ*, 469: 355, Sept. 1996. doi:10.1086/177785.

Gaidos, E. J., Henry, G. W., & Henry, S. M. Spectroscopy and Photometry of Nearby Young Solar Analogs. *AJ*, 120: 1006–1013, Aug. 2000. doi:10.1086/301488.

Gough, D. O. Solar interior structure and luminosity variations. *Sol. Phys.*, 74: 21–34, Nov. 1981. doi:10.1007/BF00151270.

Hall, J. C., Lockwood, G. W., & Skiff, B. A. The Activity and Variability of the Sun and Sun-like Stars. I. Synoptic Ca II H and K Observations. *AJ*, 133: 862–881, Mar. 2007. doi:10.1086/510356.

Harder, J. W., Fontenla, J. M., Pilewskie, P., Richard, E. C., & Woods, T. N. Trends in solar spectral irradiance variability in the visible and infrared. *Geophys. Res. Lett.*, 36: L07801, Apr. 2009. doi:10.1029/2008GL036797.

Harvey, K. L. & White, O. R. Magnetic and Radiative Variability of Solar Surface Structures. I. Image Decomposition and Magnetic-Intensity Mapping. *ApJ*, 515: 812–831, Apr. 1999. doi:10.1086/307035.

Henry, G. W., Fekel, F. C., & Hall, D. S. An Automated Search for Variability in Chromospherically Active Stars. *AJ*, 110: 2926, Dec. 1995. doi:10.1086/117740.

Holmberg, J., Nordström, B., & Andersen, J. The Geneva-Copenhagen survey of the solar neighbourhood. III. Improved distances, ages, and kinematics. *A&A*, 501: 941–947, July 2009. doi:10.1051/0004-6361/200811191.

Judge, P. G. & Egeland, R. Century-long monitoring of solar irradiance and Earth's albedo using a stable scattering target in space. *MNRAS*, 448: L90–L93, Mar. 2015. doi:10.1093/mnrasl/slv004.

Kopp, G. An assessment of the solar irradiance record for climate studies. *Journal of Space Weather and Space Climate*, 4 (27): A14, Apr. 2014. doi:10.1051/swsc/2014012.

Lee, Y. S., Beers, T. C., Allende Prieto, C., Lai, D. K., Rockosi, C. M., Morrison, H. L., Johnson, J. A., An, D., Sivarani, T., & Yanny, B. The SEGUE Stellar Parameter Pipeline. V. Estimation of Alpha-element Abundance Ratios from Low-resolution SDSS/SEGUE Stellar Spectra. *AJ*, 141: 90, Mar. 2011. doi:10.1088/0004-6256/141/3/90.

Leighton, R. B. Observations of Solar Magnetic Fields in Plage Regions. *ApJ*, 130: 366, Sept. 1959. doi:10.1086/146727.

Li, T. D., Bi, S. L., Liu, K., Tian, Z. J., & Shuai, G. Z. Stellar parameters and seismological analysis of the star 18 Scorpii. *A&A*, 546: A83, Oct. 2012. doi:10.1051/0004-6361/201219063.

Linsky, J. L. & Avrett, E. H. The Solar H and K Lines. *PASP*, 82: 169, Apr. 1970. doi:10.1086/128904.

Lockwood, G. W., Skiff, B. A., & Radick, R. R. The Photometric Variability of Sun-like Stars: Observations and Results, 1984-1995. *ApJ*, 485: 789–811, Aug. 1997.

Lockwood, G. W., Skiff, B. A., Henry, G. W., Henry, S., Radick, R. R., Baliunas, S. L., Donahue, R. A., & Soon, W. Patterns of Photometric and Chromospheric Variation among Sun-like Stars: A 20 Year Perspective. *ApJS*, 171: 260–303, July 2007. doi:10.1086/516752.

Meehl, G. A., Arblaster, J. M., & Marsh, D. R. Could a future "Grand Solar Minimum" like the Maunder Minimum stop global warming? *Geophys. Res. Lett.*, 40: 1789–1793, May 2013. doi:10.1002/grl.50361.

Meléndez, J., Ramírez, I., Karakas, A. I., Yong, D., Monroe, T. R. Bedell, M., Bergemann, M., Asplund, M., Tucci Maia, M., Bean, J., do Nascimento, Jr., J.-D., Bazot, M., Alves-Brito, A., Freitas, F. C., & Castro, M. 18 Sco: A Solar Twin Rich in Refractory and Neutron-capture Elements. Implications for Chemical Tagging. *ApJ*, 791: 14, Aug. 2014. doi:10.1088/0004-637X/791/1/14.

Noyes, R. W., Hartmann, L. W., Baliunas, S. L., Duncan, D. K., & Vaughan, A. H. Rotation, convection, and magnetic activity in lower main-sequence stars. *ApJ*, 279: 763–777, Apr. 1984. doi:10.1086/161945.

Petit, P., Dintrans, B., Solanki, S. K., Donati, J.-F., Aurière, M., Lignières, F., Morin, J., Paletou, F., Ramirez Velez, J., Catala, C., & Fares, R. Toroidal versus poloidal magnetic fields in Sun-like stars: a rotation threshold. *MNRAS*, 388: 80–88, July 2008. doi:10.1111/j.1365-2966.2008.13411.x.

Pevtsov, A. A., Virtanen, I., Mursula, K., Tlatov, A., & Bertello, L. Reconstructing solar magnetic fields from historical observations. I. Renormalized Ca K spectroheliograms and pseudo-magnetograms. *A&A*, 585: A40, Jan. 2016. doi:10.1051/0004-6361/201526620.

Porto de Mello, G. F. & da Silva, L. HR 6060: The Closest Ever Solar Twin? *ApJ*, 482: L89, June 1997. doi:10.1086/310693.

Prša, A., Harmanec, P., Torres, G., Mamajek, E., Asplund, M., Capitaine, N., Christensen-Dalsgaard, J., Depagne, É., Haberreiter, M., Hekker, S., Hilton, J., Kopp, G., Kostov, V., Kurtz, D. W. Laskar, J., Mason, B. D., Milone, E. F., Montgomery, M., Richards, M., Schmutz, W., Schou, J., & Stewart, S. G. Nominal Values for Selected Solar and Planetary Quantities: IAU 2015 Resolution B3. *AJ*, 152: 41, Aug. 2016. doi:10.3847/0004-6256/152/2/41.

Radick, R. R., Lockwood, G. W., Skiff, B. A., & Baliunas, S. L. Patterns of Variation among Sun-like Stars. *ApJS*, 118: 239–258, Sept. 1998. doi:10.1086/313135.

Sagan, C. & Mullen, G. Earth and Mars: Evolution of Atmospheres and Surface Temperatures. *Science*, 177: 52–56, July 1972. doi:10.1126/science.177.4043.52.

Schwabe, M. Sonnenbeobachtungen im Jahre 1843. Von Herrn Hofrath Schwabe in Dessau. *Astronomische Nachrichten*, 21: 233, Feb. 1844.

Skumanich, A. Time Scales for CA II Emission Decay, Rotational Braking, and Lithium Depletion. *ApJ*, 171: 565, Feb. 1972. doi:10.1086/151310.

Solanki, S. K., Krivova, N. A., & Haigh, J. D. Solar Irradiance Variability and Climate. *ARA&A*, 51: 311–351, Aug. 2013. doi:10.1146/annurev-astro-082812-141007.

Stocker, T., Qin, D., Plattner, G., Tignor, M., Allen, S., Boschung, J., Nauels, A., Xia, Y., Bex,

B., & Midgley, B. Ipcc, 2013: climate change 2013: the physical science basis. contribution of working group i to the fifth assessment report of the intergovernmental panel on climate change. 2013.

Willson, R. C. & Hudson, H. S. The sun's luminosity over a complete solar cycle. *Nature*, 351: 42–44, May 1991. doi:10.1038/351042a0.

Wilson, O. C. Flux Measurements at the Centers of Stellar H- and K-Lines. *ApJ*, 153: 221, July 1968. doi:10.1086/149652.

Wilson, O. C. Chromospheric variations in main-sequence stars. *ApJ*, 226: 379–396, Dec. 1978. doi:10.1086/156618.

Yeo, K. L., Krivova, N. A., & Solanki, S. K. Solar Cycle Variation in Solar Irradiance. *Space Sci. Rev.*, 186: 137–167, Dec. 2014. doi:10.1007/s11214-014-0061-7.

Living Around Active Stars
Proceedings IAU Symposium No. 328, 2016
D. Nandy, A. Valio & P. Petit, eds.

© International Astronomical Union 2017
doi:10.1017/S1743921317004276

The solar proxy κ^1 Cet and the planetary habitability around the young Sun

J.-D. do Nascimento, Jr.[1,2], **A. A. Vidotto**[3,4], **P. Petit**[5], **C. Folsom**[6], **G. F. Porto de Mello**[7], **S. Meibom**[2], **X. C. Abrevaya**[8], **I. Ribas**[9], **M. Castro**[1], **S. C. Marsden**[10], **J. Morin**[11], **S. V. Jeffers**[12], **E. Guinan**[13] and Bcool Collaboration

[1] Univ. Federal do Rio G. do Norte, UFRN, Dep. de Fisica, CP 1641, 59072-970, Natal, RN, Brazil: jdonascimento@fisica.ufrn.br
[2] Harvard-Smithsonian Center for Astrophysics, 60 Garden Street, Cambridge MA 02138, US
[3] Observatoire de Genève, 51 ch. des Maillettes, CH-1290, Switzerland
[4] School of Physics, Trinity College Dublin, Dublin 2, Ireland
[5] Univ. de Toulouse, UPS-OMP, IRAP, CNRS, 14 Av. E. Belin, F-31400 Toulouse, France
[6] Univ. Grenoble Alpes, IPAG, F-38000 Grenoble, France
[7] Observ. do Valongo, UFRJ, L do Pedro Antoio,43 20080-090, RJ, Brazil
[8] Inst. de Astronom y Fica del Espacio (IAFE), UBA CONICET, Buenos Aires, Argentina
[9] Inst. de Ciències de l'Espai, C. de Can Magrans, s/n, Campus UAB, 08193 Bellaterra, Spain
[10] CESCR, Univ. of Southern Queensland, Toowoomba, 4350, Australia
[11] LUPM-UMR5299, U. Montpellier, Montpellier, F-34095, France
[12] I. für Astrophysik, G.-August-Univ., D-37077, Goettingen, Germany
[13] Univ. of Villanova, Astron. Department, PA 19085 Pennsylvania, US

Abstract. Among the solar proxies, κ^1 Cet, stands out as potentially having a mass very close to solar and a young age. We report magnetic field measurements and planetary habitability consequences around this star, a proxy of the young Sun when life arose on Earth. Magnetic strength was determined from spectropolarimetric observations and we reconstruct the large-scale surface magnetic field to derive the magnetic environment, stellar winds, and particle flux permeating the interplanetary medium around κ^1 Cet. Our results show a closer magnetosphere and mass-loss rate 50 times larger than the current solar wind mass-loss rate when Life arose on Earth, resulting in a larger interaction via space weather disturbances between the stellar wind and a hypothetical young-Earth analogue, potentially affecting the habitability. Interaction of the wind from the young Sun with the planetary ancient magnetic field may have affected the young Earth and its life conditions.

Keywords. Kappa 1 Ceti, HD 20630, solar analog, magnetic field, winds, outflows, habitability

1. Introduction

Large-scale surface magnetic fields measurements of a young Sun proxy from Zeeman Doppler imaging (ZDI) techniques (Semel 1989; Donati *et al.* 2006) give us crucial information about the early Sun's magnetic activity, and today, thanks to spectropolarimetric observations we can reconstruct the magnetic field topology of the stellar photosphere and provide quantitatively the interactions between the stellar wind and the surrounding planetary system. For understanding the origin and evolution of life on Earth we need to know the evolution of the Sun itself, especially the early evolution of its radiation field, particle and magnetic properties. The irradiation from the central star is, by far, the most important source of energy in planetary atmospheres. The radiation field defines the habitable zone, a region in which orbiting planets could sustain liquid water at their surface (Huang 1960; Kopparapu *et al.* 2013). The magnetic environment (and particles)

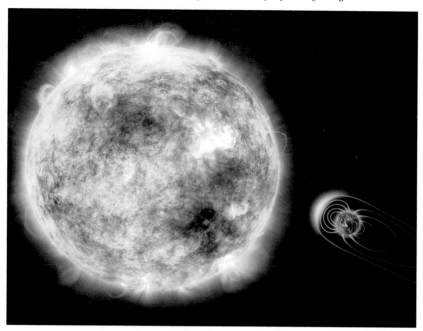

Figure 1. Artist's illustration of the young Sun-like star κ^1 Cet blotched with large starspots, a sign of its high level of magnetic activity and an Earth-like planet with magnetic field protecting its atmosphere and habitability. The physical sizes of the star and planet and distance between them are not to scale. Copyright M. Weiss/CfA-Harvard Smithsonian and do Nascimento *et al.* (2016).

define the interactions star – planet. In the case of magnetized planets, such as the Earth that developed a magnetic field at least four billion years ago (Tarduno *et al.* 2015), their magnetic fields act as obstacles to the stellar wind, deflecting it and protecting the upper planetary atmospheres and ionospheres against the direct impact of stellar wind plasmas and high-energy particles (Kulikov *et al.* 2007; Lammer *et al.* 2007). Focused on carefully selected and well-studied stellar proxies that represent key stages in the evolution of the Sun, The Sun in Time program from Dorren & Guinan (1994), Ribas *et al.* (2005), studied a small sample in the X-ray, EUV, and FUV domains. However, nothing or little has been done in this program with respect to the magnetic field properties for those stars. A characterization of a genuine young Sun proxy is a difficult task, because ages for field stars, particularly for those on the bottom of the main sequence are notoriously difficult to be derived (e.g., do Nascimento *et al.* 2014). Fortunately, stellar rotation rate for young low mass star decrease with time as they lose angular momenta. This rotation rates give a relation to determine stellar age (Kawaler 1989; Barnes 2007; Meibom *et al.* 2015). Young solar analogue stars rotate faster than the Sun and show a much higher level of magnetic activity with highly energetic flares. This behavior is driven by the dynamo mechanism, which operates in rather different regimes in these young objects.

2. κ^1 Cet spectropolarimetric observations and measurements

Find a star that faithfully represents the young Sun in terms of age, mass and metallicity is not an easy task. Kappa 1 Ceti (HD 20630, HIP 15457) is a solar proxy studied by the Sun in Time program. A nearby G5 dwarf star with V = 4.85 and age from 0.4

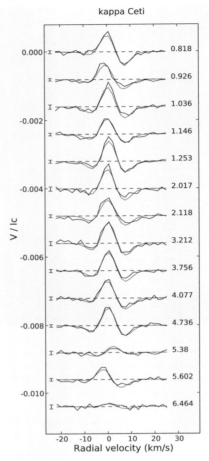

Figure 2. The normalized Stokes V LSD profiles time series of κ^1 Cet. Continuum black lines represent observed profiles and red lines correspond to synthetic profiles of our model. Successive profiles are shifted vertically for display clarity. Rotational cycle is shown on the right of each profile. 1 σ error bars for each observation are indicated on the left of each rotational phase (calculated by assuming a rotation period of 9.2 d, and a reference Julian date arbitrary set to 2456195.0). Horizontal dashed lines illustrate the zero levels of each observation (do Nascimento *et al.* 2016).

Gyr to 0.6 Gyr (Ribas *et al.* 2010) that stands out potentially having a mass very close to solar and age of the Sun when the window favorable to the origin of life opened on Earth around 3.8 Gyr ago or earlier (Mojzsis *et al.* 1996). This corresponds to the period when favorable physicochemical and geological conditions became established and after the late heavy bombardment. As to the Sun at this stage, κ^1 Cet radiation environment determined the properties and chemical composition of the close planetary atmospheres, and provide an important constraint of the role played by the Earth magnetospheric protection at the critical time at the start of the Archean epoch (Mojzsis *et al.* 1996), when life is thought to have originated on Earth. This is also the epoch when Mars lost its liquid water inventory at the end of the Noachian epoch some 3.7 Gyr ago (Jakosky *et al.* 2001). Study based on κ^1 Cet can also clarify the biological implications of the high-energy particles at this period (Cnossen *et al.* 2008) and add straight constraints limiting the possibility of lithopanspermia process on the Mars-Earth system (Abrevaya

2016). Such a studies require careful analysis based on reasonably bright stars at specific evolutionary state, and there are only a few number of bright solar analogues at this age of κ^1 Cet. The Figure 1 represents an artistic illustration of the young Sun-like star κ^1 Cet blotched with large starspots, a sign of its high level of magnetic activity. Physical sizes of the star and planet and distance between them are not to scale.

Spectropolarimetric data of κ^1 Cet were collected with the NARVAL spectropolarimeter (Aurière 2003) at the 2.0-m Bernard Lyot Telescope (TBL) of Pic du Midi Observatory as part of TBL Bcool Large Program (Marsden *et al.* 2014). NARVAL comprises a Cassegrain-mounted achromatic polarimeter and a bench-mounted cross-dispersed echelle spectrograph. In polarimetric mode, NARVAL has a spectral resolution of about 65,000 and covers the whole optical domain in one single exposure, with nearly continuous spectral coverage ranging from 370 nm to 1000 nm over 40 grating orders. The data reduction is performed through Libre-ESpRIT package based on ESPRIT (Donati *et al.* 1997). As described by do Nascimento *et al.* (2016), Stokes I and V (circularly-polarized) spectra were gathered. The resulting time-series is composed of 14 individual observations collected over 53 consecutive nights, during which more than one rotation period (assuming a rotation period of 9.2 d). As usually for cool active stars, Stokes V spectra do not display any detectable signatures in the individual spectral lines, even with a peak S/N in excess of 1,000 (at wavelengths close to 730 nm). In this situation, we take advantage of the fact that, at first order, Stokes V Zeeman signatures of different spectral lines harbor a similar shape and differ only by their amplitude, so that a multiline approach in the form of a cross-correlation technique is able to greatly improve the detectability of tiny polarized signatures. We employ here the Least-Squares-Deconvolution method (LSD, Donati *et al.* 1997; Kochukhov 2010) using a procedure similar to the one described in Marsden *et al.* (2014). Our line-list is extracted from the VALD data base (Kupka *et al.* 2000) and is computed for a set of atmospheric parameters (effective temperature and surface gravity) similar to those of κ^1 Cet. We assume a total of about 8,400 spectral lines recorded in NARVAL spectra and listed in our line mask, the final S/N of Stokes V LSD pseudo-profiles is ranging from 16,000 to 28,000, well enough to detect Zeeman signatures at all available observations (Figure 2). From the Stokes I spectra, we also determined the S_{index}, calibrated from the Mount Wilson S_{index}, to quantify the chromospheric emission changes in the Ca II H line (do Nasciemento *et al.* 2014). The complete pipeline of the S_{index} computation is described in Morgenthaler *et al.* (2012) and Wright *et al.* (2004).

3. Fundamental parameters and evolutionary status

Based on our NARVAL data we performed spectroscopic analysis of κ^1 Cet to redetermine stellar parameters as in do Nascimento *et al.* (2013) and references therein. We used excitation and ionization equilibrium of a set of 209 Fe I and several Fe II lines and an atmosphere model and mostly laboratory gf-values to compute a synthetic spectra. The best solution from this synthetic analysis was fitted to the NARVAL spectrum for the set of parameters T_{eff} = 5705 ± 50 K, [Fe/H] = +0.10 ± 0.05 dex, log g = 4.49 ± 0.10. Several photometric and spectroscopic observational campaigns were carried out to determine κ^1 Cet fundamental parameters. Ribas *et al.* (2010) determined the photometric T_{eff} of κ^1 Cet from intermediate-band Strömgren photometry, based on the 2MASS near-IR photometry and a fit of the spectral energy distribution with stellar atmosphere models. This photometric method yielded T_{eff} = 5685 ± 45 K. Ribas *et al.* (2010) also determined spectroscopic fundamental parameters of κ^1 Cet as T_{eff} = 5780 ± 30 K, log g = 4.48 ± 0.10 dex, [Fe/H] = +0.07 ± 0.04 dex. Valenti & Fischer (2005) gives T_{eff} = 5742 K, log g = 4.49 dex, [M/H] = +0.10 dex. Paletou *et al.* (2015), from high resolution

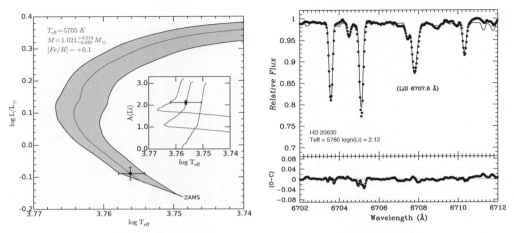

Figure 3. The **left painel**, κ^1 Cet in the Hertzsprung-Russell diagram. Luminosities and related errors have been derived from the Hipparcos parallaxes. The typical errors on Teff are from our spectroscopic analysis. The shaded zone represents the range of masses in TGEC models (the maximum and the minimum masses are indicated) limited by the 1σ observational error bars. The inside plot shows the lithium destruction for κ^1 Cet along the evolutionary tracks as a function of the effective temperature. The blue tracks represent the range of masses of TGEC models (the maximum and the minimum masses) limited by the 1σ observational error bars. The **right painel**, Spectral synthesis of the λ 6707 Li I line of κ^1 Cet based on spectroscopic set of atmospheric parameters as described by Ribas *et al.* (2010). The derived abundance logN(Li) = 2.12, in the usual scale where logN(H) = 12.00. This value is in good agreement with the determinations of Luck & Heiter (2006) and Pasquini *et al.* (1994)

NARVAL Echelle spectra (R = 65,000, S/N $\sim$ 1000) described in the Sect. 2, determined T_{eff} = 5745 $\pm$ 101 K, log g = 4.45 $\pm$ 0.09 dex, [Fe/H] = +0.08 $\pm$ 0.11. Spectroscopic T_{eff} values are hotter than photometric, and a possible explanation of this offset could be effects of high chromospheric activity and, an enhanced non-local UV radiation field resulting in a photospheric overionization (Ribas *et al.* 2010). The presented spectroscopic T_{eff} values are in agreement within the uncertainty. Finally we used our determined solution T_{eff} = 5705 $\pm$ 50 K, [Fe/H] = +0.10 $\pm$ 0.05 dex, log g = 4.49 $\pm$ 0.10. This yields a logN(Li) = 2.05, in good agreement with logN(Li) = 2.12 determined by Ribas *et al.* (2010) and presented in the Figure 3 right. To constrain the evolutionary status of κ^1 Cet, we used the spectroscopic solution within computed models with the Toulouse-Geneva stellar evolution code (do Nascimento *et al.* 2013). We used models with an initial composition from Grevesse & Noels (1993). Transport of chemicals and angular momentum due to rotation-induced mixing are computed as described in Vauclair & Théado (2003) and presented in the Figure 3 left. The angular momentum evolution follows the Kawaler (1988) prescription. We calibrated a solar model as Richard *et al.* (1996) and used this calibration to compute κ^1 Cet model. These models, together with lithium abundance measurement, result in mass of 1.02 $\pm$ 0.02 $M_{\odot}$, an age between 0.5 Gyr to 0.9 Gyr for κ^1 Cet, consistent with Güdel *et al.* (1997) estimated age of 0.75 Gyr and Marsden *et al.* (2014) estimated age of 0.82 Gyr using our data and activity-age calibration.

For rotation period P_{rot}, as in do Nascimento *et al.* (2014), we measured the average surface P_{rot} from light curves. Here, for κ^1 Cet we used MOST (Microvariability and Oscillations of Stars) (Walker *et al.* 2003) light curve modulation (Figure 4). MOST continuously observed κ^1 Cet for weeks at a time providing a P_{rot} (Walker *et al.* 2003). We extract P_{rot} from Lomb–Scargle periodogram (Scargle 1982) and a wavelet analysis of the light curve (Figure 4). The P_{rot} obtained was P_{rot} = 8.77d $\pm$ 0.8 days, three

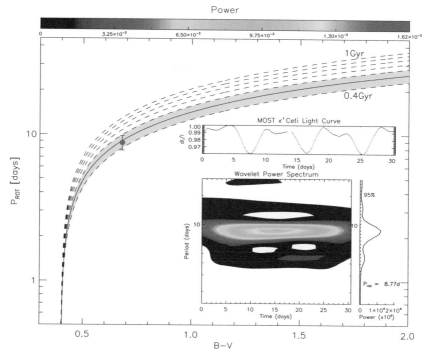

Figure 4. Colour–period diagram for κ^1 Cet based on P_{rot} measurement from MOST (Microvariability and Oscillations of Stars) light curve. Lomb–Scargle periodogram (Scargle 1982) and wavelet (inside figure) was used to derive rotation period. The predictions from gyrochronology models (Barnes 2007) of cool star are plotted for different ages.

times shorter than the solar P_{rot}. The P_{rot} we have measured from the *MOST* light curves allows us an independent (from classical isochrone) age derivation of κ^1 Cet using gyrochronology (Skumanich 1972; Barnes 2007). The gyrochronology age of κ^1 Cet that we derive range from 0.4 Gyr to 0.6 Gyr (Figure 4), consistent with the predictions from Ribas *et al.* (2010) and ages determined from evolutionary tracks.

4. κ^1 Cet Magnetic field topology

Our implementation of the ZDI algorithm is the one detailed by Donati *et al.* (2006), where the surface magnetic field is projected onto a spherical harmonics frame. From the time-series of Stokes V profiles, we used the ZDI method (Zeeman-Doppler Imaging, Semel 1989) to reconstruct the large-scale magnetic topology of the star. We assume during reconstruction a projected rotational velocity equal to 5 km/s (Valenti & Fischer 2005), a radial velocity equal to 19.1 km/s, and an inclination angle of 60 degrees (from the projected rotational velocity, radius and stellar rotation period). We truncate the spherical harmonics expansion to modes with $l \leqslant 10$ since no improvement is noticed in our model if we allow for a more complex field topology. Given the large time-span of our observations, some level of variability is expected in the surface magnetic topology. A fair amount of this intrinsic evolution is due to differential rotation, which can be taken into account in our inversion procedure assuming that the surface shear obeys a simple law of the form $\Omega(l) = \Omega_{\mathrm{eq}} - \sin^2(l)d\Omega$, where $\Omega(l)$ is the rotation rate at latitude l, Ω_{eq} is the rotation rate of the equator and $d\Omega$ is the difference of rotation rate between the

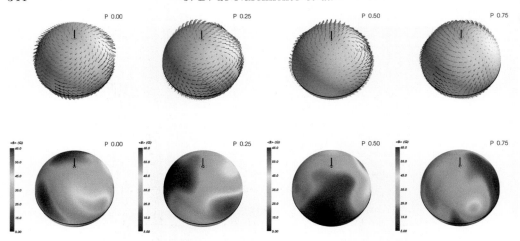

Figure 5. Large-scale magnetic topology of κ^1 Cet at different rotation phases indicated in the top right of each panel. The top row shows the inclination of field lines over stellar surface, with red and blue arrows depicting positive and negative field radial component values, respectively. The bottom row displays the field strength (do Nascimento *et al.* 2016).

pole and the equator. We optimize the two free parameters $\Omega_{\rm eq}$ and $d\Omega$ by computing a 2D grid of ZDI models spanning a range of values of these two parameters, following the approach of (Petit *et al.* 2002). By doing so, we obtain a minimal reduced χ^2 equal to 1.3 at $\Omega_{\rm eq} = 0.7$ rad/d and $d\Omega = 0.056$ rad/d. These values correspond to a surface shear roughly solar in magnitude, with an equatorial rotation period $P_{\rm rot}^{\rm eq} = 8.96$ d, while the polar region rotates in about $P_{\rm rot}^{\rm pole} = 9.74$ d. Figure 5 from top to bottom presents the inclination of field lines over the stellar surface and the resulting large-scale magnetic geometry. The surface-averaged field strength is equal to 24 G, with a maximum value of 61 G at phase 0.1. A majority (61%) of the magnetic energy is stored in the toroidal field component, showing up as several regions with field lines nearly horizontal and parallel to the equator, e.g. at phase 0.1. The dipolar component of the field contains about 47% of the magnetic energy of the *poloidal* field component, but significant energy is also seen at $\ell > 3$, where 20% of the magnetic energy is reconstructed. Axisymmetric modes display 66% of the total magnetic energy. These magnetic properties are rather typical of other young Sun-like stars previously observed and modeled with similar techniques (Petit *et al.* 2008, Folsom *et al.* 2016 and references therein).

5. Stellar wind and effects on the magnetosphere of the young Earth

We reconstruct the κ^1 Cet large-scale surface magnetic field based on spectropolarimetric observations and we derive the magnetic environment and particle flux permeating the interplanetary medium around κ^1 Cet. Stellar winds are is an important point, and we use here that one presented in Vidotto *et al.* (2012, 2015), in which we use the three-dimensional magnetohydrodynamics (MHD) numerical code BATS-R-US (Powell *et al.* 1999; Tóth *et al.* 2012) to solve the set of ideal MHD equations. In this model, as described in do Nascimento *et al.* (2016), we use, as inner boundary conditions for the stellar magnetic field, the radial component of the reconstructed surface magnetic field of κ^1 CetWe assume the wind is polytropic, with a polytropic index of $\gamma = 1.1$, and consists of a fully ionised hydrogen plasma. We further assume a stellar wind base density of 10^9 cm^{-3} and a base temperature of 2 MK. Figure 5 shows the large-scale

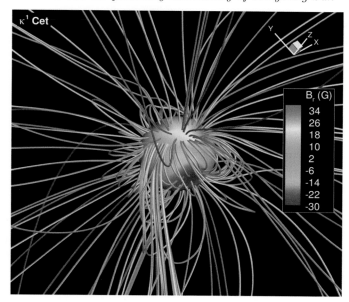

Figure 6. Large-scale magnetic field embedded in the wind of κ^1 Cet. The radial component of the observationally reconstructed surface magnetic field is shown in colour.

magnetic field embedded in the wind of κ^1 Cet. In our model, we derive a mass-loss rate of $\dot{M} = 9.7 \times 10^{-13}$ $M_\odot$ yr^{-1}, i.e., almost 50 times larger than the current solar wind mass-loss rate. It is interesting to compare our results to the empirical correlation between $\dot{M}$ and X-ray fluxes (F_X) derived by Wood *et al.* (2014). For κ^1 Cet, the X-ray luminosity is $10^{28.79}$ erg/s (Wood *et al.* 2012). Assuming a stellar radius of $0.95 R_\odot$, we derive $F_X \simeq 10^6$ erg cm^{-2} s^{-1} and, according to Wood *et al.* (2012) relation, $\dot{M}$ to be ~ 63 to 140 times the current solar wind mass-loss rate. Thus, our $\dot{M}$ derivation roughly agrees with the lower envelope of the empirical correlation of Wood *et al.* (2014) and derived mass-loss rate of Airapetian & Usmanov (2016).

5.1. *The mass-loss rate of the young Sun*

The enhanced mass-loss rate of the young solar analogue κ^1 Cet implies that the strengths of the interactions between the stellar wind and a hypothetical young-Earth analogue is larger than the current interactions between the present-day solar wind and Earth. To quantify this, we calculate the ram pressure of the wind of κ^1 Cet as $P_{\rm ram} = \rho u^2$, where ρ is the particle density and u the wind velocity (Figure 7). Pressure balance between the magnetic pressure of a hypothetical young-Earth and the ram pressure of the young Sun's wind allows us to estimate the magnetospheric size of the young-Earth:

$$\frac{r_M}{R_\oplus} = f \left(\frac{B^2_{\rm eq,\oplus}}{8\pi P_{\rm ram}} \right)^{1/6} \tag{5.1}$$

where $B_{\rm eq,\oplus}$ is the equatorial field strength of the young Earth dipolar magnetic field and $f \simeq 2^{2/6}$ is a correction factor used to account for the effects of currents (e.g., Cravens 2004). Figure 8a shows the stand-off distance of the Earth's magnetopause calculated using Eq. (5.1). Here, we assume three values for $B_{\rm eq,\oplus}$: (i) $B_{\rm eq,\oplus} = 0.31$ G, identical to the present-day magnetic field strength (e.g., Bagenal 1992); (ii) $B_{\rm eq,\oplus} = 0.15$ G, according

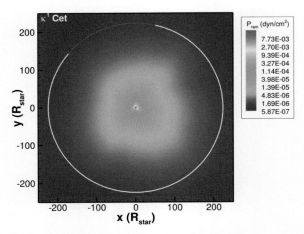

Figure 7. The ram pressure of the stellar wind of κ^1 Cet. The circle indicates the position of the orbit of a young-Earth analogue. Red portions of the orbit indicates regions of negative vertical component of the interplanetary magnetic field ($B_z < 0$).

to measurements of the Paleoarchean Earth's magnetic field (3.4Gyr ago) (Tarduno *et al.* 2010); and (iii) $B_{eq,\oplus} = 0.40$G, according to rotation-dependent dynamo model theory (see Sterenborg *et al.* 2011). Depending on the assumed field strength of the hypothetical young-Earth, the average magnetospheric sizes are (i) $4.8R_\oplus$, (ii) $3.8R_\oplus$ and (iii) $5.3R_\oplus$, respectively, indicating a size that is about 34 to 48% the magnetospheric size of the present-day Earth (about $11R_\oplus$, (Bagenal 1992).

The relative orientation of the interplanetary magnetic field with respect to the orientation of the planetary magnetic moment plays an important role in shaping the open-field-line region (polar cap) of the planet (e.g., Sterenborg *et al.* 2011). Through the polar cap, particles can be transported to/from the interplanetary space. Tarduno *et al.* (2010) discusses that the increase in polar cap area should be accompanied by an increase of the volatile losses from the exosphere, which might affect the composition of the planetary atmosphere over long timescales. In the case where the vertical component of interplanetary magnetic field B_z is parallel to the planet's magnetic moment (or anti-parallel to the planetary magnetic field at r_M), the planetary magnetosphere is in its widest open configuration and a polar cap develops. If B_z and the planet's magnetic moment are anti-parallel, there is no significant polar cap. The complex magnetic-field topology of κ^1 Cet gives rise to non-uniform directions and strengths of B_z along the planetary orbit. The red (white) semi-circle shown in Figure 7 illustrates portions of the orbital path surrounded by negative (positive) B_z. Therefore, depending on the relative orientation between B_z and the planet's magnetic moment, the colatitude of the polar cap will range from $0°$ (closed magnetosphere) to $\arcsin(R_\oplus/r_M)^{1/2}$ (widest open configuration) (e.g., Vidotto *et al.* 2013). Figure 8b shows the colatitude of the polar cap for the case where the planetary magnetic moment points towards positive z. Portions of the orbit where the planet is likely to present a closed magnetosphere (from 76 to 140 degrees in longitude) are blanked out.

6. Conclusions

We report a magnetic field detection for κ^1 Cet with an average field strength of 24 G, and maximum value of 61 G. The complex magnetic-field topology of κ^1 Cet gives rise

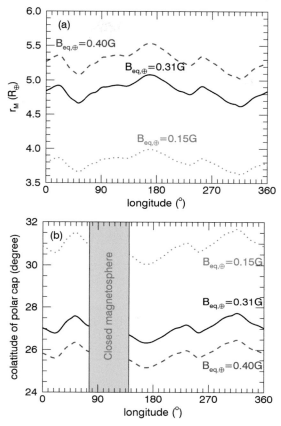

Figure 8. (a) The magnetospheric size of the young-Earth is calculated through pressure balance between the ram pressure of the young Sun's wind (Figure 7) and the magnetic pressure of the planetary magnetosphere for different equatorial dipolar field strengths. (b) The related colatitude of the polar cap, assuming that during most of the orbit, the planetary magnetic moment is parallel to the interplanetary magnetic field.

to non-uniform directions and strengths along a possible planetary orbit. Our stellar wind model for κ^1 Cet shows a mass-loss rate factor 50 times larger than the current solar wind mass-loss rate, resulting in a larger interaction between the stellar wind and a hypothetical young-Earth like planet. With $1.02\,M_\odot$, an age between $0.4\,$Gyr to $0.6\,$Gyr, κ^1 Cet is a perfect target to study habitability on Earth during the early Sun phase when life arose on Earth. An enhanced mass-loss, high-energy emissions from κ^1 Cet, supporting the extrapolation from Newkirk (1980) and Lammer *et al.* (2007) of a Sun with stronger activity 3.8 Gyr ago or earlier. Early magnetic field have affected the young Earth and its life conditions and due to the ancient magnetic field on Earth four billion years ago as measured by Tarduno *et al.* (2015), the early magnetic interaction between the stellar wind and the young-Earth planetary magnetic field may well have prevented the volatile losses from the Earth exosphere and create conditions to support life. κ^1 Cet magnetic field strength and wind mass-loss rate tell us that life at the primitive Earth surface 3.8 Gyr have been exposed to a higher radiation level, when compared to present time.

References

Abrevaya, X. C. 2017, *Living around active stars. Proceedings of IAU Symposium 328.*

Airapetian V. S. & Usmanov A. V. 2016, *ApJL*, 817, 24

Aurière, M. 2003, in Arnaud J., Meunier N. *eds., Magnetism and Activity of the Sun and Stars, EAS Pub. Ser.*, 9, 105

Bagenal, F. 1992, *Annual Review of Earth and Planetary Sciences*, 20, 289

Barnes, S. A. 2007, *ApJ*, 669, 1167

Cravens, T. E. 2004, *Physics of Solar System Plasmas*

Cnossen, I., Sanz-Forcada, J., Favata, F., Witasse, O., Zegers, T., & Arnold, N. F. 2007, *J. Geophys. Res. (Planets)*, 112, 2008

Dorren, J. D. & Guinan, E. F. 1994, *IAU 143, The Sun as a Variable Star*, ed. J. M. Pap, C. Frölich, H. S. Hudson, & S. Solanki, *Cambridge U. Press*, 206

do Nascimento, J. D., Petit, P., Castro, M. *et al.* 2014, *Magnetic Fields throughout Stellar Evolution. Proceedings of the IAU Symposium*, 302, 142

do Nascimento, J. -D., Jr., Takeda, Y., *et al.* 2013, *ApJL*, 771, 31

do Nascimento, J. -D., Jr., García, R. A., *et al.* 2014, *ApJL*, 790, 23

do Nascimento, J. -D., Jr., Vidotto, A. A., *et al.* 2016, *ApJL*, 820, 15

Donati, J.-F., Howarth, I. D., Jardine, M. M. *et al.* 2006, *MNRAS*, 370, 629

Donati, J.-F., Semel, M., Carter, B. D., *et al.* 1997, *MNRAS*, 291, 658

Folsom, C. P., Petit, P., Bouvier, J., *et al.* 2016, *MNRAS*, 457, 580

Grevesse, N. & Noels, A. 1993, *Origin and Evolution of the Elements, eds. N. Prantzos, E. Vangioni–Flam, and M. Cassé, Cambridge U. Press*, 15

Güdel, M., Guinan, E. F., & Skinner, S. L. 1997, *ApJ*, 483, 947

Huang, S.-S. 1960, *Am. Sci.*, 202, 55

Jakosky, B. M. & Phillips, R. J. 2001, *Nature*, 412, 237

Kawaler, S. D. 1988, *ApJ*, 333, 236

Kawaler, S. D. 1989, *ApJ*, 343, L65

Kopparapu, R. K., Ramirez, R., Kasting, J. F., *et al.* 2013, ApJ, 765, 131

Kochukhov, O., Makaganiuk, V., & Piskunov, N. 2010, *A&A*, 524, A5

Kulikov, Y. N., Lammer, H., *et al.* 2007, *Space Sci. Rev.*, 207, 129

Kupka, F. G., Ryabchikova, T. A., Piskunov, N. E., Stempels, H. C., & Weiss W. W. 2000, *Baltic Astronomy*, 9, 590

Lammer, H. *et al.* 2007, *Astrobiology*, 7, 185

Marsden, S. C., Petit, P., *et al.* 2014, *MNRAS*, 4444, 3517

Meibom, S., Barnes, S. A., *et al.* 2015, *Nature*, 517, 589

Mojzsis, S. J., Arrhenius, G., McKeegan, K. D., *et al.* 1996, *Nature*, 384, 55

Morgenthaler, A., Petit, P., Saar, S., Solanki, S. K. *et al.* 2012, *ApJ*, 540, 138

Newkirk, G. 1980, *Geochimica Cosmochimica Acta Suppl.*, 13, 293

Paletou, F., Böhm T., Watson V., & Trouilhet J.-F. 2015, *A&A*, 573, A67

Petit, P., Donati, J.-F., & Collier Cameron, A. 2002, *ApJL*, 771, 31

Petit, P., Dintrans, B., *et al.* 2008, *MNRAS*, 388, 80

Powell, K. G., Roe, P. L., *et al.* 1999, *J. Chem. Phys.*, 154, 284

Ribas, I., Porto de Mello, G. F., Ferreira, L. D., Hebrard, E. *et al.* 2010, *ApJ*, 714, 384

Ribas, I., Guinan, E. F., Güdel, M., & Audard, M. 2005, *ApJ*, 622, 680

Ribas, I., Guinan, E., & Güdel, Audard, M. 2005, *ApJ*, 622, 680

Richard, O., Vauclair, S., *et al.* 1996, *A&A*, 312, 1000

Scargle, J. D. 1982, *ApJ*, 263, 835

Semel, M. 1989, *A&A*, 255, 456

Skumanich, A. 1972, *ApJ*, 171, 565

Sneden, C. 1973, *ApJ*, 184, 839

Sterenborg, M. G., Cohen, O., Drake, J. J., & Gombosi, T. I. 2011, *Journal of Geophysical Research (Space Physics)*, 116, 1217

Tarduno, J. A., Cottrell, R. D., *et al.* 2010, *Science*, 327, 1238

Tarduno, J. A., Cottrell, R. D., *et al.* 2015, *Science*, 349, 521

Tóth, G., van der Holst, B., *et al. 2012*, Journal of Computational Physics, 231, 870

Vauclair, S. & Théado, S. 2003, *A&A*, 587, 777

Valenti, F. A. & Fischer, D. 2005, *ApJS*, 587, 777

Vidotto, A. A., Fares, R., Jardine, M., *et al.* 2012, *MNRAS*, 423, 3285

Vidotto, A. A., Fares, R., Jardine, M., *et al.* 2015, *MNRAS*, 449, 4117

Vidotto, A. A., Jardine, M., Morin, J., *et al.* 2013, *A&A*, 557, A67

Walker, G. A. H., *et al.* 2003, *PASP*, 115, 1023

Wood, B. E., Laming, J. M., & Karovska, M. 2012, *ApJ*, 753, 76

Wright, J. T., Marcy, G. W., Butler, R. P., Vogt, S. S. *et al.* 2004, *ApJS*, 152, 261

Wood, B. E., Müller, H.-R., Redfield, S., & Edelman, E. 2014, *ApJL*, 781, L33

Living Around Active Stars
Proceedings IAU Symposium No. 328, 2016
D. Nandy, A. Valio & P. Petit, eds.

© International Astronomical Union 2017
doi:10.1017/S1743921317004331

The Faint Young Sun and Faint Young Stars Paradox

Petrus C. Martens

Department of Physics & Astronomy,
Georgia State University,
25 Park Place, 6th Floor,
Atlanta, GA 30303, USA
email: martens@astro.gsu.edu

Abstract. The purpose of this paper is to explore a resolution for the Faint Young Sun Paradox that has been mostly rejected by the community, namely the possibility of a somewhat more massive young Sun with a large mass loss rate sustained for two to three billion years. This would make the young Sun bright enough to keep both the terrestrial and Martian oceans from freezing, and thus resolve the paradox. It is found that a large and sustained mass loss is consistent with the well observed spin-down rate of Sun-like stars, and indeed may be required for it. It is concluded that a more massive young Sun must be considered a plausible hypothesis.

Keywords. Stars: late-type, stars: rotation, stars: mass loss, stars: evolution, solar wind

1. Introduction

The young Sun started its life on the main sequence with about 70% of the luminosity of what it has now according to standard stellar evolution theory. It is still a scientific riddle how, with such a faint Sun, the young Earth could be warm enough to host liquid water in its first couple of billion years. Yet geological evidence clearly indicates there have been warm oceans from very early on (Kasting 1989), and that these oceans were a key ingredient in the development of life.

This is called the Faint Young Sun Paradox. The paradox is even more compelling for the planet Mars which we know now to have been covered with oceans for periods of hundreds of millions of years in its early life, with only half of the incoming energy flux of sunlight of the Earth.

Stellar evolution simulations dictate his paradox and it therefore applies to all G stars, and less so for K and M stars that evolve much slower. In all cases the habitable zone around the star gradually moves outwards and planets that started out balmy are expected to end up scorched. Given that it took about four billion years on planet Earth for the development from single cell organisms to multi-cellular life – and since that is the only example of evolution we have – it is a reasonable assumption that the development of multicellular intelligent life takes a very long time in general, with most G star planets not spending enough time in the habitable zone.

This paradox has been known for a long time, and one of the first to hint at a solution was well known science popularizer Carl Sagan (Sagan & Mullan 1972). Many solutions to the Faint Young Sun Paradox have been proposed over the years, and they come from very different fields. Fairly straightforward proposals are an enhanced greenhouse effect by carbon dioxide or methane, geothermal heath from an initially much warmer terrestrial core, a much smaller Earth albedo, life developing in a cold environment under a 200 meter thick ice sheet, a secular variation in the gravitational constant, etc. Most of these models have serious shortcomings: For example the greenhouse effect from methane appears to be self-limiting, and not enough CO2 is indicated by the geological record to justify a greatly enhanced greenhouse effect in the past (Kasting, 2004)

There is not enough space in a proceedings paper to review all the material discussed above, so I refer to a recent review by Feulner (2012) and a series of very enlightening presentations and papers by Dr. James Kasting (Kasting, Toon & Pollack 1988; Kasting, 2004) on the many hypotheses proposed to resolve the Faint Young Sun Paradox.

Most of the solutions proposed for the Faint Young Sun Paradox apply to Earth alone; they do not explain the presence of liquid oceans on early Mars. Also, they are not solutions for the Faint Young Stars Paradox in general. A simpler solution has been proposed in that the early Sun was more massive and hence more luminous. This necessitates a massive, sustained solar wind for the first billions of years of the Sun's evolution, a condition that most in the community find implausible.

In the current paper I will explore the hypothesis of a more massive and hence much brighter young Sun in more detail, and investigate whether this leads to logical contradictions or can be ruled out by observations. I will find, surprisingly, that a more massive young Sun is not implausible at all, and links together what we know about stellar spin-down with simulations of the same. The results of this paper do not prove that the young Sun was more massive – we would need observations that demonstrate the sustained presence of a massive solar wind. But it does show that such a hypothesis cannot be ruled out at present, and consequently that the presence of oceans on early Mars may not be a conundrum, and that the habitable zones around solar analogs as well may remain in place for the billions of years it takes for multi-cellular and intelligent life to develop.

2. Mass Loss and Luminosity

A slightly more massive Sun would be significantly more luminous: In the solar portion of the Hertzsprung-Russel diagram luminosity scales with mass to the power 4 to 5. If the Sun were more massive earlier on the Earth would be closer in as well: Because of conservation of it angular momentum the mean Sun-Earth distance varies as the inverse of the mass of the Sun, while the incoming radiation at the top of the Earth's atmosphere scales with the inverse of the square of that distance. Hence the amount of radiation the Earth receives varies as the mass of Sun to the power 6 to 7. A 30% less luminous Sun at the Zero Age Main Sequence (ZAMS) at one solar mass could be compensated for by a mere 4 to 5% mass increase going back in time from the current Sun.

A sustained solar mass loss rate of roughly 10^{-11} $M_\odot$ yr^{-1} is required to accomplish that. The current solar mass loss rate is estimated at 2-3×10^{-14} $M_\odot$ yr^{-1} for the fast wind and 10^{-15} $M_\odot$ yr^{-1} for Coronal Mass Ejections (CMEs). Interestingly the mass loss from photon emission is twice as large, around 7×10^{-14} $M_\odot$ yr^{-1}, but the latter contributes much less to angular momentum loss, because the photons are not forced to co-rotate with the magnetic field.

So the current mass loss rate, if extrapolated to the past, is insufficient to resolve the Faint Young Sun Paradox by a factor of 300 or more. Hence we must assume a much higher mass loss rate for the young Sun, sustained for several billions of years. Observations of some young solar-type stars indicate mass loss rates of roughly the right magnitude: e.g. 70 Opiuchi with a mass of 0.92 times that of the Sun, and an estimated age of 0.8 billion years, has a mass loss rate of 3×10^{-12} $M_\odot$ yr^{-1}, and κ^{-1} Cet (Do Nascimento *et al.* 2016) yields the same result from X-ray calibration.

A much more detailed analysis than the back-of-the-envelope calculation above, by Minton & Malhotra (2007), narrows down the mass loss rate constraint further. Minton

& Malhotra calculate the solar mass and hence mass loss rate required to keep the radiative equilibrium temperature of the Earth's atmosphere at 273° K, the freezing point of water, during solar evolution. It is plausibly assumed that the greenhouse effect will add about 15° K to that, as it does at present, to achieve the average atmospheric temperature we have now, that is favorable to life. The result of their analysis is, again, a required mass loss rate of about 10^{-11} M$_\odot$ yr^{-1}, but it only has to be sustained for the first 2.4 billion years.

The choice of maintaining the radiative equilibrium temperature at or above freezing is a rational one, because at a lower temperature planet Earth could flip to an equilibrium in which all of the surface is frozen over – snowball Earth – where the albedo is much higher, because of all the ice and snow. The geological record indicates that several "snowball Earth" episodes have occurred in Earth's history – in addition to the much more recent ice ages, where there is no full planetary ice coverage.

As an aside, but in response to an obvious question: How does the Earth's atmosphere escape from a "snowball Earth" state? The answer probably lies in the addition of CO2 to the atmosphere from volcanic eruptions. A slow but steady addition of CO2 by volcanism and no uptake of CO2 by the weathering of rocks and diffusion into the oceans – all covered by ice – will eventually create enough of a greenhouse effect to initiate melting at equatorial latitudes, after which, via various feedbacks, melting will proceed precipitously.

Minton & Malhotra also point out that their model for "minimum mass loss", according to the simulations of Kasting (1991), maintains solar luminosity at a high enough level to keep the atmosphere of Mars above the freezing point for the first billion years of its history – when oceans are believed to have existed on Mars. So indeed a strong early solar wind can resolve the paradox for both Earth and Mars, no separate solutions are required, much to the liking of Father William of Occam.

3. Stellar Spin Down and Mass Loss

In this section I will relate stellar mass loss rates, which are hard to observe, with the much better known stellar spin down rates in order to verify whether these can be made consistent.

It is well known that Sun-like stars spin down from rotation periods of just a few days in their first billion years to several weeks in their mid-life, e.g. 26 days for the Sun at 4.5 billion years. The loss of angular momentum is usually ascribed to the torque applied by the stellar wind that co-rotates with the star near the surface and is forced to co-rotate roughly out to the Alfvén radius where the wind outflow velocity equals the Alfvén speed.

Weber & Davis (1967) were the first to relate spin-down rates to a stellar wind model. Their model is of a purely radial magnetic field that changes polarity at the equator. Their key result that the torque applied by the wind on the star is to a good approximation given by

$$T = \omega R_A^2 \dot{M}, \tag{3.1}$$

where ω is the angular rotation rate of the star, R_A the critical Alfvén radius where the wind outflow velocity equals the local Alfvén speed, and $\dot{M}$ is the stellar mass loss rate. The calculation of Weber & Davis includes at factor 2/3 on the right hand side resulting from the azimuthal integration of the torque, which I omit here for simplicity

Physically interpreted this means that the stellar wind torque is roughly equal to the angular momentum of a stellar wind forced to co-rotate up to R_A and then let go, flowing out further preserving its angular momentum. The result of Eq. (3.1) follows from the requirement that the solution flows smoothly through the critical point in the defining equations, much like the requirement for the critical point in the thermally driven Parker wind.

The location of the critical Alfvén radius in the Weber & Davis solution is 24 solar radii, while sophisticated numerical solutions (Keppens & Goedbloed 2000, their Fig. 3) yield 7 to 14 stellar radii in the segment of their solution with open field lines – where the stellar wind comes from. Recent observations for the Sun (Velli, Tenerani & DeForest 2016) also indicate that for the Sun the Alfvén radius is of the order of 12 radii out over the polar regions. So there is broad agreement between observations, theory and simulations here.

However, it turns out that the expression of Eq. (3.1) as defining the stellar wind torque is strongly dependent upon the geometry of the stellar coronal magnetic field. While the field in Weber & Davis (1967) is purely radial (reversing at the equator), that in Keppens & Goedbloed (2000) has a much more realistic "dead zone" over the equator, where the field is closed, while open field lines spread out from the poles. The dead zone in the simulations takes on a form very similar to observed solar helmet streamers, as observed during a solar eclipse.

The torque from the stellar wind in Keppens & Goedbloed is a factor 15 to 60 smaller than that given by Eq. (3.1) (a factor 10 to 40 compared to Weber & Davis), because of the difference in magnetic field topology. The same result had already been pointed out by Priest & Pneuman (1974), based on the helmet streamer geometry of Pneuman & Kopp (1971).

I will show now that this result has important implications both for the mass loss required for the Sun to slow down from its initial rotation rate at the ZAMS, to its current rate, and for the slow down of solar rotation in the remainder of the Sun's main sequence lifetime.

First we need to know the moment of inertia of a star to be able to estimate the slow down rate for a given stellar mass loss rate. The angular momentum of a star is given by

$$L_\star = I_\star \omega = M_\star (\beta_I R_\star)^2 \omega, \tag{3.2}$$

where $L_\star$ is the angular momentum, $I_\star$ the moment of inertia, ω the rotation rate, $M_\star$ the stellar mass, $R_\star$ its radius, and $\beta_I R_\star$ the radius of gyration, with β_I the gyration constant, i.e. the fraction of the radius for the arm in the moment of inertia. Stellar evolution codes show that the value of β_I decreases a little after arrival of a star at the ZAMS, because of the production of Helium from lighter Hydrogen in the core (see Schrijver & Zwaan 2000, their Sect. 13.1). Interestingly then we would expect Sun-like stars to slowly spin up as they evolve on the main sequence if it weren't for stellar mass loss. Later, as the star evolves towards its giant phase, the moment of inertia greatly increases of course, but that is of no concern here. The typical value of β_I for a Sun-like star on the main sequence is of the order of 0.25, the value I shall use from here on, and assumed to decrease much less than the rotation rate.

The decrease in angular momentum of a star as it evolves. i.e. $\dot{L}_\star$ is of course equal to the torque applied to it by the stellar wind, i.e.

$$\dot{L}_\star = M_\star (\beta_I R_\star)^2 \dot{\omega} = f \omega R_A^2 \dot{M}, \tag{3.3}$$

where f is the efficiency factor discussed above, determined by Keppens & Goedbloed (2000) to range from 1/60 to 1/15. When we write the Alfvén radius R_A as a multiple of the stellar radius, $\alpha_A R_\star$ a very simple expression results for the stellar mass loss that is required to produce the much better observed spin-down of late type stars after arriving

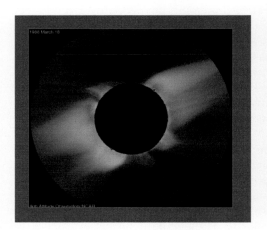 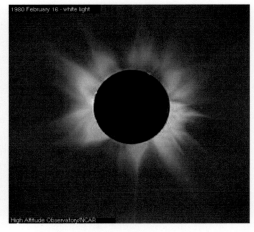

Figure 1. Helmet streamers observed during solar eclipses at solar minimum (left) and solar maximum (right). During maximum more streamers are present but their size is smaller than the ones at minimum.

on the main sequence,

$$\dot{M} = \frac{\beta_I^2}{f\alpha_A{}^2}\frac{\dot{\omega}}{\omega}M_\star. \tag{3.4}$$

The term $\frac{\dot{\omega}}{\omega}$ is simply the inverse of the e-folding time for the slow down in rotation of late type stars, which is of the order of 2-3 billion years, with not much variation between different stars (e.g. Nandy & Martens, 2007). Above I have found $\beta_I \approx 0.25$, $\alpha_A \approx 10$, and $f \approx 1/30$. Inserting these values into Eq. (3.4) we derive our main result,

$$\dot{M}_\star = -7.5 \times 10^{-12} M_\star yr^-1. \tag{3.5}$$

This represents the mass loss required to explain the observed stellar spin down rate by magnetic breaking. This mass loss rate also equals the mass loss required to resolve the Faint Young Sun Paradox, as discussed in the previous section. Indeed a very large mass loss, sustained for several billion years, is not just a possibility, but it may very well be required to explain the spin down of Sun-like stars. The analysis above also demonstrates that at its current mass loss rate of 2-3$\times 10^{-14}$ M$_\odot$ yr^{-1} our Sun will not slow down significantly for the remainder of its presence on the main sequence. This appears consistent with the observed rotation rates of older late type stars on the main sequence (Egeland, these proceedings).

4. Discussion and Conclusions

The current mass loss rate of 2-3$\times 10^{-14}$ M$_\odot$ yr^{-1} is not sufficient to slow down the Sun from an initial rotation period of 4-5 days to its current value : That would require an Alfvén radius of 170 solar radii, beyond the orbit of Venus. One might argue that the young Sun most likely had a much stronger magnetic field, which would increase its Alfvén radius. However, the simulations of Keppens & Goedbloed show that the slow down is not very sensitive to magnetic field strength: A stronger magnetic field is compensated by a larger dead zone, keeping the wind torque nearly constant.

Observations during solar eclipses even suggest a shrinking Alfvén radius with solar magnetic field. Fig. 1 shows juxtaposed helmet streamers observed during eclipses near solar minimum and solar maximum. The solar maximum image on the right shows more helmets, as expected, but of smaller size, with in particular the peak of the helmets closer to the solar surface. If, as in the simulations of Keppens & Goedbloed, the peak of the helmets approximately coincides with the Alfvén radius at that position angle, the Alfvén radius of the Sun, in its current phase of evolution, is indeed smaller at higher activity levels.

I conclude that a high mass loss rate is a reasonable hypothesis to resolve both the Faint Young Sun and the Faint Young Stars Paradoxes. Observations of mass loss rates of late type stars in the first billions of years after their arrival on the ZAMS are needed to verify this hypothesis, as well as, if possible, investigations of remaining signatures of the early solar wind.

References

Kasting, J. F. 1989, *Global and Planetary Change*, 1, 83

Kasting, J. F. 1991, *Icarus*, 94, 1

Kasting, J. F. 2004, *AGU Fall Meeting Abstracts*

Kasting, J. F., Toon, O. B., & Pollack, J. B. 1988, *Scientific American*, 258, 46

Keppens, R. & Goedbloed, J. P. 2000, *ApJ*, 530, 1036

Minton, D. A. & Malhotra, R. 2007, *ApJ*, 660, 1700

Nandy, D. & Martens, P. C. H. 2007, *Advances in Space Research*, 40, 891

do Nascimento, J.-D., Jr., Vidotto, A. A., Petit, P., *et al.* 2016, *ApJ Letters*, 820, L15

Pneuman, G. W. & Kopp, R. A. 1971, *Sol. Phys.*, 18, 258

Priest, E. R. & Pneuman, G. W. 1974, *Sol. Phys.*, 34, 231

Sagan, C. & Mullen, G. 1972, *Science*, 177, 52

Schrijver, C. J. & Zwaan, C. 2000, Solar and stellar magnetic activity / Carolus J. Schrijver, Cornelius Zwaan. New York: Cambridge University Press, 2000. (Cambridge astrophysics series; 34)

Velli, M., Tenerani, A., & DeForest, C. 2016, AAS/*Solar Physics Division Meeting*, 47, 402.05

Weber, E. J. & Davis, L., Jr. 1967, *ApJ*, 148, 217

Living Around Active Stars
Proceedings IAU Symposium No. 328, 2016
D. Nandy, A. Valio & P. Petit, eds.

© International Astronomical Union 2017
doi:10.1017/S1743921317004379

High Energy Exoplanet Transits

Joe Llama[1] and Evgenya L. Shkolnik[2]

[1] Lowell Observatory, 1400 W. Mars Hill Road, Flagstaff, AZ. 86001. USA
email: joe.llama@lowell.edu
[2] ASU School of Earth and Space Exploration, Tempe, AZ 85287. USA
email: shkolnik@asu.edu

Abstract. X-ray and ultraviolet transits of exoplanets allow us to probe the atmospheres of these worlds. High energy transits have been shown to be deeper but also more variable than in the optical. By simulating exoplanet transits using high-energy observations of the Sun, we can test the limits of our ability to accurately measure the properties of these planets in the presence of stellar activity. We use both disk-resolved images of the Solar disk spanning soft X-rays, the ultraviolet, and the optical and also disk-integrated Sun-as-a-star observations of the Lyα irradiance to simulate transits over a wide wavelength range. We find that for stars with activity levels similar to the Sun, the planet-to-star radius ratio can be overestimated by up to 50% if the planet occults an active region at high energies. We also compare our simulations to high energy transits of WASP-12b, HD 189733, 55 Cnc b, and GJ 436b.

Keywords. Sun: activity, stars: activity, stars: spots, (stars:) planetary systems

1. Introduction

High energy exoplanet transits offer the exciting opportunity to study the atmospheres of planets outside our Solar System. In particular, near-ultraviolet (NUV), and far-ultraviolet (FUV) transits probe the upper-most atmospheres of planets, where this light is absorbed. Multi-wavelength transits have been used to determine the composition of the atmosphere through transmission spectroscopy (see for example Sing *et al.* 2011; Gibson *et al.* 2012; Nikolov *et al.* 2014; Stevenson *et al.* 2014; Pont *et al.* 2013). Determining the parts-per-million change in light due to the atmosphere of an exoplanet requires an extremely accurate and precise measurement of the planet-to-star radius ratio ($R_p/R_\star$). Understanding the impact stellar activity has on these measurements is, therefore, crucial since activity will alter the depth and shape of the transit profile and the resultant ($R_p/R_\star$) measurement.

The signature of stellar activity has been seen in a many transit light curves across the entire wavelength spectrum. In the optical, bumps in the transit light curve correspond to the planet occulting a star spot on the stellar surface. By tracking the location of these spots over consecutive transits, the spin-orbit alignment of the exoplanet can be inferred and agrees well with Rossiter-Mclaughlin measurements (Winn *et al.* 2010; Sanchis-Ojeda & Winn 2011). For misaligned systems, it is possible to track changes in the phase location of the spots that correspond to latitudinal changes in the star spot distribution allowing us to recover stellar butterfly patterns and differential rotation patterns (Llama *et al.* 2012; Sanchis-Ojeda *et al.* 2013; Davenport *et al.* 2015). These observations are providing insight into stellar dynamos (Berdyugina 2005).

In the NUV, Hubble Space Telescope observations of WASP-12b have revealed an asymmetric transit when compared to the optical data (Fossati *et al.* 2010; Haswell *et al.* 2012). The NUV light curve is deeper than the optical transit, and also begins before but ends simultaneously with the optical transit. This early ingress suggests the presence of additional asymmetrically-distributed occulting material in the NUV leading the orbit of the planet.

There have been a number of explanations in the literature to explain such asymmetrical transits. One theory is that very close-in planets are heavily inflated and likely to be overflowing their Roche lobes, resulting in an accretion stream from the planet onto the star (Vidal-Madjar *et al.* 2003; Gu *et al.* 2003; Ibgui *et al.* 2010; Lai *et al.* 2010). An alternative suggestion is that the NUV early-ingress is the observational signature of the stellar wind colliding with the magnetosphere of the planet, resulting in a magnetic bow shock (Vidotto *et al.* (2010)). From the timing difference between the early-ingress and optical transits, Vidotto *et al.* (2010) were able to determine the magnetosphere stand-off distance to be $\sim 5R_p$. From this, they were able to constrain the magnetic field strength of WASP-12b to be < 24 G. Llama *et al.* (2011) modeled this idea using a Monte-Carlo Radiative Transfer Simulation and were able to fit the HST NUV light curve with a bow shock model with parameters similar to those found by Vidotto *et al.* (2010).

In the FUV, Lyα transits have been used to study both the atmospheric escape and inflated atmospheres of close-in exoplanets. Since these planets orbit much closer to their parent star, they will be subjected to increased levels of stellar irradiance which will heat the planet's exosphere and may lead to mass loss (Lammer *et al.* 2003; Lecavelier des Etangs *et al.* 2004). Deep Lyα transits have been observed using HST/STIS, hinting at the presence of extended atmospheres around HD 209458b, a hot-Jupiter orbiting a G star (Vidal-Madjar *et al.* 2003, 2004; Ben-Jaffel & Sona Hosseini 2010; Vidal-Madjar *et al.* 2004) and of HD 189733b, a hot Jupiter orbiting a more active K dwarf (Lecavelier Des Etangs *et al.* 2010; Lecavelier des Etangs *et al.* 2012; Bourrier *et al.* 2013).

Ehrenreich *et al.* (2012) used HST/STIS to observe 55 Cnc searching for the transit of 55 Cnc b, which has yet to be observed in optical photometry. Their data showed a transit depth of $\sim 7.5\%$, suggesting that this planet's atmosphere is grazing the stellar disk. FUV observations of the transit of GJ 436b have also revealed that this heavily irradiated hot Neptune orbiting an M dwarf is undergoing mass loss with the transit exhibiting a late egress and a transit depth of $\sim 25\%$, far greater than the 0.69% optical transit depth (Kulow *et al.* (2014)). A further study of this system by Ehrenreich *et al.* (2015) found an even deeper asymmetric Lyα light curve with a transit depth of $\sim 56\%$ and a cometary tail trailing the orbit of the planet.

Multi-epoch observations of HD 189733 in Lyα have shown variability in the FUV transit depth of this planet. In 2010 April there was no detectable Lyα transit; however, a $14.4 \pm 3.6\%$ transit was observed in 2011 September (Lecavelier des Etangs *et al.* 2012; Bourrier *et al.* 2013). The 2011 September transit depth is $\sim 10\times$ deeper than the optical transit and has been attributed to strong atmospheric escape from the planet which is likely correlated to the stellar wind strength of the star (Bourrier *et al.* 2013).

X-ray observations of HD 189733b's transit have been made using Chandra by Poppenhaeger *et al.* (2013). They observed a total of seven transits of HD 189733b and found a combined transit depth of $6\% - 8\%$, some $3\times$ larger than the optical transit depth. The authors note that stellar activity is clearly visible in their data, and depending on which observations they choose to combine, the resultant transit depth varies between $2.3\% - 9.4\%$.

In this proceedings, we review the work of Llama & Shkolnik (2015) and Llama & Shkolnik (2016), where high energy observations of the Sun were used as input into a transit code to simulate the impact of stellar activity on exoplanet transits.

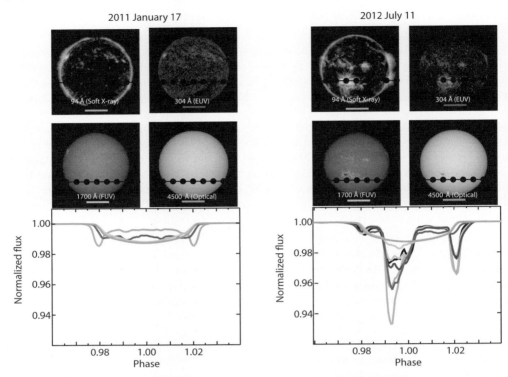

Figure 1. Top: Two sets of images of the Sun taken with NASA's SDO/AIA spanning the soft X-ray/EUV through to the optical. The left panel was taken while the Sun was quiet, while the right panel was obtained when there were a large number of active regions on the solar disk. Overplotted is the trajectory of the simulated hot Jupiter. Bottom: The simulated exoplanet transit for each of the light curves. Figure adapted from Llama & Shkolnik (2015).

2. Solar Observations

Using the Sun-as-a-star allows us to study the impact of stellar activity on exoplanet observations where we can directly observe the solar surface. NASA's Solar Dynamics Observatory (SDO) has been observing the Sun continuously since 2010. In this work, we make use of the data from the Atmospheric and Imaging Assembly (AIA) on board SDO. AIA images the solar disk in 10 wavelengths every 10 seconds, with a wavelength coverage spanning soft-Xray/EUV through to the optical. The images have a spatial resolution of 1" using a CCD of 4096 px^2. In Llama & Shkolnik (2015) we obtained images of the Sun in all ten wavelengths once every 24 hours between 2011 January 01 and 2014 September 01. In total our sample was comprised of 1300 images at each of the ten wavelengths.

Figure 1 shows images of the Sun from SDO/AIA in four wavelengths spanning the soft X-ray/EUV through to the optical on two separate dates. The short wavelength (EUV) images show how active regions on the solar disk appear as bright, extended features, whereas at longer wavelengths (FUV and optical) the active regions appear as smaller, dark regions.

3. Transit Model

Our transit model is based on that of Llama *et al.* (2013). Each image of the Sun is used as the stellar background over which we simulate the transit of a hot Jupiter with

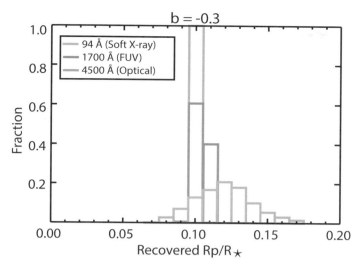

Figure 2. Histogram showing the distribution of measured $R_p/R_\star$ for the soft-Xray, FUV, and optical transits. In all cases, the input value of $R_p/R_\star$=0.1. This Figure was adapted from Llama & Shkolnik (2015).

$R_p/R_\star = 0.1$. To enable us to determine the impact of stellar activity on the simulations we assumed the planet to have the same radius in all wavelengths, i.e., the atmosphere of the planet does not change with wavelength. We also assumed the planet to be completely dark so that any region of the stellar disk occulted by the planet will have zero flux. At each step in the simulation, the position of the planet over the stellar disk is computed, and the loss in flux caused by the planet occulting that particular region of the star is summed to produce the simulated transit light curve.

4. Transit Depths as a function of wavelength

For each of our ~ 1300 images in each of the ten wavelengths, we computed the transit light curve and measured $R_p/R_\star$ using either a limb-darkened (FUV and optical) or limb-brightened (soft-Xray/EUV) transit model (see Llama & Shkolnik (2015) for more details).

Figure 2 shows the distribution of measured $R_p/R_\star$ for all our simulated light curves. As expected, the optical histogram is single peaked and centered on the input value of $R_p/R_\star = 0.1$, showing that for stars with similar activity levels as the Sun, optical star spots have little impact on our ability to accurately determine $R_p/R_\star$. At shorter wavelengths, there is a wider spread in the recovered values of $R_p/R_\star$. In the FUV, even though the active regions are dark, faculae and plage appear much brighter relative to the ambient photosphere than in the optical. Since these regions are brighter than their surroundings, the planet occults more starlight as it passes over them, resulting in a deeper transit. In the soft X-ray/EUV, active regions appear as bright extended regions, which, when occulted by the planet cause a depth increase in the transit. Since these regions are very extended relative to the optical, the recovered transit depth and hence $R_p/R_\star$, is on average, overestimated at high energies.

At short wavelengths, we find that up to 70% of the simulated transits exhibited stellar activity signatures that resulted in the measured radius of the planet to be overestimated by up to $\sim 50\%$. In the FUV, the number of impacted transits was lower, at $\sim 20\%$,

J. Llama & E. L. Shkolnik

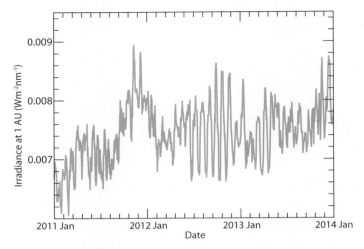

Figure 3. EVE MEGS-P Lyα irradiance. Figure adapted from Llama & Shkolnik (2016).

while in the optical we find that stars with activity signatures similar to the Sun, only 5% of light curves will show localized bumps caused by the planet occulting a star spot. For a complete description of the impact of stellar activity on the light curves we refer the reader to Sections 3 and 4 of Llama & Shkolnik (2015).

5. Lyα Transits

In Llama & Shkolnik (2016) we used observations of the Solar Lyα flux to study the impact of stellar activity of this EUV line on transit observations. The Extreme Ultraviolet Variability Experiment (EVE) instrument on board SDO measures the disk-integrated (Sun-as-a-star) EUV spectrum with a 10 s cadence within the wavelength range of 6.5 - 105 nm using the MEGS-A and MEGS-B components. Additionally, the photodiode MEGS-P monitors a 10 nm band centered around Lyα at 121.6 nm. To minimize degradation to the instruments, MEGS-P only observes for 3 hr a day at a 10 s cadence and once every 5 minutes every hour for the remainder of the day.

We obtained all the available EVE Level 2 line data which contain the Solar irradiance adjusted to 1 AU at selected emission lines, including Lyα. Figure 3 shows the Solar Lyα irradiance between 2011 January and 2014 January. There are two clear types of variability in this light curve. Firstly, the 27 day solar rotation period can clearly be seen. This modulation varies by up to 30%; however, the timescale of the variability is less problematic for exoplanet observations which are typically taken over a few hours. The more problematic issue, however, is short term variability which hampers our ability to determine the correct normalization level and accurately measure $R_p/R_\star$.

Since the Lyα data is not disk resolved, we could not use the transit model of Llama *et al.* (2013). Instead, we isolated each 3 hr section of continuous Lyα monitoring and combine it with a simulated limb-darkened transit using the code developed by Mandel & Agol (2002) with the same 10 s cadence. Again, we chose the input $R_p/R_\star = 0.1$, and the 3 hr transit was comprised of the 1 hr transit event and 1 hr of out-of-transit data on either side of the transit for normalization. To measure $R_p/R_\star$ we performed a χ^2 test with Mandel & Agol (2002) light curves of varying $R_p/R_\star$ to find the best fit.

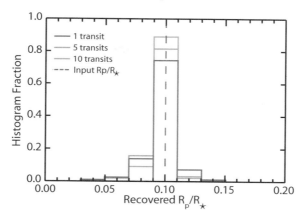

Figure 4. Histogram of recovered $R_p/R_\star$ for our Lyα simulations. For a single transit, the correct answer is measured $\sim 75\%$ of the time. Once 10 consecutive transits are averaged together, the correct answer is measured over 90% of the time. Figure adapted from Llama & Shkolnik (2016).

It is worthy of note that we used disk resolved data, as such all spatial information is lost this experiment provides the limits of probing the upper atmospheres of exoplanets at this important wavelength.

In total we simulated 1100 transits. Figure 4 shows the distribution of measured $R_p/R_\star$ for this set of simulations. The histogram for 1 transit shows that we obtain the input value of $R_p/R_\star = 0.1$ (within error of $\pm 5\%$) in $\sim 75\%$ of the simulations. In the X-ray transit work of Poppenhaeger *et al.* (2013) they obtained seven transits and combined them in an effort to combat stellar activity. In Figure 4 we show histograms for combining 5 and 10 consecutive transits. Once 10 transits are combined, the correct answer was measured over 90% of the time. This, therefore, shows that despite stellar activity changes on the surface of the star between observations, combining multiple transits together increases the likelihood of measuring the correct value of $R_p/R_\star$.

6. Summary

Accounting for stellar activity on high energy exoplanet transits is imperative in ensuring the correct value of $R_p/R_\star$ is measured. The Sun offers a unique opportunity to study the impact of stellar activity by simulating transits of a hot Jupiter over high cadence data of the solar disk. Using disk-resolved images of the Sun from NASA's SDO/AIA instrument we have simulated over 13,000 transits in wavelengths spanning soft X-rays/EUV through to the FUV, and optical.

Our simulations show that in 70% of short wavelength transits the light curves show signatures of stellar activity that would result in the derived value of $R_p/R_\star$ differing from the correct value. As expected, this number decreased at longer wavelengths.

By using disk integrated Lyα data from NASA's SDO/EVE instrument we were able to place lower limits on the impact stellar activity has on observations at this wavelength. In none of our 1000 simulated Lyα transits were we able to reproduce the observed early ingress seen in the NUV light curves of WASP-12b observed by Fossati *et al.* (2010); Haswell *et al.* (2012), adding strength to the conclusion that this asymmetry is the signature of a magnetospheric bow shock (Vidotto *et al.* 2010; Llama *et al.* 2011). In none of our simulations were we able to reproduce the large transit of 55 Cnc b observed by Ehrenreich *et al.* (2012), again strengthening the conclusion that this planet has an extended atmosphere that is grazing the stellar disk.

We conclude that it is vital to account for stellar activity in high-energy transit observations. Simultaneous observations in the optical and long baselines can help accurately determine the normalization level and disentangle planetary properties from stellar activity.

References

Ben-Jaffel, L. & Sona Hosseini, S. 2010, *ApJ*, 709, 1284
Berdyugina, S. V. 2005, *Living Reviews in Solar Physics*, 2, 8
Bourrier, V., Lecavelier des Etangs, A., Dupuy, H., *et al.* 2013, *A&A*, 551, A63
Davenport, J. R. A., Hebb, L., & Hawley, S. L. 2015, *ApJ*, 806, 212
Ehrenreich, D., Bourrier, V., Bonfils, X., *et al.* 2012, *A&A*, 547, A18
Ehrenreich, D., Bourrier, V., Wheatley, P. J., *et al.* 2015, *Nature*, 522, 459
Fossati, L., Haswell, C. A., Froning, C. S., *et al.* 2010, *ApJL*, 714, L222
Haswell, C. A., Fossati, L., Ayres, T., *et al.* 2012, *ApJ*, 760, 79
Gibson, N. P., Aigrain, S., Pont, F., *et al.* 2012, *MNRAS*, 422, 753
Gu, P.-G., Lin, D. N. C., & Bodenheimer, P. H. 2003, *ApJ*, 588, 509
Ibgui, L., Burrows, A., & Spiegel, D. S. 2010, *ApJ*, 713, 751
Kulow, J. R., France, K., Linsky, J., & Loyd, R. O. P. 2014, *ApJ*, 786, 132
Lai, D., Helling, C., & van den Heuvel, E. P. J. 2010, *ApJ*, 721, 923
Lammer, H., Selsis, F., Ribas, I., *et al.* 2003, *ApJL*, 598, L121
Lecavelier des Etangs, A., Vidal-Madjar, A., McConnell, J. C., & Hébrard, G. 2004, *A&A*, 418, L1
Lecavelier Des Etangs, A., Ehrenreich, D., Vidal-Madjar, A., *et al.* 2010, *A&A*, 514, A72
Lecavelier des Etangs, A., Bourrier, V., Wheatley, P. J., *et al.*2012, *A&A*, 543, L4
Llama, J., Wood, K., Jardine, M., *et al.* 2011, *MNRAS*, 416, L41
Llama, J., Jardine, M., Mackay, D. H., & Fares, R. 2012, *MNRAS*, 422, 72
Llama, J., Vidotto, A. A., Jardine, M., *et al.* 2013, *MNRAS*, 436, 2179
Llama, J. & Shkolnik, E. L. 2015, *ApJ*, 802, 41
Llama, J. & Shkolnik, E. L. 2016, *ApJ*, 817, 81
Mandel, K. & Agol, E. 2002, *ApJL*, 580, L171
Nikolov, N., Sing, D. K., Pont, F., *et al.* 2014, *MNRAS*, 437, 46
Pont, F., Sing, D. K., Gibson, N. P., *et al.* 2013, *MNRAS*, 432, 2917
Poppenhaeger, K., Schmitt, J. H. M. M., & Wolk, S. J. 2013, *ApJ*, 773, 62
Sanchis-Ojeda, R. & Winn, J. N. 2011, *ApJ*, 743, 61
Sanchis-Ojeda, R., Winn, J. N., Marcy, G. W., *et al.* 2013, *ApJ*, 775, 54
Sing, D. K., Pont, F., Aigrain, S., *et al.* 2011, *MNRAS*, 416, 1443
Stevenson, K. B., Bean, J. L., Seifahrt, A., *et al.* 2014, *AJ*, 147, 161
Vidal-Madjar, A., Lecavelier des Etangs, A., Désert, J.-M., *et al.* 2003, *Nature*, 422, 143
Vidal-Madjar, A., Désert, J.-M., Lecavelier des Etangs, A., *et al.* 2004, *ApJL*, 604, L69
Vidotto, A. A., Jardine, M., & Helling, C. 2010, *ApJL*, 722, L168
Winn, J. N., Johnson, J. A., Howard, A. W., *et al.* 2010, *ApJL*, 723, L223

Living Around Active Stars
Proceedings IAU Symposium No. 328, 2016
D. Nandy, A. Valio & P. Petit, eds.

© International Astronomical Union 2017
doi:10.1017/S1743921317003660

Detection of secondary eclipses of WASP-10b and Qatar-1b in the Ks band and the correlation between Ks-band temperature and stellar activity.

Patricia Cruz[1,2], David Barrado[2], Jorge Lillo-Box[3,2], Marcos Diaz[1], Mercedes López-Morales[4], Jayne Birkby[4,5] Jonathan J. Fortney[6] and Simon Hodgkin[7],

[1]Instituto de Astronomia, Geofísica e Ciências Atmosféricas, Universidade de São Paulo (IAG/USP), Brazil
email: patricia.cruz@usp.br
[2]Departamento de Astrofísica, Centro de Astrobiología (CAB/INTA-CSIC), Spain
[3]European Southern Observatory (ESO), Chile
[4]Harvard-Smithsonian Center for Astrophysics, USA
[5]NASA Sagan Fellow
[6]Department of Astronomy and Astrophysics, University of California, USA
[7]Institute of Astronomy, University of Cambridge, UK

Abstract. The Calar Alto Secondary Eclipse study was a program dedicated to observe secondary eclipses in the near-IR of two known close-orbiting exoplanets around K-dwarfs: WASP-10b and Qatar-1b. Such observations reveal hints on the orbital configuration of the system and on the thermal emission of the exoplanet, which allows the study of the brightness temperature of its atmosphere. The observations were performed at the Calar Alto Observatory (Spain). We used the OMEGA2000 instrument (Ks band) at the 3.5m telescope. The data was acquired with the telescope strongly defocused. The differential light curve was corrected from systematic effects using the Principal Component Analysis (PCA) technique. The final light curve was fitted using an occultation model to find the eclipse depth and a possible phase shift by performing a MCMC analysis. The observations have revealed a secondary eclipse of WASP-10b with depth of 0.137%, and a depth of 0.196% for Qatar-1b. The observed phase offset from expected mid-eclipse was of -0.0028 for WASP-10b, and of -0.0079 for Qatar-1b. These measured offsets led to a value for $|e \cos \omega|$ of 0.0044 for the WASP-10b system, leading to a derived eccentricity which was too small to be of any significance. For Qatar-1b, we have derived a $|e \cos \omega|$ of 0.0123, however, this last result needs to be confirmed with more data. The estimated Ks-band brightness temperatures are of 1647 K and 1885 K for WASP-10b and Qatar-1b, respectively. We also found an empirical correlation between the $\log(R'_{\rm HK})$ activity index of planet hosts and the Ks-band brightness temperature of exoplanets, considering a small number of systems.

Keywords. Planetary systems, stars, photometry

1. Introduction and data reduction

The Calar Alto Secondary Eclipse study (the CASE study) was an observational program dedicated to observe in the near-infrared, from the ground, secondary eclipses of two known close-orbiting exoplanets: WASP-10b (Christian *et al.* 2009) and Qatar-1b (Alsubai *et al.* 2011). Such observations reveal hints on the orbital configuration of the system and on the thermal emission of the exoplanet.

The observations were performed in the Ks band with the OMEGA2000 instrument mounted on the 3.5m telescope of the Calar Alto Observatory (CAHA), in Spain. The telescope was strongly defocused, resulting in a ring-shaped PSF. The data were gathered in staring mode, where we observed the target continuously without any dithering during several hours. Before and after the staring mode sequence, we also obtained in-focus images composed by five dither-point images each with the purpose of obtaining sky images for further subtraction.

The initial data reduction was performed using IRAF for the bad pixel removal, flat-fielding, and sky subtraction. We constructed a normalized sky map based on the in-focus images, which was scaled to the observed median background of each image and then subtracted. A circular aperture photometry was done in IDL for the target and for all sufficiently bright stars in the field of view. Only the stars that did not present any strong variations or other odd behavior in their light curves were used as reference for the differential photometry.

2. Secondary eclipse analysis

2.1. *The CASE study: WASP-10b*

We observed the planet-host WASP-10 during approximately 5.4 continuous hours. After the reduction described above, we used the Principal Component Analysis (PCA) technique to identify parameters related to visible trends in the differential light curve, revealing significant correlations of the normalized flux with the star's y-position at the detector, defocused seeing, airmass, and background count level. We fit for these systematics simultaneously by performing a multiple linear regression, where we modeled the trends only on the out-of-eclipse portions of the light curve, applied the model to the in-eclipse portion, and removed it from the light curve. The complete description of the reduction procedure and analysis performed can be found in Cruz *et al.* (2015).

Firstly, we binned the light curve with 127 points per bin, and performed a fit to obtain the secondary eclipse depth and detect a possible phase offset by using the occultation model from Mandel & Agol (2002), assuming no limb-darkening. In order fit for the expected phase of mid-eclipse ϕ_c, the depth of the secondary eclipse ΔF, and the out-of-eclipse baseline level F_{bl}, we performed a Markov chain Monte Carlo (MCMC) method, generating four chains of 1×10^6 simulations each. After combining the results from all chains (total of 4×10^6 simulations), the best model obtained from the MCMC analysis was the one with ΔF of $0.137\%^{+0.013\%}_{-0.019\%}$, $\Delta\phi$ of $-0.0028^{+0.0005}_{-0.0004}$, and $F_{\mathrm{bl}} = 0.99984^{+0.00008}_{-0.00010}$ (fig. 1). We also fit for the systematics and the occultation model simultaneously. For that, we applied the MCMC analysis to the unbinned light curve with 3900 individual measurements. With the baseline level fixed at $F_{\mathrm{bl}} = 1.0$, this analysis provided the following results: $\Delta F = 0.139\%^{+0.011\%}_{-0.023\%}$ and $\Delta\phi = -0.0019^{+0.0019}_{-0.0002}$.

Derived orbital configuration

The measured offset of $\Delta\phi = -0.0028$ means that the center of the secondary eclipse occured at phase $\phi_{ecl} = 0.4972$ instead of at phase $\phi = 0.5$, expected for a circular orbit. From the results, we obtain that $e \cos\omega \simeq -0.0044$. The negative sign indicates that ω is a value between 90 and 270 degrees. Considering the ω of $167.13°$, published by Christian *et al.* (2009), its orbit would have an eccentricity of $e \simeq 0.0045$. This value is fully consistent with a circular orbit, although it is in disagreement with their published eccentricity of $e \simeq 0.059$. Instead, if we use the value given by Johnson *et al.* (2009) for ω of $153.3°$, we get $e \simeq 0.0049$, also in favor of a null eccentricity and different from their result ($e \simeq 0.051$). Nevertheless, these derived eccentricities are too small to be of any significance.

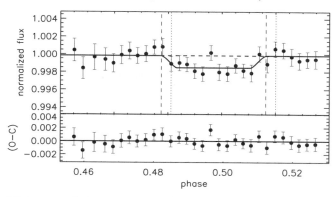

Figure 1. Binned light curve (127 points per bin) showing the secondary eclipse of WASP-10b in the Ks band. The best-fitting model (solid line) provided an eclipse depth of $\Delta F = 0.137\%$. The ingress and egress positions of the expected eclipse, considering a circular orbit, are represented by dotted lines, and the dashed lines show a phase shift of $\Delta\phi = -0.0028$. In the lower panel, the residuals are shown. (Cruz *et al.* 2015, fig. 3.)

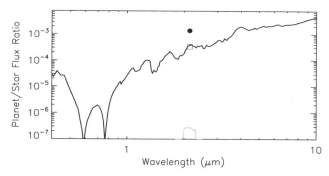

Figure 2. Model spectrum of thermal emission of WASP-10b computed without TiO/VO and considering $f = 2/3$. The measured planet-to-star flux ratio (at 2.14 μm) is represented by a filled circle (the error bars are inside the filled circle). The expected flux considering the presented model is represented by a square. The Ks-band transmission curve (gray line) is shown at the bottom of the panel at arbitrary scale. (Cruz *et al.* 2015, fig. 8.)

Estimated thermal emission

The MCMC analysis resulted in a planet-to-star flux ratio of 0.137%, which is given by the eclipse depth. Assuming both components of the system, planet and star, emit as black-bodies, we derived a Ks-band brightness temperature for WASP-10b of $T_{Ks} \simeq 1647^{+97}_{-131}$ K (Cruz *et al.* 2015). The maximum expected equilibrium temperature for the planet, assuming zero Bond albedo and instant reradiation ($A_B = 0$ and $f = 2/3$, respectively) is $T_{eq} \simeq 1224$ K. This temperature is about 25% cooler than the observed T_{Ks}.

From the comparision of the measured planet-to-star flux ratio to several atmospheric spectral models, we noticed that none of the models used was able to reproduce the high emission of WASP-10b in the Ks band. We used the models by Fortney *et al.* (2006, 2008), generated with different reradiation factors (f) and computed without TiO/VO (models without TiO/VO tend to have higher K-band fluxes). The model which produces the highest emission possible in the observed band considers an instant reradiation over the dayside ($f = 2/3$) and it is shown in figure 2.

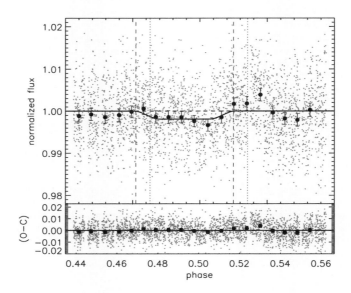

Figure 3. Detrended light curve of the secondary eclipse of Qatar-1b in the Ks band. The best-fitting model (solid line) resulted in $\Delta F = 0.186\%$, and $\Delta \phi = -0.0071$. The individual measurements are represented by small dots, and the filled circles represent the same data binned every 143 points, for better visualization. The expected ingress and egress positions, considering circular orbit, are shown as dotted vertical lines. The dashed lines represent the phase shift provided by the analysis. (Cruz *et al.* 2016, fig. 3.)

2.2. *The CASE study: Qatar-1b*

We observed the planet-host Qatar-1 during approximately 4.2 continuous hours. The data reduction followed the same procedure described earlier. The differential light curve generated was also corrected from systematic effects by using the PCA technique, presenting significant correlations of the normalized flux with star's xy-position, defocused seeing, and airmass. The details on the reduction and analysis are describde in Cruz *et al.* (2016).

The final decorrelated light curve was fitted using the occultation model from Mandel & Agol (2002), assuming no limb-darkening, by performing a MCMC analysis. We then performed the MCMC analysis of the uncorrected light curve (with 2850 individual measurements), unifying the decorrelation function and the occultation model in a joint fit, with the objective of integrating the effect of the systematics into the uncertainty estimates. The best fitting model resulted in $\Delta F = 0.186\%^{+0.022\%}_{-0.024\%}$, and $\Delta \phi = -0.0071^{+0.0006}_{-0.0010}$, with $F_{\rm bl}$ fixed at 1.0 (fig. 3).

Since the red noise present in these data is not negligible, we have used the "Prayer Bead" method (Moutou *et al.* 2004, Gillon *et al.* 2007) to assess the impact of red noise on the derived parameters and errors, which resulted in revised values for ΔF and $\Delta \phi$ of $0.196\%^{+0.071\%}_{-0.051\%}$ and $-0.0079^{+0.0162}_{-0.0043}$, respectively. It is worth mentioning that the sample of red noise we have is limited, and its timescale is comparable with the eclipse duration, which could have compromised this final analysis and led to an overestimation of the derived uncertainties.

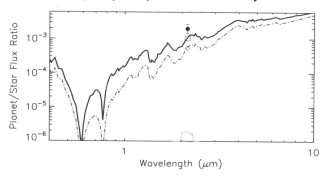

Figure 4. Model spectrum of thermal emission of Qatar-1b computed without TiO/VO and $f = 2/3$ (solid line). The measured planet-to-star flux ratio is represented by a filled circle. For comparison, a similar model with $f = 1/2$ is presented (dot-dashed line). The Ks-band transmission curve (gray line) is also shown, at arbitrary scale. (Cruz *et al.* 2016, fig. 5.)

Derived orbital configuration

The measured offset in phase led to a derived $e\cos\omega$ of $-0.0123^{+0.0252}_{-0.0067}$. Due to the large uncertainty of $\Delta\phi$, the estimated $e\cos\omega$ can lead to several combinations of e and ω, making impossible to constrain ω for eccentricities lower than 0.02 (see Cruz *et al.* 2016 for more details). Additional data are needed to further investigate the orbital configuration of this exoplanet.

Estimated thermal emission

From the ΔF derived in the analysis, and assuming that Qatar-1b and its central star emit as black-bodies, we estimated a Ks-band brightness temperature of $T_{\rm Ks} \simeq 1885^{+212}_{-168}$ K. Considering an instant reradiation ($f = 2/3$) and a zero Bond albedo ($A_B = 0$), we would expect a maximum equilibrium temperature for this exoplanet of $T_{\rm eq} \simeq 1768$ K (Cruz *et al.* 2016). Then, the estimated $T_{\rm Ks}$ of Qatar-1b is in accordance with $T_{\rm eq}$, within the error estimates.

We then compared the measured planet-to-star flux ratio with several atmospheric spectral models by Fortney *et al.* (2006, 2008). As done previously, these models were computed with different reradiation factors, and without TiO/VO. Figure 4 shows the atmospheric model that better reproduces the observed emission of Qatar-1b, with an inefficient heat redistribution ($f = 2/3$).

3. Discussion

3.1. *Presence or absence of thermal inversions*

A thermal inversion could be caused due to the presence of an opacity source in the atmosphere of an exoplanet. For instance, some studies have suggested that TiO and VO present in a high layer of the atmosphere could lead to a hot stratosphere and to a temperature inversion. According to Fortney *et al.* (2006), atmospheric models with such an inversion would show a weak emission in the near-infrared (JHK bands), however, models without this thermal inversion would have stronger emissions. From our analyses presented earlier, the atmospheric models that most approximate the planet-to-star flux ratio measured for WASP-10b and Qatar-1b are models computed without TiO/VO. This suggests that both exoplanets whould have atmospheres without a thermal inversion layer.

Table 1. Secondary eclipses (SE) in the Ks band from the literature.

Object	SE depth (%)	T_{Ks} (K)	Reference
CoRoT-1b	0.336 ± 0.042	2460^{+80}_{-160}	Rogers *et al.* 2009
CoRoT-2b	0.16 ± 0.09	1890^{+260}_{-350}	Alonso *et al.* 2010
HAT-P-1b	0.109 ± 0.025	2136^{+150}_{-170}	de Mooij *et al.* 2011
Qatar-1b	$0.196^{+0.071}_{-0.051}$	1885^{+212}_{-168}	Cruz *et al.* 2016
TrES-2b	$0.062^{+0.013}_{-0.011}$	1636^{+79}_{-88}	Croll *et al.* 2010a
TrES-3b	$0.133^{+0.018}_{-0.016}$	1731^{+56}_{-60}	Croll *et al.* 2010b
WASP-3b	0.181 ± 0.020	~ 2435	Zhao *et al.* 2012
WASP-4b	$0.185^{+0.014}_{-0.013}$	1995 ± 40	Cáceres *et al.* 2011
WASP-10b	$0.137^{+0.013}_{-0.019}$	1647^{+97}_{-131}	Cruz *et al.* 2015
WASP-12b	$0.309^{+0.013}_{-0.012}$	2988^{+45}_{-46}	Croll *et al.* 2011
WASP-19b	0.287 ± 0.020	2310 ± 60	Zhou *et al.* 2014

Fortney *et al.* (2008) and Burrows *et al.* (2008) proposed that highly irradiated exoplanets, with $F_{\mathrm{inc}} > 10^9$ *erg s^{-1} cm^2*, would present a thermal inversion layer in their atmospheres, however, at low irradiation levels, the exoplanetary atmospheres would not show temperature inversions. Both objects studied, WASP-10b and Qatar-1b, fall in the category of low irradiated exoplanets, with $F_{\mathrm{inc}} \sim 0.2 \times 10^9$ and $F_{\mathrm{inc}} \sim 0.8 \times 10^9$ *erg s^{-1} cm^2*, respectively. Nevertheless, the direct relation of the presence or absence of a thermal inversion layer with the incident radiation level could be confirmed since several exoplanets discussed in the literature do not seem to follow this trend.

3.2. *Are the stellar activity and the absence of a thermal inversion correlated?*

Knutson *et al.* (2010) investigated the correlation between the stellar activity and the presence of a thermal inversion in the atmosphere of hot Jupiters. They calculated the $\log(R'_{\mathrm{HK}})$ activity index from the CaII H and K spectral lines (at 3968 and 3933 Å, respectively) for 16 transiting planet-host stars with available secondary eclipse observations in the 3.6 and 4.5 μm IRAC bandpasses (from Spitzer). They found that those exoplanets without a thermal inversion layer orbit the most active stars ($\log(R'_{\mathrm{HK}}) > -4.9$), and those with an inversion layer are associated with quiet stars ($\log(R'_{\mathrm{HK}}) < -4.9$). In order to explain the observed correlation, these authors suggested that the higher UV flux emitted by active stars would influence the photochemistry in hot Jupiter atmospheres and destroy the absorber responsible for the inversion layer.

From the list of transiting planet host stars with measured $\log(R'_{\mathrm{HK}})$ from Knutson *et al.* (2010), we identified nine exoplanets from the literature with observed secondary eclipse in the Ks band (table 1). To discuss this correlation of stellar activity with the temperature profile (with or without inversion), we investigated the relationship of $\log(R'_{\mathrm{HK}})$ with the estimated Ks-band brightness temperature, which is shown in figure 5. From this sample, the exoplanets presenting a thermal inversion layer (diamonds) are placed in the left part of the plot, in the region of the less active or quiet stars ($\log(R'_{\mathrm{HK}}) < -4.9$). Those without an inversion layer (triangles) are placed on the right, in the region of the active stars ($\log(R'_{\mathrm{HK}}) > -4.9$).

The Ks-band brightness temperatures obtained in the CASE study are $T_{\mathrm{Ks}} \simeq 1647$ K and $T_{\mathrm{Ks}} \simeq 1885$ K for WASP-10b and Qatar-1b, respectively. The activity indices ($\log(R'_{\mathrm{HK}})$) for these objects are of -4.3 for WASP-10 (Maciejewski *et al.* 2011) and of -4.6 for Qatar-1 (Covino *et al.* 2013), placing both systems in the region of active stars and exoplanets without thermal inversion layer (fig. 5).

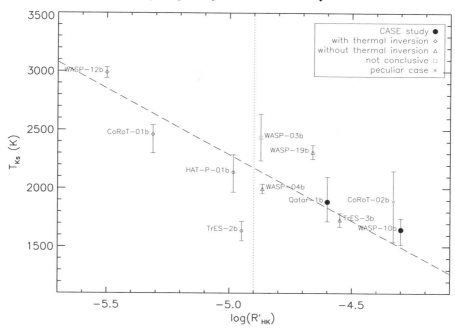

Figure 5. The $\log(R'_{\mathrm{HK}})$ activity index from Knutson *et al.* (2010) as function of the Ks-band brightness temperature from the literature. The vertical dotted line delineates the separation between active $(\log(R'_{\mathrm{HK}}) > -4.9)$ and quiet stars $(\log(R'_{\mathrm{HK}}) < -4.9)$. The dashed line shows an empirical correlation between both quantities with $dT_{\mathrm{Ks}}/d\log(R'_{\mathrm{HK}}) \simeq -1131 \pm 60$ K (Cruz 2015b).

To quantify the apparent correlation in figure 5, we performed a linear fit resulting in the following empirical relation:

$$T_{\mathrm{Ks}} = (-3367 \pm 291) + (-1131 \pm 60) \cdot \log(R'_{\mathrm{HK}}), \qquad (3.1)$$

with T_{Ks} given in K. This empirical relationship between $\log(R'_{\mathrm{HK}})$ and T_{Ks} was based on a small number of data points (Cruz 2015b). For a more precise analysis, we need to increase the sample by measuring the activity indices of the exoplanets that have the Ks-band brightness tempertature already estimated, and to observed new secondary eclipses of known close-in hot Jupiters in the Ks band. This is an interesting correlation to be further investigated and to understand better what is driving thermal inversions in the exoplanetary atmospheres.

References

Alonso, R., Deeg, H. J., Kabath, P., & Rabus, M. 2010, *AJ*, 139, 1481

Alsubai, K. A., Parley, N. R., Bramich, D. M., West, R. G., Sorensen, P. M., Collier Cameron, A., Latham, D. W., Horne, K., Anderson, D. R., Bakos, G. Á., Brown, D. J. A., Buchhave, L. A., Esquerdo, G. A., Everett, M. E., Fżrész, G., Hartman, J. D., Hellier, C., Miller, G. M., Pollacco, D., Quinn, S. N., Smith, J. C., Stefanik, R. P., & Szentgyorgyi, A. 2011, *MNRAS*, 417, 709

Burrows, A., Budaj, J., & Hubeny, I. 2008, *ApJ*, 678, 1436

Cáceres, C., Ivanov, V. D., Minniti, D., Burrows, A., Selman, F., Melo, C., Naef, D., Mason, E., & Pietrzynski, G. 2011, *A&A*, 530, A5

Christian, D. J., Gibson, N. P., Simpson, E. K., Street, R. A., Skillen, I., Pollacco, D., Collier Cameron, A., Joshi, Y. C., Keenan, F. P., Stempels, H. C., Haswell, C. A.,

Horne, K., Anderson, D. R., Bentley, S., Bouchy, F., Clarkson, W. I., Enoch, B., Hebb, L., Hébrard, G., Hellier, C., Irwin, J., Kane, S. R., Lister, T. A., Loeillet, B., Maxted, P., Mayor, M., McDonald, I., Moutou, C., Norton, A. J., Parley, N., Pont, F., Queloz, D., Ryans, R., Smalley, B., Smith, A. M. S., Todd, I., Udry, S., West, R. G., Wheatley, P. J., & Wilson, D. M. 2009, *MNRAS*, 392, 1585

Covino, E., Esposito, M., Barbieri, M., Mancini, L., Nascimbeni, V., Claudi, R., Desidera, S., Gratton, R., Lanza, A. F., Sozzetti, A., Biazzo, K., Affer, L., Gandolfi, D., Munari, U., Pagano, I., Bonomo, A. S., Collier Cameron, A., Hébrard, G., Maggio, A., Messina, S., Micela, G., Molinari, E., Pepe, F., Piotto, G., Ribas, I., Santos, N. C., Southworth, J., Shkolnik, E., Triaud, A. H. M. J., Bedin, L., Benatti, S., Boccato, C., Bonavita, M., Borsa, F., Borsato, L., Brown, D., Carolo, E., Ciceri, S., Cosentino, R., Damasso, M., Faedi, F., Martínez Fiorenzano, A. F., Latham, D. W., Lovis, C., Mordasini, C., Nikolov, N., Poretti, E., Rainer, M., Rebolo López, R., Scandariato, G., Silvotti, R., Smareglia, R., Alcalá, J. M., Cunial, A., Di Fabrizio, L., Di Mauro, M. P., Giacobbe, P., Granata, V., Harutyunyan, A., Knapic, C., Lattanzi, M., Leto, G., Lodato, G., Malavolta, L., Marzari, F., Molinaro, M., Nardiello, D., Pedani, M., Prisinzano, L., & Turrini, D. 2013, *A&A*, 554, A28

Croll, B., Albert, L., Lafreniere, D., Jayawardhana, R., & Fortney, J. J. 2010a, *ApJ*, 717, 1084

Croll, B., Jayawardhana, R., Fortney, J. J., Lafreniere, D., & Albert, L. 2010b, *ApJ*, 718, 920

Croll, B., Lafreniere, D., Albert, L., Jayawardhana, R., Fortney, J. J., & Murray, N. 2011, *AJ*, 141, 30

Cruz, P., Barrado, D., Lillo-Box, J., Diaz, M., Birkby, J., López-Morales, M., Hodgkin, S., & Fortney, J. J. 2015, *A&A*, 574, A103

Cruz, P. 2015b, *PhD Thesis*, Universidad Autónoma de Madrid

Cruz, P., Barrado, D., Lillo-Box, J., Diaz, M., Birkby, J., López-Morales, M., & Fortney, J. J. 2016, *A&A*, 595, A61

de Mooij, E. J. W., de Kok, R. J., Nefs, S. V., & Snellen, I. A. G. 2016, *A&A*, 528, A49

Fortney, J. J., Saumon, D., Marley, M. S., Lodders, K., & Freedman, R. S. 2006, *ApJ*, 642, 495

Fortney, J. J., Lodders, K., Marley, M. S., & Freedman, R. S. 2008, *ApJ*, 678, 1419

Gillon, M., Demory B.-O., Barman, T., Bonfils, X., Mazeh, T., Pont, F., Udry, S., Mayor, M., & Queloz, D. 2007, *A&A*, 471, L51

Johnson, J. A., Winn, J. N., Cabrera, N. E., & Carter, J. A. 2009, *ApJL*, 692, L100

Maciejewski, G., Raetz St., Nettelmann, N., Seeliger, M., Adam, C., Nowak, G., & Neuhäuser, R. 2011, *A&A*, 535, A7

Mandel, K. & Agol, E. 2002, *ApJ*, 580, L171

Moutou, C., Pont, F., Bouchy, F., & Mayor, M. 2004, *A&A*, 424, L31

Rogers, J., Apai, D., López-Morales, M., Sing, D., & Burrows, A. 2009, *ApJ*, 707, 1707

Zhao, M., Milburn, J., Barman, T., Hinkley, S., Swain, M. R., Wright, J., & Monnier, J. D. 2012, *ApJ*, 748, L8

Zhou, G., Bayliss, D. D. R., Kedziora-Chudczer, L., Salter, G., Tinney, C. G., & Bailey, J. 2014, *MNRAS*, 445, 2746

Living Around Active Stars
Proceedings IAU Symposium No. 328, 2016
D. Nandy, A. Valio & P. Petit, eds.

© International Astronomical Union 2017
doi:10.1017/S1743921317003696

Atmospheric Parameters and Luminosities of Nearby M Dwarfs – Estimating Habitable Exoplanet Detectability with the E-ELT

Gustavo F. Porto de Mello†, Riano E. Giribaldi, Diego Lorenzo-Oliveira[1] and Nathália M. Paes Leme

Observatório do Valongo, Universidade Federal do Rio de Janeiro, Ladeira do Pedro Antonio
43, Rio de Janeiro, RJ, CEP: 20080-090, Brazil
email: gustavo@astro.ufrj.br

Abstract. We derive T_{eff} and [Fe/H] for a sample of 72 nearby M-dwarfs with Hipparcos parallaxes and $\delta < +30$. Spectra, acquired at the Observatório do Pico dos Dias, Brazil, have R = 10,000 and S/N $\gtrsim$ 100 for nearly all targets in the $\lambda\lambda$8380-8880 range. Atmospheric parameters were derived from VJHK colors and a system of spectral line indices calibrated against sample stars with interferometric T_{eff} and [Fe/H] from detailed analysis of FGK binary companions. A PCA method of calibration yields internal errors within 70 K and 0.1 dex for T_{eff} and [Fe/H]. For 18 stars we present the first T_{eff} or [Fe/H] derivation in the literature. We compute the star's luminosities, calculate the position of their habitable zones and estimate that, were all of they to harbour rocky planets inside their HZ, 15−20 of these would be detectable by the E-ELT Planetary Camera and Spectrograph.

Keywords stars: abundances – stars: late-type – stars: low-mass – (stars:) planetary systems

1. Introduction

Low mass M dwarf stars occupy the center stage in modern searches for habitable planets in the solar neighborhood owing to: 1) their very large number densities, accounting for more than 70% of nearby stars, making them the most likely hosts of habitable planets (Henry *et al.* 1994); 2) the relative ease with which rocky planets may be discovered orbiting them, and the high occurrence rate of such planets (Cassan *et al.* 2012) at $\sim$1 planet per low mass star with period < 50 days; 3) the tantalizing observability (Turbet *et al.* 2016) of such planets in the future $\sim$40m-class telescopes, following the discovery (Anglada-Escudé *et al.* 2016) of an Earth-like, possibly habitable (Ribas *et al.* 2016) planet in the late-type M dwarf Proxima Centauri, the closest stellar system to the Sun. Exoplanetology demands the systematic and accurate knowledge of the atmospheric parameters, effective temperature T_{eff} and metallicity [Fe/H], of nearby M dwarfs. This knowledge remains wanting in the face of the high importance of these systems, yet there has been progress in their distance determinations and census completeness for the nearby population. Ongoing surveys keep enlarging the samples of probable nearby unstudied M dwarfs (Winters *et al.* 2015). We aim to take advantage of these new candidate lists to provide intermediate resolution spectroscopic data for nearby, southern M dwarfs, contributing to the systematic determination of their T_{eff} and [Fe/H]. Finally, the discovery of an Earthlike and possibly habitable planet in Proxima Centauri highlights the possibility that other nearby M dwarfs harbor equally interesting, potentially directly

† Based on observations collected at the Observatório do Pico dos Dias (OPD), operated by the Laboratório Nacional de Astrofísica, CNPq, Brazil.

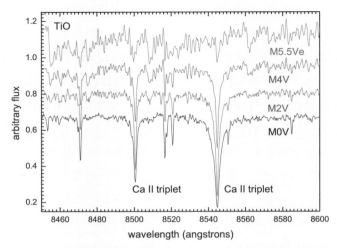

Figure 1. Sample spectra: the TiO λ8455 bandhead and two of the Ca II triplet lines are indicated.

observable planets. We use our new T_{eff} and [Fe/H] data to obtain luminosities and re-assess the dimensions of the habitable zones of nearby M dwarfs, estimating the possible fraction of their potentially habitable rocky planets that might be directly observable by the next generation of >30m telescopes.

2. Observations

Spectra were obtained at the coudé spectrograph of the 1.60m telescope of the Obser-vatório do Pico dos Dias (OPD), operated by the Laboratório Nacional de Astrofíísica (Brazil), and were mostly exposed to S/N > 100 (Fig. 1) at a spectral resolution of 0.85 Å (R = 10,000), the median of the S/N values being ∼150 and reaching out to ∼250 in the best cases. Coverage is λλ8380-8880. A total of 72 stars had spectra of suffi-cient quality, corresponding to 81% of the population of these objects within 10 parsecs, with Hipparcos parallaxes and below $\delta < +30$.

3. Determination of atmospheric parameters

We used VJHK colors and a system of 68 initial spectral line indexes to calibrate T_{eff} and [Fe/H] by means of a PCA analysis. For the T_{eff} calibration we demanded the corre-lation coefficient of the index versus T_{eff} regression to be R ⩾ 0.80 and kept 38 indices; for [Fe/H] we enforced R ⩾ 0.70 and kept 6 indices only. For the T_{eff} calibration 7 PCA components explained 90% of the variance, while 3 PCA components achieved this for [Fe/H]. The final calibrations employed only the two most significant PCA components for each atmospheric parameter. Internal uncertainties respectively for T_{eff} and [Fe/H] are 70 K and 0.11 dex. The T_{eff} values for the calibrating stars come mostly from interfer-ometric, model-free determinations taken from Maldonado *et al.* (2015) and Rojas-Ayala *et al.* (2012); the [Fe/H] values come exclusively from binary systems for which [Fe/H] is known from a spectroscopic model atmosphere analysis of the primary FGK-star (Neves *et al.* 2014). For 18 stars no published either T_{eff} or [Fe/H] was found in the literature, our results being thus the first ever atmospheric parameters provided for these objects.

4. Exoplanet Detectability

The upcoming 39m E-ELT telescope is designed, e.g., with the Planetary Camera and Spectrograph (PCS), to obtain a star-planet contrast in the 10^{-7} to 10^{-8} range at 1 μm. An inner working angle of $\sim 7\lambda/D$ corresponds to 38 *mas* at one parsec (Turbet *et al.* 2016), which is the angular separation of the Proxima Centauri planet for the equivalent irradiance the Earth receives from the Sun, $S_{\text{eff}} \sim 1$. Under the hypothesis that $\sim 5\lambda/D$ achieves direct imaging detection for a rocky, high albedo planet with one Earth-radius (Turbet *et al.* 2016), a working limit of ~ 30 *mas* is inferred. The angular separation, in *mas*, between a potentially habitable putative planet and its host star is given by the star's luminosity in solar units $L/L_\odot$ and its distance d in parsecs by $\alpha_{\text{plan}} = 960$ $[(L/L_\odot)/d]^{-1/2}$. We calculated luminosities for the M dwarf stars in our sample using our derived T_{eff} and [Fe/H], Hipparcos parallaxes and up-to-date (Mann *et al.* 2015) bolometric corrections. Observability is also set by the star's apparent magnitude and distance of the star-planet system, but under optimistic conditions the above formula allows the estimation of the detectability of hypothetical rocky planets sited inside the habitable zone (HZ) of our sample's stars by means of the PCS-E-ELT. For $S_{\text{eff}} \sim 1$ there could be 15 stars in our sample for which a rocky planet would be detectable inside the HZ. Assuming the rocky planets are at the very limit of the theoretical HZ — which is $S_{\text{eff}} \sim 0.3$ (the so called maximum greenhouse limit) (Kasting *et al.* 1993) - would improve detectability due to a larger star-planet angular separation, bringing a total of 22 possible systems of our sample under the direct imaging capabilities of the PCS-E-ELT.

5. Perspectives

Our method can be much improved by a more detailed definition of the spectral indices, isolating the sensitivity of specific groups of transitions to T_{eff} and [Fe/H] and possibly even to surface gravity log g (which we did not determine here). We also plan to extend the calibration to include more colors, such as *griz*, (V-R) and (V-I) colors (Winters *et al.* 2015) as well as WISE and Gaia colors (Mann *et al.* 2015). The inclusion of a Bayesian approach with a large quantity of colors and indices has been shown to improve considerably the T_{eff} and [Fe/H] determinations (Lorenzo-Oliveira 2016). More observations are planned at OPD: the instrumentation has been improved, with a new CCD and better efficiency at the coudé spectrograph, with which it is possible to extend the sample at least to V $\sim$ 12. Potentially all southern M dwarfs down to type $\sim$M3V-M4V, and within 10 parsecs, are accessible to the OPD spectrograph for reasonable integration times.

References

Anglada-Escudé, G., Amado, P. J., Barnes J. *et al.* 2016, *Nature*, 536, 437
Cassan, A., Kubas, D., Beaulieu, J.-P., *et al.* 2012, *Nature*, 481, 167
Henry, T. J., Kirkpatrick, J. D., & Simmons, D. A. 1994, *AJ*, 108, 1437
Kasting, J. F., Whitmire, D. P., & Reynolds, R. T. 1993, *Icarus*, 101, 108
Lorenzo-Oliveira, D. L. 2016, *PhD thesis*, Observatório do Valongo, Universidade Federal do Rio de Janeiro
Maldonado, J., Affer, L., Micela, G. *et al.* 2015, *A&A*, 577, A312
Mann, A. W., Feiden, G. A., Gaidos, E. *et al.* 2015, *ApJ*, 804, 64
Neves, V., Bonfils, X., Santos, N. C. *et al.* 2014, *A&A*, 568, A121
Ribas, I,, Bolmont, E., Selsis F. *et al.* 2016, *arXiv:1608:06813*,
Rojas-Ayala, B., Covey, K. R., Muirhead P. S. *et al.* 2012, *ApJ*, 748, 93
Turbet, M., Leconte, J., Selsis F. *et al.* 2016, *arXiv:1608:06827*,
Winters, J. G., Henry, T. J., Lurie, J. C. *et al.* 2015, *AJ*, 149, 5

Author index

Abrevaya, X. C. – 338
Airapetian, V. S. – 315
Alves, L. R. – 224
Amado, P. J. – 46

Baliunas, S. – 329
Balmaceda, L. A. – 149, 159, 233
Barbosa, M. J. – 224
Barrado, D. – 124, 363
Beaudoin, P. – 1
Berni, L. A. – 221, 224
Birkby, J. – 124
Boardman, S. – 85
Boas, J. V. V. – 224
Bolzam, M. J. A. – 230
Braga, C. R. – 130, 218
Brain, D. A. – 211
Branco, R. H. F. – 224
Browning, M. K. – 85
Brun, A. S. – 1, 77
Buccino, A. P. – 180
Busá, I. – 38

Caballero, J. A. – 46
Camacho, F. J. – 117
Cardoso, F. – 224
Carlesso, F. – 221, 224
Castilho, B. V. – 224
Castro, M. – 338
Charbonneau, P. – 1
Chicrala, A. – 127
Clarke, J. – 85
Connerney, J. E. – 211
Costa, J. E. R. – 127
Cruz, P. – 124, 363
Cruz, W. – 134
Curry, S. M. – 211

da Silva Rockenbach, M. – 127
Dallaqua, R. S. – 127
D'Angelo, C. V. – 192
de Castro, G. G. – 120
de Gouveia Dal Pino, E. M. – 61
de Mello, G. F. P. – 338, 371
de Mendonça, R. R. S. – 130, 218
Denig, W. – 93
de Oliveira e Silva, A. J. – 137
de Souza Echer, M. P. – 298
de Souza, A. M. – 230
Diaz, M. – 124, 363
do Nascimento, J.-D. – 338
Dong, C. F. – 211

Dong, Y. – 211
dos Santos, L. A. – 274

Echer, E. – 130, 218, 227, 230, 233, 298
Egeland, R. – 329
Emeriau-Viard, C. – 77
Emery, B. A. – 93
Eparvier, F. – 211
Espley, J. – 211
Esquivel, A. – 192
Estrela, R. – 152

Flamenco, E. – 180
Folsom, C. – 338
Fortney, J. – 363
Fränz, M. – 230
Freitas, F. C. – 274

Gibson, S. E. – 93
Giribaldi, R. E. – 371
Gómez, J. M. R. – 127, 149, 159, 224
Gonzalez, W. – 227
Gonzalez, W. D. – 233
Gregory, S. G. – 252
Guarnieri, F. – 224
Guerrero, G. – 30, 61, 117
Guinan, E. – 338
Gusmão, E. A. – 140

Halekas, J. – 211
Hall, J. C. – 329
Hara, T. – 211
Henry, G. W. – 329
Hewins, I. M. – 93
Hill, C. A. – 54, 101
Hodgkin, S. – 363

Jakosky, B. M. – 211
Jardine, M. – 162
Jeffers, S. V. – 338
Johnstone, C. P. – 168
Jonathan, J. B. – 363
Jouve, L. – 12

Kaufmann, P. – 134
Klopf, M. – 134
Korhonen, H. – 198
Kosovichev, A. G. – 30, 61, 117
Kovács, O. E. – 198
Kumar, R. – 12

Lago, A. D. – 127, 130, 149, 159, 218, 221, 224, 233
Landin, N. R. – 30, 146
Leblanc, F. – 211
Lee, C. O. – 211
Leitzinger, M. – 198
Leme, N. M. P. – 371
Li, Y. – 211
Lillis, R. – 211
Lillo-Box, J. – 363
Llama, J. – 356
Lopes, R. – 227
López-Morales, M. – 363
Lorenzo-Oliveira, D. – 371
Luhmann, J. G. – 204, 211

Ma, Y. J. – 211
MacKinnon, A. – 120
Maehara, H. – 22
Magalhães, F. P. – 227
Mansour, N. N. – 61, 117
Marsden, S. C. – 338
Martens, P. C. – 350
Matijevič, G. – 143
Mauas, P. J. D. – 180
McFadden, R. H. – 93
McIntosh, P. S. – 93
Meibom, S. – 338
Meléndez, J. – 274
Mendes, L. T. S. – 146
Mendes, O. – 224
Mendoza-Torres, J. E. – 240
Menezes, F. – 113
Miranda, E. – 221
Morgenthaler, J. P. – 227
Morin, J. – 338
Munakata, K. – 130

Nelson, N. J. – 264
Netto, Y. – 107
Nishimori, M. – 221, 224

Odert, P. – 198
Oliveira, A. S. – 140
Osten, R. A. – 243

Pacini, A. A. – 298
Palacios, J. – 127, 149, 159
Paula, A. S. – 224
Payne, C. – 264
Pérez-León, J. E. – 240
Petit, P. – 338
Pillitteri, I. – 290
Poppenhaeger, K. – 290, 308
Portugal, W. – 298

Pugsley, S. – 85

Quirrenbach, A. – 46

Raju, K. P. – 110
Rathbun, J. – 227
Raulin, J.-P. – 134
Reiners, A. – 46
Ribas, I. – 46, 338
Riley, P. – 204
Rockenbach, M. – 130, 233
Russell, C. T. – 204, 211

Sampaio, M. – 224
Santos, J. – 224
Savonov, G. S. – 221, 224
Schneiter, M. – 192
Schuch, N. J. – 130
Seifert, W. – 46
Selhorst, C. L. – 137, 140, 305
Shkolnik, E. L. – 356
Silva, D. – 301
Silva, L. A. – 224
Silva, M. R. – 224
Smolarkiewicz, P. K. – 30, 61, 117
Soon, W. – 329
Sorensen, C. M. – 264
Souza-Echer, M. P. – 227
Stekel, T. – 149, 159
Stekel, T. R. C. – 127
Strugarek, A. – 1
Szpigel, S. – 120, 134

Thiemann, E. – 211
Townsend, E. – 85
Tuneu, J. – 120

Valio, A. – 69, 107, 113, 152, 301, 305
Vida, K. – 198
Vidotto, A. A. – 237
Vidotto, Jr. A. A. – 338
Vieira, L. E. A. – 127, 149, 159, 221, 224, 233

Webb, D. – 93
Weber, M. A. – 85
Withers, P. – 211
Wolk, S. J. – 243, 290

Yu, L. – 282

Zaire, B. – 30, 61
Zechmeister, M. – 46
Žerjal, M. – 143
Zwitter, T. – 143

Printed in the United States
by Baker & Taylor Publisher Services

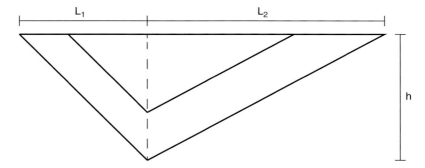

Figure 8.A.1 Geometrically similar triangles.

Starting with the width, $W = L_1 + L_2$, of the cross-section

$$\frac{W}{h} = \alpha_1 + \alpha_2. \tag{8.A.2}$$

The cross-sectional area is

$$A = \tfrac{1}{2}hW. \tag{8.A.3}$$

Eliminating h between Eqs. (8.A.2) and (8.A.3) results in

$$W = \beta_1 \sqrt{A}, \tag{8.A.4}$$

with

$$\beta_1 = \sqrt{2(\alpha_1 + \alpha_2)}. \tag{8.A.5}$$

The hydraulic radius R is the ratio of the cross-sectional area A and the wetted perimeter P,

$$R = \frac{A}{P}. \tag{8.A.6}$$

In terms of the ratios α_1 and α_2 the cross-sectional area is

$$A = \tfrac{1}{2}(\alpha_1 + \alpha_2)h^2, \tag{8.A.7}$$

and the wetted perimeter is

$$P = h\left(\sqrt{\alpha_1^2 + 1} + \sqrt{\alpha_2^2 + 1}\right). \tag{8.A.8}$$

The expression for the hydraulic radius then is

$$R = \frac{\tfrac{1}{2}h(\alpha_1 + \alpha_2)}{\left(\sqrt{\alpha_1^2 + 1} + \sqrt{\alpha_2^2 + 1}\right)}. \tag{8.A.9}$$

Eliminating W between Eqs. (8.A.2) and (8.A.3) results in

$$h = \sqrt{\frac{2}{(\alpha_1 + \alpha_2)}} \sqrt{A}. \tag{8.A.10}$$

Substituting for h from Eq. (8.A.10) in Eq. (8.A.9), it follows that

$$R = \beta_2 \sqrt{A}, \tag{8.A.11}$$

with

$$\beta_2 = \frac{\frac{1}{\sqrt{2}} \sqrt{\alpha_1 + \alpha_2}}{\left(\sqrt{\alpha_1^2 + 1} + \sqrt{\alpha_2^2 + 1} \right)}. \tag{8.A.12}$$

The coefficients β_1 and β_2 are referred to as *shape factors*.

8.B Linear Stability Analysis

When removed from equilibrium, the rate of change of cross-sectional area is given by Eq. (8.7) in the main text, i.e.,

$$\frac{dA}{dt} = \frac{M'}{L_e} \left[\left(\frac{\hat{u}}{\hat{u}_{eq}} \right)^n - 1 \right]. \tag{8.B.1}$$

The right-hand side of this equation is linearized by defining

$$\hat{u} = \hat{u}_{eq} + \Delta\hat{u}. \tag{8.B.2}$$

From Eq. (8.B.2) it follows that

$$\left(\frac{\hat{u}}{\hat{u}_{eq}} \right)^n = \left(1 + \frac{\Delta\hat{u}}{\hat{u}_{eq}} \right)^n. \tag{8.B.3}$$

For small deviations from equilibrium, i.e., $\Delta\hat{u}/\hat{u}_{eq} \ll 1$,

$$\left(1 + \frac{\Delta\hat{u}}{\hat{u}_{eq}} \right)^n \cong 1 - n + n\frac{\hat{u}}{\hat{u}_{eq}}, \tag{8.B.4}$$

and thus

$$\left(\frac{\hat{u}}{\hat{u}_{eq}} \right)^n - 1 \cong \frac{n(\hat{u} - \hat{u}_{eq})}{\hat{u}_{eq}}. \tag{8.B.5}$$

Furthermore,

$$(\hat{u} - \hat{u}_{eq}) \cong \frac{\partial\hat{u}}{\partial A}(A - A_{eq}), \tag{8.B.6}$$

where $\partial \hat{u} / \partial A$ is evaluated at the equilibrium, $A = A_{eq}$. Using Eqs. (8.B.5) and (8.B.6), Eq. (8.B.1) is written as

$$\frac{d(A - A_{eq})}{dt} = \lambda (A - A_{eq}), \tag{8.B.7}$$

with

$$\lambda = \frac{nM}{\hat{u}_{eq} L} \frac{\partial \hat{u}}{\partial A}. \tag{8.B.8}$$

From Eq. (8.B.7)

$$(A - A_{eq}) = (A - A_{eq})_0 e^{\lambda t}. \tag{8.B.9}$$

In Eq. (8.B.9), $(A - A_{eq})_0$ is the initial perturbation. The perturbation increases for λ is positive and decreases for λ is negative. It follows from Eq. (8.B.8) that the sign of λ is determined by the sign of $\partial \hat{u} / \partial A$, evaluated at $A = A_{eq}$. Referring to the Escoffier Diagram, Fig. 8.1 in the main text, it follows that with $\partial \hat{u} / \partial A$ positive, the equilibrium $A = A_{eq_1}$ is unstable and with $\partial \hat{u} / \partial A$ negative, the equilibrium $A = A_{eq_2}$ is stable.

9

Cross-Sectional Stability of a Double Inlet System, Assuming a Uniformly Varying Basin Water Level

9.1 Introduction

Instead of one inlet, many back-barrier lagoons, bays and inland seas are connected to the ocean by multiple inlets. Examples are the Ría Formosa in south Portugal (Salles et al., 2005), the Dutch, German and Danish Wadden Sea (Ehlers, 1988) and the Venice Lagoon in Italy (Tambroni and Seminara, 2006); see Fig. 1.1. This chapter concerns the interaction of these inlets, with emphasis on cross-sectional stability.

Depending on the hydraulic efficiency, inlets connecting the same back-barrier lagoon capture a smaller or larger part of the tidal prism. The tidal prism is the volume of water entering and leaving the back-barrier lagoon during a tidal cycle. The fraction of the tidal prism entering and leaving an individual inlet is the prime parameter determining the cross-sectional stability of that inlet. If this fraction is too small, the inlet closes. In this respect the opening of a new inlet is of interest. Potentially, the new inlet could lead to a decrease of the tidal prisms of the already existing inlet(s) and chances are that some of these inlets close. In that case, it has to be decided to either close the new inlet or leave it open.

An example of a recently opened inlet is the Breach at Old Inlet on Fire Island, NY (Fig. 2.1). The inlet was opened in October 2012 during hurricane Sandy. Together with the already existing Fire Island Inlet, the Breach at Old Inlet connects Great South Bay to the ocean. To determine the behavior of the new inlet and its effect on Fire Island Inlet, both inlets are being monitored (National Park Service, 2012). As of March 2015, the Breach at Old Inlet was still open. Based on the monitoring results it might be possible to arrive at a rational decision to either close the inlet or leave it open.

In this chapter, as a first step to determine the cross-sectional stability of tidal inlets connecting the same back-barrier lagoon to the ocean, the method used to determine the cross-sectional stability of a single inlet system (Chapter 8) is expanded and applied to a double inlet system.

9.2 Escoffier Stability Model for a Double Inlet System

The method to evaluate the cross-sectional stability of a double inlet system is an extension of the Escoffier Stability Model for a single inlet presented in Section 8.2.1. In the following, the various steps in the stability analysis are described using the Texel-Vlie inlet system as an example (Brouwer, 2013; Brouwer et al., 2008; van de Kreeke et al., 2008). This double inlet system is part of the Dutch Wadden Sea and is shown in Fig. 9.1.

9.2.1 Schematization

For the purposes of illustrating the stability analysis, the Texel-Vlie inlet system is schematized to a basin with two inlets (Fig. 9.2). The schematized system is symmetric with parameter values presented in Table 9.1. Similar to the single inlet, inlets are prismatic with diverging parts on both ends (Fig. 6.A.1).

Parameter values are the same for both inlets. Assumed is a triangular cross-section with a surface width to maximum depth ratio of 115, resulting in a shape

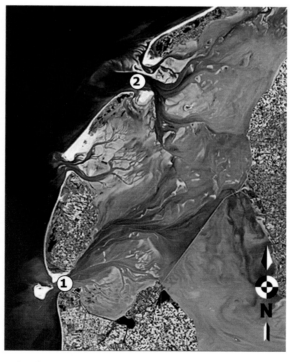

Figure 9.1 Texel-Vlie inlet system; (1) is Texel Inlet and (2) is Vlie Inlet (USGS and ESA, 2011).

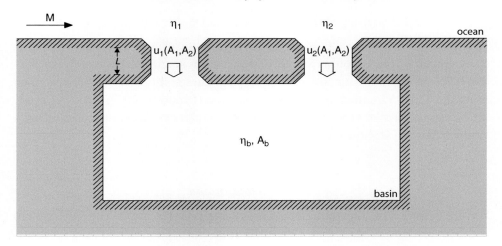

Figure 9.2 Schematization of Texel-Vlie inlet system.

Table 9.1 *Parameter values for the schematized Texel-Vlie inlet system.*

Parameter	Symbol	Dimension	Value
Inlet length	L	m	5,000
Friction factor	F	[-]	4×10^{-3}
Entrance/exit loss coefficient	m	[-]	0
Shape factor	β_2	[-]	0.065
Basin surface area	A_b	m^2	14×10^8
Tidal amplitude	$\hat{\eta}_0$	m	0.8
Tidal frequency	σ	rad s^{-1}	1.4×10^{-4}
Equilibrium velocity	$\hat{u}_{eq}$	m s^{-1}	1.0

factor $\beta_2 = 0.065$ (Appendix 8.A). The ocean tidal amplitude pertains to mean tide conditions.

9.2.2 Equilibrium Velocity

Similar to Eq. (8.1) for a single inlet system, the condition for equilibrium of a double inlet system is

$$\hat{u}_j(A_1, A_2) = \hat{u}_{eq}, \qquad j = 1, 2. \tag{9.1}$$

In this equation $j = 1$ refers to the Texel Inlet and $j = 2$ refers to the Vlie Inlet. The velocity amplitude $\hat{u}_j$ is a function of the cross-sectional areas, A, of both inlets. In analogy with the closure curve $\hat{u}(A)$ for the single inlet system (Fig. 8.1), the surfaces $\hat{u}_j(A_1, A_2)$ are referred to as closure surfaces. The velocity $\hat{u}_{eq}$ is the equilibrium velocity defined in Section 5.2.4. Depending on the cross-sectional area – tidal prism relationship, values of $\hat{u}_{eq}$ follow from Eqs. (5.12) or (5.13). Taking Eq. (5.13) as an example, the equilibrium velocity is

$$\hat{u}_{eq} = \frac{\pi}{C_l T}, \tag{9.2}$$

where T is tidal period and C_l is the coefficient in the cross-sectional area – tidal prism relationship given by Eq. (5.2). For the inlets of the Dutch Wadden Sea with P in Eq. (5.2) referring to mean tide conditions, $C_l = 6.8 \times 10^{-5}$ m^{-1} (Section 5.2.3). From Eq. (9.2), with $T = 44,712$ s, it follows that $\hat{u}_{eq} = 1.03$ m s^{-1}.

9.2.3 Governing Equations

The momentum equation for the individual inlets is Eq. (6.1). Assuming evolving cross-sections remain geometrically similar, the hydraulic radius, R, in this equation is taken proportional to the square root of the cross-sectional area resulting in

$$\frac{L_j}{g} \frac{du_j}{dt} + \left(\frac{m}{2g} + \frac{FL}{g\beta_2\sqrt{A_j}} \right) u_j |u_j| = \eta_0 - \eta_b, \qquad j = 1, 2. \tag{9.3}$$

Assuming a uniformly fluctuating basin water level, the continuity equation is

$$A_1 u_1 + A_2 u_2 = A_b \frac{d\eta_b}{dt}. \tag{9.4}$$

In these equations, L is the length of the inlet, g is gravity acceleration, u is the cross-sectionally averaged velocity, t is time, m is an entrance/exit loss coefficient, F is a bottom friction factor, β_2 is a shape factor (Appendix 8.A), A_b is the basin surface area, η_0 is the ocean tide and η_b is the basin tide. In Eqs. (9.3) and (9.4), variations in cross-sectional area with tidal stage are neglected. The ocean tide is simple harmonic with the same amplitude and phase for both inlets, i.e.,

$$\eta_0 = \hat{\eta}_0 \sin \sigma t. \tag{9.5}$$

Using the terminology introduced in Chapter 1, the system of equations constitutes a process-based exploratory model.

9.2.4 Closure Surface

Given the cross-sectional areas and the parameter values for the schematized Texel-Vlie inlet system in Table 9.1, the governing equations are solved for the basin water level and inlet velocities. For this a finite difference solution can be used as described in (Brouwer, 2006). In the present application a more efficient semi-analytical solution, described in Appendix 2A of (Brouwer, 2013), is used. The semi-analytical solution leads to inlet velocities that are sinusoidal.

Using the results of the semi-analytical solution, the closure surfaces, $\hat{u}_1(A_1, A_2)$ for Inlet 1 and $\hat{u}_2(A_1, A_2)$ for Inlet 2 are presented in, respectively, Figs. 9.3a and 9.3b. Referring to Fig. 9.3a, for a constant value of A_2, the variation in $\hat{u}_1$ with A_1 resembles the closure curve in the Escoffier Diagram presented in Fig. 8.1. For a constant value of A_1, $\hat{u}_1$ monotonically decreases with increasing values of A_2. Similarly referring to Fig. 9.3b, keeping A_1 constant, the variation in $\hat{u}_2$ with A_2 resembles the closure curve in the Escoffier Diagram. For constant values of A_2, $\hat{u}_2$ monotonically decreases with increasing values of A_1. The solid black line in Fig. 9.3a is the intersection of the plane $\hat{u}_1 = \hat{u}_{eq}$ with the closure surface of Inlet 1. Similarly, the solid black line in Fig. 9.3b is the intersection of the plane $\hat{u}_2 = \hat{u}_{eq}$ and the closure surface of Inlet 2.

9.2.5 Equilibrium Velocity Curves

The equilibrium velocity curve for Inlet 1 is the projection of the solid black line in Fig. 9.3a in the (A_1, A_2)-plane. It is the locus of (A_1, A_2)-values for

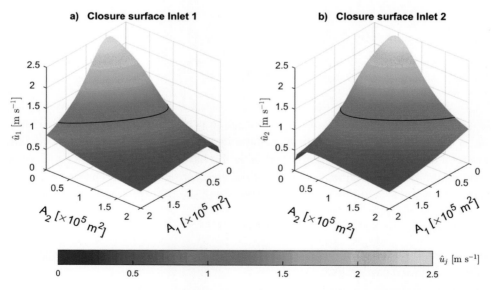

Figure 9.3 Closure surface for a) Inlet 1 (Texel Inlet) and b) Inlet 2 (Vlie Inlet).

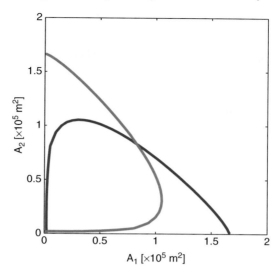

Figure 9.4 Equilibrium velocity curves for Texel Inlet (blue) and Vlie Inlet (red). (From: Brouwer, 2013)

which $\hat{u}_1 = \hat{u}_{eq}$. Similarly, projecting the solid black line in Fig. 9.3b in the (A_1, A_2)-plane results in the equilibrium velocity curve for Inlet 2, representing the locus of (A_1, A_2)-values for which $\hat{u}_2 = \hat{u}_{eq}$. Both equilibrium velocity curves are presented in Fig. 9.4. Because the double inlet system is symmetric, the equilibrium velocity curves for the two inlets are symmetric with respect to the line $A_1 = A_2$. The two intersections of the equilibrium velocity curves represent combinations of (A_1, A_2) for which the velocity amplitudes in both inlets equal the equilibrium velocity and both inlets are in equilibrium with the hydrodynamic environment. The two sets of equilibrium cross-sectional areas are $(A_1 A_2) \cong (600 \text{ m}^2, 600 \text{ m}^2)$ and $(A_1 A_2) \cong (8,100 \text{ m}^2, 8,100 \text{ m}^2)$.

9.2.6 Flow Diagram

The flow diagram for the double inlet system is the counterpart of the Escoffier Diagram for the single inlet system. A flow diagram consists of the two equilibrium velocity curves together with a vector plot. The vectors represent the rate of change of the cross-sectional areas after a perturbation and are defined as

$$\frac{d\vec{A}}{dt} = \frac{dA_1}{dt}\vec{e}_1 + \frac{dA_2}{dt}\vec{e}_2, \tag{9.6}$$

where $\vec{e}_1$ and $\vec{e}_2$ are the unit vectors in, respectively, the A_1 and A_2 direction. Referring to Eq. (8.7), the rate of change of the cross-sectional areas is related to the ratio of the velocity amplitudes $\hat{u}_j$ and the equilibrium velocity $\hat{u}_{eq}$ by

$$\frac{dA_j}{dt} = \frac{M'}{L_{e_j}} \left[\left(\frac{\hat{u}_j}{\hat{u}_{eq}} \right)^n - 1 \right], \qquad j = 1, 2, \qquad (9.7)$$

where L_{e_j} is the length of the entrance section of Inlet j. The entrance section is the seaward part of the inlet where, during a storm event, an excess volume of sand is deposited. M' is the fraction of the long-term average longshore sand transport entering the inlet on the flood and leaving on the ebb. The direction of the vectors in the flow diagram shows whether, after a perturbation, the system returns to the original equilibrium or moves away from it. The direction of the vectors thus determines whether the equilibrium is stable or unstable. Because it is only the direction and not the magnitude that is of interest, vectors are given a unit length. In the present example values of M' and L_{e_j} are the same for both inlets. Therefore, the ratio M'/L_{e_i} does not play a role in the direction of the vectors.

The flow diagram of the schematized Texel-Vlie inlet system, presented in Fig. 9.5, shows two equilibriums. Details in the vicinity of the two equilibriums are presented in Figs. 9.6a and 9.6b. For the equilibrium in Fig. 9.6a holds that, regardless of the direction of the perturbation, the cross-sectional areas move away from the original equilibrium; the equilibrium is unstable. For the equilibrium in Fig. 9.6b, a distinction is made between perturbations whereby both cross-sectional areas increase or decrease and perturbations whereby one cross-sectional

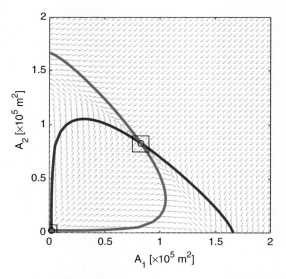

Figure 9.5 Flow diagram for the schematized Texel-Vlie inlet system. Blue is the equilibrium velocity curve of the Texel Inlet and red is the equilibrium velocity curve of the Vlie Inlet. Circles denote a set of equilibrium cross-sectional areas.

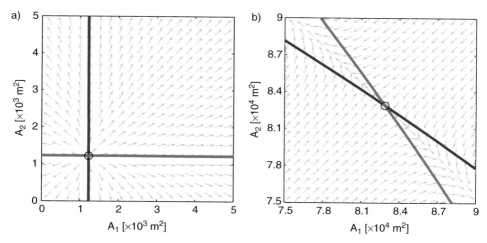

Figure 9.6 Zoomed-in flow diagram for the Texel-Vlie inlet system around equilibriums a) $(A_1, A_2) \cong (600\,\mathrm{m}^2, 600\,\mathrm{m}^2)$ and b) $(A_1, A_2) \cong (8{,}100\,\mathrm{m}^2,$ $8{,}100\,\mathrm{m}^2)$. Blue is the equilibrium velocity curve of the Texel Inlet and red is the equilibrium velocity curve of the Vlie Inlet.

area increases and the other decreases. When both cross-sectional areas increase or decrease, they return to the original equilibrium. When the perturbations are in opposite direction, cross-sectional areas move further away from the original equilibrium. Defining a stable equilibrium as one where, regardless of the direction of the perturbation, the cross-sectional areas return to the original equilibrium, the equilibrium in Fig. 9.6b is unstable.

In summary, the Escoffier Stability Model, with the hydrodynamics described by Eqs. (9.3) and (9.4) and the same sinusoidal ocean tide at both inlets given by Eq. (9.5), does not lead to a set of stable equilibrium areas for the Texel-Vlie inlet system.

9.3 Conditions for a Set of Stable Cross-Sectional Areas

In addition to the Texel-Vlie inlet system, the Escoffier Stability Model described in the preceding sections has been applied to the Pass Cavallo-Matagorda inlet system in Texas (van de Kreeke, 1985) and the Big Marco-Capri inlet system in Florida (van de Kreeke, 1990b). Similar to the Texel-Vlie inlet system, the equilibrium velocity curves of these double inlet systems show two intersections, representing unstable equilibriums. The absence of stable equilibriums for the Texel-Vlie, Pass Cavallo-Matagorda and Big Marco-Capri inlet systems when applying the Escoffier Stability Model is somewhat surprising, given the fact that these double inlet systems have existed for long periods of time and are considered stable.

The lack of stable equilibriums when using the Escoffier Stability Model for a double inlet system was first investigated in van de Kreeke (1990a). Similar to the preceding examples, he considered a symmetric, friction-dominated double inlet system, forced with equal ocean tides at both inlets. Based on the results of this study, he postulated four possible configurations for the two equilibrium velocity curves. Three of the configurations showed two intersections of the equilibrium velocity curves, none of them representing stable equilibriums. The fourth configuration showed four intersections, three of them representing unstable equilibriums and the fourth representing a stable equilibrium. From this it was concluded that for a stable equilibrium to exist, the equilibrium velocity curves must have four intersections. To investigate if Eqs. (9.3) and (9.4) with the boundary condition, Eq. (9.5), lead to four intersections, Eq. (9.3) was simplified by neglecting the inertia and entrance/exit loss term. As shown in Section 6.3.1, these terms are usually small compared to the bottom friction term. Using an analytical solution it was then shown that the reduced set of equations at best leads to two intersections, excluding the presence of a stable equilibrium.

Using the Escoffier Stability Model, with Eqs. (9.3) and (9.4) describing the hydrodynamics but with ocean tides that have different amplitudes for the two inlets, Brouwer et al. (2012) investigated the cross-sectional stability of the Faro-Armona double inlet system in Portugal. The equilibrium velocity curves showed four intersections, representing three unstable and one stable equilibrium. The Faro and Armona Inlets are somewhat exceptional in that the inlets are relatively short and deep. As a result, the hydrodynamic balance is between pressure gradient and entrance/exit losses rather than pressure gradient and bottom friction. How this affects the equilibrium and stability of the inlets is further discussed in Brouwer et al. (2012).

A relatively simple explanation why the Escoffier Stability Model does not lead to stable cross-sectional areas is given in Roos et al. (2013). The Escoffier Stability Model includes the assumption of a uniformly fluctuating basin level and the same ocean tide at both inlets. This implies that at all times the pressure gradients over the two inlets are the same. As shown in Roos et al. (2013) this condition prevents stable equilibrium cross-sectional areas. To explain this, they distinguish between a destabilizing and stabilizing mechanism. The first one is associated with the bottom friction in the inlets and the second one with the pressure gradients over the inlets. The way these mechanisms operate is demonstrated for a perturbation whereby the cross-sectional area of Inlet 1 is increased and the cross-sectional area of Inlet 2 is decreased. Increasing the cross-sectional area of Inlet 1 leads to a decrease in bottom friction in Inlet 1 and decreasing the cross-sectional area of Inlet 2 leads to an increase in bottom friction in Inlet 2 (see Eq. (9.3)). To explain the destabilizing mechanism, it is first assumed that after the perturbation the pressure gradients over the inlets remain the same. Assuming a friction-dominated system, the decrease in bottom friction leads to an increase in velocity and erosion in Inlet 1. Similarly, the increase in bottom friction leads to a decrease in velocity and accretion in Inlet 2. As a result, both inlets move away from equilibrium. To explain a possible return to equilibrium, the assumption of the pressure gradients over the inlets remaining the same after the perturbation is removed. For Inlet 1 to return to equilibrium, the pressure gradient needs to decrease, resulting in a decrease in velocity, counteracting the increase in velocity resulting from the decrease in bottom friction. Similarly for Inlet 2, the pressure gradient needs to increase, resulting in an increase in velocity, counteracting the decrease in velocity resulting from the increase in bottom friction. Decreasing the pressure gradient for Inlet 1 and at the same time increasing the pressure gradient over Inlet 2 is not possible when the system is identically forced at the ocean side and at the same time the water level in the basin is forced to fluctuate uniformly. As a consequence, the Escoffier Stability Model as described above does not lead to stable double inlet systems.

It follows that a necessary, but not necessarily sufficient, condition for a set of stable cross-sectional areas is that the formulation of the hydrodynamics allows for different pressure gradients across the inlets. In practice this implies allowing for spatial variation in basin water level. This excludes the assumption of a uniformly fluctuating basin water level.

9.4 Basin with Topographic High

9.4.1 Schematization

One way to introduce spatial variations in the basin water level is by dividing the basin into two sub-basins separated by a barrier (van de Kreeke et al., 2008; de Swart and Volp, 2012). To allow for exchange, the barrier has an opening in the form of a tidal inlet. Each sub-basin is connected to the ocean by an inlet. The ocean tides off the inlets are the same. The barrier may be viewed as a topographic high which in nature could be a tidal flat. It is there where the tides entering the inlets meet, resulting in relatively low velocities and persistent sedimentation. The tidal flats, depending on extent and height, allow a certain degree of exchange between the sub-basins. For little exchange, corresponding to a small opening in the barrier, the sub-basins act independently and the double inlet system reverts to two single inlet systems. For a large opening in the barrier, the sub-basins act as a single basin with the water levels the same in both sub-basins and fluctuating uniformly.

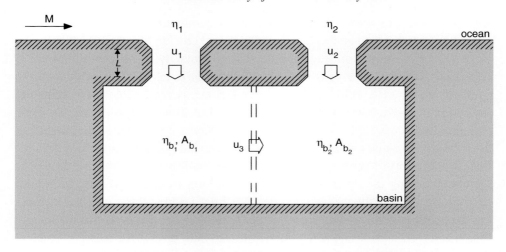

Figure 9.7 Schematization of a double inlet system with topographic high.

The tidal inlet schematization when including a topographic high is presented in Fig. 9.7. The schematization is based on the Texel-Vie inlet system. Parameter values for this system are given in Table 9.1. The topographic high divides the basin in two sub-basins with equal surface areas of 7×10^8 m^2. Assumed is a rectangular opening over the topographic high. Using Google Earth images of the Wadden Sea, the opening over the topographic high is given a width of b = 25,000 m.

9.4.2 Governing Equations

The governing equations are those used in van de Kreeke et al. (2008). The momentum equation for the flow in the inlets is Eq. (9.3). Because the inlets are relatively long, entrance/exit losses are small compared to bottom friction losses and are neglected. This result in the momentum equation

$$\frac{L_j}{g}\frac{du_j}{dt} + \frac{F_j L_j}{g\beta_2\sqrt{A_j}}u_j|u_j| = \eta_0 - \eta_{b_j}, \qquad j = 1, 2, \tag{9.8}$$

with subscript j referring to the inlets and the sub-basins. The momentum equation for the opening over the topographic high is similar to that for the inlets and reads

$$\frac{L_3}{g}\frac{du_3}{dt} + \frac{F_3 L_3 b}{g A_3}u_3|u_3| = \eta_{b_1} - \eta_{b_2}. \tag{9.9}$$

Subscript 3 refers to the opening over the topographic high. With the opening relatively wide, the hydraulic radius to a good approximation is A_3/b.

Water levels in the sub-basins are assumed to fluctuate uniformly, resulting in the continuity conditions

$$A_{b_1} \frac{\eta_{b_1}}{dt} = A_1 u_1 - A_3 u_3, \tag{9.10}$$

and

$$A_{b_2} \frac{\eta_{b_2}}{dt} = A_2 u_2 + A_3 u_3. \tag{9.11}$$

Water levels prescribed at the ocean entrance of the inlets are the same, and are given by Eq. (9.5). The equations can be solved with a finite difference method (Brouwer, 2006) or the semi-analytical method described in Appendix 2A of Brouwer (2013) and used in the present application. The semi-analytical solution leads to inlet velocities that are sinusoidal.

In Herman (2007), an exploratory model based on a similar set of equations as Eqs. (9.8)–(9.11) is used to describe the hydrodynamics of a tidal inlet system consisting of three inlets and three basins. The system of equations is solved using a finite difference method. For the same tidal inlet, a process-based simulation model is deployed, in which the hydrodynamics is described by the shallow water wave equations, eliminating the assumption of uniformly fluctuating basin water levels. The shallow water wave equations are solved numerically using the modeling system Delft3D (Lesser et al., 2004). Comparison of the results obtained from the exploratory and simulation model suggests that the exploratory model based on Eqs. (9.8)–(9.11) is capable of reproducing the water transport between basins and between basins and ocean.

9.4.3 Flow Diagrams

Using Eqs. (9.8)–(9.11) and an equilibrium velocity of 1 m s^{-1}, flow diagrams for the double inlet system are calculated for openings in the barrier with cross-sectional areas $A_3 = 25,000$ m^2, $A_3 = 50,000$ m^2 and $A_3 = 75,000$ m^2. For the relatively large opening of 75,000 m^2, the flow diagram in Fig. 9.8a shows two unstable and no stable equilibriums. The sub-basins act as one with water levels fluctuating uniformly. As a result, the pressure gradients across both inlets are always the same, thereby excluding sets of stable equilibrium cross-sectional areas. The flow diagram resembles that presented in Fig. 9.5 for a basin without a topographic high.

For the relatively small opening of 25,000 m^2, the flow diagram in Fig. 9.8c shows three unstable and one stable equilibrium. The cross-sectional areas of the stable equilibrium are $A_1 = A_2 = 83,000$ m^2. The stability of this equilibrium is explained by introducing a perturbation, whereby A_1 is increased and A_2 is decreased by 10,000 m^2, resulting in $A_1 = 93,000$ m^2 and $A_2 = 73,000$ m^2.

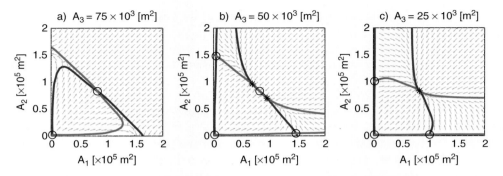

Figure 9.8 Flow diagrams for the schematized Texel-Vlie inlet system for an opening over the topographic high of a) 75,000 m², b) 50,000 m² and c) 25,000 m². Blue is the equilibrium velocity curve of the Texel Inlet and red is the equilibrium velocity curve of the Vlie Inlet. Circles are unstable sets and stars are stable sets of cross-sectional areas. For parameter values reference is made to Table 9.1.

Calculating the water level differences across the two inlets at times of maximum flow velocity for $(A_1, A_2) = (83,000 \text{ m}^2; 83,000 \text{ m}^2)$, results in $|\eta_0 - \eta_{b_1}| = |\eta_0 - \eta_{b_2}| = 0.071$ m. Carrying out the same calculations for the perturbed system with $(A_1, A_2) = (93,000 \text{ m}^2; 73,000 \text{ m}^2)$ results in $|\eta_0 - \eta_{b_1}| = 0.069$ m and $|\eta_0 - \eta_{b_2}| = 0.073$ m. The perturbation causes the water level difference across Inlet 1 to decrease and across Inlet 2 to increase. The decrease in the water level difference across Inlet 1 leads to a decrease in the maximum velocity from $\hat{u}_1 = \hat{u}_{eq} = 1.03$ m s^{-1} to $\hat{u}_1 = 0.99$ m s^{-1}. The increase in water level difference across Inlet 2 leads to an increase in the maximum velocity from $\hat{u}_2 = \hat{u}_{eq} = 1.03$ m s^{-1} to $\hat{u}_2 = 1.07$ m s^{-1}. As a result, Inlet 1 erodes and Inlet 2 accretes and both inlets return to the original equilibrium. The foregoing is in agreement with the statement at the end of Section 9.3 that, a necessary condition to have stable inlets is that the formulation of the hydrodynamics allows for different pressure gradients across the two inlets.

For very small openings the double inlet system approaches two single inlet systems. Assuming a complete separation of the two sub-basins and using the data in Table 9.1, the stability analysis for a single inlet system described in Chapter 8 results in stable cross-sectional areas, $A_1 = A_2 = 72,000 \text{ m}^2$. This is slightly smaller than the stable cross-sectional areas of 83,000 m² with the opening of 25,000 m², i.e., for this opening there is still some interaction between the sub-basins.

The flow diagram for the mid-size opening of 50,000 m² is presented in Fig. 9.8b. There are four unstable and two stable equilibriums. Although not investigated in detail, as done for the small opening, it seems reasonable to assume that also for the midsize opening pressure gradients across the two inlets can be different for the two inlets, resulting in stable equilibriums. Why there is more than one set of stable equilibriums is not clear and needs further investigation.

Contrary to the equations used by de Swart and Volp (2012), Eqs. (9.8)–(9.11) do not account for hypsometric effects. Hypsometric effects refer to the dependence of the inlet cross-sectional areas and back-barrier lagoon surface areas on the tidal stage. As shown in de Swart and Volp (2012), the conclusions when including hypsometric effects remain qualitatively the same, i.e., depending on the size of the opening over the topographic high there can be stable equilibriums. The hypsometry does affect the size of the equilibrium cross-sectional areas.

10

Cross-Sectional Stability of a Double Inlet System, Assuming a Spatially Varying Basin Water Level

10.1 Introduction

For inlets where the dynamic balance is between bottom friction and pressure gradient, stability requires spatial variations in basin water level and/or different ocean tides (Section 9.3). In a primitive way, for the basin water level this was demonstrated by introducing a topographic high. The topographic high divides the basin into two sub-basins with different uniformly fluctuating water levels. In the present chapter, instead of a topographic high, spatial variations in water level are introduced describing the hydrodynamics by the shallow water wave equations. Using these equations, the cross-sectional stability of a double inlet system is investigated. As in Chapter 9, the double inlet system resembles the Texel-Vlie inlet system but with a slightly different schematization. In the various experiments, special attention is given to the effect of basin depth, basin geometry, Coriolis acceleration and radiation damping on the spatial variation in basin water level and the cross-sectional stability. Cross-sectional stability of the inlets is determined using the Escoffier Stability Model for a double inlet system described in Section 9.2.

10.2 Schematization

The double inlet system consists of four rectangular compartments of length L_j, width B_j and uniform depth h_j ($j = 0$–3); see Fig. 10.1. Compartment 0 represents the ocean with an open boundary at $x = -L_0$. Compartments 1 and 2 are the inlets. Inlets have equal lengths and a rectangular cross-section. Compartment 3 is the tidal basin. The double inlet system, consisting of Compartments 1, 2 and 3, is symmetrically aligned with respect to the central axis of the ocean compartment. The two inlets are at equal distance, $\Delta y / 2$, from the central axis.

10.3 Governing Equations and Boundary Conditions

In each compartment, conservation of momentum and mass is expressed by the depth-averaged shallow water wave equations, including the Coriolis acceleration.

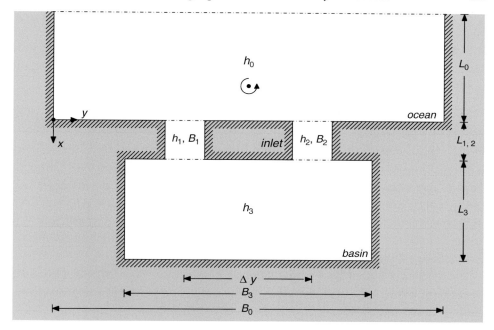

Figure 10.1 Schematized Texel-Vlie inlet system. Subscript 1 refers to Texel Inlet, subscript 2 refers to Vlie Inlet and subscript 3 refers to the basin. Furthermore, h is mean water depth and B is compartment width.

To allow for a semi-analytical solution of the equations, the advective terms are neglected and the bottom friction term is linearized. The mean water depth in all compartments is assumed to be large compared to the water level amplitude. With these constraints, the shallow water wave equations are

$$\frac{\partial u_j}{\partial t} - f v_j + \frac{r_j u_j}{h_j} = -g \frac{\partial \eta_j}{\partial x}, \tag{10.1}$$

$$\frac{\partial v_j}{\partial t} + f u_j + \frac{r_j v_j}{h_j} = -g \frac{\partial \eta_j}{\partial y}, \tag{10.2}$$

$$\frac{\partial \eta_j}{\partial t} + h_j \left[\frac{\partial u_j}{\partial x} + \frac{\partial v_j}{\partial y} \right] = 0, \tag{10.3}$$

where subscript j refers to Compartment j, with $j = 0\text{--}3$. In Eqs. (10.1)–(10.3), $u_j(x, y, t)$ and $v_j(x, y, t)$ are the depth-averaged velocities in, respectively, the x- and y-direction, $\eta_j(x, y, t)$ is the free surface elevation with respect to mean sea level, g is the gravity acceleration, h_j is the mean water depth and f is the Coriolis parameter. For the coordinate system reference is made to Fig. 10.1. The Coriolis parameter is given by

$$f = 2\Omega \sin\varphi, \tag{10.4}$$

with $\Omega = 7.292 \times 10^{-5}$ rad s^{-1} is the angular frequency of the Earth's rotation and φ is latitude. The coefficient r_j is a linear bottom friction coefficient. Using the Lorentz linearization (Lorentz, 1926; Zimmerman, 1982),

$$r_j = \frac{8FU_j}{3\pi}. \tag{10.5}$$

In Eq. (10.5), F is the nonlinear bottom friction coefficient and U_j is the velocity scale of Compartment j, defined as the velocity amplitude averaged over Compartment j. To ensure that the velocity scale agrees with the solution, the linear friction coefficients are obtained using an iterative procedure. For details, see Appendix A.4 in Brouwer et al. (2013). Even though the nonlinear bottom friction coefficient F is assumed to be the same for each compartment, the linear bottom friction coefficient differs due to the different velocity scales.

The system is forced by a single damped Kelvin wave with angular frequency σ, entering through the open boundary of the ocean compartment and propagating in the positive x-direction

$$\eta(x, y, t) = \Re \left\{ Z e^{-\frac{y}{\gamma L_r}} e^{i(\sigma t - \gamma k_0 x)} \right\}, \tag{10.6}$$

with L_r the Rossby radius of deformation for the ocean compartment defined by

$$L_r = \frac{\sqrt{g h_0}}{f}, \tag{10.7}$$

γ is a complex friction factor for the ocean compartment defined by

$$\gamma = \sqrt{1 - \frac{i r_0}{\sigma h_0}} \tag{10.8}$$

and k_0 is the wave number without bottom friction for the ocean compartment

$$k_0 = \frac{\sigma}{\sqrt{g h_0}}. \tag{10.9}$$

Z is the amplitude of the wave at $(x, y) = (-L_0, 0)$.

For the special case with the Coriolis acceleration equal to zero and introducing

$$i\gamma k_0 = \mu + ik, \tag{10.10}$$

Eq. (10.6) can be written as

$$\eta(x, y, t) = Z e^{-\mu x} \cos(\omega t - kx). \tag{10.11}$$

Eq. (10.11) represents a damped shallow water wave traveling in the positive x-direction, where μ is a damping factor and k is the wave number when accounting for bottom friction. Given k_0, σ, h_0 and r_0, μ and k follow from Eq. (10.10) (Ippen, 1966; van de Kreeke, 1998).

Using the terminology introduced in Chapter 1, the system of Eqs. (10.1), (10.2) and (10.3) with the ocean forcing given by Eq. (10.6) constitutes a process-based exploratory model. In applying the model, a water level amplitude is prescribed halfway between the two inlets. For the situation without inlets, the corresponding amplitude Z of the incoming Kelvin wave is then determined by trial and error. Due to the Coriolis acceleration, the Kelvin wave travels along the coast past the two inlets, causing the water levels off the inlets to be different. The Kelvin wave, along with other waves generated within the model domain, leaves the ocean compartment without reflecting at the open boundary.

At the closed boundaries, a no-normal flow condition,

$$u = 0 \quad \text{and} \quad v = 0, \tag{10.12}$$

is imposed. Continuity of elevation and normal flux is required across the interfaces between the ocean and inlet compartments, formulated as

$$\eta_0 = \eta_1 \quad \text{and} \quad \eta_0 = \eta_2, \tag{10.13}$$

and

$$h_0 u_0 = h_1 u_1 \quad \text{and} \quad h_0 u_0 = h_2 u_2. \tag{10.14}$$

Similarly, for the interfaces between basin and inlets holds

$$\eta_3 = \eta_1 \quad \text{and} \quad \eta_3 = \eta_2, \tag{10.15}$$

and

$$h_3 u_3 = h_1 u_1 \quad \text{and} \quad h_3 u_3 = h_2 u_2. \tag{10.16}$$

10.4 Solution Method

Following Brouwer et al. (2013), in each compartment the solution for η_j, u_j and v_j is written as the truncated sum of analytical wave solutions in an infinite channel. This involves incoming and reflected Kelvin waves as well as Poincaré waves generated at closed boundaries. The unknown amplitudes in the solution are determined from the matching conditions at the interfaces, Eqs. (10.13)–(10.16), and the no-normal flow condition, Eq. (10.12). For this a collocation method is used (Boyd, 2001; Roos and Schuttelaars, 2011), i.e., a large number of discrete points are defined along the interfaces and closed boundaries and the amplitude of the Kelvin and Poincaré waves are calculated such that the matching and boundary conditions in these points are satisfied. The solution leads to sinusoidal inlet velocities.

The process-based exploratory model is computationally efficient, which makes it possible to carry out large numbers of simulations in a short period of time. This makes it eminently suitable to calculate inlet velocities for a large number of combinations of cross-sectional areas (A_1, A_2), needed to construct the flow diagrams discussed in Section 9.2.6.

10.5 Effect of Spatial Variations in Basin Water Level on Cross-Sectional Stability

10.5.1 Spatial Variations in Basin Water Level

In the following a qualitative description of the effects of basin depth, Coriolis acceleration, radiation damping and basin geometry on the spatial variations of water levels is presented (Brouwer et al., 2013).

Basin depth. Finite basin depth affects the tidal propagation and dissipation. The tidal wave is short compared to the same wave in deep water, resulting in differences in surface elevation over relatively short distances.

Coriolis acceleration. The Coriolis acceleration leads to asymmetry and thus spatial variation in the basin surface elevation. In addition, the Coriolis acceleration allows prescribing the ocean forcing by a Kelvin wave traveling along the coastal boundary, resulting in different ocean tides off the inlets.

Radiation damping. In earlier studies, the forcing of the system has been prescribed by water levels, usually in the form of a series of tidal components. Formulating the boundary conditions in this way does not account for the effect of the inlets on the forcing. This effect is included by prescribing the boundary forcing by an incoming wave. The resulting water level variations off the inlets induce oscillating flow in each inlet which triggers co-oscillations in the basin. In turn, these co-oscillations result in waves radiating away into the ocean, influencing the surface elevation off the inlets. This mechanism is referred to as radiation damping. The effect of radiation damping on the surface elevations off the inlets decreases with decreasing values of the ratio of ocean and inlet depth (Buchwald, 1971).

Basin geometry. Short and wide basins, as opposed to long and narrow basins, make a difference in the spatial variation of basin water level.

The effect of basin water depth, Coriolis acceleration, radiation damping and basin geometry on the basin level and inlet stability is further investigated using the hydrodynamics model described in Sections 10.3 and 10.4 and the Escoffier Stability Model introduced in Section 9.2. In the experiments the water level amplitude described halfway between the inlets is $Z_{char} = 0.8$ m.

10.5.2 Comparison with Earlier Stability Analysis

As a first step, the process-based exploratory model is used to evaluate the stability of the schematized Texel-Vlie inlet system described in Section 9.2. Parameter values are presented in Table 10.1 and are close to the same as those presented in Table 9.1. To compare with the stability analysis of the Texel-Vlie inlet system in Section 9.2, the basin depth is given a large value $h_3 = 1,000$ m, resulting in a close to uniformly fluctuating basin water level. In the stability analysis the inlet cross-sections are assumed to be geometrically similar, resulting in $h_j = 0.07\sqrt{A_j}$ with $j = 1,2$ (Appendix 8.A), with subscript 1 referring to the Texel Inlet and subscript 2 referring to the Vlie Inlet. To minimize radiation damping, the ocean compartment is given a large depth, $h_0 = 1,000$ m. The basin dimensions are 30×40 km. With the Coriolis parameter equal to zero, the Rossby radius of deformation

Table 10.1 *Parameter values for schematized Texel-Vlie inlet system.*

Parameter	Symbol	Dimension	Value
Ocean compartment (0)[a]			
Length	L_0	km	50
Width	B_0	km	100
Inlet compartment (1)[b]			
Length	L_1	km	6
Inlet compartment (2)[b]			
Length	L_2	km	6
Basin compartment (3)[a]			
Length	L_3	km	30
Width	B_3	km	40
General			
Distance between inlets	Δy	km	10
Ocean water level amplitude	$\hat{\eta}_0$	m	0.8
Tidal wave frequency (M$_2$)	σ	rad s^{-1}	1.4×10^{-4}
Bottom friction factor	F	–	2.5×10^{-3}
Equilibrium velocity	$\hat{u}_{eq}$	m s^{-1}	1.0
Shape factor[b]	β_2	–	0.07

[a]Compartment depth varies depending on experiment.
[b]Inlet cross-sections are rectangular with a width to depth ratio of 200, resulting in a shape factor with a value of 0.07 (Appendix 8.A).

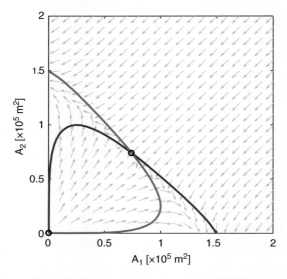

Figure 10.2 Flow diagram for the schematized Texel-Vlie inlet system; basin depth $h_3 = 1,000$ m, ocean depth $h_0 = 1,000$ m, basin dimensions 30×40 km, Coriolis parameter $f = 0$. Circles represent sets of unstable cross-sectional areas. Blue is equilibrium velocity curve for Inlet 1 and red is equilibrium velocity curve for Inlet 2. (From: Brouwer et al., 2013)

L_r approaches infinity and the prescribed Kelvin wave reduces to a damped shallow water wave traveling perpendicular to the coast. This results in the same tides off both inlets. The resulting flow diagram is presented in Fig. 10.2. The same as in Fig. 9.6, the flow diagram shows two unstable and no stable equilibriums. Minor differences are attributed to differences in the shape factor and the equilibrium velocity; compare parameter values in Tables 9.1 and 10.1.

10.5.3 *Effects of Basin Depth, Coriolis Acceleration, Radiation Damping and Basin Geometry*

To demonstrate the effects of basin depth, Coriolis acceleration, radiation damping and basin geometry on the stability of a double inlet system, four experiments are carried out (see Fig. 10.3). The schematized Texel-Vlie inlet system with the parameter values presented in Table 10.1 is used as the inlet configuration.

Basin Depth

To investigate the effect of basin depth on cross-sectional stability, the basin is given a shallow depth $h_3 = 5$ m. With the Coriolis parameter $f = 0$, the incoming wave is a damped shallow water wave propagating perpendicular to

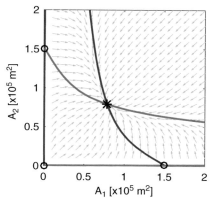

(a) **Effect of basin depth:** Basin depth $h_3 = 5$ m, ocean depth $h_0 = 1,000$ m, basin dimensions 30×40 km, Coriolis parameter $f = 0$.

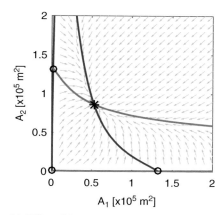

(b) **Effect of Coriolis acceleration:** Basin depth $h_3 = 5$ m, ocean depth $h_0 = 1000$ m, basin dimensions 30×40 km, Coriolis parameter $f = 1.164 \times 10^{-4}$ rad s^{-1}.

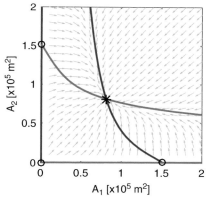

(c) **Effect of radiation damping:** Basin depth $h_3 = 5$ m, ocean depth $h_0 = 20$ m, basin dimensions 30×40 km, Coriolis parameter $f = 0$.

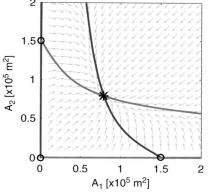

(d) **Effect of basin geometry:** Basin depth $h_3 = 5$ m, ocean depth $h_0 = 1000$ m, basin dimensions 60×20 km, Coriolis parameter $f = 0$.

Figure 10.3 Flow diagrams for the schematized Texel-Vlie inlet system displaying the effect of a) basin depth, b) Coriolis acceleration, c) radiation damping and d) basin geometry. Circles are unstable sets and stars are stable sets of cross-sectional areas. Blue is equilibrium velocity curve for Inlet 1 and red is equilibrium velocity curve for Inlet 2.

the coast. This results in equal tides at the inlets. The water depth in the ocean compartment is $h_0 = 1,000$ m. This is relatively large compared to the depth of the inlets, thus limiting the effect of radiation damping on the offshore water levels. The basin dimensions are 30×40 km. The flow diagram, presented in Fig. 10.3a, shows four equilibriums, one of which is stable and the other three are unstable.

Coriolis Acceleration

To investigate the effect of the Coriolis acceleration on the cross-sectional stability, the Coriolis parameter is given a value $f = 1.164 \times 10^{-4}$ rad s^{-1}, corresponding to the latitude of the Texel-Vlie inlet system. Other parameters remain the same as in the experiment with the finite basin depth, i.e., $h_0 = 1,000$ m and $h_3 = 5$ m. The basin dimensions are 30×40 km. As a result of the Coriolis acceleration, the incoming wave is a Kelvin wave traveling along the coast, resulting in different amplitudes and phases of the ocean tides at the inlets. In addition to the water levels off the inlets, the Coriolis acceleration affects the basin water levels, leading to asymmetry in the spatial pattern of the basin tidal amplitudes. The flow diagram when including the Coriolis acceleration is presented in Fig. 10.3b. The flow diagram differs little from that presented in Fig. 10.3a. This suggests that the effect of the Coriolis acceleration on the cross-sectional stability is of secondary importance compared to that of basin depth. As a result of the Coriolis acceleration, the problem becomes asymmetric and the stable cross-sectional areas differ slightly for the two inlets.

Radiation Damping

The effect of radiation damping on cross-sectional stability is investigated by decreasing the depth of the ocean compartment to $h_0 = 20$ m, thereby increasing the effect of radiation damping. The basin depth remains $h_3 = 5$ m. The basin dimensions are 30×40 km. The Coriolis parameter $f = 0$ and the incoming wave is a damped shallow water wave traveling perpendicular to the coast, resulting in the same tides off both inlets. The flow diagram is presented in Fig. 10.3c and shows three unstable and one stable equilibrium. The equilibrium cross-sectional areas differ little from those in Fig. 10.3a, showing the marginal effect of radiation damping on the cross-sectional stability compared to that of reducing the basin depth.

Basin Geometry

To demonstrate the effect of basin geometry on cross-sectional stability, the basin is given a length $L_3 = 60$ km and a width $B_3 = 20$ km (as opposed to 30×40 km in the previous experiments). The ocean depth $h_0 = 1,000$ m, the Coriolis parameter $f = 0$ and the basin depth $h_3 = 5$ m. The incoming wave travels perpendicular to the coast. This results in the same tides off both inlets. The flow diagram, presented in Fig. 10.3d, is qualitatively the same as that in Fig. 10.3a with four equilibriums, one of which is stable and the other three are unstable. However, the stable cross-sectional areas are now considerably smaller.

In summary, allowing for a spatially varying basin water level by introducing a finite basin depth results in three sets of unstable and one set of stable equilibrium cross-sectional areas. The same result is found when, in addition to finite basin depth, Coriolis acceleration, radiation damping and different basin geometry is included. Accounting for Coriolis acceleration and radiation damping slightly affects the size of the stable equilibrium cross-sectional areas. Changes in basin geometry significantly affect the size of the stable equilibrium cross-sectional areas.

10.6 Multiple Inlets

In a recent study, the stability analysis for the double inlet system was expanded to include multiple inlets (Roos et al., 2013). In search for multiple stable inlets, their simulations start with a large number of open inlets along a barrier island coast. During the simulations some inlets close, while others remain open competing for the remaining tidal prism. The final result is a barrier island coast with a number of stable inlets. It was found that for a given length of barrier island coast there is a maximum number of stable inlets. Furthermore, simulations revealed that the number of stable inlets increases with increasing values of tidal range and basin surface area and decreasing values of longshore sand transport.

11

Morphodynamic Modeling of Tidal Inlets Using a Process-Based Simulation Model

11.1 Introduction

Morphodynamic models come in two categories. Based on their architecture a distinction is made between process-based models and empirical models. In the present chapter the focus is on process-based modeling. Process-based models start with small-scale physics and integrate the results over the larger timescales. Because at this time the state of the art of process-based modeling limits the time period over which can be integrated, most model applications focus on problems with timescales of months to decades. Examples are the adaptation of the inlet cross-section after a storm, the formation of an ebb delta after the opening of a new inlet, migration and breaching of channels and spit formation (Nahon et al., 2012; Tung et al., 2012; van der Wegen et al., 2010). To address problems with larger timescales, parallel to the process-based models, empirical models have been developed. They are the subject of Chapter 12. A common characteristic of process-based and empirical morphodynamic models is the feedback between morphology (bathymetry) and hydrodynamics; the hydrodynamics causes a change in bathymetry and in turn this affects the hydrodynamics.

11.2 Model Concept and Formulation

Process-based morphodynamic models consist of a series of computational modules as shown in Fig. 11.1. Starting with a known bathymetry and water level boundary conditions, the hydrodynamic equations (Lesser et al., 2004) are solved in the Flow module. Given the wave boundary conditions, wave transformation including wave height and direction is calculated in the Waves module. To account for tidal stage and wave-current interaction, information on water levels and current velocities is transferred from the Flow to the Waves module. Radiation stresses are calculated and the results are transferred to the Flow module. Exchange between the two modules takes place at specified time intervals.

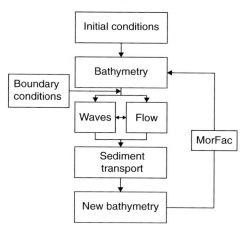

Figure 11.1 Example of an elementary feedback loop.

In general this interval is much larger than the computational time step used in either of the modules. Using the information on currents and waves from the Flow and Waves modules, sediment transport is calculated in the Sediment Transport module using selected sediment transport formulae. With the results of the sand transport calculations the bathymetry is updated in the Bathymetry module.

The computer time for one complete cycle varies with the application and is measured in seconds to minutes. Even though short, for calculations extending over a large time period, e.g., one year, this would still lead to lengthy computer times. Computer time is reduced by taking advantage of the difference in timescales of the hydrodynamic and morphodynamic processes. For this, the depth changes after one cycle are multiplied by a morphological factor MorFac (Roelvink and Reniers, 2010). As an example, assuming MorFac is 30 and the computer time of one complete cycle is one minute, the change in bathymetry over a period of 30 minutes is obtained, while wave, current and transport fields only have to be calculated once during that one-minute period. The wave height and direction, and current and transport fields, only have to be calculated once during that 30-minute period. Selecting the value of MorFac is not trivial, as too large a value can lead to the wrong morphology.

As examples of process-based morphodynamic modeling, two sets of experiments presented in Tung et al. (2012) are discussed. The purpose of the first set of experiments is to simulate the development of the inlet morphology after the opening of a new inlet. The purpose of the second set of experiments is to investigate the cross-sectional area – tidal prism relationship for a set of geologically and hydrodynamically similar inlets.

11.3 Morphology of a Newly Opened Inlet

In the calculations, the depth-averaged version of the Delft3D online modeling system is used (Lesser et al., 2004). Following Murray (2003), this system can be characterized as a process-based simulation model. In the present application, in the Flow module the water motion is described by the shallow water wave equations, excluding the Coriolis term. The Waves module uses the Swan wave model (Booij et al., 1999). At intervals of 30 minutes, information on water levels and current velocities is transferred from the Flow to the Waves module. In the Sediment Transport module the sediment transport formulation of van Rijn (1993) is used, which includes both tide- and wave-induced sediment transport. The formulation accounts for bottom sediment transport and suspended load transport. The model includes dry bank and flat erosion. Carrying out experiments in which the morphological factor MorFac was varied systematically, it was found that for values of 50 and less the model results are unaffected by the upscaling. A value of 40 was selected for the numerical experiments in this study. The horizontal eddy viscosity coefficient and eddy diffusion coefficient are $0.1 \ \mathrm{m^2 \ s^{-1}}$. The Chézy friction coefficient is the same for the entire computational domain and equal to $65 \ \mathrm{m^{1/2} \ s^{-1}}$. Bottom sediment consists of a single fraction of non-cohesive sand with a density of $2{,}650 \ \mathrm{kg \ m^{-3}}$ and a median grain diameter of $250 \ \mu\mathrm{m}$.

The idealized tidal inlet system consists of a rectangular basin connected to the ocean by an inlet. Initially, the tidal basin and inlet have a uniform depth; the cross-section of the inlet is trapezoidal. Inside the surf zone the bathymetry has a concave equilibrium profile (Dean, 1991). Outside the surf zone a gentle profile with a slope 1:200 to a water depth of SWL -13 m (still water level) is used. The elevation of the barrier islands on both sides of the inlet channel is set at SWL $+3$ m. To allow for an unrestricted widening and/or migration of the inlet channel the barrier islands are defined as erodible barriers.

The computational grid is rectangular with a resolution of 30 m in the inlet region, gradually increasing to 200 m in the offshore and basin area. A simple harmonic tide is prescribed at the ocean boundary, located 3 km offshore. The lateral boundaries are open, non-reflective Neumann boundaries where a zero alongshore water level gradient is prescribed (Roelvink and Walstra, 2004). Waves in the form of a JONSWAP spectrum are prescribed at the offshore boundary. The spectrum is characterized by a significant wave height, peak period and mean wave direction.

To study the development of the morphology of a newly opened inlet, experiments were carried out with two different forcings: one with tide only and the other with tides and waves. Parameter values used in the experiments are listed in Table 11.1. With waves approaching the shore at an oblique angle, the major difference between the two experiments is the absence of longshore sand transport for tide only. For the experiment with waves, the calculated longshore sand

Table 11.1 *Morphology of a newly opened inlet; parameter values used in the experiments.*

	Parameter	Value
Basin	Surface area A_b	15 km^2
	Initial depth H_0	2 m
Inlet	Length L	500 m
	Initial depth H_0	2 m
	Shape cross-section	trapezoidal
Tide	Amplitude $\hat{\eta}_0$	0.5 m
	Period T	12 hours
Waves	Significant wave height H_s	1.5 m
	Peak frequency f_p	7 s
	Direction from north ϑ	25°
Sand	Median diameter d_{50}	250 μm
	Density ρ_s	2650 kg m^{-3}
MorFac	MF	40

transports is 0.5×10^6 m^3 year^{-1}. Except for the waves and the initial cross-sectional area of the inlet (850 m^2 for the experiment with tide only and 250 m^2 for the experiment with waves and tide), parameter values are the same for both experiments. The main interest here is to determine whether the morphology approaches equilibrium and, if so, what the equilibrium conditions are. To highlight the difference in response, results of the two experiments are presented in parallel.

The initial bathymetry and the bathymetry after 10,000 days for tide only and 800 days for waves and tide are presented in Fig. 11.2. The reason for the relatively short simulation time for the experiment with tide and waves is the much faster morphological response in comparison to the experiment with tide only. Comparing Figs. 11.2a and 11.2b, for tide only the bathymetry remains symmetric with the inlet perpendicular to the shore. Comparing Figs. 11.2c and 11.2d, when adding waves, and thus longshore sand transport, the bathymetry becomes asymmetric with the inlet taking on a NE–SW direction. The P/M ratio in the experiment is approximately 50, corresponding to poor to fair location stability (Section 3.5). In agreement with this low value, the inlet has shifted in the downdrift direction.

Changes in the cross-section at the gorge for tide only and tide and waves are presented in Figs. 11.3a and 11.3b, respectively. For tide only, the cross-section maintains its position. The width at mean sea level remains the same, but the depth increases resulting in a change from the original trapezoidal to a V-shaped cross-section. For tide and waves, the cross-section shifts in the downdrift direction. The width at mean sea level increases and so does the depth. Furthermore, there is a tendency to form a dual channel system.

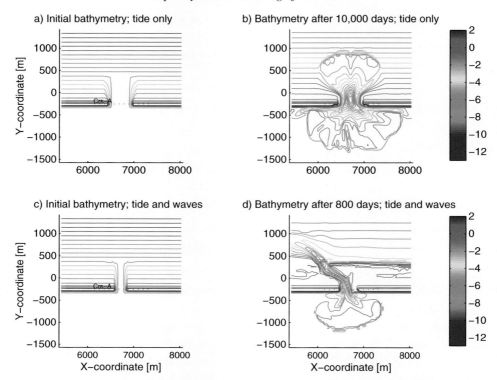

Figure 11.2 a) Initial bathymetry for the experiment with tide only, b) bathymetry for tide only after 10,000 days, c) initial bathymetry for the experiment with tide and waves, and d) bathymetry for tide and waves after 800 days (figures a and b reprinted from Tung et al., 2012, copyright 2016, with permission from Elsevier). For parameter values reference is made to Table 11.1.

Tidal prism P, velocity amplitude $\hat{u}$ and cross-sectional area A as a function of time are presented in Fig. 11.4a for tide only and in Fig. 11.4b for tide and waves. The tidal prism was calculated by integrating the cross-sectionally averaged velocity over the ebb cycle. With the known tidal prism, the velocity amplitude follows from Eq. 5.6. For tide only, the velocity amplitude approaches a value of 0.5 m s^{-1}. This is considerably larger than the critical velocity of erosion of 0.27 m s^{-1} for the sand used in the experiments; the sand bottom constitutes a 'live bed' where sand carried in by the flood is removed during ebb. For tide and waves, the equilibrium velocity is even larger and approaches a value of 0.8 m s^{-1}. This is attributed to the wave-induced longshore sand transport (Section 5.2.4). Similar to the velocity amplitude, the cross-sectional areas for both tide only and tide and waves approach an equilibrium value. The larger equilibrium cross-sectional area corresponds with the smaller equilibrium velocities. This agrees with the Escoffier Diagram in Section 8.2.2.

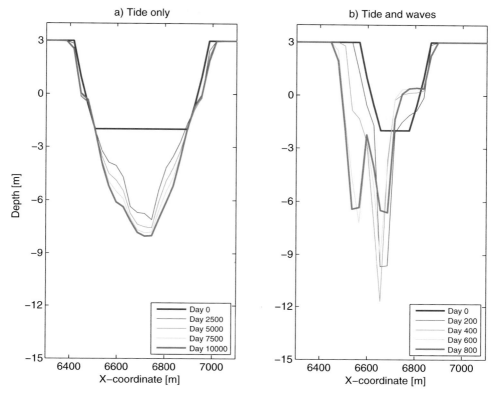

Figure 11.3 Changes in cross-section at the gorge for a) tide only and b) tide and waves. For parameter values and coordinate system, reference is made to Table 11.1 and Fig. 11.2, respectively.

To investigate the dependence of the equilibrium cross-sectional areas and equilibrium velocity on the value of the initial cross-sectional area, numerical experiments were carried out with initial cross-sectional areas that are smaller and larger than the estimated equilibrium values. The velocity amplitudes for initial cross-sectional areas of 250 m^2 and 2,400 m^2 are presented in Fig. 11.5. Regardless of the initial value, the cross-sectional areas approach the same equilibrium value and the same holds for the velocity amplitude. Additional examples are presented in Tung et al. (2012). The cluster of points near the equilibrium is an indication that the equilibrium is dynamic rather than static.

11.4 Cross-Sectional Area – Tidal Prism Relationship

Using the depth-averaged version of the Delft3D online modeling system, the cross-sectional area – tidal prism relationship for inlets at equilibrium is determined

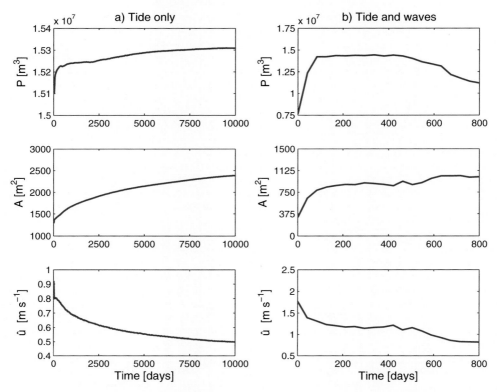

Figure 11.4 Tidal prism, inlet cross-sectional area and velocity amplitude for a) tide only and b) tide and waves. For parameter values reference is made to Table 11.1.

for a set of five geologically and hydrodynamically similar inlets. Experiments were carried out for different values of ocean tidal amplitude, a_0, and basin surface area, A_b, as summarized in Table 11.2. The remaining parameter values are kept constant, i.e., tidal period is 12 hr, offshore significant wave height is 0.7 m, peak wave period is 6 s, wave direction is 25° from the north, median grain diameter is 250 μm, grain density is 2,650 kg m^{-3}, initial depth of the basin and the inlet is 2 m. The equilibrium cross-sectional areas and corresponding velocity amplitudes are calculated using initial cross-sectional areas that are smaller and larger than the estimated equilibrium cross-sectional area. The selected vales are 250 m^2 and 2,400 m^2. The resulting equilibrium cross-sectional areas together with the corresponding tidal prisms are presented in the last two columns of Table 11.2. The longshore sand transport calculated with the model is the same for all experiments and equals 50,000 m^3 year^{-1}.

Table 11.2 *Results of model experiments used to determine the cross-sectional area – tidal prism relationship (Tung et al., 2012).*

Experiment	a_0 [m]	A_b [m^2]	A [m^2]	P [m^3]	$\hat{u}$ [m s^{-1}]
1	0.50	15×10^6	1,380	15.3×10^6	0.81
2	0.75	15×10^6	2,400	22.7×10^6	0.69
3	1.00	15×10^6	3,450	29.8×10^6	0.63
4	0.60	30×10^6	3,560	37.5×10^6	0.77
5	0.30	15×10^6	1,080	9.3×10^6	0.63

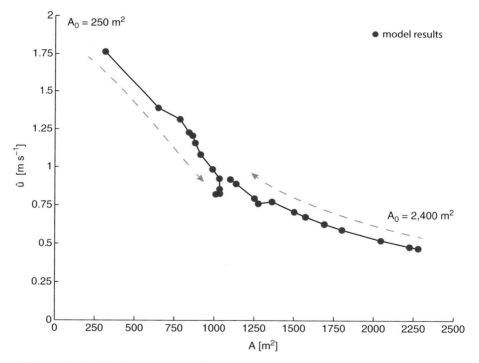

Figure 11.5 Velocity amplitude $\hat{u}$ versus cross-sectional area A for two values of the initial cross-sectional area A_0. For parameter values, reference is made to Table 11.1.

Values of equilibrium cross-sectional areas and tidal prisms are plotted in Fig. 11.6. Using the correlation functions, Eqs. (5.1) and (5.2), a best fit results in, respectively,

$$A = 2.33 \times 10^{-4} P^{0.95} \qquad \text{with} \quad r^2 = 0.96 \tag{11.1}$$

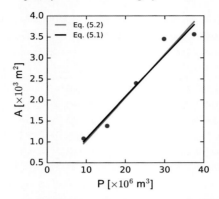

Figure 11.6 The cross-sectional area – tidal prism relationship for five experiments in Table 11.2.

and

$$A = 1.03 \times 10^{-4} P \qquad \text{with} \quad r^2 = 0.94. \qquad (11.2)$$

Similar to the inlets of the Dutch Wadden Sea and the North Island of New Zealand, the equilibrium cross-sectional areas calculated with the morphodynamic model show good correlation with the tidal prisms. It is noted that the coefficient in Eq. (11.2) is considerably larger than for the inlets of the Dutch Wadden Sea and the North Island of New Zealand (Section 5.2). Referring to Eq. (5.7), and with k values approximately the same, this is attributed to the small longshore sand transport of 50,000 m^3 year^{-1} in the morphodynamic model.

11.5 Limitations of Process-Based Morphodynamic Models

The examples presented in Sections 11.3 and 11.4 illustrate the potential of modeling morphological processes in tidal inlets. Although the results look promising, process-based morphodynamic models are still in a developmental state. In particular, there are fundamental problems associated with the upscaling of the small-scale processes to larger space and timescales resulting from (1) limited knowledge of sediment transport processes, (2) the statistical nature of the wave forcing, (3) the propagation of numerical errors when integrating over large timescales (larger than decades) and (4) computer time. The formulation of the sediment transport by tidal currents and waves is still largely empirical. Wave forcing is only known in a statistical sense, whereas indications are that morphological changes are sensitive to the time history of the forcing (Southgate, 1993). With regards to the propagation of numerical errors, there is the problem that small changes in initial conditions can ultimately lead to widely different results (Ridderinkhof and Zimmerman, 1992).

As a result of the strongly varying bathymetry, and to obtain sufficient spatial resolution, the space step in the calculations has to be small. As a consequence, to maintain computational stability, the time step has to be small. For large simulation periods, this leads to lengthy computing time. Fortunately, with the development of faster computers this becomes less and less of a problem.

In view of the aforementioned shortcomings, the general belief is that, at this time, upscaling of small-scale processes to larger timescales can only be done over short periods, after which everything diverges and collapses; indeed, this behavior has been seen in many cases when a model that was calibrated over a period of some years was continued for a much longer period (Roelvink and Reniers, 2010). Recently, a number of improvements have been introduced that allow the process-based morphodynamic models to be run over longer periods and reduce computer time. A key issue in this has been the strategy to bridge the gap between short-term hydrodynamic processes, with timescales of hours to days, and morphological changes, with timescales of months and longer. A number of strategies for this are presented in Roelvink (2006). The most promising seems to be the morphological factor or online approach used in the present application. In spite of progress, it is concluded that at this time process-based morphodynamic models should only be used to diagnose, as opposed to predict, the morphological behavior of inlets. For the use of process-based morphodynamic models in a predictive mode, additional research and development is needed. As part of this research and development effort, process-based morphodynamic models have been applied to schematized tidal inlets. Examples are found in van der Wegen et al. (2010), Nahon et al. (2012), and Tung et al. (2012). In van der Wegen et al. (2010), special attention is given to morphological changes covering timescales of centuries.

12

Morphodynamic Modeling of Tidal Inlets Using an Empirical Model

12.1 Introduction

Process-based models are a valuable tool when the relevant timescales are measured in months to years (Section 11.5). For timescales of decades to centuries, recourse is often taken to empirical models, also referred to as long-term behavior models or aggregate models. Empirical models start with the premise that, after a perturbation, the morphology tends towards an equilibrium state. The equilibrium is defined by equations of state. Examples are the relationship between inlet cross-sectional area and tidal prism (Eqs. (5.1) and (5.2)) and the relationship between ebb delta volume and tidal prism (Eq. (5.14)). Examples of perturbations are changes in the inlet morphology resulting from a storm and changes resulting from such engineering activities as dredging and basin reduction. The objective of empirical modeling is to predict the transition from the old to the new equilibrium.

12.2 Modeling Concepts

In applying empirical modeling to tidal inlets, the morphology is divided into a number of large-scale geomorphic elements, e.g., inlet, ebb delta and flood delta. Each of these elements is viewed on an aggregated scale and characterized by either a sand or a water volume. When the morphology as a whole tends to an equilibrium, so do the individual elements.

Examples of empirical models are presented in Kraus (2000), van de Kreeke (2006) and Stive et al. (1998). Kraus (2000) applied an empirical model to simulate the ebb delta development after the opening of Ocean City Inlet (MD). van de Kreeke (2006) used a similar empirical model to explain the transition from the old to the new equilibrium of the Frisian Inlet (The Netherlands) after basin reduction. In both studies, the sand transport entering and leaving an element is prescribed in terms of the ratio between the actual and equilibrium sand or water volume of the element. The model by Stive et al. (1998) is fundamentally different from the models used by Kraus (2000) and van de Kreeke (2006) in that the sand transport entering and leaving an element is formulated as a diffusive transport.

12.3 Ebb Delta Development at Ocean City Inlet

12.3.1 Ocean City Inlet

Ocean City Inlet (MD) was opened in 1933 by a hurricane. Since that time, the development of the ebb delta has been observed at regular intervals. Starting at the updrift side of the inlet, the ebb delta is divided in three geomorphic features: an ebb tidal shoal, a bypassing bar and an attachment bar; see Fig. 12.1. Sand is transported from the updrift coast to the ebb tidal shoal and from there continues its path via the bypassing bar and attachment bar to the downdrift coast. Little information is available on the hydrodynamic setting of the inlet.

12.3.2 Schematization and Model Formulation

The empirical model, referred to in Kraus (2000) as the reservoir model, consists of three elements, each representing one of the three ebb delta elements. In the model the elements are connected in series. Each element is characterized by a sand volume, V. The elements together with sand transport pathways are presented in Fig. 12.2.

Sand transport entering an element equals the sand transport leaving the neighboring updrift element. Referring to Fig. 12.2, the sand transport entering Element 1 (ebb tidal shoal) is taken as being equal to the longshore sand transport M,

$$S_{e_1} = M, \tag{12.1}$$

where the subscript e stands for entering and the subscript 1 refers to Element 1. The sand transport leaving Element 1 is taken as being proportional to the product of the sand transport entering and the ratio of the actual and equilibrium sand volume

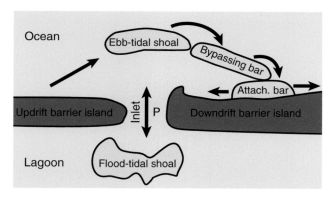

Figure 12.1 Geomorphic elements and sand transport pathways at Ocean City Inlet (MD) (adapted from Kraus, 2000, with permission from ASCE).

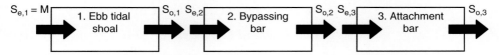

Figure 12.2 Empirical model for Ocean City Inlet; for definition of symbols reference is made to the text.

$$S_{l_1} = M \frac{V_1}{V_{1_{eq}}}, \tag{12.2}$$

where the subscript l stands for leaving, V_1 is the actual sand volume and $V_{1_{eq}}$ is the equilibrium sand volume.

The sand transport entering Element 2 (bypassing bar) equals the sand transport leaving Element 1,

$$S_{e_2} = S_{l_1}. \tag{12.3}$$

The sand transport leaving Element 2 is taken as being equal to the product of the sand transport entering and the ratio of the actual and equilibrium sand volume,

$$S_{l_2} = M \frac{V_1}{V_{1_{eq}}} \frac{V_2}{V_{2_{eq}}}. \tag{12.4}$$

Similarly, the sand transport entering Element 3 (attachment bar) is

$$S_{e_3} = S_{l_2}, \tag{12.5}$$

and the sand transport leaving Element 3 is

$$S_{l_3} = M \frac{V_1}{V_{1_{eq}}} \frac{V_2}{V_{2_{eq}}} \frac{V_3}{V_{3_{eq}}}. \tag{12.6}$$

With the sand transports defined by Eqs. (12.1) and (12.2), the sand conservation equation for Element 1 is

$$\frac{dV_1}{dt} = M \left(1 - \frac{V_1}{V_{1_{eq}}} \right). \tag{12.7}$$

Similarly, using Eqs. (12.3) and (12.4) and Eqs. (12.5) and (12.6), the sand conservation equation for Element 2 and Element 3 are, respectively,

$$\frac{dV_2}{dt} = M \frac{V_1}{V_{1_{eq}}} \left(1 - \frac{V_2}{V_{2_{eq}}} \right), \tag{12.8}$$

and

$$\frac{dV_3}{dt} = M \frac{V_1}{V_{1_{eq}}} \frac{V_2}{V_{2_{eq}}} \left(1 - \frac{V_3}{V_{3_{eq}}} \right). \tag{12.9}$$

12.3.3 Model Results

With $V_1 = V_2 = V_3 = 0$ at $t = 0$, the solutions to Eqs. (12.7), (12.8) and (12.9) are, respectively,

$$V_1 = V_{1_{eq}} \left(1 - e^{-\alpha t}\right), \quad \text{with } \alpha = \frac{M}{V_{1_{eq}}}, \tag{12.10}$$

$$V_2 = V_{2_{eq}} \left(1 - e^{-\beta t'}\right), \quad \text{with } \beta = \frac{M}{V_{2_{eq}}}, \quad \text{and } t' = t - \frac{V_1}{M}, \tag{12.11}$$

$$V_3 = V_{3_{eq}} \left(1 - e^{-\gamma t''}\right), \quad \text{with } \gamma = \frac{M}{V_{3_{eq}}}, \quad \text{and } t'' = t' - \frac{V_2}{M}. \tag{12.12}$$

The reader can verify this by substituting the solutions in Eqs. (12.7)–(12.9). The inverses of the coefficients α, β and γ represent timescales. Here, α^{-1} is the adaptation timescale of the ebb tidal shoal, β^{-1} is the adaptation timescale of the bypassing bar in the absence of the ebb tidal shoal and γ^{-1} is the adaptation timescale of the attachment bar in the absence of the ebb tidal shoal and the bypassing bar. Timescales are independent of those of the downdrift elements, implying that there is no feedback.

For Ocean City Inlet, parameter values presented in Kraus (2000) are $M = 0.15 \times 10^6$ m^3 year^{-1}, $V_{1_{eq}} = 3 \times 10^6$ m^3, $V_{2_{eq}} = 7 \times 10^6$ m^3 and $V_{3_{eq}} = 0.5 \times 10^6$ m^3. It is not clear how the values of the equilibrium volumes were determined. The development of the sand volumes of the different elements as calculated from Eqs. (12.10)–(12.12), together with observations, are plotted in Fig. 12.3. The timescale for the ebb tidal shoal is $\alpha^{-1} = 20$ years, for the bypassing bar is $\beta^{-1} = 46$ years and for the attachment bar is $\gamma^{-1} = 3$ years. The model results and

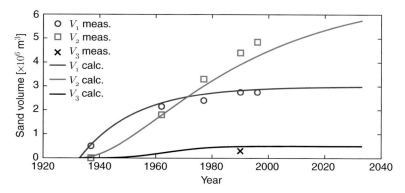

Figure 12.3 Modeled development of ebb tidal shoal, bypassing bar and attachment bar at Ocean City Inlet together with observations; subscript 1 refers to ebb tidal shoal, subscript 2 refers to bypassing bar and subscript 3 refers to attachment bar (adapted from Kraus, 2000, with permission from ASCE).

observations show reasonable agreement. For an in-depth discussion of the results the reader is referred to Kraus (2000).

12.4 Adaptation of the Frisian Inlet after Basin Reduction

12.4.1 Frisian Inlet

The Frisian Inlet is one of the tidal inlets of the Dutch Wadden Sea, located between the island of Schiermonnikoog and the Engelsmanplaat shoal (Fig. 12.4). Major morphological elements are the ebb delta, the inlet (Zoutkamperlaag) and the tidal flats. The water motion in the inlet is governed by the tide, as freshwater inflow is insignificant. Tides are semi-diurnal with an offshore mean tidal range of 2.25 m. The offshore mean annual significant wave height is 1.13 m. The dominant wave direction is from the northwest. Estimates of the gross annual longshore sand transport, M, vary between 0.5×10^5 and 1×10^5 m^3. The sand transport is primarily in a west–east direction.

In 1969, the basin surface area of the inlet was reduced by approximately 30 percent, resulting in a decrease in the tidal prism from 325×10^6 to 225×10^6 m^3. Prior

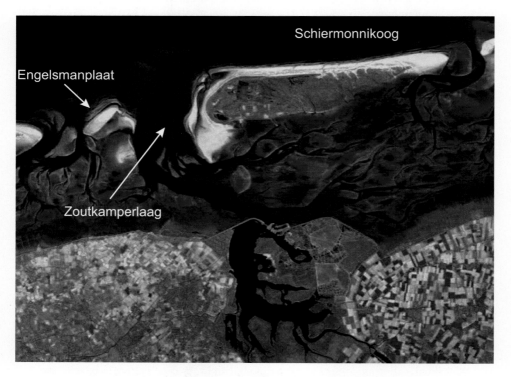

Figure 12.4 Frisian Inlet, The Netherlands (Esri et al., 2016).

to basin reduction, the morphology was in equilibrium with an ebb delta (sand) volume $V_{s_0} = 132 \times 10^6$ m^3 and an inlet (water) volume $V_{w_0} = 171 \times 10^6$ m^3. Following the basin reduction, measurements every four years over an 18-year period showed that by the end of this period the delta volume had decreased by 21×10^6 m^3 and the inlet volume had decreased by 31×10^6 m^3 (Biegel and Hoekstra, 1995). The inlet volume decreased monotonically while the changes in delta volume showed a more random character (see Fig. 12.6 discussed in Section 12.4.3). For example, during the period 1975–1979, instead of a decrease, an increase in delta volume was observed. A possible reason for this random character could be the difficulty in defining the extent of the (moving) delta. During the 18-year period, the tidal prism remained constant at 225×10^6 m^3 and most likely will stay close to this value for the remainder of the adaptation period. Both the observed ebb delta volumes and inlet volumes are used to calibrate the empirical model.

The ebb delta volume and the inlet volume are expected to ultimately reach new equilibrium values. The new equilibrium volume of the ebb delta is estimated using the relationship between ebb delta sand volume and tidal prism for inlets in the Wadden Sea presented in Louters and Gerritsen (1994). With $P = 225 \times 10^6$ m^3, this resulted in an equilibrium sand volume of the delta $V_{s_{eq}} = 83 \times 10^6$ m^3. The equilibrium water volume of the inlet was estimated using the two cross-sectional area – tidal prism relationships for the inlets of the Wadden Sea (Section 5.2.3). Both relationships resulted in an equilibrium cross-sectional area of the gorge of the inlet of approximately $A = 15,300$ m^2. From this, and making certain assumptions for the variation in cross-sectional area over the inlet, the equilibrium water volume of the inlet is estimated at $V_{w_{eq}} = 118 \times 10^6$ m^3 (van de Kreeke, 2006). After the basin reduction, the ebb delta has a surplus and the inlet has a shortage of sand. When the new equilibrium is reached, the delta and inlet combined have gained a volume of sand equal to 4×10^6 m^3. This gain comes at the expense of the downdrift coast.

12.4.2 Schematization and Model Formulation

To model the adaptation of the morphology after basin reduction, the tidal inlet is represented by two of the three major elements, the ebb delta and the inlet. The third element, the tidal flats, is not included as its response time is much larger than that of the ebb delta and inlet (Biegel and Hoekstra, 1995). As shown in Fig. 12.5, elements are connected in series. Contrary to the model for Ocean City Inlet there is feedback, in this case between the ebb delta and the inlet. The ebb delta is characterized by the sand volume V_s and the inlet by the water volume V_w. A fraction M_1 of the longshore sand transport enters the delta and the remaining fraction M_2 enters the inlet. S_1 is the sand transport from the inlet to the delta resulting from

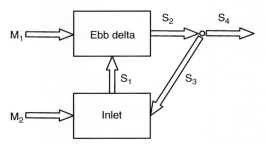

Figure 12.5 Empirical model for the Frisian Inlet; for definition of the symbols reference is made to the text.

the ebb tidal current. S_2 is the wave-induced sand transport leaving the delta in the direction of the downdrift coast. Before reaching the downdrift coast, a part of this sand volume, S_3, is diverted to the inlet. The fraction reaching the downdrift coast is $S_4 = S_2 - S_3$. Because the interest is in timescales ranging from decades to centuries, sand transports are taken as annually averaged transports.

The sand transport from the inlet to the delta is taken as being proportional to the ratio of the equilibrium and the actual water volume of the inlet to a power n:

$$S_1 = \kappa_1 \left(\frac{V_{w_{eq}}}{V_w} \right)^n , \tag{12.13}$$

where $V_{w_{eq}}$ is the equilibrium water volume of the inlet, n is an empirical constant and κ_1 is a proportionality constant. The sand transport from the delta towards the downdrift coast is taken as being proportional to the ratio of the actual and equilibrium sand volume of the delta to a power m:

$$S_2 = \kappa_2 \left(\frac{V_s}{V_{S_{eq}}} \right)^m , \tag{12.14}$$

where $V_{S_{eq}}$ is the equilibrium sand volume of the ebb delta, m is an empirical constant and κ_2 is a proportionality constant. A difference with the reservoir model (Section 12.3) is that the sand transport leaving an element is solely dependent on the morphological state of that element and is independent of the sand transport entering the element.

The transport diverted from the delta to the inlet, S_3, is taken as a fraction α of S_2, i.e.,

$$S_3 = \alpha S_2, \tag{12.15}$$

with $0 \leq \alpha \leq 1$. Through this feedback mechanism, the adaptation of the inlet depends on that of the ebb delta.

The transport to the downdrift coast is

$$S_4 = (1 - \alpha)S_2. \tag{12.16}$$

When at equilibrium, the sand transport leaving equals the sand transport entering an element. For the inlet element,

$$M_2 - S_{1_{eq}} + \alpha S_{2_{eq}} = 0, \tag{12.17}$$

and for the delta element

$$M_1 + S_{1_{eq}} - S_{2_{eq}} = 0, \tag{12.18}$$

with the subscript *eq* referring to equilibrium conditions. Solving for $S_{1_{eq}}$ and $S_{2_{eq}}$ from Eqs. (12.17) and (12.18) results in

$$S_{1_{eq}} = \frac{\alpha M_1 + M_2}{1 - \alpha}, \tag{12.19}$$

and

$$S_{2_{eq}} = \frac{M_1 + M_2}{1 - \alpha}. \tag{12.20}$$

Using Eq. (12.13), it follows that $S_{1_{eq}} = \kappa_1$ and using Eq. (12.14), $S_{2_{eq}} = \kappa_2$. It then follows from Eqs. (12.19) and (12.20) that

$$\kappa_1 = \frac{\alpha M_1 + M_2}{1 - \alpha}, \tag{12.21}$$

and

$$\kappa_2 = \frac{M_1 + M_2}{1 - \alpha}. \tag{12.22}$$

Substituting for κ_1 from Eq. (12.21) in Eq. (12.13) results in

$$S_1 = \frac{\alpha M_1 + M_2}{1 - \alpha} \left(\frac{V_{w_{eq}}}{V_w} \right)^n. \tag{12.23}$$

Substituting for κ_2 from Eq. (12.22) in Eq. (12.14) results in

$$S_2 = \frac{M_1 + M_2}{1 - \alpha} \left(\frac{V_s}{V_{s_{eq}}} \right)^m. \tag{12.24}$$

Using the expressions for the transports S_1, S_2 and S_3 and assuming that excess volumes of sand entering an element are deposited on the bed and excess volumes of sand leaving an element are eroded from the bed, the conservation of sand equation for the delta is

$$\frac{dV_s}{dt} + \frac{M_1 + M_2}{1 - \alpha} \left(\frac{V_s}{V_{s_{eq}}} \right)^m - \frac{\alpha M_1 + M_2}{1 - \alpha} \left(\frac{V_{w_{eq}}}{V_w} \right)^n = M_1, \tag{12.25}$$

and for the inlet is

$$\frac{dV_w}{dt} + \frac{\alpha(M_1 + M_2)}{1 - \alpha}\left(\frac{V_s}{V_{s_{eq}}}\right)^m - \frac{\alpha M_1 + M_2}{1 - \alpha}\left(\frac{V_{w_{eq}}}{V_w}\right)^n = -M_2. \qquad (12.26)$$

The two coupled differential equations (12.25) and (12.26) describe the transition of the delta and the inlet from the old to the new equilibrium.

12.4.3 Model Results

Eqs. (12.25) and (12.26) are solved numerically for V_s and V_w with the observed initial conditions $V_{w_0} = 171 \times 10^6$ m^3 and $V_{s_0} = 132 \times 10^6$ m^3 and the equilibrium sand and water volumes are $V_{s_{eq}} = 83 \times 10^6$ m^3 and $V_{w_{eq}} = 118 \times 10^6$ m^3 (Section 12.4.1). Values of the parameters M_1, M_2, m, n and α are determined by matching calculated and observed values of the inlet water volume V_w and the delta sand volumes V_s. In view of the aberrant behavior of the delta volume around year 10, emphasis in the calibration has been on matching observed and calculated inlet volumes. With the tidal prism $P = 225 \times 10^6$ m^3 and the longshore sand transport $M < 1 \times 10^6$ m^3 year^{-1} the ratio $P/M > 225$. Referring to Section 3.3.2, it follows that for the Frisian Inlet the sand bypassing mode is tidal flow bypassing. Therefore, the fraction of the longshore sand transport M_2 entering the inlet is assumed to be larger than the fraction of the longshore sand transport M_1 entering the delta. With the constraints that 0.5×10^6 m^3 year$^{-1} < M < 1 \times 10^6$ m^3 year^{-1} and $0 \le \alpha < 1$, a good match between observed and calculated values of V_s and V_w during the first 18 years after basin reduction was found for $M_1 = 0.25 \times 10^6$ m^3 year^{-1}, $M_2 = 0.75 \times 10^6$ m^3 year^{-1}, $m = 0.9$, $n = 3$ and $\alpha = 0.65$; see Figs. 12.6a and 12.6b.

Using Eqs. (12.25) and (12.26) and the parameter values determined by matching observed and calculated values of V_s and V_w for the 18-year observational period, the volume deficits $V_s - V_{s_{eq}}$ and $V_w - V_{w_{eq}}$ are calculated for a period of 200 years. The results are plotted in Fig. 12.7. The trajectory towards equilibrium for the inlet shows a monotonic decrease in the water volume. Initially, the delta volume decreases but then shows an overshoot before approaching equilibrium. This is further discussed in Section 12.4.5.

For the same 200-year period, sand transport values are presented in Fig. 12.8. The transport from inlet to delta (S_1) increases monotonically to an equilibrium value of $S_{1_{eq}} = 2.61 \times 10^6$ m^3 year^{-1}. The transport from the delta (S_2) shows a slight overshoot; it decreases to reach a minimum approximately 60 years after basin reduction and then increases to the equilibrium value of $S_{2_{eq}} = 2.86 \times 10^6$ m^3 year^{-1}. Because S_3 and S_4 are directly related to S_2, they both show a similar overshoot.

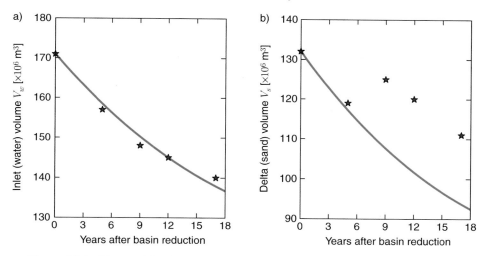

Figure 12.6 Observed (star) and numerically calculated (solid red line) values of a) the water volume of the inlet V_w and b) the sand volume of the delta V_s.

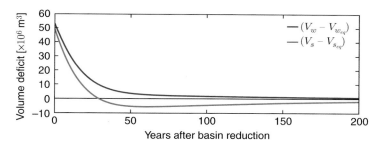

Figure 12.7 Changes in delta (sand) volume deficit and inlet (water) volume deficit for a 30 percent basin reduction using the numerical solution to Eqs. (12.25) and (12.26); for definition of the symbols reference is made to the text.

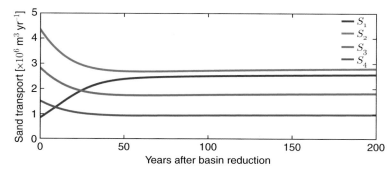

Figure 12.8 Sand transport after basin reduction; for definition of the symbols reference is made to the text.

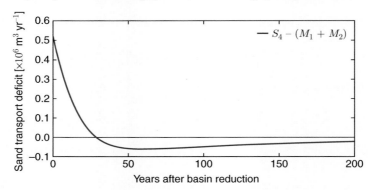

Figure 12.9 Difference between the sand transport to the downdrift coast and the longshore sand transport; for definition of the symbols reference is made to the text.

Because of its effect on the downdrift beaches, the transport S_4 is of particular interest. Referring to Fig. 12.9, initially this transport is considerably larger than the longshore sand transport of 1.0×10^6 m^3 year^{-1}, potentially resulting in accretion of the downdrift beaches. After approximately 40 years the value of S_4 becomes, less than the longshore sand transport and at that time erosion of the downdrift beaches should be expected.

12.4.4 Analytical Solution; Local and System Timescales

To explain the numerically calculated evolution of the delta and inlet and to identify the relevant timescales, an approximate analytical solution to Eqs. (12.25) and (12.26) is presented. For this, the quotients $(V_s / V_{s_{eq}})^m$ and $(V_{w_{eq}} / V_w)^n$ are linearized by imposing the condition that the morphological state is close to equilibrium, i.e.,

$$\frac{|V_{s_o} - V_{s_{eq}}|}{V_{s_{eq}}} \ll 1, \quad \text{and} \quad \frac{|V_{w_0} - V_{w_{eq}}|}{V_{w_{eq}}} \ll 1. \tag{12.27}$$

In that case, to a good approximation

$$\left(\frac{V_s}{V_{s_{eq}}}\right)^m = 1 + m\left(\frac{V_s - V_{s_{eq}}}{V_{s_{eq}}}\right), \tag{12.28}$$

and

$$\left(\frac{V_{w_{eq}}}{V_w}\right)^n = 1 - n\left(\frac{V_w - V_{w_{eq}}}{V_{w_{eq}}}\right). \tag{12.29}$$

For a detailed description of the method used to derive Eqs. (12.28) and (12.29) reference is made to Section 8.3.

To satisfy the conditions expressed by Eq. (12.27), a hypothetical case is considered in which the basin surface area of the Frisian Inlet is reduced by slightly less than 10 percent, resulting in a reduction in tidal prism from 325×10^6 m^3 to 300×10^6 m^3. The water volume of the inlet and the sand volume of the delta before basin reduction are $V_{wo} = 171 \times 10^6$ m^3 and $V_{so} = 132 \times 10^6$ m^3, respectively. Using the cross-sectional area – tidal prism relationship for the inlets of the Dutch Wadden Sea (Section 5.2.3), it follows that with a tidal prism of 300×10^6 m^3, the equilibrium inlet cross-sectional area after basin reduction is 20,400 m^2. Making certain assumptions for the variation in cross-sectional area over the inlet, the corresponding water volume of the inlet $V_{w_{eq}} = 158 \times 10^6$ m^3 (van de Kreeke, 2006). The equilibrium sand volume of the ebb delta follows from the relationship between ebb delta sand volume and tidal prism for tidal inlets in the Wadden Sea (Louters and Gerritsen, 1994). With the tidal prism of 300×10^6 m^3, this result in an equilibrium sand volume is $V_{s_{eq}} = 119 \times 10^6$ m^3. Substitution of the different sand and water volumes in Eq. (12.27) shows that the condition imposed by this equation is satisfied.

Starting with Eqs. (12.28) and (12.29) and introducing the timescales,

$$\tau_s = \frac{V_{s_{eq}}(1-\alpha)}{m(M_1 + M_2)}, \quad \text{and} \quad \tau_w = \frac{V_{w_{eq}}(1-\alpha)}{n(\alpha M_1 + M_2)}. \tag{12.30}$$

Eqs. (12.25) and (12.26) are written as, respectively,

$$\frac{d(V_s - V_{s_{eq}})}{dt} + \frac{(V_s - V_{s_{eq}})}{\tau_s} + \frac{(V_w - V_{w_{eq}})}{\tau_w} = 0, \tag{12.31}$$

and

$$\frac{d(V_w - V_{w_{eq}})}{dt} + \frac{(V_w - V_{w_{eq}})}{\tau_w} + \alpha \frac{(V_s - V_{s_{eq}})}{\tau_s} = 0. \tag{12.32}$$

It follows from Eq. (12.31) that when the inlet is in equilibrium, the delta adapts exponentially with the timescale τ_s. Similarly, when the delta is in equilibrium, it follows from Eq. (12.32) that the inlet adapts exponentially with the timescale τ_w. As these timescales do not involve the interaction of the two elements, they are referred to as local timescales.

Defining $V_s - V_{s_{eq}} = y_1$ and $V_w - V_{w_{eq}} = y_2$, Eqs. (12.31) and (12.32) in matrix form are

$$\begin{bmatrix} \frac{dy_1}{dt} \\ \frac{dy_2}{dt} \end{bmatrix} = \begin{bmatrix} -\frac{1}{\tau_s} & -\frac{1}{\tau_w} \\ -\frac{\alpha}{\tau_s} & -\frac{1}{\tau_w} \end{bmatrix} \begin{bmatrix} y_1 \\ y_2 \end{bmatrix}. \tag{12.33}$$

The solution to Eq. (12.33) is

$$\begin{pmatrix} y_1 \\ y_2 \end{pmatrix} = C_1 \begin{pmatrix} 1 \\ x_1 \end{pmatrix} e^{\lambda_1 t} + C_2 \begin{pmatrix} 1 \\ x_2 \end{pmatrix} e^{\lambda_2 t}, \tag{12.34}$$

in which λ_1 and λ_2 are the eigenvalues of the square matrix and $\left(\begin{smallmatrix}1\\x_1\end{smallmatrix}\right)$ and $\left(\begin{smallmatrix}1\\x_2\end{smallmatrix}\right)$ are the corresponding eigenvectors. The coefficients C_1 and C_2 are determined by the initial conditions. In terms of the timescales τ_s and τ_w and the feedback coefficient α, the eigenvalues are

$$\lambda_1 = \frac{-\left(\frac{1}{\tau_s} + \frac{1}{\tau_w}\right) + \sqrt{\left(\frac{1}{\tau_s} + \frac{1}{\tau_w}\right)^2 + \frac{4(\alpha-1)}{\tau_s\tau_w}}}{2}, \tag{12.35}$$

$$\lambda_2 = \frac{-\left(\frac{1}{\tau_s} + \frac{1}{\tau_w}\right) - \sqrt{\left(\frac{1}{\tau_s} + \frac{1}{\tau_w}\right)^2 + \frac{4(\alpha-1)}{\tau_s\tau_w}}}{2}. \tag{12.36}$$

With $0 \le \alpha < 1$, both λ_1 and λ_2 are negative with $|\lambda_1| < |\lambda_2|$. The reciprocals of the absolute value of the eigenvalues, $\tau_1 = 1/|\lambda_1|$ and $\tau_2 = 1/|\lambda_2|$ are referred to as the system timescales. As opposed to the local timescales, τ_s and τ_w, the system timescales account for the interaction of the two elements. For additional interpretation of the system timescales, reference is made to the last paragraph of this section. In terms of the system timescales, the solution for y_1 and y_2 is

$$\begin{pmatrix}y_1\\y_2\end{pmatrix} = C_1 \begin{pmatrix}1\\x_1\end{pmatrix} e^{-\frac{t}{\tau_1}} + C_2 \begin{pmatrix}1\\x_2\end{pmatrix} e^{-\frac{t}{\tau_2}}. \tag{12.37}$$

The components of the eigenvectors are

$$x_1 = -\tau_w \left(\frac{1}{\tau_s} + \lambda_1\right), \quad \text{and} \quad x_2 = -\tau_w \left(\frac{1}{\tau_s} + \lambda_2\right). \tag{12.38}$$

Substituting for λ_1 and λ_2 from, respectively, Eqs. (12.35) and (12.36) in Eq. (12.38) results in the expressions for the components of the eigenvectors

$$x_{1,2} = \frac{-\left(\frac{1}{\tau_s} - \frac{1}{\tau_w}\right) \mp \sqrt{\left(\frac{1}{\tau_s} + \frac{1}{\tau_w}\right)^2 - 4(1-\alpha)\left(\frac{1}{\tau_s}\frac{1}{\tau_w}\right)}}{\frac{2}{\tau_w}}, \tag{12.39}$$

with the negative sign referring to x_1 and the positive sign referring to x_2. This equation can be written as

$$x_{1,2} = \frac{-\left(\frac{1}{\tau_s} - \frac{1}{\tau_w}\right) \mp \sqrt{\left(\frac{1}{\tau_s} - \frac{1}{\tau_w}\right)^2 + 4\alpha\left(\frac{1}{\tau_s}\frac{1}{\tau_w}\right)}}{\frac{2}{\tau_w}}. \tag{12.40}$$

Because $0 \le \alpha < 1$, the square root in Eq. (12.40) is always larger than $|1/\tau_s - 1/\tau_w|$. It follows that the components of the eigenvectors, x_1 and x_2, have opposite signs.

With the initial conditions for the sand volume of the delta, $V_s = V_{s0}$ and for the water volume of the inlet, $V_w = V_{w0}$, the expressions for C_1 and C_2 are

$$C_1 = \frac{(V_{s0} - V_{s_{eq}})x_2 - (V_{w0} - V_{w_{eq}})}{x_2 - x_1},$$ (12.41)

and

$$C_2 = \frac{-(V_{s0} - V_{s_{eq}})x_1 + (V_{w0} - V_{w_{eq}})}{x_2 - x_1}.$$ (12.42)

To illustrate the adaptation of the delta volume and the inlet volume, the parameter values for M_1, M_2, m, n and α in Section 12.4.3 for the basin reduction of 30 percent are assumed to also be valid for the basin reduction of 10 percent. With these values and $V_{s_{eq}} = 119 \times 10^6$ m^3 and $V_{w_{eq}} = 158 \times 10^6$ m^3, the values of the local timescales are $\tau_s = 46$ years and $\tau_w = 20$ years. The eigenvalues are $\lambda_1 = -0.0056$ year^{-1} and $\lambda_2 = -0.0656$ year^{-1}, resulting in the system timescales $\tau_1 = 1/|\lambda_1| = 118$ years and $\tau_2 = 1/|\lambda_2| = 15$ years. Values of the eigenvector are $x_1 = -0.38$ and $x_2 = 0.83$. With the initial deficits, $V_{s0} - V_{s_{eq}} = 13 \times 10^6$ m^3 and $V_{w0} - V_{w_{eq}} = 13 \times 10^6$ m^3, the values of the coefficients C_1 and C_2 are -1.9×10^6 m^3 and 14.9×10^6 m^3, respectively. Using the results of the analytical solution, the adaptation of the delta volume and inlet volume are presented in Fig. 12.10. Comparing with Fig. 12.7, it follows that the analytically calculated adaptation for the 10 percent basin reduction is qualitatively the same as the numerically calculated adaptation for the 30 percent basin reduction. The inlet (water) volume shows a monotonic decrease until it reaches the equilibrium value. Initially, the delta (sand) volume decreases but then overshoots before returning to the equilibrium value.

Referring to Eq. (12.37), the fractions of the initial sand and water volume deficits adjusting with the shorter system timescale τ_2 are C_2 and $C_2 x_2$, respectively. With C_2 and x_2 positive, both volumes have the same sign, i.e., the sand volume of the delta and the water volume of the inlet are both too large. This

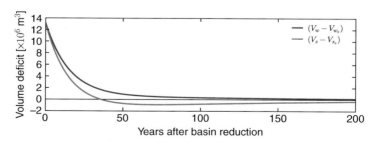

Figure 12.10 Changes in delta (sand) volume deficit and inlet (water) volume deficit for a 10 percent basin reduction using the analytical solution, Eq. (12.37).

results in an internal distribution of sand, whereby the shortage of sand in the inlet is partially compensated by a supply of sand from the delta. The fractions of the initial sand and water volume deficits adjusting with the longer timescale, τ_1, are C_1 and $C_1 x_1$, respectively. With C_1 and x_1 negative, the sand volume and water volume have opposite signs, i.e., the sand volume of the delta is too small and the water volume of the inlet is too large. Compensation of the shortage of sand in both the delta and inlet has to come from outside, which explains the longer timescale.

12.4.5 *Bumps and Overshoots*

With the eigenvalues λ_1 and λ_2 negative, both the delta and the inlet evolve towards an equilibrium state. As explained in Kragtwijk et al. (2004), this evolution is not necessarily monotonic. Depending on the interaction of the two elements, the initial response of an element may be away from its equilibrium, referred to as bump behavior. As shown in Figs. 12.7 and 12.10, it is also possible for an element to overshoot its equilibrium. These two contrasting situations are illustrated for the delta element in Fig. 12.11. In the case of a bump, even though there is already a surplus, sand is added to the delta. In the case of an overshoot, even though the delta element has reached equilibrium, additional sand is withdrawn.

The criterion for a bump is that the sign of the initial rate of change of the volume deficit and the sign of the initial volume deficit are the same. For the 10 percent basin reduction, the terms in Eq. (12.37) with the shortest timescale are positive. As a result, the rates of change of the initial inlet and delta volumes are negative. Because the initial volume deficits for both inlet and delta are positive, neither the evolution of the delta nor that of the inlet shows a bump. This agrees with the results in Figs. 12.7 and 12.10.

The criterion for an overshoot is that the initial volume deficit and the volume deficit close to equilibrium have different signs. For the 10 percent basin reduction, both the initial delta and inlet volume deficits are positive. Referring to the terms

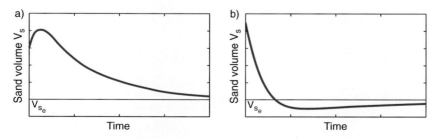

Figure 12.11 Schematic of a bump a) and an overshoot b) for a delta element. V_s is sand volume of the delta and V_{s_e} is equilibrium sand volume of the delta.

with the longest timescale in Eq. (12.37), it follows that the volume deficit for the delta is negative and for the inlet is positive. The evolution of the delta sand volume will show an overshoot and the evolution of the inlet water volume will be monotonic. This agrees with the results in Figs. 12.7 and 12.10.

12.5 Adaptation of an Inlet-Delta System Using a Diffusive Transport Formulation

A modeling approach, using a diffusive transport between the elements, was introduced by Di Silvio (1989) and later expanded by Stive et al. (1998). The principles are illustrated using a schematized inlet-ebb delta system (Fig. 12.12). Similar to the examples presented in Sections 12.3 and 12.4, it is assumed that after a perturbation the inlet and delta return to an equilibrium state. In Fig. 12.12, Element 1 represents the inlet with a water volume V_w and Element 2 represents the delta with a sand volume V_s. Sand enters and leaves Element 1 through exchange with Element 2. Sand enters and leaves Element 2 through exchange with Element 1 and the outside world (ocean). The diffusive sand transport between Elements 1 and 2 is

$$S_{12} = \delta(c_1 - c_2). \qquad (12.43)$$

Similarly, the transport between Element 2 and the outside world is

$$S_{20} = \delta(c_2 - c_E). \qquad (12.44)$$

In these equations, c_1 and c_2 are the local volume concentrations of suspended sand in Element 1 and 2, respectively, c_E is the volume concentration of suspended sand in the outside world and δ is a diffusion coefficient. The concentration c_E is assumed to remain constant during the transition period. To simplify the algebra, the diffusion coefficient δ is taken the same for Element 1 and 2. Transports are annually averaged values. The diffusive type transport results in a feedback between the ebb delta and the inlet.

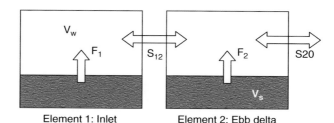

Element 1: Inlet   Element 2: Ebb delta

Figure 12.12 Two-element system with diffusive transport; for definition of the symbols see text.

The volume flux of sand from the bottom to the water column in Element 1 is

$$F_1 = w_s A(c_{1_e} - c_1), \tag{12.45}$$

in which c_{1_e} is the local equilibrium sand concentration in the water column. The volume flux of sand from the bottom to the water column in Element 2 is

$$F_2 = w_s A(c_{2_e} - c_2), \tag{12.46}$$

in which c_{2_e} is the local equilibrium sand concentration in the water column. In Eqs. (12.45) and (12.46), w_s is an exchange coefficient and A is the bottom surface area. To simplify the algebra, the bottom surface areas are taken the same for both elements.

The conservation of suspended sand equation for Element 1 is

$$\frac{dV_{sus}}{dt} = F_1 - S_{12} = 0, \tag{12.47}$$

and for Element 2 is

$$\frac{dV_{sus}}{dt} = F_2 + S_{12} - S_{20} = 0. \tag{12.48}$$

In these equations, V_{sus} is the suspended sand volume in the element. Assuming that on an annual average base the volume of suspended sand in the water column remains constant, $dV_{sus}/dt = 0$. It then follows from Eq. (12.47), with S_{12} given by Eq. (12.43) and F_1 given by (12.45), that

$$w_s A(c_{1_e} - c_1) - \delta(c_1 - c_2) = 0. \tag{12.49}$$

Similarly, it follows from Eq. (12.48), with S_{12} given by Eq. (12.43), S_{20} given by Eq. (12.44) and F_2 given by (12.46), that

$$w_s A(c_{2_e} - c_2) + \delta(c_1 - c_2) - \delta(c_2 - c_E) = 0. \tag{12.50}$$

Solving for c_1 and c_2 from Eqs. (12.49) and (12.50) results in

$$c_1 = \frac{w_s A(w_s A + 2\delta)c_{1_e} + w_s A\delta c_{2_e} + \delta^2 c_E}{(w_s A + 2\delta)(w_s A + \delta) - \delta^2}, \tag{12.51}$$

and

$$c_2 = \frac{w_s A\delta c_{1_e} + w_s A(w_s A + \delta)c_{2_e} + \delta(w_s A + \delta)c_E}{(w_s A + 2\delta)(w_s A + \delta) - \delta^2}. \tag{12.52}$$

For the inlet, the local and the overall equilibrium concentrations are related by the ratio of the equilibrium water volume and the actual water volume. For the delta, these concentrations are related by the ratio of the actual sand volume and the equilibrium sand volume (van Goor, 2003), i.e.,

$$c_{1_e} = c_E \left(\frac{V_{w_e}}{V_w}\right)^n, \tag{12.53}$$

$$c_{2_e} = c_E \left(\frac{V_s}{V_{se}} \right)^m .$$ (12.54)

In these equations, V_w is the actual water volume and V_{we} is equilibrium water volume of Element 1. V_s is the actual sand volume and V_{se} is the equilibrium sand volume of Element 2. The exponents n and m are empirical coefficients. When the system is at equilibrium, $c_1 = c_{1_e} = c_E$ and $c_2 = c_{2_e} = c_E$.

The rate of change of the water volume V_w and the sand volume V_s are related to the volume flux of sand from the bottom to the water column by, respectively,

$$\frac{dV_s}{dt} = -F_2 = -w_s A (c_{2_e} - c_2),$$ (12.55)

and

$$\frac{dV_w}{dt} = F_1 = w_s A (c_{1_e} - c_1).$$ (12.56)

Using Eqs. (12.51)–(12.54), it follows from Eq. (12.55) that

$$\frac{dV_s}{dt} + (\mu_1 + 2\mu_2) c_E \left(\frac{V_s}{V_{se}} \right)^m - \mu_2 c_E \left(\frac{V_{we}}{V_w} \right)^n = (\mu_1 + \mu_2) c_E,$$ (12.57)

with the coefficients μ_1 and μ_2 defined as

$$\mu_1 = \frac{w_s A \delta^2}{((w_s A + 2\delta)(w_s A + \delta) - \delta^2)},$$ (12.58)

and

$$\mu_2 = \frac{(w_s A)^2 \delta}{((w_s A + 2\delta)(w_s A + \delta) - \delta^2)}.$$ (12.59)

Using Eqs. (12.51)–(12.54), it follows from Eq. (12.56) that

$$\frac{dV_w}{dt} + \mu_2 c_E \left(\frac{V_s}{V_{se}} \right)^m - (\mu_1 + \mu_2) c_E \left(\frac{V_{we}}{V_w} \right)^n = -\mu_1 c_E.$$ (12.60)

Similar to Eqs.(12.25) and (12.26), the nature of the solution to Eqs. (12.57) and (12.60) can be investigated by assuming a small perturbation. This allows linearizing the nonlinear terms in these equations, resulting in

$$\frac{d(V_s - V_{se})}{dt} + (\mu_1 + 2\mu_2) \frac{m c_E}{V_{se}} (V_s - V_{se}) + \mu_2 \frac{n c_E}{V_{we}} (V_w - V_{we}) = 0,$$ (12.61)

and

$$\frac{d(V_w - V_{we})}{dt} + \mu_2 \frac{m c_E}{V_{se}} (V_s - V_{se}) + (\mu_1 + \mu_2) \frac{n c_E}{V_{we}} (V_w - V_{we}) = 0.$$ (12.62)

Except for the coefficients, these equations are the same as Eqs. (12.31) and (12.32), respectively. In view of this, the response of the inlet-ebb delta system to a perturbation using a diffusive transport formulation is similar to the response using the model described in Section 12.4.4.

Applications of the empirical model using diffusive transport are found in van Goor (2003) and Kragtwijk et al. (2004). van Goor (2003) deals with the impact of sea level rise on the morphological equilibrium state of tidal inlets, with application to two inlets in the Dutch Wadden Sea: the Ameland Inlet described in Section 4.8 and the Eyerlandse Gat Inlet. In Kragtwijk et al. (2004) the focus is on the evolution of the Dutch Wadden Sea inlets after closure of the Zuiderzee. In this study, special attention is given to the Texel and Vlie Inlets, described in Chapter 9, as they are most affected by the closure.

12.6 Limitations of Empirical Modeling

Empirical modeling of tidal inlets is directed at the transition of the morphology from a perturbed to an equilibrium state. A requirement is that the equilibrium state of the elements of the tidal inlet is known. The equilibrium state is expressed in terms of a relationship between the sand or water volume of the element and the tidal prism. The formulation of the exchange of sediment between the various morphological elements is empirical and involves a large number of parameters. Many of these lack a physical base. As a result, models need extensive calibration, requiring data over a relatively long period of time. Because of the lack of a physical base, parameter values found for one tidal inlet do not necessarily apply to another tidal inlet (Wang et al., 2008).

The assumption of the inlet system tending towards an equilibrium is not always valid. An example is the reduction of the cross-sectional area of an inlet below a certain value where, instead of tending towards an equilibrium, the inlet closes.

Provided they are properly calibrated over a sufficient long time period, empirical models can provide valuable information over the transition from a perturbed to an equilibrium state and the timescales involved. They are a useful addition to process-based morphodynamic models until some of the shortcomings in these models, preventing application to problems with timescales of decades and centuries, have been resolved.

13

River Flow and Entrance Stability

13.1 Introduction

In the preceding chapters, the emphasis is on tidal inlets where tidal currents are dominant and river flow is of secondary importance. Sand is carried towards the inlets by longshore sand transport and cross-shore sand transport is small. Inlets are open at all times. The few that closed did so gradually through spit formation, thereby increasing the inlet length and decreasing the inlet velocity. Examples are Captain Sam's Inlet and Mason Inlet, both located in South Carolina and described, respectively, in Sections 4.4 and 4.5.

A different category of inlets is where river flow is dominant and the tide is of secondary importance in keeping the inlet open. Inlets in this category are found in Vietnam (Lam, 2009; Tung, 2011), South Africa (Cooper, 2001; Whitfield, 1992) and Australia (Baldock et al., 2008; Hinwood and McLean, 2015b; Hinwood et al., 2012; Morris and Turner, 2010; Ranasinghe and Pattiaratchi, 2003). Many of these inlets connect to small lagoons and have a small tidal range, resulting in a small tidal prism. In addition to longshore sand transport, cross-shore sand transport plays an important role in carrying the sand towards the entrance. The river flow shows strong inter-annual variations with periods of high alternating with periods of low river flow. The height and the period between peak flows have an important bearing on whether the inlets stay open or close. In this respect, a distinction is made between inlets that remain open at all times and inlets that are open only seasonally or intermittently.

Even though only open part of the year, many seasonally and intermittently closed inlets are used extensively as harbors for small fishing boats and as recreational areas for swimming and boating. Closure presents a three-fold problem. Firstly, ocean access for boats that use the back-barrier lagoon as a harbor is limited to when the inlet is open. Secondly, the water quality in the lagoon could deteriorate during the months of inlet closure. Thirdly, flooding of low-lying land may disrupt land use and access and lead to land siltation. Consequently, there is an interest in keeping the inlet permanently open.

In the following sections, the effect of river flow on entrance stability is discussed for three inlets, Thuan An Inlet on the central coast of Vietnam, Wilson Inlet in southwestern Australia and Lake Conjola Inlet on the southeast coast of Australia. Thuan An is an inlet that, in spite of the seasonal character of the river flow, remains open at all times. The seasonal variations in river flow at Wilson Inlet cause this inlet to be open only part of the year. The river flow at Lake Conjola Inlet is highly irregular, resulting in an intermittently open entrance.

Prior to dealing with the three selected inlets, the effect of river flow on the basin tide and inlet velocity is discussed. For this, the Öszoy–Mehta Solution (Section 6.4) is expanded to include river flow.

13.2 Effect of River Flow on Basin Tide and Inlet Velocity

River flow results in a basin water level set-up, a net ebb velocity and a lowering of the amplitudes of the basin tide and the inlet velocity. The mean inlet velocity resulting from variations of depth with tidal stage discussed in (Section 7.4.2) is usually small compared to the mean velocity generated by the river flow.

The effect of the river flow on basin tide and inlet velocity is demonstrated using an expanded version of a lumped parameter model that includes river flow. The dynamic equation for the expanded lumped parameter model is the same as for the lumped parameter model without river flow and is given by Eq. (6.7), repeated here as Eq. (13.1):

$$K_2^2 \frac{du^*}{dt^*} + \frac{1}{K_1^2} u^* |u^*| = \eta_0^* - \eta_b^*.$$ (13.1)

In this equation, density gradients resulting from the river flow are neglected. The variables u^*, η_b^*, η_0^* and t^* are the non-dimensional velocity, basin water level, ocean tide and time, respectively, defined by Eq. (6.5). Velocities are positive in the flood direction. The coefficients K_1 and K_2 are defined by Eq. (6.9). Continuity is expressed by Eq. (6.8) with a term added to account for river flow, resulting in

$$u^* = \frac{d\eta_b^*}{dt^*} - Q^*,$$ (13.2)

with the non-dimensional river discharge Q^* defined as

$$Q^* = \frac{Q}{\sigma \hat{\eta}_0 A_b}.$$ (13.3)

In this equation, Q is the river discharge, $\sigma \hat{\eta}_0 A_b$ is the tidal discharge scale and A_b is the basin surface area. Assumed is a sinusoidal ocean tide with frequency σ and amplitude $\hat{\eta}_0$.

A semi-analytical solution to Eqs. (13.1) and (13.2) forced by a sinusoidal ocean tide is presented in Appendix 13.A and Escoffier and Walton (1979). The solution presented in Appendix 13.A is an expanded version of the Öszoy–Mehta Solution presented in Section 6.4. In both solutions the basin water level and inlet velocity consist of a time-dependent and a time-independent part. For the basin water level,

$$\eta_b^* = \tilde{\eta}_b^* + \langle \eta_b^* \rangle, \tag{13.4}$$

where $\tilde{\eta}_b^*$ is the basin tide and $\langle \eta_b^* \rangle$ is the mean basin level, both measured with respect to mean sea level. Similarly, for the inlet velocity

$$u^* = \tilde{u}^* + \langle u^* \rangle. \tag{13.5}$$

In this equation, $\tilde{u}^*$ is the tidal velocity and $\langle u^* \rangle$ is the mean velocity resulting from river flow.

With the trial solution

$$\tilde{\eta}_b^* = \hat{\tilde{\eta}}_b^* \sin(t^* - \alpha), \tag{13.6}$$

it follows from Eq. (13.2) that

$$\tilde{u}^* = \hat{\tilde{u}}^* \cos(t^* - \alpha). \tag{13.7}$$

Referring to Appendix 13.A, the tidal velocity amplitude $\hat{\tilde{u}}^*$ and basin tidal amplitude $\hat{\tilde{\eta}}_b^*$ are

$$\hat{\tilde{u}}^* = \hat{\tilde{\eta}}_b^* = \sqrt{\frac{\sqrt{\left(1 - K_2^2\right)^4 + \frac{4k_{10}^2}{K_1^4}} - \left(1 - K_2^2\right)^2}{\frac{2k_{10}^2}{K_1^4}}}, \tag{13.8}$$

and the phase α given by

$$\alpha = \tan^{-1}\left(\frac{k_{10}\hat{\tilde{u}}^*}{K_1^2 \left(1 - K_2^2\right)}\right). \tag{13.9}$$

The expressions for the mean basin level and mean inlet velocity are, respectively,

$$\langle \eta_b^* \rangle = -\frac{1}{K_1^2} k_{00} \hat{\tilde{u}}^{*2}, \tag{13.10}$$

and

$$\langle u^* \rangle = -Q^*. \tag{13.11}$$

The coefficients k_{10} and k_{00} depend on the ratio $\langle u^* \rangle / \hat{\bar{u}}^*$, which makes the solution implicit. Also, as shown in Appendix 13.A, the solution is limited to values of $\langle u^* \rangle / \hat{\bar{u}}^* \leq 1$. For zero river flow the expression for k_{10} reduces to that for the Öszoy–Mehta Solution, i.e., $k_{10} = 8/3\pi$ (compare Eqs. (6.25) and (13.8)) and $k_{00} = 0$.

Escoffier and Walton (1979) use a slightly different approach to solve Eqs. (13.2) and (13.3). The difference with the expanded Öszoy–Mehta Solution is the treatment of the nonlinear bottom friction term. However, after rearranging some of the expressions in Escoffier and Walton (1979), the resulting solution is similar to the expanded Öszoy–Mehta Solution. The solution for the amplitude of the basin tide $\hat{\eta}_b^*$ and velocity $\hat{\bar{u}}^*$ is the same as Eq. (13.8), with k_{10} replaced by a coefficient e. The solution for the basin water level set-up $\langle \eta_b^* \rangle$ is the same as Eq. (13.10) with k_{00} replaced by a coefficient f. The coefficients e and f are functions of the ratio $\langle u^* \rangle / \hat{\bar{u}}^*$ that are numerically calculated and presented in Fig. 1 of Escoffier and Walton (1979). Just as for the expanded Öszoy–Mehta Solution, the solution requires iteration and is limited to values of $\langle u^* \rangle / \hat{\bar{u}}^* \leq 1$.

To demonstrate the effect of the river flow on the basin water level and the inlet velocity, the expanded Öszoy–Mehta Solution is applied to the representative inlet. Parameter values for this inlet are given in Table 6.2. The results of the computations for the basin water level set-up and the amplitudes of the basin tide and the tidal velocity are presented in Fig. 13.1. The basin water level set-up increases and the amplitudes of the basin tide and the tidal velocity decrease with increasing river discharge.

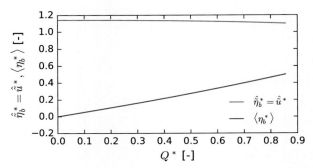

Figure 13.1 Basin water level set-up $\langle \eta_b^* \rangle$, amplitude of the basin tide $\hat{\eta}_b^*$ and tidal velocity $\hat{\bar{u}}^*$ as a function of the river discharge Q^*. For parameter values, reference is made to Table 6.2.

13.3 Effect of River Flow on Cross-Sectional Stability of Selected Inlets

13.3.1 Thuan An Inlet: A Permanently Open Inlet

Thuan An Inlet is located on the central coast of Vietnam in a tropical monsoon region (Lam, 2009; Tung, 2011). The inlet opened in 1897 as a result of a breach in the barrier island. It was briefly closed in 1903–1904 but has been open ever since. Ocean tides are semi-diurnal with a tidal range of 0.41 m. The inlet connects the Tam Giang Lagoon to the Gulf of Tonkin. The part of the lagoon that is served by Thuan An Inlet has an estimated surface area of 104 km^2. Reported values of the tidal prism range from 28×10^6 m^3 to 47×10^6 m^3. The monsoon regime exerts its influence on the tidal inlet through the river flow. Most of this is concentrated in the period from September to December. Little is known about its magnitude, other than that the river flow exiting through Thuan An Inlet during the extreme flood of 1999 was estimated at 12,000 m^3 s^{-1}. Observed velocities in the inlet are never higher than 0.5 m s^{-1} (it is not clear whether this included periods of river flow). The mean annual offshore significant wave height is 1 m.

Estimates of the longshore sand transport at Thuan An Inlet vary between 0.4×10^6 m^3 year^{-1} and 1.7×10^6 m^3 year^{-1}. With the reported values of the tidal prism, this result in P/M values ranging from 16 to 117. Taking an average value of 65 would imply that the location stability for tide alone is fair (Tables 3.1 and 3.2).

Entrance cross-sectional areas of Thuan An Inlet range from 2,900 to 4,000 m^2. Observations show that during the rainy season the inlet becomes wider and deeper. For example, as a result of the severe flood of November 1999 the inlet widened to 350 m and reached a maximum depth of 12 m. During the subsequent dry season, the inlet became narrower and shallower, a result of longshore sediment transport entering the inlet.

The cross-sectional stability of Thuan An Inlet is determined by both tide and river flow. To determine their relative importance, the cross-sectional stability for tide alone and tide and river flow is evaluated using the Escoffier Diagram introduced in Section 8.2. Parameter values for the inlet are presented in Table 13.1. The closure curves in the diagram are calculated using the expanded Öszoy–Mehta Solution. Because with river flow it is not possible to define a tidal prism, the velocity amplitude is taken as the maximum ebb velocity instead of the velocity $\hat{u}$ defined by Eq. (5.6). Assumed is a trapezoidal cross-section with side slopes 1:3 and a ratio of surface width to bottom width of 130, resulting in a shape factor $\beta_2 = 0.17$. Evolving cross-sections remain geometrically similar. Accounting for the moderate to large longshore sand transport, the equilibrium velocity $\hat{u}_{eq}$ is estimated to be on the high side of the values presented in Section 5.2.4 and taken equal to 1 m s^{-1}.

Table 13.1 *Parameter values for Thuan An Inlet used to calculate the closure curve.*

Parameter	Symbol	Dimension	Value
Inlet length	L	m	4,000
Friction factor	F	–	3×10^{-3}
Entrance/exit loss coefficient	m	–	1
Shape factor	β_2	–	0.17
Basin surface area	A_b	m^2	1.04×10^8
Tidal amplitude	$\hat{\eta}_0$	m	0.2
Tidal frequency	σ	rad s^{-1}	1.4×10^{-4}
Equilibrium velocity	$\hat{u}_{eq}$	m s^{-1}	1

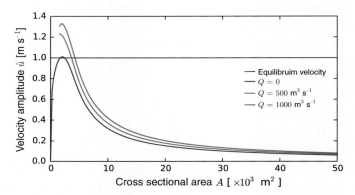

Figure 13.2 Escoffier Diagram for Thuan An Inlet with $Q = 0$, $Q = 500$ m^3 s^{-1} and $Q = 1,000$ m^3 s^{-1}. For parameter values, see Table 13.1.

The Escoffier Diagram with river flows of $Q = 0$, $Q = 500$ m^3 s^{-1} and $Q = 1,000$ m^3 s^{-1} is presented in Fig. 13.2. From this, it is concluded that for tide alone the stable equilibrium cross-sectional area is approximately 2,400 m^2. The unstable equilibrium cross-sectional area is only slightly less, implying that the degree of cross-sectional stability is marginal and chances are that in the absence of river flow the inlet closes. The stable equilibriums are considerably larger when including river discharge and are 3,600 m^2 for a river discharge of 500 m^3 s^{-1} and 4,200 m^2 for a river discharge of 1,000 m^3 s^{-1}.

During periods of river discharge, the inlet cross-sectional area increases beyond that for tide alone, thus increasing the entrance stability. When river flow diminishes, the inlet cross-sectional area decreases and tends towards the stable cross-sectional area for tide alone. In spite of the marginal stability for tide alone,

Thuan An Inlet is open year round. This is attributed to the longshore sand transport not being large enough to close the inlet before the next flood arrives.

13.3.2 Wilson Inlet: A Seasonally Open Inlet

Wilson Inlet is a small inlet on the southwest coast of Australia. Tides are diurnal with a spring tidal range of 0.8 m. The surface area of the back-barrier lagoon is approximately 25 km^2. The climate is characterized by a clearly defined wet and dry season, resulting in a seasonally varying river flow. The wet season coincides with the Australian winter (July–October). The mean annual river discharge is 207×10^6 m^3, of which 80 percent enters during the wet season. At the beginning of the wet season, the inlet is artificially opened by dredging a small channel through the entrance bar. Subsequently, it is kept open by river flow. Following the wet season, the inlet closes for a period of six to seven months due to the formation of a sand bar across the entrance. The coast has little exposure to wind waves limiting the longshore sand transport to an estimated 10,000 m^3 year^{-1}.

To study the closure of the inlet, a field study was carried out (Ranasinghe and Pattiaratchi, 1999). Field observations at the end of the wet season included bathymetric surveys and wave and current measurements. During the observational period the inlet was open with river flow being insignificant. Wave measurements showed waves to be dominantly swell that approached the coast perpendicularly. Longshore currents were observed to be weak. Current measurements showed maximum flood velocities of 1 m s^{-1} and maximum ebb velocities of 0.8 m s^{-1}. The predominant swell conditions, the weak and inconsistent longshore currents and the absence of spit formation suggest that for Wilson Inlet, onshore transport of sand is the main transport process responsible for inlet closure.

This was further investigated using a morphodynamics model (Ranasinghe et al., 1999). The overall structure of the morphodynamics model is similar to that presented in Chapter 11, with separate modules for flow, waves, sand transport and bathymetric changes. Applying the model to Wilson Inlet, it was found that, in the absence of river flow, longshore sand transport alone would not close the inlet, whereas swell-generated onshore sand transport alone would. This confirms the conclusion from the field observations that onshore sand transport, rather than longshore sand transport, is responsible for the seasonal closure of Wilson Inlet.

13.3.3 Lake Conjola Inlet: An Intermittently Open Inlet

Lake Conjola Inlet is a small inlet on the southeastern coast of Australia. Tides are semi-diurnal and at the upper end of micro-tidal. River flow is intermittent. The coast is characterized by high energy waves. Longshore sand transport is variable

both in magnitude and direction. Onshore sand transport may be at least as important as longshore sand transport in carrying sand towards the inlet (McLean and
Hinwood, 2000).

In the absence of river flow, Lake Conjola Inlet has an inadequate tidal prism
to remain open and continually moves to a closed state. Two means of entrance
restrictions have been observed: 1) gradual closure through inward tidal transport
of sand and 2) a sudden restriction associated with coastal storms, resulting in an
entrance bar (Hinwood and McLean, 2001). Summarizing, entrance conditions are
sensitive to coastal storms and fluvial events with tidal flows providing the energy
producing gradual entrance change following these large perturbations.

The dependence of the entrance condition on fluvial and storm events, rather
than on a trend towards a regime state, is illustrated using the two diagrams in
Fig. 13.3. The diagrams show non-dimensional values of basin tidal amplitude $\hat{\eta}_b*$
(Fig. 13.3a) and basin water level set-up $\langle \hat{\eta}_b^* \rangle$ (Fig. 13.3b) as a function of non-
dimensional river discharge Q^* and repletion coefficient K_1. Neglecting inertia,

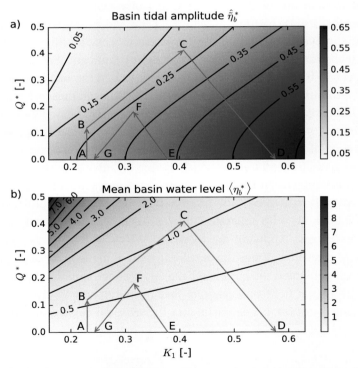

Figure 13.3 a) Basin tidal amplitude and b) mean basin water level as a function
of Q^* and K_1 with trajectories (in red) for the response of Lake Conjola Inlet
to a fluvial event (ABCD) and a storm event (EFG) (adapted from Hinwood and
McLean, 2001).

the diagrams are constructed using Eqs. (13.1) and (13.2) with $K_2 = 0$. The equations are solved numerically (van de Kreeke, 1967). By calculating Q^* and K_1 for typical fluvial and storm events and including them in the diagram, the effect of these events on basin tidal amplitude and basin water level set-up may be seen as trajectories in the diagrams.

In case of a fluvial event, Q^* initially increases but the entrance cross-sectional area has insufficient time to substantially increase, hence the value of K_1 does not change. The trajectory is shown as AB, which shows an increased mean basin level and some attenuation of the tidal amplitude. When river flow increases, it continues to increase Q^* and enlarges the entrance, thereby increasing K_1. The corresponding trajectory is BC. At C the mean basin water level is raised and the basin tidal amplitude is slightly reduced. The attenuation of the tidal amplitude is a result of the increased river flow which is only partially offset by a decrease in the resistance parameter. When the river flow diminishes, Q^* decreases and, with the entrance cross-sectional area still increasing, the value K_1 increases. This results in trajectory CD. While at D, the mean basin water level approaches the mean sea level and the basin tidal amplitude has increased compared to the value before the river flow event.

In the case of a storm event, the steep waves carry sand into the inlet, decreasing the value of K_1. Storms are usually accompanied by some rainfall, resulting in an increase in Q^*. A typical trajectory for such an event is EF. Because of the increased river flow and the increase in the resistance parameter, at F the basin tidal amplitude has decreased. After the river flow decreases, the entrance constriction remains. Following trajectory FG, as a result of the decreasing river flow, the tidal amplitude increases and the mean basin water level decreases. When at G, the basin tidal amplitude is smaller than at the beginning of the event and the mean tide level is the same, approaching mean sea level.

13.4 A Morphodynamic Model for the Long-Term Evolution of an Inlet

To improve the understanding of the long-term evolution of an inlet in the presence of tide and river flow, Hinwood et al. (2012) and Hinwood and McLean (2015a) developed a process-based exploratory morphodynamics model. The model consists of a hydrodynamics, a sediment transport and a bathymetry module. Schematization of the tidal inlet is shown in Figs. 6.1 and 6.2. The hydrodynamics is described by Eq. (7.1) and includes the variation of depth with tidal stage. With the water level in the inlet approximated by the basin tide, Eq. (7.1) is written as

$$\frac{L}{g}\frac{du}{dt} + \left(\frac{m}{2g} + \frac{FL}{g(h + \eta_b)}\right)u|u| = \eta_0 - \eta_b. \tag{13.12}$$

Assumed is a rectangular inlet cross-section with a tidally averaged depth h; L is inlet length, g is gravity acceleration, u is cross-sectionally averaged velocity, positive in the (tidal) flood direction, t is time, m is entrance/exit loss coefficient, F is a friction factor, η_0 is ocean tide and η_b is basin tide. Continuity is described by Eq. (7.2), with a term added to account for the river flow and, as in Eq. (13.12), the water level in the inlet approximated by the basin tide, resulting in

$$b(h + \eta_b)u = A_b \frac{d\eta_b}{dt} - Q, \qquad (13.13)$$

where b is inlet width, A_b is basin surface area and Q is river discharge.

The ocean tide is a simple harmonic

$$\eta_0 = \hat{\eta}_0 \sin(\sigma t), \qquad (13.14)$$

where $\hat{\eta}_0$ is ocean tidal amplitude and σ is the tidal frequency. In the sediment transport module, sediment transport is in the form of suspended load with C_0 and C_b the sediment concentrations in the ocean and the basin, respectively. The concentration in the inlet follows from

$$C = k \left[\left(\frac{u}{u_c} \right)^2 - 1 \right], \quad \text{for } u \geq u_c,$$
$$C = 0, \qquad\qquad \text{for } u < u_c. \qquad (13.15)$$

In this equation, C is the sediment concentration in the inlet, u_c is a threshold velocity for both pickup and deposition and k is an empirical constant. During flood, the sand transport into the inlet is $b(h + \eta_0)|u|C_0$ and the transport out of the inlet is $b(h + \eta_0)|u|C$. During ebb, the sand transport into the inlet is $b(h + \eta_b)|u|C_b$ and the transport out of the inlet is $b(h + \eta_b)|u|C$.

The bathymetry module consists of the conservation of sediment equation,

$$L\frac{dh}{dt} = (h + \eta_0)|u|(C - C_0), \qquad \text{for flood}, \qquad (13.16)$$

and

$$L\frac{dh}{dt} = -(h + \eta_0)|u|(C - C_b), \qquad \text{for ebb}. \qquad (13.17)$$

For flood, there is erosion for $C_0 < C$ and deposition for $C_0 > C$. For ebb, there is erosion for $C_b < C$ and deposition for $C_b > C$.

Except for the wave module, computations follow the feedback loop described in Section 11.2. Starting with an initial value for the tidally averaged depth, h, Eqs. (13.12) and (13.13) with the boundary condition Eq. (13.14) are solved numerically for $u(t)$ and $\eta_b(t)$. Concentrations are then calculated from Eq. (13.15) and used in Eqs. (13.16) and (13.17) to calculate the change in depth. Some parameter values, notably C_0 and C_b, are based on observations in intermittently closed inlets on Australia's southeast coast. Others, like k, F and u_c, are taken from the literature.

The relatively simple formulation of the model makes it possible to carry out many calculations in a short period of time. As an example, for a given ocean tide and river flow, calculations were made for a large number of initial inlet depths. Runs were made for a sufficiently long duration to show that after some time the depth reaches an equilibrium value. Results are plotted in a $\frac{\pi}{2}Q^* - h^*$ diagram (Fig. 13.4), with the non-dimensional river flow, Q^*, given by Eq. (13.3) and the non-dimensional water depth $h^* = h/\hat{\eta}_0$. The expression $\frac{\pi}{2}Q^*$ represents the non-dimensional river flow used by Hinwood and McLean (2015a). The left panel in Fig. 13.4 shows the starting positions, which are evenly distributed over the $\frac{\pi}{2}Q^* - h^*$ plane. The right panel shows the final position of each run. Final positions, representing the equilibrium depths, are clustered in two narrow bands, referred to as tidal flow and river flow attractors. The equilibrium depths depend on the value of river flow but are independent of the initial depth. An initially shallow inlet will evolve towards the attractor by increasing its depth and an initially deep inlet will evolve towards the attractor by decreasing its depth. Even though the model shows the attractors to cluster in two bands, a physical explanation is still lacking.

For zero river flow the model described in this section leads to similar results as the Escoffier Stability Model, and inlets approach an equilibrium depth that is independent of the initial depth. Because of the different sand transport formulations, the values of the equilibrium depths are expected to be different for the two models (Hinwood and McLean, 2015b).

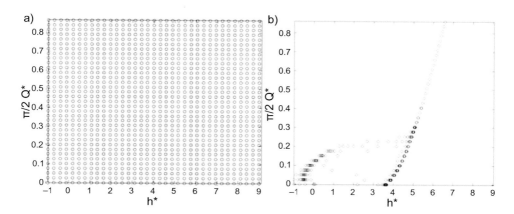

Figure 13.4 Evolution of inlet depths. a) The starting conditions for the model runs and b) the final depth positions showing clustering in two narrow bands (reprinted from Hinwood and McLean, 2015b, copyright 2016, with permission from Elsevier).

13.A Öszoy–Mehta Solution Including River Flow

The governing equations are Eqs. (13.1) and (13.2) in the main text. Assumed is a sinusoidal ocean tide

$$\eta_0^* = \sin(t^*),\qquad(13.A.1)$$

with the non-dimensional variables defined by Eq. (6.5).

The solution for the basin water level, η_b^*, is written as the sum of the basin tide $\tilde{\eta}_b^*$ and a mean basin water level $\langle\eta_b^*\rangle$,

$$\eta_b^* = \tilde{\eta}_b^* + \langle\eta_b^*\rangle.\qquad(13.A.2)$$

Similarly, the inlet velocity is written as the sum of a tidal velocity $\tilde{u}^*$ and a mean velocity $\langle u^*\rangle$,

$$u^* = \tilde{u}^* + \langle u^*\rangle.\qquad(13.A.3)$$

Substituting for η_b^* and u^* from, respectively, Eqs. (13.A.2) and (13.A.3) in Eq. (13.2) and collecting time-independent and time-dependent terms results in, respectively,

$$\langle u^*\rangle = -Q^*,\qquad(13.A.4)$$

and

$$\tilde{u}^* = \frac{d\tilde{\eta}_b^*}{dt^*}.\qquad(13.A.5)$$

With trial solutions for the basin tide and tidal velocity of, respectively,

$$\tilde{\eta}_b^* = \hat{\tilde{\eta}}_b^* \sin(t^* - \alpha),\qquad(13.A.6)$$

and

$$\tilde{u}^* = \hat{\tilde{u}}^* \cos(t^* - \beta),\qquad(13.A.7)$$

it follows from Eq. (13.A.5) that

$$\hat{\tilde{\eta}}_b^* = \hat{\tilde{u}}^* \qquad \text{and} \qquad \beta = \alpha.\qquad(13.A.8)$$

To arrive at a solution of Eqs. (13.1) and (13.2), the product $u^*|u^*|$ is expanded in a Fourier series. Following Dronkers (1964, 1968), the Fourier expansion is carried out by substituting the expression for u^*, Eq. (13.A.3) with $\tilde{u}^*$ given by Eq. (13.A.7), in $u^*|u^*|$. Retaining only the constant and first harmonic in the series expansion results in

$$u^*|u^*| \cong k_{00}\hat{\tilde{u}}^{*2} + k_{10}\hat{\tilde{u}}^{*2} \cos(t^* - \beta),\qquad(13.A.9)$$

with

$$k_{00} = \frac{1}{4}(2 + 2\gamma)\left(2 - \frac{4\gamma}{\pi}\right) + \frac{2}{3\pi}\sin(2\gamma),\qquad(13.A.10)$$

$$k_{10} = \frac{3}{\pi}\sin(\gamma) + \frac{1}{3\pi}\sin(3\gamma) + \left(2 - \frac{4\gamma}{\pi}\right)\cos(\gamma), \quad (13.A.11)$$

and

$$\cos(\gamma) = \frac{\langle u^* \rangle}{\hat{u}^*}. \quad (13.A.12)$$

For zero river flow with $\gamma = \pi/2$, $k_{00} = 0$ and $k_{10} = 8/(3\pi)$. With $\langle u^* \rangle$ negative, γ is in the second quadrant. Eq. (13.A.12) requires that $\langle u^* \rangle < \hat{u}^*$. It follows from Eqs. (13.A.10), (13.A.11) and (13.A.12) that both k_{00} and k_{10} are functions of the unknown $\hat{u}^*$.

Using the Fourier expansion of $u^*|u^*|$, and substituting for η_0^*, η_b^* and u^*, from, respectively, Eqs. (13.A.1), (13.A.6) and (13.A.7) in Eq. (13.1) and using Eqs. (13.A.6), (13.A.7) and (13.A.8), it follows that:

$$\left(1 - K_2^2\right)\hat{u}^*\sin(t^* - \alpha) + \frac{k_{00} + k_{10}\cos(t^* - \alpha)}{K_1^2}\hat{u}^{*2} = \sin(t^*) - \langle \eta_b^* \rangle. \quad (13.A.13)$$

Collecting time-dependent terms results in an equation for the tidal velocity amplitude:

$$\left(1 - K_2^2\right)\hat{u}^*\sin(t^* - \alpha) + \frac{k_{10}}{K_1^2}\hat{u}^{*2}\cos(t^* - \alpha) = \sin(t^*). \quad (13.A.14)$$

Collecting time-independent terms results in the equation for the mean basin level:

$$\langle \eta_b^* \rangle = -\frac{k_{00}}{K_1^2}\hat{u}^{*2}. \quad (13.A.15)$$

For zero river flow, with $k_{00} = 0$, the basin water level set-up is zero.

Except for the factor $8/3\pi$, which is replaced by k_{10}, Eq. (13.A.14) is the same as Eq. (6.20). Using the same method used to solve Eq. (6.20), the solution to Eq. (13.A.14) is

$$\hat{u}^* = \sqrt{\frac{\left[\left(1 - K_2^2\right)^4 + \frac{4k_{10}^2}{K_1^4}\right]^{1/2} - \left(1 - K_2^2\right)^2}{\frac{2k_{10}^2}{K_1^4}}}, \quad (13.A.16)$$

and

$$\alpha = \tan^{-1}\left(\frac{k_{10}\hat{u}^*}{K_1^2\left(1 - K_2^2\right)}\right). \quad (13.A.17)$$

With k_{10} a function of the velocity amplitude, Eq. (13.A.16) is implicit in $\hat{u}^*$ and has to be solved iteratively. Once $\hat{u}^*$ is known, values of $\langle \eta_b^* \rangle$, $\hat{\eta}_b^*$ and α follow from Eqs. (13.A.15), (13.A.8) and (13.A.17), respectively. For zero river flow, with $k_{10} = 8/3\pi$, Eqs. (13.A.16) and (13.A.17) are the same as Eqs. (6.25) and (6.26), respectively.

14

Engineering of Tidal Inlets

14.1 Introduction

Back-barrier lagoons are home to recreational marinas and fishing ports. With a few exceptions, most vessels using these facilities are relatively small, with lengths in the 5–30 m range and a maximum draft of 5 m. To access the lagoon, vessels need to navigate the ebb delta channel and the inlet. This requires that both channel and inlet are relatively stable, have sufficient depth and an alignment relative to the wave direction that allows safe access and passage. Not many natural inlets satisfy these requirements and measures are needed to remedy the shortcomings. A distinction is made between soft and hard measures. Soft measures include the opening of a new inlet, inlet relocation, dredging and artificial sand bypassing. Hard measures are jetty construction and weir-jetty systems. In addition to providing boating access, inlets play a role in maintaining the water quality of the back-barrier lagoon; they serve as conduits for the exchange of lagoon and ocean water.

14.2 Artificial Opening of a New Inlet

The objectives of the artificial opening of a new inlet are to provide passage for vessels to the back-barrier lagoon and/or to improve water quality. With regards to the passage of vessels, design requirements include sufficient channel depth, width, alignment and stability. Improving water quality requires sufficient exchange, implying a large enough tidal prism. Examples of inlets that were artificially opened, but with different objectives, are Bakers Haulover Inlet (FL), Faro-Olhão Inlet (Portugal) and Packery Channel (TX). Bakers Haulover Inlet (Fig. 14.1a) was opened in 1925 for the prime purpose of improving water quality in the northern part of Biscayne Bay (Dombrowsky and Mehta, 1993). Faro-Olhão Inlet (Fig. 14.1b) was opened in 1929 to improve navigational access to the city of Faro

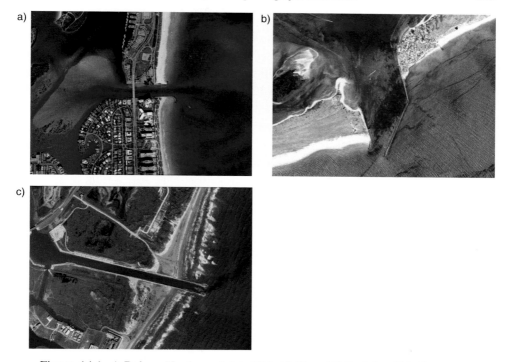

Figure 14.1 a) Bakers Haulover Inlet (FL), b) Faro-Olhão Inlet (Ría Formosa, Portugal) and c) Packery Channel (TX) (Source: Google Earth).

(Pacheco et al., 2011). Packery Channel (Fig. 14.1c) was opened in 2006 with the objectives to facilitate recreational fishing and boating and to improve the exchange between the Gulf of Mexico and Corpus Christi Bay (Williams et al., 2007).

In the following, some of the elements that go into the design of a new inlet are discussed. A first step is to determine the equilibrium entrance cross-sectional area and its stability. For this the Escoffier Diagram, described in some detail in Section 8.2.2, is used. Assuming the back-barrier lagoon is small and deep enough to allow for a uniformly fluctuating basin water level, the closure curve in the diagram can be calculated using the Öszoy–Mehta or the Keulegan Solution (Sections 6.4 and 6.5). For back-barrier lagoons that are large and shallow, more advanced models should be used. This implies an increase in computer time. To limit computer time and cost, a solution is to run the more advanced models in combination with a simpler, less computationally extensive model. The simple model could then be used to narrow down the parameter range of interest and reduce the computing time of the advanced model.

Calculating the equilibrium velocity curve in the Escoffier Diagram requires that the $A-P$ relationship for the location of the new inlet is known. If not available, the next best option is to look for relationships for similar coasts, i.e., coasts with the same offshore wave characteristics (wave height, period and direction), same tide characteristics (semi-diurnal, diurnal or mixed tide) and the same grain characteristics (density and size). Depending on the selected $A–P$ relationship, the equilibrium velocity is given by Eq. (5.12) or Eq. (5.13). The equilibrium velocity is usually close to 1 m s^{-1}, with values slightly larger for coasts with a large longshore sand transport and somewhat smaller for coasts with a small longshore sand transport.

With the known stable cross-sectional area determined from the Escoffier Diagram, the tidal prism follows from the cross-sectional area – tidal prism relationship. With the known tidal prism, P, and assuming the gross longshore sand transport M is known, the P/M ratio is calculated. Using Tables 3.1 and 3.2, this gives the degree of location stability and the bypassing mode for the new inlet. To assure sufficient stability of the inlet and the channels on the ebb delta, a P/M ratio larger than 80 is recommended.

After opening a new inlet, ebb and flood deltas develop. The deltas continue to grow by capturing part of the longshore sand transport, until they reach equilibrium. An example is Ocean City Inlet (MD), for which the development of the ebb delta was observed and calculated using the empirical model discussed in Section 12.3. Without the new inlet, the sand stored in the deltas would have been available to the downdrift beach. Given the tidal prism, an estimate of the ebb delta volume, and the potential sand loss to the downdrift beach, follows from Eq. (5.14). In particular, for inlets with a large tidal prism the sand volume of the deltas is substantial, resulting in potentially large erosion of the downdrift beaches. To partly compensate for the sand loss to the downdrift beach, it is recommended that sand from opening the inlet is placed on the downdrift beach or at the future location of the ebb delta.

An important step in determining the effect of a new inlet on the downdrift shore is development of a sand management plan. The sand management plan includes the sediment budget discussed in Section 3.2. For a given control volume, it delineates the sources and sinks of sediment, the sediment transport, sediment transport pathways and volume changes. The sediment budget should be constructed for both the adaptation/transition period and the period after the morphology of the tidal inlet has reached equilibrium. A helpful tool in constructing the sediment budget, especially for the adaptation/transition period, is the empirical models described in Chapter 12. For most inlets, the major source of sand is the longshore sand transport. Major sinks are the back-barrier lagoon and the offshore. The sediment budget should result in an estimate of the volume of sand that bypasses the inlet at any given period and reaches the downdrift coast. Given this volume the resulting shoreline changes can be calculated with the line models described in Kamphuis

(2006) and Bakker (2013). Examples of inlet sand management plans for eight inlets along the southeast coast of Florida are presented in Dombrowsky and Mehta (1993).

Results of preliminary studies should indicate whether inlet cross-sectional area meets navigational demands, a sufficient volume of sand is bypassed to the down-drift beach and cross-sectional and location stability are large enough to limit maintenance dredging. If any of these requirements is not met, measures described in Sections 14.3–14.7 can be contemplated.

14.3 Relocation of an Existing Inlet

Inlets with a relatively small P/M ratio have a tendency to migrate and form a sand spit across the entrance. Spit formation increases the inlet length and leads to a lowering of the inlet velocity and shoaling. To improve navigability and/or water exchange at these inlets, one measure is to relocate the inlet. Examples are Captain Sam's Inlet and Mason Inlet, discussed in Sections 4.4 and 4.5, respectively. Another well-documented example is Ancão Inlet (Pacheco et al., 2007; Vila-Concejo et al., 2003). Ancão Inlet is part of the Ría Formosa barrier island system in Portugal. The inlet has a history of eastward migration and breaching. In late 1996, the inlet was in the last stage of its migration cycle, highly sinuous and infilling. To assure navigability and to provide sufficient water exchange, in the middle of 1997, a new inlet was opened 3.5 km to the west of the closed inlet. After a relatively rapid adjustment, the relocated inlet started to resemble a natural inlet with well-developed ebb and flood deltas. After moving a short distance to the west it resumed its migration towards the east. The inlet in 2004 (Fig. 14.2a) still showed the classic features of a natural inlet, but in 2011 (Fig. 14.2b) had deteriorated to the extent that artificial breaching again was considered. The effect of the inlet relocation was felt up to 4 km updrift of the opening position, showing an

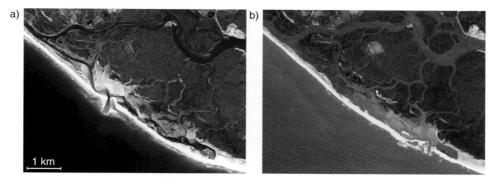

Figure 14.2 Ancão Inlet in a) 2004 and b) 2011 (Source: Google Earth).

increase in beach volume (Ferreira, 2011). Erosion after relocation was observed downdrift of the relocation position (Matias et al., 2009). Similar the dredging of shoals and shallows (Section 14.4) and sand bypassing (Section 14.5), relocation of a migrating inlet is a repeat process. Partly based on the studies carried out at Ancão Inlet, recommendations for inlet relocation are presented in Vila-Concejo et al. (2004).

14.4 Dredging

Dredging at inlets includes maintenance dredging and sand bypassing. Maintenance dredging refers to the removal of shoals and shallows from inlet and navigation channels. Sand bypassing involves dredging sand from a sand trap or impoundment basin and moving it to the downdrift coast. Sand traps or impoundment basins are pre-dredged depressions where sand is collected by natural processes. They are usually located in protected waters that allow dredges to operate in a quiet wave environment.

Dredges used at inlets are hydraulic suction dredges and hopper dredges (Vlasblom, 2003). In case of a hydraulic suction dredge, sand is transported through a floating or submerged pipeline onto the downdrift beach or it is offloaded on a split hull barge and dumped on the foreshore. Hopper dredges carry the sand to the downdrift beaches where they dump it on the foreshore or pump it onto the beach by a floating pipe line or by rainbowing. For details of these methods reference is made to the US Army Corps of Engineers' *Shore Protection Manual* (1984).

An example of an inlet that is maintained by dredging is Wiggins Pass (FL). Using a hydraulic suction dredge, the sand is directly placed on the downdrift beach via a pipeline (Dabees et al., 2011). Examples of inlets with sand traps or impoundment basins from which the sand is transported to the downdrift beaches are Sebastian Inlet (FL), Jupiter Inlet (FL), Masonboro Inlet (NC) and Channel Islands Harbor (CA).

14.5 Sand Bypassing Plants

Bypassing plants are designed to transfer sand accumulated at the updrift jetty directly to the downdrift beaches, avoiding the need for an impoundment basin and dredging. The oldest sand bypassing plant is at South Lake Worth Inlet (FL) and was constructed in 1937. The plant uses slurry pumps that operate from a platform on the updrift jetty. A submerged discharge line carries the sand–water mixture to the downdrift beach. After upgrading the pump in 1948, the design capacity was estimated at 70,000 m^3 year^{-1}. In the late fifties, a similar plant was installed at Lake Worth Inlet, located approximately 25 km to the north (Fig. 14.3). The plant was designed to bypass 230,000 m^3 year^{-1}. The sand–water mixture is carried

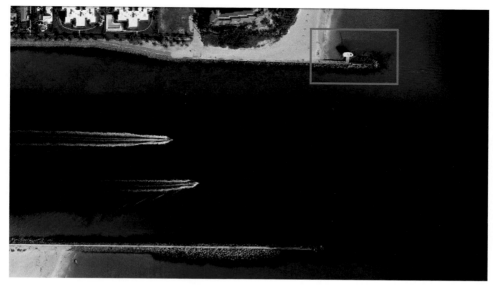

Figure 14.3 Sand bypassing plant at Lake Worth Inlet (FL) (Source: Google Earth).

through a submerged discharge line to the downdrift beach. A detailed description of the plant is given in (Zurmuhlen, 1957). A problem with fixed bypassing plants, as constructed at South Lake Worth Inlet and Lake Worth Inlet, is that sand does not always accumulate at the same location relative to the jetty. This, and the limited length of the dredge arm, means that the intake cannot always reach the accumulated sand. This makes the plants inefficient.

To deal with the natural variability of the location of the accumulated sand fillet, at Nerang River Entrance (Queensland, Australia) a 500 m long pier was constructed perpendicular to the beach close to the updrift jetty (Fig. 14.4). Along the outer 300 m, ten jet pumps were installed. As opposed to conventional slurry pumps, jet pumps do not have moving parts. At Nerang River Entrance the slurry from the jet pumps is discharged in a buffering hopper and from there is discharged to the downdrift beach. Operation started in 1986. The design bypass rate is 600,000 m^3 year^{-1}. A detailed description of the plant is given in US Waterway Experiment Station (1989).

In 1990, a moving bypassing system was implemented at Indian River Inlet (DE). The system uses a single jet pump. To account for the natural variability of the location of the accumulated sand, the jet pump must cover a relatively large area. This is accomplished by employing the jet pump from a crawler crane moving along the beach in the swash zone. Discharge of the jet pump is through a line to a booster pump and from there a discharge line crosses the inlet via an existing bridge. The discharge line extends up to a maximum of 500 m on the beach on the

Figure 14.4 Sand bypassing plant at Nerang River Entrance (Queensland, Australia) (Source: Google Earth).

downdrift beach. The system was designed to bypass $90,000 \text{ m}^3 \text{ year}^{-1}$. For details of the operation and performance, reference is made to Clausner et al. (1991).

14.6 Jetties; Jetty Length and Orientation

Jetties are implemented to prevent the longshore sand transport from entering the navigation channel. In addition, they confine the current, thereby reducing maintenance dredging (Kraus, 2005). Inlets have one or two jetties. The more common configuration is that of two parallel jetties. Examples are presented in Figs. 14.1a and 14.1c. Where sand traps and wave damping beaches are located within the confines of the jetties, the configuration is more as shown in Fig. 14.1b.

To prevent sand from entering the navigation channel, jetties should be sand tight and preferably extend beyond the seaward limit of the longshore sand transport. When shorter, some of the longshore sand transport will enter the navigation channel. This sand has to be removed by dredging, which in some cases might be cheaper than extending the jetty. For the seaward limit of the longshore sand transport, reference is made to Komar (1998) and Kamphuis (2006).

After the sand fillet at the updrift side of a jetty has sufficiently developed, some sand travels seaward along the jetty and at the jetty tip enters the navigation channel. To slow this process, sometimes spur jetties are added (Fig. 14.5). The spur jetty is typically perpendicular to the jetty but could also be at an acute

Figure 14.5 Jetty with spur jetty at Fort Pierce Inlet(FL) (Source: Google Earth).

angle. The basic function of a spur jetty is to divert the sand, keeping it away from the jetty and navigation channel. For spur jetty design considerations including location, elevation and length, reference is made to Seabergh and Kraus (2003).

Depending on length and orientation, jetties contribute to the safe entrance of vessels into the inlet. Preferably, jetties should extend beyond the breaker zone. The location of the breaker zone depends on wave height, wave period, bottom slope and tidal stage. As a simple rule, waves break when the wave height approaches the water depth. As a result, on any given day, the distance between the shore and the breaker zone differs. The relevant breaker zone is where waves break for wave conditions at which the design vessel should still be able to enter the inlet. Summarizing, jetties should preferably extend beyond the seaward limit of the longshore sand transport and beyond the relevant breaker zone. In addition to jetty length, jetty orientation is important. To maintain rudder control, orientation of the jetties should be such that, under rough conditions, vessels can enter with waves coming in at the aft quarter as opposed to having to deal with a following sea.

Jetties interrupt the natural sand transport pathways (Section 3.3), leading to adverse effects on the downdrift beaches. Therefore, a central element of inlet restoration and improvement is the bypassing of the sand accumulated at the updrift side of the jetties. Preferably, bypassing is accomplished by natural processes, whereby sand is transferred to the downdrift beaches by waves and tide. If this is not possible, artificial means are used to bypass the sand. Examples are dredging

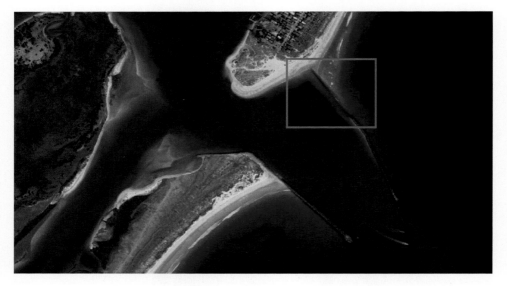

Figure 14.6 Weir-jetty system at Masonboro Inlet (NC) (Source: Google Earth).

(Section 14.4) and bypassing plants (Section 14.5). In the mid-sixties the weir-jetty system was introduced to bypass sand.

14.7 Weir-Jetty Systems

Jetties interrupt the flow of sand. A weir-jetty system allows the transport of sand across the jetty. The weir section is a depressed region in the jetty. The sand that is transported over the weir is collected in an impoundment basin or sand trap. The location of the impoundment basin or sand trap in the lee of the jetty allows small dredges to transfer the sand to the downdrift beaches. As an example, the weir-jetty system at Masonboro Inlet (NC) is shown in Fig. 14.6. An additional benefit of the weir-jetty system is that the seaward transport of sand along the outside of the jetty is reduced. As a result, the jetty may not need to extend as far as without a weir. Another benefit is that flood currents enter the inlet over the weir, bypassing the navigation channel. This leads to flood currents in the navigation channel being weaker than ebb currents, thereby reducing the volume of sand that enters the inlet.

The first weir-jetty systems were built in the mid-sixties at Hillsboro Inlet (FL) and Masonboro Inlet (NC). They were followed in the seventies and eighties by weir-jetty systems at Ponce de Leon Inlet (FL), Murrells Inlet (NC) and the mouth of the Colorado River (TX). A review of these projects, including their performance and recommendations for weir location, elevation and length, and location and size of the deposition basin, is presented in Seabergh and Kraus (2003).

References

Bakker, W.T. (1968). "A mathematical theory about sand waves and its application on the Dutch Wadden Isle of Vlieland". *Shore and Beach* 36(2): 4–14.

— (2013). *Coastal Dynamics*. World Scientific Publishing Co., p. 114.

Baldock, T.E., F. Weir, and M.G. Hughes (2008). "Morphodynamic evolution of a coastal lagoon entrance during swash overwash". *Geomorphology* 95(3–4): 398–411. DOI: 10.1016/j.geomorph.2007.07.001.

Batten, B.K., N.C. Kraus, and L. Lin (2007). "Long-term inlet stability of a multiple inlet system, Pass Cavallo, Texas". In: *Coastal Sediments '07*. American Society of Civil Engineers, pp. 1515–1528. DOI: 10.1061/40926(239)117.

Beets, D.J. and A.J.F. van der Spek (2000). "The Holocene evolution of the barrier and the back-barrier basins of Belgium and the Netherlands as a function of late Weichselian morphology, relative sea-level rise and sediment supply". *Netherlands Journal of Geosciences* 79(1): 3–16.

Biegel, E.J. and P. Hoekstra (1995). "Morphological response characteristics of the Zoutkamperlaag (the Netherlands) to a sudden reduction in basin area". *The International Association of Sedimentologists* 24: 85–99.

Booij, N., R.C. Ris, and L.H. Holthuijsen (1999). "A third-generation model for coastal regions: 1. Model description and validation". *Journal of Geophysical Research* 104(C4): 7649–7666. DOI: 10.1029/98JC02622.

Boon, J.D. and R.J. Byrne (1981). "On basin hypsometry and the morphodynamic response of coastal inlet systems". *Marine Geology* 40(1–2): 27–48. DOI: 10.1016/0025-3227(81)90041-4.

Boyd, J.P. (2001). *Chebyshev and Fourier Spectral Methods*. 2nd edn. New York: Dover Publications, 665 pp.

Brouwer, R.L. (2006). "Equilibrium and stability of a double inlet system". MA thesis. Delft University of Technology.

Brouwer, R.L. (2013). "Cross-sectional stability of double inlet systems". PhD thesis. Delft University of Technology.

Brouwer, R.L., H.M. Schuttelaars, and P.C. Roos (2013). "Modelling the influence of spatially varying hydrodynamics on the cross-sectional stability of double inlet systems". *Ocean Dynamics* 63(11): 1263–1278. DOI: 10.1007/s10236-013-0657-6.

Brouwer, R.L., H.M. Schuttelaars, J. van de Kreeke, and T.J. Zitman (2008). "Effects of amplitude differences on equilibrium and stability of a two-inlet bay system". In: *Proceedings of the 5th IAHR Symposium on River, Coastal and Estuarine Morphodynamics (RCEM) 2007*. Ed. by C. Dohmen-Janssen and S. Hulscher. Vol. 1. Leiden: Taylor & Francis/Balkema, pp. 33–39.

Brouwer, R.L., J. van de Kreeke, and H.M. Schuttelaars (2012). "Entrance/exit losses and cross-sectional stability of double inlet systems". *Estuarine, Coastal and Shelf Science* 107: 69–80. DOI: `10.1016/j.ecss.2012.04.033`.

Brown, E.I. (1928). "Inlets on sandy coasts". In: *Proceedings of the American Society of Civil Engineers*. Vol. 54. 1, pp. 505–553.

Bruun, P. (1981). *Port Engineering*. Houston, Texas: Gulf Publishing Company, 436 pp.

Bruun, P. and F. Gerritsen (1959). "Natural bypassing of sand at coastal inlets". *Journal of the Waterways and Harbours Division* 85: 75–107.

— (1960). *Stability of Coastal Inlets*. North Holland Publishing Co.

Bruun, P., A.J. Mehta, and I.G. Johnsson (1978). *Stability of Tidal Inlets: Theory and Engineering*. Elsevier Scientific Publishing Co., 510 pp.

Buchwald, V.T. (1971). "The diffraction of tides by a narrow channel". *Journal of Fluid Mechanics* 46(3): 501–511. DOI: `10.1017/S0022112071000661`.

Buijsman, M.C. and H. Ridderinkhof (2007). "Long-term ferry-ADCP observations of tidal currents in the Marsdiep Inlet". *Journal of Sea Research* 57(4): 237–256. DOI: `10.1016/j.seares.2006.11.004`.

Byrne, R.J., P.A. Bullock, and D.G. Tyler (1975). "Response characteristics of a tidal inlet". In: *Estuarine Research*. Ed. by L.E. Cronin. Vol. 2. Academic Press, pp. 267–276.

Byrne, R.J., J.T. DeAlteris, and Bullock (1974). "Channel stability in tidal inlets: a case study". In: *Proceedings of the 14th International Conference on Coastal Engineering*. Ed. by M.P. O'Brien. Vol. 1. ASCE, pp. 1585–1604.

Byrnes, M.R., J.L. Bakker, and N.C. Kraus (2003). "Coastal sediment budgets for Grays Harbor, Washington". In: *Coastal Sediments '03*, pp. 1–10.

Cheung, K.F., F. Gerritsen, and J. Cleveringa (2007). "Morphodynamics and sand bypassing at Ameland inlet, the Netherlands". *Journal of Coastal Research* 23(1): 106–118. DOI: `10.2112/04-0403.1`.

Clausner, J.E., J.A. Gebert, A.T. Rambo, and K.D. Watson (1991). "Sand bypassing at Indian River Inlet, Delaware". In: *Coastal Sediments '91*, pp. 1177–1191.

Cleary, W.J. and D.M. FitzGerald (2003). "Tidal inlet response to natural sedimentation processes and dredging-induced tidal prism changes: Mason Inlet, North Carolina". *Journal of Coastal Research* 19(4): 1018–1025.

Cooper, J.A.G. (2001). "Geomorphological variability among microtidal estuaries from the wave-dominated South African coast". *Geomorphology* 40(1–2): 99–122. DOI: `10.1016/S0169-555X(01)00039-3`.

Dabees, M.A., B.D. Moore, and K.K. Humiston (2011). "Evaluation of inlets channel migration and management practices in southwest Florida". In: *Coastal Sediments '11*, pp. 484–496.

Davis, R.A. (1994). *Geology of Holocene Barrier Island Systems*. Berlin Heidelberg: Springer, 464 pp. DOI: `10.1007/978-3-642-78360-9`.

Davis, R.A. and D.M. FitzGerald (2004). *Beaches and Coasts*. Blackwell Publishing Ltd., 419 pp.

Davis, R.A. and M.O. Hayes (1984). "What is a wave-dominated coast?" *Marine Geology* 60: 313–329. DOI: `10.1016/S0070-4571(08)70152-3`.

Davis, R.A., A.C. Hine, and M.J. Bland (1987). "Midnight pass, Florida: inlet instability due to man-related activities in Little Sarasota Bay". In: *Coastal Sediments '87*. Ed. by N.C. Kraus. ASCE, pp. 2062–2077.

DeAlteris, J.T. and R.J. Byrne (1975). "The recent history of Wachapreague Inlet, Virginia". In: *Estuarine Research*. Ed. by L.E. Cronin. Vol. 2. New York: Academic Press, pp. 167–183.

Dean, R.G. (1988). "Sediment interaction at modified coastal inlets; processes and policies". In: *Hydrodynamics and Sediment Dynamics of Tidal Inlets*. Ed. by D.G. Aubrey and L. Weishar. Vol. 29. Lecture Notes on Coastal and Estuarine Studies. New York: Springer-Verlag, pp. 412–439. DOI: 10.1029/LN029p0412.

— (1991). "Equilibrium beach profiles: characteristics and applications". *Journal of Coastal Research* 7(1): 53–84.

Dean, R.G. and R.A. Dalrymple (2002). *Coastal Processes with Engineering Applications*. Cambridge University Press, 475 pp.

Dean, R.G. and P.A. Work (1993). "Interaction of navigational entrances with adjacent shorelines". *Journal of Coastal Research Special Issue* 18: 91–110.

Dean-Rosati, J. (2005). "Concepts in sediment budgets". *Journal of Coastal Research* 21(2): 307–322. DOI: 10.2112/02-475A.1.

de Swart, H.E. and N.D. Volp (2012). "Effects of hypsometry on the morphodynamic stability of single and multiple tidal inlet systems". *Journal of Sea Research* 74: 35–44. DOI: 10.1016/j.seares.2012.05.008.

de Swart, H.E. and J.T.F. Zimmerman (2009). "Morphodynamics of tidal inlet systems". *Annual Review of Fluid Mechanics* 41: 203–229. DOI: 10.1146/annurev.fluid.010908.165159.

Dieckmann, R., M. Osterhun, and H.W. Partenscky (1988). "A comparison between German and North American tidal inlets". In: *Proceedings of the 21st International Conference on Coastal Engineering*. Ed. by B.L. Edge. Vol. 1. ASCE, pp. 2681–2691.

Dillingh, D. (2013). *Kenmerkende waarden Kustwateren en Grote Rivieren*. Report 1207509-000-ZKS-0010. Delft, The Netherlands: Deltares.

DiLorenzo, J.L. (1988). "The overtide and filtering response of small inlet/bay systems". In: *Hydrodynamics and Sediment Dynamics of Tidal Inlets*. Ed. by D.G. Aubrey and L. Weishar. Vol. 29. Lecture Notes on Coastal and Estuarine Studies. New York: Springer New York, pp. 24–53. DOI: 10.1007/978-1-4757-4057-8_2.

Di Silvio, G. (1989). "Modelling of the morphological evolution of tidal lagoons and their equilibrium configuration". In: *Proceedings of the 13th Congress of the IAHR*, pp. 169–175.

Dombrowsky, M.R. and A.J. Mehta (1993). "Inlets and management practices: southeast coast of Florida". *Journal of Coastal Research* (SI 18): 29–57.

Dronkers, J. (1986). "Tidal asymmetry and estuarine morphology". *Netherlands Journal of Sea Research* 20(2–3): 117–131. DOI: 10.1016/0077-7579(86)90036-0.

Dronkers, J., (2005). *Dynamics of coastal systems*. World Scientific Publishing Co., 25: 519 pp.

Dronkers, J.J. (1964). *Tidal Computations in Rivers and Coastal Waters*. Amsterdam: North Holland Publishing Co., 518 pp.

— (1968). "Discussion of 'Water-level fluctuations and flow in tidal inlets' by J. van de Kreeke". *Journal of the Waterways and Harbors Division* 94(WW3): 376–377.

— (1975). "Tidal theory and computations". In: *Advances in Hydroscience*. Vol. 10. Academic Press Inc., pp. 145–230.

Ehlers, J. (1988). *The Morphodynamics of the Wadden Sea*. Rotterdam: Balkema, 397 pp.

El-Ashry, M.T and Wanless, H.R. (1965). Birth and early growth of a tidal delta. *Journal of Geology* 73: 404–406.

Escoffier, F.F. (1940). "The stability of tidal inlets". *Shore and Beach* 8(4): 111–114.

Escoffier, F.F. and T.L. Walton (1979). "Inlet stability solutions for tributary inflow". *Journal of the Waterway, Port, Coastal and Ocean Division* 105(WW4): 341–355.

Esri, DigitalGlobe, GeoEye, Earthstar Geographics, CNES/Airbus DS, USDA, USGS, AEX, Getmapping, Aerogrid, IGN, IGP, swisstopo, and the GIS User Community (2016). *ArcGIS Online*. URL: www.arcgis.com/home/webmap/viewer.html?useExisting=1.

Ferreira, Ó. (2011). "Morphodynamic impact of inlet relocation to the updrift coast: Ancão peninsula (Ría Formosa, Portugal)". In: *Coastal Sediments '11*, pp. 497–504.

FitzGerald, D.M. (1984). "Interactions between the ebb-delta and landward shoreline: Price Inlet, South Carolina". *Journal of Sedimentary Research* 54(4): 1301–1318. DOI: 10.1306/212F85C6-2B24-11D7-8648000102C1865D.

— (1988). "Shoreline erosional-depositional processes associated with tidal inlets". In: *Hydrodynamics and Sediment Dynamics of Tidal Inlets*. Ed. by D.G. Aubrey and L. Weisher. Vol. 29. Lecture Notes on Coastal and Estuarine Studies. New York: Springer-Verlag, pp. 186–225. DOI: 10.1029/LN029p0186.

— (1996). "Geomorphic variability and morphological and sedimentological controls on tidal inlets". *Journal of Coastal Research* (SI 23): 47–71. DOI: /10.2307/25736068.

FitzGerald, D.M., N.C. Kraus, and E.B. Hands (2000). *Natural Mechanisms of Sediment Bypassing at Tidal Inlets*. ERDC/CHL CHETN-IV-30. US Army Corps of Engineers, 10 pp.

FitzGerald, D.M. and D. Nummedal (1983). "Response characteristics of an ebb-dominated tidal inlet channel". *Journal of Sedimentary Research* 53(3): 833–845.

Fitzgerald, D.M., S. Penland, and D. Nummedal (1984). "Control of barrier island shape by inlet sediment bypassing: East Frisian Islands, West Germany". *Marine Geology* 60(1–4): 355–376. DOI: 10.1016/0025-3227(84)90157-9.

Friedrichs, C.T. and D.G. Aubrey (1988). "Non-linear distortion in shallow well-mixed estuaries; a syntheses". *Estuarine Coastal and Shelf Science* 27(5): 521–545. DOI: 10.1016/0272-7714(88)90082-0.

Friedrichs, C.T., D.G. Aubrey, G.S. Giese, and P.E. Speer (1993). "Hydrodynamical modeling of a multiple-inlet estuary/barrier system: insight into tidal inlet formation and stability". In: *Formation and Evolution of Multiple Tidal Inlets*. Ed. by D.G. Aubrey and G.S. Giese. Vol. 44, pp. 95–112. DOI: 10.1029/CE044p0095.

Fry, V.A. and D.G. Aubrey (1990). "Tidal velocity asymmetries and bedload transport in shallow embayments". *Estuarine, Coastal and Shelf Science* 30(5): 453–473. DOI: 10.1016/0272-7714(90)90067-2.

Gaudiano, D.J. and T.W. Kana (2001). "Shoal bypassing in mixed energy inlets: geomorphic variables and empirical predictions for nine South Carolina inlets". *Journal of Coastal Research* 17(2): 280–291. DOI: 10.2307/4300178.

Glaeser, J.D. (1978). "Global distribution of barrier islands in terms of tectonic setting". *Journal of Geology* 86(3): 283–297.

Goor, M. van (2003). "Impact of sea-level rise on the morphological equilibrium state of tidal inlets". *Marine Geology* 202(3–4): 211–227. DOI: 10.1016/S0025-32 27(03)00262-7.

Hanisch, J. (1981). "Sand transport in the tidal inlet between Wangerooge and Spiekeroog (W. Germany)". In: *Holocene Marine Sedimentation in the North Sea Basin: Special Publication 5 of the IAS*. Ed. by S.-D. Nio, R.T.E. Schüttenhelm, and T.C.E. van Weering. Wiley, pp. 175–185. DOI: 10.1002/9781444303759.ch13.

Hayes, M.O. (1977). "Development of Kiawah Island, SC". In: *Coastal Sediments '77*. ASCE, pp. 828–847.

— (1979). "Barrier island morphology as a function of tidal and wave regime". In: *Barrier Islands*. Ed. by S.P. Leatherman. New York: Springer-Verlag.

— (1980). "General morphology and sediment patterns in tidal inlets". *Sedimentary Geology* 26(1–3): 139–156. DOI: 10.1016/0037-0738(80)90009-3.

— (1994). "The Georgia Bight barrier system". In: *Geology of Holocene Barrier Island Systems*. Ed. by R.A. Davis. Springer-Verlag, pp. 233–304.

Heath, R.A. (1975). "Stability of some New Zealand coastal inlets". *New Zealand Journal of Marine and Freshwater Research* 9(4): 449–457. DOI: 10.1080/00288330.1975.9515580.

Herman, A. (2007). "Numerical modelling of water transport processes in partially-connected tidal basins". *Coastal Engineering* 54(4): 297–320. DOI: 10.1016/j.coastaleng.2006.10.003.

Hicks, M.D. and T.M. Hume (1996). "Morphology and size of ebb tidal deltas at natural inlets on open-sea and pocket-beach coasts, North Island, New Zealand". *Journal of Coastal Research* 12(1): 47–63.

— (1997). "Determining sand volumes and bathymetric change on an ebb delta". *Journal of Coastal Research* 13(2): 407–416.

Hicks, M.D., T.M. Hume, A. Swales, and M.O. Green (1999). "Magnitudes, spacial extent, time scales and causes of shoreline change adjacent to an ebb tidal delta, Katikati Inlet, New Zealand". *Journal of Coastal Research* 15(1): 220–240.

Hine, A.C. (1975). "Bedform distribution and migration patterns on tidal deltas in the Chatham Harbor estuary, Cape Cod, Massachusetts". In: *Estuarine Research: Geology and engineering*. Ed. by L.E. Cronin. Vol. 2. Acad, pp. 235–252.

Hinwood, J.B. and E.J. McLean (2001). "Monitoring and modeling tidal regime changes following inlet scour". *Journal of Coastal Research* (SI 34): 449–458.

— (2015a). "Estuaries, tidal inlets, Escoffier, O'Brien and geomorphic attractors". In: *Proceedings of the 36th IAHR World Congress*, pp. 2344–2356.

— (2015b). "Predicting the dynamics of intermittently closed/open estuaries using attractors". *Coastal Engineering* 99: 64–72. ISSN: 0378-3839. DOI: 10.1016/j.coastaleng.2015.02.008.

Hinwood, J.B., E.J. McLean, and B.C. Wilson (2012). "Non-linear dynamics and attractors for the entrance state of a tidal estuary". *Coastal Engineering* 61: 20–26. ISSN: 0378-3839. DOI: 10.1016/j.coastaleng.2011.11.007.

Hume, T.M. and C.E. Herdendorf (1992). "Factors controlling tidal inlet characteristics on low drift coasts". *Journal of Coastal Research* 8(2): 355–375.

Ippen, A.T. (1966). *Estuary and Coastline Hydrodynamics*. McGraw-Hill Book Co., pp. 505–510.

Israel, C.G. and D.W. Dunsbergen (1999). "Cyclic morphological development of the Ameland Inlet". In: *Proceedings of the 1st IAHR Symposium on River, Coastal and Estuarine Morphodynamics (RCEM) 1999*. Vol. 2, pp. 705–714.

Jarrett, J.T. (1976). *Tidal Prism-Inlet Area Relationships*. GITI Report 3. Vicksburg, MS: U.S. Army Coastal Engineering Research Center.

Jelgersma, S. (1983). "The Bergen Inlet, transgressive and regressive Holocene shoreline deposits in the northwestern Netherlands". *Geologie en Mijnbouw* 62(3): 471–486.

Kamphuis, J.W. (2006). *Introduction to Coastal Engineering and Management*. World Scientific Publishing Co., pp. 280–297.

Kana, T.W. (1989). "Erosion and beach restoration at Seabrook Island, South Carolina". *Shore and Beach* 57(3): 3–18.

Kana, T.W. and J.E. Mason (1988). "Evolution of an ebb tidal delta after an inlet relocation". In: *Hydrodynamics and Sediment Dynamics of Tidal Inlets*. Ed. by D.G. Aubrey and L. Weishar. Vol. 29. Lecture Notes on Coastal and Estuarine Studies. New York: Springer-Verlag, pp. 382–412. DOI: 10.1029/LN029p0382.

Kana, T.W. and P.A. McKee (2003). "Relocation of Captain Sams Inlet – 20 years later". In: *Coastal Sediments '03*. ASCE, pp. 1–13.

Keulegan, G.H. (1951). *Third Progress Report on Tidal Flow in Entrances: Water Level Fluctuations of Basins in Communication with Seas*. Report No. 1146. Washington, DC: National Bureau of Standards, 28 pp.

— (1967). *Tidal Flow in Entrances; Water-Level Fluctuations of Basins in Communication with Seas*. Technical Bulletin No. 14. Vicksburg, MS: U.S. Army Engineer Waterways Experiment Station, 100 pp.

King, C.A.M. (1972). *Beaches and Coasts*. Butler and Tanner Ltd., 570 pp.

Kjerfve, B. (1986). "Comparative oceanography of coastal lagoons". In: *Estuarine Variability*. Ed. by D.A. Wolfe. Academic Press, pp. 63–81.

Komar, P.D. (1998). *Beach Processes and Sedimentation*. Prentice Hall, p. 276.

Kragtwijk, N.G., T.J. Zitman, M.J.F. Stive, and Z.B. Wang (2004). "Morphological response of tidal basins to human interventions". *Coastal Engineering* 51(3): 207–221. ISSN: 0378-3839. DOI: 10.1016/j.coastaleng.2003.12.008.

Kraus, N.C. (1998). "Adaptation of the Frisian Inlet to a reduction in basin area". In: *Proceedings of the 25th International Conference on Coastal Engineering*. Ed. by B.L. Edge. Vol. 3. ASCE, pp. 3265–3278.

Kraus, N.C. (2000). "Reservoir model of ebb-tidal shoal evolution and sand bypassing". *Journal of Waterway, Port, Coastal and Ocean Engineering* 126(6): 305–313.

— (2005). "Coastal inlet functional design: anticipating morphological response". In: *Coastal Dynamics '05*. Ed. by A. Sanchez-Arcilla. ASCE, pp. 1–13. DOI: 10.1061/40855(214)108.

Kraus, N.C., L. Lin, B.K. Batten, and G.L. Brown (2006). *Matagorda Shipping Channel, Texas: Jetty Stability Study*. ERDC/CHL TR-06-7. US Army Corps of Engineers, Engineering Research and Development Center.

Kraus, N.C. and J.D. Rosati (1999). "Estimating uncertainty in coastal inlet sediment budgets". In: *Proceedings of the 12th Annual National Conference on Beach Preservation Technology*. Florida Shore and Beach Preservation Association, pp. 287–302.

Kreeke, J. van de (1967). "Water-level fluctuations and flow in tidal inlets". *Journal of the Waterways, Harbor and Coastal Engineering Division* 93(WW4): 97–106.

— (1985). "Stability of tidal inlets – Pass Cavallo, Texas". *Estuarine, Coastal and Shelf Science* 21(1): 33–43. DOI: 10.1016/0272-7714(85)90004-6.

— (1990a). "Can multiple tidal inlets be stable?" *Estuarine, Coastal and Shelf Science* 30(3): 261–273. DOI: 10.1016/0272-7714(90)90051-R.

— (1990b). "Stability analysis of a two-inlet bay system". *Coastal Engineering* 14(6): 481–497. DOI: 10.1016/0378-3839(90)90031-Q.

— (1992). "Stability of tidal inlets; Escoffier's analysis". *Shore and Beach* 60(1): 9–12.

— (1996). "Morphological changes on a decadal time scale in tidal inlets: modeling approaches". *Journal of Coastal Research* (SI 23): 73–81.

— (1998). "Adaptation of the Frisian Inlet to a reduction in basin area with special reference to the cross-sectional area". In: *Proceedings of the 8th Conference on Physics of Estuaries and Coastal Seas (PECS) 1996*. Ed. by J. Dronkers and M.B.A.M. Scheffers, pp. 355–362.

— (2004). "Equilibrium and cross-sectional stability of tidal inlets: application to the Frisian Inlet before and after basin reduction". *Coastal Engineering* 51(5–6): 337–350. DOI: 10.1016/j.coastaleng.2004.05.002.

— (2006). "An aggregate model for the adaptation of the morphology and sand bypassing after basin reduction of the Frisian Inlet". *Coastal Engineering* 53(2–3): 255–263. DOI: 10.1016/j.coastaleng.2005.10.013.

Kreeke, J. van de, R.L. Brouwer, T.J. Zitman, and H.M. Schuttelaars (2008). "The effect of a topographic high on the morphological stability of a two-inlet bay system". *Coastal Engineering* 55(4): 319–332. DOI: 10.1016/j.coastaleng.2007.11.010.

Kreeke, J. van de and D.W. Dunsbergen (2000). "Tidal asymmetry and sediment transport in the Frisian Inlet". In: *Proceedings of the 9th Conference on Physics of Estuaries and Coastal Seas (PECS) 1998*. Ed. by T. Yanagi. Tokyo: Terra Scientific Publishing Company, pp. 139–159.

Kreeke, J. van de and A. Hibma (2005). "Observations on silt and sand transport in the throat section of the Frisian Inlet". *Coastal Engineering* 52(2): 159–175. DOI: 10.1016/j.coastaleng.2004.10.002.

Kreeke, J. van de and K. Robaczewska (1993). "Tide-induced residual transport of coarse sediment; application to the Ems estuary". *Netherlands Journal of Sea Research* 31(3): 209–220. DOI: 10.1016/0077-7579(93)90022-K.

Lam, N.T. (2009). "Hydrodynamics and morphodynamics of seasonally forced tidal inlet systems". PhD thesis. Delft University of Technology, 142 pp.

LeConte, L.J. (1905). "Discussion of 'Notes on the improvement of river and harbour outlets in the United States' by D.A. Watts". *Transactions of the American Society of Civil Engineers* LV(2): 306–308.

LeProvost, C. (1991). "Generation of overtides and compound tides (review)". In: *Tidal Hydrodynamics*. Ed. by B.B. Parker, pp. 269–295.

Lesser, G.R., J.A. Roelvink, J.A.T.M. van Kester, and G.S. Stelling (2004). "Development and validation of a three-dimensional morphological model". *Coastal Engineering* 51(8–9): 883–915. ISSN: 0378-3839. DOI: http://dx.doi.org/10.1016/j.coastaleng.2004.07.014.

Lorentz, H.A. (1926). *Verslag Staatscommissie Zuiderzee 1918–1926*. In Dutch. Den Haag.

Louters, T. and F. Gerritsen (1994). *The Riddle of the Sands; A Tidal System's Answer to a Rising Sea Level*. Report RIKZ-94.040. Ministry of Transport, Public Works and Water Management, Directorate-General of Public Works and Water Managements, National Institute for Coastal and Marine Management, 69 pp.

Marino, J.N. and A.J. Mehta (1988). "Sediment trapping at Florida's east coast inlets". In: *Hydrodynamics and Sediment Dynamics of Tidal Inlets*. Ed. by D.G. Aubrey and L. Weishar. Vol. 29. Lecture Notes on Coastal and Estuarine Studies. New York: Springer-Verlag, pp. 284–296. DOI: 10.1029/LN029p0284.

Matias, A., A. Vila-Concejo, Ó Ferreira, and J.M.A. Dias (2009). "Sediment dynamics of barriers with frequent overwash". *Journal of Coastal Research* 25(3): 768–780.

McLean, E.J. and J.B. Hinwood (2000). "Modelling entrance resistance in estuaries". In: *Proceedings of the 27th International Conference on Coastal Engineering*. Ed. by B.L. Edge. Vol. 4. ASCE, pp. 3446–3457. DOI: 10.1061/40549(276)268.

Mehta, A.J. and E. Özsoy (1978). "Inlet hydraulics". In: *Stability of Tidal Inlets: Theory and Engineering*. Ed. by P. Bruun. Amsterdam, The Netherlands: Elsevier Scientific Publishing Co., pp. 83–161.

Morris, B.D., M.A. Davidson and D.A. Huntley (2004). "Estimates of the seasonal morphological evolution of the Barra Nova Inlet using video techniques". *Continental Shelf Research* 24(2): 263–278. DOI: 10.1016/j.csr.2003.09.009.

Morris, B.D. and I.L. Turner (2010). "Morphodynamics of intermittently open-closed coastal lagoon entrances: new insights and conceptual model". *Marine Geology* 271(1–2): 55–66. DOI: 10.1016/j.margeo.2010.01.009.

Murray, A.B. (2003). "Contrasting the goals, strategies, and predictions associated with simplified numerical models and detailed simulations". In: *Prediction in*

Geomorphology. Ed. by P.R. Wilcock and R.M. Iverson. Vol. 135. Geophysical Monograph Series. AGU. Washington DC, pp. 151–165. DOI: 10.1029/GM135.

Nahon, A., X. Bertin, A.B. Fortunato, and A. Oliveira (2012). "Process-based 2DH morphodynamic modeling of tidal inlets: A comparison with empirical classifications and theories". *Marine Geology* 291–294: 1–11. DOI: 10.1016/j.margeo.2011.10.001.

NASA, GSFC, MITI, ERSDAC, JAROS, and U.S./Japan ASTER Science Team (2003). *Venice, Italy*. URL: http://earthobservatory.nasa.gov/IOTD/view.php?id=3827.

National Park Service (2012). *Post-Hurricane Sandy: Old Inlet Breach on Fire Islands*. URL: www.nps.gov/fiis/naturescience/post-hurricane-sandy-breaches.htm.

O'Brien, M.P. (1931). "Estuary tidal prism related to entrance areas". *Civil Engineering* 1(8): 738–739.

— (1969). "Equilibrium flow areas of inlets on sandy coasts". *Journal of the Waterways and Harbors Division* 95(WW1): 43–52.

O'Brien, M.P. and R.G. Dean (1972). "Hydraulics and sedimentary stability of coastal inlets". In: *Proceedings of the 14th International Conference on Coastal Engineering*. Ed. by M.P. O'Brien. Vol. 2, pp. 761–780.

Pacheco, A., Ó. Ferreira, and J.J. Williams (2011). "Long-term morphological impacts of the opening of a new inlet on a multiple inlet system". *Earth Surface Processes and Landforms* 36(13): 1726–1735. DOI: 10.1002/esp.2193.

Pacheco, A., A. Vila-Concejo, Ó. Ferreira, and J.A. Dias (2007). "Present hydrodynamics of Ancão Inlet, 10 years after its relocation". In: *Coastal Sediments '07*. Ed. by N.C. Kraus and J. Dean-Rosati. ASCE, pp. 1557–1570. DOI: 10.1061/40926(239)120.

— (2008). "Assessment of tidal inlet evolution and stability using sediment budget computations and hydraulic parameter analysis". *Marine Geology* 247(1–2): 104–127. DOI: 10.1016/j.margeo.2007.07.003.

Pingree, R.D. and D.K. Griffiths (1979). "Sand transport paths around the British Isles resulting from M2 and M4-tidal interactions". *Journal of the Marine Biological Association of the United Kingdom*, 59(2): 497–513. DOI: 10.1017/S0025315400042806.

Powell, M.A., R.J. Thieke, and A.J. Mehta (2006). "Morphodynamic relationships for ebb and flood delta volumes at Florida's tidal entrances". *Ocean Dynamics* 56(3): 295–307. DOI: 10.1007/s10236-006-0064-3.

Ranasinghe, R. and C. Pattiaratchi (1999). "The seasonal closure of tidal inlets: Wilson Inlet – a case study". *Coastal Engineering* 37(1): 37–56. DOI: 10.1016/S0378-3839(99)00007-1.

— (2003). "The seasonal closure of tidal inlets: causes and effects". *Coastal Engineering Journal* 45(4): 601–627. DOI: 10.1142/S0578563403000919.

Ranasinghe, R., C. Pattiaratchi, and G. Masselink (1999). "A morphodynamic model to simulate the seasonal closure of tidal inlets". *Coastal Engineering* 37(1): 1–36. DOI: 10.1016/S0378-3839(99)00008-3.

Reid, R.O. and B.R. Bodine (1969). "Numerical model for storm surges in Galveston Bay". *Journal of the Waterways and Harbors Division* 94(WW1): 35–59.

Ridderinkhof, H. and J.T.F. Zimmerman (1992). "Chaotic stirring in a tidal system". *Science* 258(5085): 1107–1111. DOI: 10.1126/science.258.5085.1107.

Ridderinkhof, W., H.E. de Swart, M. van der Vegt, N.C. Alebregtse, and P. Hoekstra (2014). "Geometry of tidal inlet systems: a key factor for the net sediment transport in tidal inlets". *Journal of Geophysical Research: Oceans* 119(10): 6988–7006. DOI: 10.1002/2014JC010226.Received.

Rijn, L. van (1993). *Principles of Sediment Transport in Rivers, Estuaries and Coastal Seas*. Part I. Aqua Publications, 500 pp.

Roelvink, J.A. (2006). "Coastal morphodynamic evolution techniques". *Coastal Engineering* 53(2–3): 277–287. DOI: 10.1016/j.coastaleng.2005.10.015.

Roelvink, J. A. and Reniers, A. J. H. M., (2010). *A Guide to Modeling Coastal Morphology*. World Scientific Publishing Co., 12: 274 pp. DOI: 10.1142/7712.

Roelvink, J.A. and D.J.R. Walstra (2004). "Keeping it simple by using complex models". In: *Proceedings of Advances in Hydro-Science and -Engineering*. Vol. 6, pp. 1–11.

Roos, P.C. and H.M. Schuttelaars (2011). "Influence of topography on tide propagation and amplification in semi-enclosed basins". *Ocean Dynamics* 61(1): 21–38. DOI: 10.1007/s10236-010-0340-0.

Roos, P.C., H.M. Schuttelaars, and R.L. Brouwer (2013). "Observations on barrier island length explained using an exploratory morphodynamic model". *Geophysical Research Letters* 40(16): 4338–4343. DOI: 10.1002/grl.50843.

Rosati, J.D. and N.C. Kraus (1999). *Formulation of Sediment Budgets at Inlets*. Coastal Engineering Technical Note CETN IV-15. Vicksburg (MS): US Army Engineer Research, Development Center, Coastal, and Hydraulics Laboratory, 20 pp.

Salles, P., G. Voulgaris, and D.G. Aubrey (2005). "Contribution of nonlinear mechanisms in the persistence of multiple tidal inlet systems". *Estuarine, Coastal and Shelf Science* 65(3): 475–491. DOI: 10.1016/j.ecss.2005.06.018.

Seabergh, W.C. (2002). "Hydrodynamics of Tidal Inlets". In: *Coastal Engineering Manual*. 1110-2-1100. Washington, DC: US Army Corps of Engineers. Chap. II-6.

Seabergh, W.C. and N.C. Kraus (2003). "Progress in management of sediment bypassing at coastal inlets; natural bypassing, weir jetties, jetty spurs, and engineering aids in design". *Coastal Engineering Journal* 45(4): 533–563. DOI: 10.1142/S0578563403000944.

Serrano, D., E. Ramírez-Félix, and A. Valle-Levinson (2013). "Tidal hydrodynamics in a two-inlet coastal lagoon in the Gulf of California". *Continental Shelf Research* 63:1–12. DOI: 10.1016/j.csr.2013.04.038.

Sha, L.P. (1989). "Variation in ebb-delta morphologies along the West and East Frisian Islands, the Netherlands and Germany". *Marine Geology* 89(1–2): 11–28. DOI: 10.1016/0025-3227(89)90025-X.

Shigemura, T. (1980). "Tidal prism - throat area relationships of the bays of Japan". *Shore and Beach* 48(3): 30–35.

Sorensen, R.M. (1977). *Procedures for Preliminary Analysis of Tidal Inlet Hydraulics and Stability*. Coastal Engineering Technical Aid No. 77-8. CERC.

Soulsby, R. (1997). *Dynamics of Marine Sands: A Manual for Practical Applications*. Thomas Telfordt, 249 pp.

Southgate, H.N. (1993). "The effect of wave event sequencing on long-term beach response". In: *Proceedings of Large Scale Coastal Behavior '93*. Ed. by J.H. List. US Geological Survey Open-File Report 93-381. St. Petersburg, FL, 238 pp.

Spanhoff, R., E. Biegel, J. van de Graaff, and P. Hoekstra (1997). "Shoreface nourishments at Terschelling, the Netherlands: Feeder berm or breaker berm?" In: *Coastal Dynamics '97*. New York: ASCE, pp. 863–872.

Stauble, D.K. (1993). "An overview of southeast Florida inlet morphogdynamics". *Journal of Coastal Research* (SI 18): 1–27.

Stive, M.J.F., Z.B. Wang, M. Capobianco, P. Ruol, and M.C. Buijsman (1998). "Morphodynamics of a tidal lagoon and the adjacent coast". In: *Proceedings of the 8th Conference
on Physics of Estuaries and Coastal Seas (PECS) 1996*. Ed. by J. Dronkers and
M.B.A.M. Scheffers, pp. 355–362.

Stommel, H.M. and H.G. Farmer (1952). *On the nature of estuarine circulation*. Woods
Hole Oceanographic Institution, 38 pp. DOI: 10.1575/1912/2032.

Suprijo, T. and A. Mano (2004). "Dimensionless parameters to describe topographical
equilibrium of coastal inlets". In: *Proceedings of the 29th International Conference on Coastal Engineering*. Ed. by J. McKee Smith. Vol. 3, pp. 2531–2543. DOI:
10.1142/9789812701916_0204.

Tambroni, N. and G. Seminara (2006). "Are inlets responsible for the morphological degradation of Venice Lagoon?" *Journal of Geophysical Research* 111(F03013): 1–19.
DOI: 10.1029/2005JF000334.

Townend, I. (2005). "An examination of empirical stability relationships for UK estuaries".
Journal of Coastal Research 21(5): 1042–1063. DOI: 10.2112/03-0066R.1.

Tung, T.T. (2011). "Morphodynamics of seasonally closed coastal inlets at the central coast
of Vietnam". PhD thesis. Delft University of Technology, 192 pp.

Tung, T.T., J. van de Kreeke, M.J.F. Stive, and D.-J.R. Walstra (2012). "Cross-sectional
stability of tidal inlets: a comparison between numerical and empirical approaches".
Coastal Engineering 60: 21–29. DOI: 10.1016/j.coastaleng.2011.08.
005.

US Army Corps of Engineers (1984). *Shore Protection Manual*. Vol. II, 656 pp.

USGS and ESA (2011). *Reclaimed Lands*. URL: www.esa.int/spaceinimages/
Images/2011/11/Reclaimed_lands.

van der Spek, A.J.F. and D.J. Beets (1992). "Mid-Holocene evolution of a tidal basin in the
western Netherlands: a model for future changes in the northern Netherlands under
conditions of accelerated sea-level rise?" *Sedimentary Geology* 80(3–4): 185–197.
DOI: 10.1016/0037-0738(92)90040-X.

Veen, J. van (1936). "Onderzoekingen in de Hoofden in verband met de gesteldheid der
Nederlandsche kust". PhD thesis. Leiden University.

Vennell, R. (2006). "ADCP measurements of momentum balance and dynamic topography
in a constricted tidal channel". *Journal of Physical Oceanography* 36(2): 177–188.
DOI: 10.1175/JPO2836.1.

Vila-Concejo, A., Ó. Ferreira, A. Matias, and J.M.A. Dias (2003). "The first two years of
an inlet: sedimentary dynamics". *Continental Shelf Research* 23(14–15): 1425–1445.
DOI: 10.1016/S0278-4343(03)00142-0.

Vila-Concejo, A., Ó. Ferreira, B.D. Morris, A. Matias, and J.M.A. Dias (2004). "Lessons
from inlet relocation: example from southern Portugal". *Coastal Engineering* 51(10):
967–990. DOI: 10.1016/j.coastaleng.2004.07.019.

Vlasblom, W.J. (2003). *Designing Dredging Equipment*. University lecture notes. Delft
University of Technology, 323 pp.

Walton, T.L. (2004a). "Escoffier curves and inlet stability". *Journal of Waterway, Port, Coastal, and Ocean Engineering* 130(1): 54–57. DOI: 10.1061/
(ASCE)0733-950X(2004)130:1(54).

— (2004b). "Linear systems analysis approach to inlet-bay systems". *Ocean Engineering*
31(3–4): 513–522. DOI: 10.1016/j.oceaneng.2003.07.002.

Walton, T.L. and W.D. Adams (1976). "Capacity of inlet outer bars to store sand". In:
Proceedings of the 15th International Conference on Coastal Engineering. Vol. 1.
ASCE, pp. 1919–1937.

Walton, T.L. and F.F. Escoffier (1981). "Linearized solution to inlet equation with inertia". *Journal of the Waterway, Port, Coastal and Ocean Division, ASCE* 107(WW3): 191–195.

Wang, Z.B., H.J. de Vriend, M.J.F. Stive, and I.H. Townend (2008). "On the parameter setting of semi-empirical long-term morphological models for estuaries and tidal lagoons". In: *Proceedings of the 5th IAHR Symposium on River, Coastal and Estuarine Morphodynamics (RCEM) 2007*, Ed. by C. Dohmen-Janssen and S. Hulscher. Vol. 1. Leiden: Taylor & Francis/Balkema, pp. 103–111.

Wegen, M. van der, A. Dastgheib, and J.A. Roelvink (2010). "Morphodynamic modeling of tidal channel evolution in comparison to empirical PA relationship". *Coastal Engineering* 57(9): 827–837. DOI: http://dx.doi.org/10.1016/j.coastaleng.2010.04.003.

Welsh, J.M. and W.J. Cleary (2007). "Evolution of a relocated tidal inlet: Mason Inlet, NC". In: *Coastal Sediments '07*. Ed. by N.C. Kraus and J. Dean-Rosati. ASCE, pp. 1543–1555. DOI: 10.1061/40926(239)119.

Whitfield, A.K. (1992). "A characterization of southern African estuarine systems". *Journal of Aquatic Science* 18(1–2): 89–103. DOI: 10.1080/10183469.1992.9631327.

Williams, D.D., N.C. Kraus, and L.M. Anderson (2007). "Morphologic response to a new inlet, Packery Channel, Corpus Christi, Texas". In: *Coastal Sediments '07*. Ed. by N.C. Kraus and J. Dean-Rosati. ASCE, pp. 1529–1542. DOI: 10.1061/40926(239)118.

Winton, T.C. and A.J. Mehta (1981). "Dynamic model for closure of small inlets due to storm-induced littoral drift". In: *Proceedings of the 19th Congress of IAHR*. Vol. 2. 2. New Delhi, pp. 153–159.

Zimmerman, J.T.F. (1982). "On the Lorentz linearization of a quadratically damped forced oscillator". *Physics Letters A* 89A(3): 123–124. DOI: 10.1016/0375-9601(82)90871-4.

Zurmuhlen, F.H. (1957). "The sand transfer plant at Lake Worth Inlet". In: *Proceedings of the 6th International Conference on Coastal Engineering*. Ed. by J.W. Johnson, pp. 457–462.

Index

Attachment bar, 121–123
Attractor, 149

Back-barrier lagoons
    Matagorda Bay, 80, 81
    Ría Formosa, 86, 153
    Tam Giang Lagoon, 143
    Venice Lagoon, 86
    Wadden Sea, 6, 31, 35, 38–40, 76, 80, 86, 87, 89,
        96, 118, 124, 125, 131, 138
Barrier island coasts
    Adriatic coast of Italy, 2, 3
    Algarve coast of Portugal, 2, 3
    Fire Island coast, 7, 86
    North Island of New Zealand, 2, 24, 30, 38, 39, 41,
        118
    Ría Formosa barrier island coast, 15, 155
    US East Coast, 2
    US Gulf Coast, 2
    Wadden Sea coast, 2, 3, 24, 67, 68
Beach erosion, 1, 27
Bottom friction, 46, 49, 53, 55, 89, 94–96, 100–103,
    105, 142
Breaker zone, 159
Breakwater, 13
Bump, 134
Bypassing bar, 121–123

Cape Hatteras, 7
Channel margin linear bar, 10, 16, 24
Chézy friction coefficient, 112
Closure curve, 76, 77, 81, 89, 90, 143, 144, 153
Closure depth, 14, 22
Colorado River, 160
Coriolis acceleration, 100, 102–109

Damped progressive wave, 22
Diffusion equation, 22
Double inlet systems

Big Marco-Capri inlet system, 93
Faro-Armona inlet system, 94
Pass Cavallo-Matagorda inlet system, 93
Texel-Vlie inlet system, 87, 88, 90, 92, 93, 98, 100,
    101, 105–108
Downdrift beach, 16, 17, 24, 27, 130, 154–160
Downdrift coast, 1, 8, 13, 15–17, 22, 121, 125–127,
    130, 154, 156
Dredging, 1, 13, 15, 28, 120, 145, 152, 155, 156, 158,
    159

Ebb delta, 2, 4–11, 13–17, 19, 21, 24, 27–30, 32, 42,
    110, 120, 121, 124–126, 131, 135, 137, 154
Ebb delta channel, 16, 152
Ebb delta platform, 21, 25, 32, 41
Ebb delta volume, 26, 27, 29, 30, 32, 34, 41–43, 120,
    125, 131, 154
Ebb tidal shoal, 121, 123
Empirical model, 5, 110, 120, 121, 125, 138, 154
Empirical relationships
    Cross-sectional area – tidal prism relationship, 34,
        38–40, 89, 111, 115, 118, 131, 154
    Ebb delta volume – tidal prism relationship, 41
    Flood delta volume – tidal prism relationship, 43
Entrance/exit losses, 45, 46, 49, 62, 82, 88, 89, 94,
    144, 148
Equilibrium, 3, 7, 14, 35–37, 40, 41, 75, 78, 80, 84,
    88, 91–95, 97–99, 106–108, 112–115, 120, 144,
    154
Equilibrium cross-sectional area, 34, 75–78, 80–82,
    85, 91–94, 97–99, 108, 109, 114–118, 125, 144,
    153
Equilibrium depth, 149
Equilibrium velocity, 4, 40, 46, 75–77, 81, 88, 89, 91,
    97, 105, 106, 114, 115, 143, 144, 154
Equilibrium velocity curve, 76, 77, 90–94, 98, 106,
    107, 153
Escoffier Diagram, 4, 76, 77, 81, 82, 143, 144, 153

Escoffier Stability Model, 4, 75, 76, 80, 87, 93–95, 105, 149
Even–odd analysis, 21, 22

First-order equations, 62, 63, 65, 66, 73
First-order solution, 63, 72
Flood delta, 2, 4, 6–8, 11, 13–16, 120
Flood delta volume, 43
Flow diagram, 4, 91–93, 97, 98, 104, 106–108
Forcing frequency, 47

Geologic and hydrodynamic similarity, 37, 39, 111, 116
Geometric similarity, 36, 37, 76, 81, 82, 89, 105, 143
Geomorphology, 6, 42
Gorge, 9, 32, 34, 113, 115, 125
Great diurnal amplitude, 80
Great South Bay, 7, 86
Gulf of Tonkin, 143

Helmholtz frequency, 64
Hurricane Sandy, 7, 86
Hypsometry, 58, 61, 99

Intermittently open inlet, 5, 145

Kelvin wave, 102–104, 106, 108
Keulegan repletion factor, 47
Keulegan Solution, 53, 57, 81, 153

Leading-order equations, 62, 63, 73
Leading-order solution, 70, 72
Longshore current, 13, 14, 145
Lumped parameter model, 4, 46, 47, 140

Marginal flood channels, 10, 24
Mean basin level, 4, 5, 61, 62, 66, 67, 72, 141, 147, 151
Mean inlet velocity, 4, 61, 62, 66, 140, 141
MorFac, 111, 113
Morphology, 1–8, 10, 11, 34, 110–113, 120, 125, 138, 154
Multiple inlets, 2, 54, 86, 109

Natural frequency, 53
Neap tide, 34

Öszoy–Mehta Solution, 49, 53, 55, 56, 65, 71, 76, 140, 142, 143, 150
Overshoot, 128, 134, 135
Overtides, 4, 44, 50, 61, 62, 66, 69–72

Permanently open inlet, 5, 143
Perturbation analysis, 61, 73
Poincaré wave, 103, 104
Pressure gradient, 94, 95, 100
Process-based exploratory model, 2, 4, 89, 97, 103–105

Process-based simulation model, 2, 3, 5, 97, 110, 112

Radiation damping, 100, 104–109
Representative inlet, 4, 44, 55–58, 68, 70, 72, 76, 77, 142
Resonance, 44, 49
River flow, 5, 37, 38, 139–151
Rossby radius, 102, 105

Sand bypassing, 4, 5, 16–20, 24–26, 28, 30, 128, 152, 156
  Bar bypassing, 16–19
  Ebb delta breaching, 17–20, 26
  Spit formation and breaching, 8, 17–19, 27–29, 110, 139, 145
  Tidal flow bypassing, 16–20, 25, 30, 31, 33, 128
Sand bypassing plant, 156–158
Sand transport
  Cross-shore sand transport, 2, 19, 139
  Diffusive sand transport, 120, 135, 138
  Longshore sand transport, 4, 6, 9, 10, 12, 13, 17–19, 21, 22, 24–33, 35–40, 78, 92, 109, 113, 114, 116, 118, 121, 124, 125, 128, 130, 139, 143, 145, 146, 154, 158, 159
  Sand transport pathways, 1, 4, 13, 14, 29, 121, 159
Sea level rise, 8, 12
Seasonally open inlet, 5, 145
Sediment budget, 14, 15, 154
Shallow water wave equations, 4, 97, 100, 101, 112
Shape factor, 55, 76, 81, 84, 87–89, 105
Spring tide, 17, 24, 25, 27, 29–31, 34, 35, 38, 40
Spur jetty, 158, 159
Stability
  Cross-sectional stability, 4, 5, 75–77, 80, 82, 86, 87, 94, 97, 100, 104–106, 108, 143, 144, 153
  Entrance stability, 5, 139, 140, 144
  (Linear) stability analysis, 4, 77, 84, 87, 105, 109
  Location stability, 4, 18, 19, 24–26, 28–31, 33, 113, 143, 152, 154, 155
Swash bar, 10, 13, 16, 25
Swash platform, 10, 16

Terminal lobe, 10, 16
Throat, 9, 17, 19, 24, 25, 27, 29, 30, 32, 34, 35, 40
Tidal asymmetry, 4, 69, 70, 72
Tidal current, 1, 7, 11, 14, 16, 21, 32, 41, 78, 118, 126, 139
Tidal discharge scale, 141
Tidal inlets, cuts and passes
  Ameland Inlet, 6, 8, 18, 21, 24, 31, 32, 38, 138
  Ancão Inlet, 8, 155, 156
  Aveiro Inlet, 18
  Bakers Haulover Inlet, 7, 152, 153
  Barra Nova Inlet, 8
  Bergen Inlet, 12
  Big Pass, 18
  Breach at Old Inlet, 86
  Breach Inlet, 18, 24–26

Cape Canaveral Inlet, 21, 22
Captain Sam's Inlet, 18, 27, 139, 155
Channel Islands Harbor, 156
Eyerlandse Gat Inlet, 38, 138
Faro-Olhão Inlet, 15, 152, 153
Figueira da Foz, 18
Fire Island Inlet, 86
Fort Pierce Inlet, 159
Frisian Inlet, 5, 6, 37, 38, 67, 68, 80, 120, 124, 126, 128, 131
Government Cut, 7
Indian River Inlet, 157
John's Pass, 18
Jupiter Inlet, 156
Katikati Inlet, 18, 24, 30, 31
Lake Conjola Inlet, 140, 145, 146
Lake Worth Inlet, 7, 156, 157
Longboat Pass, 18
Maketu, 39
Mangawhai North, 39
Mangawhai South, 39
Mason Inlet, 8, 18, 24, 28, 29, 139, 155
Masonboro Inlet, 160
Midnight Pass, 19
Murrells Inlet, 160
Nerang River Entrance, 157, 158
Ngunguru, 39
Ocean City Inlet, 5, 120–123, 125, 154
Ohiwa, 39
Oregon Inlet, 18
Packery Channel, 152, 153
Pass Cavallo, 4, 80–82
Pataua, 39
Ponce de Leon Inlet, 18, 160
Price Inlet, 8, 18, 24, 25, 34

Puhoi, 39
Sebastian Inlet, 156
South Lake Worth Inlet, 7, 156, 157
Tairua, 39
Tauranga, 39
Texel Inlet, 18, 37, 38, 58, 68, 87, 89–93, 98, 101, 105
Thuan An Inlet, 140, 143–145
Vlie Inlet, 18, 21, 38, 87, 89–93, 98, 101, 105
Wachapreague Inlet, 8, 18, 24, 29, 30, 34
Whananaki, 39
Whangamata, 39
Whangapoua, 39
Whangarei, 39
Whangateau, 39
Whitianga, 39
Wilson Inlet, 19, 140, 145
Tidal prism, 2–4, 6, 9, 11, 17–19, 24–26, 28–34, 36–43, 76, 78–81, 86, 109, 114, 116–118, 120, 124, 125, 128, 131, 138, 139, 143, 146, 152, 154
Tidal range, 1, 24–26, 28–31, 37, 38, 53, 54, 109, 124, 139, 143, 145
Timescales, 1, 2, 8, 16, 19, 21, 32, 46, 73, 78, 80, 110, 111, 118–120, 123, 126, 130–132, 134, 135, 138
  Adaptation timescale, 4, 78, 80, 123
  Local timescale, 130–133
  System timescale, 130, 132, 133
Topographic high, 4, 95–100

Velocity scale, 46, 73, 102

Washover, 6, 7, 17
Wave number, 22, 102, 103
Weir-jetty system, 160